Notes in Computer Science 13245

ng Editors

rd Goos
Karlsruhe Institute of Technology, Karlsruhe, Germany
Juris Hartmanis
Cornell University, Ithaca, NY, USA

More information about this series at https://link.springer.com/bookserie

Arnab Bhattacharya · Janice Lee Mong Li ·
Divyakant Agrawal · P. Krishna Reddy ·
Mukesh Mohania · Anirban Mondal ·
Vikram Goyal · Rage Uday Kiran (Eds.)

Database Systems for Advanced Applications

27th International Conference, DASFAA 2022
Virtual Event, April 11–14, 2022
Proceedings, Part I

 Springer

Editors

Arnab Bhattacharya
Indian Institute of Technology Kanpur
Kanpur, India

Divyakant Agrawal
University of California, Santa Barbara
Santa Barbara, CA, USA

Mukesh Mohania
Indraprastha Institute of Information
Technology Delhi
New Delhi, India

Vikram Goyal
Indraprastha Institute of Information
Technology Delhi
New Delhi, India

Janice Lee Mong Li
National University of Singapore
Singapore, Singapore

P. Krishna Reddy ⓘ
IIIT Hyderabad
Hyderabad, India

Anirban Mondal
Ashoka University
Sonepat, Haryana, India

Rage Uday Kiran
University of Aizu
Aizu, Japan

ISSN 0302-9743   ISSN 1611-3349 (electronic)
Lecture Notes in Computer Science
ISBN 978-3-031-00122-2   ISBN 978-3-031-00123-9 (eBook)
https://doi.org/10.1007/978-3-031-00123-9

General Chairs' Preface

On behalf of the Organizing Committee, it is our great pleasure to welcome you to the proceedings of the 27th International Conference on Database Systems for Advanced Applications (DASFAA 2022), which was held during April 11–14, 2022, in Hyderabad, India. The conference has returned to India for the second time after a gap of 14 years, moving from New Delhi in 2008 to Hyderabad in 2022. DASFAA has long established itself as one of the leading international conferences in database systems. We were expecting to welcome you in person and give you a feel of our renowned Indian hospitality. However, unfortunately, given the Omicron wave of COVID-19 and the pandemic circumstances, we had to move the conference to a fully online mode.

Our gratitude goes first and foremost to the researchers, who submitted their work to the DASFAA 2022 main conference, workshops, and the data mining contest. We thank them for their efforts in submitting the papers, as well as in preparing high-quality online presentation videos. It is our distinct honor that five eminent keynote speakers graced the conference: Sunita Sarawagi of IIT Bombay, India, Guoliang Li of Tsinghua University, China, Gautam Das of the University of Texas at Arlington, Ioana Manolescu of Inria and Institut Polytechnique de Paris, and Tirthankar Lahiri of the Oracle Corporation. Each of them is a leader of international renown in their respective areas, and their participation significantly enhanced the conference. The conference program was further enriched with a panel, five high–quality tutorials, and six workshops on cutting-edge topics.

We would like to express our sincere gratitude to the contributions of the Senior Program Committee (SPC) members, Program Committee (PC) members, and anonymous reviewers, led by the PC chairs, Arnab Bhattacharya (IIT Kanpur), Lee Mong Li Janice (National University of Singapore), and Divyakant Agrawal (University of California, Santa Barbara). It is through their untiring efforts that the conference had an excellent technical program. We are also thankful to the other chairs and Organizing Committee members: industry track chairs, Prasad M. Deshpande (Google), Daxin Jiang (Microsoft), and Rajasekar Krishnamurthy (Adobe); demo track chairs, Rajeev Gupta (Microsoft), Koichi Takeda (Nagoya University), and Ladjel Bellatreche (ENSMA); workshop chairs, Maya Ramanath (IIT Delhi), Wookey Lee (Inha University), and Sanjay Kumar Madria (Missouri Institute of Technology); tutorial chairs, P. Sreenivasa Kumar (IIT Madras), Jixue Liu (University of South Australia), and Takahiro Hara (Osaka university); panel chairs, Jayant Haritsa (Indian Institute of Science), Reynold Cheng (University of Hong Kong), and Georgia Koutrika (Athena Research Center); Ph.D. consortium chairs, Vikram Pudi (IIIT Hyderabad), Srinath Srinivasa (IIIT Bangalore), and Philippe Fournier-Viger (Harbin Institute of Technology); publicity chairs, Raj Sharma (Goldman Sachs), Jamshid Bagherzadeh Mohasefi (Urmia University), and Nazha Selmaoui-Folcher (University of New Caledonia); publication chairs, Vikram Goyal (IIIT Delhi), and R. Uday Kiran (University of Aizu); and registration/local arrangement chairs, Lini Thomas (IIIT Hyderabad), Satish Narayana Srirama (University of Hyderabad), Manish Singh (IIT Hyderabad), P. Radha Krishna (NIT Warangal), Sonali Agrawal (IIIT Allahabad), and V. Ravi (IDRBT).

We appreciate the hosting organization IIIT Hyderabad, which is celebrating its silver jubilee in 2022. We thank the researchers at the Data Sciences and Analytics Center (DSAC) and the Kohli Center on Intelligent Systems (KCIS) at IIIT Hyderabad for their support. We also thank the administration and staff of IIIT Hyderabad for their help. We thank Google for the sponsorship. We feel indebted to the DASFAA Steering Committee for its continuing guidance.

Finally, our sincere thanks go to all the participants and volunteers. There would be no conference without them. We hope all of you enjoy these DASFAA 2022 proceedings.

February 2022   P. Krishna Reddy
                                                        Mukesh Mohania
                                                         Anirban Mondal

Program Chairs' Preface

It is our great pleasure to present the proceedings of the 27th International Conference on Database Systems for Advanced Applications (DASFAA 2022). DASFAA is a premier international forum for exchanging original research results and practical developments in the field of databases.

For the research track, we received 488 research submissions from across the world. We performed an initial screening of all submissions, leading to the desk rejection of 88 submissions due to violations of double-blind and page limit guidelines. For submissions entering the double-blind review process, each paper received at least three reviews from Program Committee (PC) members. Further, an assigned Senior Program Committee (SPC) member also led a discussion of the paper and reviews with the PC members. The PC co-chairs then considered the recommendations and meta-reviews from SPC members in making the final decisions. As a result, 72 submissions were accepted as full papers (acceptance ratio of 18%), and 76 submissions were accepted as short papers (acceptance ratio of 19%). For the industry track, 13 papers were accepted out of 36 submissions. Nine papers were accepted out of 16 submissions for the demo track. For the Ph.D. consortium, two papers were accepted out of three submissions. Four short research papers and one industry paper were withdrawn. The review process was supported by Microsoft's Conference Management Toolkit (CMT).

The conference was conducted in an online environment, with accepted papers presented via a pre-recorded video presentation with a live Q&A session. The conference program also featured five keynotes from distinguished researchers in the community, a panel, five high–quality tutorials, and six workshops on cutting-edge topics.

We wish to extend our sincere thanks to all SPC members, PC members, and external reviewers for their hard work in providing us with thoughtful and comprehensive reviews and recommendations. We especially thank the authors who submitted their papers to the conference. We hope that the readers of the proceedings find the content interesting, rewarding, and beneficial to their research.

March 2022

Arnab Bhattacharya
Janice Lee Mong Li
Divyakant Agrawal
Prasad M. Deshpande
Daxin Jiang
Rajasekar Krishnamurthy
Rajeev Gupta
Koichi Takeda
Ladjel Bellatreche
Vikram Pudi
Srinath Srinivasa
Philippe Fournier-Viger

Organization

DASFAA 2022 was organized by IIIT Hyderabad, Hyderabad, Telangana, India.

Steering Committee Chair

Lei Chen   Hong Kong University of Science and
Technology, Hong Kong

Honorary Chairs

P. J. Narayanan   IIIT Hyderabad, India
S. Sudarshan   IIT Bombay, India
Masaru Kitsuregawa   University of Tokyo, Japan

Steering Committee Vice Chair

Stephane Bressan   National University of Singapore, Singapore

Steering Committee Treasurer

Yasushi Sakurai   Osaka University, Japan

Steering Committee Secretary

Kyuseok Shim   Seoul National University, South Korea

General Chairs

P. Krishna Reddy   IIIT Hyderabad, India
Mukesh Mohania   IIIT Delhi, India
Anirban Mondal   Ashoka University, India

Program Committee Chairs

Arnab Bhattacharya   IIT Kanpur, India
Lee Mong Li Janice   National University of Singapore, Singapore
Divyakant Agrawal   University of California, Santa Barbara, USA

Steering Committee

Zhiyong Peng	Wuhan University, China
Zhanhuai Li	Northwestern Polytechnical University, China
Krishna Reddy	IIIT Hyderabad, India
Yunmook Nah	Dankook University, South Korea
Wenjia Zhang	University of New South Wales, Australia
Zi Huang	University of Queensland, Australia
Guoliang Li	Tsinghua University, China
Sourav Bhowmick	Nanyang Technological University, Singapore
Atsuyuki Morishima	University of Tsukaba, Japan
Sang-Won Lee	Sungkyunkwan University, South Korea
Yang-Sae Moon	Kangwon National University, South Korea

Industry Track Chairs

Prasad M. Deshpande	Google, India
Daxin Jiang	Microsoft, China
Rajasekar Krishnamurthy	Adobe, USA

Demo Track Chairs

Rajeev Gupta	Microsoft, India
Koichi Takeda	Nagoya University, Japan
Ladjel Bellatreche	ENSMA, France

PhD Consortium Chairs

Vikram Pudi	IIIT Hyderabad, India
Srinath Srinivasa	IIIT Bangalore, India
Philippe Fournier-Viger	Harbin Institute of Technology, China

Panel Chairs

Jayant Haritsa	Indian Institute of Science, India
Reynold Cheng	University of Hong Kong, China
Georgia Koutrika	Athena Research Center, Greece

Sponsorship Chair

P. Krishna Reddy	IIIT Hyderabad, India

Publication Chairs

Vikram Goel   IIIT Delhi, India
R. Uday Kiran   University of Aizu, Japan

Workshop Chairs

Maya Ramanath   IIT Delhi, India
Wookey Lee   Inha University, South Korea
Sanjay Kumar Madria   Missouri Institute of Technology, USA

Tutorial Chairs

P. Sreenivasa Kumar   IIT Madras, India
Jixue Liu   University of South Australia, Australia
Takahiro Hara   Osaka University, Japan

Publicity Chairs

Raj Sharma   Goldman Sachs, India
Jamshid Bagherzadeh Mohasefi   Urmia University, Iran
Nazha Selmaoui-Folcher   University of New Caledonia, New Caledonia

Organizing Committee

Lini Thomas   IIIT Hyderabad, India
Satish Narayana Srirama   University of Hyderabad, India
Manish Singh   IIT Hyderabad, India
P. Radha Krishna   NIT Warangal, India
Sonali Agrawal   IIIT Allahabad, India
V. Ravi   IDRBT, India

Senior Program Committee

Avigdor Gal   Technion - Israel Institute of Technology, Israel
Baihua Zheng   Singapore Management University, Singapore
Bin Cui   Peking University, China
Bin Yang   Aalborg University, Denmark
Bingsheng He   National University of Singapore, Singapore
Chang-Tien Lu   Virginia Tech, USA
Chee-Yong Chan   National University of Singapore, Singapore
Gautam Shroff   Tata Consultancy Services Ltd., India
Hong Gao   Harbin Institute of Technology, China

Jeffrey Xu Yu	Chinese University of Hong Kong, China
Jianliang Xu	Hong Kong Baptist University, China
Jianyong Wang	Tsinghua University, China
Kamalakar Karlapalem	IIIT Hyderabad, India
Kian-Lee Tan	National University of Singapore, Singapore
Kyuseok Shim	Seoul National University, South Korea
Ling Liu	Georgia Institute of Technology, USA
Lipika Dey	Tata Consultancy Services Ltd., India
Mario Nascimento	University of Alberta, Canada
Maya Ramanath	IIT Delhi, India
Mohamed Mokbel	University of Minnesota, Twin Cities, USA
Niloy Ganguly	IIT Kharagpur, India
Sayan Ranu	IIT Delhi, India
Sourav S. Bhowmick	Nanyang Technological University, Singapore
Srikanta Bedathur	IIT Delhi, India
Srinath Srinivasa	IIIT Bangalore, India
Stephane Bressan	National University of Singapore, Singapore
Tok W. Ling	National University of Singapore, Singapore
Vana Kalogeraki	Athens University of Economics and Business, Greece
Vassilis J. Tsotras	University of California, Riverside, USA
Vikram Pudi	IIIT Hyderabad, India
Vincent Tseng	National Yang Ming Chiao Tung University, Taiwan
Wang-Chien Lee	Pennsylvania State University, USA
Wei-Shinn Ku	Auburn University, USA
Wenjie Zhang	University of New South Wales, Australia
Wynne Hsu	National University of Singapore, Singapore
Xiaofang Zhou	Hong Kong University of Science and Technology, China
Xiaokui Xiao	National University of Singapore, Singapore
Xiaoyong Du	Renmin University of China, China
Yoshiharu Ishikawa	Nagoya University, Japan
Yufei Tao	Chinese University of Hong Kong, China

Program Committee

Abhijnan Chakraborty	IIT Delhi, India
Ahmed Eldawy	University of California, Riverside, USA
Akshar Kaul	IBM Research, India
Alberto Abell	Universitat Politecnica de Catalunya, Spain
An Liu	Soochow University, China
Andrea Cali	Birkbeck, University of London, UK

Andreas Züfle	George Mason University, USA
Antonio Corral	University of Almeria, Spain
Atsuhiro Takasu	National Institute of Informatics, Japan
Bin Wang	Northeastern University, China
Bin Yao	Shanghai Jiao Tong University, China
Bo Jin	Dalian University of Technology, China
Bolong Zheng	Huazhong University of Science and Technology, China
Chandramani Chaudhary	National Institute of Technology, Trichy, India
Changdong Wang	Sun Yat-sen University, China
Chaokun Wang	Tsinghua University, China
Cheng Long	Nanyang Technological University, Singapore
Chenjuan Guo	Aalborg University, Denmark
Cheqing Jin	East China Normal University, China
Chih-Ya Shen	National Tsing Hua University, Taiwan
Chittaranjan Hota	BITS Pilani, India
Chi-Yin Chow	Social Mind Analytics (Research and Technology) Limited, Hong Kong
Chowdhury Farhan Ahmed	University of Dhaka, Bangladesh
Christos Doulkeridis	University of Pireaus, Greece
Chuan Xiao	Osaka University and Nagoya University, Japan
Cindy Chen	University of Massachusetts Lowell, USA
Cuiping Li	Renmin University of China, China
Dan He	University of Queensland, Australia
Demetrios Zeinalipour-Yazti	University of Cyprus, Cyprus
De-Nian Yang	Academia Sinica, Taiwan
Dhaval Patel	IBM TJ Watson Research Center, USA
Dieter Pfoser	George Mason University, USA
Dimitrios Kotzinos	University of Cergy-Pontoise, France
Fan Zhang	Guangzhou University, China
Ge Yu	Northeast University, China
Goce Trajcevski	Iowa State University, USA
Guoren Wang	Beijing Institute of Technology, China
Haibo Hu	Hong Kong Polytechnic University, China
Haruo Yokota	Tokyo Institute of Technology, Japan
Hiroaki Shiokawa	University of Tsukuba, Japan
Hongzhi Wang	Harbin Institute of Technology, China
Hongzhi Yin	University of Queensland, Australia
Hrishikesh R. Terdalkar	IIT Kanpur, India
Hua Lu	Roskilde University, Denmark
Hui Li	Xidian University, China
Ioannis Konstantinou	University of Thessaly, Greece

Iouliana Litou	Athens University of Economics and Business, Greece
Jagat Sesh Challa	BITS Pilani, India
Ja-Hwung Su	Cheng Shiu University, Taiwan
Jiali Mao	East China Normal University, China,
Jia-Ling Koh	National Taiwan Normal University, Taiwan
Jian Dai	Alibaba Group, China
Jianqiu Xu	Nanjing University of Aeronautics and Astronautics, China
Jianxin Li	Deakin University, Australia
Jiawei Jiang	ETH Zurich, Switzerland
Jilian Zhang	Jinan University, China
Jin Wang	Megagon Labs, USA
Jinfei Liu	Zhejiang University, China
Jing Tang	Hong Kong University of Science and Technology, China
Jinho Kim	Kangwon National University, South Korea
Jithin Vachery	National University of Singapore, Singapore
Ju Fan	Renmin University of China, China
Jun Miyazaki	Tokyo Institute of Technology, Japan
Junjie Yao	East China Normal University, China
Jun-Ki Min	Korea University of Technology and Education, South Korea
Kai Zeng	Alibaba Group, China
Karthik Ramachandra	Microsoft Azure SQL, India
Kento Sugiura	Nagoya University, Japan
Kesheng Wu	Lawrence Berkeley National Laboratory, USA
Kjetil Nørvåg	Norwegian University of Science and Technology, Norway
Kostas Stefanidis	Tempere University, Finland
Kripabandhu Ghosh	Indian Institute of Science Education and Research Kolkata, India
Kristian Torp	Aalborg University, Denmark
Kyoung-Sook Kim	Artificial Intelligence Research Center, Japan
Ladjel Bellatreche	ENSMA, France
Lars Dannecker	SAP, Germany
Lee Roy Ka Wei	Singapore University of Technology and Design, Singapore
Lei Cao	Massachusetts Institute of Technology, USA
Leong Hou U.	University of Macau, China
Lijun Chang	University of Sydney, Australia
Lina Yao	University of New South Wales Australia
Lini Thomas	IIIT Hyderabad, India

Liping Wang	East China Normal University, China
Long Yuan	Nanjing University of Science and Technology, China
Lu-An Tang	NEC Labs America, USA
Makoto Onizuka	Osaka University, Japan
Manish Kesarwani	IBM Research, India
Manish Singh	IIT Hyderabad, India
Manolis Koubarakis	University of Athens, Greece
Marco Mesiti	University of Milan, Italy
Markus Schneider	University of Florida, USA
Meihui Zhang	Beijing Institute of Technology, China
Meng-Fen Chiang	University of Auckland, New Zealand
Mirella M. Moro	Universidade Federal de Minas Gerais, Brazil
Mizuho Iwaihara	Waseda University, Japan
Navneet Goyal	BITS Pilani, India
Neil Zhenqiang Gong	Iowa State University, USA
Nikos Ntarmos	Huawei Technologies R&D (UK) Ltd., UK
Nobutaka Suzuki	University of Tsukuba, Japan
Norio Katayama	National Institute of Informatics, Japan
Noseong Park	George Mason University, USA
Olivier Ruas	Inria, France
Oscar Romero	Universitat Politècnica de Catalunya, Spain
Oswald C.	IIT Kanpur, India
Panagiotis Bouros	Johannes Gutenberg University Mainz, Germany
Parth Nagarkar	New Mexico State University, USA
Peer Kroger	Christian-Albrecht University of Kiel, Germany
Peifeng Yin	Pinterest, USA
Peng Wang	Fudan University, China
Pengpeng Zhao	Soochow University, China
Ping Lu	Beihang University, China
Pinghui Wang	Xi'an Jiaotong University, China
Poonam Goyal	BITS Pilani, India
Qiang Yin	Shanghai Jiao Tong University, China
Qiang Zhu	University of Michigan – Dearborn, USA
Qingqing Ye	Hong Kong Polytechnic University, China
Rafael Berlanga Llavori	Universitat Jaume I, Spain
Rage Uday Kiran	University of Aizu, Japan
Raghava Mutharaju	IIIT Delhi, India
Ravindranath C. Jampani	Oracle Labs, India
Rui Chen	Samsung Research America, USA
Rui Zhou	Swinburne University of Technology, Australia
Ruiyuan Li	Xidian University, China

Sabrina De Capitani di Vimercati	Università degli Studi di Milano, Italy
Saiful Islam	Griffith University, Australia
Sanghyun Park	Yonsei University, South Korea
Sanjay Kumar Madria	Missouri University of Science and Technology, USA
Saptarshi Ghosh	IIT Kharagpur, India
Sebastian Link	University of Auckland, New Zealand
Shaoxu Song	Tsinghua University, China
Sharma Chakravarthy	University of Texas at Arlington, USA
Shiyu Yang	Guangzhou University, China
Shubhadip Mitra	Tata Consultancy Services Ltd., India
Shubhangi Agarwal	IIT Kanpur, India
Shuhao Zhang	Singapore University of Technology and Design, Singapore
Sibo Wang	Chinese University of Hong Kong, China
Silviu Maniu	Université Paris-Saclay, France
Sivaselvan B.	IIIT Kancheepuram, India
Stephane Bressan	National University of Singapore, Singapore
Subhajit Sidhanta	IIT Bhilai, India
Sungwon Jung	Sogang University, South Korea
Tanmoy Chakraborty	Indraprastha Institute of Information Technology Delhi, India
Theodoros Chondrogiannis	University of Konstanz, Germany
Tien Tuan Anh Dinh	Singapore University of Technology and Design, Singapore
Ting Deng	Beihang University, China
Tirtharaj Dash	BITS Pilani, India
Toshiyuki Amagasa	University of Tsukuba, Japan
Tsz Nam (Edison) Chan	Hong Kong Baptist University, China
Venkata M. Viswanath Gunturi	IIT Ropar, India
Verena Kantere	National Technical University of Athens, Greece
Vijaya Saradhi V.	IIT Guwahati, India
Vikram Goyal	IIIT Delhi, India
Wei Wang	Hong Kong University of Science and Technology (Guangzhou), China
Weiwei Sun	Fudan University, China
Weixiong Rao	Tongji University, China
Wen Hua	University of Queensland, Australia
Wenchao Zhou	Georgetown University, USA
Wentao Zhang	Peking University, China
Werner Nutt	Free University of Bozen-Bolzano, Italy
Wolf-Tilo Balke	TU Braunschweig, Germany

Wookey Lee	Inha University, South Korea
Woong-Kee Loh	Gacheon University, South Korea
Xiang Lian	Kent State University, USA
Xiang Zhao	National University of Defence Technology, China
Xiangmin Zhou	RMIT University, Australia
Xiao Pan	Shijiazhuang Tiedao University, China
Xiao Qin	Amazon Web Services, USA
Xiaochun Yang	Northeastern University, China
Xiaofei Zhang	University of Memphis, USA
Xiaofeng Gao	Shanghai Jiao Tong University, China
Xiaowang Zhang	Tianjin University, China
Xiaoyang Wang	Zhejiang Gongshang University, China
Xin Cao	University of New South Wales, Australia
Xin Huang	Hong Kong Baptist University, China
Xin Wang	Tianjin University, China
Xu Xie	Peking University, China
Xuequn Shang	Northwestern Polytechnical University, China
Xupeng Miao	Peking University, China
Yan Shi	Shanghai Jiao Tong University, China
Yan Zhang	Peking University, China
Yang Cao	Kyoto University, Japan
Yang Chen	Fudan University, China
Yanghua Xiao	Fudan University, China
Yang-Sae Moon	Kangwon National University, South Korea
Yannis Manolopoulos	Aristotle University of Thessaloniki, Greece
Yi Yu	National Institute of Informatics, Japan
Yingxia Shao	Beijing University of Posts and Telecommunication, China
Yixiang Fang	Chinese University of Hong Kong, China
Yong Tang	South China Normal University, China
Yongxin Tong	Beihang University, China
Yoshiharu Ishikawa	Nagoya University, Japan
Yu Huang	National Yang Ming Chiao Tung University, Taiwan
Yu Suzuki	Gifu University, Japan
Yu Yang	City University of Hong Kong, China
Yuanchun Zhou	Computer Network Information Center, China
Yuanyuan Zhu	Wuhan University, China
Yun Peng	Hong Kong Baptist University, China
Yuqing Zhu	California State University, Los Angeles, USA
Zeke Wang	Zhejiang University, China

Zhaojing Luo	National University of Singapore, Singapore
Zhenying He	Fudan University, China
Zhi Yang	Peking University, China
Zhixu Li	Soochow University, China
Zhiyong Peng	Wuhan University, China
Zhongnan Zhang	Xiamen University, China

Industry Track Program Committee

Karthik Ramachandra	Microsoft, India
Akshar Kaul	IBM Research, India
Sriram Lakshminarasimhan	Google Research, India
Rajat Venkatesh	LinkedIn, India
Prasan Roy	Sclera, India
Zhicheng Dou	Renmin University of China, China
Huang Hu	Microsoft, China
Shan Li	LinkedIn, USA
Bin Gao	Facebook, USA
Haocheng Wu	Facebook, USA
Shivakumar Vaithyanathan	Adobe, USA
Abdul Quamar	IBM Research, USA
Pedro Bizarro	Feedzai, Portugal
Xi Yin	International Digital Economy Academy, China
Xiangyu Niu	Facebook

Demo Track Program Committee

Ahmed Awad	University of Tartu, Estonia
Beethika Tripathi	Microsoft, India
Carlos Ordonez	University of Houston, USA
Djamal Benslimane	Université Claude Bernard Lyon 1, France
Nabila Berkani	Ecole Nationale Supérieure d'Informatique, Algeria
Philippe Fournier-Viger	Shenzhen University, China
Ranganath Kondapally	Microsoft, India
Soumia Benkrid	Ecole Nationale Supérieure d'Informatique, Algeria

Sponsoring Institutions

Google, India

INTERNATIONAL INSTITUTE OF
INFORMATION TECHNOLOGY
H Y D E R A B A D

IIIT Hyderabad, India

Contents – Part I

Database Queries

Approximate Continuous Top-K Queries over Memory Limitation-Based
Streaming Data 3
   Rui Zhu, Liu Meng, Bin Wang, Xiaochun Yang, and Xiufeng Xia

Cross-Model Conjunctive Queries over Relation and Tree-Structured Data 21
   Yuxing Chen, Valter Uotila, Jiaheng Lu, Zhen Hua Liu,
   and Souripriya Das

Leveraging Search History for Improving Person-Job Fit 38
   Yupeng Hou, Xingyu Pan, Wayne Xin Zhao, Shuqing Bian, Yang Song,
   Tao Zhang, and Ji-Rong Wen

Efficient In-Memory Evaluation of Reachability Graph Pattern Queries
on Data Graphs 55
   Xiaoying Wu, Dimitri Theodoratos, Dimitrios Skoutas, and Michael Lan

Revisiting Approximate Query Processing and Bootstrap Error Estimation
on GPU 72
   Hang Zhao, Hanbing Zhang, Yinan Jing, Kai Zhang, Zhenying He,
   and X Sean Wang

μ-join: Efficient Join with Versioned Dimension Tables 88
   Mika Takata, Kazuo Goda, and Masaru Kitsuregawa

Learning-Based Optimization for Online Approximate Query Processing 96
   Wenyuan Bi, Hanbing Zhang, Yinan Jing, Zhenying He, Kai Zhang,
   and X. Sean Wang

Knowledge Bases

Triple-as-Node Knowledge Graph and Its Embeddings 107
   Xin Lv, Jiaxin Shi, Shulin Cao, Lei Hou, and Juanzi Li

LeKAN: Extracting Long-tail Relations via Layer-Enhanced
Knowledge-Aggregation Networks 122
   Xiaokai Liu, Feng Zhao, Xiangyu Gui, and Hai Jin

TRHyTE: Temporal Knowledge Graph Embedding Based
on Temporal-Relational Hyperplanes 137
  Lin Yuan, Zhixu Li, Jianfeng Qu, Tingyi Zhang, An Liu, Lei Zhao,
  and Zhigang Chen

ExKGR: Explainable Multi-hop Reasoning for Evolving Knowledge Graph 153
  Cheng Yan, Feng Zhao, and Hai Jin

Improving Core Path Reasoning for the Weakly Supervised Knowledge
Base Question Answering 162
  Nan Hu, Sheng Bi, Guilin Qi, Meng Wang, Yuncheng Hua,
  and Shirong Shen

Counterfactual-Guided and Curiosity-Driven Multi-hop Reasoning
over Knowledge Graph 171
  Dan Shi, Anchen Li, and Bo Yang

Visualizable or Non-visualizable? Exploring the Visualizability
of Concepts in Multi-modal Knowledge Graph 180
  Xueyao Jiang, Ailisi Li, Jiaqing Liang, Bang Liu, Rui Xie, Wei Wu,
  Zhixu Li, and Yanghua Xiao

Spatio-Temporal Data

JS-STDGN: A Spatial-Temporal Dynamic Graph Network Using
JS-Graph for Traffic Prediction 191
  Pengfei Li, Junhua Fang, Pingfu Chao, Pengpeng Zhao, An Liu,
  and Lei Zhao

When Multitask Learning Make a Difference: Spatio-Temporal Joint
Prediction for Cellular Trajectories 207
  Yuan Xu, Jiajie Xu, Junhua Fang, An Liu, and Lei Zhao

Efficient Retrieval of Top-k Weighted Spatial Triangles 224
  Ryosuke Taniguchi, Daichi Amagata, and Takahiro Hara

DIOT: Detecting Implicit Obstacles from Trajectories 232
  Yifan Lei, Qiang Huang, Mohan Kankanhalli, and Anthony Tung

Exploring Sub-skeleton Trajectories for Interpretable Recognition of Sign
Language 241
  Joachim Gudmundsson, Martin P. Seybold, and John Pfeifer

Significant Engagement Community Search on Temporal Networks 250
  Yifei Zhang, Longlong Lin, Pingpeng Yuan, and Hai Jin

Influence Computation for Indoor Spatial Objects 259
    Yue Li, Guojie Ma, Shiyu Yang, Liping Wang, and Jiujing Zhang

A Localization System for GPS-free Navigation Scenarios 268
    *Jiazhi Ni, Xin Zhang, Beihong Jin, Fusang Zhang, Xin Li, Qiang Huang,
    Pengsen Wang, Xiang Li, Ning Xiao, Youchen Wang, and Chang Liu*

Systems

HEM: A Hardware-Aware Event Matching Algorithm for Content-Based
Pub/Sub Systems 277
    Wanghua Shi and Shiyou Qian

RotorcRaft: Scalable Follower-Driven Raft on RDMA 293
    Xuecheng Qi, Huiqi Hu, Xing Wei, and Aoying Zhou

Efficient Matrix Computation for SGD-Based Algorithms on Apache Spark 309
    Baokun Han, Zihao Chen, Chen Xu, and Aoying Zhou

Parallel Pivoted Subgraph Filtering with Partial Coding Trees on GPU 325
    Yang Wang, Yu Gu, and Chuanwen Li

TxChain: Scaling Sharded Decentralized Ledger via Chained Transaction
Sequences 333
    Zheng Xu, Rui Jiang, Peng Zhang, Tun Lu, and Ning Gu

Zebra: An Efficient, RDMA-Enabled Distributed Persistent Memory File
System 341
    Jingyu Wang, Shengan Zheng, Ziyi Lin, Yuting Chen, and Linpeng Huang

Data Security

ADAPT: Adversarial Domain Adaptation with Purifier Training
for Cross-Domain Credit Risk Forecasting 353
    *Guanxiong Zeng, Jianfeng Chi, Rui Ma, Jinghua Feng, Xiang Ao,
    and Hao Yang*

Poisoning Attacks on Fair Machine Learning 370
    Minh-Hao Van, Wei Du, Xintao Wu, and Aidong Lu

Bi-Level Selection via Meta Gradient for Graph-Based Fraud Detection 387
    *Linfeng Dong, Yang Liu, Xiang Ao, Jianfeng Chi, Jinghua Feng,
    Hao Yang, and Qing He*

Contrastive Learning for Insider Threat Detection 395
M. S. Vinay, Shuhan Yuan, and Xintao Wu

Metadata Privacy Preservation for Blockchain-Based Healthcare Systems 404
Lixin Liu, Xinyu Li, Man Ho Au, Zhuoya Fan, and Xiaofeng Meng

Blockchain-Based Encrypted Image Storage and Search in Cloud
Computing 413
Yingying Li, Jianfeng Ma, Yinbin Miao, Ximeng Liu, and Qi Jiang

Applications of Algorithms

Improving Information Cascade Modeling by Social Topology and Dual
Role User Dependency 425
Baichuan Liu, Deqing Yang, Yuchen Shi, and Yueyi Wang

Discovering Bursting Patterns over Streaming Graphs 441
Qianzhen Zhang, Deke Guo, and Xiang Zhao

Mining Negative Sequential Rules from Negative Sequential Patterns 459
Chuanhou Sun, Xiaoqi Jiang, Xiangjun Dong, Tiantian Xu, Long Zhao,
Zhao Li, and Yuhai Zhao

CrossIndex: Memory-Friendly and Session-Aware Index for Supporting
Crossfilter in Interactive Data Exploration 476
Tianyu Xia, Hanbing Zhang, Yinan Jing, Zhenying He, Kai Zhang,
and X. Sean Wang

GHStore: A High Performance Global Hash Based Key-Value Store 493
Jiaoyang Li, Yinliang Yue, and Weiping Wang

Hierarchical Bitmap Indexing for Range Queries on Multidimensional
Arrays 509
Luboš Krčál, Shen-Shyang Ho, and Jan Holub

Membership Algorithm for Single-Occurrence Regular Expressions
with Shuffle and Counting 526
Xiaofan Wang

(p, n)-core: Core Decomposition in Signed Networks 543
Junghoon Kim and Sungsu Lim

TROP: Task Ranking Optimization Problem on Crowdsourcing Service
Platform 552
Jiale Zhang, Haozhen Lu, Xiaofeng Gao, Ailun Song, and Guihai Chen

HATree: A Hotness-Aware Tree Index with In-Node Hotspot Cache
for NVM/DRAM-Based Hybrid Memory Architecture 560
 Gaocong Liu, Yongping Luo, and Peiquan Jin

A Novel Null-Invariant Temporal Measure to Discover Partial Periodic
Patterns in Non-uniform Temporal Databases 569
 R. Uday Kiran, Vipul Chhabra, Saideep Chennupati, P. Krishna Reddy,
 Minh-Son Dao, and Koji Zettsu

Utilizing Expert Knowledge and Contextual Information
for Sample-Limited Causal Graph Construction 578
 Xuwu Wang, Xueyao Jiang, Sihang Jiang, Zhixu Li, and Yanghua Xiao

A Two-Phase Approach for Recognizing Tables with Complex Structures 587
 Huichao Li, Lingze Zeng, Weiyu Zhang, Jianing Zhang, Ju Fan,
 and Meihui Zhang

Towards Unification of Statistical Reasoning, OLAP and Association Rule
Mining: Semantics and Pragmatics 596
 Rahul Sharma, Minakshi Kaushik, Sijo Arakkal Peious, Mahtab Shahin,
 Amrendra Singh Yadav, and Dirk Draheim

A Dynamic Heterogeneous Graph Perception Network with Time-Based
Mini-Batch for Information Diffusion Prediction 604
 Wei Fan, Meng Liu, and Yong Liu

Graphs

Cascade Enhanced Graph Convolutional Network for Information
Diffusion Prediction 615
 Ding Wang, Lingwei Wei, Chunyuan Yuan, Yinan Bao, Wei Zhou,
 Xian Zhu, and Songlin Hu

Diversify Search Results Through Graph Attentive Document Interaction 632
 Xianghong Xu, Kai Ouyang, Yin Zheng, Yanxiong Lu, Hai-Tao Zheng,
 and Hong-Gee Kim

On Glocal Explainability of Graph Neural Networks 648
 Ge Lv, Lei Chen, and Caleb Chen Cao

Temporal Network Embedding with Motif Structural Features 665
 Zhi Qiao, Wei Li, and Yunchun Li

Learning Robust Representation Through Graph Adversarial Contrastive
Learning 682
Jiayan Guo, Shangyang Li, Yue Zhao, and Yan Zhang

What Affects the Performance of Models? Sensitivity Analysis
of Knowledge Graph Embedding 698
Han Yang, Leilei Zhang, Fenglong Su, and Jinhui Pang

CollaborateCas: Popularity Prediction of Information Cascades Based
on Collaborative Graph Attention Networks 714
*Xianren Zhang, Jiaxing Shang, Xueqi Jia, Dajiang Liu, Fei Hao,
and Zhiqing Zhang*

Contrastive Disentangled Graph Convolutional Network
for Weakly-Supervised Classification 722
*Xiaokai Chu, Jiashu Zhao, Xinxin Fan, Di Yao, Zhihua Zhu, Lixin Zou,
Dawei Yin, and Jingping Bi*

CSGNN: Improving Graph Neural Networks with Contrastive
Semi-supervised Learning 731
Yumeng Song, Yu Gu, Xiaohua Li, Chuanwen Li, and Ge Yu

IncreGNN: Incremental Graph Neural Network Learning by Considering
Node and Parameter Importance 739
Di Wei, Yu Gu, Yumeng Song, Zhen Song, Fangfang Li, and Ge Yu

Representation Learning in Heterogeneous Information Networks Based
on Hyper Adjacency Matrix 747
Bin Yang and Yitong Wang

Author Index 757

Contents – Part II

Recommendation Systems

MDKE: Multi-level Disentangled Knowledge-Based Embedding
for Recommender Systems 3
   Haolin Zhou, Qingmin Liu, Xiaofeng Gao, and Guihai Chen

M^3-IB: A Memory-Augment Multi-modal Information Bottleneck Model
for Next-Item Recommendation 19
   Yingpeng Du, Hongzhi Liu, and Zhonghai Wu

Fully Utilizing Neighbors for Session-Based Recommendation with Graph
Neural Networks 36
   Xingyu Zhang and Chaofeng Sha

Inter- and Intra-Domain Relation-Aware Heterogeneous Graph
Convolutional Networks for Cross-Domain Recommendation 53
   *Ke Wang, Yanmin Zhu, Haobing Liu, Tianzi Zang, Chunyang Wang,
and Kuan Liu*

Enhancing Graph Convolution Network for Novel Recommendation 69
   Xuan Ma, Tieyun Qian, Yile Liang, Ke Sun, Hang Yun, and Mi Zhang

Knowledge-Enhanced Multi-task Learning for Course Recommendation 85
   Qimin Ban, Wen Wu, Wenxin Hu, Hui Lin, Wei Zheng, and Liang He

Learning Social Influence from Network Structure for Recommender
Systems 102
   Ting Bai, Yanlong Huang, and Bin Wu

PMAR: Multi-aspect Recommendation Based on Psychological Gap 118
   Liye Shi, Wen Wu, Yu Ji, Luping Feng, and Liang He

Meta-path Enhanced Lightweight Graph Neural Network for Social
Recommendation 134
   Hang Miao, Anchen Li, and Bo Yang

Intention Adaptive Graph Neural Network for Category-Aware
Session-Based Recommendation 150
   *Chuan Cui, Qi Shen, Shixuan Zhu, Yitong Pang, Yiming Zhang,
Hanning Gao, and Zhihua Wei*

Multi-view Multi-behavior Contrastive Learning in Recommendation 166
  Yiqing Wu, Ruobing Xie, Yongchun Zhu, Xiang Ao, Xin Chen, Xu Zhang,
  Fuzhen Zhuang, Leyu Lin, and Qing He

Joint Locality Preservation and Adaptive Combination for Graph
Collaborative Filtering 183
  Zhiqiang Guo, Chaoyang Wang, Zhi Li, Jianjun Li, and Guohui Li

Gated Hypergraph Neural Network for Scene-Aware Recommendation 199
  Tianchi Yang, Luhao Zhang, Chuan Shi, Cheng Yang, Siyong Xu,
  Ruiyu Fang, Maodi Hu, Huaijun Liu, Tao Li, and Dong Wang

Hyperbolic Personalized Tag Recommendation 216
  Weibin Zhao, Aoran Zhang, Lin Shang, Yonghong Yu, Li Zhang,
  Can Wang, Jiajun Chen, and Hongzhi Yin

Diffusion-Based Graph Contrastive Learning for Recommendation
with Implicit Feedback 232
  Lingzi Zhang, Yong Liu, Xin Zhou, Chunyan Miao, Guoxin Wang,
  and Haihong Tang

Multi-behavior Recommendation with Two-Level Graph Attentional
Networks 248
  Yunhe Wei, Huifang Ma, Yike Wang, Zhixin Li, and Liang Chang

Collaborative Filtering for Recommendation in Geometric Algebra 256
  Longcan Wu, Daling Wang, Shi Feng, Kaisong Song, Yifei Zhang,
  and Ge Yu

Graph Neural Networks with Dynamic and Static Representations
for Social Recommendation 264
  Junfa Lin, Siyuan Chen, and Jiahai Wang

Toward Paper Recommendation by Jointly Exploiting Diversity
and Dynamics in Heterogeneous Information Networks 272
  Jie Wang, Jinya Zhou, Zhen Wu, and Xigang Sun

Enhancing Session-Based Recommendation with Global Context
Information and Knowledge Graph 281
  Xiaohui Zhang, Huifang Ma, Zihao Gao, Zhixin Li, and Liang Chang

GISDCN: A Graph-Based Interpolation Sequential Recommender
with Deformable Convolutional Network 289
  Yalei Zang, Yi Liu, Weitong Chen, Bohan Li, Aoran Li, Lin Yue,
  and Weihua Ma

Deep Graph Mutual Learning for Cross-domain Recommendation 298
Yifan Wang, Yongkang Li, Shuai Li, Weiping Song, Jiangke Fan,
Shan Gao, Ling Ma, Bing Cheng, Xunliang Cai, Sheng Wang,
and Ming Zhang

Core Interests Focused Self-attention for Sequential Recommendation 306
Zhengyang Ai, Shupeng Wang, Siyu Jia, and Shu Guo

SAER: Sentiment-Opinion Alignment Explainable Recommendation 315
Xiaoning Zong, Yong Liu, Yonghui Xu, Yixin Zhang, Zhiqi Shen,
Yonghua Yang, and Lizhen Cui

Toward Auto-Learning Hyperparameters for Deep Learning-Based
Recommender Systems 323
Bo Sun, Di Wu, Mingsheng Shang, and Yi He

GELibRec: Third-Party Libraries Recommendation Using Graph Neural
Network 332
Chengming Zou and Zhenfeng Fan

Applications of Machine Learning

Hierarchical Attention Factorization Machine for CTR Prediction 343
Lianjie Long, Yunfei Yin, and Faliang Huang

MCRF: Enhancing CTR Prediction Models via Multi-channel Feature
Refinement Framework 359
Fangye Wang, Hansu Gu, Dongsheng Li, Tun Lu, Peng Zhang,
and Ning Gu

CaSS: A Channel-Aware Self-supervised Representation Learning
Framework for Multivariate Time Series Classification 375
Yijiang Chen, Xiangdong Zhou, Zhen Xing, Zhidun Liu, and Minyang Xu

Temporal Knowledge Graph Entity Alignment via Representation Learning 391
Xiuting Song, Luyi Bai, Rongke Liu, and Han Zhang

Similarity-Aware Collaborative Learning for Patient Outcome Prediction 407
Fuqiang Yu, Lizhen Cui, Yiming Cao, Ning Liu, Weiming Huang,
and Yonghui Xu

Semi-supervised Graph Learning with Few Labeled Nodes 423
Cong Zhang, Ting Bai, and Bin Wu

Human Mobility Identification by Deep Behavior Relevant Location
Representation 439
    Tao Sun, Fei Wang, Zhao Zhang, Lin Wu, and Yongjun Xu

Heterogeneous Federated Learning via Grouped Sequential-to-Parallel
Training 455
    Shenglai Zeng, Zonghang Li, Hongfang Yu, Yihong He, Zenglin Xu,
    Dusit Niyato, and Han Yu

Transportation-Mode Aware Travel Time Estimation via Meta-learning 472
    Yu Fan, Jiajie Xu, Rui Zhou, and Chengfei Liu

A Deep Reinforcement Learning Based Dynamic Pricing Algorithm
in Ride-Hailing 489
    Bing Shi, Zhi Cao, and Yikai Luo

Peripheral Instance Augmentation for End-to-End Anomaly Detection
Using Weighted Adversarial Learning 506
    Weixian Zong, Fang Zhou, Martin Pavlovski, and Weining Qian

HieNet: Bidirectional Hierarchy Framework for Automated ICD Coding 523
    Shi Wang, Daniel Tang, Luchen Zhang, Huilin Li, and Ding Han

Efficient Consensus Motif Discovery of All Lengths in Multiple Time Series . . . 540
    Mingming Zhang, Peng Wang, and Wei Wang

LiteWSC: A Lightweight Framework for Web-Scale Spectral Clustering 556
    Geping Yang, Sucheng Deng, Yiyang Yang, Zhiguo Gong, Xiang Chen,
    and Zhifeng Hao

Dual Confidence Learning Network for Open-World Time Series
Classification 574
    Junwei Lv, Ying He, Xuegang Hu, Desheng Cai, Yuqi Chu, and Jun Hu

Port Container Throughput Prediction Based on Variational AutoEncoder 590
    Jingze Li, Shengmin Shi, Tongbing Chen, Yu Tian, Yihua Ding,
    Yiyong Xiao, and Weiwei Sun

Data Source Selection in Federated Learning: A Submodular Optimization
Approach 606
    Ruisheng Zhang, Yansheng Wang, Zimu Zhou, Ziyao Ren, Yongxin Tong,
    and Ke Xu

MetisRL: A Reinforcement Learning Approach for Dynamic Routing
in Data Center Networks 615
   Yuanning Gao, Xiaofeng Gao, and Guihai Chen

CLZT: A Contrastive Learning Based Framework for Zero-Shot Text
Classification 623
   Kun Li, Meng Lin, Songlin Hu, and Ruixuan Li

InDISP: An Interpretable Model for Dynamic Illness Severity Prediction 631
   Xinyu Ma, Meng Wang, Xing Liu, Yifan Yang, Yefeng Zheng, and Sen Wang

Learning Evolving Concepts with Online Class Posterior Probability 639
   *Junming Shao, Kai Wang, Jianyun Lu, Zhili Qin, Qiming Wangyang,
   and Qinli Yang*

Robust Dynamic Pricing in Online Markets with Reinforcement Learning 648
   Bolei Zhang and Fu Xiao

Multi-memory Enhanced Separation Network for Indoor Temperature
Prediction 656
   *Zhewen Duan, Xiuwen Yi, Peng Li, Dekang Qi, Yexin Li, Haorun Xu,
   Yanyong Huang, Junbo Zhang, and Yu Zheng*

An Interpretable Time Series Classification Approach Based on Feature
Clustering 664
   Fan Qiao, Peng Wang, Wei Wang, and Binjie Wang

Generative Adversarial Imitation Learning to Search in Branch-and-Bound
Algorithms 673
   Qi Wang, Suzanne V. Blackley, and Chunlei Tang

A Trace Ratio Maximization Method for Parameter Free Multiple Kernel
Clustering 681
   Yan Chen, Lei Wang, Liang Du, and Lei Duan

Supervised Multi-view Latent Space Learning by Jointly Preserving
Similarities Across Views and Samples 689
   *Xiaoyang Li, Martin Pavlovski, Fang Zhou, Qiwen Dong, Weining Qian,
   and Zoran Obradovic*

Market-Aware Dynamic Person-Job Fit with Hierarchical Reinforcement
Learning 697
   *Bin Fu, Hongzhi Liu, Hui Zhao, Yao Zhu, Yang Song, Tao Zhang,
   and Zhonghai Wu*

TEALED: A Multi-Step Workload Forecasting Approach Using
Time-Sensitive EMD and Auto LSTM Encoder-Decoder 706
   Xiuqi Huang, Yunlong Cheng, Xiaofeng Gao, and Guihai Chen

Author Index 715

Contents – Part III

Text and Image Processing

Emotion-Aware Multimodal Pre-training for Image-Grounded Emotional
Response Generation 3
   Zhiliang Tian, Zhihua Wen, Zhenghao Wu, Yiping Song, Jintao Tang,
   Dongsheng Li, and Nevin L. Zhang

Information Networks Based Multi-semantic Data Embedding for Entity
Resolution 20
   Chenchen Sun, Derong Shen, and Tiezheng Nie

Semantic-Based Data Augmentation for Math Word Problems 36
   Ailisi Li, Yanghua Xiao, Jiaqing Liang, and Yunwen Chen

Empowering Transformer with Hybrid Matching Knowledge for Entity
Matching 52
   Wenzhou Dou, Derong Shen, Tiezheng Nie, Yue Kou, Chenchen Sun,
   Hang Cui, and Ge Yu

Tracking the Evolution: Discovering and Visualizing the Evolution
of Literature 68
   Siyuan Wu and Leong Hou U

Incorporating Commonsense Knowledge into Story Ending Generation
via Heterogeneous Graph Networks 85
   Jiaan Wang, Beiqi Zou, Zhixu Li, Jianfeng Qu, Pengpeng Zhao, An Liu,
   and Lei Zhao

Open-Domain Dialogue Generation Grounded with Dynamic Multi-form
Knowledge Fusion 101
   Feifei Xu, Shanlin Zhou, Yunpu Ma, Xinpeng Wang, Wenkai Zhang,
   and Zhisong Li

KdTNet: Medical Image Report Generation via Knowledge-Driven
Transformer 117
   Yiming Cao, Lizhen Cui, Fuqiang Yu, Lei Zhang, Zhen Li, Ning Liu,
   and Yonghui Xu

Fake Restaurant Review Detection Using Deep Neural Networks
with Hybrid Feature Fusion Method 133
   Yifei Jian, Xingshu Chen, and Haizhou Wang

Aligning Internal Regularity and External Influence of Multi-granularity
for Temporal Knowledge Graph Embedding 149
   Tingyi Zhang, Zhixu Li, Jiaan Wang, Jianfeng Qu, Lin Yuan, An Liu,
   Lei Zhao, and Zhigang Chen

AdCSE: An Adversarial Method for Contrastive Learning of Sentence
Embeddings 165
   Renhao Li, Lei Duan, Guicai Xie, Shan Xiao, and Weipeng Jiang

HRG: A Hybrid Retrieval and Generation Model in Multi-turn Dialogue 181
   Deji Zhao, Xinyi Liu, Bo Ning, and Chengfei Liu

FALCON: A Faithful Contrastive Framework for Response Generation
in TableQA Systems 197
   Shineng Fang, Jiangjie Chen, Xinyao Shen, Yunwen Chen,
   and Yanghua Xiao

Tipster: A Topic-Guided Language Model for Topic-Aware Text
Segmentation 213
   Zheng Gong, Shiwei Tong, Han Wu, Qi Liu, Hanqing Tao, Wei Huang,
   and Runlong Yu

SimEmotion: A Simple Knowledgeable Prompt Tuning Method for Image
Emotion Classification 222
   Sinuo Deng, Ge Shi, Lifang Wu, Lehao Xing, Wenjin Hu, Heng Zhang,
   and Ye Xiang

Predicting Rumor Veracity on Social Media with Graph Structured
Multi-task Learning 230
   Yudong Liu, Xiaoyu Yang, Xi Zhang, Zhihao Tang, Zongyi Chen,
   and Zheng Liwen

Knowing What I Don't Know: A Generation Assisted Rejection
Framework in Knowledge Base Question Answering 238
   Junyang Huang, Xuantao Lu, Jiaqing Liang, Qiaoben Bao,
   Chen Huang, Yanghua Xiao, Bang Liu, and Yunwen Chen

Medical Image Fusion Based on Pixel-Level Nonlocal Self-similarity
Prior and Optimization 247
   Rui Zhu, Xiongfei Li, Yu Wang, and Xiaoli Zhang

Knowledge-Enhanced Interactive Matching Network for Multi-turn
Response Selection in Medical Dialogue Systems 255
   Ying Zhu, Shi Feng, Daling Wang, Yifei Zhang, and Donghong Han

KAAS: A Keyword-Aware Attention Abstractive Summarization Model
for Scientific Articles 263
Shuaimin Li and Jungang Xu

E-Commerce Knowledge Extraction via Multi-modal Machine Reading
Comprehension 272
Chaoyu Bai

PERM: Pre-training Question Embeddings via Relation Map for Improving
Knowledge Tracing 281
Wentao Wang, Huifang Ma, Yan Zhao, Fanyi Yang, and Liang Chang

A Three-Stage Curriculum Learning Framework with Hierarchical Label
Smoothing for Fine-Grained Entity Typing 289
*Bo Xu, Zhengqi Zhang, Chaofeng Sha, Ming Du, Hui Song,
and Hongya Wang*

PromptMNER: Prompt-Based Entity-Related Visual Clue Extraction
and Integration for Multimodal Named Entity Recognition 297
*Xuwu Wang, Junfeng Tian, Min Gui, Zhixu Li, Jiabo Ye, Ming Yan,
and Yanghua Xiao*

TaskSum: Task-Driven Extractive Text Summarization for Long News
Documents Based on Reinforcement Learning 306
*Moming Tang, Dawei Cheng, Cen Chen, Yuqi Liang, Yifeng Luo,
and Weining Qian*

Concurrent Transformer for Spatial-Temporal Graph Modeling 314
*Yi Xie, Yun Xiong, Yangyong Zhu, Philip S. Yu, Cheng Jin, Qiang Wang,
and Haihong Li*

Towards Personalized Review Generation with Gated Multi-source Fusion
Network 322
*Hongtao Liu, Wenjun Wang, Hongyan Xu, Qiyao Peng, Pengfei Jiao,
and Yueheng Sun*

Definition-Augmented Jointly Training Framework for Intention Phrase
Mining 331
*Denghao Ma, Yueguo Chen, Changyu Wang, Hongbin Pei, Yitao Zhai,
Gang Zheng, and Qi Chen*

Modeling Uncertainty in Neural Relation Extraction 340
Yu Hong, Yanghua Xiao, Wei Wang, and Yunwen Chen

Industry Papers

A Joint Framework for Explainable Recommendation with Knowledge
Reasoning and Graph Representation 351
  *Luhao Zhang, Ruiyu Fang, Tianchi Yang, Maodi Hu, Tao Li, Chuan Shi,
  and Dong Wang*

XDM: Improving Sequential Deep Matching with Unclicked User
Behaviors for Recommender System 364
  *Fuyu Lv, Mengxue Li, Tonglei Guo, Changlong Yu, Fei Sun, Taiwei Jin,
  and Wilfred Ng*

Mitigating Popularity Bias in Recommendation via Counterfactual
Inference 377
  Ming He, Changshu Li, Xinlei Hu, Xin Chen, and Jiwen Wang

Efficient Dual-Process Cognitive Recommender Balancing Accuracy
and Diversity 389
  *Yixu Gao, Kun Shao, Zhijian Duan, Zhongyu Wei, Dong Li, Bin Wang,
  Mengchen Zhao, and Jianye Hao*

Learning and Fusing Multiple User Interest Representations for Sequential
Recommendation 401
  Ming He, Tianshuo Han, and Tianyu Ding

Query-Document Topic Mismatch Detection 413
  *Sahil Chelaramani, Ankush Chatterjee, Sonam Damani,
  Kedhar Nath Narahari, Meghana Joshi, Manish Gupta,
  and Puneet Agrawal*

Beyond QA: 'Heuristic QA' Strategies in JIMI 425
  *Shuangyong Song, Bo Zou, Jianghua Lin, Xiaoguang Yu,
  and Xiaodong He*

SQLG+: Efficient *k*-hop Query Processing on RDBMS 430
  *Li Zeng, Jinhua Zhou, Shijun Qin, Haoran Cai, Rongqian Zhao,
  and Xin Chen*

Modeling Long-Range Travelling Times with Big Railway Data 443
  *Wenya Sun, Tobias Grubenmann, Reynold Cheng, Ben Kao,
  and Waiki Ching*

Multi-scale Time Based Stock Appreciation Ranking Prediction via Price
Co-movement Discrimination 455
  *Ruyao Xu, Dawei Cheng, Cen Chen, Siqiang Luo, Yifeng Luo,
  and Weining Qian*

RShield: A Refined Shield for Complex Multi-step Attack Detection
Based on Temporal Graph Network 468
    Weiyong Yang, Peng Gao, Hao Huang, Xingshen Wei, Wei Liu,
    Shishun Zhu, and Wang Luo

Inter-and-Intra Domain Attention Relational Inference for Rack
Temperature Prediction in Data Center 481
    Fang Shen, Zhan Li, Bing Pan, Ziwei Zhang, Jialong Wang,
    Wendy Zhao, Xin Wang, and Wenwu Zhu

DEMO Papers

An Interactive Data Imputation System 495
    Yangyang Wu, Xiaoye Miao, Yuchen Peng, Lu Chen, Yunjun Gao,
    and Jianwei Yin

FoodChain: A Food Delivery Platform Based on Blockchain for Keeping
Data Privacy 500
    Rodrigo Folha, Valéria Times, Arthur Carvalho, André Araújo,
    Henrique Couto, and Flaviano Viana

A Scalable Lightweight RDF Knowledge Retrieval System 505
    Yuming Lin, Chuangxin Fang, Youjia Jiang, and You Li

CO-AutoML: An Optimizable Automated Machine Learning System 509
    Chunnan Wang, Hongzhi Wang, Bo Xu, Xintong Song, Xiangyu Shi,
    Yuhao Bao, and Bo Zheng

OIIKM: A System for Discovering Implied Knowledge from Spatial
Datasets Using Ontology 514
    Liang Chang, Long Wang, Xuguang Bao, and Tianlong Gu

IDMBS: An Interactive System to Find Interesting Co-location Patterns
Using SVM 518
    Liang Chang, Yuxiang Zhang, Xuguang Bao, and Tianlong Gu

SeTS^3: A Secure Trajectory Similarity Search System 522
    Yiping Teng, Fanyou Zhao, Jiayv Liu, Mengfan Zhang, Jihang Duan,
    and Zhan Shi

Data-Based Insights for the Masses: Scaling Natural Language Querying
to Middleware Data 527
    Kausik Lakkaraju, Vinamra Palaiya, Sai Teja Paladi,
    Chinmayi Appajigowda, Biplav Srivastava, and Lokesh Johri

Identifying Relevant Sentences for Travel Blogs from Wikipedia Articles 532
Arnav Kapoor and Manish Gupta

PhD Constorium

Neuro-Symbolic XAI: Application to Drug Repurposing for Rare Diseases 539
Martin Drancé

Leveraging Non-negative Matrix Factorization for Document
Summarization 544
Alka Khurana

Author Index 549

Database Queries

Approximate Continuous Top-K Queries over Memory Limitation-Based Streaming Data

Rui Zhu[1]([✉]), Liu Meng[1], Bin Wang[2], Xiaochun Yang[2], and Xiufeng Xia[1]

[1] School of Computer Science, Shenyang Aerospace University, Shenyang, China
{zhurui,xiufengxia}@mail.sau.edu.cn
[2] College of Computer Science and Engineering, Northeastern University,
Shenyang, China
{binwang,yangxc}@mail.neu.edu.cn

Abstract. Continuous top-k query over sliding window is a fundamental problem over data stream. It retrieves k objects with the highest scores when the window slides. Existing efforts include exact-based algorithms and approximate-based algorithms. Their common idea is maintaining a small subset of objects in the window. When the window slides, query results could be found from this set as much as possible. However, the space cost of all existing efforts is high, i.e., linear to the scale of objects in the window, cannot work under memory limitation-based streaming data, i.e., a general environment in real applications.

In this paper, we define a novel query named $\rho-$approximate continuous top-k query, which returns error-bounded answers to the system. Here, ρ is a threshold, used for bounding the score ratio between approximate and exact results. In order to support $\rho-$approximate continuous top-k query, we propose a novel framework named $\rho-$TOPK. It can self-adaptively adjust ρ based on the distribution of streaming data, and achieve the goal of supporting $\rho-$approximate continuous top-k query over memory limitation-based streaming data. Theoretical analysis indicates that even in the worsst case, both running cost and space cost of $\rho-$TOPK are all unrelated with data scale.

Keywords: Data stream · Continuous top-k query · Approximate · Memory limitation

1 Introduction

Continuous top-k query over sliding window is a fundamental problem in the domain of streaming data management, yet has been deeply studied over 15 years [1]. Formally, a continuous top-k query q, expressed by the tuple $q\langle n, F, k, s\rangle$, monitors the window W, which returns k objects with the highest scores to the system whenever the window slides. The query window can be

A. Bhattacharya et al. (Eds.): DASFAA 2022, LNCS 13245, pp. 3–20, 2022.
https://doi.org/10.1007/978-3-031-00123-9_1

time- or count-based [2–5]. For simplicity, in this paper, we only focus on count-based windows. However, our techniques also can be applied to answer top-k queries over time-based sliding window. Under this setting, n refers to the number of objects contained in the window, s refers to the number of objects arriving in/expiring from the window whenever the window slides. In other words, a continuous top-k query q monitors a window containing n objects. Whenever the window slides, s object arrive in the window, and another s objects expire from the window, q returns k objects $\{o_1, o_2, \cdots, o_k\}$ from the query window that have the highest scores $\{F(o_1), F(o_2), \cdots, F(o_k)\}$.

Based on whether *exact* query results are required, the state-of-the-art efforts can be divided into *exact*-based and *approximate*-based algorithms [6–9]. Their common idea is selecting a set of objects in the window as candidates, incrementally maintaining them. When the window slides, query results could be found from these candidates as much as possible. However, as discussed in Sect. 2, the space cost of all exact algorithms is high, i.e., $\mathcal{O}(n)$ in the worst cases. Some approximate algorithms are proposed for reducing space cost [10]. They are allowed to return deviation-bounded results to the system via introducing a user defined "error threshold". Compared with exact algorithms, they can use fewer candidates for supporting approximate top-k search. However, such threshold is usually set in advance, which is *inflexible* and cannot be self-adaptively adjusted based on the changing of streaming data distribution. Obviously, if the threshold is large, they cannot provide users with high quality query results. If the threshold is small, the space cost of such algorithms is still high. Therefore, it is difficult for them to use as low as space cost to support query processing in the premising of providing users with high quality query results. Note, in real applications, memory resource of streaming data management system is usually limited. Using linear space cost for supporting a signal query is usually unacceptable especially under memory limitation-based streaming data, a general environment in real applications.

For example, continuous top-k search could be applied in stock recommendation system, which can recommend stocks for users via monitoring the 10 most significant transactions within the last one hour. However, the importance of transactions could be evaluated based on different score functions defined by different users, the system has to submit thousands of continuous queries with different score functions so as to satisfy requirements of different users. Therefore, the system has to allocate limited memory for each query, and it is unacceptable for an algorithm to use high space cost for supporting one signal query. As another example, continuous top-k search also could be applied in road monitoring system. A RSU (short for Road Side Unit) timely takes photo for vehicles passing it, and then analyses content of photos via machine learning, reports 5 vehicles with the highest probability that may violate traffic regulations within the last 10 min to the system. As memory size of each RSU is limited, size of each photo is usually large, each RSU only can store a small number of photos other than maintaining a large number of photos. Therefore, existing efforts cannot effectively work under such environment.

Table 1. Frequent notations

$F(o)$	The score of the object o
$T(o)$	The arrived order of o
$o.d$	The dominate number of o
C	The candidate set
M	The maximal number of objects the system can maintain
$|C|$	The number of objects in C
SK_ρ	The $\rho-$skyband set
C_ρ	The set of objects in the $\rho-$MSET

In order to solve the above problems, in this paper, we propose a novel framework named $\rho-$TOPK to support approximate continuous top-k search over memory limitation-based streaming data with ρ being an error threshold ranging from 0 to 1. Let M be the maximal number of objects the system can maintain, C be the candidate set maintained by $\rho-$TOPK. When $|C| < M$, we use exact algorithm to support query processing. Here, $|C|$ refers to the number of candidates in C. Otherwise, we compute a suitable ρ based on objects in C, and then form the set $\rho-$MSET C_ρ (short for $\rho-$Minimum-SET),i.e., a subset of C with minimal scale based on ρ, for supporting approximate query processing. Let $\{o_1, o_2, \cdots, o_k\}$ be exact query results, $\{a_1, a_2, \cdots, a_k\}$ be approximate query results. $\forall i$, $\frac{F(a_i)}{F(o_i)} \geq \rho$ should be guaranteed (Table 1).

Challenges. In order to make $\rho-$TOPK effectively work, we should overcome the above challenges. Firstly, as distribution of streaming data is timely changed, it is difficult to find a suitable ρ. Secondly, even ρ could be found, it is also difficult to form, as small as possible, a set to support approximate top-k search. To deal with the above challenges, the contributions of this paper are as follows.

- (i) We propose a novel query named ρ-approximate continuous top-k query. Here, ρ is a threshold. We use it for bounding the score ratio between exact and approximate results;
- (ii) We propose a baseline algorithm and an optimization algorithm for supporting $\rho-$selection. The key of $\rho-$selection is partitioning objects in C into a group of partitions, and then evaluating ρ based on score/arrived order relationship among objects in these partitions. We can guarantee that the running and space cost of $\rho-$selection is bounded by $\mathcal{O}(M \log k)$ and $\mathcal{O}(\log M)$ respectively.
- (iii) We propose the $\rho-$MSET construction algorithm, where we use a *min-heap* for evaluating which objects could be selected as elements of $\rho-$MSET. Theoretical analysis shows that the running and space cost of forming C_ρ are bounded by $\mathcal{O}(M \log k)$ and $\mathcal{O}(\log M)$ respectively. Overall, the incremental maintenance cost and space cost of $\rho-$TOPK are bounded by $\mathcal{O}(\log k + \frac{M}{s})$ and $\mathcal{O}(M)$ respectively.

The rest of this paper is as follows. Section 2 reviews the related work and proposes the problem definition. Section 3 explains the framework $\rho-$TOPK. Section 4 evaluates the performance of $\rho-$TOPK. Section 5 concludes this paper.

2   Preliminary

In this section, we first review some important existing results about continuous top-k queries over sliding window. Next, we introduce the problem definition. Lastly, we will explain the algorithm S-Merge, an exact algorithm that can support continuous top-k query over data stream.

2.1   Related Works

Continuous top-k query and its variant over the sliding window is an important query in the domain of streaming data management, which has been well studied over 15 years. Based on whether approximate query results are allowed, existing algorithms could be divided into two types: *exact algorithms* and *approximate algorithms*.

Exact Algorithms. could be further divided into *multi-pass* based algorithms and *one-pass* based algorithms. *one-pass* based algorithms use domination relationship among objects to support query processing. Specially, given two objects o and o' in the window, if o arrives *no earlier* than o', and $F(o) > F(o')$, we say o dominates o'. If o' is dominated by less than k objects, we regard o' as a k-skyband object. Otherwise, o' is regarded as a non-$k-$skyband object [7]. As k objects with scores higher than $F(o')$ arrive no earlier than o', o' cannot become a query result object before it expires from the window. Therefore, maintaining all k-skyband objects in the window is enough to support query processing. However, as shown in [1], its performance is sensitive to both the window size n and data distribution. In the worst case, its space cost is $\mathcal{O}(n)$. The algorithm minTopK improves k-skyband-based algorithm via using a natural property of sidling window, i.e., s. When $s \gg k$, since a set of s objects flow into (or expire from) the window at the same time, only k objects with the highest scores among them have chance to become query results. In other words, the larger the s, the lower the space cost of minTopK [3]. However, when s approaches to k, the space cost of minTopK is still high, i.e., also $\mathcal{O}(n)$ in the worst cases. SMA [7] and SAP [1] are two classical *multi-pass* based algorithms. As stated in [1], they should restore all objects in the window with space cost $\mathcal{O}(n)$. Therefore, all exact algorithms cannot be applied for supporting continuous top-k search over memory limitation-based streaming data.

Approximate Algorithms. Zhu et al. [8] proposed an efficient framework, named PABF, to support approximate continuous top-k query over sliding window. Compared with exact algorithms, a $k-$skyband object o could be discarded if: (i)existing k objects have scores no smaller than $F(o)+\varepsilon$ with ε being an error threshold defined by users; (ii)the probability of o becoming a query result object in the future is smaller than another threshold δ. Accordingly, it only maintains a subset of $k-$skyband objects, and can return deviation-bounded results to the system. However, a serious issue of PABF is both ε and δ are set in advance. In real applications, it is difficult to find a suitable ε(and δ) especially when the distribution of streaming data is timely changed. KRESIMIR [10] et al. proposed a probabilistic k-skyband based algorithm. Compared with maintaining all k-skyband objects, it evaluates the probability of each newly arrived object o becoming a query result object. If the probability is smaller than a threshold, o could be deleted directly. Similar with that of PABF, it is difficult to find a suitable probabilistic threshold.

Discussion. In summary, the space cost of *exact* based algorithms is high, which all cannot work under memory limited environment. *Approximate*-based algorithms usually cannot find a suitable error threshold, leading that: (i) their corresponding space cost is still high; or (ii) quality of query results is low. Therefore, an approximate algorithm that can effectively work under memory limitation-based environment in the premise of returning high quality approximate results is desired.

2.2   Problem Definition

A continuous top-k query q, expressed by the tuple $q\langle n, F, k, s\rangle$, monitors the window W, which returns k objects with the highest scores to the system whenever the window slides. A unique property of s is naturally partitioning the window W into a group of partitions. Take an example in Fig. 1(a). The current window W_0 consists of 15 objects, which is partitioned into $\{s_0, s_2, \cdots, s_4\}$, i.e., $s_0 = \{67, 42, 69\}$, $s_1 = \{94, 76, 57\}$, and etc. Query results are $\{94, 92\}$. When the window slides to W_1, objects in s_0 expire from the window, and objects in the new partition s_5 flow into the window. As no object with score larger than 92 flow into the window, query results are still $\{94, 92\}$. In the following, we will formally explain the ρ-approximate continuous top-k query.

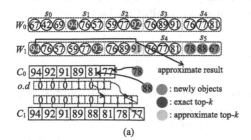

ρ	SK_ρ	C_ρ
0.80	{88,78}	{88,78}
0.90	{88,78}	{91,88,78}
0.95	{88,78}	{**94**,91,88,78}
0.99	{94,92,91,89,88,81,78}	{94,92,91,89,88,81,78}

(a)   (b)

Fig. 1. $k = 2$, $s = 3$, $M = 6$, $\rho = 0.95$

Definition 1. ρ-**Approximate Continuous Top-**k **Query.** *A ρ-approximate continuous top-k query, denoted as $\langle \rho, n, s, k, F \rangle$, monitors the window. When the window slides, q returns k objects $\{a_1, a_2, \cdots, a_k\}$ to the system. $\forall i$, it is satisfied that $\frac{F(a_i)}{F(o_i)} \geq \rho$.*

Back to the example in Fig. 1(a). Let ρ be 0.95. Under W_0, any object with score no smaller than 94*0.95 and 92*0.95 could be used as the top-1 and top-2 approximate query results respectively. For example, $\{92, 91\}$ could be used as approximate top-2 results under W_0 and W_1. Note, approximate top-2 result set is not unique. For example, $\{92, 88\}$ also could be regarded as approximate top-2 results under W_1.

2.3   The Algorithm S-Merge

This section explains an efficiently continuous top-k algorithm over data stream named S-Merge [1], which is a sub-step of our proposed algorithm in Sect. 3.2. Let $W\{s_0, s_1, \cdots, s_{m-1}\}$ be the set of objects in the window, SK be the k-skyband object set in the window, and s_m be the set of newly arrived objects. As s objects flow into the window at the same time, we scan objects in s_m one by one, select k objects with the highest scores as k-skyband objects, sort them based on their scores, and then initialize their dominate numbers. The others could be deleted directly. For example, if an object $o \in s_m$ has the i-th highest score $(i < k)$, it is dominated by $i - 1$ objects in s_m. Its dominate number is set to $i - 1$, i.e., $o.d = i - 1$. Next, we merge these k objects into C. As objects in both C and SK_m are sorted in ascending order by their scores, *merge sort* could be applied. During the merging, we update dominate number of objects in C. In Fig. 1(a). When the window slides from W_0 to W_1, $\{78, 88, 67\}$ flow into the window. As $k=2$, $\{78, 88\}$ are k-skyband objects. The dominate number of 78 and 88 are set to 1 and 0 respectively. Next, we merge $\{78, 88\}$ into C. During the merge, as $78 > 77$, the dominate number of 77 adds 2, i.e., it must be dominated by $\{78, 88\}$.

3   The Framework ρ−TOPK

In this section, we propose a novel framework named ρ−TOPK for supporting approximate continuous top-k query over memory limitation-based streaming data.

3.1   The ρ-MSET

Definition 2. ρ-**dominate.** *Given any two object o and o' in the window W, if $T(o) \geq T(o')$ and $F(o) \geq \rho F(o')$, we say o ρ−dominates o' with ρ satisfying $0 < \rho \leq 1$. Here, $T(o)$ refers to the arrived order of o.*

Definition 3. k_ρ-**skyband.** *Let o be an object in the window W. If o is ρ−dominated by less than k objects, we call o as a k_ρ-skyband object.*

Take an example in Fig. 1(a)–(b). Let ρ be 0.95. As $T(94) < T(92)$ and $94 * 0.95 < 92$, we say 92 can ρ-dominate 94. Furthermore, as 94 is ρ–dominated by $\{92, 91\}$, it is a non-k_ρ-skyband object. 92 is ρ–dominated by $\{91, 89\}$, which is also a non-k_ρ-skyband object. All k_ρ-skyband objects in W_1 from the k_ρ-skyband set SK_ρ, which is $\{88, 78\}$. Note, compared with k–skyband object set that can support exact continuous top-k search, we cannot use SK_ρ to support approximate query processing. The reason behind it is, for each non-k_ρ-skyband object o, objects that can ρ–dominate it also may be ρ–dominated by another k objects, these objects are still not contained in SK_ρ. Thus, sometimes, we cannot find enough high quality objects from SK_ρ for supporting approximate top-k search.

Back to the example in Fig. 1(a)–(b). After the window slides to W_1, the top-2 objects in SK_ρ turn to $\{88, 78\}$, and the exact top-2 objects in the window are $\{94, 92\}$. As $\frac{88}{94} < 0.95$ and $\frac{78}{92} < 0.95$, $\{88, 78\}$ could not be used as an approximate query result. The reason behind it is 94 is ρ–dominated by $\{92, 91\}$, 92 is ρ–dominated by $\{91, 89\}$, both $\{94, 92\}$ are out of SK_ρ. Also, $\{91, 92\}$ are out of SK_ρ due to the reason that they are ρ–dominated by 2 objects. Thus, we cannot find enough high quality objects from SK_ρ for supporting approximate top-k search. Based on the above observation, we propose the concept of ρ-MSET (short for ρ–Minimum-SET). Our goal is to form, as small as, an object set to support approximate top-k search. Here, O_W refers to the set of objects in the window.

Definition 4. ρ-MSET. *The ρ-MSET C_ρ is a subset of objects in the window W satisfying that:*

- *(i) $\forall o \in C_\rho$, it is ρ-dominated by less than k objects in C_ρ;*
- *(ii) $\forall o' \in O_W - C_\rho$, at least k objects in C_ρ ρ–dominate o'.*

According to Definition 4, SK_ρ is a subset of C_ρ, some non k_ρ–skyband objects are contained in C_ρ. We can guarantee that, $\forall o \notin C_\rho$, at least k objects in C_ρ ρ-dominate it. Besides, $\forall o' \in C_\rho$, it is ρ–dominated by less than another k objects in C_ρ. Therefore, C_ρ could be regarded as the set with minimal scale that can support approximate top-k search.

Discussion. It is significant to find a suitable ρ based on both k–skyband scale and maximal number of objects the system can maintain, i.e., M. Figure 1(b) shows the ρ-MSET under different ρ with M being 6. If $\rho = 0.99$, as $|C_{0.99}| = 7$, $C_{0.99}$ cannot be used as the error threshold. Here, $C_{0.99}$ corresponds to C_ρ with ρ being 0.99. If $\rho = 0.95$, $|C_{0.95}| = 4 < M$. It can be used as the threshold. Also, the quality of query results is relatively high. If $\rho = 0.90$, $|C_{0.90}|$ is reduced, but the quality of query results also turns to low. If $\rho = 0.80$, $|C_{0.8}| \ll 6$, but it cannot provide users with high quality query results any longer.

3.2   The Incremental Maintenance Algorithms

The Solution Overview. Let $W\langle n, s\rangle$ be the query window, M be the maximal number of objects that the system can maintain, and C be the candidate set. In order to make the quality of query results as high as possible, we use the algorithm S-Merge to support *exact* top-k search when $|C| \leq M$ (line 1–2 in algorithm 1). At the moment $|C|$ achieves to M, we first find a suitable ρ (line 4–10), and then construct ρ–MSET C_ρ (line 11–16), use C_ρ for substituting C. From then on, we repeatedly use S-Merge for supporting query processing. Along with the window slides, when $|C|$ achieves to M again, we re-invoke the ρ–MSET algorithm, and then re-form C. In the following, we first discuss the ρ selection.

Algorithm 1: The Baseline Algorithm

Input: The Query Window W, the Maximal Memory M, a set of s newly arrived objects s_m

Output: The Updated Result set R_0, the updated candidate set C

1 $C \leftarrow$ S-merge(s_m);
2 **if** $|C| \leq M$ **then**
3 $\quad$ return C;
4 $C \leftarrow$ SORT(C);
5 $\mathcal{P} \leftarrow$ formPartition(C);
6 $H \leftarrow \emptyset$;
7 **for** i *from m to 1* **do**
8 $\quad$ $H \leftarrow$ formMinHeap(H, P_{2i});
9 $\quad$ **for** j *from k to 1* **do**
10 $\quad\quad$ $o_{2i-1}^j.\rho \leftarrow \frac{F(H_{min})}{F(o_{2i-1}^j)}$;

11 $\rho \leftarrow$ computeMedian(C);
12 $H \leftarrow$ formMinHeap(P_{2m});
13 **for** i *from $|C| - k$ to 1* **do**
14 $\quad$ **if** $\rho F(o_i) > F(H_{min})$ **then**
15 $\quad\quad$ $C_\rho \leftarrow C_\rho \cup o_i$;
16 $\quad\quad$ $H \leftarrow$ updateHeap(H, H_{min}, o_i);

17 $C \leftarrow C_\rho$;
18 return C;

The ρ Selection Algorithm. Let C be the set of objects sorted in ascending order by their arrived order. We partition objects in C into $\{P_1, P_2, \cdots, P_{2m}\}$ satisfying:

- (i) $\forall o_i \in P_i, o_j \in P_j$, if $i < j$, $T(o_i) \leq T(o_j)$;
- (ii) $|P_i| = k$ with $|P_i|$ being the number of objects in P_i;

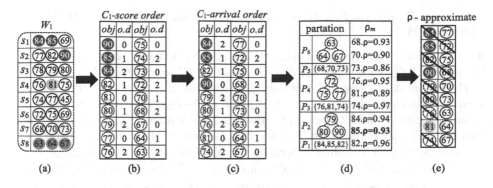

Fig. 2. Running example of $\rho-$MSET construction

Here, we assume that $\frac{|C|}{2k}$ is an integer. However, our proposed techniques also could be efficiently work under other cases. We call partitions $\{P_1, P_3, \cdots, P_{2m-1}\}$ as *s-partitions*(short for substituted partitions), call $\{P_2, P_4, \cdots, P_{2m}\}$ as *e-partitions* (short for evaluation partitions). Our goal is to compute ρ via comparing score relationship among objects in *s-partitions* and *e-partitions*. In the initialization phase, we first associate each object o in *s-partitions* with a variable, denoted as $o.\rho$ with initial value 1. Next, we form a min-heap H based on objects' scores in the last partition P_{2m}. After initialization, we reversely scan objects in each partition, update H based on objects located at *e-partitions*, compute $o.\rho$ corresponding to each object o in *s-partitions*.

Specially, $\forall o_{2m-1}^i \in P_{2m-1}(i \in [1, k])$, $o_{2m-1}^i.\rho$ is set to $\frac{F(H_{min})}{F(o_{2m-1}^i)}$. The intuition behind it is if ρ is set to $o_{2m-1}^i.\rho$ and all elements in H are contained in C_ρ, there are at least k objects in C_ρ $\rho-$dominating o_{2m-1}^i, and we can safely exclude o_{2m-1}^i from C_ρ. Here, o_{2m-1}^i refers to the i-th object in the partition P_{2m-1}. After scanning objects in P_{2m-1}, we reversely scan objects in P_{2m-2}. As P_{2m-2} is an e-partition, for each object in it, if its score is higher than $F(H_{min})$, we insert it into H, and then remove H_{min} from H. After scanning objects in P_{2m-2}, we reversely scan objects in P_{2m-3}, $\forall o_{2m-3}^i \in P_{2m-3}(i \in [1, k])$, $o_{2m-3}^i.\rho$ is set to $\frac{F(H_{min})}{F(o_{2m-3}^i)}$. From then on, we repeat the above operations until all objects in C are accessed. Lastly, we set ρ to the median of $\{o_1^1.\rho, o_1^2.\rho, \cdots, o_{2m-1}^k.\rho\}$.

Figure 2 shows a running example of $\rho-$MSET construction with $n = 24$, $k = 3$, $M = 18$ and $s = 3$. When the window slides to W_1, $\{63, 64, 67\}$ are merged into C. Here, Fig. 2(b)-(c) shows $k-$skyband objects in W_1 sorted by their scores and arrived orders respectively. As $|C|$ achieves to $M(=18)$, we should compute ρ. We first sort objects in C based on their arrived order, and then form partitions $\{P_1, P_2, \cdots, P_6\}$. Among them, s-partitions are $\{P_1, P_3, P_5\}$. The remaining partitions are e-partitions. After partitioning, we initialize H based on P_6. Next, we scan objects P_5, compute $o.\rho$ of each object $o \in P_5$, i.e., $68.\rho = 0.93$, $70.\rho = 0.90$, $73.\rho = 0.86$. From then on, we update H based on objects in P_4, compute $o.\rho$ of each object $o \in P_3$, update H based on objects in P_2, compute $o.\rho$ of each object $o \in P_1$ in turn. Finally, ρ is set to the median of $o.\rho$ of these 9 objects, which is 0.93.

The ρ−MSET Construction Algorithm. Once ρ is found, we now construct the ρ−MSET C_ρ. Firstly, we still form a min-heap H' to maintain objects in P_{2m} at the beginning. These k objects are also inserted into C_ρ. Next, we reversely scan the remaining objects. For each object $o \in C - P_{2m}$, if $\rho F(o) > F(H_{min})$, it means we cannot find k objects in C_ρ that can ρ−dominate o. Thus, o is inserted into both C_ρ and H respectively. Also, H_{min} is removed from H. The algorithm is terminated after scanning objects in C. As is discussed in Theorem 1, it is guaranteed that $\frac{|C_\rho|}{|C|} \le \frac{3}{4}$. Lastly, C is updated to C_ρ. As shown in Fig. 2(e), we reversely scan objects in these 6 partitions, and then select objects for forming C_ρ. For example, when processing P_5, as $63 < 73 * 0.93$, 73 is inserted into both $C_{0.93}$ and H, $F(H_{min})$ is updated to 64. Also, as $64 < 70 * 0.93$, 70 is inserted into $C_{0.93}$ and H, where H is updated to $\{67, 70, 73\}$, $F(H_{min})$ is updated to 68. As $67 > 68 * 0.93$, it is out of $C_{0.93}$. Lastly, 10 objects are selected as elements of C_ρ.

Theorem 1. *Let $\{o_1, o_2, \ldots, o_r\}$ be a set of objects in $\cup_{i=1}^{i=m} P_{2i-1}$, $\mathcal{RH}\{o_1^1.\rho,$ $o_1^2.\rho, \cdots, o_{2m-1}^k.\rho\}$ be $o.\rho$ of objects located at s-partitions. If ρ is set to the median of elements in $\mathcal{RH}$, $|C_\rho|$ is bounded by $\frac{3|C|}{4}$.*

Proof. Let o be an object contained in a s-partition $P_{2i-1}(1 \le i \le m)$, $F(H_{min})$ be the $k - th$ highest score among objects contained in C_ρ with arrived order no smaller than o, and $F(H'_{min})$ be the $k - th$ highest score among objects contained in e-partitions with partition ID larger than $2i - 1$. We first form another set C'_ρ. It contains: (i) all objects in e-partitions; (ii) each object o in s-partition $P_{2i-1}(i \in [1, m])$ with score larger than $\frac{F(H'_{min})}{\rho}$. Obviously, if ρ is set to the median of elements in $\mathcal{RH}$, half of objects located at s-partitions are excluded from C'_ρ, and $\frac{|C'_\rho|}{|C|} \le \frac{3}{4}$ is guaranteed. Compared with C'_ρ, as $F(H'_{min}) \le F(H_{min})$, more objects located at s-partitions could be excluded from C_ρ based on $F(H_{min})$. In addition, the ρ−MSET construction algorithm allows objects in e-partitions be excluded from C_ρ. Therefore, $|C'_\rho| \ge |C_\rho|$, and $\frac{|C_\rho|}{|C|} \le \frac{3}{4}$ is guaranteed.

3.3   The Optimization Incremental Maintenance Algorithms

In this section, we improve algorithms discussed in Sect. 3.2 as follows. Firstly, we propose a novel algorithm named DSORT(short for Domination-based SORT) to sort objects in C in descending order by their arrived order. Compared with using traditional sort algorithms, DSORT fully utilizes domination relationship among objects to speed up the sort. Secondly, recalling Sect. 3.2, we compute $o.\rho$ for each object o located at *s-partitions*. The corresponding space cost is $\mathcal{O}(|C|)$. It is obviously unacceptable under memory limitation-based environment. In this section, we propose a novel algorithm named QGRS(short for Group Queue-based Rho Selection), which could use $\mathcal{O}(\log M)$ for ρ-selection.

The Algorithm DSORT. It is based on Theorem 2. Let C be the set of candidates sorted in descending order by their scores(before sorting). We first construct k empty sets $\{S_0, S_1, \cdots, S_{k-1}\}$. Next, we reversely scan objects in C, insert objects in C into these k sets based on their dominate number,i.e., an object o with dominate number $i(o.d = i)$ is inserted into S_i.

Theorem 2. *Let o and o' be two objects contained in the window W. If $o.d = o'.d$ and $F(o) < F(o')$, o must arrive later than that of o'.*

Proof. We first explain a useful fact: let o_1, o_2 and o_3 be 3 objects. If o_1 dominates o_2, and o_2 dominates o_3, o_1 must dominate o_3. We go on proving this theorem. Let $\{o_1, o_2, \cdots, o_d\}$ be the set of objects that dominate o, and $\{o'_1, o'_2, \cdots, o'_d\}$ be the set of objects that dominate o', i.e., $o.d = o'.d$. Theorem 1 can be proved by contradiction, where we assume that o arrives earlier than o'. As $F(o) < F(o')$, if o arrives no later than o', o mufst be dominated by o'. As o' is dominated by d objects, and these objects are also dominating o, the dominate number of o must be larger than d. Thus, our assumption is invalid and hence o must arrive later than that of o'.

Let S_{iu}, S_{iv} be two elements contained in the set S_i, with S_{iu} (or S_{iv}) being the u-th(or v-th) inserted element. As we *reversely* scan objects in C, and objects in C are sorted in descending order by their score before sorting, if $u < v$, both $T(S_{iu}) > T(S_{iv})$ and $F(S_{iu}) < F(S_{iv})$ (see Theorem 2) are guaranteed. In other words, after $\{S_0, S_1, \cdots, S_{k-1}\}$ are formed, objects in S_i must be sorted in *descending order* by their scores(also sorted in *ascending order* by their arrived order). Based on the above properties, we apply *merge sort* for speeding up the sorting.

Specially, given $\{S_0, S_1, \cdots, S_{k-1}\}$, we construct a min-heap SH to organize these k sets based on their scale,i.e., top of SH is the set with minimal scale. Then we select two sets,i.e., S_i and S_j, with minimal size, merge S_j into S_i via *merge sort*, remove S_j from SH. From then on, we repeat the above operations until all sets are merged into one set S. At that moment, objects in S must be sorted in ascending order by their arrived order. Back to the example in Fig. 2(b)–(c). We first construct three sets S_0, S_1 and S_2. Next, we reversely scan objects in C_1 with descending order by their score, insert them into the right set based on their dominate number, i.e.,$S_0 = \{90, 81, 77, 75, 73, 67\}$, $S_1 = \{85, 82, 80, 70, 64\}$ and $S_2 = \{84, 79, 76, 74, 72, 68, 63\}$. As $|S_1| < |S_0| < |S_2|$, we first merge S_1 into S_0. Next, we merge S_2 into S_0. At the same time, objects in S_0 must be sorted by their arrival order.

The Optimization ρ Selection. We first apply DSORT for sorting objects in the candidate set C based on their arrived order. Next, we still partition objects in C into a group of partitions $\{P_1, P_2, \cdots, P_{2m}\}$, where $\{P_1, P_3, \cdots, P_{2m-1}\}$ and $\{P_2, P_4, \cdots, P_{2m}\}$ are also regarded as *s-partitions* and *e-partitions* respectively. From then on, we reversely scan objects in these $2m$ partitions, use different strategies to access objects contained in them. Formally, if we meet an *e-partition* P_i,i.e., $i = \{2m, 2m-2, 2m-4, \cdots, 2\}$, we update H as the manner discussed before. Otherwise (meet a *s-partition* P_j), we compute $o.\rho$ of each object $o \in P_j$, and then input it into the function QGRS (short for Group Queue-based Rho Selection). Here, GQRS is used for computing ρ based on $o.\rho$ of each object o located at *s-partitions*. Note, compared with the algorithm discussed in Sect. 3.2, we do not retain $o.\rho$ for o. By contrast, once $o.\rho$ is computed, we input it into the function QGRS, and then discard it at once. When all objects in C are scanned, we compute ρ.

We now explain the algorithm GQRS. Let C be the candidate set. We first initialize a group of u empty queues $\{Q_1, Q_2, \cdots, Q_u\}$ with u being $\lceil \log \frac{|C|}{2} \rceil$. Maximal size of each queue is set to 3. In implementation, these queues are regarded as *static variables*. In other words, they are always maintained until all objects in C are processed, and it is unnecessary to repeatedly initialize them whenever GQRS is invoked. From then on, the algorithm is run based on whether these queues are full.

Algorithm 2: The Algorithm GQRS

Input: DOUBLE ρ_o, INT state
Output: The parameter ρ
1   Queue $\{Q_1, Q_2, \cdots, Q_u\}$;
2   **if** *state=FINISH* **then**
3      $\rho \leftarrow$ getMedian(q_u);
4   return ρ;
5   $q_0 \leftarrow q_0 \cup \rho_o$, Int $i \leftarrow 0$;
6   **while** q_i *is full* **do**
7   DOUBLE $\rho_m \leftarrow$ getMedian(q_i);
8      $q_{i+1} \leftarrow q_{i+1} \cup \rho_m$, $q_i \leftarrow \emptyset$, $i \leftarrow i+1$;
9   return 0;

Specially, we first insert $o.\rho$ into Q_1. If Q_1 is not full, the algorithm is terminated. Otherwise, we compute the median, i.e., denoted as ρ_m, of elements in Q_1. Next, we delete all elements in Q_1, and then insert ρ_m into Q_2. From then on, we repeat the above operations to process these queues until meeting a non-full queue. For example, after ρ_m is inserted into Q_2, if Q_2 is not full, the algorithm is terminated. Otherwise, ρ_m is updated to the median of elements in Q_2, and then insert ρ_m into Q_3. In particularly, when executing GQRS, if the state is FINISH, it means all objects in C are accessed. At that moment, we compute the median based on elements in Q_u, use it as the final ρ.

Take an example in Fig. 3. When $|C|$ achieves to 54 ($M = 54$), DSORT is applied, and then ρ−election is invoked. Next, we partition objects in C into 18 partitions, where e-partition sets and s-partition sets are shown in Fig. 3(a) respectively. Secondly, we form a group of $\lceil \log 0.5 \times 54 \rceil (=3)$ queues,i.e., Q_1, Q_2 and Q_3. Also, we form a min-heap H based on objects in P_{18}. From then on, we reversely scan objects in these 17 partitions. After $12 \in P_{17}$ is processed, Q_1

turns to full, we insert the median of elements in Q_1 into Q_2, which is $\rho_m = 0.15$. After $36 \in P_{13}$ is processed, Q_2 turns to full, we insert the median of elements in Q_2 into Q_3. After $88 \in P_1$ is processed, the algorithm is terminated. The corresponding ρ is set to 0.87.

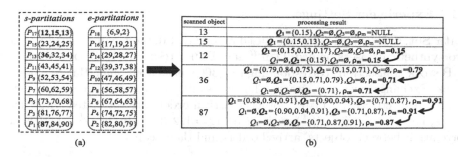

(a)   (b)

Fig. 3. Running example of optimization ρ Selection $(k = 3, M = 54)$

The Cost Analysis. We first analyze the running cost of forming $\rho-$MSET under optimization algorithms. It includes: DSORT, $\rho-$selection and $\rho-$MSET construction. For DSORT, the cost of forming these k sets is $\mathcal{O}(M)$. As the cost of merge sort is linear to the data scale, and the scale of each min-heap is k, the total merging cost is bounded by $\mathcal{O}(M \log k)$. For optimization ρ selection, whenever 3^{i-1} objects are processed, the queue q_i is updated one time. Therefore, the total cost of maintaining these queues is bounded by $\mathcal{O}(\sum_{i=0}^{i=\lceil \log \frac{|C|}{2} \rceil} \frac{|C|}{3^{i-1}})$, which is $\mathcal{O}(M)$. As the cost of maintaining min-heap is bounded by $\mathcal{O}(M \log k)$, the total cost of computing ρ is bounded by $\mathcal{O}(M \log k + M)$, which is $\mathcal{O}(M \log k)$. Similarly, for the cost of $\rho-$MSET construction, the cost of maintaining min-heap is $\mathcal{O}(M \log k)$, the corresponding running cost is bounded by $\mathcal{O}(M \log k)$. Therefore, the cost of forming $\rho-$MSET is $\mathcal{O}(M \log k)$.

We now analyze the incremental maintenance cost. When a set of s objects arrive in the window, the cost of constructing SK_m is bounded by $\mathcal{O}(s \log k)$, and the cost of merging is bounded by $\mathcal{O}(|C| + k)$. When we construct $\rho-$MSET C_ρ, the corresponding running cost is bounded by $\mathcal{O}(M \log k)$. As is discussed in Theorem 1, under baseline algorithm, once C_ρ is formed, as $\frac{|C_\rho|}{M} \leq \frac{3}{4}$ is guaranteed, it implies $\rho-$MSET construction is invoked whenever at least $\frac{sM}{4k}$ objects flow into the window, and we can amortize this part of running cost to $\frac{sM}{4k}$ objects. Under improving algorithm, based on the median searching algorithm, the selected ρ also could guarantee that $M - |C_\rho|$ is $\mathcal{O}(M)$. Therefore, the amortized incremental maintenance cost of each object is bounded by $\mathcal{O}(\frac{M \log k}{\frac{sM}{k}} + \frac{|C|}{s} + \log k)$, which is $\mathcal{O}(\log k + \frac{M}{s})(|C|$ is bounded by $M)$. For the space cost, as $|C|$ is bounded by M, the cost of $\rho-$selection is bounded by $\mathcal{O}(\log M)$, the overall space cost is bounded by $\mathcal{O}(M)$.

4 The Experiment

In this section, we conduct extensive experiments to demonstrate the efficiency of the ρ-TOPK. In the following, we first explain the settings of our experiments.

4.1 Experiment Settings

Data Set. The experiments are based on one real dataset named STOCK, and two synthetic datasets named TIME^R and TIME^U respectively. STOCK refers to 1GB stock transactions corresponding to 2,300 stocks from ShangHai/ShenZhen Stock Exchange over two years. The two synthetic datasets are TIME^R and TIME^U respectively. In TIME^R, objects' arrival orders are correlated with their scores, i.e., $F(o) = sin(\frac{\pi \times o.t}{10^6})$ with $o.t$ being $= 1, 2, 3 \cdots$. In TIME^U, there is no correlation between objects' arrival orders and their scores.

Parameter Setting. In our experiment, we evaluate algorithms differences via the following metrics, which are *response time, candidate amount*, and *score ratio* as the main performance indicators. Here, response time refers to the total running time that we spend after processing all objects. Candidate amount refers to the average number of candidates we should maintain. Score ratio refers to the degree of deviation between approximation results and accurate results. Besides, four parameters are considered, which are window size n, the parameters s, k, and M. Parameter Settings are shown in Table 2 with the default values bolded.

Table 2. Parameter setting

Parameter	Value
n	100 KB, 200 KB, 500 KB, **1 MB**, 5 MB, 10 MB
k	5, 10, 20, **50**, 100, 500
s	$0.1\%, 0.5\%, \mathbf{1\%}, 5\%, 10\%, 20\%(\times n)$
M	2, **4**, 8, 16, 32$(\times k)$

In addition to ρ-TOPK, we also implement minTopK [3], PABF [8] and PA [10] for answering continuous top-k query as competitors. minTopK is a representative exact approach. The others are approximate algorithms. Note, these 3 algorithms do not consider relationship between k-skyband object amount and memory size, in implementation, if the candidate set scale under these 3 algorithms is larger than M, we remove the object with minimal arrived order from the candidate set directly. All the algorithms are implemented with C++, and all the experiments are conducted on a CPU i7, running Microsoft Windows 10.

4.2 The Performance Evaluation

Effect of the Parameter ρ. First of all, we evaluate the effectiveness of ρ–TOPK under different ρ. Compared with using a variable ρ, we set ρ in advance. Whenever C achieves to M, we use this ρ for forming C_ρ. As is shown in Table 3, the running time of ρ–TOPK is reduced with the decreasing of ρ. The reason behind it is the higher the ρ, the more k–skybands we should maintain. However, after ρ is reduced to 0.9, the running time of ρ–TOPK is not drastically changed. That is because when ρ is smaller than 0.9, the candidate set scale is small enough. In this case, the mainly running cost is spent in processing newly arrived objects. Furthermore, the error ratio of ρ–TOPK is increasing with the decreasing of ρ. The reason is the lower the ρ, the more k–skybands are deleted, and the higher the error ratio it is. Above all, when ρ is small enough, its running time is no longer drastically changed, but it may significantly the quality of query results. It is important to find a suitable ρ based on M and scale k–skyband set.

Table 3. Effect of the parameter ρ

Dataset		The parameter (ρ)										
		0.80	0.82	0.84	0.86	0.88	0.90	0.92	0.94	0.96	0.98	0.99
STOCK	Running time	20.89	21.33	22.02	22.92	23.03	23.52	24.32	25.66	27.78	28.01	29.85
	Accuracy	0.86	0.87	0.88	0.89	0.91	0.92	0.94	0.96	0.98	0.99	0.99
$Time^U$	Running time	18.77	19.13	18.75	19.32	20.09	21.91	23.38	25.01	26.33	27.93	28.66
	Accuracy	0.84	0.86	0.87	0.89	0.90	0.92	0.93	0.95	0.97	0.99	0.99
$Time^R$	Running time	20.31	21.20	21.44	22.06	23.11	24.77	25.82	28.11	30.28	34.30	35.49
	Accuracy	0.86	0.86	0.88	0.89	0.91	0.92	0.94	0.96	0.97	0.98	0.99

Running Time Comparison. We first evaluate the performance of different algorithms under different n. The other parameters are default values. From Fig. 4(a) to Fig. 4(c), we observe that ρ–TOPK performs the best in terms of running time. For example, it averagely consumes only 40% of minTopKs running time and 70% of PABFs running time. This is no surprise since ρ–TOPK only maintains a subset of k–skybands to support query processing. Compared with that of PABF and PA, ρ–TOPK could use, as small as running cost, to construct C_ρ. Thus, it performs better than theirs. Next, we evaluate the performance of different algorithms under different k. The other parameters are default values. From Fig. 4(d)-4(f), we find that ρ–TOPK performs the best of all. Beside the reasons discussed before, another reason is ρ–TOPK is not sensitive to the distribution of streaming data. For example, under TIME^U, the running time of other algorithms is much higher than of ρ–TOPK, where the candidate set scale under this data set is M in many cases. Intuitively, the larger the candidate set scale, the higher the running cost. Thus, it implies that $\rho-$TOPK is not sensitive to data distribution. Thirdly, we evaluate the performance of these algorithms under different s. The other parameters are default values. From Fig. 4(g) to Fig. 4(i), $\rho-$TOPK still performs the best of all. We also find that, with the increasing of s, the running time difference among these 4 algorithms all turn to small. The reason behind it is, the larger the s, the more objects could be filtered. When the $k-$skyband set is small enough, $\rho-$TOPK is equivalent to minTopK. However, $\rho-$TOPK is more stable than others (Table 4).

Table 4. Candidate set size under different window length (MB)

Algorithm	STOCK						$TIME^U$						$TIME^R$					
	0.1	0.2	0.5	1	5	10	0.1	0.2	0.5	1	5	10	0.1	0.2	0.5	1	5	10
$\rho-$TOPK	71	86	103	124	133	140	69	82	92	119	130	141	73	89	109	125	137	142
minTopK	119	144	174	200	200	200	116	137	154	200	200	200	122	149	183	200	200	200
PA	93	113	135	162	174	179	91	114	125	155	167	174	96	117	143	154	169	179
PABF	83	101	121	146	156	165	82	102	112	140	153	167	88	109	127	145	158	171

Candidates Amount Comparison. In the following, we compare the candidate set scale under different algorithms. For the limitation of space, we only compare candidate set scale under different n. The other parameters are default values. We find that, candidate set scale are all increasing with the increasing of n, but the candidate set scale under $\rho-$TOPK is the smallest of all. The reason behind it is when $|C|$ achieves to M, we should form $\rho-$MSET, whose scale is roughly half of M in most cases. Therefore, the candidate amount under $\rho-$TOPK is the smallest of all (Fig. 5).

Score Ratio Comparison. We find that the score ratio under $\rho-$TOPK is the highest of all. Also, when M is small, the score ratio under $\rho-$TOPK is also acceptable. The reason behind it is, $\rho-$TOPK could self-adaptively adjust ρ based on k-skyband set scale and M. In this way, approximate results quality under $\rho-$TOPK is the highest of all.

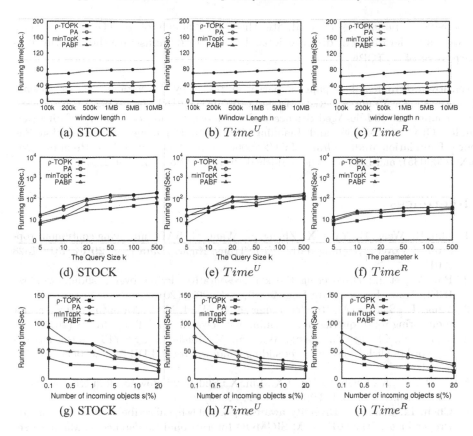

Fig. 4. Running time comparison of different algorithms under different data sets.

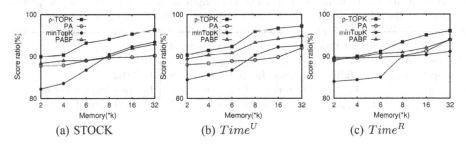

Fig. 5. Score rate of different algorithms under different data sets.

5 Conclusion

This paper proposes a novel framework named $\rho-$TOPK for supporting $\rho-$approximate continuous top-k query over data stream. It can self-adaptively adjust ρ based on distribution of streaming data, and supporting $\rho-$approximate top-k query under memory limitation-based data stream. We have conducted extensive experiments to evaluate the performance of $\rho-$TOPK on several datasets under different distributions. The results demonstrate the superior performance of $\rho-$TOPK.

Acknowledgment. This paper is partly supported by the National Natural Science Foundation for Young Scientists of China (61702344, 61802268, 62102271, 62072088), the Young and Middle-Aged Science and Technology Innovation Talent of Shenyang under Grant RC200439, and Liaoning Provincial Department of Education Science Foundation under Grant JYT2020066, Ten Thousand Talent Program (No. ZX20200035), and Liaoning Distinguished Professor (No. XLYC1902057).

References

1. Zhu, R., Wang, B., Yang, X., Zheng, B., Wang, G.: SAP: improving continuous top-k queries over streaming data. IEEE Trans. Knowl. Data Eng. **29**(6), 1310–1328 (2017)
2. Bai, M., et al.: Discovering the k representative skyline over a sliding window. IEEE Trans. Knowl. Data Eng. **28**(8), 2041–2056 (2016)
3. Yang, D., Shastri, A., Rundensteiner, E.A., Ward, M.O.: An optimal strategy for monitoring top-k queries in streaming windows. In: EDBT, pp. 57–68 (2011)
4. Wang, X., Zhang, Y., Zhang, W., Lin, X., Huang, Z.: SKYPE: top-k spatial-keyword publish/subscribe over sliding window. Proc. VLDB Endow. **9**(7), 588–599 (2016)
5. Jin, C., Yi, K., Chen, L., Yu, J.X., Lin, X.: Sliding-window top-k queries on uncertain streams. VLDB J. **19**(3), 411–435 (2010)
6. Chen, L., Cong, G.: Diversity-aware top-k publish/subscribe for text stream. In: Proceedings of the 2015 ACM SIGMOD International Conference on Management of Data, Melbourne, Victoria, Australia, 31 May–4 June 2015, pp. 347–362 (2015)
7. Mouratidis, K., Bakiras, S., Papadias, D.: Continuous monitoring of top-k queries over sliding windows. In: SIGMOD Conference, pp. 635–646 (2006)
8. Zhu, R., Wang, B., Luo, S., Yang, X., Wang, G.: Approximate continuous top-k query over sliding window. J. Comput. Sci. Technol. **32**(1), 93–109 (2017)
9. Shen, Z., Cheema, M.A., Lin, X., Zhang, W., Wang, H.: Efficiently monitoring top-k pairs over sliding windows. In: ICDE, pp. 798–809 (2012)
10. Pripuzic, K., Zarko, I.P., Aberer, K.: Time- and space-efficient sliding window top-k query processing. ACM Trans. Database Syst. **40**(1), 1:1–1:44 (2015)

Cross-Model Conjunctive Queries over Relation and Tree-Structured Data

Yuxing Chen[1], Valter Uotila[1], Jiaheng Lu[1(✉)], Zhen Hua Liu[2],
and Souripriya Das[2]

[1] University of Helsinki, Helsinki, Finland
{yuxing.chen,valter.uotila,jiaheng.lu}@helsinki.fi
[2] Oracle, Redwood City, USA
{zhen.liu,souripriya.das}@oracle.com

Abstract. Conjunctive queries are the most basic and central class of
database queries. With the continued growth of demands to manage
and process the massive volume of different types of data, there is little
research to study the conjunctive queries between relation and tree data.
In this paper, we study Cross-Model Conjunctive Queries (CMCQs) over
relation and tree-structured data (XML and JSON). To efficiently process CMCQs with bounded intermediate results we first encode tree
nodes with position information. With tree node original label values
and encoded position values, it allows our proposed algorithm *CMJoin*
to join relations and tree data simultaneously, avoiding massive intermediate results. *CMJoin* achieves worst-case optimality in terms of the total
result of label values and encoded position values. Experimental results
demonstrate the efficiency and scalability of the proposed techniques to
answer a CMCQ in terms of running time and intermediate result size.

Keywords: Cross-model join · Worst-case optimal · Relation and tree
data

1 Introduction

Conjunctive queries are the most fundamental and widely used database queries
[2]. They correspond to `project-select-join` queries in the relational algebra.
They also correspond to non-recursive datalog rules [7]

$$R_0(u_0) \leftarrow R_1(u_1) \wedge R_2(u_2) \wedge \ldots \wedge R_n(u_n), \tag{1}$$

where R_i is a relation name of the underlying database, R_0 is the output relation,
and each argument u_i is a list of $|u_i|$ variables, where $|u_i|$ is the arity of the
corresponding relation. The same variable can occur multiple times in one or
more argument lists.

It turns out that traditional database engines are not optimal to answer conjunctive queries, as all pair-join engines may produce unnecessary intermediate

A. Bhattacharya et al. (Eds.): DASFAA 2022, LNCS 13245, pp. 21–37, 2022.
https://doi.org/10.1007/978-3-031-00123-9_2

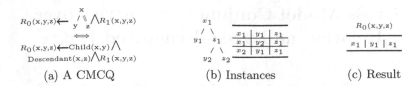

(a) A CMCQ   (b) Instances   (c) Result

Fig. 1. An example of a CMCQ

results on many join queries [26]. For example, consider a typical triangle conjunctive query $R_0(a, b, c) \leftarrow R_1(a, b) \wedge R_2(b, c) \wedge R_3(a, c)$, where the size of input relations $|R_1| = |R_2| = |R_3| = N$. The worst-case size bound of the output table $|R_0|$ yields $\mathcal{O}(N^{\frac{3}{2}})$. But any pairwise relational algebra plan takes at least $\Omega(N^2)$, which is asymptotically worse than the optimal engines. To solve this problem, recent algorithms (e.g. *NPRR* [26], *LeapFrog* [31], *Joen* [8]) were discovered to achieve the optimal asymptotic bound for conjunctive queries.

Conjunctive queries over trees have recently attracted attention [13], as trees are a clean abstraction of HTML, XML, JSON, and LDAP. The tree structures in conjunctive queries are represented using node label relations and axis relations such as *Child* and *Descendant*. For example, the *XPath* query $A[B]//C$ is equivalent to the conjunctive query:

$$R(z) \leftarrow Label(x, \text{``}A\text{''}) \wedge Child(x, y) \wedge Label(y, \text{``}B\text{''})$$
$$\wedge Descendant(x, z) \wedge Label(z, \text{``}C\text{''}). \tag{2}$$

Conjunctive queries with trees have been studied extensively. For example, see [13] on their complexity, [3] on their expressive power, and [4, 13] on the satisfiability problem. While conjunctive queries with relations or trees have been studied separately in the literature, hybrid conjunctive queries have gained less attention.

This paper embarks on the study of a Cross-Model Conjunctive Query (CMCQ) over both relations and trees. Figure 1 depicts a CMCQ. CMCQs emerge in modern data management and analysis, which often demands a hybrid evaluation with data organized in different formats and models, e.g. data lake [16], multi-model databases [22], polystores [9], and computational linguistics [32].

The number of applications that we have hinted at above motivates the study of CMCQs, and the main contributions of this paper are as follows:

1. This paper embarks on the study of the cross-model conjunctive query (CMCQ) and formally defines the problem of CMCQ processing, which integrates both relational conjunctive query and tree conjunctive pattern.
2. We propose *CMJoin*-algorithm to process relations and encoded tree data efficiently. *CMJoin* produces the worst-case optimal join result in terms of the label values as well as the encoded information values. In some cases, *CMJoin* is the worst-case optimal join in the absence of encoded information.

$$R_1(r_a,r_b) \bowtie R_2(r_a,r_c) \bowtie$$
$$R_0(r_a,r_b,r_c) \longleftarrow \text{Descendant}(t_a, t_b) \bowtie \text{Descendant}(t_a, t_c) \bowtie$$
$$\text{Label}(t_a,r_a) \bowtie \text{Label}(t_b,r_b) \bowtie \text{Label}(t_c,r_c)$$

(a) Conjunctive query form

$$R_0(a,b,c) \longleftarrow \overset{a}{\underset{b \quad c}{\diagup\diagdown}} \bowtie R_1(a,b) \bowtie R_2(a,c)$$

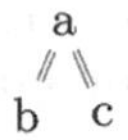

(b) Simplified expression

Fig. 2. Complete and simplified expressions of CMCQ

3. Experiments on real-life and benchmark datasets show the effectiveness and efficiency of the algorithm in terms of running time and intermediate result size.

The remainder of the paper is organized as follows. In Sect. 2 we provide preliminaries of approaches. We then extend the worst-case optimal algorithm for CMCQs in Sect. 3. We evaluate our approaches empirically in Sect. 4. We review related works in Sect. 5. Section 6 concludes the paper.

2 Preliminary

Cross-model Conjunctive Query. Let $\mathbb{R}$ be a database schema and $R_1, \ldots, R_n$ be relation names in $\mathbb{R}$. A rule-based conjunctive query over $\mathbb{R}$ is an expression of the form $R_0(u_0) \leftarrow R_1(u_1) \wedge R_2(u_2) \wedge \ldots \wedge R_n(u_n)$, where $n \geq 0$, R_0 is a relation not in $\mathbb{R}$. Let $u_0, u_1, \ldots, u_n$ be free tuples, i.e. they may be either variables or constants. Each variable occurring in u_0 must also occur at least once in $u_1, \ldots, u_n$.

Let T be a tree pattern with two binary axis relations: `Child` and `Descendant`. The axis relations `Child` and `Descendant` are defined in the normal way [13]. In general, a cross-model conjunctive query contains three components: (i) the relational expression $\tau_1 := \exists r_1, \ldots, r_k \colon R_1(u_1) \wedge R_2(u_2) \wedge \ldots \wedge R_n(u_n)$, where $r_1, \ldots, r_k$ are all the variables in the relations $R_1, \ldots, R_n$; (ii) the tree expression $\tau_2 := \exists t_1, \ldots, t_k \colon Child(v_1) \wedge \ldots \wedge Descendant(v_n)$, where $t_1, \ldots, t_k$ are all the node variables occurring in v_i, for $i \geq 1$ and each v_i is a binary tuple (t_{i_1}, t_{i_2}); and (iii) the cross-model label expression $\tau_3 := \exists r_1, \ldots, r_k, t_1, \ldots, t_k \colon label_1(t_{i_1}, r_{j_1}) \wedge \ldots \wedge label_n(t_{i_n}, r_{j_n})$, where Σ denotes a labeling alphabet. Given any node $t \in T$, $label(t, s)$ means that the label of the node t is $s \in \Sigma$. The label relations bridge the expressions of relations and trees by the equivalence between the label values of the tree nodes and the values of relations.

By combining the three components together, we define a cross-model conjunctive query with the calculus of form $\{e_1, \ldots, e_m \mid \tau_1 \wedge \tau_2 \wedge \tau_3\}$, where the variables $e_1, \ldots, e_m$ are the return elements which occur at least once in relations.

Figure 2a shows an example of a cross-model conjunctive query, which includes two relations and one tree pattern. For the purpose of expression simplicity, we do not explicitly distinguish between the variable of trees (e.g. t_a) and that of relations (e.g. r_a), but simply write them with one symbol (i.e. a) if $label(t_a, r_a)$ holds. We omit the label relation when it is clear from the context. Figure 2b shows a simplified representation of a query.

Revisiting Relational Size Bound. We review the size bound for the relational model, which Asterias, Grohe, and Marx (AGM) [2] developed. The AGM bound is computed with linear programming (LP). Formally, given a relational schema $\mathbb{R}$, for every table $R \in \mathbb{R}$ let A_R be the set of attributes of R and $\mathcal{A} = \cup_R A_R$. Then the worst-case size bound is precisely the optimal solution for the following LP:

$$
\begin{aligned}
\text{maximize} \quad & \Sigma_r^{\mathcal{A}} x_r \\
\text{subject to} \quad & \Sigma_r^{A_R} x_r \leq 1 \quad \text{for all } R \in \mathbb{R}, \\
& 0 \leq x_r \leq 1 \quad \text{for all } r \in \mathcal{A}.
\end{aligned}
\tag{3}
$$

Let ρ denote the optimal solution of the above LP. Then the size bound of the query is N^ρ, where N denotes the maximal size of each table. The AGM bound can be proved as a special case of the discrete version of the well-known Loomis-Whitney inequality [20] in geometry. Interested readers may refer to the details of the proof in [2]. We present these results informally and refer the readers to Ngo et al. [27] for a complete survey.

For example, we consider a typical triangle conjunctive query $R_0(a, b, c) \leftarrow R_1(a, b) \wedge R_2(b, c) \wedge R_3(a, c)$ that we introduced in Sect. 1. Then the three LP inequalities corresponding to three relations include $x_a + x_b \leq 1$, $x_b + x_c \leq 1$, and $x_a + x_c \leq 1$. Therefore, the maximal value of $x_a + x_b + x_c$ is $3/2$, meaning that the size bound is $\mathcal{O}(N^{\frac{3}{2}})$. Interestingly, the similar case for CMCQ in Fig. 3, the query $Q = a[b]/c \bowtie R(b, c)$ has also the size bound $\mathcal{O}(N^{\frac{3}{2}})$.

3   Approach

In this section, we tackle the challenges in designing a worst-case optimal algorithm for CMCQs over relational and tree data. We briefly review the existing relational worst-case optimal join algorithms. We represent these results informally and refer the readers to Ngo et al. [27] for a complete survey. The first algorithm to have a running time matching these worst-case size bounds is the NPRR algorithm [26]. An important property in NPRR is to estimate the intermediate join size and avoid producing a case that is larger than the worst-case bound. In fact, for any join query, its execution time can be upper bounded by the AGM [2]. Interestingly, *LeapFrog* [31] and *Joen* [8] completely abandon a query plan and propose to deal with one attribute at a time with multiple relations simultaneously.

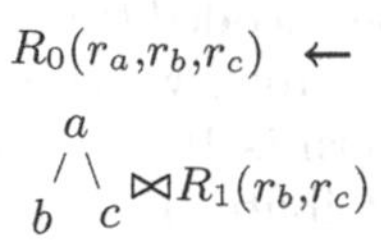

$$R_0(r_a, r_b, r_c) \leftarrow$$

$$\begin{array}{c} a \\ / \ \backslash \\ b \quad c \end{array} \bowtie R_1(r_b, r_c)$$

r_b	r_c
b_0	c_0
b_0	c_1
b_1	c_0
b_1	c_1

a_0

p_0

$b_0 \quad b_1 \quad b_2 \quad b_3 \quad c_0 \quad c_1 \quad c_2 \quad c_3$

$p_1 \quad p_2 \quad p_3 \quad p_4 \quad p_5 \quad p_6 \quad p_7 \quad p_8$

(a) Tree query Q (b) Table D_{R_1} (c) Encode Tree D_T

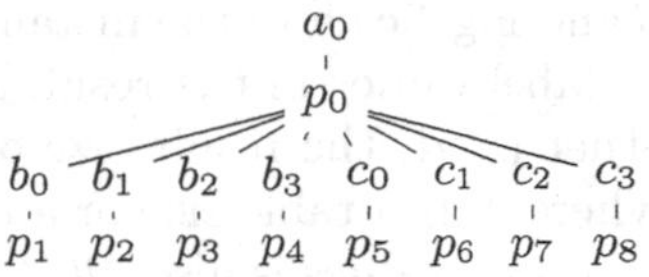

Fig. 3. A CMCQ (a) and its table instance (b) and tree instance (c).

3.1 Tree and Relational Data Representation

To answer a tree pattern query, a positional representation of occurrences of tree elements and string values in the tree database is widely used, which extends the classic inverted index data structure in information retrieval. There existed two common ways to encode an instance tree, i.e. Dewey encoding [23] and containment encoding [6]. These decodings are necessary as they allow us to partially join tree patterns to avoid the undesired intermediate result. After encoding, each attribute j in the query node can be represented as a **node table** in form of $t_j(r_j, p_j)$, where r_j and p_j are the label value and position value, respectively. Check an example from a encoded tree instance in Fig. 3c. The position value can be added in $\mathcal{O}(N)$ by one scan of the original tree. Note that we use Dewey coding in our implementation but the following algorithm is not limited to such representation. Any representation scheme which captures the structure of trees such as a region encoding [6] and an extended Dewey encoding [23] can all be applied in the algorithm.

All the data in relational are label data, and all relation tables and node tables will be expressed by the Trie index structure, which is commonly applied in the relational worst-case optimal algorithms (e.g. [1,31]). The Trie structure can be accomplished using standard data structures (notably, balanced trees with $\mathcal{O}(\log n)$ look-up time or nested hashed tables with $\mathcal{O}(1)$ look-up time).

3.2 Challenges

In our context, tree data and twig pattern matching do make the situation more complex. Firstly, directly materializing tree pattern matching may yield asymptotically more intermediate results. If we ignore the pattern, we may lose some bound constraints. Secondly, since tree data are representing both label and position values, position value joining may require more computation cost for pattern matching while we do not need position values in our final result.

Example 1. Recall that a triangle relational join query $Q = R_1(r_a, r_b) \bowtie R_2(r_a, r_c) \bowtie R_3(r_b, r_c)$ has size bound $\mathcal{O}(N^{\frac{3}{2}})$. Figure 3 depicts an example of a CMCQ Q with the table $R_1(r_b, r_c)$ and twig query $a[b]/c$ to return result $R_0(r_a, r_b, r_c)$, which also has size bound $\mathcal{O}(N^{\frac{3}{2}})$ since the PC paths a/b and a/c are equivalent to the constraints $x_a + x_b \leq 1$ and $x_a + x_c \leq 1$, respectively.

Figure 3b and Fig. 3c show the instance table D_{R_1} and the encoded tree D_T. The number of label values in the result $R_0(r_a, r_b, r_c)$ is only 4 rows which is $\mathcal{O}(N)$. On the other hand, the result size of only the tree pattern is 16 rows which is $\mathcal{O}(N^2)$, where N is a table size or a node size for each attribute. The final result with the position values is also $\mathcal{O}(N^2)$. Here, $\mathcal{O}(N^2)$ is from the matching result of the position values of the attributes t_b and t_c.

EmptyHeaded [1] applied the existing worst-case optimal algorithms to process the graph edge pattern matching. We may also attempt to solve relation-tree joins by representing the trees as relations with the node-position and the node-label tables and then reformulating the cross-model conjunctive query as a relational conjunctive query. However, as Example 1 illustrated, such a method can not guarantee the worst-case optimality as extra computation is required for position value matching in a tree.

3.3   Cross-Model Join (CMJoin) Algorithm

In this part, we discuss the algorithm to process both relational and tree data. As the position values are excluded in the result set while being required for the tree pattern matching, our algorithm carefully deals with it during the join. We propose an efficient cross-model join algorithm called *CMJoin* (cross-model join). In certain cases, it guarantees runtime optimality. We discover the join result size under three scenarios: with all node position values, with only branch node position value, and without position value.

Lemma 1. *Given relational tables $\mathcal{R}$ and pattern queries $\mathcal{T}$, let S_r, S_p, and S_p' be the sets of all relation attributes, all position attributes, and only branch node position attributes, respectively. Then it holds that*

$$\rho_1(S_r \cup S_p) \geq \rho_2(S_r \cup S_p') \geq \rho_3(S_r). \tag{4}$$

Proof. $Q(S_r)$ is the projection result from $Q(S_r \cup S_p')$ by removing all position values, and $Q(S_r \cup S_p')$ is the projection result from $Q(S_r \cup S_p)$ by removing non-branch position values. Therefore, the result size holds $\rho_1(S_r \cup S_p) \geq \rho_2(S_r \cup S_p') \geq \rho_3(S_r)$.

Example 2. Recall the CMCQ Q in Fig. 3a, which is $Q = R_1(r_b, r_c) \bowtie a[b]/c$. Nodes a, b, c in the tree pattern can be represented as node tables (r_a, p_a), (r_b, p_b), and (r_c, p_c), respectively. So we have $Q(S_r \cup S_p) = R(r_a, r_b, r_c, p_a, p_b, p_c)$, $Q(S_r \cup S_p') = R(r_a, r_b, r_c, p_a)$, and $Q(S_r) = R(r_a, r_b, r_c)$. By the LP constraint bound for the relations and PC-paths, we achieve $\mathcal{O}(N^2)$, $\mathcal{O}(N^{\frac{3}{2}})$, and $\mathcal{O}(N^{\frac{3}{2}})$ for the size bounds $\rho_1(S_r \cup S_p)$, $\rho_2(S_r \cup S_p')$, and $\rho_3(S_r)$, respectively.

We elaborate *CMJoin* Algorithm 1 more in the following. In the case of $\rho_1(S_r \cup S_p) = \rho_3(S_r)$, *CMJoin* executes a generic relational worst-case optimal join algorithm [1,8] as the extra position values do not affect the worst-case final result. In other cases, *CMJoin* computes the path result of the tree pattern first.

Algorithm 1: *CMJoin*

Input: Relational tables $\mathcal{R}$, pattern queries $\mathcal{T}$

1   $\mathcal{R}' \leftarrow \emptyset$ `// Tree intermediate result`

2   **if** $\rho_1(S_r \cup S_p) \leq \rho_3(S_r)$ **then** // **Theorem 1 condition (1)**

3   **foreach** $N \in \mathcal{T}$ **do**

4         $\lfloor$ $\mathcal{R}' \leftarrow \mathcal{R}' \cup C_N(r_N, p_N)$ `// Nodes as tables`

5   **else**

6      $\mathcal{P} \leftarrow \mathcal{T}.getPaths()$

7   **foreach** $P \in \mathcal{P}$ **do**

8         $R_P(S_r \cup S_p) \leftarrow$ path result of P `// Paths as tables`

9         $R'_P(S_r \cup S'_p) \leftarrow$ project out non-branch position values of $R_P(S_r \cup S_p)$

10       $\lfloor$ $\mathcal{R}' \leftarrow \mathcal{R}' \cup \{R'_P(S_r \cup S'_p)\}$

11   $Q(S_r \cup S'_p) \leftarrow generic\_join(\mathcal{R} \cup \mathcal{R}')$

12   $Q(S_r) \leftarrow$ project out all position values $Q(S_r \cup S'_p)$

Output: Join results $Q(S_r)$

In this case, we project out all position values of a non-branch node for the query tree pattern. Then, we keep the position values of the only branch node so that we still can match the whole part of the tree pattern.

Theorem 1. *Assume we have relations $\mathcal{R}$ and pattern queries $\mathcal{T}$. If either*

(1) $\rho_1(S_r \cup S_p) \leq \rho_3(S_r)$ or
(2) (i) $\rho_1(S_r \cup S'_p) \leq \rho_3(S_r)$ and (ii) for each path P in $\mathcal{T}$ let S''_r and S''_p be the set of label and position attributes for P so that $\rho_4(S''_r \cup S''_p) \leq \rho_3(S_r)$.

Then, CMJoin is worst-case optimal to $\rho_3(S_r)$.

Proof. (1) Since the join result of the only label value $\rho_3(S_r)$ is the projection of $\rho_1(S_r \cup S_p)$, we can compute $Q(S_r \cup S_p)$ first, then project out all the position value in linear of $\mathcal{O}(N^{\rho_1(S_r \cup S_p)})$. Since $\rho_1(S_r \cup S_p) \leq \rho_3(S_r)$, we can estimate that the result size is limited by $\mathcal{O}(N^{\rho_3(S_r)})$.

(2) $\rho_1(S''_r \cup S''_p) \leq \rho_3(S_r)$ means that each path result with label and position values are under worst-case result of $\rho_3(S_r)$. We may first compute the path result and then project out all the non-branch position values. The inequality $\rho_2(r, p') \leq \rho_3(r)$ means that the join result containing all branch position values has a worst-case result size which is still under $\rho_3(S_r)$. Then by considering those position values as relational attribute values and by a generic relation join [1, 26], *CMJoin* is worst-case optimal to $\rho_3(S_r)$. $\square$

Example 3. Recall the CMCQ query $Q = R_1(r_b, r_c) \bowtie a[b]/c$ in Fig. 3a. Since $\rho_1(S_r \cup S_p) > \rho_3(r)$, directly computing all label and position values may generate asymptotically bigger result ($\mathcal{O}(N^2)$ in this case). So we can compute path results of a/b and a/c, which are (r_a, p_a, r_b, p_b) and (r_a, p_a, r_c, p_c) and in $\mathcal{O}(N)$. Then we obtain only branch node results (r_a, p_a, r_b) and (r_a, p_a, r_c). By joining these project-out results with relation R_1 using a generic worst-case optimal algorithm, we can guarantee that the size bound is $\mathcal{O}(N^{\frac{3}{2}})$.

Table 1. Intermediate result size (10^6) and running time (S) for queries. "/" and "–" indicate "timeout" (≥ 10 min) and "out of memory". We measure the intermediate size by accumulating all intermediate and final join results.

Query	Intermediate result size (10^6)					Running time (second)				
	PG	SJ	VJ	EH	CMJoin	PG	SJ	VJ	EH	CMJoin
Q1	7.87x	2.60x	2.00x	1.68x	0.15	18.02x	1.39x	1.51x	1.66x	3.22
Q2	/	–	3.75x	4.83x	0.08	/	–	4.52x	129x	1.96
Q3	86.0x	62.6x	3.63x	4.61x	0.08	21.3x	4.27x	1.99x	4.28x	3.06
Q4	/	1.96x	1.75x	1.64x	0.24	/	2.34x	2.63x	1.82x	3.55
Q5	/	–	1.86x	1.77x	0.22	/	–	4.75x	39.8x	3.11
Q6	/	2.24x	2.00x	1.85x	0.21	/	6.10x	3.30x	2.89x	3.00
Q7	133x	106x	–	35.1x	0.29	4.82x	9.05x	–	7.18x	8.36
Q8	350x	279.8x	–	/	0.11	4.36x	5.61x	–	/	13.8
Q9	8.87x	8.34x	–	2.01x	4.62	1.12x	2.13x	–	1.48x	35.0
Q10	110x	440x	4.86x	/	0.07	2.91x	12.7x	1.22x	/	5.62
Q11	110x	440x	4.86x	/	0.07	2.11x	10.5x	0.88x	/	6.84
Q12	110x	440x	4.86x	/	0.07	2.68x	9.99x	1.06x	/	7.25
Q13	1.04x	1.22x	1.22x	1.07x	43.2	1.37x	4.81x	4.79x	1.31x	34.2
Q14	19.7x	2.56x	2.56x	3.90x	0.39	2.04x	3.82x	3.79x	2.14x	2.73
Q15	14.2x	1.85x	1.85x	17.0x	0.54	1.68x	3.53x	3.54x	2.01x	2.87
Q16	1.24x	1.24x	6.81x	2.15x	0.37	2.88x	1.32x	1.96x	1.02x	12.9
Q17	1.59x	7.84x	2.28x	1.31x	0.32	7.03x	3.58x	3.38x	2.10x	5.08
Q18	1.59x	7.13x	1.59x	1.64x	0.32	14.1x	5.21x	5.02x	2.06x	2.98
Q19	/	5.47x	6.62x	1.77x	0.45	/	1.41x	1.94x	0.89x	14.7
Q20	7.80x	25.1x	7.30x	4.19x	0.10	12.1x	6.77x	6.39x	3.17x	3.37
Q21	12.0x	36.1x	18.4x	14.7x	0.10	14.6x	8.82x	8.75x	10.9x	2.89
Q22	1.00x	18.5x	18.5x	0.96x	0.57	1.16x	5.22x	4.31x	2.35x	12.7
Q23	18.5x	18.5x	18.5x	1.61x	0.57	1.92x	2.17x	1.83x	1.02x	15.8
Q24	14.3x	3.02x	4.02x	0.96x	0.57	>9kx	>11kx	>12kx	0.18x	0.01
AVG	5.46x	5.90x	1.92x	1.90x	2.24	4.37x	5.34x	3.33x	3.46x	8.54

4   Evaluation

In this section, we experimentally evaluate the performance of the proposed algorithms and *CMJoin* with four real-life and benchmark data sets. We comprehensively evaluate *CMJoin* against state-of-the-art systems and algorithms concerning efficiency, scalability, and intermediate cost.

4.1   Evaluation Setup

Datasets and Query Design. Table 2 provides the statistics of datasets and designed CMCQs. These diverse datasets differ from each other in terms of the tree structure, data skewness, data size, and data model varieties. Accordingly,

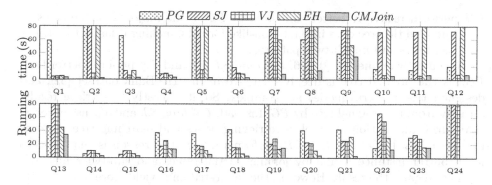

Fig. 4. Efficiency: runtime performance for all queries by *PG*, *SJ*, *VJ*, *EH*, and *CMJoin*. The performance time of more than 80 s is cut for better presentation.

we designed 24 CMCQs to evaluate the efficiency, scalability, and cost performance of the *CMJoin* in various real-world scenarios.

Comparison Systems and Algorithms. *CMJoin* is compared with two types of state-of-the-art cross-model solutions. The first solution is to use one query to retrieve a result without changing the nature of models [25,34]. We implemented queries in PostgreSQL (*PG*), that supports cross-model joins. This enables the usage of the *PG*'s default query optimizer.

The second solution is to encode and retrieve tree nodes in a relational engine [1,5,29,35]. We implemented two algorithms, i.e. structure join (*SJ*) (pattern matching first, then matching the between values) and value join (*VJ*) (label value matching first, then matching the position values). Also, we compared to a worst-case optimal relational engine called EmptyHeaded (*EH*) [1].

Experiment Setting. We conducted all experiments on a 64-bit Windows machine with a 4-core Intel i7-4790 CPU at 3.6 GHz, 16 GB RAM, and 500 GB HDD. We implemented all solutions, including *CMJoin* and the compared algorithms, in-memory processing by Python 3. We measured the computation time of joining as the main metric excluding the time used for compilation, data loading, index presorting, and representation/index creation for all the systems and algorithms. We employed the Dewey encoding [23] in all experiments. The join order of attributes is greedily chosen based on the frequency of attributes. We measured the intermediate cost metric by accumulating all intermediate and final join results. For *PG* we accumulated all sub-query intermediate results. We repeated five experiments excluding the lowest and the highest measure and calculated the average of the results. Between each measurement of queries, we wiped caches and re-loaded the data to avoid intermediate results.

Efficiency. Figure 4 shows the evaluation of the efficiency. In general, *CMJoin* is 3.33–13.43 times faster in average than other solutions as shown in Table 1. These numbers are conservative as we exclude the "out of memory" (OOM) and "time out" (TO) results from the average calculation. Algorithms *SJ*, *VJ*, and

EH perform relatively better compared to *PG* in the majority of the cases as they encoded the tree data into relation-like formats, making it faster to retrieve the tree nodes and match twig patterns.

Specifically in queries *Q*1–*Q*6, *CMJoin*, *SJ*, *VJ*, and *EH* perform better than *PG*, as the original tree is deeply recursive in the TreeBank dataset [32], and designed tree pattern queries are complex. So, it is costly to retrieve results directly from the original tree by *PG*. Instead, *CMJoin*, *SJ*, and *VJ* use encoded structural information to excel in retrieving nodes and matching tree patterns in such cases. In *Q*2 and *Q*5, *EH* performs worse. The reason is that it seeks for a better instance bound by joining partial tables and sub-twigs first and then aggregates the result. However, the separated joins yield more intermediate results in such cases in this dataset. In *Q*1 and *Q*4, which deal with a single table, *SJ* and *VJ* perform relatively close as no table joining occurs in these cases. However, in *Q*2–*Q*3 and *Q*5–*Q*6, *SJ* performs worse as joining two tables first leads to huge intermediate results in this dataset.

In contrast to the above, *SJ* outperforms *VJ* in *Q*7–*Q*9. The reason is that in the Xmark dataset [30], the tree data are flat and with fewer matching results in twig queries. The data in tables are also less skew. Therefore, *SJ* operates table joins and twig matching separately yielding relatively low results. Instead, *VJ* considers tree pattern matching later yielding too many intermediate results (see details of *Q*7 in Fig. 5 and Fig. 6) when joining label values between two models with non-uniform data. *PG*, which implements queries in a similar way of *SJ*, performs satisfactorily as well. The above comparisons show that compared solutions, which can achieve superiority only in some cases and can not adapt well to dataset dynamics.

Queries *Q*10–*Q*12 have more complex tree pattern nodes involved. In these cases *VJ* filters more values and produces fewer intermediate results. Thus it outperforms *SJ* ($\sim$10$\times$) and *PG* ($\sim$2$\times$). For queries, *Q*10–*Q*12 *EH* also yields huge intermediate results with more connections in attributes. The comparison between *Q*7–*Q*9 and *Q*10–*Q*12 indicates that the solutions can not adapt well to query dynamics.

Considering queries *Q*13–*Q*15 and *Q*22–*Q*24, *PG* performs relatively well since it involves only JSON and relational data. *PG* performs well in JSON retrieving because JSON documents have a simple structure. In *Q*14–*Q*15 most of the solutions perform reasonably well when the result size is small but *SJ*, and *VJ* still suffer from a large result size in *Q*13. With only JSON data, *SJ* and *VJ* perform similarly, as they both treat a simple JSON tree as one relation. In contrast in *Q*16–*Q*21, it involves XML, JSON and relational data from the UniBench dataset [34]. *CMJoin*, *SJ*, *VJ*, and *EH* perform better than *PG*. This is again because employing the encoding technique in trees accelerates node retrieval and matching tree patterns. Also, *CMJoin*, *SJ*, *VJ*, and *EH* are able to treat all the data models together instead of achieving results separately from each model by queries in *PG*.

Though compared systems and algorithms possess advantages of processing and matching data, they straightforwardly join without bounding intermediate

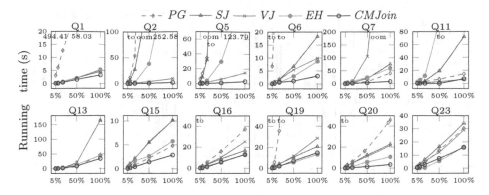

Fig. 5. Scalability: runtime performance of *PG*, *SJ*, *VJ*, *EH*, and *CMJoin*. The x-axis is the percentage of data size. "oom" and "to" stand for "out of memory" error and "timeout" (≥10 min), respectively.

results, thus achieving sub-optimal performance during joining. *CMJoin* is the clear winner against other solutions, as it can wisely join between models and between data to avoid unnecessary quadratic intermediate results.

Scalability. Figure 5 shows the scalability evaluation. In most queries, *CMJoin* performs flatter scaling as data size increases because *CMJoin* is designed to control the unnecessary intermediate output.

As discussed, *CMJoin*, *SJ*, *VJ*, and *EH* outperform *PG* in most of the queries, as the encoding method of the algorithms speeds up the twig pattern matching especially when the documents or queries are complex. However, *PG* scales better when involving simpler documents (e.g. in *Q*15 and *Q*23) or simpler queries (e.g. in *Q*7). Compared to processing XML tree pattern queries, *PG* processes JSON data more efficiently.

Interestingly in *Q*2, *SJ* and *PG* join two relational tables separately from twig matching, generating quadratic intermediate results, thus leading to the OOM and TO, respectively. In *Q*7 *VJ* joins tables with node values without considering tree pattern structural matching and outputs an unwanted non-linear increase of intermediate results, thus leading to OOM in larger data size. Likewise evaluating *EH* between *Q*2 and *Q*7, it can not adapt well with different datasets. Performing differently in diverse datasets between *SJ/PG* and *VJ/EH* indicates that they can not smartly adapt to dataset dynamics. While increasing twig queries in *Q*11 compared to *Q*7, *VJ* filters more results and thus decreases the join cost and time in *Q*11. The comparison between *SJ/EH* and *VJ* shows dramatically different performance in the same dataset with different queries that indicates they can not smartly adapt to query dynamics.

In *Q*11, both *CMJoin* and *VJ* perform efficiently as they can filter out most of the values at the beginning. In this case, *CMJoin* runs slightly slower than *VJ*, which is reasonable as *CMJoin* maintains a tree structure whereas *VJ* keeps only tuple results. Overall, *CMJoin* judiciously joins between models and controls

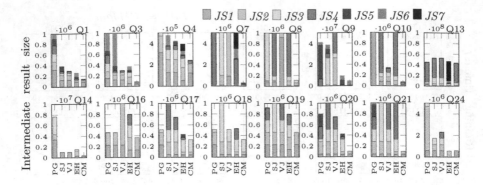

Fig. 6. Cost: intermediate result size. CM: *CMJoin*. JS: Joining step.

unwanted massive intermediate results. The evaluation shows that it performs efficiently and stably in dynamical datasets, with various queries and it also scales well.

Cost Analysis. Table 1 presents the intermediate result sizes showing that *CMJoin* outputs 5.46×, 5.90×, 1.92×, and 1.90× less intermediate results on average than *PG*, *SJ*, *VJ*, and *EH*, respectively. Figure 6 depicts more detailed intermediate results for each joining step. In general, *CMJoin* generates less intermediate results due to its designed algorithmic process, worst-case optimality, as well as join order selections. In contrast, *PG*, *SJ*, and *VJ* can easily yield too many (often quadratic) intermediate results during joining in different datasets or queries. This is because they have no technique to avoid undesired massive intermediate results.

PG and *SJ* suffer when the twig matching becomes complex in datasets (e.g. Q3 and Q10), while *VJ* suffers in the opposite case of simpler twig pattern matching (e.g. Q7 and Q16). More specifically in Q3, *PG* and *SJ* output significant intermediate results by joining of two relational tables. In turn, *VJ* controls intermediate results utilizing the values of common attributes and tags between two models. On the other hand, in Q7, Q9, and Q16 *VJ* does not consider structural matching at first yielding unnecessary quadratic intermediate results. The above two-side examples indicate that solutions considering only one model at a time or joining values first without twig matching produce an undesired significant intermediate result.

EH suffers when the queries and attributes are more connected which leads to larger intermediate results during join procedures. The reason is that *EH* seeks a better instance bound so that it follows the query plan based on the GHD decomposition [1]. Our proposed method, *CMJoin*, by wisely joining between models, avoids an unnecessary massive intermediate output from un-joined attributes.

Summary. We summarize evaluations of *CMJoin* as follows:

1. Extensive experiments on diverse datasets and queries show that averagely *CMJoin* achieves up to 13.43× faster runtime performance and produces up to 5.46× less intermediate results compared to other solutions.
2. With skew data *CMJoin* avoids undesired huge intermediate results by wisely joining data between models. With uniform data *CMJoin* filters out more values by joining one attribute at a time between all models.
3. With more tables, twigs, or common attributes involved *CMJoin* seems to perform more efficiently and scale better.

5   Related Work

Worst-Case Size Bounds and Optimal Algorithms. Recently, Grohe and Marx [15] and Atserias, Grohe, and Marx [2] estimated size bounds for conjunctive joins using the fractional edge cover. That allows us to compute the worst-case size bound by linear programming. Based on this worst-case bound, several worst-case optimal algorithms have been proposed (e.g. *NPRR* [26], *LeapFrog* [31], *Joen* [8]). Ngo et al. [26] constructed the first algorithm whose running time is worst-case optimal for all natural join queries. Veldhuizen [31] proposed an optimal algorithm called *LeapFrog* which is efficient in practice to implement. Ciucanu et al. [8] proposed an optimal algorithm *Joen* which joins each attribute at a time via an improved tree representation. Besides, there exist research works on applying functional dependencies (FDs) for size bound estimation. The initiated study with FDs is from Gottlob, Lee, Valiant, and Valiant (GLVV) [14], which introduces an upper bound called GLVV-bound based on a solution of a linear program on polymatroids. The follow-up study by Gogacz et al. [11] provided a precise characterization of the worst-case bound with information-theoretic entropy. Khamis et al. [19] provided a worst-case optimal algorithm for any query where the GLVV-bound is tight. See an excellent survey on the development of worst-case bound theory [27].

Multi-model Data Management. As more businesses realized that data, in all forms and sizes, are critical to making the best possible decisions, we see a continuing growth of demands to manage and process massive volumes of different types of data [21]. The data are represented in various models and formats: structured, semi-structured, and unstructured. A traditional database typically handles only one data model. It is promising to develop a multi-model database to manage and process multiple data models against a single unified backend while meeting the increasing requirements for scalability and performance [21,24]. Yet, it is challenging to process and optimize cross-model queries.

Previous work applied naive or no optimizations on (relational and tree) CMCQs. There exist two kinds of solutions. The first is to use one query to retrieve the result from the system without changing the nature of the model [25,34]. The second is to encode and retrieve the tree data into a relational engine [1,5,29,35]. Even though the second solution accelerates twig matching, they both may suffer from generating large, unnecessary intermediate results. These solutions or optimizations did not consider cross-model worst-case optimality.

Table 2. Dataset statistics and designed queries ($m = 10^6$, $k = 10^3$).

Dataset	Statistics	Query	Relational table	XML or JSON path query	LP	#Result
D1:TreeBank[32] (Linguistic data)	Zipfian Tables: 1m rows XML: 2.4m nodes	Q1	R1(NP,VP)	S[NP]/VP//PP[IN]//NNP	N^3	7.6k
		Q2	R1(NP,VP) R2(NP,PP)		N^3	4.6k
		Q3	R1(NP,VP) R3(NP,NNP)		N^3	<0.1k
		Q4	R1(NP,VP)	S[NP]/VP//PP[IN]//NNP	N^3	1.4k
		Q5	R1(NP,VP) R2(NP,PP)	S/VP/PP/IN	N^2	0.8k
		Q6	R1(NP,VP) R3(NP,NNP)		N^3	<0.1k
D2:Xmark[30] (Auction data)	Normal Tables: 1m rows XML: 1.6m nodes	Q7	R1(incategory,quantity,email)	T7=Item[incategory]/quantity	N^2	91k
		Q8		T8=Item[incategory][localtion][quantity]//email	N^3	0.4k
		Q9	R2(item,incategory,email)	T9=Item[location]//email	N^3	2.4m
		Q10	R3(item,quantity,email)	T7, T8	N^3	0.7k
		Q11		T7, T9	N^3	0.7k
		Q12		T7, T8, T9	N^3	0.7k
D3:UniBench[34] (E-commerce)	Uniform Tables: 1m rows JSON: 2m-4m nodes	Q13	R1(asin,productID,orderID)	$.[orderID,personID]	N^3	37.0m
		Q14	R2(personID,lastname)	$.[orderID,personID,orderline[productID]]	N^3	<0.1k
		Q15	R3(productID,product.info)	$.[personID,orderline[productID, asin]]	N^3	0.1k
		Q16		OrderLine[asin]/price	N^3	1.1k
		Q17		T17=Invoice[orderID]/orderline[asin]/price	N^3	<0.1k
D3:UniBench[34] (E-commerce)	Uniform Tables: 1m rows JSON: 4m nodes XML: 1.4m nodes	Q18	R1(asin,orderID) R2(personID,lastname)	T18=Invoice[orderID]//asin	N^3	<0.1k
		Q19	$.[orderID,personID,orderline[asin]]	orderline/asin, orderline/price	N^3	1.1k
		Q20		Invoice(I)/orderID, I/orderline(O)/asin, I/O/price	N^3	<0.1k
		Q21		T17, T18	N^3	<0.1k
D4:MIMIC-III[18] (Clinical data)	Uniform Tables:0.5-10m rows JSON: 10m nodes	Q22	R1(RowID,ICUstayID,ItemID,CGID), R2(RowID,SubjectID,ICUstayID,ItemID)	T22=$.[RowID,SubjectID,HADMID]	N	<0.1k
		Q23	R1,R3(SubjectID,ItemID)	T22	$N^{\frac{3}{2}}$	<0.1k
		Q24	R1,R2,R4(RowID,SubjectID,HADMID)	T22, T23=$.[RowID,ICUstayID,ItemID,CGID], T24=$.[RowID,SubjectID,HADMID,ICUstayID,ItemID,CGID]	N	<0.1k

Some advances are already in development to process graph patterns [1,17,28]. In contrast to previous work, this paper initiates the study on the worst-case bound for cross-model conjunctive queries with both relation and tree structure data.

Join Order. In this paper, we do not focus on more complex query plan optimization. A better query plan [10,12] may lead to a better bound for some instances [1] by combining the worst-case optimal algorithm and non-cyclic join optimal algorithm (i.e. Yannakakis [33]). We leave this as the future work to continue optimizing CMCQs.

6   Conclusion and Future Work

In this paper, we studied the problems to find the worst-case size bound and optimal algorithm for cross-model conjunctive queries with relation and tree structured data. We provide the optimized algorithm, i.e. CMJoin, to compute the worst-case bound and the worst-case optimal algorithm for cross-model joins. Our experimental results demonstrate the superiority of proposal algorithms against state-of-the-art systems and algorithms in terms of efficiency, scalability, and intermediate cost. Exciting follow-ups will focus on adding graph structured data into our problem setting and designing a more general cross-model algorithm involving three data models, i.e. relation, tree and graph.

Acknowledgment. This paper is partially supported by Finnish Academy Project 310321 and Oracle ERO gift funding.

References

1. Aberger, C.R., Tu, S., Olukotun, K., Ré, C.: EmptyHeaded: a relational engine for graph processing. In: SIGMOD Conference, pp. 431–446. ACM (2016)
2. Atserias, A., Grohe, M., Marx, D.: Size bounds and query plans for relational joins. In: FOCS, pp. 739–748. IEEE Computer Society (2008)
3. Benedikt, M., Fan, W., Kuper, G.: Structural properties of XPath fragments. Theoret. Comput. Sci. **336**(1), 3–31 (2005). Database Theory
4. Björklund, H., Martens, W., Schwentick, T.: Conjunctive query containment over trees. In: Proceedings of the 11th International Conference on Database Programming Languages. DBPL 2007 (2007)
5. Bousalem, Z., Cherti, I.: XMap: a novel approach to store and retrieve XML document in relational databases. JSW **10**(12), 1389–1401 (2015)
6. Bruno, N., Koudas, N., Srivastava, D.: Holistic twig joins: optimal XML pattern matching. In: SIGMOD Conference, pp. 310–321. ACM (2002)
7. Chaudhuri, S., Vardi, M.Y.: On the equivalence of recursive and nonrecursive datalog programs. In: PODS, pp. 55–66. ACM Press (1992)
8. Ciucanu, R., Olteanu, D.: Worst-case optimal join at a time. Technical report, Oxford (2015)
9. Duggan, J., et al.: The BigDAWG polystore system. SIGMOD Rec. **44**(2), 11–16 (2015)

10. Fischl, W., Gottlob, G., Pichler, R.: General and fractional hypertree decompositions: hard and easy cases. In: PODS, pp. 17–32. ACM (2018)
11. Gogacz, T., Toruńczyk, S.: Entropy bounds for conjunctive queries with functional dependencies. arXiv preprint arXiv:1512.01808 (2015)
12. Gottlob, G., Grohe, M., Musliu, N., Samer, M., Scarcello, F.: Hypertree decompositions: structure, algorithms, and applications. In: Kratsch, D. (ed.) WG 2005. LNCS, vol. 3787, pp. 1–15. Springer, Heidelberg (2005). https://doi.org/10.1007/11604686_1
13. Gottlob, G., Koch, C., Schulz, K.U.: Conjunctive queries over trees. J. ACM 53(2), 238–272 (2006)
14. Gottlob, G., Lee, S.T., Valiant, G., Valiant, P.: Size and treewidth bounds for conjunctive queries. J. ACM 59(3), 16:1–16:35 (2012)
15. Grohe, M., Marx, D.: Constraint solving via fractional edge covers. ACM Trans. Algorithms 11(1) (2014). https://doi.org/10.1145/2636918
16. Hai, R., Geisler, S., Quix, C.: Constance: an intelligent data lake system. In: SIGMOD Conference, pp. 2097–2100. ACM (2016)
17. Hogan, A., Riveros, C., Rojas, C., Soto, A.: A worst-case optimal join algorithm for SPARQL. In: Ghidini, C., et al. (eds.) ISWC 2019. LNCS, vol. 11778, pp. 258–275. Springer, Cham (2019). https://doi.org/10.1007/978-3-030-30793-6_15
18. Johnson, A.E., et al.: MIMIC-III, a freely accessible critical care database. Sci. Data 3, 160035 (2016)
19. Khamis, M.A., Ngo, H.Q., Suciu, D.: Computing join queries with functional dependencies. In: PODS, pp. 327–342. ACM (2016)
20. Loomis, L.H., Whitney, H.: An inequality related to the isoperimetric inequality. Bull. Am. Math. Soc. 55(10), 961–962 (1949)
21. Lu, J., Holubová, I.: Multi-model data management: what's new and what's next? In: EDBT, pp. 602–605. OpenProceedings.org (2017)
22. Lu, J., Holubová, I.: Multi-model databases: a new journey to handle the variety of data. ACM Comput. Surv. 52(3), 55:1-55:38 (2019)
23. Lu, J., Ling, T.W., Chan, C.Y., Chen, T.: From region encoding to extended dewey: on efficient processing of XML twig pattern matching. In: VLDB, pp. 193–204. ACM (2005)
24. Lu, J., Liu, Z.H., Xu, P., Zhang, C.: UDBMS: road to unification for multi-model data management. CoRR abs/1612.08050 (2016)
25. Nassiri, H., Machkour, M., Hachimi, M.: One query to retrieve XML and relational data. Procedia Comput. Sci. 134, 340–345 (2018). FNC/MobiSPC
26. Ngo, H.Q., Porat, E., Ré, C., Rudra, A.: Worst-case optimal join algorithms. J. ACM 65(3), 16:1-16:40 (2018)
27. Ngo, H.Q., Ré, C., Rudra, A.: Skew strikes back: new developments in the theory of join algorithms. SIGMOD Rec. 42(4), 5–16 (2013)
28. Nguyen, D.T., et al.: Join processing for graph patterns: an old dog with new tricks. In: GRADES@SIGMOD/PODS, pp. 2:1–2:8. ACM (2015)
29. Qtaish, A., Ahmad, K.: XAncestor: an efficient mapping approach for storing and querying XML documents in relational database using path-based technique. Knowl.-Based Syst. 114, 167–192 (2016)
30. Schmidt, A., Waas, F., Kersten, M.L., Carey, M.J., Manolescu, I., Busse, R.: XMark: a benchmark for XML data management. In: VLDB, pp. 974–985. Morgan Kaufmann (2002)
31. Veldhuizen, T.L.: Leapfrog triejoin: a simple, worst-case optimal join algorithm. arXiv preprint arXiv:1210.0481 (2012)

32. Xue, N., Xia, F., Chiou, F.D., Palmer, M.: The Penn Chinese TreeBank: phrase structure annotation of a large corpus. Nat. Lang. Eng. $11(2)$, 207–238 (2005)
33. Yannakakis, M.: Algorithms for acyclic database schemes. In: VLDB, pp. 82–94. IEEE Computer Society (1981)
34. Zhang, C., Lu, J., Xu, P., Chen, Y.: UniBench: a benchmark for multi-model database management systems. In: Nambiar, R., Poess, M. (eds.) TPCTC 2018. LNCS, vol. 11135, pp. 7–23. Springer, Cham (2019). https://doi.org/10.1007/978-3-030-11404-6_2
35. Zhu, H., Yu, H., Fan, G., Sun, H.: Mini-XML: an efficient mapping approach between XML and relational database. In: ICIS, pp. 839–843. IEEE Computer Society (2017)

Leveraging Search History for Improving Person-Job Fit

Yupeng Hou[1], Xingyu Pan[2], Wayne Xin Zhao[1,4(✉)], Shuqing Bian[2],
Yang Song[3], Tao Zhang[3], and Ji-Rong Wen[1,2,4]

[1] Gaoling School of Artificial Intelligence, Renmin University of China,
Beijing, China
{houyupeng,jrwen}@ruc.edu.cn, batmanfly@gmail.com
[2] School of Information, Renmin University of China, Beijing, China
bianshuqing@ruc.edu.cn
[3] BOSS Zhipin, Beijing, China
{songyang,kylen.zhang}@kanzhun.com
[4] Beijing Key Laboratory of Big Data Management and Analysis Methods,
Beijing, China

Abstract. As the core technique of online recruitment platforms, *person-job fit* can improve hiring efficiency by accurately matching job positions with qualified candidates. However, existing studies mainly focus on the *recommendation* scenario, while neglecting another important channel for linking positions with job seekers, *i.e., search*. Intuitively, search history contains rich user behavior in job seeking, reflecting important evidence for job intention of users.

In this paper, we present a novel **Search History enhanced Person-Job Fit** model, named as **SHPJF**. To utilize both text content from jobs/resumes and search histories from users, we propose two components with different purposes. For *text matching component*, we design a BERT-based text encoder for capturing the semantic interaction between resumes and job descriptions. For *intention modeling component*, we design two kinds of intention modeling approaches based on the Transformer architecture, either based on the click sequence or query text sequence. To capture underlying job intentions, we further propose an intention clustering technique to identify and summarize the major intentions from search logs. Extensive experiments on a large real-world recruitment dataset have demonstrated the effectiveness of our approach.

Keywords: Person-job fit · Self-attention · User intention

1 Introduction

With the rapid development of Web techniques, online recruitment has become prevalent to match qualified candidates with suitable job positions or vacancies

---

Y. Hou—Work done during internship at BOSS Zhipin NLP Center.

© The Author(s), under exclusive license to Springer Nature Switzerland AG 2022
A. Bhattacharya et al. (Eds.): DASFAA 2022, LNCS 13245, pp. 38–54, 2022.
https://doi.org/10.1007/978-3-031-00123-9_3

through the online service. As the core technique of online recruitment, it is key to develop an effective algorithm for the task of *person-job fit* [24], given the huge increase in both job seekers and job positions.

On online recruitment platforms, businesses publish job descriptions introducing the requirement for the positions, while job seekers upload their resumes stating their skills and the expectation about jobs. A job seeker can browse the job listings, send applications to interested positions and further schedule the interviews. Since both job descriptions and candidate resumes are presented in natural language, person-job fit is typically casted as a text-matching task [24], and a number of studies [1, 2, 15] aim to learn effective text representation models for capturing the semantic compatibility between job and resume contents.

Existing studies mainly focus on the *recommendation* scenario, where the system provides recommended job positions and a job seeker provides implicit or explicit feedback to these recommendations through behaviors such as clicking or making an application. However, another important channel of *search* has seldom been considered in previous studies: a user can also actively issue queries and click interested jobs like in general-purpose search engine. Although on online recruitment platforms *recommendation* takes a higher volume of user behaviors, *search* also serves an important complementary function. According to the statistics on the collected dataset, search takes a considerable proportion of 19% volume of user behaviors against the major proportion of 81% by recommendation. Intuitively, search history contains rich user behavior in job seeking, reflecting important evidence for job intention of users. In particular, it will be more important to consider when resume text is not well-written (noisy, short or informal), or when the intention of the job seeker is unclear (*e.g.*, low-skilled job seekers tend to consider a diverse range of job categories).

Considering the importance of search history data, we aim to leverage these valuable user data for improving the task of person-job fit. However, there are two major challenges to effectively utilize search history data. First, search history is usually presented in the form of short queries together with clicked (or applied) jobs, and it is unclear how to capture actual job needs from search logs. Second, the semantics reflected in search history can be redundant or diverse, and it is difficult to find out the underlying major intentions for job seekers.

To this end, in this paper, we propose a novel **Search History** enhanced **Person-Job Fit** model, named as **SHPJF**. It jointly utilizes text content from job descriptions/resumes and search histories from users. On one hand, we design a BERT-based text encoder for capturing the semantic interaction between resumes and job descriptions, called *text matching component*. On the other hand, we leverage the search history to model the intention of a candidate, called *intention modeling component*. In intention modeling component, we design two kinds of intention modeling approaches based on the Transformer architecture. These two approaches consider modeling the click sequence and query text sequence, respectively. To capture underlying job intentions, we further propose an intention clustering technique to identify and summarize the major intentions from search logs. Finally, we integrate the above two components together.

To the best of our knowledge, it is the first time that search history data has been leveraged to improve the task of person-job fit. To evaluate our approach, we construct a large real-world dataset from a popular online recruitment platform. Extensive experiments have shown that our approach is more effective than a number of competitive baseline methods.

2   Related Work

Person-Job Fit (PJF) has been extensively studied in the literature as an important task in recruitment data mining [9,17]. The early research efforts of person-job fit can be dated back to Malinowski *et al.* [14], who built a bilateral person-job recommendation system using expectation maximization algorithm. Then, some works casted this problem as a collaborative filtering task [12,22].

The major limitation of early methods lies in the ignorance of the semantic information of job/resume documents. Thus, recent research mostly treats person-job fit as a text-matching task. Several models are designed to extract expressive representations from resume and job description documents via Convolutional Neural Network (CNN) [24], Recurrent Neural Network (RNN) [15] and self-attention mechanisms [1,13]. Besides, rich features of candidates and jobs are also leveraged, such as historical matched documents [19], multi-level labels [11], multi-behavioral sequences [5] or other general contextual features [8].

Most previous work considers person-job interactions from the *recommendation* scenario as implicit or explicit signals. As a comparison, we would like to extend this body of research by leveraging search history of users on online recruitment platforms, which is a valuable resource to understand the job intentions of users. We are also aware that there are some general studies that jointly consider search and recommendation [20,21]. However, to our knowledge, we are the first to leverage search history for improving person-job fit task.

3   Problem Definition

Assume there are a set of job positions $\mathcal{J} = \{j_1, j_2, \ldots, j_M\}$ and a set of candidates $\mathcal{U} = \{u_1, u_2, \ldots, u_N\}$, where M and N represent the total number of job positions and candidates respectively. Each job j (job ID) is associated with a text of job description t_j, and each candidate u (user ID) is associated with a text of resume r_u. And, an observed match set (recommendations with explicit feedback) $\mathcal{D} = \{\langle u, j, y_{u,j} \rangle | u \in \mathcal{U}, j \in \mathcal{J}\}$ is given, where $y_{u,j}$ is the binary label indicating the match result (1 for *success* and 0 for *failure*) for user u and job j.

Different from existing settings [1,15,24], we assume that the search history of each user u is also available. On recruitment platforms, a user can issue queries, and then she clicks or applies for some jobs for interviews during the session of this query. For user u, a L-length search history is formally denoted as $\mathcal{H}_u = \{\langle q_1, j_1 \rangle, \ldots, \langle q_L, j_L \rangle\}$, where q_i denotes a query (a short word sequence describing the desired position) and j_i denotes the job that user u clicks or applies for this job during the query session for q_i. The tuples in $\mathcal{H}_u$ are sorted in temporal order. Note that queries in one session might be the same, *i.e.*, a user applies for multiple jobs under the same query content. Besides, different sessions might correspond to the same query content, *i.e.*, a user issued the same query at a different time. Based on the observed match set $\mathcal{D}$ and the search history $\mathcal{H}$, our task is to learn a predictive function $f(u, j) \in [0, 1]$ to predict the confident score that a user $u \in \mathcal{U}$ will accept job position $j \in \mathcal{J}$ each other.

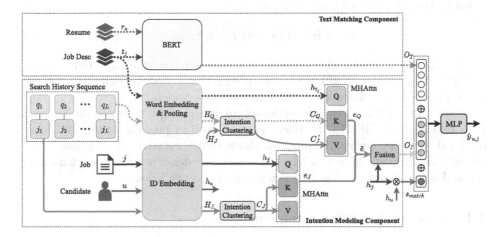

Fig. 1. The overall architecture of SHPJF.

4   The Proposed Approach

In this section, we present the proposed **Search History** enhanced **Person-Job Fit** model, named as **SHPJF**. It considers two kinds of data signals to develop the matching model, either text content from resumes/job descriptions or search histories issued by users. On one hand, we design a BERT-based text encoder for capturing the semantic interaction between resumes and job descriptions, called *text matching component*. On the other hand, we leverage the search history to model the intention of a candidate, called *intention modeling component*. Based on the Transformer architecture, we design two kinds of intention modeling approaches, either with job ID sequence or with query text sequence. Figure 1 presents the overall architecture of our proposed approach.

4.1   Text Matching Component

On online recruitment platforms, a major criterion for person-job fit is that a candidate's resume should well match the job description in text contents. It is not easy to learn a good semantic match model for this task, as candidate resumes and job descriptions are likely to be described or presented in different ways [24].

Recent years, self-attention mechanisms (*e.g.*, Transformer [18]) and its extensions on pre-trained model (*e.g.*, BERT [4]) have made great progress in various natural language processing tasks. Therefore, we adopt a standard BERT model to develop the text matching component. Given a pair of resume r_u and job description t_j, we firstly concatenate (1) a special symbol [CLS], (2) resume document, (3) a special symbol [SEP], and (4) job description document, in order and derive the input sequence for BERT. Then the concatenated sequence is fed to the BERT model, so we have

$$o_T = \text{BERT}([CLS]; r_u; [SEP]; t_j]), \tag{1}$$

where o_T is the final hidden vector corresponding to the first input token ([CLS]). The self-attention mechanism in BERT can effectively characterize the semantic interaction between resume and job description.

Note that another possible architecture is to adopt a two-tower encoder with two separate BERTs [1]. However, as shown in previous studies in dense retrieval [7], the currently adopted single-tower encoder (also called *cross encoder*) is more effective than the two-tower encoder, since it can capture fine-grained semantic interaction between the texts on two sides.

4.2   Intention Modeling Component

Above, text-based matching component mainly measures the matching degree between job and candidate based on semantic compatibility in content. As introduced in Sect. 3, search history data is also available in our setting. Since search history contains the issued queries and the applied (or clicked) jobs, it provides important evidence for modeling user's intention. To learn such intention, we design a two-stream intention aggregation approach, where both job ID sequence and query text sequence are modeled to capture the underlying intention semantics. In particular, different queries are likely to correspond to the same intention, *e.g.*, "*layer*" and "*case source*". Considering this issue, we further design an intention clustering technique. Finally, the derived representations in two streams are combined as the final intention representation.

Intention Modeling with Job ID Sequence. Recall that the search history for user u is given as $\mathcal{H}_u = \{\langle q_1, j_1 \rangle, \langle q_2, j_2 \rangle, \ldots, \langle q_L, j_L \rangle\}$ in Sect. 3. We first consider extracting the intention semantic from the sequence of clicked or applied job IDs, namely $j_1 \rightarrow j_2 \ldots \rightarrow j_L$, where these jobs reflect the job preference of user u. We first embed the job IDs in the search history:

$$h_1, h_2, \ldots, h_L = \text{IDEmb}(j_1, j_2, \ldots, j_L), \tag{2}$$
$$H_J = [h_1; h_2; \ldots; h_L], \tag{3}$$

where h_j denotes the embedding for job j with the IDEmb layer. Here, we apply a look-up operation to obtain job ID embeddings from the IDEmb layer. Note that IDs of the recommended job and those in the search history are embedded into the same semantic space (by the same look-up table IDEmb in Eqn. (2)).

Intention Clustering. As one candidate may have diverse job intentions, the embeddings of interested jobs from search history are clustered into several *intentions*. Formally, we have

$$C_J = P_J H_J, \tag{4}$$
$$P_J = \text{softmax}(W_1 H_J^\top + b_1), \tag{5}$$

where P_J is a probability assignment matrix that gives the probabilities of a job (from search history) into each cluster, and W_1 and b_1 are learnable parameter matrix and bias respectively. Here, $C_J \in \mathbb{R}^{k \times d_j}$ is an intention embedding matrix, where each of the k intentions is mapped into a b-dimensional vector.

Job-specific Intention Representation. After obtaining the representations of k intentions (C_J) for user u, we next learn the intention representation of user u given a recommended job position j. The basic idea is to consider the relevance of job j with each of the learned intention representations. Intuitively, a job tends to be accepted by a user if it highly matches with some specific intention of the user. We adopt the self-attention architecture [18] to characterize this idea by attending job j to each intention embedding. To be specific, we adopt multi-head attention operation [18]:

$$\text{MHAttn}(Q, K, V) = [head_1, \ldots, head_h] W^O, \tag{6}$$
$$head_i = \text{softmax}\left(Q W_i^Q (K W_i^K)^\top / \sqrt{D}\right) V W_i^V, \tag{7}$$

where W_i^Q, W_i^K, W_i^V and W^O are parameter matrices, and $\frac{1}{\sqrt{D}}$ is the scaling factor. With the above attention operation, we specially designed *queries* (job embedding), *keys* (intention embedding) and *values* (intention embedding) as:

$$e_J = \text{MHAttn}(h_j, C_J, C_J), \tag{8}$$

where h_j is the embedding for job j obtained from the IDEmb layer, and $e_J \in \mathbb{R}^{d_j}$ denotes the learned intention representation based on job ID sequence. By mapping job j (as a *query*) against a sequence of intentions (as *keys*), each intention is assigned with a relevance score. Those highly relevant intentions (as *values*) will receive a larger attention weight. Since the recommended job j has attended to each intention, e_J encodes important evidence for measuring the match degree between job j and user u.

Intention Modeling with Query Text Sequence. To learn user intention, another kind of important data signal from search history is the *query text*, *i.e.*, $q_1 \rightarrow q_2 \ldots \rightarrow q_L$. Each query is a short sequence of words, reflecting the user intention in job seeking. We follow the similar approach as modeling job ID sequence to model query text sequence. Firstly, we apply the look-up operation and the average pooling to represent each query in low-dimensional space:

$$\widetilde{h}_1, \widetilde{h}_2, \ldots, \widetilde{h}_L = \text{Pooling}(\text{WordEmb}(q_1, q_2, \ldots, q_L)), \tag{9}$$

$$H_Q = [\widetilde{h}_1, \widetilde{h}_2, \ldots, \widetilde{h}_L], \tag{10}$$

where $\text{WordEmb}(\cdot)$ is a learnable embedding layer, and $\text{Pooling}(\cdot)$ is an average pooling layer that aggregates several vectors into a single representation.

Intention Clustering. By considering both query text and job ID, we can derive a more comprehensive intention learning approach by following Eqn. (4) and (5):

$$C_Q = P_Q H_Q, \tag{11}$$

$$C'_J = P_Q H_J, \tag{12}$$

$$P_Q = \text{softmax}(W_2[H_Q; H_J]^\top + b_2), \tag{13}$$

where $C_Q \in \mathbb{R}^{k \times d_j}$ and $C_J \in \mathbb{R}^{k \times d_j}$ denote the learned intention representations (k intentions) based on query text and job ID, respectively, P_Q is a probability assignment matrix of the probabilities of jobs in the learned intentions (clusters), and W_2 and b_2 are learnable parameter matrix and bias respectively.

Job-specific Intention Representation. After obtaining the representations of k intentions (C'_J and C_Q) for user u, we next learn the intention representation of user u given a recommended job position j. Following Eqn. (8), we still adopt the multi-head attention to learn the intention representation:

$$e_Q = \text{MHAttn}(h_{t_j}, C_Q, C_J), \tag{14}$$

where the representation of job description for j (denoted by t_j) to be matched, denoted as h_{t_j}, is used as *query*, the clustered query representations are used as *key*, and the clustered ID representations are used as *value*. Here the relevance of job j to intentions is defined by the similarity between job description representation h_{t_j} and the clustered query text representations C_Q. Intentions associated with highly relevant queries are assigned with high attention weights. The derived e_Q can represent the learned intention through query text. A key point is that we still adopt C_J as *values*, so that e_Q and e_J can be subsequently fused.

Intention Representation Fusion. To combine the above two kinds intention-based representations, we apply a weighted linear combination:

$$\tilde{e} = \lambda e_J + (1 - \lambda)e_Q, \tag{15}$$

where e_J (Eqn. (8)) and e_Q (Eqn. (14)) are the learned intention representations, and λ is a tuning coefficient. If a recommended job well matches the job seeking intention of a user, it indicates that such a person-job pair should be more likely to be successful. Following [15], we measure the match degree by fusion:

$$o_I = \text{MLP}\big([\tilde{e}; h_j; \tilde{e} - h_j; \tilde{e} \circ h_j]\big), \tag{16}$$

where o_I encodes the match information about this person-job pair based on intention, h_j is the embedding for job j, $\tilde{e}$ is the combined intention representation (Eqn. (15)), MLP is multilayer perceptron stacked with fully connected layers, and "$\circ$" denotes the hadamard product operation.

4.3   Prediction and Optimization

With the text matching and intention modeling component, we finally integrate the two match representations to predict the confident score of a person-job pair:

$$\hat{y}_{u,j} = \sigma(\text{MLP}\big([o_T; o_I; s_{match}]\big)), \tag{17}$$

where $\hat{y}_{u,j} \in [0,1]$ indicates the matching degree between candidate u and job j, o_T and o_I are the learned match representations in Eqns. (1) and (16), respectively. In addition to two match representations, we also incorporate a simple match score based on the user embedding and job embedding, *i.e.*, $s_{match} = h_j^\top \cdot h_u$, where h_u and h_j are obtained from the IDEmb($\cdot$) layer.

We adopt binary cross entropy loss to optimize our model,

$$\mathcal{L} = \sum_{\langle u,j,y_{u,j}\rangle \in \mathcal{D}} - [y_{u,j} \cdot \log \hat{y}_{u,j} + (1 - y_{u,j}) \cdot \log(1 - \hat{y}_{u,j})], \tag{18}$$

where we iterate the training dataset and compute the accumulate loss.

Learning. In our model, various kinds of embeddings (IDEmb layer in Eqn. (2), WordEmb layer in Eqn. (9)) and involved component parameters are the model parameters. Note that each MHAttn (Eqns. (8) and (14)) and intention clustering layers (Eqns. (5) and (13)) are with different parameters. In order to avoid overfitting, we adopt the dropout strategy with a rate of 0.2. More implementation details can be found in Sect. 5.1.

Time Complexity. For online service, it is more important to analyze online time complexity for a given impression list of q job positions. Our text matching component requires a time of $\mathcal{O}(lm^2 d_w + lmd_w^2)$, where d_w is the token embedding dimension, m is the truncated length of input tokens and l is the number of BERT layers. While the cross-encoder architecture can be efficiently accelerated with an approximately learned dual-encoder (*e.g.*, distillation [10]). We can also accelerate our text matching component with the two-tower architecture described in Sect. 4.1. In this way, the representations of resumes and job descriptions can be calculated offline, and we only need to perform an inner product operation online.

As for our intention modeling component, we adopt a lazy-update technique to pre-calculate each user's intention representations C_J (Eqn. (4)), C'_J

(Eqn. (12)) and C_Q (Eqn. (11)). We update user's search history and update C_J, C'_J and C_Q offline every several hours. Thus, the complexity mainly depends on the calculation of multi-head attention mechanism (Eqn. (8) and Eqn. (14)). Suppose k is the number of clusters and d_j is dimension of clustered intention representations. The complexity of one single pass of intention modeling component is $\mathcal{O}(kd_j^2)$. Suppose MLP in Eqns. (16) and (17) are both s layers, the complexity of intention representation fusion (Eqn. (16)) is $\mathcal{O}(s \cdot d_j^2)$. For one pair of candidate and job, the complexity of prediction (Eqn. (17)) is $\mathcal{O}(s \cdot (d_w + d_j)^2)$. The overall complexity of online serving for a session of n jobs is $\mathcal{O}\left(nlm^2 d_w + nlmd_w^2 + nkd_j^2 + ns(d_w + d_j)^2\right)$. Such a time complexity can be further reduced with parallel efficiency optimization.

Table 1. Statistics of the datasets. $\overline{L}$ denotes the average length of search history per candidate. $\overline{|q|}$ denotes the average number of words per query.

#candidates	#jobs	#positive	#negative	$\overline{L}$	$\overline{|q|}$
53, 566	307, 738	257, 922	2, 109, 876	16.55	1.50

5 Experiments

In this section, we conduct extensive experiments to verify the effectiveness of our model. In what follows, we first set up the experiments, and then present and analyze the evaluation results.

5.1 Experimental Setup

Datasets. We evaluate our model on a real-world dataset provided by a popular online recruiting platform. The records of our dataset are collected from the real online logs between November 3 and November 12 in 2020. We anonymize all the records by removing identity information to protect the privacy of candidates. Our dataset and code are available at https://github.com/RUCAIBox/SHPJF.

There are three kinds of user behavior in our dataset, called *Accept*, *Apply*, and *Exposure*. *Accept* means that a candidate and a company reach an agreement on an offline interview. *Apply* means that a candidate applies for a job. Generally, it means that the candidate shows a clear intention to the applied job [11]. *Exposure* means that a job position has been exposed to the candidate but the candidate may not perform further behavior. To construct the evaluation dataset, all the job-user pairs with *Accept* are considered to be *positive instances*, while those pairs with *Exposure* but without further behavior are considered to be *negative instances*. Since the number of *Exposure* is huge so that for each positive instance we pair it with several negative instances from the same exposure of a recommendation list. The ratio of positive and negative instances is approximately 1 : 8. Note that all the instances are from the *recommendation*

Table 2. Performance comparisons of different methods. The improvement of our model over the best baseline is significant at the level of 0.01 with paired t-test.

Method	GAUC	R@1	R@5	MRR
PJFNN	0.5313	0.1412	0.5192	0.4025
BPJFNN	0.5343	0.1391	0.5217	0.4008
APJFNN	0.5323	0.1403	0.5185	0.4000
BERT	0.5449	0.1515	0.5297	0.4129
MV-CoN	0.5463	0.1554	0.5307	0.4165
SHPJF (ours)	**0.5785**	**0.1630**	**0.5516**	**0.4297**

channel, we remove the ones which also appear in the *search* channel for the same user to avoid data leakage.

We sort the records of the selected instances by timestamp. Records of the last two days are used as validation set and test set respectively, while the others are used for training. For a user, we also obtain her search history (queries with clicked jobs). The number of browsing and clicking behaviors is very large and noisy. Therefore, we only keep jobs with the status of *Apply*. We truncate the search history before the timestamp of validation set. The statistics of the processed data are summarized in Table 1.

Baselines. We compare our model with the following representative methods:

- **PJFNN** [24] is a convolutional neural network (CNN) based method, resumes and job descriptions are encoded independently by hierarchical CNNs.
- **BPJFNN** [15] leverages bidirectional LSTM to derive the resume and job description representations.
- **APJFNN** [15] proposes to use hierarchical RNNs and co-attention techniques to process the job positions and resumes.
- **BERT** [4] is a broadly used pre-trained model for learning text representations. Here, we adopt the released pre-trained BERT-Base-Uncased model. Then it was fine-tuned on our dataset as a sentence pair classification task.
- **MV-CoN** [1] is a BERT-based multi-view co-teaching network that is able to learn from sparse, noisy interaction data for job-resume matching.

We finally report metrics on the test set with models that gain the highest performance on the validation set.

Implementation Details. Text-matching module is initialized via the BERT-Base-Uncased[1]. The dimensions of word embeddings and ID embeddings are 128 and 16 respectively. The dimension of the hidden state is 64. Hyperparameters of baselines are tuned in the recommended range from the original papers. We select combination coefficient λ as 0.6, number of clusters k as 4 and number of heads h as 1. The Adam optimizer is used to learn our model, and the

---

[1] https://github.com/huggingface/transformers.

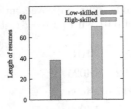

(a) Average resume length.

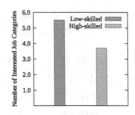

(b) Average #interested categories.

Fig. 2. Comparison of low-skilled and high-skilled candidates. Each job position belongs to one of the 950+ pre-defined job categories.

Table 3. Performance comparisons of different methods on low-skilled and high-skilled candidates. APJFNN is taken as the base model to compute the RelaImpr.

Groups	Low-skilled Candidates		High-skilled Candidates	
Method	GAUC	RelaImpr	GAUC	RelaImpr
PJFNN	0.5295	−4.53%	0.5318	−2.75%
BPJFNN	0.5399	+29.13%	0.5326	−0.31%
APJFNN	0.5309	0.00%	0.5327	0.00%
BERT	0.5381	+23.30%	0.5470	+43.73%
MV-CoN	0.5396	+28.16%	0.5484	+48.01%
SHPJF	**0.5689**	**+122.98%**	**0.5814**	**+148.93%**

learning rate is tuned in $\{0.01, 0.001, 0.00001\}$. The dropout ratio is tuned in $\{0, 0.1, 0.2, 0.3, 0.5\}$. We adopt early stopping with patience of 5 epochs.

Evaluation Metrics. Given a candidate, we tend to rank the more appropriate job positions higher in the recommended list. Thus, we adopt Grouped AUC (GAUC), Recall@K (R@$\{1, 5\}$) and MRR to evaluate our models. As the traditional AUC metric doesn't treat different users differently, we use GAUC [23, 25], which averages AUC scores over users.

5.2 The Overall Comparison

Table 2 presents the performance comparison between our model and the baselines on person-job fit. Overall, the three methods PJFNN, BPJFNN and APJFNN tend to have similar performance on our dataset. Furthermore, the two BERT-based methods (*i.e.,* BERT and MV-CoN) seem to perform better leveraging the excellent modeling capacities of pre-trained models. Different from previous studies on person-job fit, our negative instances are more strong, *i.e.,* they are from the same impression list with the positive instance. So, the text match model should be capable of identifying fine-grained variations in semantic representations. This may explain why BERT-based methods are better than traditional neural network-based methods. Another observation is that MV-CoN is slightly better than BERT by further considering multi-view learning.

As a comparison, our model achieves the best performance on all the metrics. In specific, SHPJF can improve the best baseline's GAUC result by 3.22% and 5.89% absolutely and relatively, respectively. These results demonstrate the effectiveness of our approach. In particular, we incorporate a special intention component that learns user intention representations from search history. Such a component is able to enhance the base text matching component (a cross encoder based on BERT), which is the key to performance improvement.

5.3 Evaluation in Different Skill Groups

According to the employability in the labor market, candidates can be divided into *low-skilled* and *high-skilled* candidates. Furthermore, candidates who applied for jobs with less training participation and high task flexibility can be classified as low-skilled candidates [3,16]. It is usually more difficult for low-skilled candidates to find suitable job positions. Therefore, we would like to examine the performance improvement *w.r.t.* different groups.

In our recruitment platform, domain experts manually annotate low-skilled candidates in order to provide specific requirement strategies, so that each user in our dataset will be associated with a label indicating that whether she/he is a low-skilled candidate. Figure 2 shows the average resume lengths and the average number of interested job categories in the two groups. As we can see, low-skilled candidates have a shorter resume in text and a more diverse of interested job categories. These characteristics make it more difficult to apply text-based matching algorithms in finding suitable job positions for low-skilled candidates. As a comparison, our method incorporates search history to enhance the matching model, which is expected to yield larger improvement on low-skilled candidates.

To examine this, we follow [23] and introduce *RelaImpr* as a metric of the relative improvements over the base model. Here, we adopt APJFNN as the base model and check how the other methods improve over it in different groups. As shown in Table 3, it is more difficult to recommend suitable jobs for low-skilled candidates (lower performance). The improvement of BERT and MV-CoN is actually very small, which means that BERT-based approaches mainly improve the performance of high-skilled candidates. As a comparison, our method yields substantial improvement in low-skilled candidates, which further indicates the necessity of leveraging search history data.

5.4 Ablation Study

The major technical contribution of our approach lies in the intention modeling component. It involves several parts and we now analyze how each part contributes to the final performance.

We consider the following four variants of our approach for comparison: **(A)** BERT is a BERT-based text matching model (same as the text matching component in Sect. 4.1); **(B)** BERT$_{GRU}$ replaces the intention modeling component

Table 4. Ablation study of the variants for our model.

Variants	GAUC	R@1	R@5	MRR
BERT	0.5449	0.1515	0.5297	0.4129
BERT$_{GRU}$	0.5557	0.1546	0.5334	0.4196
BERT$_{query}$	0.5572	0.1558	0.5342	0.4201
SHPJF w/o Q	0.5697	0.1599	0.5456	0.4270
SHPJF w/o J	0.5715	**0.1634**	0.5456	0.4286
SHPJF w/o C	0.5738	0.1581	0.5443	0.4237
SHPJF	**0.5785**	0.1630	**0.5516**	**0.4297**

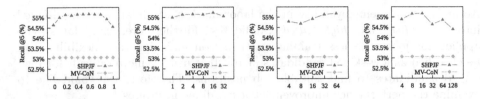

(a) Coefficient λ.　(b) The number of (c) Length of search (d) ID embedding
clusters k.　　history sequence.　dimensionality.

Fig. 3. Performance tuning of our model on different hyper-parameters

with GRU4Rec [6] to encode the job ID sequence; **(C)** BERT$_{query}$ replaces the intention modeling component (Sect. 4.2) with a BERT-based model to encode the concatenated query text sequence; **(D)** SHPJF w/o Q removes the part of intention modeling with query text sequence; **(E)** SHPJF w/o J removes the part of intention modeling with job ID sequence; **(F)** SHPJF w/o C removes the intention clustering (Eqns. (4), (12) and (11)).

In Table 4, we can see that the performance order can be summarized as: BERT < BERT$_{GRU}$ $\simeq$ BERT$_{query}$ < SHPJF w/o Q < SHPJF w/o J $\simeq$ SHPJF w/o C < SHPJF. These results indicate that all the parts are useful to improve the final performance. Besides, GRU4Rec doesn't perform well in our experiments. We observe that user behavior sequences in person-job fit differ from those in traditional sequential recommendation scenarios [6]. Existing research also mainly treats person-job fit as a text-matching task but doesn't directly leverage user behavior sequences. We can see that the carefully designed Transformer architecture performs better in modeling search logs (*e.g.*, BERT$_{GRU}$ < SHPJF w/o Q, BERT$_{query}$ < SHPJF w/o J). In particular, intention modeling with query text sequences brings more improvement than only with job ID sequences, as query texts directly reflect users' intention in job seeking.

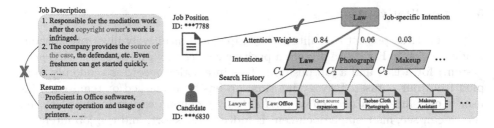

Fig. 4. Case study: a matched pair between a low-skilled candidate and a job position.

5.5 Performance Tuning

In this part, we examine the robustness of our model and perform a detailed parameter analysis. For simplicity, we only incorporate the best baseline MV-CoN from Table 2 as a comparison.

Varying the Combination Coefficient λ***.*** We use a coefficient λ to combine the two involved intention representations in Eqn. (15). Here, we vary the combination coefficient λ in the range of 0 and 1 with a gap of 0.1. When $\lambda = 1$ and 0, the model degenerates to SHPJF w/o Q and SHPJF w/o J described in Sect. 5.4 respectively. As Fig. 3a presented, our model achieves the best performance when $\lambda = 0.6$. With a selection between 0.4 and 0.8, the performance is relatively stable. The results indicate both intention modeling methods are important for our task, which can complement each other in performance.

Varying the Number of Clusters. In the intention modeling component, we introduce k to denote the number of clusters. The larger k is, the more fine-grained the learned intentions will be. Here, we vary the number of clusters in the set $\{1, 2, 4, 8, 16, 32\}$. As shown in Fig. 3b, our model achieves the best performance when $k = 4$ and $k = 16$. Overall, the performance is relatively stable with different values for k.

Varying the Length of Search History. In our work, we leverage search history to enhance person-job fit. It is intuitive that the length of search history will affect the final performance. Here, we vary the (maximum) sequence length of search history in a selection of $\{4, 8, 16, 32, 64\}$. The tuning results are presented in Fig. 3c. From Fig. 3c, we can see that our model gradually improves with the increase of the sequence length. It shows that using more search history will boost the match performance.

Varying the ID Embedding Dimensionality. We vary the embedding dimensionality d_j in a selection of $\{4, 8, 16, 32, 64, 128\}$ to examine how the performance changes. As shown in Fig. 3d, we find a small embedding dimensionality ($d_j = 8$ or 16) can lead to a good performance, which can be more efficient in the industrial deployment.

5.6 Qualitative Analysis

In this part, we present a qualitative example, and intuitively illustrate how our model works. In Fig. 4, we present a positive case (*i.e.,* with the status *Accept* in system) for a user and a job position. This interaction record is randomly sampled from our dataset. Privacy information has been masked or removed. This user is classified as a *low-skilled candidate* by a domain expert in our platform. As we can see, his resume document is indeed very short, and the job intention is not clearly stated. Given this case, it is difficult to match the candidate with the current position, as the job description has few overlapping words with the resume (semantically different). Therefore, previous text-based matching algorithms [1,15,24] would fail in this matched case.

However, by checking the candidate's search history, we find that he has issued queries about several job intentions, like *"Makeup Assistant"*, *"Photograph"*, *"Lawyer"*, etc. These intentions cannot be extracted or inferred from his resume. While our intention modeling component is effective to derive meaningful clusters about intentions. Then, the intention clusters with at least one of the following conditions will be assigned high attentions weights: **(1)** contain similar jobs in search history as the job position to be matched. **(2)** contain query words highly related or similar to words in the job description, such as *"source of case - Case source"* (marked in red) and *"copyright owner - Law/Lawyer"* (marked in purple). Once the correct intention cluster has been identified (with a large attention weight), we can indeed derive a job-specific intention representation (see Sect. 4.2). In this way, we can correctly match this job-candidate pair even if their documents are semantically different.

This example shows that leveraging search history is able to improve the performance for person-job fit. It also intuitively explains why our method can yield performance improvement on low-skilled candidates (see Table 3).

6 Conclusion

In this paper, we presented the first study that leveraged search history data for improving the task of person-job fit. Our approach developed a basic component based on BERT for capturing the semantic interaction between resumes and job descriptions. As the major technical contribution, we designed a novel intention modeling component that was able to learn and identify the intention of job seekers. It modeled two kinds of sequences, either the click sequence or query text sequence. Furthermore, an intention clustering technique is proposed to accurately capture underlying job intentions. Extensive experiments on a large recruitment data have shown the demonstrated the effectiveness of our approach.

Besides search history, there are other kinds of side information for person-job fit. As future work, we will consider developing a more general approach to leverage various kinds of side information in a unified way.

Acknowledgements. This work was partially supported by the National Natural Science Foundation of China under Grant No. 61872369 and 61832017, Beijing Outstanding Young Scientist Program under Grant No. BJJWZYJH012019100020098.

References

1. Bian, S., et al.: Learning to match jobs with resumes from sparse interaction data using multi-view co-teaching network. In: CIKM (2020)
2. Bian, S., Zhao, W.X., Song, Y., Zhang, T., Wen, J.R.: Domain adaptation for person-job fit with transferable deep global match network. In: EMNLP (2019)
3. De Grip, A., Van Loo, J., Sanders, J.: The industry employability index: Taking account of supply and demand characteristics. Int. Lab Rev. **143**, 211 (2004)
4. Devlin, J., Chang, M.W., Lee, K., Toutanova, K.: Bert: Pre-training of deep bidirectional transformers for language understanding. In: NAACL (2019)
5. Fu, B., Liu, H., Zhu, Y., Song, Y., Zhang, T., Wu, Z.: Beyond matching: Modeling two-sided multi-behavioral sequences for dynamic person-job fit. In: DASFAA (2021)
6. Hidasi, B., Karatzoglou, A., Baltrunas, L., Tikk, D.: Session-based recommendations with recurrent neural networks. In: ICLR (2016)
7. Humeau, S., Shuster, K., Lachaux, M., Weston, J.: Poly-encoders: architectures and pre-training strategies for fast and accurate multi-sentence scoring. In: ICLR (2020)
8. Jiang, J., Ye, S., Wang, W., Xu, J., Luo, X.: Learning effective representations for person-job fit by feature fusion. In: CIKM (2020)
9. Kenthapadi, K., Le, B., Venkataraman, G.: Personalized job recommendation system at LinkedIn: practical challenges and lessons learned. In: RecSys (2017)
10. Lan, Z., Chen, M., Goodman, S., Gimpel, K., Sharma, P., Soricut, R.: ALBERT: a lite BERT for self-supervised learning of language representations. In: ICLR (2020)
11. Le, R., Hu, W., Song, Y., Zhang, T., Zhao, D., Yan, R.: Towards effective and interpretable person-job fitting. In: CIKM (2019)
12. Lu, Y., Helou, S.E., Gillet, D.: A recommender system for job seeking and recruiting website. In: WWW (2013)
13. Luo, Y., Zhang, H., Wen, Y., Zhang, X.: ResumeGAN: an optimized deep representation learning framework for talent-job fit via adversarial learning. In: CIKM (2019)
14. Malinowski, J., Keim, T., Wendt, O., Weitzel, T.: Matching people and jobs: a bilateral recommendation approach. In: HICSS (2006)
15. Qin, C., Zhu, H., Xu, T., Zhu, C., Jiang, L., Chen, E., Xiong, H.: Enhancing person-job fit for talent recruitment: an ability-aware neural network approach. In: SIGIR (2018)
16. Sanders, J., De Grip, A.: Training, task flexibility and the employability of low-skilled workers. Int. J. Manpower **25**, 1–21 (2004)
17. Shalaby, W., et al.: Help me find a job: a graph-based approach for job recommendation at scale. In: BigData (2017)
18. Vaswani, A., et al.: Attention is all you need. In: NeurIPS (2017)
19. Yan, R., Le, R., Song, Y., Zhang, T., Zhang, X., Zhao, D.: Interview choice reveals your preference on the market: To improve job-resume matching through profiling memories. In: SIGKDD (2019)
20. Zamani, H., Croft, W.B.: Joint modeling and optimization of search and recommendation. In: DESIRES (2018)

21. Zamani, H., Croft, W.B.: Learning a joint search and recommendation model from user-item interactions. In: WSDM (2020)
22. Zhang, Y., Yang, C., Niu, Z.: A research of job recommendation system based on collaborative filtering. In: ISCID (2014)
23. Zhou, G., et al.: Deep interest network for click-through rate prediction. In: SIGKDD (2018)
24. Zhu, C., et al.: Person-job fit: Adapting the right talent for the right job with joint representation learning. In: TMIS (2018)
25. Zhu, H., et al.: Optimized cost per click in Taobao display advertising. In: KDD (2017)

Efficient In-Memory Evaluation of Reachability Graph Pattern Queries on Data Graphs

Xiaoying Wu[1], Dimitri Theodoratos[2(✉)], Dimitrios Skoutas[3], and Michael Lan[2]

[1] Wuhan University, Wuhan, China
`xiaoying.wu@whu.edu.cn`
[2] New Jersey Institute of Technology, Newark, USA
`{dth,mll22}@njit.edu`
[3] R.C. Athena, Athens, Greece
`dskoutas@imis.athena-innovation.gr`

Abstract. Graphs are a widely used data model in modern data-intensive applications. Graph pattern matching is a fundamental operation for the exploration and analysis of large data graphs. In this paper, we present a novel approach for efficiently finding homomorphic matches of graph pattern queries, where pattern edges denote reachability relationships between nodes in the data graph. We first propose the concept of query reachability graph to compactly encode all the possible homomorphisms from a query pattern to the data graph. Then, we design a graph traversal-based filtering method to prune nodes from the data graph which violate reachability conditions induced by the pattern edges. We use the pruned data graph to generate a refined query reachability graph which serves as a compact search space for the pattern query answer. Finally, we design a multiway join algorithm to enumerate answer tuples from the query reachability graph without generating an excessive number of redundant intermediate results (a drawback of previous approaches). We experimentally verify the efficiency of our approach and demonstrate that it outperforms by far existing approaches and a recent graph DBMS on evaluating reachability graph pattern queries.

Keywords: Graph pattern matching · Edge-to-path homomorphism · Multi-way join

1 Introduction

Graphs model complex relationships among entities in a plethora of applications. A fundamental operation for querying, exploring and analysing graphs is graph matching, which identifies the matches of a query pattern in the data graph. Graph matching is crucial in many application domains, such as social

The research of the first author was supported by the National Natural Science Foundation of China under Grant No. 61872276.

A. Bhattacharya et al. (Eds.): DASFAA 2022, LNCS 13245, pp. 55–71, 2022.
https://doi.org/10.1007/978-3-031-00123-9_4

network mining, biological network analysis, cheminformatics, fraud detection, and network traffic monitoring.

Existing graph matching approaches can be characterized by: (a) the *type of edges* the patterns have and (b) the *type of morphism* used to map the patterns to the data structure. An edge in a pattern can be either a *child* edge, which is mapped to an *edge* in the data graph (*edge-to-edge* mapping), or a *descendant* edge, which is mapped to a *path* in the data graph (*edge-to-path* mapping). The morphism determines how a pattern is mapped to the data graph: a homomorphism is a mapping from pattern nodes to data graph nodes, while an isomorphism is a one-to-one function from pattern nodes to data graph nodes. Research has initially considered edge-to-edge mappings [13] for matching patterns, with several algorithms being proposed [1,2,7,11]. However, isomorphisms are very restricted [6] as they cannot represent reachability (transitive) relationships. More recently, *homomorphisms* for matching *patterns with descendant edges* have been considered [5,6,8,14,15]. By allowing edge-to-path mappings, homomorphisms can extract matches "hidden" deep within large graphs, e.g., transitive subClassOf, partOf, influence or other relationships, which are missed by isomorphisms.

Existing graph pattern matching algorithms can be broadly classified into the following two approaches: the *join-based* approach (*JM*) [1,5,9] and the *tree-based* approach (*TM*) [2,3,7,15]. Given a graph pattern query Q, *JM* first decomposes Q into a set of binary (reachability) relationships between pairs of nodes. The query is then evaluated by matching each binary relationship against the data graph and joining together these individual matches. Unlike *JM*, *TM* first decomposes or transforms Q into one or more tree patterns using various methods, and then uses them as the basic processing unit. Both *JM* and *TM* suffer from a potentially exploding number of intermediate results which can be substantially larger than the final output size of the query, thus spending a prohibitive amount of time on examining false positives. As a consequence, they display limited scalability. Our experimental results also reveal that query engines of existing graph DBMS are unable to handle reachability graph pattern queries efficiently.

In this paper, we address the problem of evaluating graph pattern queries with descendant edges (edge-to-path mapping) using homomorphisms over a data graph. This is a general setting for graph pattern matching. We develop a new graph pattern matching framework, which consists of two phases: (a) the *summarization phase*, where a query-dependent summary graph is built on-the-fly, serving as a compact search space for the given query, and (b) the *enumeration phase*, where query solutions are produced using the summary graph.

Contribution. Our main contributions are summarized as follows:

- We propose the concept of *query reachability graph* to encode all possible homomorphisms from a query pattern to the data graph. By losslessly summarizing the occurrences of a given pattern, a query reachability graph represents results more succinctly. A query reachability graph can be efficiently built *on-the-fly* and does not have to persist on disk.

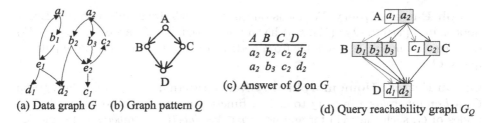

(a) Data graph G (b) Graph pattern Q

A	B	C	D
a_2	b_2	c_2	d_2
a_2	b_3	c_2	d_2

(c) Answer of Q on G

(d) Query reachability graph G_Q

Fig. 1. A data graph G, a query Q, and query reachability graph G_Q.

- We develop a node filtering method which traverses the data graph to efficiently prune nodes violating reachability constraints imposed by query edges. Using this filtering method, we build a refined query reachability graph to further reduce the query answer search space.
- We develop a novel algorithm for enumerating the results of graph pattern queries. In order to compute the results, our algorithm performs multiway joins by intersecting node lists and node adjacency lists in the query reachability graph. Unlike both *JM* and *TM*, it avoids generating a potentially exploding number of intermediate results and has small memory footprint. We integrate the above techniques to design a graph pattern matching algorithm, called *GM*.
- We implement *GM* and experimentally verify its time efficiency and scalability for evaluating reachability graph pattern queries on real datasets. We compare *GM* with both the *JM* and *TM* approaches as well as the query engine of a recent graph DBMS. The results show that *GM* can efficiently evaluate graph pattern queries with varied structural characteristics and with tens of nodes on data graphs, and that it outperforms by a wide margin both *JM* and *TM* and the graph query engines.

2   Preliminaries and Problem Definition

We focus on directed, connected, and node-labeled graphs. However, the techniques presented in this paper can be readily extended to handle other cases, such as undirected/disconnected graphs and graphs with multiple labels on nodes/edges.

Let $G = (V, E)$ be a directed node-labeled graph, where V denotes the set of nodes and E denotes the set of edges (ordered pairs of nodes) of G. Let $\mathcal{L}$ be a finite set of node labels. Each node $v \in V$ has a label $label(v) \in \mathcal{L}$ associated with it. Given a label $a \in \mathcal{L}$, the inverted list I_a is the list of nodes on G whose label is a. Figure 1(a) shows a data graph G. Subscripts are used in the labels of nodes in G to distinguish between nodes with the same label (e.g., nodes a_1 and a_2 whose label is a). The inverted list I_a of label a is $\{a_1, a_2\}$.

A node u is said to *reach* node v in G, denoted $u \prec v$, if there exists a path from u to v in G. Clearly, if $(u, v) \in E$, $u \prec v$. Abusing tree notation, we refer to v as a *child* of u (or u as a *parent* of v) if $(u, v) \in E$, and v as a *descendant* of u (or u as an *ancestor* of v) if $u \prec v$.

Graph Pattern Query. We focus on queries which are graph patterns. Every node x in a pattern $Q = (V_Q, E_Q)$ has a label $label(x) \in \mathcal{L}$. An edge $(x, y) \in E_Q$ denotes that node y is reachable from node x. Figure 1(b) shows a graph pattern query Q.

Graph Pattern Homomorphism. Given a pattern query Q and a data graph G, a *homomorphism* from Q to G is a function m mapping the nodes of Q to nodes of G, such that: (1) for any node $x \in V_Q$, $label(x) = label(m(x))$; and (2) for any edge $(x, y) \in E_Q$, $m(x) \prec m(y)$ in G.

Query Answer. An *occurrence* of Q on G is a tuple indexed by the nodes of Q whose values are the images of the nodes in Q under a homomorphism from Q to G. The *answer* of Q on G is a relation whose schema is the set of nodes of Q, and whose instance is the set of occurrences of Q under all possible homomorphisms from Q to G.

If x is a node in Q labeled by label a, the *occurrence set of x on G* is a subset L_x of the inverted list I_a containing only those nodes that occur in the answer of Q on G for x (that is, nodes that occur in the column x of the answer).

Let $(q_i, q_j) \in E_Q$, and v_i and v_j be two nodes in G, such that $label(q_i) = label(v_i)$ and $label(q_j) = label(v_j)$. The pair (v_i, v_j) is called a *match* of the query edge (q_i, q_j) if $v_i \prec v_j$ in G. The set of all the matches of an edge of Q is called *match set* of this edge. For instance, the match set of edge (A, B) of Q is $\{(a_1, b_1), (a_2, b_2), (a_2, b_3)\}$.

The answer of Q on G is the set of occurrences of Q on G. In the example of Fig. 1(c) shows the answer of Q on G.

Problem Statement. Given a large directed graph G and a pattern query Q, our goal is to efficiently find the answer of Q on G.

3   Query Reachability Graph

Given a graph pattern query Q, the data graph G constitutes the entire search space for Q. However, many nodes and edges in G might be irrelevant for Q. This motivates us to define the concept of *query reachability graph* of Q, which serves as a, typically, much more compact search space for Q on G.

Definition 1 (Query Reachability Graph). *A query reachability graph G_Q of Q on G is a k-partite graph such that: (i) Every node in G_Q is incident to at least one edge; (ii) Every node $q \in V_Q$ is associated with an independent node set n_q which is a subset of the inverted list $I_{label(q)}$ of G; (iii) There exists an edge (v_x, v_y) in G_Q from a node $v_x \in n_x$ to a node $v_y \in n_y$ if and only if (v_x, v_y) is a match of the query edge $(x, y) \in E_Q$.*

Figure 1(d) shows a query reachability graph G_Q of Q over G. One can see that G_Q has the same structure as Q. The independent node set in G_Q for query node B is $\{b_1, b_2, b_3\}$.

Let $G_T = (V_T, E_T)$ denote the transitive closure of G: $V_T = V$, and G_T has an edge from node v_x to node v_y if and only if there exists a path from v_x to

v_y in G. Edges in G_Q correspond to edges in G_T. Thus, G_Q contains all the reachability information between nodes of G relevant to query Q. Moreover, G_Q encodes the answer of Q on G as shown by the following proposition.

Proposition 1. *Let G_Q be a query reachability graph of a pattern query Q over a data graph G. If there exists a homomorphism from Q to G that maps node $x \in V_Q$ to node $v_x \in V_G$, then there is a homomorphism from Q to G_Q which maps x to node $v_x \in V_{G_Q}$.*

According to Proposition 1, G_Q is complete. That is, it encodes all the homomorphisms from Q to G. Thus, it can serve as a search space for the answer of Q on G. Using the inverted lists of the query nodes as input and reachability information between data graph nodes, G_Q can be constructed by computing the matches of every query edge.

Note that several recent subgraph matching algorithms also use query related auxiliary data structures to represent the query answer search space [3,7]. These auxiliary data structures are designed to support searching for subgraph isomorphisms. Unlike G_Q, they are subgraphs of the data graph, hence they do not contain reachability information between data nodes, and consequently, they cannot be used for computing edge-to-path homomorphic matches.

While G_Q is typically much smaller than G, it can still be quite large for efficiently computing the query answer. The reason is that it may contain redundant nodes (and their incident edges). Redundant nodes are nodes which are not part of any occurrence of query Q. Obtaining a query reachability graph which is minimal (that is, one which does not have redundant nodes) is an NP-hard problem even for isomorphisms and edge-to-edge mapping [2]. Therefore, we focus on constructing a refined G_Q whose independent node sets contain much less redundant nodes while maintaining its ability of serving as a search space for the answer of Q on G. We do so by providing next an efficient data graph node filtering algorithm.

4 A Graph Traversal Filtering Algorithm

Existing data node filtering methods are either simply based on query node labels [1,9], or use a BFS tree of the query to filter out data nodes violating children or parent structural constraints of the tree [2,7]. They are unable to prune nodes violating ancestor/descendant structural constraints of Q.

We present next a node filtering method based on graph traversal to efficiently prune nodes violating ancestor/descendant structural constraints of Q. Our filtering method is inspired by the node pre-filtering technique [4,15]. In contrast to that technique which is restricted to directed acyclic graphs (dags), our method is designed for filtering general graphs.

Graph Node Filtering Algorithm. Algorithm 1 shows our graph node filtering algorithm, referred to as *GraphFilter*. Let p, q denote nodes in the query Q, and u, v denote nodes in the data graph G. The algorithm uses the following data structures: (1) We associate each query node q with three n-wide bit-vectors

---

Algorithm 1. Algorithm *GraphFilter* (query Q, data graph G).

---

1. **for** $(q \in V_Q)$ **do**
2.   Set bits corresponding to descendants of q in QBitVecDes[q] to 1;
3.   Set bits corresponding to ancestors of q in QBitVecAnc[q] to 1;
4.   Set the bit corresponding to q in QBit[q] to 1;
5.   cis[q] := $\emptyset$; /*cis is the vector of candidate independent node sets*/
6. **for** $(u \in V_G)$ **do**
7.   **for** (each $p \in V_Q$ where label(u)==label(p)) **do**
8.   bitVecDes[u] := bitOR(bitVecDes[u], QBit[p]);
9.   bitVecAnc[u] := bitOR(bitVecAnc[u], QBit[p]);
10.   Set satDes[u][p] to *false* for each $p \in V_Q$;
11. Compute the strongly connected components (SCC) graph G_s of G;
12. BUPCheck();
13. TDWCheck();
14. **return** The vector cis;

Procedure BUPCheck()

1. **for** (each C_s of G_s in a reverse topological order) **do**
2.   **repeat**
3.   **for** (each node $u \in C_s$) **do**
4.   **for** (each child node v of u) **do**
5.   bitVecDes[u] := bitOR(bitVecDes[u], bitVecDes[v]);
6.   **until** (bit-vectors in bitVecDes have no changes)
7.   **for** (each node $u \in C_s$ and each $p \in V_Q$ where label(u)==label(p)) **do**
8.   **if** (QBitVecDes[p] == bitAND(bitVecDes[u], QBitVecDes[p])) **then**
9.   satDes[u][p]:=*true*;
10.   **else**
11.   bitVecDes[u] := bitAND(bitVecDes[u], $\sim$QBit[p]);

Procedure TDWCheck()

1. **for** (each C_s of G_s in a topological order) **do**
2.   **repeat**
3.   **for** (each node $u \in C_s$) **do**
4.   **for** (each parent node v of u) **do**
5.   bitVecAnc[u] := bitOR(bitVecAnc[u], bitVecAnc[v]);
6.   **until** (bit-vectors in bitVecAnc have no changes)
7.   **for** (each node $u \in C_s$ and each $p \in V_Q$ where label(u)==label(p)) **do**
8.   **if** (QBitVecAnc[p] == bitAND(bitVecDes[u], QBitVecAnc[p])) **then**
9.   **if** (satDes[u][p]) **then**
10.   add u to cis[p];
11.   **else**
12.   bitVecAnc[u] := bitAND(bitVecAnc[u], $\sim$QBit[p]);

QBit[q], QBitVecDes[q] and QBitVecAnc[q], where n is the number of nodes of Q. The last two bit-vectors, QBitVecDes[q] and QBitVecAnc[q], encode the descendant and ancestor query nodes of the query node q, respectively. (2) For each data node u, we use two n-wide bit-vectors bitVecDes[u] and bitVecAnc[u] to record whether u has ancestors or descendants matching a particular query node. We denote the strongly connected component (SCC) graph of G as $G_s = (V_s, E_s)$. An element C_s in V_s represents a SCC in G_s. $(C_s, C_s') \in E_s$ iff there exists $u \in C_s$ and $v \in C_s'$ such that $(u, v) \in E_G$.

Algorithm *GraphFilter* first initializes the bit vectors associated with query nodes and graph nodes (lines 1–10). The SCC graph G_s of G is then computed using Tarjan's algorithm [12] (line 11). Then, *GraphFilter* conducts two graph traversals on G_s (bottom-up and top-down) implemented by procedures BUPCheck and TDWCheck, respectively. During the graph traversal, it updates the bit-vectors of nodes of G and identifies nodes whose bit-vectors are inconsistent with the vectors assigned to their corresponding query nodes (line 12–13). After the traversals, a vector of candidate independent sets, indexed by query nodes, is returned for constructing a refined query reachability graph (line 14). We describe below the graph traversal procedures in more detail.

Procedure BUPCheck (bottom-up traversal) processes the nodes C_s of G_s, in reverse topological order, as follows: the bit vectors of bitVecDes for nodes in C_s are repeatedly updated until the *fixpoint* is reached (lines 2–6). Specifically, for each node u in C_s, BUPCheck consolidates the bit-vectors of its child nodes with its own bitVecDes[u] using a bitOR operation (lines 4–5). Once the bit vectors become stable, BUPCheck proceeds to check whether bitVecDes[u] contains QBitVecDes[p], for each node u in C_s and each query node p that has the same label as u, using a bitAND operation (line 7–8). If this is the case, we say that u *downward matches* p, and set satDes[u][p] to *true* (line 9), indicating that u has descendants in the data graph which downward match descendants of p in the query. Otherwise, the bit for p in bitVecDes[u] is reset to 0 (line 11).

Procedure TDWCheck (top-down traversal) processes each node C_s of G_s in a forward topological order. The bit-vectors in bitVecAnc of the nodes C_s of G_s are repeatedly consolidated with the bit-vectors of their parents, until no more changes can be applied to them (lines 2–6). Then, nodes in C_s are checked for upward matching with nodes of Q (lines 9–10). We say that a data node u *upward matches* a query node p if bitVecAnc[u] contains QBitVecAnc[p]. All the nodes satisfying both the upward and the downward matching conditions are added to their corresponding candidate independent node sets (lines 8–10).

Example. We apply Algorithm *GraphFilter* to the graph pattern query Q and the data graph G of Fig. 1. After the bottom-up processing by BUPCheck, the algorithm identifies two redundant nodes a_1 and c_1 since they violate the downward matching condition. Then, the top-down processing by TDWCheck identifies a redundant node d_1 violating the upward matching condition. The redundant nodes are not added to the candidate independent node sets in vector *cis*. Based on the refined candidate independent node sets, we can build a smaller reachability graph for Q on G. In Fig. 1(c), red nodes denote redundant nodes pruned from the original reachability graph G_Q. The refined G_Q contains only the rest of the nodes.

Complexity. The preprocessing (lines 1–11) takes time $O(|V_G| + |E_G| + |V_Q|^2 + |V_G||V_Q|)$. The loop (lines 2–6 of BUPCheck and TDWCheck) is repeated $O(|V_G|)$

Algorithm 2. Algorithm *MJoin*(query Q, query eachability graph G_Q).

1. Pick an order $q_1, \ldots, q_n$ for the nodes of Q, where $n = |V_Q|$;
2. Let t be a tuple where $t[i]$ is initialized to be *null* for $i \in [1, n]$;
3. Enumerate$(1, t)$;

Function Enumerate (index i, tuple t)

1. **if** $(i = n + 1)$ **then**
2.   **return** t;
3. $N_i := \{q_j \mid (q_i, q_j) \in E_Q \text{ or } (q_j, q_i) \in E_Q, j \in [1, i-1]\}$
4. $C_i := S_i$;   /*S_i denotes the node set of q_i in G_Q*/
5. **for** (every $q_j \in N_i$) **do**
6. $C_i^j := \{v_i \in C_i \mid (v_i, t[j]) \in E(G_Q) \text{ or } (t[j], v_i) \in E(G_Q)\}$;
7. $C_i := C_i \cap C_i^j$;
8. **for** (every node $v_i \in C_i$) **do**
9. $t[i] := v_i$;
10.   Enumerate$(i + 1, t)$;

---

times. Each iteration takes $O(|E_G||V_Q|)$ time to update bit-vectors of nodes in G. Checking possible (downward and upward) matching between nodes of G and Q takes time $O(|V_G||V_Q|^2)$. Putting these together, Algorithm *GraphFilter* takes $O(|V_G||E_G||V_Q| + |V_G||V_Q|^2)$ time in the worst case.

5 A Join-Based Query Occurrence Enumeration Algorithm

In this section, we present a novel multi-way join algorithm called *MJoin* (Algorithm 2) to compute the query answer. Given query Q and data graph G, let R_e denote the relation containing the matches of each query edge e on G. Conceptually, *MJoin* produces occurrences of Q by joining multiple R_es at the same time. Given G_Q as input, the multi-way join operator of *MJoin* works by multi-way intersecting node adjacency lists of G_Q.

Algorithm *MJoin* first picks a linear order of query nodes to match. Then, it performs a recursive backtracking search to find candidate matches for query nodes, considering query nodes one at a time by the given order, before returning any result tuples.

More concretely, at a given recursive step i, function *Enumerate* first determines query nodes that have been considered in a previous recursive step and are adjacent to the current node q_i. These nodes are collected in the set N_i. Let S_i denote the node set of every $q_i \in V_Q$ in G_Q. To get the candidate occurrences of q_i, for each $q_j \in N_i$, *Enumerate* intersects S_i with the forward adjacency list of $t[j]$ in G_Q when $(q_i, q_j) \in E_Q$, or with the backward adjacency list of $t[j]$ when $(q_j, q_i) \in E_Q$ (lines 5–7). Next, *Enumerate* iterates over the candidate occurrences (line 8). In every iteration, a candidate occurrence is assigned to $t[i]$ (line 9) and *Enumerate* proceeds to the next recursive step (line 10). In the final recursive step, when $i = n + 1$, tuple t contains one specific occurrence for all the query nodes and is returned as a result of Q (line 2).

Example 1. In our running example, let G_Q be the refined query reachability graph, i.e., the graph of Fig. 1(c) without the red nodes and their incident edges. Assume the matching order of Q is A, B, C, D. When $i = 1$, Algorithm *MJoin* first assigns a_2 to tuple $t[1]$, then it recursively calls *Enumerate*$(2, t)$. The intersection of a_2's adjacency list with the node set of B is $\{b_2, b_3\}$. Node b_2 is assigned to $t[2]$ first. Similarly, at $i = 3$, c_2 is the only node in the intersection and is assigned to $t[3]$. At $i = 4$, since the intersection of adjacency lists of b_2 and c_2 with the node set of D is $\{d_2\}$, *MJoin* assigns d_2 to $t[4]$, and returns a tuple $t = \{a_2, b_2, c_2, d_2\}$. Then, *MJoin* backtracks, assigns the next node b_3 from the intersection result for node B to $t[2]$ and proceeds in the same way. It returns another tuple $t = \{a_2, b_3, c_2, d_2\}$.

6   Experimental Evaluation

We now present a thorough evaluation of our proposed approach for edge-to-path based homomorphic graph pattern matching.

6.1   Setup

We implemented the join-based approach (*JM*) and the tree-based approach (*TM*) for finding homomorphisms of graph pattern queries on data graphs. Our approach is abbreviated as *GM*. Our implementation was coded in Java.

For *JM*, we first compute the occurrences for each edge of the input query on the data graph, then find an optimized left-deep join plan through dynamic programming, and finally use the plan to evaluate the query as a sequence of binary joins. For *TM*, we first transform the graph pattern query into a tree query, evaluate the tree query using a tree pattern evaluation algorithm, and filter out occurrences of the tree query that violate the reachability relationships specified by the missing edges of the original query. For the *TM* approach, we implemented the recent tree pattern evaluation algorithm described in [14], which has been shown to outperform other existing algorithms. In our implementation, we applied our filtering algorithm *GraphFilter* (Sect. 4) to both approaches, *JM* and *TM*.

The above three graph matching algorithms were implemented using a recent efficient scheme, called *Bloom Filter Labeling* (BFL) [10], for reachability checking which was shown to greatly outperform most existing schemes [10].

In addition to pure algorithms, we also compare with a recent graph query engine[1] *GraphflowDB* [9], referred to here as *GF*.

All the experiments reported were performed on a workstation running Ubuntu 16.04 with 32GB memory and 8 cores of Intel(R) Xeon(R) processor (3.5GHz). The Java virtual machine memory size was set to 16GB.

Datasets. We ran experiments on real-world graph datasets from the Stanford Large Network Dataset Collection which have been used extensively in previous

---

[1] https://github.com/queryproc/optimizing-subgraph-queries-combining-binary-and-worst-case-optimal-joins.

Table 1. Key statistics of the graph datasets used.

Domain	Dataset	# of nodes	# of edges	# of labels	Avg #incident edges
Biology	Yeast	3.1K	12K	71	8.05
	Human	4.6K	86K	44	36.9
	HPRD	9.4K	35K	307	7.4
Social	Epinions	76K	509K	20	6.87
	DBLP	317K	1049K	20	6.62
Communication networks	Email	265K	420K	20	2.6

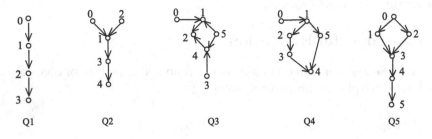

Fig. 2. Templates of graph pattern queries used for evaluation (each pattern edge denotes a descendant relationship between the associated nodes).

works [7,9,11]. Table 1 lists the properties of six datasets. Its last column displays the average number of incident edges (both incoming and outgoing) per node. The datasets have different structural properties and come from a variety of application domains: biology, social networks, and communication networks. We also ran experiments on other datasets. The results are similar and omitted due to space limitations.

Queries. For each of the three datasets from the biology domain, we used a query set of 10 queries selected from 200 randomly generated queries used in [11]. For the rest of the datasets, we used a query set of 5 queries. The templates of the 5 queries are shown in Fig. 2. These are representative queries commonly used in existing work [5,9]. The number associated with each node of a query template denotes the node id. Query instances are generated by assigning labels to nodes.

Metrics. We measured the runtime of individual queries in a query set in seconds (sec). This includes the preprocessing time (i.e., the time spent on filtering vertices and building auxiliary data structures) and the enumeration time (i.e., the time spent on enumerating results). The number of occurrences for a given query on a data graph can be quite large. Following usual practice [7,11], we terminated the evaluation of a query after finding 10^7 matches covering as much of the search space as time allowed. We stopped the execution of a query if it did not complete within 10 min, so that the experiments could be completed in a reasonable amount of time. We refer to these queries as *unsolved*.

Table 2. Performance of *JM, TM* and *GM* for evaluating large queries.

Dataset	Alg	Time out	Out of memory	Solved queries	Avg. time of solved queries (sec.)
Human	*JM*	1	7	2	1.51
	TM	3	0	7	16.7
	GM	0	0	**10**	**0.53**
HPRD	*JM*	2	4	4	1.86
	TM	1	0	9	134.21
	GM	0	0	**10**	**0.58**
Yeast	*JM*	5	3	2	0.14
	TM	3	0	7	20.8
	GM	0	0	**10**	**0.34**

6.2 Performance Results

Random Graph Patterns. We have measured the performance of *JM, TM* and *GM* for evaluating 10 random queries over Human, HPRD and Yeast. These three datasets are selected because have been used by recent contributions on graph pattern matching [7,11] and have different structural properties. For instance, Human is very dense. Because of its higher average degree and fewer distinct labels, graph matching on it is harder. In contrast Yeast is sparse. The number of nodes in the 10 queries for each data graph range from 4 to 20 for Human, and from 4 to 32 for HPRD and Yeast. In Table 2, we record, for each algorithm, the number of unsolved queries in two categories: time out and out of memory. We also record the number of solved queries as well as the average runtime of solved queries for each algorithm.

We observe that *GM* has the best performance overall among the three algorithms. It is able to solve all the given queries. In contrast, *JM* is only able to solve the first 2 or 4 queries on each data graph, and the number of nodes of solved queries are no more than 8. *TM* solves more and larger queries than *JM*, but it is up to two orders of magnitude slower than *GM*.

A large percentage of the failures of *JM* is due to an out-of-memory error, since it generates a large number of intermediate results during the query evaluation. Another cause of the inefficiency of *JM* is due to the join plan selection. As described in [5], in order to select an optimized join plan, *JM* uses dynamic programming to exhaustively enumerate left-deep tree query plans. For queries with more than 10 nodes, the number of enumerated query plans can be huge. For example, for a query with 24 nodes on HPRD, *JM* enumerates 2,384,971 query plans in total.

Most of the failures of *TM* is due to a time out error. Recall that *TM* works by evaluating a tree query of the original graph query. For each tuple of the tree query, it checks the missing edges for satisfaction. Hence, its performance is severely affected when the number of solutions of the tree query is very large.

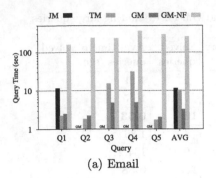 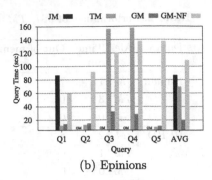

(a) Email	(b) Epinions

Fig. 3. Evaluation time of *JM, TM* and *GM* on Email and Epinions.

Designed Graph Patterns. Figure 3 shows the elapsed time of *JM, TM* and *GM* on evaluating five queries instantiated from the five templates in Fig. 2 on Email and Epinions. The average time of queries solved by each algorithm is also shown in the figure.

GM again has the best performance among the three algorithms. *JM* only solves Q_1 for both Email and Epinions, and is unable to finish for the rest of the cases due to out-of-memory issues. *TM* solves all the queries, but its average time performance is around 2 times slower than *GM*. In particular, for Q_3 and Q_4, it is outperformed by *GM* by up to 5 times. The results are consistent with the random query results shown above.

We observe from the results (not shown due to space limitations) that the time for constructing the query reachability graph dominates the total query evaluation time of *GM*. Its percentage of the total query evaluation time is about 87% and 98.6% on Email and Epinions, respectively. The query reachability graph is the core component used for the multi-way join phase of *GM*. Clearly, a smaller query reachability graph requires a smaller construction time but it also reduces the cost of multi-way join operations. This demonstrates the importance of designing an effective filtering strategy.

Effect of Node Filtering. Figure 3 shows also the time performance of *GM* on evaluating the queries on Email and Epinions without the node filtering procedure. This approach is abbreviated as *GM-NF*. We observe that the node filtering is highly effective, offering *GM* a speedup of up to 143 and 11 over *GM-NF* on Email and Epinions, respectively. The average filtering time is very small, around 5.6% and 0.4% of the query time on Email and Epinions, respectively. The average percentage of nodes pruned from the input inverted lists of the query nodes is about 79.5% and 68.4% on Email and Epinions, respectively. By pruning redundant nodes, the node filtering procedure greatly reduces the number of nodes accessed during query evaluation.

Scalability. We evaluated the scalability of the algorithms as the data set size grows. For this experiment, we recorded the elapsed query time on increasingly larger randomly chosen subsets of the DBLP data. Figure 4 shows the results

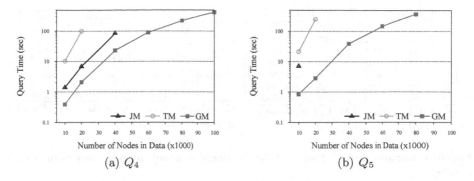

Fig. 4. Elapsed time of Q_4 and Q_5 on increasingly larger subsets of DBLP.

of the three algorithms evaluating instantiations of the query templates Q_4 and Q_5 shown in Fig. 2. The other queries gave similar results in our experiments. We restrict the comparison on DBLP subsets of up to 10^5 nodes due to *TM* and *JM*'s inability to handle larger data graphs. As expected, the execution time for all algorithms increases when the total number of graph nodes increases. *GM* scales smoothly compared to both *TM* and *JM* for evaluating the two queries.

6.3   Comparison with Graph DB Systems

We compared the performance of *GM* with the recent *GraphFlow* [9] (abbreviated as *GF*) graph DBMS on evaluating reachability graph pattern queries.

Setup. Before evaluating queries, *GM* builds a reachability index on the input data graph G using the BFL (Bloom Filter Labeling) algorithm [10].

The engine *GF* was designed to process graph pattern queries whose edges are mapped with homomorphisms to edges in the data graph (therefore, it does not need a reachability index). As *GM* is able to match edges to paths in data graph, we designed an indirect way for *GF* to evaluate reachability graph pattern queries: first generate the transitive closure G' on the input graph G and then use G' as the input data graph to *GF*. In the experiments, we used the Floyd-Warshall algorithm to compute the transitive closure of the data graph. As a join-based method, *GF* enumerates and optimizes join plans based on a cost model. In order to estimate join plan costs, for each input data graph, *GF* constructs a catalog containing entries on cardinality information for subgraphs.

Comparison Results. Figure 5 presents the results on Email data graph. Specifically, Table 5(a) shows the time for constructing the BFL reachability index, the graph transitive closure and the catalog on Email graphs with different numbers of labels and nodes. As we can see, the time for constructing BFL indices remains very small for graphs of different sizes (it is 0.38 sec. for the original Email graph with 265K nodes). In contrast, the transitive closure construction time grows very fast as the number of graph nodes increases. For a graph with 3k nodes, the transitive closure construction takes more than one

#lbs	#nodes	BFL(sec.)	TC(sec.)	CAT(sec.)
5	1k	0.01	22.95	5.52
10	1k	0.01	22.67	10.84
15	1k	0.01	23.07	55.97
20	1k	0.01	23.58	323.92
20	2k	0.01	207.93	outOfMemory
20	3k	0.02	765.65	outOfMemory
20	5k	0.03	4042.62	outOfMemory

(a) Building time of BFL, transitive closure (TC) and catalog (CAT).

Query	Alg.	#lbs=5	#lbs=10	#lbs=15	#lbs=20
Q_2	GF	0.27	0.12	0.09	0.09
	GM	1.12	0.1	0.01	0.01
Q_4	GF	2.69	0.26	0.38	0.39
	GM	13.84	0.31	0.03	0.03
Q_5	GF	0.70	0.25	0.20	0.36
	GM	4.34	0.11	0.07	0.01

(b) Query time (seconds) on Email graphs of 1K nodes.

Fig. 5. Comparison of GM and GF for reachability graph pattern queries on Email graphs.

hour. We observe also that the catalog construction is affected enormously by the growing cardinality of the label and node sets of the graph.

Because the time for building transitive closures and catalogs on large-sized graphs is prohibitive large, we used only 1k-sized Email graphs (with different numbers of labels) to compare the query time of GM and GF. We evaluated instantiations of the five query templates of Fig. 2. For each query, GM and GF enumerate all the matchings of the query.

Table 5(b) shows only the query time of GM and GF for Q_2, Q_4, and Q_5. The results for the other queries are similar. We observe that GF performs better than GM on the Email graph with 5 labels. However, GM greatly outperforms GF when the number of labels increases from 10 to 20. Note that in reporting the query times of GF, the transitive closure and the catalog construction times and ignored since otherwise GF underperforms GM by several orders of magnitude if feasible at all.

Summary. Overall, GM is much more efficient than GF on evaluating reachability graph pattern queries. To determine node reachability in graphs, it does not need to compute the graph transitive closure. Instead, it uses a reachability index (BFL) which can be computed efficiently. Also, unlike GF which relies on statistics (i.e., catalogs) that are prohibitively expensive to compute, GM uses a query reachability graph that can be built efficiently on-the-fly during query processing and does not have to be materialized on disk.

Neo4j is the most popular graph DBMS and EmptyHeaded [1] is one of the most efficient graph database systems. We did not directly compare GM with Neo4j and EmptyHeaded because it is expected that it outperforms both on evaluating reachability graph pattern queries since the evaluation results of [9] show that GF largely outperforms Neo4j and EmptyHeaded.

7   Related Work

We review related work on graph pattern query evaluation algorithms. Our discussion focuses on in-memory algorithms designed for the evaluation of a single query.

Isomorphic Mapping Algorithms. The majority of algorithms for isomorphic mapping adopt a backtracking method [13], which recursively extends partial matches by mapping query nodes to data nodes to find the query answer. Many of the earlier algorithms directly explore the input data graph to enumerate all results. Several recent ones [1,2,7] adopt the preprocessing-enumeration framework, which performs preprocessing to find, for each query node, the set of possible data nodes (called candidates), and builds an auxiliary data structure (ADS) to maintain edges between candidates. Unlike ADS, our query reachability graph is not a subgraph of the data graph; instead, it is a summary of the matches of a given query and is used by our multi-way join algorithm to efficiently enumerate query matches.

An alternative approach to the backtracking method is *JM*, the join-based approach, which converts graph pattern matching to a sequence of binary joins. This is the method used by [5,9] and in database management systems such as PostgreSQL, MonetDB and Neo4j. In this paper, we adopt homomorphisms which can map edges to paths. This general framework is not constraint by the restrictions of isomorphisms.

Homomorphic Mapping Algorithms. Homomorphisms for mapping graph patterns similar to those considered in this paper were introduced in [6] (called p-hom), which however did not address the problem of efficiently computing graph pattern matches; instead, it uses the notion of p-hom to resolve a graph similarity problem between two graphs.

Cheng et al. [5] proposed an algorithm called R-Join, which is a join-based algorithm. An important challenge for join-based algorithms is finding a good join order. To optimize the join order, R-Join uses dynamic programming to exhaustively enumerate left-deep tree query plans. Due to the large number of potential query plans, R-Join is efficient only for small queries (less than 10 nodes). As is typical with join-based algorithms, R-Join suffers from the problem of numerous intermediate results. As a consequence, its performance degrades rapidly when the graph becomes larger [8].

A graph pattern matching algorithm called DagStackD was developed in [4]. DagStackD implements a tree-based approach. Given a graph pattern query Q, DagStackD first finds a spanning tree Q_T of Q, then evaluates Q_T and filters out tuples that violate the reachability relationships specified by the edges of Q missing in Q_T. To evaluate Q_T, a tree pattern evaluation algorithm is presented. This algorithm decomposes the tree query into a set of root-to-leaf paths, evaluates each query path, and merge-joins their results to generate the tree-pattern query answer. Several pattern matching algorithm designed specifically for evaluating tree patterns on graphs have been proposed [8,14]. Among them, TPQ-3Hop [8] is designed on top of a hop-based reachability indexing scheme. The one presented in [14] leverages simulation to compute the query answer without producing any redundant intermediate results.

Unlike existing algorithms that follow the tree-based or the join-based approach, our graph pattern matching approach is holistic in the sense that it does not decompose the given query into subpatterns. Instead, it tries to match the query against the input graph as a whole.

8 Conclusion

We have addressed the problem of efficiently evaluating graph patterns using homomorphisms over a large data graph. By allowing *edge-to-path* mappings, homomorphisms can extract matches "hidden" deep within large graphs which might be missed by *edge-to-edge* mappings of subgraph isomorphisms. We have introduced the concept of query reachability graph to compactly encode the pattern matching search space. To further reduce the search space, we have developed a filtering method to prune data nodes violating reachability constraints imposed by query edges. We have also designed a novel join-based query occurrence enumeration algorithm which leverages multi-way joins realized as intersections of adjacency lists and node sets of the query reachability graph. We have verified experimentally the efficiency and scalability of our approach and showed that it largely outperforms state-of-the-art approaches.

We are currently working on extending the proposed approach to handle more general hybrid graph patterns (which consist of both child-edges and descendant-edges) evaluation over large graph data. We are also investigating alternative node filtering strategies.

References

1. Aberger, C.R., Tu, S., Olukotun, K., Ré, C.: Emptyheaded: a relational engine for graph processing. In: SIGMOD, pp. 431–446 (2016)
2. Bhattarai, B., Liu, H., Huang, H.H.: CECI: compact embedding cluster index for scalable subgraph matching. In: SIGMOD, pp. 1447–1462 (2019)
3. Bi, F., Chang, L., Lin, X., Qin, L., Zhang, W.: Efficient subgraph matching by postponing cartesian products. In: SIGMOD, pp. 1199–1214 (2016)
4. Chen, L., Gupta, A., Kurul, M.E.: Stack-based algorithms for pattern matching on DAGs. In: VLDB, pp. 493–504 (2005)
5. Cheng, J., Yu, J.X., Yu, P.S.: Graph pattern matching: a join/semi join approach. IEEE Trans. Knowl. Data Eng. **23**(7), 1006–1021 (2011)
6. Fan, W., Li, J., Ma, S., Wang, H., Wu, Y.: Graph homomorphism revisited for graph matching. PVLDB **3**(1), 1161–1172 (2010)
7. Han, M., Kim, H., Gu, G., Park, K., Han, W.: Efficient subgraph matching: Harmonizing dynamic programming, adaptive matching order, and failing set together. In: SIGMOD, pp. 1429–1446 (2019)
8. Liang, R., Zhuge, H., Jiang, X., Zeng, Q., He, X.: Scaling hop-based reachability indexing for fast graph pattern query processing. IEEE Trans. Knowl. Data Eng. **26**(11), 2803–2817 (2014)
9. Mhedhbi, A., Kankanamge, C., Salihoglu, S.: Optimizing one-time and continuous subgraph queries using worst-case optimal joins. ACM Trans. Database Syst. **46**(2), 6:1-6:45 (2021)
10. Su, J., Zhu, Q., Wei, H., Yu, J.X.: Reachability querying: can it be even faster? IEEE Trans. Knowl. Data Eng. **29**(3), 683–697 (2017)

11. Sun, S., Luo, Q.: In-memory subgraph matching: an in-depth study. In: SIGMOD, pp. 1083–1098 (2020)
12. Tarjan, R.E.: Depth-first search and linear graph algorithms. SIAM J. Comput. **1**(2), 146–160 (1972)
13. Ullmann, J.R.: An algorithm for subgraph isomorphism. J. ACM **23**(1), 31–42 (1976)
14. Wu, X., Theodoratos, D., Skoutas, D., Lan, M.: Leveraging double simulation to efficiently evaluate hybrid patterns on data graphs. In: WISE, pp. 255–269 (2020)
15. Zeng, Q., Zhuge, H.: Comments on "stack-based algorithms for pattern matching on DAGs.". PVLDB **5**(7), 668–679 (2012)

Revisiting Approximate Query Processing and Bootstrap Error Estimation on GPU

Hang Zhao[1,2], Hanbing Zhang[1,2], Yinan Jing[1,2(✉)], Kai Zhang[1,2],
Zhenying He[1,2], and X Sean Wang[1,2,3]

[1] School of Computer Science, Fudan University, Shanghai, China
{zhaoh19,hbzhang17,jingyn,zhangk,zhenying,xywangcs}@fudan.edu.cn
[2] Shanghai Key Laboratory of Data Science, Shanghai, China
[3] Shanghai Institute of Intelligent Electronics and Systems, Shanghai, China

Abstract. Sampling-based Approximate Query Processing (AQP) is one of the promising approaches for timely and cost-effective analytics over big data. There are mainly two methods to estimate errors of approximate query results, namely analytical method and bootstrap method. Although the bootstrap method is much more general than the first method, it is rarely used in the existing AQP system due to its high computation overhead. In this paper, we propose to use the powerful GPU and a series of advanced optimization mechanisms to accelerate bootstrap, thus make it feasible to address the essential err r estimation problem for AQP by utilizing bootstrap. Besides, since modern GPUs have bigger and bigger memory capacity, we can store samples in the GPU memory and use GPU to accelerate the execution of AQP queries in addition to using GPU to accelerate the bootstrap-based error estimation. Extensive experiments on the SSB benchmark show that our GPU-accelerated method is at most about two orders of magnitude faster than the CPU method.

Keywords: Big data analytics · Approximate query processing · Bootstrap · GPU

1 Introduction

Online Analytical Processing (OLAP) is the core function of the data management and analysis system. With the development of online commerce, Internet of Things and scientific research, a large amount of data is created every moment. These massive amounts of data play a vital role in the decision-making of governments, enterprises and scientific research. However, the latency of analysis on big data is very high, which is not conducive to OLAP. To solve this problem, the researchers propose approximate query processing (AQP) [11,17,21,22]. Sampling is one of the most commonly used techniques in AQP. By leveraging on samples, the AQP system can provide an approximate answer with an accuracy guarantee, such as confidence interval under a given confidence level. Confidence

A. Bhattacharya et al. (Eds.): DASFAA 2022, LNCS 13245, pp. 72–87, 2022.
https://doi.org/10.1007/978-3-031-00123-9_5

interval help users decide whether to accept approximate result. This affiliated process is also called error estimation.

In the past research, there are two ways to calculate the confidence interval, namely *analytical* method [18] and *bootstrap* method [2]. The analytical method is based on the central-limit theorem (CLT). CLT tells us the normalized sum of sample tends toward a normal distribution. The analytical method is efficient, but the application scope of this method is very limited. It only support COUNT, SUM, AVG, and VARIANCE. Bootstrap is completely based on simulation and is suitable for almost all functions. In contrast to the analytical method, bootstrap does not require any confidence interval formula derivation, thus freeing users from complex mathematical theorems. Of course, its shortcomings are also obvious. A large number of simulations cause an explosion of computation, which usually requires hundreds or thousands of bootstrap trials. Due to the high computation overhead of bootstrap, bootstrap method is rarely used in the existing AQP system. Although, some optimization mechanisms have been proposed for bootstrap on CPU [13] and on GPU [5,7,9], it still cannot meet the real-time response requirements of interactive analysis scenarios. In this paper, we propose two approximate query processing models on GPU, namely *coprocessor model* and *main processor model*. The difference between these two models is the role played by GPU for AQP. As for the coprocessor model, we first use the CPU to process the query on the original sample and then use the GPU to perform the resampling of bootstrap. Finally, we send the multiple resampled samples from GPU to CPU through the PCIE and calculate the confidence interval according to these samples. With the development of modern GPUs, GPU can not only be used as a coprocessor of CPU, but also it can be used as a main processor for AQP. As for the main processor model, since the modern GPU has a bigger memory capacity, we can store samples in the GPU memory to avoid the bandwidth bottleneck of PCIE between CPU and GPU. Therefore, in the main processor model, both the query execution and the bootstrap can be performed on the GPU. Furthermore, we propose some advanced optimization mechanisms for the new bottlenecks encountered in the main processor model to further improve the performance.

In summary, the main contributions of this paper are as follows.

- We propose two approximate query processing models on GPU, called *coprocessor model* and *main processor model* according to the different roles played by GPU. The main processor model has higher performance compared to the coprocessor model if GPU has engouh memory to store samples.
- We further propose two advanced optimization mechanisms for main processor model, called *one-step calculation* and *count sampling*. Experiments shows that *count sampling* has better optimization effect.
- We extend the main processor model from bootstrap to the whole process of AQP, providing fast query execution and error estimation capabilities. Experiments shows our method can achieve nearly two orders of magnitude performance improvement compared to the method on CPU.

Algorithm 1: Bootstrap Algorithm

Input: R: sample relation
Output: $result$: result of b bootstrap trials
1 Result $result = \{\}$;
2 **for** $i = 1 \rightarrow b$ **do**
3   let $R* = \{\}$;
4   **for** $j = 1 \rightarrow R.length$ **do**
5   let idx = rand() * R.length ;
6         $R*[j] = R[idx]$;
7   add $q(R*)$ to $result$;
8 sort($result$) ;
9 return $result$;

This paper is organized as follows. Section 2 introduces the research background. Section 3 describes two approximate query processing models and bootstrap on GPU. Section 4 introduces two advanced optimization mechanisms. Section 5 describes our experiments. Finally, we conclude this paper in Sect. 6.

2   Preliminary

2.1   AQP and Bootstrap

AQP provides an approximate answer mainly through sampling techniques. To make it credible, AQP introduces error estimation. There are two methods of error estimation: analytical method and bootstrap method. The bootstrap method is widely used in statistics because of its simplicity and high applicability.

Bootstrap was first proposed by Bradley Efron [2] in 1979, and detailed description and application instructions were given in [3]. Since bootstrap was proposed, 40 years have passed. Although it has strong applicability, bootstrap is computationally intensive. Because of the limitation of computing power in the past, the response time was very long. Below we introduce how to use the classic bootstrap for the error estimation of database AQP queries.

First, we have an original sample table R and an aggregate query $q(R)$ on this sample table. We want to know the distribution of the query results, so as to derive the confidence interval of the query results and other important information. In many cases, the analysis and derivation of these confidence limits is not easy. However, using bootstrap can easily derive the distribution of $q(R)$. Bootstrap can be divided into b bootstrap trials. Each trial resamples the original sample table R to generate a new sample table $R*$, and then executes the query $q(R*)$ to obtain b query results. After that, we sort the b query results and take the quantile corresponding to the confidence level to obtain the confidence interval. The specific algorithm is shown in Algorithm 1.

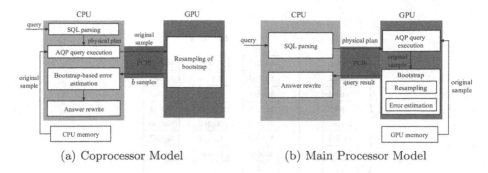

(a) Coprocessor Model   (b) Main Processor Model

Fig. 1. Two approximate query processing models with GPU

It can be seen that bootstrap needs to execute bootstrap trial b times. A bootstrap trial includes two phases: the re-sampling phase and the query execution phase. Resampling phase has the largest computation overhead. Each re-sampling needs to generate the same amount of re-sampled samples as the original samples. Computation overhead in the query execution phase is small, because it runs on the sample whose size is small.

2.2   Two Approximate Query Processing Models with GPU

The main reason we use GPU is to solve the problem of time-consuming error estimation in AQP. According to the relationship between GPU and CPU, we proposed two models: coprocessor model and main processor model. As shown in Fig. 1(a), in the coprocessor model, the original samples are mainly located in the CPU memory. After the CPU performs the AQP query execution, the original samples are transmitted to the GPU side through the PCIE bus. After GPU resampling, b new samples are generated and transmitted to the CPU through the PCIE bus, and the CPU performs the final bootstrap-based error estimation. As shown in Fig. 1(b), in the main processor model, the original samples are naturally stored in GPU memory. All data is processed on the GPU side without going through the PCIE bus. Finally GPU passes the result to CPU.

These two models have their own advantages and disadvantages. If we use the GPU as the coprocessor of the CPU, the biggest problem is the need to repeatedly transfer data between the CPU and the GPU. This price is unbearable, because as we all know, the interconnection bandwidth of CPU and GPU is much lower than memory bandwidth. But the coprocessor model also has advantages. Its implementation is simple and the changes to bootstrap are small. The main processor model fully implements AQP and bootstrap on the GPU side. His advantage is that it completely avoids the interconnection bottleneck between CPU and GPU, and greatly improves the calculation speed of AQP and bootstrap. Its disadvantage is that it needs to introduce new dependencies to achieve query execution on GPU.

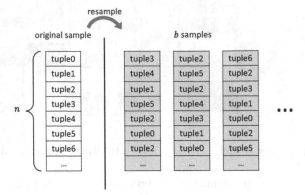

Fig. 2. Resampling of boostrap: generate b new samples from original sample

3   AQP and Bootstrap-Based Error Estimation on GPU

3.1   Coprocessor Model

In the database field, past work [4, 6, 8, 16] focused on using GPU as the coprocessor of the CPU to help the CPU run some computationally intensive tasks, which we call the coprocessor model. We introduced this model in the second section.

So, how to realize the parallelization of resampling? As shown in the Fig. 2, bootstrap needs to perform resampling b times, and each resampling includes n uniform samples. In the past parallelized bootstrap research [5, 7, 9] let b threads perform a resample separately. The shortcomings of these parallelization studies are obvious, that is, the upper limit of the degree of parallelism is the number of bootstrap trials b, which is far lower than the highest degree of parallelism that the GPU can achieve, and a lot of computing power is wasted. We found that the n times of uniform sampling within one resampling are also independent, and can be completely divided into n small tasks to achieve fine-grained parallelism. The advantage of this is that the parallel capability of the GPU can be fully utilized, and the cost needs to be more Threads, but GPUs are inherently suitable for multi-threading, and increasing the number of threads will not cause additional overhead.

In the specific implementation, we need a total of $b * n$ threads, so we need to set appropriate blocknum and threadnum to cover $b * n$ in the CUDA kernel function. The task of each thread is to solve two subscripts, and then construct a sample from the original sample array to the resampled sample array based on these two subscripts, and then copy the corresponding data. It can be seen that the implementation logic is very simple, which is the advantage of the coprocessor model.

Table 1. Crystal primitives [15]

Primitive	Description
BlockLoad	Copies a tile of items from global memory to shared memory. Uses vector instructions to load full tiles
BlockLoadSel	Selectively load a tile of items from global memory to shared memory based on a bitmap
BlockStore	Copies a tile of items in shared memory to device memory
BlockPred	Applies a predicate to a tile of items and stores the result in a bitmap array
BlockScan	Co-operatively computes prefix sum across the block. Also returns sum of all entries
BlockShuffle	Uses the thread offsets along with a bitmap to locally rearrange a tile to create a contiguous array of matched entries
BlockLookup	Returns matching entries from a hash table for a tile of keys
BlockAggregate	Uses hierarchical reduction to compute local aggregate for a tile of items

3.2 Main Processor Model

Although the coprocessor model works well in our experiments, it still has a lot of room for improvement. The coprocessor model needs to first transfer the query-related columns to the global memory of the GPU. Then, each time the GPU resamples, it sends the resampled samples back to the CPU memory. Using PCIE bus with such a high flow rate requires several seconds of transmission. What is even more intolerable is that if the amount of original sample data is doubled, the total cost will increase b times.

In order to solve the problem of the interconnection bottleneck between the CPU and GPU in the coprocessor mode, we proposed the main processor model. The main processor model uses the GPU as the main processor for AQP and error estimation, and uses the CPU as a coprocessor to process SQL Parsing. The main processor model performs queries on the sample table on the GPU side, and does not need to repeatedly transmit samples in the PCIE pipeline. In this way, we have eliminated the interconnected IO bottleneck. However, this method requires the GPU to have query execution capabilities that are not inferior to the CPU. There have been many research results on this issue that can meet our needs, such as YDB [19], HippogriffDB [10], Omnisci [14], etc. We found that the latest research result Crystal [15] is the best in performance. The Crystal library is a CUDA function library that provides a variety of function primitives (kernel functions), as shown in the Table 1. By combining these functions, we can achieve the main functions defined by the SQL syntax. Using GPU as the main processor has far-reaching significance. As far as we know, no one has used GPU as the main processor for AQP. The memory size of modern GPUs is sufficient to accommodate all samples in the calculation process. In order to avoid transferring samples through the PCIE pipeline during online query, we can import the samples offline in advance into the GPU memory.

Algorithm 2: Main Processor Model

Input: R: sample relation
Output: $result$: result of b bootstrap trials
1 Result $result = \{\}$;
2 **for** $i = 1 \rightarrow b$ **do**
3 | Resample R to generate resampled samples R^* ;
4 |__ Execute query on R^*, and add the result to $result$;
5 load back $result$ from GPU memory to CPU memory ;
6 sort($result$) ;
7 return $result$;

We introduced Crystal to solve the query execution problem on the GPU. The main difficulty lies in embedding Crystal in bootstrap. Crystal only provides a series of function primitives, and we need to program them to combine them into equivalent SQL statements. This process is very difficult. In the experiments in Chap. 5, we programmed 13 SQL queries. But in the system, it should be implemented automatically by the system. The system can obtain the physical plan programmed by CUDA by scanning the logical plan of the query. This process is usually called "code generation" [1]. The main processor model algorithm is shown in the Algorithm 2.

4 Advanced Optimization

4.1 One-Step Calculation

Bootstrap can be divided into two phases: resampling and query execution. In the previous algorithm, we regarded it as two phases, strictly following the logic of re-sampling first, and then querying the re-sampled samples. However, previous studies have shown that many aggregate functions can be calculated in a stream, that is, we do not need to execute the query after the resampling, but calculate the aggregate value of the query at the same time as the resampling. We refer to the classic bootstrap as the two-step algorithm, and the new optimized version as the one-step algorithm. As shown in the Fig. 3, we use a bootstrap trial of the simplest SUM calculation to demonstrate one-step optimization.

4.2 Count Sampling

We analyze the main processor model proposed in Sect. 3 and get a very important finding: when resampling, we copy the selected tuples to the new samples, which is unnecessary. We can add a column to the original sample table to indicate the number of times the tuple was sampled in a bootstrap trial. Then, the new resampling algorithm becomes: we uniformly sample a tuple and then add one to the count of the tuple in the new column. The new algorithm is shown in the Fig. 4.

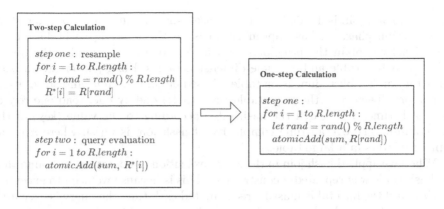

Fig. 3. One-step calculation optimization

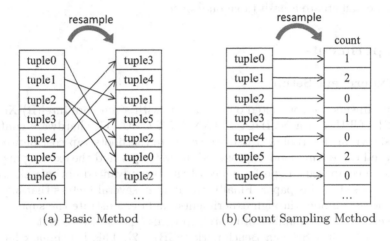

(a) Basic Method   (b) Count Sampling Method

Fig. 4. Count sampling optimizations

Although the Count Sampling optimization did not reduce the time complexity of resampling, considering the cost of each data copy, we can still save a considerable amount of time. In our experiments, the Count Sampling optimization saves about 1/4 time. Count Sampling optimization also greatly reduces the space complexity. Originally, the entire resampled sample needs to be stored, but now only an integer array is stored, which saves a lot of space when the number of elements is the same.

4.3   One-Time Hashtable Building

Many of the benchmark queries used in this paper contain join statements. In the real workloads, there are also many join statements in the queries. There are many ways to implement join. The most commonly used is the hash join method. When we use the Crystal library, we need to manually write the hash join code. Hash join is divided into two phases: the building hashtable phase and the detection phase. A hash optimization is for the building hashtable phase. First, briefly explain the principle of hash join. In the star model, the object of join is a fact table and multiple dimension tables. Hash join creates a hash index for small tables (dimension tables) based on join columns and stores them in memory. Then scan the large table (fact table) one by one, read one row of data each time, and search the hash table according to the value (key) of the join column. The overall time complexity of hash join is $\mathcal{O}(n)$, where n is the number of rows in a large table.

When we applied hash join in the query execution of bootstrap, we found that the hash table was repeatedly constructed. This is because we need to execute b queries, and the hashtable must be recalculated each time the query is executed. Since we only sample the fact table, the dimension table does not need to be changed, so the hash table does not need to be changed. When we calculate b queries, we can create a hash table only once.

5 Experiments

5.1 Experiment Setup

Our experiments are conducted on a machine with a ten-core Intel(R) Xeon(R) Gold 5215 CPU and a NVIDIA TITAN RTX GPU. The hardware configuration is shown in the Table 2. First, we compare the bootstrap cost between the GPU-based computing models proposed in this paper and the existing methods. Second, we compared the performance of the two advanced optimization mechanism proposed in this paper. Finally, we analyze several factors that affect the main processor model through experiments, and demonstrate the superiority of GPU over CPU. The number of bootstrap trials b is 100 by default.

We use the Star Schema Benchmark (SSB) [12]. This benchmark has been widely used in various data analysis studies. SSB is simplified to a star model, which is more in line with OLAP application scenarios. SSB has a fact table LINEORDER and four dimension tables DATE, SUPPLIER, CUSTOMER, PART, which are organized in a star schema and connected by foreign keys. When generating the SSB dataset, you can control the size of the dataset by specifying a scale factor. In our experiments, the scale factor is 100. This will generate a dataset with a total size of approximately 60 GB, in which the fact table has approximately 600 million tuples. Then we sample the fact table uniformly with a probability of 0.01, and obtain the sample fact table with a size of about 600 MB.

A total of 13 queries are provided in the SSB benchmark, to evaluate performance, we use all 13 queries in SSB. These queries are divided into four categories and queries in the same category has the same template with different selectivities. These queries cover project, selection, join, sort and aggregations.

Table 2. Hardware configuration

Operating system	Ubuntu Linux 18.04.5 64 bit
CPU	Intel(R) Xeon(R) Gold 5215 CPU @ 2.50 GHz, 10 cores
Momory	64 GB
GPU	NVIDIA TITAN RTX
GPU memory	24 GB
CUDA cores number	4608

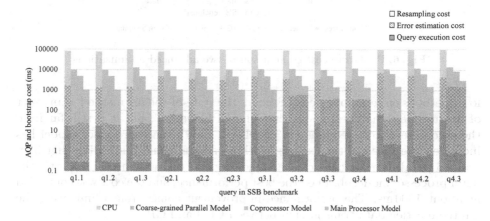

Fig. 5. Bootstrap cost comparison of four methods

5.2 Performance Comparison

Performance Comparison of Four Methods. We compare the bootstrap cost of the CPU and GPU. This is the most time-consuming part. Its cost is hundreds of times that of AQP query execution. The performance improvement in this part comes from the new method proposed in this paper, so it is very critical.

We implemented and compared four methods. The first method is based on CPU, which is the most widely used method and the baseline of our experiments. The second method is the state-of-the-art method called Coarse-grained Parallel Model [7], which parallelize the bootstrap via dividing the b times bootstrap trials roughly to b threads. The third method is the coprocessor model proposed in Sect. 3.1. The fourth method is the main processor model proposed in Sect. 3.2. Our parameters are: the number of bootstrap trials $b = 100$, and the size of the sample table LO_LEN = 6001171.

Figure 5 shows the bootstrap cost of four methods. The abscissa represents 13 kinds of queries defined in SSB, denoted as $q1.1, q1.2, ..., q4.3$. The ordinate values are too different, so we use a logarithmic scale. The results show that, first of all, Resampling is the most time-consuming phase in bootstrap. Error estimation and AQP query execution does not cost much time. So we mainly compare the

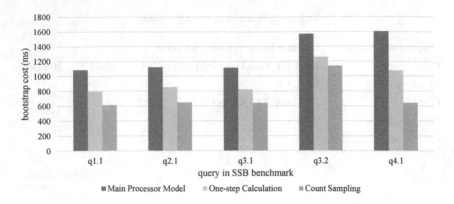

Fig. 6. Bootstrap cost comparison of two advanced optimizations

four methods on resampling cost. Second, the state-of-the-art method is an order of magnitude (ten times) faster than the CPU method, but it is still far from the second-level cost requirement of OLAP. Third, the coprocessor model has increased the speed of the existing work by several times, indicating that fine-grained parallelization can achieve higher benefits on modern GPUs. Forth, the main processor model obtained the best performance, and the cost was controlled at about 1000 ms. The performance improvement comes from Eliminating the interconnection data transmission between CPU and GPU.

Performance Comparison on Advanced Optimization. We compare two optimization mechanisms with the main processor model, One-step Calculation and Count Sampling proposed in Sect. 4. We selected 5 representative queries in SSB: $q1.1, q2.1, q3.1, q3.2, q4.1$. The tables they connect are different, the predicates and grouping columns are also different, and the aggregate functions used are all SUM. The results are shown in Fig. 6. The cost spent in One-step calculation is reduced by an average of 1/4 compared to the main processor model. Compared with the main processor, the cost spent on Count Sampling is reduced by an average of 1/2.

We can see that the optimization effect of One-step Calculation is not as good as Count Sampling. This is an uncommon sense result, because One-step Calculation directly eliminates the intermediate step of saving the resampled samples and directly calculates the final result. And Count Sampling still retains the resampling results, but greatly compresses the volume. We further analyzed and found that the SQL engine we used has vector computing capabilities implemented by GPU. Count Sampling only optimizes the sampling phase and does not involve query execution phase, so vector calculations can be used. One-step Calculation integrates the two phases and can not use vector calculations.

5.3   Factor Analysis

Effect of the Number of Columns Involved in Query. We explore the effect of the number of columns involved in the query on the bootstrap cost.

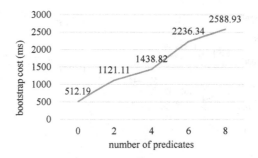

Fig. 7. Effect of the number of columns involved in query on bootstrap cost

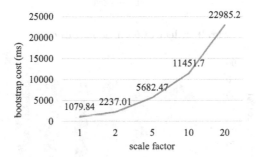

Fig. 8. Effect of original sample size on bootstrap cost

Because we use columnar storage, a column is only loaded when needed, and each column of data will generate resampled samples when resampling. Therefore, we believe that the number of columns involved in the query is positively related to bootstrap cost.

We chose the SSB query q1.1 as the test query, and changed the number of columns involved in the query by modifying the number of predicates. When there is no predicate, the query involves only two columns. After that, we keep increasing the number of predicates (2, 4, 6, 8), and record bootstrap cost. While increasing the number of predicates, we control the predicate selectivity at 50% to reduce the influence of selectivity. The results are shown in Fig. 7. The results are consistent with our expectations. As the number of predicates grows, bootstrap cost grows linearly. This shows that the number of columns involved in the query has a significant impact on bootstrap cost.

Effect of Original Sample Size. We explore the effect of original sample size on bootstrap cost. We believe that the original sample size affects the size of the re-sampling sample, which in turn affects bootstrap cost. We change the size of the original sample, expressed by the SSB scale factor, set to 1, 2, 5, 10, 20, and compare their bootstrap cost. The results are shown in Fig. 8, and it can be seen that the two are roughly proportional. This is in line with our expectations. At the same time, we should pay attention to that, because of the GPU memory

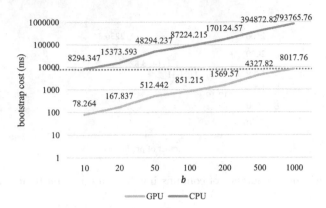

Fig. 9. Effect of b on bootstrap cost

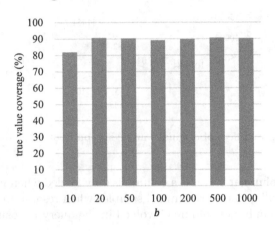

Fig. 10. True value coverage of the confidence interval calculated by different b

size limitation, the original sample cannot be too large. In this query, the scale factor can be up to 28.

Effect of the Number of Bootstrap Trials. We explore the effect of the number of bootstrap trials (i.e., b) on bootstrap cost and confidence interval accuracy. Obviously, the larger the b, the longer the bootstrap cost, and the higher the accuracy of the confidence interval. But we want to know the specific relationship between bootstrap cost and b, so that we can preset b according to the cost that users can tolerate. We also want to know the specific relationship between the accuracy of the confidence interval and b, so that we can set b according to the accuracy that the user wants, and control the convergence of the confidence interval.

We choose the SSB query $q1.1$ as the test query. We gradually increase b, and record the bootstrap cost. The results are shown in Fig. 9. It can be seen that the bootstrap cost is roughly proportional to b. In order to quantify the

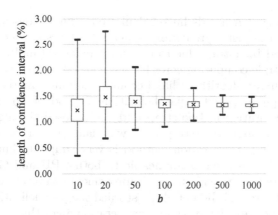

Fig. 11. The length of confidence intervals calculated 100 times for eah different b

relationship between the accuracy of the confidence interval and b, we designed a new experiment: We set b from 10 to 1000. For each value, we repeat bootstrap 100 times, and each time we can get a 90% confidence interval. We calculate the true value coverage and the length of the confidence interval, as shown in Figs. 10 and 11. When b increases, the coverage rate is roughly 90%, which does not change significantly. For the length of the confidence interval, its average value does not change significantly, but its distribution converges as b increases, which indicate that the confidence interval is getting more and more accurate as b increases.

From the Figs. 10 and 11, we can get the difference in the performance of the bootstrap between CPU and GPU. We limit the bootstrap cost to 10 s. As shown by the dotted line in the Fig. 9, the CPU can only execute bootstrap trials 10 times, and the GPU can execute bootstrap trials 1000 times. According to the number of bootstrap trials, we can find the corresponding accuracy of the confidence interval in the Fig. 11. It can be seen that under the same cost requirements, the accuracy of confidence interval calculated by GPU is 5 times that of CPU.

6   Related Work

There are many studies on the serialization and parallelization of bootstrap. For serialized bootstrap, Chris Jermaine proposed Tuple Augmentation (TA) [13]. His point is that when we try to reduce bootstrap cost, even if the user may specify thousands of bootstrap trials, we should avoid running the basic query multiple times in a certain way. Kai Zeng proposed Analytical Bootstrap Method (ABM) [20]. The method is to establish a probabilistic multi-set relationship model to represent all possible simulation data sets generated by bootstrap trials, and then expand the relationship algebra on this basis. Through theoretical derivation, predict the results of thousands of bootstrap trials. The advantage of this method is that it avoids huge calculations, but it does not conduct real bootstrap trials. It is just a prediction, which introduces new errors.

For parallelized bootstrap, due to the development of multi-core CPUs and GPUs, using multiple computing cores for parallel computing, accelerating computationally intensive algorithms has become a common method for data scientists. G. Guo [5] proposed a parallelization method of bootstrap. He gave the matrix representation of bootstrap and decomposed the algorithm coarse-grained, that is, a single bootstrap trial was handed over to a single thread for processing. This parallelization method is very simple to implement, but it is very effective. This scheme is applicable to both CPU and GPU. In experiments, several to ten times performance improvement can be obtained depending on the number of cores. M. Iida et al. [7] studied the parallelization acceleration of bootstrap to calculate the maximum expectation algorithm. MS Lee et al. [9] studied the use of bootstrap to calculate the parallelization acceleration of a fully homomorphic encryption algorithm. Their parallelization methods are similar to G Guo, and they have made different optimizations for specific statistics. The above three studies have improved the performance of bootstrap to a certain extent.

7 Conclusion

This paper studies how to use GPU to accelerate approximate query processing, focusing on the error estimation part of approximate query processing. We proposed the GPU-based bootstrap algorithm which is a fine-grained parallelization method to make full use of the parallel capabilities of the GPU. Further, we migrated the approximate query processing to the GPU as a whole, and imported the samples into the GPU memory in advance to minimize the interconnection IO overhead. We call it main processor model. Experimental results show that the error estimation speed of GPU is at most about two orders of magnitude faster than that of the CPU method. This paper also analyzes and derives the important factors that affect the performance of the main processor model.

Acknowledgement. This work is funded by the NSFC (No. 61732004 and No. 62072113), the National Key R&D Program of China (No. 2018YFB1004404) and the Zhejiang Lab (No. 2021PE0AC01).

References

1. Armbrust, M., et al.: Spark SQL: relational data processing in spark. In: SIGMOD (2015)
2. Efron, B.: Bootstrap methods: another look at the jackknife. In: Kotz, S., Johnson, N.L. (eds.) Breakthroughs in Statistics. Springer Series in Statistics (Perspectives in Statistics), pp. 569–593. Springer, New York (1992). https://doi.org/10.1007/978-1-4612-4380-9_41
3. Efron, B., Tibshirani, R.J.: An Introduction to the Bootstrap. CRC Press, Boca Raton (1994)

4. Govindaraju, N., Gray, J., Kumar, R., Manocha, D.: GPUteraSort: high performance graphics co-processor sorting for large database management. In: SIGMOD (2006)
5. Guo, G.: Parallel statistical computing for statistical inference. J. Statist. Theory Pract. **6**(3), 536–565 (2012)
6. He, B., et al.: Relational joins on graphics processors. In: SIGMOD (2008)
7. Iida, M., Miyata, Y., Shiohama, T.: Bootstrap estimation and model selection for multivariate normal mixtures using parallel computing with graphics processing units. Commun. Statist. Simul. Comput. **47**(5), 1326–1342 (2018)
8. Kaldewey, T., Lohman, G., Mueller, R., Volk, P.: GPU join processing revisited. In: Proceedings of the Eighth International Workshop on Data Management on New Hardware. pp, 55–62 (2012)
9. Lee, M.S., Lee, Y., Cheon, J.H., Paek, Y.: Accelerating bootstrapping in FHEW using GPUs. In: ASAP (2015)
10. Li, J., Tseng, H.W., Lin, C., Papakonstantinou, Y., Swanson, S.: HippogriffDB: Balancing I/O and GPU bandwidth in big data analytics. Proc. VLDB Endow. **9**(14), 1647–1658 (2016)
11. Mozafari, B.: Approximate query engines: commercial challenges and research opportunities. In: SIGMOD (2017)
12. O'Neil, P.E., O'Neil, E.J., Chen, X.: The star schema benchmark (SSB). PAT **200**, 50 (2007)
13. Pol, A., Jermaine, C.: Relational confidence bounds are easy with the bootstrap. In: SIGMOD (2005)
14. Root, C., Mostak, T.: MapD: a GPU-powered big data analytics and visualization platform. In: ACM SIGGRAPH 2016 Talks, pp. 1–2 (2016)
15. Shanbhag, A., Madden, S., Yu, X.: A study of the fundamental performance characteristics of GPUs and CPUs for database analytics. In: SIGMOD (2020)
16. Sitaridi, E.A., Ross, K.A.: Optimizing select conditions on GPUs. In: Proceedings of the Ninth International Workshop on Data Management on New Hardware, pp. 1–8 (2013)
17. Wu, Z., Jing, Y., He, Z., Guo, C., Wang, X.S.: POLYTOPE: a flexible sampling system for answering exploratory queries. World Wide Web **23**(1), 1 22 (2019). https://doi.org/10.1007/s11280-019-00685-x
18. Yan, Y., Chen, L.J., Zhang, Z.: Error-bounded sampling for analytics on big sparse data. Proc. VLDB Endow. **7**(13), 1508–1519 (2014)
19. Yuan, Y., Lee, R., Zhang, X · The Yin and Yang of processing data warehousing queries on GPU devices. Proc. VLDB Endow. **6**(10), 817–828 (2013)
20. Zeng, K., Gao, S., Mozafari, B., Zaniolo, C.: The analytical bootstrap: a new method for fast error estimation in approximate query processing. In: SIGMOD (2014)
21. Zhang, H., et al.: An agile sample maintenance approach for agile analytics. In: ICDE (2020)
22. Zhang, Y., Zhang, H., He, Z., Jing, Y., Zhang, K., Wang, X.S.: Parrot: a progressive analysis system on large text collections. Data Sci. Eng. **6**(1), 1–19 (2021)

μ-join: Efficient Join with Versioned Dimension Tables

Mika Takata[✉], Kazuo Goda, and Masaru Kitsuregawa

The University of Tokyo, 4-6-1 Komaba, Meguro-ku, Tokyo, Japan
{mtakata,kgoda,kitsure}@tkl.iis.u-tokyo.ac.jp

Abstract. The star schema is composed of two types of tables; fact tables record business process, whereas dimension tables store description of business resources and contexts that are referenced from the fact tables. Analytical business queries join both tables to provide and verify business findings. Business resources and contexts are not necessarily constant; the dimension table may be updated at times. Versioning preserves every version of a dimension record and allows a fact record to reference its associated version of the dimension record correctly. However, major existing versioning practices (utilizing a binary join operator and a union operator) cause processing redundancy in queries joining a fact table and a dimension table. This paper proposes μ-join, an extended join operator that directly accepts a fact table and an arbitrary number of dimension tables, and presents that this operator reduces the redundancy and speeds up fact-dimension joins queries. Our experiment demonstrates that μ-join offers speedup using the synthetic dataset (up to 71.7%).

Keywords: Relational database · Join · Dimension table

1 Introduction

Versioning dimension tables is a widely employed practice to offer referential integrity in the star schema [12]. The star schema is a popular approach for organizing business data into relational database. The schema is composed of two types of tables. Fact tables record business process, whereas dimension tables store description of business resources and contexts. Analytical business queries join both tables to provide and verify business findings. Business resources and contexts are not necessarily constant. Suppose that database records a company's sales history. Daily sales events are recorded into a fact table, whereas product names and prices are stored in a dimension table. The company may update its business resources or contexts, for example, by launching new products, discontinuing existing products, and changing products names or prices. These updates are not necessarily frequent, but they may actually happen; the updates of business resources and contexts are described in the dimension table.

A. Bhattacharya et al. (Eds.): DASFAA 2022, LNCS 13245, pp. 88–95, 2022.
https://doi.org/10.1007/978-3-031-00123-9_6

Versioning preserves every version of a dimension record and allows a fact record to reference its associated version of the dimension record correctly. Existing versioning practices using a binary join and a union can answer queries joining a fact table and a versioned dimension table. One major practice joins a fact table and multiple versions of a dimension table one by one and then eliminates redundant tuples. Another practice merges multiple versions of a dimension table, joins it with the fact table, and then eliminates redundant tuples if necessary. These practices do not necessarily offer optimized processing, but rather induce inefficient processing.

This paper proposes, μ-join, an extended join operator, which can directly join a fact table and multiple versions of a dimension table by considering the join condition on a join key and a version compatibility. The μ-join operator allows database engines to evaluate the join conditions at an early phase and to reduce the version-related complication. Thus, the database engine potentially improves the join performance. This paper presents our experiments to clarify the benefit of μ-join in terms of query response time using a synthetic dataset. To our knowledge, similar attempts have not been reported in literature.

The remainder of this paper is organized as follows. Section 2 describes versioned dimension tables and their problem. Section 3 defines a version-aware fact-dimension join operation and presents the μ-join operator. Section 4 presents the experimental study. Section 5 shows related work and Sect. 6 concludes the paper.

2 Join with Multiple Versions of Dimension Tables

Let us exhibit another example of a versioned dimension table, which has motivated this study. Figure 1 presents an example database of Japanese public healthcare insurance claims. Japan has employed the universal service policy; all the certified healthcare services necessary to the citizens are basically covered by the public healthcare insurance system. When providing an insured patient with healthcare services, healthcare service providers (e.g., hospitals) are supposed to submit an insurance claim to request the compensation of the expense to the public healthcare insurers. Let us think about the design of database for managing the insurance claims. One typical solution is to record insurance claims in a fact table (S in the figure). Each claim contains the description of diagnosed diseases, which are coded in the standardized format. Such descriptive information of disease codes is to be stored in a separate dimension table. The disease code system does not remain unchanged. Rather, it may change at times. Figure 1 presents a partial portion of the real Japanese disease code standard, indicating that the disease code system changed twice from June, 2020 to October, 2021. For example, the disease code indicating Apparent strabismus changed from 8831256 to 3789010 after January, 2021. When healthcare researchers analyze the database, they need to interpret each insurance claim by referencing an appropriate disease description record according to the recorded disease code and the claiming date. Thus, the database must keep all historical

(R₁) Disease description (effective June to December, 2020)

Disease code	Description
49005	Dysentery
8840201	Botulism poisoning
-	-
3685004	Color-blindness
7843001	Wernicke's aphasia
8831489	Sensory aphasia
-	-
-	-
8831256	Apparent strabismus
2749004	Goat
8847201	Heavy chain disease
-	-

(R₂) Disease description (effective January to May, 2021)

Disease code	Description
49005	Dysentery
8840201	Botulism poisoning
8850385	Shrimp allergy
3685004	Color-blindness
-	-
8831489	Sensory aphasia
-	-
-	-
3789010	Apparent strabismus
2749004	Goat
8847201	Heavy chain disease
-	-

(R₃) Disease description (effective June to October, 2021)

Disease code	Description
49005	Dysentery
8840201	Botulism poisoning
8850385	Shrimp allergy
3685016	Color-blindness
-	-
8831489	Sensory aphasia
8850617	COVID-19 sequela
8850619	COVID-19 history
3789010	Apparent strabismus
2749004	Goat
8847201	Heavy chain disease
3685001	Acquired color blindness

(S) Healthcare insurance claims

Patient ID	Disease code	Medical institution ID	Medical service	Service date	Billing month
1	8831256	88147900	Fundus examination	2020-12-22	2020-12
1	3789010	88147900	Rehabilitation	2021-01-03	2021-01
2	7843001	88147900	Injection	2020-12-29	2020-12
3	8850385	88147900	Special disease treatment	2021-01-10	2021-01
3	3685016	88147900	Community comprehensive care	2021-06-10	2021-06
...	...	...	...	...	...
7	8831489	88147900	MRI imaging	2021-05-02	2021-05
8	8850617	88147900	Special disease treatment	2021-06-10	2021-06
9	8840201	88147900	Emergency home visit	2021-05-31	2021-05
10	8840201	88147900	Intravenous injection	2020-10-11	2020-10
10	8850617	88147900	Primary care	2021-06-12	2021-06
...	...	...	...	...	...

Fig. 1. An example of a fact table and versioned dimension tables

updates of the dimension tables (R_1, R_2, and R_3 in the figure) for answering such queries. This complication is not limited to disease codes. Insurance claims contain more information such as medical treatments and medicinal drugs, the descriptive information of which change every month, in order to reflect the evolution of medical science and industry.

Versioning preserves every version of a dimension record and allows a fact record to reference its associated version of the dimension record correctly. Unfortunately, most current database management systems do not support such dimension versioning explicitly. Instead, implementing the dimension versioning on the relational schema is a major engineering approach. They can be roughly grouped into three policies. *Snapshot versioning* stores every entire version of a dimension table as a separate table. *Differential versioning* stores only the difference of every version from the first version as a separate table. *Incremental versioning* stores only the difference of every version from the previous version as a separate table. Snapshot versioning is relatively simple, but it consumes redundant space and may cause additional processing due to tuple redundancy. Incremental versioning only consumes smaller space, but it is complicated, needing additional processing to reproduce a tuple of a specific version.

Analytical queries often performed across a fact table and such a versioned dimension table. Those queries must be written to guarantee that each fact record can reference its associated version of the dimension record. One typical practice uses a binary join operation to join a fact table and every version of a dimension table, merges the join results, and then eliminates non-effective and redundant tuples. This practice often causes inefficient processing due to redundant work and elimination work. Another practice merges all possible versions of a dimension table, joins the merged dimension table, and then eliminates non-effective and redundant tuples. This practice do not necessarily offer optimized processing, but it is likely to induce inefficient processing.

3   The μ-join Operator

In this section, we introduce a new database operator, named μ-join. The μ-join operator extends the existing binary join and it can directly join a fact table and multiple versions of a dimension table by considering the join condition on a join key and a version compatibility. Being implementing this new operator, database engines can efficiently perform version-aware fact-dimension join queries.

Version-Aware Fact-Dimension Join Operation. First, we define a versioned dimension table and version-aware fact-dimension join operation.

Definition 1 (Versioned dimension table). Let $\check{R}$ be a versioned dimension table, having $<\dot{R}>$ versions, where the k-th version $\check{R}_k$ is defined as $\check{R}_k := \{r | r \in \check{R} \wedge e_k(r)\}$, where $e_k(r)$ is true if and only if a tuple r is effective at a point of a version number k $(0 \leq k < <\dot{R}>)$.

Definition 2 (Snapshot versioning). When snapshot versioning is deployed, each version $\check{R}_k$ is directly stored as a separate table $R_k^{(s)}$ as $R_k^{(s)} \leftarrow \check{R}_k$.

Definition 3 (Differential versioning). When differential versioning is deployed, the first version $\check{R}_0$ is directly stored as a separated table $R_0^{(d)}$ and the difference of each version $\check{R}_k$ from the first version $\check{R}_0$ is stored as a separate table $R_k^{(d)}$ in the database as $R_0^{(d)} \leftarrow \check{R}_0$, $R_k^{(d)} \leftarrow \check{R}_k \ominus \check{R}_0 (k > 0)$, where the operator $\ominus$ produces the difference of a left-hand table from a right-hand table as a table[1].

Definition 4 (Incremental versioning). When incremental versioning is deployed, the first version $\check{R}_0$ is directly stored as a separate table $R_0^{(i)}$ and the difference of each version $\check{R}_k$ from the first version $\check{R}_0$ is stored as a separate table $R_k^{(i)}$ in the database as $R_0^{(i)} \leftarrow \check{R}_0$, $R_k^{(i)} \leftarrow \check{R}_k \ominus \check{R}_{k-1} (k > 0)$.

Next, we define version-aware fact-dimension join operation that can be performed over a fact table S and a versioned dimension table $\check{R}$.

---

[1] Assuming that a record given in a right-hand table is deleted in a left-hand table, the operator $\ominus$ returns the deleted record with the deleted flag. This is analogous to the minus operator in the normal arithmetic system. The similar technique is widely deployed in database logging [8]. In our implementation, the deleted flag is stored in a separated attribute.

Definition 5 (Version-aware fact-dimension join operation). A version-aware fact-dimension join operation $S \bowtie_{v,Y=X} \check{R}$ of a fact table S and a versioned dimension table $\check{R}$ is defined as $S \bowtie_{v,Y=X} \check{R} := \bigcup_{0 \leq k < <\check{R}>} \{s \cup r | s \in S \land r \in \check{R}_k \land s_Y = r_X \land v(s) = k\}$, where an attribute set Y is a foreign key of S, another attribute set X is a primary key of $\check{R}$ and a function $v(s)$ return a version number indicating a version of the dimension table $\check{R}$ with which a tuple s of the fact table S is associated. The equation $Y = X$ denotes a join key condition and the equation $v(s) = k$ denotes a version compatibility condition.

μ-**join Operator.** The version-aware fact-dimension join operation presented in Definition 5 can be expressed by a combination of multiple binary join operations, a union operation, and a selection operation. Thus, it can be performed on conventional database engines. However, such practices are likely to cause processing inefficiency as discussed in Sect. 2.

This paper proposes μ-*join*, a new database operator to be implemented in database engines. The μ-join operator extends the existing binary join. This operator directly accepts a fact table S and multiple versions of a dimension table $\check{R}_0, \check{R}_1, \cdots, \check{R}_{<\check{R}>-1}$ to join them by considering the join condition on a join key and a version compatibility according to Definition 5. The native implementation of the μ-join operator allows the database engine to evaluate the join key condition $s_Y = r_X$ and the version compatibility condition $v(s) = k$ at an early phase to reduce the redundant tuple processing that is likely to be imposed by the binary-join practice.

In addition, the database engine is allowed to build a *version index* data structure to identify *target* relational tables (storing a versioned dimension table) with which each incoming fact tuple s joins. This is beneficial for differential versioning and incremental versioning, which induces complexity in the identification of target relational tables. Suppose a fact tuple s has an apparent version k, meaning that the tuple s is supposed to join with $\check{R}_k$. In snapshot versioning, obviously the target dimension tuple with which the tuple s joins is stored in the dimension table $R_k^{(s)}$. However, the target dimension tuple with which the tuple s joins is stored in any of $R_0^{(d)}$ and $R_k^{(d)}$ in differential versioning or $R_0^{(i)}, \cdots, R_k^{(i)}$ in incremental versioning. The version index data structure accepts a primary key value x (for the attribute X) and an apparent version number k as input, and returns a version number indicating a target relational table storing a versioned dimension table with which a fact tuple having x and k is associated. The database engine is allowed to efficiently identify the target relational table storing a versioned dimension table for each incoming fact tuple, thus improving the efficiency of the join processing.

In the hash-based implementation of μ-join operator, hash tables are built from target relational tables storing a versioned dimension table in the first phase in similar to hash binary joins. At the same time, the version index data structure is built from the target relational tables. Building the version index data structure does not incur major performance overhead because dimension tables are often much smaller than fact tables and the majority part of the version index building is shared with the normal hash table building. In the second phase, a fact table is scanned; each fact tuple probes in the version index data structure to identify the target table, and then probes in the identified target table. Thus, the μ-join operator offers an opportunity for database engines to improve the processing efficiency for the version-aware fact-dimension join operation. Also, μ-join is so intuitive that it can encapsulate the logic complication of version management in the database engine. That would potentially relieve the database designer's effort.

4   Evaluation

We present our experiment to clarify the benefit of the μ-join operator in terms of query response time. We revised the TPC-H data generator (dbgen) so that it could produce versioned dimension tables. The revised generator was enabled to produce multiple versions of dimension tables, where every version of a dimension table contained different tuple content[2] from its previous version, according to n_v ($n_v \geq 1$) that specifies the number of versions to be generated for each dimension table. The first version is identical to the original dimension table according to the TPC-H specification. In addition, the revised generator appended a version attribute, for each fact table, indicating a dimension version with which each fact tuple is supposed to join. For a given fact table, we divided the date attribute (e.g., O_ORDERDATE) space into n_v periods and determined a version of each fact tuple based the period with which the date value is associated. We generated the TPC-H dataset with versioned dimension tables organized with three version management policies.

For comparison, we tested the query execution according to the conventional binary-join practices and the new μ-join operator. **Binary join (naive)** performs the conventional binary-join operator to join a fact table and each version of a dimension table, merges the join results, and eliminates non-effective or redundant tuples. **Binary join (pre-dedup 1)** merges all possible versions of a dimension table, performs the conventional binary-join operator to join a fact table and the merged dimension table, and eliminates non-effective or redundant tuples. **Binary join (pre-dedup 2)** merges all possible versions of a dimension table, eliminates redundant dimension tuples as much as possible, performs the conventional binary-join operator to join a fact table and the merged dimension table, and eliminates non-effective or redundant tuples. **μ-join** performs the μ-join operator to directory join a fact tuple and all possible versions of a dimension table, without generating any non-effective or redundant tuples. We implemented these execution cases in the same code base from scratch and measured the average execution time of five trials for each test. We only employed the on-core hash join algorithm (accommodating an entire hash table in memory) rather than the nested-loop join algorithm without any selection predicates. The query execution was single-threaded and conducted on Linux with two Intel

---

[2] We only updated a non-key attribute for each dimension table. For example, we generated a new version of PART by shifting the value of P_RETAILPRICE randomly within the $\pm 5\%$ range from its previous version.

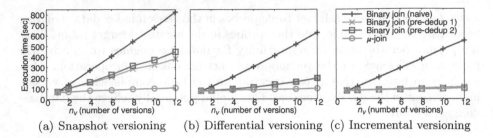

(a) Snapshot versioning   (b) Differential versioning   (c) Incremental versioning

Fig. 2. Evaluation on different numbers of versions

Xeon two processors (56 processing cores in total running as 2.60 GHz) and 96GB DRAM with 31TB RAID-6 storage by twenty-four disks.

Figure 2 presents the execution performance of a query joining LINEITEM and versioned PART on different numbers of dimension versions. Figure 2(a) shows μ-join speeds up a fact-dimension (w/two versions) join up to 31.7% from the best case of binary-join practices on snapshot versioning. As more versions are involved, μ-join achieves higher speedups up to 71.7% for twelve versions. Figure 2(b) shows μ-join consistently achieves speedups up to 2.89% for two versions and 48.9% for twelve versions on differential versioning. Figure 2(c) shows μ-join performs comparably with the best case of binary-join practices on incremental versioning. These results indicate that μ-join achieves significant speedups by evaluating the join key condition and the version compatibility at an early phase and by using the version index data structure to reduce the redundant tuple processing.

5   Related Work

Various join operator was explored for improving analytical query performance on database such as hash join, merge join, and nested loop join [1,2,13,15,16]. As a data structure, the columnar-based system was explored for statistical process to enhance analytical query performance [9,14,17]. To evolve analytical efficiency on database, many operations were implemented such as min, max, and group-by [7]. Stored procedures were explored to incorporate a user's specific requirements into database systems [6,7,18]. Sql-like operators were extended by the standard SQL to allow analysts easily do certain types of analytics [3–5,10]. Cohort analysis operation was attempted to integrate complicated processing into an operator based on requirements from real world [11]. Although such operators have been studied, join across a fact table and multiple versions of a dimension table has not been studied yet in literature to our best knowledge.

6   Conclusion

This paper proposes μ-join, an extended join operator that directly accepts a fact table and an arbitrary number of dimension tables, and presents that this operator speeds up fact-dimension joins queries. Our experiment demonstrates that μ-join offers speedup up to 71.7% from the best case of binary-join practices for twelve versions using the synthetic dataset. There remain open problems. The query language and the query optimization should be studied to allow users to intuitively exploit μ-join on the standard query interface. We would like to consider the problems in future research.

References

1. Barber, R., et al.: Memory-efficient hash joins. PVLDB **8**(4), 353–364 (2014)
2. Bloom, B.H.: Space/time trade-offs in hash coding with allowable errors. Commun. ACM **13**(7), 422–426 (1970)
3. Bosc, P., Dubois, D., Pivert, O., Prade, H.: Flexible queries in relational databases - the example of the division operator. Theor. Comput. Sci. **171**(1–2), 281–302 (1997)
4. Bosc, P., Pivert, O.: SQLF: a relational database language for fuzzy querying. IEEE Trans. Fuzzy Syst. **3**(1), 1–17 (1995)
5. Chatziantoniou, D., Ross, K.A.: Querying multiple features of groups in relational databases. In: VLDB, vol. 96, pp. 295–306 (1996)
6. Eisenberg, A.: New standard for stored procedures in SQL. SIGMOD Rec. **25**(4), 81–88 (1996)
7. Gray, J., et al.: Data cube: a relational aggregation operator generalizing group-by, cross-tab, and sub totals. Data Min. Knowl. Discov. **1**(1), 29–53 (1997)
8. Gray, J., Reuter, A.: Transaction Processing: Concepts and Techniques. Morgan Kaufmann, Burlington (1993)
9. Gupta, A., et al.: Amazon redshift and the case for simpler data warehouses. In: SIGMOD, pp. 1917–1923 (2015)
10. Hosain, S., Jamil, H.: Algebraic operator support for semantic data fusion in extended SQL. In: ICCTIS, pp. 1–6. IEEE (2010)
11. Jiang, D., et al.: Cohort query processing. PVLDB **10**(1), 1–12 (2016)
12. Kimball, R., Ross, M.: The Data Warehouse Toolkit: The Complete Guide to Dimensional Modeling. Wiley, Hoboken (2011)
13. Kitsuregawa, M., Tanaka, H., Moto-Oka, T.: Application of hash to data base machine and its architecture. New Gener. Comput. **1**(1), 63–74 (1983)
14. Manegold, S., Boncz, P.A., Kersten, M.L.: Optimizing database architecture for the new bottleneck: memory access. VLDB J. **9**(3), 231–246 (2000)
15. Patel, J.M., Carey, M.J., Vernon, M.K.: Accurate modeling of the hybrid hash join algorithm. In: SIGMETRICS, pp. 56–66 (1994)
16. Shapiro, L.D.: Join processing in database systems with large main memories. ACM Trans. Database Syst. **11**(3), 239–264 (1986)
17. Stonebraker, M., et al.: C-store: a column-oriented DBMS. In: Making Databases Work: The Pragmatic Wisdom of Michael Stonebraker, pp. 491–518. ACM/Morgan & Claypool (2019)
18. Yu, X., et al.: PushdownDB: accelerating a DBMS using S3 computation. In: ICDE, pp. 1802–1805. IEEE (2020)

Learning-Based Optimization for Online Approximate Query Processing

Wenyuan Bi[1,2], Hanbing Zhang[1,2], Yinan Jing[1,2(✉)], Zhenying He[1,2],
Kai Zhang[1,2], and X. Sean Wang[1,2,3]

[1] School of Computer Science, Fudan University, Shanghai, China
{wybi19,hbzhang17,jingyn,zhenying,zhangk,xywangCS}@fudan.edu.cn
[2] Shanghai Key Laboratory of Data Science, Shanghai, China
[3] Shanghai Institute of Intelligent Electronics and Systems, Shanghai, China

Abstract. Approximate query processing (AQP) technique speeds up query execution by reducing the amount of data that needs to be processed, while sacrificing the accuracy of the query result to some extent. AQP is essentially a trade-off between the accuracy of the query result and the query latency. However, the heuristic AQP optimization and error control mechanism used by the existing AQP system fails to meet the accuracy requirements of users. This paper proposes a deep learning-based error prediction model to guide AQP query optimization. By using this model, we can estimate the errors of candidate query plans and select the appropriate plans that can meet the accuracy requirement with high probability. Extensive experiments show that the AQP system proposed in this paper can outperform the state-of-the-art online sampling-based AQP approach.

Keywords: Approximate query processing · OLAP · Error prediction

1 Introduction

For interactive data exploration (IDE), it is necessary to answer the query as soon as possible to ensure the user's concentration. Since in big data analytics, many decisions can be made on the big picture of the data, we can use sampling-based approximate query processing (AQP) system to speed up query processing at the cost of accuracy. There are mainly two types of sampling-based AQP methods, namely *offline sampling* and *online sampling*. Offline sampling-based AQP systems [1,7] usually perform queries on pre-computed samples to speed up query processing. The advantage of offline sampling-based AQP methods is that they can quickly return query results. However, offline methods cannot give the estimated error bound of the query before the query execution is completed, so it is difficult to meet the user's accuracy requirement. In addition, offline methods only work well when a priori knowledge of the workload is given. Online sampling-based AQP methods perform sampling at the runtime. The representative of this type of AQP method is Quickr [3]. It incorporates the sampler

© The Author(s), under exclusive license to Springer Nature Switzerland AG 2022
A. Bhattacharya et al. (Eds.): DASFAA 2022, LNCS 13245, pp. 96–103, 2022.
https://doi.org/10.1007/978-3-031-00123-9_7

as a logical operator into the query plan and uses a rule-based error control mechanism to guide the sampler push-down optimization. However, for complex queries and skewed data distribution in real datasets, the assumptions in Quickr are not valid that makes the accuracy of the approximate query results will be low, and the accuracy requirements specified by the user will not be met.

In this paper, instead of using the heuristic rule-based optimization strategy like Quickr [3], we propose a learning-based AQP query optimization method, which uses a priori error prediction model to guide the AQP query optimization. Additionally, we combine the model with the SparkSQL-based AQP system together. Given a query and the accuracy requirements specified by users, by leveraging the error prediction model, the system can select query plans whose error can meet the user's requirements from all candidate query plans. In this way, it can better balance the accuracy requirements put forward by users with the acceleration effect of the AQP system. In summary, we make the following contributions in this paper:

- We design and implement an error prediction model based on deep learning, which provides a prior error guarantee for an approximate query and improves the usability of the AQP system.
- We integrate the above-mentioned error prediction model into the SparkSQL framework to implement a distributed AQP system. The system can return approximate query results that meet the user's accuracy requirements with a higher probability.
- We conduct extensive experiments on the TPC-H dataset. The experiment results show that the AQP system proposed in this paper can outperform the state-of-the-art online sampling-based AQP approach.

2   Approximate Query Optimization

By extending the standard SQL with an enhanced grammar *"ERROR WITHIN a% AT CONFIDENCE b%"*, users can easily pose an AQP query by specifying the expected error bounds. Within the query engine, a sampler is incorporated into the query plan to support the AQP. As shown in Fig. 1, if there is an aggregation operator in the query, we first insert a sampler below it. The logical state of the sampler is maintained, including the sampling rate and the set of columns that may require stratified sampling. To further optimize the AQP query, we can push down the sampler through the *filter* operator, *join* operator, and *projection* operator. Thus, a bunch of candidate query plans are generated as shown in Fig. 1. Unfortunately, not all the candidate query plans can meet the accuracy/error requirement specified by users because the sampler push-down may violate the *equivalence rules* of the relational algebra.

Instead of the strong equivalence principle, we make a **weak equivalence rule** for the AQP query optimization, especially for the probabilistic sampler push-down, i.e., if the error of the approximate query result meets the user's specified accuracy requirements, we say this push-down can comply with the

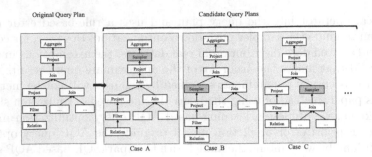

Fig. 1. Incorporating sampler into the query plan and generating candidate plans

weak equivalence rule. In order to ensure that the query accuracy can meet the needs of users, we propose an error prediction model to provide a priori error prediction. The workflow of approximate query optimization with a learned error estimator is shown in Fig. 2. The error estimator can be trained offline. In the online phase, given a query q and an error threshold δ, the AQP engine will generate a set of candidate query plans V by pushing down the sampler. Then, the AQP engine uses the error prediction model m to predict errors of all candidate plans in V and filter out the query plans that do not meet the error requirements. Thus, the AQP engine selects a set of legitimate candidate query plans V' with an error less than the given threshold δ. Finally, we choose the least expensive plan from the remaining query plans to execute. In General, we choose the query plan with the sampler pushed down the farthest.

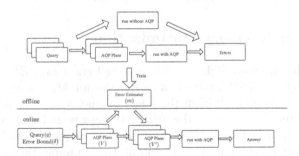

Fig. 2. Approximate query optimization with a learned error estimator

3   Deep Learning-Based Error Prediction Model

In this paper, we design an error prediction model based on deep learning, which is mainly composed of three parts: *feature encoding, tree-structured representation extraction* and *error prediction.*

The feature encoding part is responsible for encoding the features of each node in the query plan into sparse feature vectors and compressing them to obtain a dense representation. We abstract each node in the query plan into four features: the sampler features contained in the node, the filter features, the features of the relation contained, and the topological features of the current node. Among these four features, only the filter feature may not have a fixed dimension because it contains a compound predicate. Therefore, we treat the compound predicate as a sequence of atomic predicates and use LSTM [2] to encode this sequence. Then, we use the final hidden state to represent the encoding of the compound predicate with a fixed length. At present, we restrict the predicates only to be connected with "AND" boolean logic. The encoding of the other three features is a simple one-hot or bitmap type of encoding. After a simple concatenating of the encoding of these four features, we use a fully connected neural network to learn a more compact encoding representation.

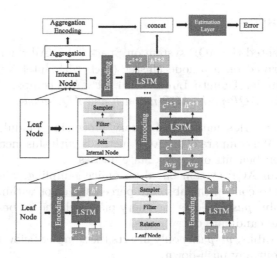

Fig. 3. Tree-structured representation extraction and error prediction

In order to further utilize the recursive topological properties of the tree-structured query plan, the representation extraction network will capture the information contained in each sub-plan from the bottom to up. Therefore, the features output by the topmost root node other than the aggregation operator in the query plan can represent the characterization of the entire query plan. The tree-structured representation extraction network is implemented using LSTM [2]. At the same time, in order to further utilize the information of the sub-tree, we need to design a mechanism that can save and use the output of the sub-tree. Considering that the internal node representing the join operator has two sub-trees, as shown in Fig. 3, we can use the *Hidden State* and *Cell State* of the left and right sub-trees in the LSTM to initialize the corresponding state of the current node.

As shown in Fig. 3, the error prediction network uses a fully connected neural network. The input is the concatenating of the tree-structured query plan feature vector and the aggregation operator encoding, which is the concatenation of one-hot vectors of the aggregate column and aggregate function. The output is the absolute percentage error of the current approximate query plan.

As a regression problem, the error prediction model needs to predict the absolute percentage error (APE) of the approximate query plan. This paper chooses the mean absolute error (MAE) as the loss function of the model. The loss function is formalized by Formula 1, where n is the batch size, $Error_{true}$ is the actual APE label of a query plan, and $Error'$ is the APE predicted by the error prediction model.

$$MAE = \frac{1}{n} \sum_{i=1}^{n} |Error_{true} - Error'| \tag{1}$$

4   Experiment

We have implemented the AQP system on Spark 2.4.5. All the following experiments are performed on a 5-node cluster (each with Intel Xeon Silver 4208, 128 GB RAM) under Ubuntu Linux 18.04 LTS. We compared five methods: *SparkSQL*, *Quickr*, *AQP-push*, *AQP-notpush* and *AQP-model*.

- *SparkSQL*[1] is Spark's native SQL query processing module and does not support AQP. We compare other AQP methods with this method to evaluate the acceleration benefits of approximate queries.
- *Quickr* [3] is an AQP system based on online sampling, which uses a rule-based method to control the absolute percentage error within 10%.
- *AQP-push* (abbr. *push*) selects the query plan with the deepest sampler position among the candidate query plans.
- *AQP-notpush* (abbr. *notpush*) only inserts the sampler below the aggregation operator without any push-down.
- *AQP-model* (abbr. *model*) is the AQP system based on the error prediction model proposed in this paper. This method selects a query plan that can meet the user's error requirements from all candidate query plans by using the error prediction model and has the lowest sampler position at the same time. If there is no query plan satisfied, it will fall back to a precise query without online sampling.

We use the Mean Absolute Percentage Error (MAPE) of a certain approximate processing method to evaluate the error control ability. The MAPE is calculated by Formula 2, where n is the number of test queries, agg_{true} is the exact result of the query, and agg' is the approximate result of the query.

$$MAPE = \frac{1}{n} \sum_{i=1}^{n} \left| \frac{agg_{true} - agg'}{agg_{true}} \right| \times 100\% \tag{2}$$

---

[1] https://spark.apache.org/sql/.

In addition, we also use Mean Speedup (MS) to evaluate the acceleration performance of these methods. The MS is calculated by Formula 3, where n is the number of test queries, $T_{SparkSQL}$ is the query execution time without approximate processing, T_{AQP} is the execution time of a certain AQP method.

$$MS = \frac{1}{n} \sum_{i=1}^{n} \frac{T_{SparkSQL}}{T_{AQP}} \tag{3}$$

In the following experiments, we use the TPC-H benchmark and generate a 10 GB TPC-H dataset. In order to train the error prediction model, the experiment also generated the corresponding query workload. According to the join graph of TPC-H data, for the two-table joined queries, at the number of predicates specified as 1 to 3, 2000 queries are generated respectively. For three-table joined queries, the number of predicates is also specified in the case of 1 to 3, and 1000 queries are generated respectively. In order to further test the overall performance of the approximate query processing system, we also randomly generated 100 SQL queries for testing. In the following experiments, the accuracy requirement is specified as less than 10% by default.

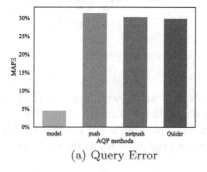
(a) Query Error

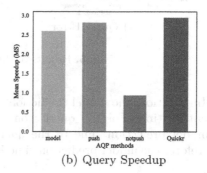
(b) Query Speedup

Fig. 4. Results of two-table joined queries

Figure 4(a) shows the MAPE of different methods under the two-table joined query. As shown in Fig. 4(a), only the method based on the error prediction model has a MAPE within 5%, and the MAPE of other methods is about 30%. The performance of the Quickr method for the two-table joined queries shows that the rule-based AQP query optimization method cannot effectively deal with the queries that contain join operators. Figure 4(b) shows the mean speedup of the two-table joined queries. We can find that the mean speedup of the notpush method without sampler push-down is still around 1, while the other methods are close to 2.5. This means that if the sampler is not pushed down, no acceleration gains can be obtained, and the other methods of pushing down can speed up the execution of the query. At the same time, we observe that the speedup of the model method is less than the counterparts of push and Quickr. This is because the model method tends to choose a less aggressive sampler push-down optimization in order to ensure that the error meets user requirements.

Figure 5(a) shows the error of different methods under three-table joined queries. As shown in Fig. 5(a), the model method can control the MAPE around 2.5%, while other methods generally exceed 13%. It can be seen from Fig. 5(b) that the mean speedup of the notpush method without sampler push-down is still around 1, while other approximation methods can reach 11. Compared with the experiment results of the two-table joined queries, we can observe that as the query contains more join operators, the AQP method can obtain more speedup benefits.

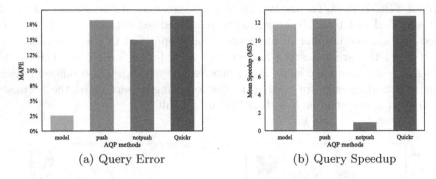

(a) Query Error   (b) Query Speedup

Fig. 5. Results of three-table joined queries

In addition, the model prediction time is very short compared to the query execution time. For the queries on the two or three joined tables, the ratio of model prediction time to the query execution time is only 1.3% to 1.9%. Therefore, the model prediction time is almost negligible for complex queries.

5   Related Work

The sampling-based AQP systems can be divided into two categories: offline sampling-based and online sampling-based systems. There are many studies on offline sampling-based systems in recent years [1,7], but these offline AQP systems need to depend on the prior workload knowledge. The AQP systems based on online sampling [3,5,10] can boost unknown query workload. For example, Quickr [3] system treats the sampling as a logical operator and incorporates it into the query plan. It uses a heuristic rule-based sampler push-down strategy to optimize AQP. However, such a heuristic optimization needs to depend on a strong assumption, and it cannot work well if the assumption does not hold.

In recent years, the application of deep learning techniques to data-driven problems has proven feasible. In the database field, there are also many attempts to apply deep learning to query optimization [4,6,8,9]. To the best of our knowledge, there is no existing work for error prediction of approximate queries by using deep learning.

6   Conclusion

In this paper, we propose a deep learning-based error prediction model, which can provide a prior error guarantee for the AQP system. We also implemented an online approximate query processing system based on SparkSQL and integrated the above-mentioned error prediction model into the system. Compared with the rule-based AQP optimization method, the learning-based method has a more robust performance on query accuracy. Experimental results also prove the superiority of the error prediction model and the AQP system proposed in this paper.

Acknowledgement. This work is funded by the NSFC (No. 61732004 and No. 62072113), the National Key R&D Program of China (No. 2018YFB1004404) and the Zhejiang Lab (No. 2021PE0AC01).

References

1. Agarwal, S., Mozafari, B., Panda, A., Milner, H., Madden, S., Stoica, I.: BlinkDB: queries with bounded errors and bounded response times on very large data. In: Eurosys (2013)
2. Hochreiter, S., Schmidhuber, J.: Long short-term memory. Neural Comput. **9**(8), 1735–1780 (1997)
3. Kandula, S., et al.: Quickr: Lazily approximating complex AdHoc queries in bigdata clusters. In: SIGMOD (2016)
4. Lakshmi, S., Zhou, S.: Selectivity estimation in extensible databases-a neural network approach. In: VLDB, vol. 98, pp. 24–27 (1998)
5. Li, F., Wu, B., Yi, K., Zhao, Z.: Wander join: Online aggregation via random walks. In: SIGMOD 2016. pp. 615–629. ACM (2016)
6. Marcus, R.C., Papaemmanouil, O.: Plan-structured deep neural network models for query performance prediction. Proc. VLDB Endow. **12**(11), 1733–1746 (2019)
7. Park, Y., Mozafari, B., Sorenson, J., Wang, J.: VerdictDB: universalizing approximate query processing. In: SIGMOD (2018)
8. Sun, J., Li, G.: An end-to-end learning-based cost estimator. Proc. VLDB Endow. **13**(3), 307–319 (2019)
9. Wang, W., Zhang, M., Chen, G., Jagadish, H., Ooi, B.C., Tan, K.L.: Database meets deep learning: challenges and opportunities. ACM SIGMOD Rec. **45**(2), 17–22 (2016)
10. Zhang, Y., Zhang, H., He, Z., Jing, Y., Zhang, K., Wang, X.S.: Parrot: a progressive analysis system on large text collections. Data Sci. Eng. **6**(1), 1–19 (2021)

Knowledge Bases

Triple-as-Node Knowledge Graph and Its Embeddings

Xin Lv[1,2], Jiaxin Shi[1,2], Shulin Cao[1,2], Lei Hou[1,2(✉)], and Juanzi Li[1,2]

[1] Department of Computer Science and Technology, BNRist, Beijing, China
[2] KIRC, Institute for Artificial Intelligence, Tsinghua University, Beijing 100084, China
lv-x18@mails.tsinghua.edu.cn, houlei@tsinghua.edu.cn

Abstract. Knowledge Graphs (KGs) aim at semantically representing the world's knowledge in the form of machine-readable graphs composed of subject-relation-object triples (**facts**). However, most previous KGs only consider the relationship between individual entities, ignoring connections between facts and entities, which are commonly used to depict useful information about the properties of facts. To this end, we formally introduce **FactKG**, a new KG form which incorporates fact nodes and extends relations from entity-level to fact-level. This new structure challenges some previous KG techniques. One of the key challenges to FactKG is how to learn compatible representation of entities and facts. In this paper, we mainly focus on the embedding task of FactKG. We contribute a benchmark **WD16K** with additional fact-relevant relations, and a framework **FactE**, which can represent facts, entities and relations in the same space via attention. Experiments demonstrate that FactE not only significantly outperforms state-of-the-art models but also brings remarkable benefits for disambiguation of 1-N relations, revealing its potential usefulness.

Keywords: Knowledge graph · Representation learning · Fact embedding

1 Introduction

Knowledge Graphs (KGs) mostly represent the world's knowledge in a structured way, taking entities (e.g., *Albert Einstein*) as nodes and their relations (e.g., *spouse*) as edges. Triples (**facts**), which consist of two entities and their relation, e.g., (*Albert Einstein*, *spouse*, *Elsa Einstein*), are the core form to store knowledge. As a complementary external resource to natural language text, KGs have proven to benefit lots of NLP tasks, such as language modeling [28], language inference [5], machine reading comprehension [34], etc.

Most of existing KGs, which we call Entity KGs (EntKGs) in this paper, only record the relationship between entities. However, in the real world, some relations happen between facts and entities. For example, *Albert Einstein* married *Mileva Marić* and *Elsa Einstein* in *Switzerland* and *Germany*, respectively. The place of marriage describes an attribute of the marriage fact. Some EntKGs like Freebase [3] connect the places directly to *Albert Einstein*, missing the correspondence between marriage facts and marriage places. Wikidata [32], one of the largest collaboratively edited knowledge

© The Author(s), under exclusive license to Springer Nature Switzerland AG 2022
A. Bhattacharya et al. (Eds.): DASFAA 2022, LNCS 13245, pp. 107–121, 2022.
https://doi.org/10.1007/978-3-031-00123-9_8

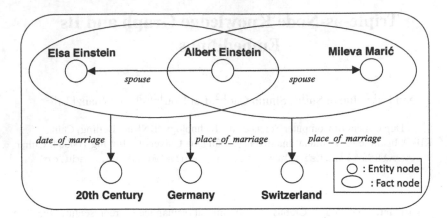

Fig. 1. An example of FactKG. Besides typical entity nodes, we also regard triples (facts) as nodes and thus we can represent the relations of facts.

graphs, records some connections between facts and entities. But it has not been well studied yet.

In this paper, we formally organize relations between facts and entities and introduce a new KG form, named as **FactKG**, by extending EntKGs. In FactKG, we establish fact nodes for typical triples and incorporate qualifiers (e.g., (*place_of_marriage*, *Germany*)) for every fact. Figure 1 gives an example depicting two marriage facts of *Albert Einstein* and their corresponding places in this new KG form. With fact nodes, we can not only incorporate more knowledge that is absent in previous KGs, but also disambiguate the objects of 1-N relations (e.g., *spouse*) by specifying fact qualifiers, which will benefit downstream tasks such as KG-based question answering [27].

Due to its novel structure, FactKG presents challenges to previous KG techniques, and the key is how to learn the representation of facts, entities and relations jointly. In this paper, we mainly focus on Knowledge Graph Embedding (KGE), a fundamental task of KG area, which aims to encode entities and relations in terms of triple constraints. Although KGE has been widely studied in recent years and many representative models have been proposed [4,8], they are limited to EntKG form. It is of course straightforward to apply these KGE methods to FactKG by simply allocating separate embeddings for fact nodes, but this will lose rich information of interactions between a fact and its internal components (i.e., subject, relation, and object) and the model also suffers from more sparsity issue. Besides, FactKG embedding task is not supported by existing KGE datasets (e.g., FB15K-237) as they lack triples with fact nodes.

To tackle these problems, we firstly establish a new KGE dataset, named **WD16K**, as a benchmark of FactKG embedding task, and then propose a novel embedding framework, named **FactE**, to jointly represent entities, relations, and facts.

For the dataset part, we extract our data from Wikidata as it provides high-quality connections between facts and entities. By aligning entities of FB15K-237 to Wikidata, extracting and filtering relevant entity-to-entity (E-E) and fact-to-entity (F-E) triples[1], we obtain our WD16K dataset consisting of 15,874 entities, 176,748 E-E triples, and 28,202 F-E triples.

For the model part, we propose a novel framework named FactE to jointly represent facts, entities and relations. Specifically, we use two strategies (i.e., ATT and FUSE) to learn the fact embedding based on the representation of its internal components. After that, we treat facts as special entities and use typical knowledge embedding methods for training. Our framework consists of three learning tasks, i.e., E-E triple prediction, F-E triple prediction and qualifier-restricted entity-to-entity (Q-E) prediction, the last of which takes qualifiers as additional input of E-E to help disambiguation of 1-N relations.

In experiments, we evaluate our model on the task of link prediction for both E-E and F-E triples. For a fair comparison with baselines, we convert our WD16K to EntKG form to apply previous KGE models. It is demonstrated that the FactKG form and FactE model bring significant improvements over baselines, especially for F-E predictions. Besides, given one or more qualifiers, 1-N relations can be well disambiguated using our form and model, indicating their superiority and potential usefulness.

2   Related Work

2.1   KGE Datasets

Most existing KGE datasets are subsets of real world knowledge bases, such as (1) FB15K [4] and FB15K-237 [30] from Freebase [3]; (2) WN18 [4] and WN18RR [8] from WordNet [22]; (3) YAGO3-10 [29] from YAGO3 [21]. However, these datasets only consider relations between entities. Some other datasets incorporates temporal information for facts, such as ICEWS [17] and GDELT [18]. However, these datasets only focus on temporal information and neglect other useful fact attributes. In this paper, we first propose the fact-level KG form FactKG and construct a dataset WD16K.

2.2   KGE Techniques

KGE techniques for typical KGs can be divided into three categories: (1) translation based models [4,20,29], which aim to learn embeddings by representing relations as translations from head to tail entities; (2) tensor-factorization based models [1,25,26], which aim to decompose relational data into low-rank matrices for representation learning and (3) non-linear models [2,8,23,24], which typically take entities and/or relations into deep neural networks and compute a semantic matching score. Besides, recently some KGE models are proposed specifically for temporal KGs [7,11,14]. Despite achieving good performance, these KGE models are limited to the typical KG form and cannot handle fact nodes well.

---

[1] Fact-to-fact triples representing the relations between two facts (e.g., cause and effect) are not provided in Wikidata. We leave it for future work.

2.3 Event KGs and Representations

Fact is similar to *Event* sometimes (e.g., (*Albert Einstein, spouse, Elsa Einstein*) describes a marriage event). Some event-related KGs are built by considering event entities (e.g., *World War 2*) [12] or actions (e.g., *watching movies*) [19] as nodes and their temporal, spatial or causal relations as edges. In these KG, facts are treated as nodes and do not have internal components, which are different from our triple-formed fact nodes. Some work [9, 10] extract triples from news as events and represent them for event-related tasks (e.g., event similarity). Their event is triple-formed. However, similar to EntKGs, they do not consider connections between events and entities. Facts and events can also be represented via n-ary [6, 33], which lists the relevant knowledge as key-value pairs. However, the n-ary form misses the triplet structure, which may break the information transformation [13].

3 Problem Formulation

FactKG. FactKG consists of entities, relations, and triples, denoted by $\mathcal{KG} = \{\mathcal{E}, \mathcal{R}, \mathcal{T}\}$. The main difference to EntKG is that besides typical entity-to-entity triples, dubbed as **facts**, FactKG contains a new type of triples, i.e., fact-to-entity triples, representing relations between a fact and an entity. Formally, we have $\mathcal{T} = (\mathcal{T}_{\text{E-E}}, \mathcal{T}_{\text{F-E}})$, where $\mathcal{T}_{\text{E-E}} = \{(h, r, t)|h, t \in \mathcal{E}, r \in \mathcal{R}\}$ and $\mathcal{T}_{\text{F-E}} = \{((h, r, t), r_q, t_q)|(h, r, t) \in \mathcal{T}_{\text{E-E}}, r_q \in \mathcal{R}, t_q \in \mathcal{E}\}$. (r_q, t_q) is denoted as the **qualifier** for a fact-to-entity triple $((h, r, t), r_q, t_q)$, where r_q and t_q are qualifier relation and qualifier entity respectively.

FactKG Embedding. Previous KGE task aims to learn a vector $\mathbf{e} \in \mathbb{R}^k$ for each entity $e \in \mathcal{E}$ and a vector $\mathbf{r} \in \mathbb{R}^k$ for each relation $r \in \mathcal{R}$, using all triples in $\mathcal{T}_{\text{E-E}}$ as constraints. For FactKG embedding task, besides typical entity and relation embeddings, we also learn a vector $\mathbf{f} \in \mathbb{R}^k$ for each fact $f \in \mathcal{T}_{\text{E-E}}$. The aim is to jointly learn $\mathbf{e}, \mathbf{r}, \mathbf{f}$ under the constraints of both $\mathcal{T}_{\text{E-E}}$ and $\mathcal{T}_{\text{F-E}}$. In Table 1, we formally compare EntKGs and FactKG.

Table 1. Comparison between EntKGs and FactKG.

	EntKGs	FactKG
Components	$\mathcal{E}, \mathcal{R}, \mathcal{T}_{\text{E-E}}$	$\mathcal{E}, \mathcal{R}, \mathcal{T}_{\text{E-E}}, \mathcal{T}_{\text{F-E}}$
Embeddings	$\mathbf{e}, \mathbf{r}$	$\mathbf{e}, \mathbf{r}, \mathbf{f}$
Constraints	$\mathcal{T}_{\text{E-E}}$	$\mathcal{T}_{\text{E-E}}, \mathcal{T}_{\text{F-E}}$

4 Our Model

Our FactE model consists of three learning tasks: 1) Entity-to-entity (E-E) prediction, which is commonly used in previous KGE methods. 2) Fact-to-entity (F-E) prediction, predicting the qualifier entity of F-E triples. 3) Qualifier-restricted entity-to-entity (Q-E) prediction, taking qualifiers as additional input of E-E to help disambiguation of

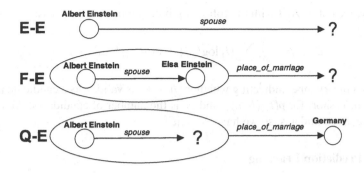

Fig. 2. Examples of our three basic tasks.

1-N relations. Figure 2 shows examples of these three tasks. Formally, we denote them as

$$\text{E-E} : (h, r, ?) \rightarrow t,$$
$$\text{F-E} : ((h, r, t), r_q, ?) \rightarrow t_q, \qquad (1)$$
$$\text{Q-E} : (h, r, ?, (r_q^1, t_q^1), \cdots, (r_q^{m}, t_q^{m})) \rightarrow t.$$

For the above three learning tasks, there are three corresponding loss functions $\mathcal{L}_{\text{E-E}}$, $\mathcal{L}_{\text{F-E}}$ and $\mathcal{L}_{\text{Q-E}}$. They share the same knowledge graph embedding and parameters. The overall objective is the combination them:

$$\mathcal{L} = \mathcal{L}_{\text{E-E}} + \mathcal{L}_{\text{F-E}} + \mathcal{L}_{\text{Q-E}}. \qquad (2)$$

We will introduce $\mathcal{L}_{\text{E-E}}$, $\mathcal{L}_{\text{F-E}}$ and $\mathcal{L}_{\text{Q-E}}$ in Sects. 4.1, 4.2, 4.3 respectively.

4.1 E-E Prediction Learning

The E-E prediction has been well studied by previous work. Given a triple (h, r, t), they compute the probability of t being matched with (h, r), denoted as $p(t|(h, r))$, using different models in Sect. 2.2, and then learn embeddings and parameters by maximizing the probability of the correct tail.

In this paper, without loss of generality, we choose ConvE [8] to demonstrate E-E prediction. First, we encode the interaction between h and r via a multi-layer convolutional network:

$$\text{Enc}_{\text{E-E}}(h, r) = g(\text{vec}(g([\overline{\mathbf{e}}_h; \overline{\mathbf{r}}_r] * \omega))\mathbf{W}), \qquad (3)$$

where $\mathbf{e}_h, \mathbf{r}_r \in \mathbb{R}^k$ are embeddings for entity h and relation r, $\overline{\mathbf{e}}_h$ and $\overline{\mathbf{r}}_r$ denote a 2D reshaping of $\mathbf{e}_h$ and $\mathbf{r}_r$, ω is the convolution kernel, g is a ReLU function to provide nonlinearity, $\text{vec}(\cdot)$ denotes the flattened vector of a tensor, and the matrix $\mathbf{W}$ projects the vector into a k-dimensional space, i.e., $\text{Enc}_{\text{E-E}}(h, r) \in \mathbb{R}^k$.

Next, we compute the probability of t being matched with (h, r) by

$$p(t|(h, r)) = \text{Sigmoid}(\text{Enc}_{\text{E-E}}(h, r) \cdot \mathbf{e}_t). \qquad (4)$$

Finally, we optimize it with the following binary cross-entropy loss:

$$\mathcal{L}_{\text{E-E}} = -\frac{1}{N} \sum_{i=1}^{N} (l_i \log(p_i) + (1 - l_i) \log(1 - p_i)), \tag{5}$$

where l_i is a binary label indicating whether (h, r, t_i) is valid, t_i is a candidate tail entity for (h, r), p_i is short for $p(t_i|(h, r))$, and N is the number of candidates. We regard all entities in $\mathcal{E}$ as candidates, so we have $N = |\mathcal{E}|$.

4.2   F-E Prediction Learning

Given an F-E triple $((h, r, t), r_q, t_q)$, similar to E-E prediction, we first encode inputs into a vector $\text{Enc}_{\text{F-E}}(h, r, t, r_q) \in \mathbb{R}^k$, then compute the probability like Eq. 4, and finally learn parameters with a binary cross-entropy loss like Eq. 5.

In the input encoding stage, we aim to capture the interactions among (h, r, t, r_q), where (h, r, t) forms a fact f whose embedding is $\mathbf{f}$. By feeding the fact embedding $\mathbf{f}$ and the qualifier relation embedding $\mathbf{r}_{r_q}$ into ConvE, we can get the F-E encoding model:

$$\text{Enc}_{\text{F-E}}(h, r, t, r_q) = g(\text{vec}(g([\overline{\mathbf{f}}; \overline{\mathbf{r}}_{r_q}] * \omega))\mathbf{W}). \tag{6}$$

As the meaning of a fact is highly determined by its subject, relation, and object, we model the fact embedding as a composition of its three components:

$$\mathbf{f} = c(\mathbf{e}_h, \mathbf{r}_r, \mathbf{e}_t). \tag{7}$$

By doing so, we can easily get the embedding of any fact once its components are given. Compared with allocating separate vectors for fact nodes like entity nodes, this strategy can significantly reduce parameters and avoid sparsity. In this paper, we propose two different strategies for the combination function c.

The first strategy is **FUSE**, in which we simply concatenate the embeddings of h, r, and t, and then fuse them via a neural network:

$$\mathbf{f} = \text{MLP}([\mathbf{e}_h; \mathbf{r}_r; \mathbf{e}_t]), \tag{8}$$

where $[\mathbf{e}_h; \mathbf{r}_r; \mathbf{e}_t] \in \mathbb{R}^{3k}$ represents vector concatenation, and MLP is a stack of Linear layers whose final output dimension is k.

The second strategy is **ATT**. As different qualifier relations may focus on different fact components, we incorporate r_q as a hint for fact embedding in our second strategy Specifically, we take $\mathbf{r}_{r_q}$ as the query, take component embeddings as the values, and use their weighted average as the fact embedding:

$$\begin{aligned} \mathbf{a} &= \text{Softmax}(\text{MLP}(\mathbf{r}_{r_q}) \cdot [\mathbf{e}_h, \mathbf{r}_r, \mathbf{e}_t]), \\ \mathbf{f} &= [\mathbf{e}_h, \mathbf{r}_r, \mathbf{e}_t] \cdot \mathbf{a}, \end{aligned} \tag{9}$$

where $[\mathbf{e}_h, \mathbf{r}_r, \mathbf{e}_t] \in \mathbb{R}^{k \times 3}$ represents the stack of three vectors, MLP is a linear mapping from k-dimension to k, and $\mathbf{a} \in \mathbb{R}^3$ denotes attention weights.

In the end, we can obtain $p(t_q|(h, r, t, r_q))$ by replacing $\text{Enc}_{\text{E-E}}$ and $\mathbf{e}_t$ of Eq. 4 with $\text{Enc}_{\text{F-E}}$ and $\mathbf{e}_{t_q}$. Similar to Eq. 5, we can obtain the loss function of F-E prediction, denoted as $\mathcal{L}_{\text{F-E}}$.

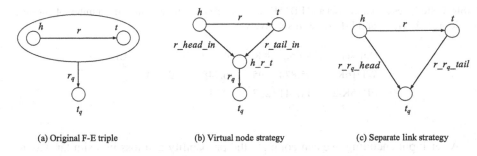

(a) Original F-E triple   (b) Virtual node strategy   (c) Separate link strategy

Fig. 3. Two strategies for converting WD16K to EntKGs.

4.3 Q-E Prediction Learning

Given an E-E triple with m qualifiers $((h, r, t), (r_q^1, t_q^1), \cdots, (r_q^m, t_q^m))$, the Q-E prediction aims to utilize qualifiers to enhance E-E prediction. Similar to E-E prediction, it can be divided into input encoding, probability computation, and loss computation.

In the input encoding stage, we encode all qualifiers into two vectors $\mathbf{q}_h$ and $\mathbf{q}_r$ to provide additional information for h and r respectively. By adding the original and additional vectors together and then feeding into ConvE, we get the Q-E encoding model:

$$
\begin{aligned}
\text{Enc}_{\text{Q-E}}(h, r, (r_q^1, t_q^1), \cdots, (r_q^m, t_q^m)) = \\
g(\text{vec}(g([\overline{(\mathbf{e}_h + \mathbf{q}_h)}; \overline{(\mathbf{r}_r + \mathbf{q}_r)}] * \omega))\mathbf{W}).
\end{aligned}
\tag{10}
$$

$\mathbf{q}_h$ and $\mathbf{q}_r$ can be computed by summing up every qualifier attentively. Taking $\mathbf{q}_h$ for exmaple:

$$
\begin{aligned}
\mathbf{a} = \text{Softmax}(\text{MLP}(\mathbf{e}_h) \cdot [\mathbf{q}_h^1, \cdots, \mathbf{q}_h^m]), \\
\mathbf{q}_h = [\mathbf{q}_h^1, \cdots, \mathbf{q}_h^m] \cdot \mathbf{a},
\end{aligned}
\tag{11}
$$

where MLP is a linear mapping from k-dimension to k, $\mathbf{a} \in \mathbb{R}^m$ denotes attention weights, and $\mathbf{q}_h^i$ is the vector of the i-th qualifier. Similarly, $\mathbf{q}_r$ can be computed using $\mathbf{r}_r$ as query. In this paper, we propose two strategies to compute $\mathbf{q}_h^i$ and $\mathbf{q}_r^i$.

The first strategy is **FUSE**. Taking $\mathbf{q}_h^i$ as an example, we fuse each (r_q^i, t_q^i) pair via a neural network:

$$
\mathbf{q}_h^i = \text{MLP}([\mathbf{r}_{r_q^i}; \mathbf{e}_{t_q^i}]),
\tag{12}
$$

where MLP is implemented as a linear layer from $2k$-dimension to k. $\mathbf{q}_r^i$ is the same as $\mathbf{q}_h^i$ in this strategy.

The second strategy is **ATT**. Take $\mathbf{q}_h^i$ for example, we encode each qualifier pair by regarding $\mathbf{e}_h$ as the attention key:

$$
\begin{aligned}
\mathbf{a} = \text{Softmax}(\text{MLP}(\mathbf{e}_h) \cdot [\mathbf{r}_{r_q^i}, \mathbf{e}_{t_q^i}]), \\
\mathbf{q}_h^i = [\mathbf{r}_{r_q^i}, \mathbf{e}_{t_q^i}] \cdot \mathbf{a},
\end{aligned}
\tag{13}
$$

where MLP is a linear mapping from k-dimension to k, and $\mathbf{a} \in \mathbb{R}^2$ denotes attention weights. Similarly, $\mathbf{q}_r^i$ can be computed using $\mathbf{r}_r$ as key.

Table 2. Statistics of our dataset and FB15K-237. The four columns denote the number of entities, relations, E-E triples and F-E triples respectively.

Dataset	# Ent.	# R.	# Trip.(E-E)	# Trip.(F-E)
WD16K	15,874	195	176,748	28,202
FB15K-237	14,541	237	310,116	–

After input encoding, we can compute the probability and loss in a similar way to Eq. 4 and Eq. 5. The final Q-E loss is denoted as $\mathcal{L}_{Q\text{-}E}$. It is worth mentioning that given a head entity and a 1-N relation, positive targets in Q-E prediction are less than those in E-E due to the qualifier constraints.

5 Dataset

5.1 Dataset Construction

We construct our dataset by merging FB15K-237 [30] and Wikidata [32]. FB15K-237 is one of the most widely-used KGE datasets, providing 15K entities and their dense connections, but lacking fact-level relations. Wikidata is one of the largest collaboratively edited KGs, providing billions of entities, relations, and most importantly, fact-to-entity connections. As lots of entities in Wikidata are annotated with Freebase IDs, we can easily align entities of FB15K-237 to Wikidata, and then extract relevant entity-level and fact-level triples to form our dataset. As a result, our dataset has a comparable scale and shares most entities with FB15K-237. Following FB15K, we name it WD16K[2].

The detailed steps for building WD16K are:

(1) We align entities in FB15K-237 to Wikidata[3] using Freebase IDs. Almost all entities (98.7%) in FB15K-237 can be found in Wikidata and these 14,353 entities make up our initial entity set $\mathcal{E}$.

(2) For every fact-relevant triple $((h, r, t), r_q, t_q)$ in Wikidata whose h and t are in $\mathcal{E}$, we collect qualifier entities that are not in $\mathcal{E}$ as an additional entity set $\mathcal{E}'$.

(3) We filter out low-frequency entities from $\mathcal{E}'$ and then add remainings to $\mathcal{E}$. Specifically, for $e \in \mathcal{E}'$, we let $\mathcal{S}_e = \{(e, r, t)|t \in \mathcal{E}\} \bigcup \{(h, r, e)|h \in \mathcal{E}\}$. We only keep those $|\mathcal{S}_e| \geq 10$, which give additional 1,521 entities into $\mathcal{E}$.

(4) We extract (h, r, t) whose $h, t \in \mathcal{E}$ from Wikidata as our E-E triples $\mathcal{T}_{E\text{-}E}$, and extract $((h, r, t), r_q, t_q)$ whose $(h, r, t) \in \mathcal{T}_{E\text{-}E}, t_q \in \mathcal{E}$ as our F-E triples $\mathcal{T}_{F\text{-}E}$.

Final statistics of our dataset WD16K is listed in Table 2. We shuffle it and use 90%/5%/5% as our training/validation/test set.

---

[2] Our codes and dataset can be available in http://github.com/davidlvxin/facte.

[3] We use the 20190506 snapshot of Wikidata.

5.2 Conversion Strategies

In order to verify the effectiveness of our model, we need to compare with some baseline models. One intuitive way is to use a conversion strategy to convert WD16K to the EntKG form and run some previous KGE models on it. In this section, we propose two strategies *virtual node* and *seperate link* for conversion.

Virtual Node Strategy. For an F-E triple $((h, r, t), r_q, t_q)$, we create a new virtual node h_r_t representing (h, r, t). Besides, we also add two new relations r_head_in and r_tail_in, which connect h and t with the virtual node h_r_t respectively. Through these two relations, h_r_t gets all information about the fact (h, r, t). Finally, the F-E triple $((h, r, t), r_q, t_q)$ is converted to (h_r_t, r_q, t_q), allowing previous KGE models to run on it. Figure 3(b) shows the conversion result of the F-E triplet in Fig. 3(a). However, compared with WD16K, such a conversion introduces additional nodes and relations, making the knowledge graph more sparse, which is not conducive to representation learning.

Separate Link Strategy. For an F-E triple $((h, r, t), r_q, t_q)$, we remove r_q and create two new relations $r\_r_q\_head$ and $r\_r_q\_tail$, which connect h and t with entity t_q respectively. Therefore, we have two new E-E triples $(h, r\_r_q\_head, t_q)$ and $(t, r\_r_q\_tail, t_q)$ instead of the original F-E triple. Figure 3(c) shows the conversion result of the F-E triplet in Fig. 3(a). It's worth noting that, similar to this strategy, some EntKGs (e.g., Freebase) also directly connect qualifier entity t_q to h or t. This strategy also has some disadvantages, e.g., the conversion incorporates new fine-grained relations, most of which are 1-N relations and will lose some information between facts and qualifier entity.

6 Experiments

We evaluate our model with link prediction task on WD16K and compare with previous KGE models on the two conversions.

6.1 Experimental Setup

Baselines: In our experiments, we select the following knowledge graph embedding models for comparison: TransE [4], DistMult [35], ComplEx [31], SimplE [15], ConvE [8], RotatE [29] and TuckER [1]. Since our model has two strategies in both F-E and Q-E learning, there are four variants of our model, i.e., FactE(FUSE+FUSE), FactE(ATT+FUSE), FactE(FUSE+ATT) and FactE (ATT+ATT), where words before and after the plus sign represent strategies of F-E and Q-E learning respectively.

Evaluation Protocol: Unlike most previous link prediction experiments, which only consider E-E prediction, we also perform experiments on F-E prediction and Q-E prediction. Specifically, since Q-E prediction is different from previous KGE tasks, it is mainly used for disambiguation of 1-N relation predictions. We put its experimental results in Sect. 6.3 for discussion. For every E-E triple (h, r, t) in the test set, take predicting tail entity t as an example, KGE models need to give a descending order of the probability that every entity is the correct tail entity. It is worth noting that we only predict the tail entity for F-E triples because it is not practical to predict a triple (fact) based on the tail entity and relation. Besides, we use the "Filter" settings proposed by TransE [4] in our experiments. We use two evaluation metrics: (1) the mean reciprocal rank of all correct entities (MRR) and (2) the proportion of correct entities that rank no larger than N (Hits@N). A good KGE model should achieve high MRR and Hits@N.

Implementation Details: Following ConvE [8], we use Adam [16] as the optimizer and label smoothing to lessen overfitting. We select the best hyperparameters via grid search according to Hits@10 on the validation dataset. The ranges of the hyperparameters for the grid search are the same as ConvE [8]. For every E-E triple (h, r, t), we also add a reverse triple $(t, r\_reverse, h)$ for head entity prediction in E-E and Q-E learning.

For virtual node strategy, some virtual nodes may not appear in the training set and have no embedding vectors. For every F-E triple $((h, r, t), r_q, t_q)$ in the test set, we will add (h, r_head_in, h_r_t) and (t, r_tail_in, h_r_t) to a finetune set $\mathcal{T}_f$. Baseline models will also be trained on $\mathcal{T}_f$ before testing on F-E triples. As shown in Fig. 3(c), for separate link strategy, we can use both $(h, r\_r_q\_head, ?)$ and $(t, r\_r_q\_tail, ?)$ for an F-E triple prediction. In our experiments, we use the average of the above two prediction scores for final prediction.

6.2 Link Prediction Results

Table 3 shows the link prediction results on WD16K. For E-E prediction, our model FactE (ATT+ATT) outperforms previous state-of-the-art models on all metrics (except for Hits@10 where TuckER does better). For F-E prediction, the advantages of our model are more obvious. FactE(ATT+ATT) performs much better than all previous models on every metric, especially on Hits@1 where our model gains significant improvements compared with the best baseline TuckER (which is about 45.1% relative improvement). This indicates that both E-E and F-E triples can be represented well in our model.

Effects of FactE Strategies: We study how the strategies in F-E and Q-E learning affect the performance. As shown in Table 3, ATT strategy in F-E learning performs much better than FUSE, which indicates that qualifier relations are helpful for fact representation in F-E learning. Besides, ATT strategy in Q-E learning also achieves slight improvement for F-E prediction.

Analyses of Conversion Strategies: From Table 3, we can learn that these two strategies bring different results for baseline models. Specifically, models using virtual node strategy perform better on F-E prediction, while models using separate link strategy achieve a better performance on E-E prediction. We think such a difference is caused by their inherent structures. For virtual node strategy, the newly created nodes can well represent fact information, which brings good performance on F-E prediction. However, too many extra nodes also make models more difficult to learn, which limits the performance on E-E prediction. For separate link strategy, it introduces a small number

Table 3. Link prediction results on E-E and F-E triples. All metrics are multiplied by 100. The best score is in **bold**. For E-E prediction, the results are the average scores of predicting the head and tail entities. For F-E prediction, the results are scores of predicting tail entities. Virtual Node and Separate Link denote results on datasets converted from WD16K using virtual node and separate link strategy respectively. FactKG denotes results on WD16K.

	Model	E-E				F-E			
		MRR	Hits@10	Hits@3	Hits@1	MRR	Hits@10	Hits@3	Hits@1
Virtual Node	TransE	23.8	39.2	27.1	15.7	24.9	45.3	27.7	14.9
	DistMult	12.3	22.3	13.0	7.2	13.4	24.5	14.2	7.9
	ComplEx	16.7	30.7	18.3	9.8	23.0	45.7	24.6	12.5
	SimplE	17.2	31.5	19.6	9.7	24.4	47.0	26.9	13.4
	ConvE	37.7	51.9	41.8	29.9	53.0	81.0	65.5	36.5
	RotatE	33.8	52.3	39.3	23.6	60.2	85.0	69.0	46.6
	TuckER	37.8	54.5	42.2	28.6	66.1	85.3	73.1	55.7
Separate Link	TransE	26.8	43.8	31.2	17.6	31.9	53.1	39.8	30.2
	DistMult	12.3	24.9	13.5	6.0	11.7	29.8	14.2	8.2
	ComplEx	17.0	31.8	19.7	9.5	19.4	40.6	24.6	16.1
	SimplE	17.8	33.0	21.2	9.4	17.3	38.7	23.3	14.7
	ConvE	39.1	54.1	42.9	30.9	37.0	57.7	42.3	25.5
	RotatE	35.7	54.3	41.5	25.4	36.8	60.3	46.3	35.6
	TuckER	39.8	**56.2**	43.9	31.0	36.9	58.0	43.9	31.0
FactKG	FactE(FUSE+FUSE)	36.8	51.4	41.0	28.9	67.3	84.4	72.5	58.7
	FactE(ATT+FUSE)	39.2	54.5	43.5	30.9	77.2	91.1	82.7	69.3
	FactE(FUSE+ATT)	36.8	51.5	40.8	28.8	68.9	84.8	73.4	60.7
	FactE(ATT+ATT)	**40.5**	55.0	**44.6**	**32.6**	**78.1**	**91.2**	**82.9**	**70.8**

of fine-grained relations and maintains the denseness of KGs, which is beneficial to E-E prediction learning. However, most added relations are 1-N relations and cannot contain all information about facts, which results in poor performance on F-E prediction. In summary, the KGs converted using both virtual node and separate link strategy have some problems, and FactKG is a better format.

6.3 Analysis

To further analyze the performance and strengths of our FactE, we carried out extensive experiments to explore the following questions:

Q1: How does the Q-E learning affect the overall performance?

A1: As shown in Table 4, with Q-E learning, even though additional qualifiers are provided only during training, both E-E and F-E predictions gain consistent improvements. We think the most important reason is that 1-N relations are handled better with the help of Q-E learning, making the learned embeddings more discriminative.

Q2: For 1-N relations, can we predict the expected object given qualifier information?

A2: As mentioned before, disambiguating 1-N relations with qualifier hints is one important application of FactKG. To demonstrate it, we pick up E-E triples that have a

Table 4. Influence of Q-E learning.

	Model	MRR	Hits@10	Hits@3	Hits@1
E-E	w/o Q-E	38.3	53.2	42.5	30.3
	FactE	**40.5**	**55.0**	**44.6**	**32.6**
F-E	w/o Q-E	77.3	90.4	82.1	69.4
	FactE	**78.1**	**91.2**	**82.9**	**70.8**

Table 5. Object disambiguation results of 1-N relations given a random qualifier (top section) or all qualifiers (bottom section) as additional information.

	Model	MRR	Hits@10	Hits@3	Hits@1
Random	ConvE	45.2	60.3	49.1	37.1
	FactE	**68.1**	**81.6**	**72.2**	**61.2**
All	ConvE	39.8	59.7	45.5	29.2
	FactE	**80.9**	**92.2**	**85.9**	**74.3**

1-N relation and at least one qualifier from the test set, extract all qualifiers for each of them, and conduct Q-E prediction by providing a random or all qualifier hints.

As shown in Table 5, our FactE achieves significant improvements over previous KGE models such as ConvE, which actually cannot utilize additional hints and thus gives the prediction in the E-E way. When given more qualifier hints, our FactE can attentively select the most informative ones from them and produce more accurate predictions. However, the performance of ConvE drops as the increase of qualifiers, more qualifiers mean more additional restrictions, the harder it is for the ConvE to find the correct answers.

Q3: What if we change the input encoder Enc of FactE from ConvE to other KGE models?

A3: We try another SOTA model TuckER, as our input encoder Enc, and list results in Table 6. We can see that the performance drops a lot using TuckER. We think it is because the theoretical basis of TuckER, i.e., third-order tensor factorization, is not suitable for our FactE well, since there exist extra restrictions in our embeddings (i.e., our fact embedding is computed using entity and relation embeddings). In fact, we have compared with other KGE models including ConvE, RotatE and TuckER in experiments, FactE can achieve the best results using ConvE as input encoder Enc.

6.4   Case Study

In Table 7, we present two object disambiguation examples mentioned in Q2 of Sect. 6.3. *award received* and *nominated for* are two typical 1-N relations in these two examples. Qualifiers are provided as additional information for every triple query. From Table 7, we can learn that FactE can make good use of qualifier information and predict accurate entities. ConvE can also predict the correct entities for a triple query without qual-

Table 6. Results of replacing ConvE part of our model by TuckER.

	Model	MRR	Hits@10	Hits@3	Hits@1
E-E	Ours(TuckER)	37.9	**55.4**	41.8	28.9
	Ours(ConvE)	**40.5**	55.0	**44.6**	**32.6**
F-E	Ours(TuckER)	63.2	83.1	70.0	52.3
	Ours(ConvE)	**78.1**	**91.2**	**82.9**	**70.8**

Table 7. Case study of two object disambiguation examples of 1-N relations given qualifiers as additional information. For every triple query, we display the predicted top 5 entities of FactE (left section) and ConvE (right section). The correct entity of every query is in **bold**.

	FactE	ConvE
	Triple query: *(Jennifer Lawrence, award received, ?)*	Qualifier: *(for work, The Hunger Games)*
1	**MTV Movie Award for Best Fight**	MTV Movie Award for Best Kiss
2	Saturn Award for Best Actress	MTV Movie Award for Best On-Screen Duo
3	MTV Movie Award for Best Kiss	Independent Spirit Award for Best Female Lead
4	MTV Movie Award for Best Jaw Dropping Moment	**MTV Movie Award for Best Fight**
5	MTV Movie Award for Best Female Performance	MTV Movie Award for Best Villain
	Triple query: *(Romeo and Juliet, nominated for, ?)*	Qualifier: *(nominee, Danilo Donati)*
1	**Academy Award for Best Costume Design**	Academy Award for Best Cinematography
2	Italian	Tony Award for Best Costume Design
3	Academy Award for Best Cinematography	Quebec
4	costume designer	Flash Gordon
5	Latin	**Academy Award for Best Costume Design**

ifier constraints. Take the first triple query of Table 7 as an example, the first, third and fourth entities predicted by ConvE are correct if we do not give qualifier constraints. However, ConvE has no ability to disambiguate these entities if we provide qualifiers as additional information.

7 Conclusion

In this paper, we formally organize relations between facts and entities and introduce a new KG form named **FactKG**, by extending EntKGs. We also construct a dataset named WD16K for FactKG embedding task, and then propose a novel model named FactE for FactKG representation. Since most previous KGE models cannot be applied to our dataset, we use two conversion strategies to convert WD16K to EntKGs and run baseline models on them for comparison. Experimental results show that FactE outperforms previous baseline models. Besides, our model can give more accurate results for 1-N predictions by using qualifier relations and entities as context information. Our future work might consider more complex KGs with events, i.e., incorporating event nodes composed of more complex internal structure and event-level relations.

Acknowledgments. This work is supported by the National Key Research and Development Program of China (2020AAA0106501), the grants from the Institute for Guo Qiang, Tsinghua University (2019GQB0003) and Beijing Academy of Artificial Intelligence, and the NSFC Youth Project (62006136).

References

1. Balažević, I., Allen, C., Hospedales, T.M.: Tucker: Tensor factorization for knowledge graph completion. In: EMNLP (2019)
2. Balažević, I., Allen, C., Hospedales, T.M.: Hypernetwork knowledge graph embeddings. In: Tetko, I.V., Kůrková, V., Karpov, P., Theis, F. (eds.) ICANN 2019. LNCS, vol. 11731, pp. 553–565. Springer, Cham (2019). https://doi.org/10.1007/978-3-030-30493-5_52
3. Bollacker, K., Evans, C., Paritosh, P., Sturge, T., Taylor, J.: Freebase: a collaboratively created graph database for structuring human knowledge. In: SIGMOD (2008)
4. Bordes, A., Usunier, N., Garcia-Duran, A., Weston, J., Yakhnenko, O.: Translating embeddings for modeling multi-relational data. In: NIPS (2013)
5. Chen, Q., Zhu, X., Ling, Z.H., Inkpen, D., Wei, S.: Neural natural language inference models enhanced with external knowledge. In: ACL (2018)
6. Codd, E.F.: A relational model of data for large shared data banks. In: Broy, M., Denert, E. (eds.) Software Pioneers, pp. 263–294. Springer, Heidelberg (2002). https://doi.org/10.1007/978-3-642-59412-0_16
7. Dasgupta, S.S., Ray, S.N., Talukdar, P.P.: Hyte: Hyperplane-based temporally aware knowledge graph embedding. In: EMNLP (2018)
8. Dettmers, T., Minervini, P., Stenetorp, P., Riedel, S.: Convolutional 2d knowledge graph embeddings. In: AAAI (2018)
9. Ding, X., Liao, K., Liu, T., Li, Z.Y., Duan, J.: Event representation learning enhanced with external commonsense knowledge. In: EMNLP (2019)
10. Ding, X., Zhang, Y., Liu, T., Duan, J.: Knowledge-driven event embedding for stock prediction. In: COLING (2016)
11. García-Durán, A., Dumancic, S., Niepert, M.: Learning sequence encoders for temporal knowledge graph completion. In: EMNLP (2018)
12. Gottschalk, S., Demidova, E.: Eventkg: A multilingual event-centric temporal knowledge graph. In: ESWC (2018)
13. Guan, S., Jin, X., Wang, Y., Cheng, X.: Link prediction on n-ary relational data. In: WWW (2019)
14. Jiang, T., Liu, T., Ge, T., Sha, L., Chang, B., Li, S., Sui, Z.: Towards time-aware knowledge graph completion. In: COLING (2016)
15. Kazemi, S.M., Poole, D.: Simple embedding for link prediction in knowledge graphs. In: NeurIPS (2018)
16. Kingma, D.P., Ba, J.: Adam: a method for stochastic optimization. arXiv preprint arXiv:1412.6980 (2014)
17. Lautenschlager, J., Shellman, S., Ward, M.: Icews event aggregations. Harvard Dataverse 3 (2015)
18. Leetaru, K., Schrodt, P.A.: Gdelt: Global data on events, location, and tone, 1979–2012. In: ISA annual convention, vol. 2, pp. 1–49. Citeseer (2013)
19. Li, Z., Ding, X., Liu, T.: Constructing narrative event evolutionary graph for script event prediction. In: IJCAI (2018)
20. Lin, Y., Liu, Z., Sun, M., Liu, Y., Zhu, X.: Learning entity and relation embeddings for knowledge graph completion. In: AAAI (2015)

21. Mahdisoltani, F., Biega, J.A., Suchanek, F.M.: Yago3: A knowledge base from multilingual wikipedias. In: CIDR (2014)
22. Miller, G.A.: Wordnet: a lexical database for English. Commun. ACM **38**(11), 39–41 (1995)
23. Nguyen, D.Q., Nguyen, T.D., Nguyen, D.Q., Phung, D.: A novel embedding model for knowledge base completion based on convolutional neural network. In: NAACL-HLT (2018)
24. Nguyen, D.Q., Vu, T., Nguyen, T.D., Nguyen, D.Q., Phung, D.: A capsule network-based embedding model for knowledge graph completion and search personalization. In: NAACL (2019)
25. Nickel, M., Rosasco, L., Poggio, T.A.: Holographic embeddings of knowledge graphs. In: AAAI (2015)
26. Nickel, M., Tresp, V., Kriegel, H.P.: A three-way model for collective learning on multi-relational data. In: ICML (2011)
27. Saha, A., Ansari, G.A., Laddha, A., Sankaranarayanan, K., Chakrabarti, S.: Complex program induction for querying knowledge bases in the absence of gold programs. Trans. ACL **7**, 185–200 (2019)
28. Sun, Y., et al.: Ernie: Enhanced representation through knowledge integration. arXiv preprint arXiv:1904.09223 (2019)
29. Sun, Z., Deng, Z.H., Nie, J.Y., Tang, J.: Rotate: Knowledge graph embedding by relational rotation in complex space. In: ICLR (2019)
30. Toutanova, K., Chen, D., Pantel, P., Poon, H., Choudhury, P., Gamon, M.: Representing text for joint embedding of text and knowledge bases. In: EMNLP (2015)
31. Trouillon, T., Welbl, J., Riedel, S., Gaussier, É., Bouchard, G.: Complex embeddings for simple link prediction. In: ICML (2016)
32. Vrandečić, D., Krötzsch, M.: Wikidata: a free collaborative knowledge base (2014)
33. Wen, J., Li, J., Mao, Y., Chen, S., Zhang, R.: On the representation and embedding of knowledge bases beyond binary relations. In: IJCAI (2016)
34. Yang, A., et al.: Enhancing pre-trained language representations with rich knowledge for machine reading comprehension. In: ACL (2019)
35. Yang, B., Yih, W.t., He, X., Gao, J., Deng, L.: Embedding entities and relations for learning and inference in knowledge bases. In: ICLR (2015)

LeKAN: Extracting Long-tail Relations via Layer-Enhanced Knowledge-Aggregation Networks

Xiaokai Liu[1,3], Feng Zhao[1,2(✉)], Xiangyu Gui[1,2], and Hai Jin[1,2]

[1] National Engineering Research Center for Big Data Technology and System, Services Computing Technology and System Lab, Cluster and Grid Computing Lab, Wuhan, China
[2] School of Computer Science and Technology, Huazhong University of Science and Technology, Wuhan, China
{zhaof,guixy,hjin}@hust.edu.cn
[3] School of Cyber Science and Engineering, Huazhong University of Science and Technology, Wuhan, China
liuxk@hust.edu.cn

Abstract. Long-tailed relation extraction is a crucial task in the information extraction field for extracting the long-tailed, imbalanced relation between two annotated entities based on related context. Although many works have been devoted to distinguishing valid instances from noisy data and have achieved promising performance, such studies still have critical defects: works based on nonhierarchical relations ignore the correlations among the relations, and those based on hierarchical relations neglect the hierarchy of the relation structure, which is unbalanced and causes difficulty in extracting data-poor classes. In this paper, a novel layer-enhanced knowledge aggregation network, named *LeKAN*, is presented to classify the relations between two annotated entities from text, especially long-tailed relations, which are very common in various corpora. Inspired by the election mechanism, we aggregate the ancestors of long-tailed relation classes into new relation representations to prevent the long-tailed relations from being ignored. Specifically, we use GraphSAGE to learn the relational knowledge from an existing knowledge graph via class embedding. Moreover, we aggregate the acquired relational knowledge into the *LeKAN* by layer-enhanced knowledge-aggregating attention mechanism. Comprehensive experimental results demonstrate that the new method yields considerable improvement over other relation extraction methods on a large-scale benchmark dataset with a long-tailed distribution.

Keywords: Natural language processing · Information extraction · Long-tailed relation extraction · Knowledge-aggregation network

1 Introduction

Relation extraction (RE) is an essential task in the NLP field for extracting the relation between two annotated entities based on the context, especially

A. Bhattacharya et al. (Eds.): DASFAA 2022, LNCS 13245, pp. 122–136, 2022.
https://doi.org/10.1007/978-3-031-00123-9_9

long-tailed, imbalanced relations, which are very common in real-world settings. Long-tailed relations cannot be ignored because they contain rich semantic information. However, it is extremely difficult to extract long-tailed relation classes at the tail of the class distribution because few data is available. There are only a few works which have attempted to dig into the problem of long-tail RE, such as the explanation-based approach [1] and the approach utilizing external knowledge (logic rules) [2]. These works have conducted beneficial studies on the extraction of long-tail relations.

As an emerging technology and an effective solution to help improve the ability of machines to understand the human world, *knowledge graphs* (KGs) can provide higher-quality support for quantitative information retrieval, question answering, recommender systems, search engines, and other natural language processing applications [3,4]. However, the construction of a large-scale knowledge graph system containing massive amounts of knowledge relies on large-scale structured training data. RE, with the purpose of extracting the relation between two named entities based on the given context, is a fundamental task in building large-scale KGs. It is also a crucial technique in automatic KG construction. Using RE, we can accumulatively extract new relation facts to expand the built KG. However, RE model performances quickly degrade when extracting long-tailed relations because many long-tailed relations suffer from data insufficiency. These difficulties make the extraction of long-tailed relations a very difficult problem.

Long-tail relations cannot be ignored because they contain rich semantic correlations. Moreover, long-tailed, imbalanced data is very common in reality. In this work, we followed previous work to employ a widely used corpus, the New York Times (NYT-10) dataset[1] [5], to verify the advantages of our method in long tailed relation extraction. To have comprehensive understanding of the long-tailed distribution in this dataset, we analyze the distribution of the relation classes in NYT-10, as shown in Fig. 1. In this figure, long-tailed relation instances account for less than 4% of the total data, while short-headed instances account for more than 96% of the data. Furthermore, the short-headed relation classes account for less than 20% of the dataset, while the long-tailed relation classes account for more than 80%. Therefore, research on long-tailed RE is significant, and methods that can extract long-tailed relations with high accuracy are urgently needed.

The task of extracting relations from long-tailed distribution context is very difficult because few examples are available to train the models, leading to insufficient relation representation and poor classifier learning. Therefore, this situation motivates us to identify methods that can transfer knowledge between relations and alleviate the imbalance inherent in the hierarchical relational structure. To tackle these problems, a layer-enhanced knowledge aggregation network, named *LeKAN*, is proposed. To transfer knowledge between relations, the conventional method considers only the transfer of knowledge between relation instances in the same branch, e.g., */people/deceased_person/place_of_death*

---

[1] http://iesl.cs.umass.edu/riedel/ecml/.

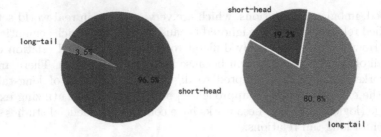

Fig. 1. Proportions of instances and classes of short-headed and long-tailed relations without NA labels in the NYT-10 dataset

and $/people/deceased\_person/place\_of\_birth$, while ignoring the fact that relation instances in different branches may also have similar semantics; e.g., both $/film/film\_festival/location$ and $/broadcast/content/location$ share the base-level relation class $/*/*/location$. LeKAN can aggregate the relational knowledge between two relations regardless of whether they are in the same branch, and the extraction of head relations provides evidence for the prediction of long-tailed relations. To alleviate the imbalance of the hierarchical relation structure, we propose a tree-based adjustment strategy to build the distributed relational representation. By pruning the long branches and extending the short branches of the network, all relation nodes are held in the same layer. Moreover, GraphSAGE with embedded KG information can sample the relevant information of the 1-step and 2-step neighbor nodes, which helps alleviate the imbalance problem of the hierarchical relation structure. Various baselines experiments were conducted on NYT-10, which demonstrate that the proposed LeKAN achieves best results in extracting the long-tailed relation. Furthermore, by leveraging the aggregated rational knowledge in different branches and levels, our proposed model can transfer relational knowledge more efficiently than existing approaches.

The remainder of the paper is organized as follows. In sect. 2, we discuss the latest progress on the long tail problem in various fields, such as relation extraction, computer vision, that can inspire this work. In Sect. 3, we mainly introduce the theory and interpretability of our proposed LeKAN method. Then, our experimental results are reported in Sect. 4. Finally, we conclude the our work and briefly introduce the work to be done in the future in Sect. 5.

2   Related Works

Relation extraction is the cornerstone of automatic construction of large-scale KGs. Early relation extraction mainly depends on the supervision model. Quantities of labeled data is required for relation extraction via conventional supervised models [6,7]. Such a process of tagging large-scale raw datasets is extremely time consuming and difficult to perform. Hence, [8] *proposed the use of distant supervision* (DS) to automatically annotate data. However, DS unavoidably introduces the incorrect labeling problem. To address such an issue caused by DS, [5,9] proposed multi-instance learning mechanisms, [10] proposed a sentence-level framework via negative training and [11] achieved promising performance by adopting DS to construct extensive datasets and alleviate the noisy label problem. Recently, [12] proposed a probabilistic approach to improve the DS relation extraction. However, these works ignored the long-tailed problem or failed to improve the effect of long-tailed RE.

The two intuitive solutions to solve the classification problem of long-tailed distribution are resampling [13–15] and reweighing [16,17]. The essence of these methods is to leverage the dataset with given distribution to violently hack the unknown distribution during the process of model training, i.e., to make change of the point weights, strengthen the tail category learning, and offset the long-tailed effect. Moreover, multi-instance learning [18] and transfer learning [19] can be employed to tackle the long-tail relevance problem. These methods have achieved good results in various computer vision tasks.

Only a few works have attempted to solve the problem of long-tailed RE [1,2,20,21]. The studies by [1,2] treated each class in isolation. Such a way of dealing with different classes of relations naturally ignores the rich semantic correlations between the classes, which are equally important. [20] proposed a hierarchical attention scheme for RE and achieved better performance than nonhierarchical schemes. [21] applied transfer knowledge between instances in the vertical direction (same branches) and leveraged implicit and explicit class embedding from Knowledge Graphs and *Graph Convolutional Networks* (GCNs) instead of learning hyper-parameter spaces using the data-driven mechanism, where similar classes may have different hyper parameters; thus, they impeded the generalization of long-tailed relations. These works conducted beneficial explorations into the long-tailed relation extraction.

Previous solutions to address the long-tail problem have mainly focused on entity hierarchies and the transfer of relational knowledge between instances in the vertical direction. Unlike them, our methods leverage GraphSAGE to learn knowledge and transfer knowledge in both vertical and horizontal directions using a relational aggregator. To alleviate the imbalance inherent in hierarchical relation structures, we also propose a method to build a layer-enhanced hierarchical relational tree to ensure that all relational branches have identical heights. Compared with the existing RE methods, our models can leverage relation correlations to better classify the given long-tailed instances by transferring knowledge from their related layers.

3   Methodology

In this section, we introduce the methodology of the layer-enhanced knowledge attention network for RE. First, we start with the relevant definitions of RE.

3.1   Framework

We follow the general definition and notations of the knowledge graph by defining the KG as a set of $\mathcal{G}$. Furthermore, $\mathcal{G} = \{\mathcal{E}, \mathcal{R}, \mathcal{F}\}$. The $\mathcal{F}$ indicates triple fact $(h, r, t) \in \mathcal{F}$, the $\mathcal{E}$ indicates entities predefined in KG, and $\mathcal{R}$ indicates

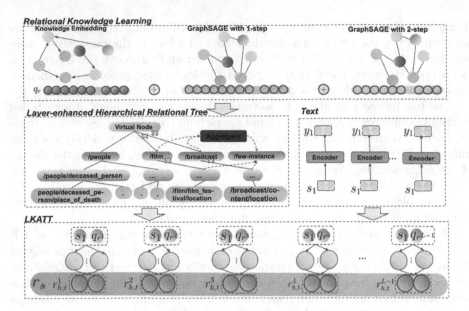

Fig. 2. The architecture of *LeKAN*

relations between such entities. The facts $\mathcal{F}$ indicate that the class of relation $r \in \mathcal{R}$ between two given entities ($h \in \mathcal{E}$ and $t \in \mathcal{E}$) is r. We adopt the multi-instance learning settings and generate multiple entity-pair bags by splitting the instances with identical entity pairs that mention h_i and t_i into the same bags $\mathcal{S}_{h_1,t_1}, \mathcal{S}_{h_2,t_2},...$. Each instance in entity-pair bags is represented as a word sequence $s = \{w_1, w_2, ...\}$.

In Fig. 2, we demonstrate the overall architecture of the *LeKAN*. There are mainly four parts in *LeKAN* as follows.

Instance Encoder: The instance encoder aims to encode the sentence semantics into a continuous low-dimensional vector. Designated an instance s with the tagged entity pair, we can use the models with neural network architecture to encode it.

Relational Knowledge Learning: Considering the pretrained KG embeddings (e.g., TransE [22]) as nonhierarchical relational knowledge, we use GraphSage to learn hierarchical relational knowledge from the aggregated relational KG. In addition, we combine the GraphSAGE with generic message-passing inference, we can acquire the relational representation for the relation classes. We concatenate the outputs of the GraphSAGE sampling neighbors with different steps and the embeddings learned from knowledge graph to construct the final distributed relational embeddings.

LKATT: Given the hierarchical relation structure of a KG, the relational knowledge aggregator automatically aggregates the parent relations of the long-tailed relation into a new relation. For example, we can aggregate two long-tailed relations under different branches, e.g., /film/film_festival/location and /broadcast/content/location, to a new relation: few_instance_location. Under the guidance of *layer-enhanced knowledge attention* (LKATT), *LeKAN* aims to select the instance with abundant information that exactly matches the relevant relation but to ignore its branch.

3.2   Instance Encoder

Given an instance $s = \{w_1, ..., w_n\}$ containing two entities, we leverage the instance encoder to encode the sentence into a continuous low-dimensional vector. The instance encoder consists of two parts: the embedding layer, which maps the words in the context into vectors, and the encoder layer, which encodes the vectors.

Embedding Layer: To better identify the synaptic and semantic meanings of the sentences. We leverage the neural networks in embedding layer to transform discrete words in specific instance into vector space. Here, we use a pretrained skip-gram model [23] to map each word w_i in the instance to a continuous vector space. Moreover, we adopt position embedding following [11]. Then, we embed the relative distances of every word in the instance from marked entities into two d_p-dimensional continuous vectors. Finally, we gather all input embeddings in the instance and concatenate all of them together. By doing so, we get a sequence of instance embedding, which is ready to be fed into the encoding layer.

Encoding Layer: In encoding layer, we also employ neural networks to encode the outputs of the embedding layer, whose input is a given instance. In this study, we employ vanilla CNNs [11] and PCNNs [24] as the instance encoder.

3.3   Distributed Relational Representation via Transfer Learning

To get distributed relational representations, we need to have pretrained KG embeddings obtained by instanced encoder and define a predefined class relation hierarchy according to the structure of KG. Then, we build the distributed relational representation. First, we use the nonhierarchical relational knowledge from the KGs. Second, we build a layer-enhanced hierarchical relational tree to learn hierarchical relational knowledge. Third, we apply GraphSAGEs with 1-step and 2-step to learn the hierarchical relational knowledge from the layer-enhanced hierarchical relational tree, and obtain a distributed relational representation.

Building a Layer-Enhanced Hierarchical Relational Tree. Given KG $\mathcal{G}$ (e.g., NYT) consisting base-level relations, we extract the set R of it to generate the corresponding layer-enhanced hierarchical relational tree set R^H. The relations in high-level sets (e.g., /location) have the same instances as their child relations (e.g., /people/deceased_person), which indicates that high-level relations are more general and common than low-level relations. The relation hierarchies are separated into tree-structured subgraphs of R^0, which is the set of all relations. The generation of subgraphs can be recursively completed to obtain the relation sets $\{R^0, R^1..., R^i, R^L\}$ and others. Then, we must adjust the hierarchy relation tree to ensure that all leaf nodes have identical heights. We propose two layer-enhanced methods to transform an imbalanced relational tree into a balanced tree: pruning and completion. The pruning method can remove layers from a relation branch, while the completion method can add layers to a relation branch. For long relational branches, we can use the pruning method to reduce their heights; for short relational branches, we can use the completion method to increase their heights. This approach can also prevent overfitting and improve the convergence speed of the network.

Learning Relational Knowledge via GraphSAGE. Because of the missing one-multiple relations in KGs, GraphSAGEs are necessary that they sample 1-step and 2-step neighbors from the hierarchical features. Given the pretrained relation embedding $v_d^{TransE} \in KGs$ via *TransE*, we use the mean aggregator to form a hierarchical representation of the i-th label:

$$h_v^k = \sigma(W^i \cdot Mean(h_v^{k-1} \bigcup h_u^{k-1}, \forall u \in \mathcal{N}(v))) \tag{1}$$

where $W^i \in \mathbb{R}^{q^i}, i = 1, 2, h_u^0 = v_d$. The convolutional aggregator concatenates the parent layer representation h_v^{k-1} of the node with the aggregated neighborhood representation $h_{N(v)}^k$. Finally, we concatenate the pretrained v_i^{TransE} and output vectors $v_i^{GSN_1}, v_i^{GSN_2}$ of the GraphSAGEs to form the hierarchical class embeddings:

$$q_r = v_i^{TransE} || v_i^{GSN_1} || v_i^{GSN_2} \tag{2}$$

where $q_r \in \mathbb{R}^{d+q^1+q^2}$.

3.4   LKATT

Conventional hierarchical RE models treat the top-level relation nodes as independent nodes, which hinders the transfer of knowledge among the base-level relational nodes of different branches and prevents the long-tailed nodes from being selected. We design a relational aggregator to solve these problems. The relational aggregator is guided by the following principles: 1) If two base-level relational nodes are semantically similar, their top-level relational nodes are aggregated. 2) Even if the basic relation nodes of two rare instances have different semantics, their top-level relations can be aggregated. Experiments show that the aggregation of relation nodes can enhance the performance of classifying long-tailed classes, and the decoupling of top-level relation nodes likely has potential effects.

In general, the output layer of the neural network will learn parameters of the specific label optimized by the given loss function. Because, the parameter space of different classes is different, it naturally leads to the fact that long-tail relations can be exposed to only a few training examples during training. Instead, our approach considers more correlations of the long-tailed relations by making the ancestor nodes of semantically similar relation nodes share parameters and concatenating the sentence to the corresponding class embeddings.

First, we acquire the instance embeddings $\{s_1, s_2, ..., s_m\}$ using the instance encoder with the entity pair (h, t) and the corresponding bag of instances $S_{h,t} = \{s_1, s_2, ..., s_m\}$. Second, we split the class embeddings into different clusters according to their types (i.e., according to their levels in the layer-enhanced hierarchical relational tree), e.g., $q_r^i, i \in \{0, 1, ..., L\}$. Third, we aggregate the semantically similar relation nodes and adopt $q_r^i, i \neq N$ (we assign an another node N as root node in the tree) as a layer-enhanced attention query vector. Finally, we use the LKATT mechanism to process the vector to get its relation representation $r_{h,t}$. For each relation r, we can build the corresponding hierarchical chain of latent relations $(r^0, ..., r^{(N-1)})$ using a layer-enhanced hierarchical relational tree, where $r^{(i-1)}$ is the subrelation of r^i. We can calculate the attention weight for $s_i \in S_{h,t} = \{s_1, s_2, ..., s_m\}$ as follows:

$$e_k^i = tanh(W_s[s_k; q_r^i]) + b_s \tag{3}$$

$$a_k^i = \frac{\exp(e_k^i)}{\sum_{j=1}^m \exp(e_j)} \tag{4}$$

where $[x_1; x_2]$ denotes the vertical concatenation of x_1 and x_2, W_s is the weight matrix, and b_s is the bias. The converged nodes share parameters. Then, we can compute the attention scores on each layer of the layer-enhanced hierarchical relational tree to acquire the relational representations.

$$r_{h,t}^i = ATT(q_{r^i}, s_1, s_2, ..., s_m) \tag{5}$$

The global representation is defined as follows:

$$r_{h,r} = Concat(r_{h,t}^0, ..., r_{h,t}^{L-1}) \tag{6}$$

The conditional probability is computed by the global representation $r_{h,t}$: $\mathcal{P}(r|h, t, S_{h,t})$:

$$\mathcal{P}(r|h, t, S_{h,t}) = \frac{\exp(o_{\hat{r}})}{\sum_{\hat{r} \in \mathcal{R}} \exp(o_{\hat{r}})} \tag{7}$$

where o contains the scores of all relations. And o is calculated via a linear layer:

$$o = Ar_{h,t} \tag{8}$$

where A is the discriminative matrix.

4   Experiments

In this section, we evaluate all models using the proposed evaluation scheme. This method evaluates the models by comparing the relational facts found in the context with those in a large-scale KG, such as *DBpedia*, and adopts an approximate

accuracy measure except the manual evaluation. To show the advantages of our
methods, we plot the precision-recall curves for all methods for the evaluation.
In order to validate whether the effect of our model for the long-tailed RE is
superior to other proposed methods, we follow the same evaluation criteria as
before [20,21] by reporting the Precision@N. We report the evaluation results
in Figure 3 and Figure 4. The dataset and baseline code are from GitHub[2].

4.1   Experimental Setting

Datasets. We evaluate the performance of our method on the most commonly
used long-tail RE dataset in recent long-tail RE work: the NYT-10 dataset [21,
25–28]. There are 52 common classes and a NA class in it. The NA relation
denotes that the relation between the given instances is not labeled. The dataset
contains rich semantic information, which has been split into a training set with
522611 sentences and a testing set with 17448 sentences. There are 281270 entity
pairs and 18252 relational facts in the training set and 96678 entity pairs and
1950 relational facts in the testing set. We follow the convention of truncating
sentences that contain more than 120 words into 120 words for the dataset.

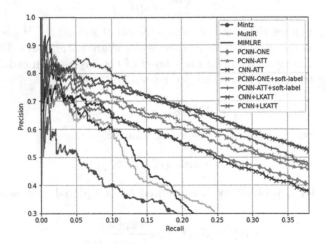

Fig. 3. P-R curves for various models

Comparison Models. For the baseline model comparison, we utilize both
neural network models and feature-based models. We report the evaluation
results of the neural networks with methods based on various attention schemes:
+**LKATT** is our layer-enhanced knowledge-aggregating attention method;
+**ONE** is a typical multi instance learning based neural model [24]. The **soft
label** is the model with attention schemes using the soft-labeling method to alleviate the effects of the noise problem [27]. In addition, we compare our model

---

[2] https://github.com/thunlp/OpenNRE

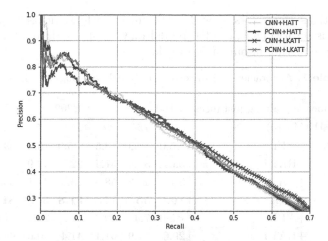

Fig. 4. P-R curves for various models with attention mechanism

with various feature-based models, including **MIML, MultiR** [9], and **Mintz** [8] [29]. To effectively evaluate the effect of our method on the long-tailed RE task, we also compare it with **HATT** [20] and **KATT** [21].

Hyperparameter Settings and Reproducibility. In order to prove that our model is superior to other baseline models and fairly compare its performance with that of baseline methods, we keep almost all experimental parameters identical to the previous work and pretrain the sentence encoder of the neural networks [25]. During the training process, a dropout layer is adopted before the output layer to prevent overfitting.

4.2   Overview of the Evaluation Results

As shown in Figs. 3 and 4, our method using a novel denoising scheme and additional auxiliary information achieved the best performance. The results also demonstrate that *LeKAN* can leverage the rich correlation between relations to improve its RE performance. We anticipate to enhance the performance of our model by integrating some novel mechanisms such as meta-learning.

To prove the advantages and performance improvement of the proposed methods for long-tail relation RE, we follow the convention to extract subsets of the test dataset, where the training instances of all relations are less than 100 and 200. We use the Hits@K metric to evaluate the long-tail RE. Then, the RE models will recommend the relations in the first K candidate classes for each entity pair. Since extracting long-tail relations is extremely difficult in existing models, we choose K from the set {10, 15, 20}. We report the macroaverage accuracy of Hits@K for all subsets. The results shown in Table 1 demonstrate that our new method is outperforms the attention mechanism based methods, even the most sophisticated HATT and KATT. Although our LKATT method achieves better results than the ordinary ATT, HATT, and KATT methods on long-tail relations, the results show that the current achievements in long-tail RE remain unsatisfactory. Thus, the RE model may require additional information. We will explore further in our future work.

Table 1. Accuracies (%) in terms of Hits@K on long-tail classes

Number of training instances		<100			<200		
Hits@K(Macro)		*10*	*15*	*20*	*10*	*15*	*20*
CNN	+ATT	<5.0	<5.0	18.5	<5.0	16.2	33.3
	+HATT	5.6	31.5	57.4	22.7	43.9	65.1
	+KATT	9.1	41.3	58.5	23.3	44.1	65.4
	+LKATT	**16.7**	**55.6**	**77.7**	**31.8**	**63.6**	**81.8**
PCNN	+ATT	<5.0	7.4	40.7	17.2	24.2	51.5
	+HATT	29.6	51.9	61.1	41.4	60.6	68.2
	+KATT	35.3	62.4	65.1	43.2	61.3	69.2
	+LKATT	29.6	61.1	**77.8**	42.4	**68.2**	**81.8**

4.3 Ablation Study

To have a comprehensive understanding of the contributions and impact of different techniques in the proposed method, we design ablation tests. We demonstrate the evaluation results of ablation study in Table 2. **+LKATT** is the proposed method; **w/o aggregation** is the method where node aggregation is not implemented; **w/o KG** is the method where the nodes are initialized with random embeddings, so it is natural that there is no relational knowledge obtained from KGs; and **w/o GraphSage** is the method without GraphSage, which denotes no structured relational knowledge. By analyzing the results in Table 2, we can draw the conclusion that the performance of our method to extract long-tailed relations is slightly degraded without KG, and the performance is significantly degraded after node aggregation or GraphSage is removed. This degradation is reasonable because GraphSAGEs consider the distances between neighbors, and node aggregation can prevent relation classes with few examples in the training set from being ignored.

Table 2. Accuracies (%) in terms of Hits@K on relations with fewer than 100/200 training instances

Number of training instances	<100			<200		
Hits@K(Macro)	*10*	*15*	*20*	*10*	*15*	*20*
+LKATT	**29.6**	**61.1**	**78.8**	**42.4**	**68.2**	**81.8**
w/o/ hier	16.7	44.4	44.4	31.8	54.5	54.5
w/o Aggregation	5.6	44.4	50.0	22.7	54.5	59.1
w/o/ KG	24.1	33.3	72.2	37.9	45.6	77.3
w/o/ GraphSage	18.5	44.4	72.2	33.3	54.5	77.3

4.4   Visualization of Class Embeddings

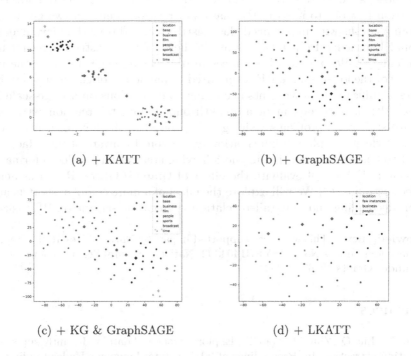

Fig. 5. Dimension reduction visualization of relation class embedding

Here, we demonstrate the rationality of our class embedding work through a visualization tool t-SNE [30]. This visualization work on relation embedding deeply shows how KG and GraphSAGE embeddings positively affect the extraction of long tail relations. In Fig. 5, the square points represent the top-level relations of the relation clusters. Figure 5(a) demonstrates that the GCN combines relations that are under the same branch and ignores semantically similar relations on different branches; Fig. 5(b) and Fig. 5(c) show that GraphSAGE can help with knowledge transfer between semantically similar long-tailed relations by aggregating the corresponding knowledge. However, if there is no KG, outliers will occur; Fig. 5(d) shows that the long-tailed relation can be emphasized by aggregating the ancestral relation nodes with few instances to a new relation, which helps to aggregate the ancestral nodes with fewer instances to prevent the corresponding base-level relation from being ignored. However, when we embed the features of ancestral relation nodes into a high-dimensional continuous vector space, the classification of long-tailed relations relies more on the representations of the base-level relations, which is a problem. In the near future, we will tackle this issue by integrating more information, for instance, relation information, or by decoupling the relations with fewer semantic correlations than other relations from their branches.

5   Conclusion

We propose a novel KG- and GraphSAGE-based layer-enhanced knowledge aggregation network to identify the classes of relations between two given entities from a corpus with imbalanced class distribution. This method leverages the relational knowledge from relations at the head of their distribution and uses semantically similar relational instances in different branches to boost the performance of the low-resource RE. Compared to previous works, the new method achieves significant improvements according to evaluations on a large-scale RE dataset. Although we have made a breakthrough in long tail relation extraction, there are still many problems waiting to be solved in the field of construction of the knowledge graph and information extraction. Be aware of these facts, we decided to conduct exploration in the following areas of long-tailed information extraction.: (1) We will evaluate the effect of GraphSAGE on RE tasks across knowledge graphs. (2) We will explore the effect of a more complex short-headed relation decoupling and long-tailed relation aggregation scheme on RE tasks.

Acknowledgment. This work was supported in part by National Key R&D Program of China under Grants No. 2018YFB1404302, National Natural Science Foundation of China under Grants No.62072203.

References

1. Gui, Y., Liu, Q., Zhu, M., Gao, Z.: Exploring long tail data in distantly supervised relation extraction. In: Proceedings of 2016 Natural Language Understanding and Intelligent Applications, pp. 514–522 (2016)
2. Lei, K., Chen, D., et al.: Cooperative denoising for distantly supervised relation extraction. In: Proceedings of the 27th International Conference on Computational Linguistics, pp. 426–436 (2018)
3. Huang, Y., Zhao, F., Gui, X., Jin, H.: Path-enhanced explainable recommendation with knowledge graphs. World Wide Web **24**(5), 1769–1789 (2021). https://doi. org/10.1007/s11280-021-00912-4
4. Song, S., Huang, Y., Lu, H.: Rumor detection on social media with out-in-degree graph convolutional networks. In: Proceedings of the 2021 IEEE Conference on Systems, Man, and Cybernetics, pp. 2395–2400 (2021)
5. Riedel, S., Yao, L., McCallum, A.: Modeling relations and their mentions without labeled text. In: Proceedings of the 2010 European conference on Machine learning and knowledge discovery in databases: Part III, pp. 148–163 (2010)
6. Zelenko, D., Aone, C., Richardella, A.: Kernel methods for relation extraction. J. Mach. Learn. Res. **3**(Feb), 1083–1106 (2003)
7. Zhou, G., Su, J., Zhang, J., Zhang, M.: Exploring various knowledge in relation extraction. In: Proceedings of the 43rd Annual Meeting of the Association for Computational Linguistics, pp. 427–434 (2005)
8. Mintz, M., Bills, S., Snow, R., Jurafsky, D.: Distant supervision for relation extraction without labeled data. In: Proceedings of the Joint Conference of the 47th Annual Meeting of the ACL and the 4th International Joint Conference on Natural Language Processing of the AFNLP, pp. 1003–1011 (2009)

9. Hoffmann, R., Zhang, C., Ling, X., Zettlemoyer, L., Weld, D.S.: Knowledge-based weak supervision for information extraction of overlapping relations. In: Proceedings of the 49th Annual Meeting of the Association for Computational Linguistics: Human Language Technologies, pp. 541–550 (2011)

10. Ma, R., Gui, T., Li, L., Zhang, Q., Zhou, Y., Huang, X.: Sent: sentence-level distant relation extraction via negative training. arXiv preprint arXiv:2106.11566 (2021)

11. Zeng, D., Liu, K., Lai, S., Zhou, G., Zhao, J.: Relation classification via convolutional deep neural network. In: Proceedings of the 25th International Conference on Computational Linguistics: Technical Papers, pp. 2335–2344 (2014)

12. Christopoulou, F., Miwa, M., Ananiadou, S.: Distantly supervised relation extraction with sentence reconstruction and knowledge base priors. arXiv preprint arXiv:2104.08225 (2021)

13. Kang, B., et al.: Decoupling representation and classifier for long-tailed recognition. arXiv preprint arXiv:1910.09217 (2019)

14. Zhou, B., Cui, Q., Wei, X.S., Chen, Z.M.: BBN: Bilateral-branch network with cumulative learning for long-tailed visual recognition. In: Proceedings of the 2020 IEEE Conference on Computer Vision and Pattern Recognition, pp. 9719–9728 (2020)

15. Wang, Y., Gan, W., Yang, J., Wu, W., Yan, J.: Dynamic curriculum learning for imbalanced data classification. In: Proceedings of the 2019 IEEE International Conference on Computer Vision, pp. 5017–5026 (2019)

16. Cui, Y., Jia, M., Lin, T.Y., Song, Y., Belongie, S.: Class-balanced loss based on effective number of samples. In: Proceedings of the 2019 IEEE Conference on Computer Vision and Pattern Recognition, pp. 9268–9277 (2019)

17. Jamal, M.A., Brown, M., Yang, M.H., Wang, L., Gong, B.: Rethinking class-balanced methods for long-tailed visual recognition from a domain adaptation perspective. In: Proceedings of the 2020 IEEE Conference on Computer Vision and Pattern Recognition, pp. 7610–7619 (2020)

18. Jiang, X., Wang, Q., Li, P., Wang, B.: Relation extraction with multi-instance multi-label convolutional neural networks. In: Proceedings of the 26th International Conference on Computational Linguistics: Technical Papers, pp. 1471–1480 (2016)

19. Zhuang, F., Qi, Z., Duan, K., Xi, D., Zhu, Y., Zhu, H., Xiong, H., He, Q.: A comprehensive survey on transfer learning. Proc. IEEE 109(1), 43–76 (2020)

20. Han, X., Yu, P., Liu, Z., Sun, M., Li, P.: Hierarchical relation extraction with coarse-to-fine grained attention. In: Proceedings of the 2018 Conference on Empirical Methods in Natural Language Processing, pp. 2236–2245 (2018)

21. Zhang, N., et al.: Long-tail relation extraction via knowledge graph embeddings and graph convolution networks. In: Proceedings of the 2019 Conference of the North American Chapter of the Association for Computational Linguistics: Human Language Technologies, vol. 1, pp. 3016–3025 (2019)

22. Bordes, A., Usunier, N., Garcia-Duran, A., Weston, J., Yakhnenko, O.: Translating embeddings for modeling multi-relational data. In: Proceedings of the 26th International Conference on Neural Information Processing Systems, Vol. 2, pp. 2787–2795 (2013)

23. Mikolov, T., Chen, K., Corrado, G., Dean, J.: Efficient estimation of word representations in vector space. arXiv preprint arXiv:1301.3781 (2013)

24. Zeng, D., Liu, K., Chen, Y., Zhao, J.: Distant supervision for relation extraction via piecewise convolutional neural networks. In: Proceedings of the 2015 Conference on Empirical Methods in Natural Language Processing, pp. 1753–1762 (2015)

25. Lin, Y., Shen, S., Liu, Z., Luan, H., Sun, M.: Neural relation extraction with selective attention over instances. In: Proceedings of the 54th Annual Meeting of the Association for Computational Linguistics, (Volume 1: Long Papers), pp. 2124–2133 (2016)
26. Han, X., Liu, Z., Sun, M.: Neural knowledge acquisition via mutual attention between knowledge graph and text. In: Proceedings of the 2018 AAAI Conference on Artificial Intelligence. vol. 32(1), pp. 4832–4839 (2018)
27. Liu, T., Wang, K., Chang, B., Sui, Z.: A soft-label method for noise-tolerant distantly supervised relation extraction. In: Proceedings of the 2017 Conference on Empirical Methods in Natural Language Processing, pp. 1790–1795 (2017)
28. Feng, J., Huang, M., Zhao, L., Yang, Y., Zhu, X.: Reinforcement learning for relation classification from noisy data. In: Proceedings of the 2018 AAAI Conference on Artificial Intelligence, vol. 32, no. 1, pp. 5779–5786 (2018)
29. Surdeanu, M., Tibshirani, J., Nallapati, R., Manning, C.D.: Multi-instance multi-label learning for relation extraction. In: Proceedings of the 2012 Joint Conference on Empirical Methods in Natural Language Processing and Computational Natural Language Learning, pp. 455–465 (2012)
30. Laurens, V.D.M., Hinton, G.: Visualizing data using t-SNE. J. Mach. Learn. Res. 9(2605), 2579–2605 (2008)

TRHyTE: Temporal Knowledge Graph Embedding Based on Temporal-Relational Hyperplanes

Lin Yuan[1], Zhixu Li[2], Jianfeng Qu[1(✉)], Tingyi Zhang[1], An Liu[1], Lei Zhao[1], and Zhigang Chen[3,4]

[1] School of Computer Science and Technology, Soochow University, Suzhou, China
{lyuan,tyzhang1}@stu.suda.edu.cn, {jfqu,anliu,zhaol}@suda.edu.cn
[2] Shanghai Key Lab of Data Science, School of Computer Science, Fudan University, Shanghai, China
zhixuli@fudan.edu.cn
[3] State Key Laboratory of Cognitive Intelligence, iFLYTEK, Hefei, China
zgchen@iflytek.com
[4] iFLYTEK Research, Suzhou, China

Abstract. Temporal Knowledge Graph Embedding (TKGE) aims at encoding evolving facts with high-dimensional vectorial representations. Although a representative hyperplane-based TKGE approach, namely HyTE, has achieved remarkable performance, it still suffers from several problems including (i) ignorance of latent temporal properties and diversity of relations; (ii) neglect of temporal dependency between adjacent hyperplanes; (iii) inefficient static random negative sampling method; (iv) incomplete testing on partial time information. To address these issues, we propose **TRHyTE**, a novel Temporal-Relational Hyperplane based TKGE model, which defines three typical properties, including interval, open-interval, and instantaneousness temporal, for relations and correspondingly constructs three relational sub-KGs, supporting distinguishing learning for facts. Within each sub-KG, TRHyTE transforms entities into relation space first, and then explicitly projects transformed entities and relations into temporal-relational hyperplanes to learn time-relation-aware embeddings. Moreover, Gate Recurrent Unit is leveraged to simulate TKG evolution so as to capture temporal dependency between adjacent hyperplanes. Additionally, we develop a dynamic negative samples mechanism for robust training. In testing phase, an *expand-and-best-merge* strategy is crafted to realize a complete testing on all valid time intervals. Extensive experiments on two well-known benchmarks verify the effectiveness of our proposals.

Keywords: Temporal knowledge graph · Knowledge Graph Embedding · Temporal property · Hyperplane projection

---

L. Yuan and Z. Li—Equal contribution.

A. Bhattacharya et al. (Eds.): DASFAA 2022, LNCS 13245, pp. 137–152, 2022.
https://doi.org/10.1007/978-3-031-00123-9_10

1 Introduction

Knowledge graphs (KGs) are to record real-world knowledge in the form of large-scale and multi-relational directed graph structures. Traditional KGs only store static factual knowledge using triplet (h, r, t), where h and t are head entity and tail entity respectively, and r denotes the relation between h and t, which however ignores the valid timeliness of facts. Recent years have witnessed the emergence of temporal knowledge graphs (TKGs), which store factual knowledge in the form of quadruple (h, r, t, T), where T denotes the temporal scope with the fact knowledge (h, r, t). The temporal scope could be either a specific time (e.g. *(Obama, <wasBornIn>, America, 1961-08-04)*), or a time interval (e.g., *(Obama, <presidentOf>, USA, 2008-2017)*), as given in the upper part of Fig. 1(a). These temporal scopes are increasingly available on several large KGs such as YAGO3 [15] and Wikidata [4].

KG embedding on static KG has been studied extensively, which learns high-dimensional vectorial representations for entities and relations in KGs [1, 4, 19, 21]. With the rise of TKGs, the research on TKG embedding has also aroused a lot of attention in recent years [18, 26]. Conceivably, considering temporal scopes during representation learning might yields better KG embeddings. Recent attempts such as [6, 14, 26] encode temporal information directly; [8, 10] leverage temporal order and regular repeats of relations as constraint terms to guide model learning; [7, 11, 27] consider that TKGs keeps evolving all the time and apply sequence model to capture long-term dependency or keep entity and relation embeddings changing through time; [5, 13] represent facts by tensors and apply tensor decomposition operations. Despite some progress achieved, many existing works do not explicitly incorporate time in the learned embeddings to make them temporally aware, and some approaches lack of interpretability especially for tensor decomposition-based models.

To address the above problems, another class of TKG embedding models, HyTE [3] as a typical representative, propose to encode time as hyperplanes and learn time aware embeddings by performing projected translation for facts on time valid hyperplanes explicitly. However, it still has four main restrictions: (1) overlooking latent temporal properties of relations (e.g. relation *<wasBornIn>* indicates knowledge occurs instantaneously while *<playsFor>* describes knowledge lasts for a period of time). Obviously, it is necessary to take full advantage of these properties and discriminatively treat relations with different properties, instead of equal treatments, so as to implement more fine-grained embedding learning. Additionally, the diversity of relations is also ignored. Different relations express different semantic meanings of entities, but it oversimplistically builds entity and relation embedding in the same semantic space; (2) encoding time as hyperplanes to segregate temporal space into different time zones and facts valid at different time zones are learned separately, which ignores the temporal dependency between adjacent hyperplanes and is contrary to the evolutionary nature of TKGs; (3) randomly replacing head or tail entity in a golden quadruple $(h, r, t, [t_s, t_e])$ to construct negative samples for training, which may generate easily distinguishable negative samples, thus prevents the model from obtaining a stronger ability to identify between positive and negative quadruples. More seriously, as existing TKGs are far from completed, random sampling may even introduce many false negative samples; (4) projecting a quadruple $(h, r, t, [t_s, t_e])$, where $[t_s, t_e]$ valid at multiple hyperplanes only onto single hyperplane associated with t_s when testing in experiments, which only considers testing on partial temporal information and is insufficient to reflect the complete test effect on the test quadruple.

To resolve these drawbacks, we propose a novel hyperplane-based TKG embedding model, namely TRHyTE, which projects facts into so-called temporal-relational hyperplanes to learn time-relation aware embeddings. More specifically, for the first challenge, considering that relations usually carry latent temporal information of triples, we summarize three main temporal properties for them: interval, open-interval, and instantaneousness properties. We attach relations and their corresponding quadruples with these temporal properties and cluster quadruples into three relational sub-KGs, which allow our model to learn distinct embeddings and make precise predictions during testing with the help of extra relation-level information. Beyond that, multiple relations also take on tasks of describing different aspects of semantic meaning for entities. Moreover, the head and tail entity within a triple mostly contain different types and attributes. Thus for every sub-KG, following TransD [9], we first transform entity embeddings into entity-relation pair specific relation vector space before projecting triples into valid hyperplanes. For the second deficiency, we utilize Gate Recurrent Unit (GRU) [2] to maintain dependency between adjacent hyperplanes so as to capture the evolution of TKGs. For the third drawback, instead of static negative sampling, we dynamically generate confusing negative samples with high score values and meanwhile avoid false negative samples during training for a more robust model learning. For the fourth problem regarding partial testing manner of HyTE, we present an *expand-and-best-merge* strategy to project a quadruple $(h, r, t, [t_s, t_e])$ into all valid hyperplanes firstly, and then merges multiple results into one in testing phase for a more comprehensive and fair evaluation.

We summarize our main contributions as follows:

We propose a novel hyperplane-based TKG embedding model, where three temporal properties for relations are defined to construct temporal-relational hyperplanes.

- We take evolving nature of TKGs into account and adopt GRU to deal with dependency between adjacent hyperplanes.
- We develop a dynamic negative sampling mechanism to generate variational deceptive negative samples dynamically and meanwhile reduce the generation of false negative samples during training iterations. To the best of our knowledge, this is the first work to incorporate dynamic negative sampling into TKGE models.
- We propose *expand-and-best-merge* strategy to address the shortcut of the incompleteness testing manner of HyTE.
- Extensive experiments on two well-established TKG benchmarks for link prediction tasks verify the effectiveness of our approach.

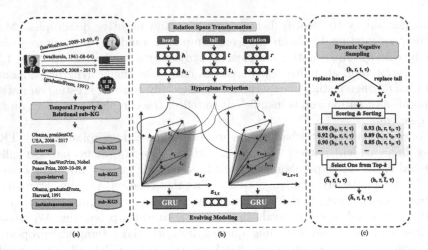

Fig. 1. An illustration of TRHyTE. Panel(a) shows the formation of three relational sub-KGs based on temporal property of relations; Panel(b) takes a quadruple from sub-KG1 as an example: $(head, relation, tail, [\tau_s, \tau_e])$, where $[\tau_s, \tau_e]$ valid at $w_{1,\tau}$ and $w_{1,\tau+1}$, it demonstrates relational space transformation, hyperplane projection, and evolving modeling; Panel(c) describes our dynamic negative sampling mechanism.

2   Notations

Temporal knowledge graphs can be defined as $G = (\mathcal{E}, \mathcal{R}, \Omega)$, where $\mathcal{E}$ is entity set, $\mathcal{R}$ is relation set and Ω is a set of quadruples in the form of $(h, r, t, [t_s, t_e])$, where $h, t \in \mathcal{E}$ and $r \in \mathcal{R}$, $[t_s, t_e]$ represents time annotation and t_s, t_e denote start time and end time, respectively. Time annotations of quadruples can mainly represent three types of facts: (1) facts with interval time, where $t_s < t_e$; (2) facts with open interval time, where $t_e = \infty$ (since) or $t_s = \infty$ (until); and (3) facts with instantaneous time, where $t_s = t_e$. A TKG can also be represented as a sequence of static KG snapshots as $G = \{G_1, G_2, \cdots, G_\tau, \cdots, G_T\}$, where $G_\tau = (\mathcal{E}, \mathcal{R}, \Omega_\tau)$, it comprises quadruples valid at time step τ. Given a quadruple $(h, r, t, [t_s, t_e])$, it is embraced in G_τ if $t_s \leq \tau \leq t_e$.

3   Our Model

In this section, we present a detailed description of TRHyTE. Figure 1 demonstrates the overview of TRHyTE. In the rest of this section, we start with temporal-relational hyperplane projection, followed by evolving modeling, and then we introduce the dynamic negative sampling mechanism, finally, we demonstrate our *expand-and-best-merge* strategy.

3.1   Temporal-Relational Hyperplane Projection

Temporal Property. We classify time annotations into three types in Sect. 2, further we observe that the type of time annotation is generally fixed regarding a particular relation within a quadruple. Relations like $<wasBornIn>$ and

<playsFor> mostly associate with instantaneous and interval time annotations, respectively. Some relations exist in quadruples with multiple types of time annotations, such as *<owns>* and *<created>*, which correlate with both open interval and interval time annotations, but the former takes a larger percentage. Accordingly, we define three temporal properties for relations: Interval property p_1 like *<playsFor>*, describing a relation lasts for a period of time; Open-interval property p_2 like *<hasWonPrize>*, representing a relation stays true since sometime or until sometime; Instantaneousness property p_3 like *<wasBornIn>*, indicating an instantaneous relation that happens at a specific time point.

Relational Sub-KG. We attach all relations and their corresponding quadruples with temporal properties in order, which depends on the distribution of time annotation types of the quadruples they exist in. In conjunction with the examples in the last subsection, we attach instantaneousness and interval property for relation *<wasBornIn>* and *<playsFor>*, respectively, and we attach open-interval and interval properties in order for relation *<owns>* and *<created>*. Next, we aggregate quadruples with interval, open-interval, and instantaneousness properties to form three relational sub-KGs, respectively, thus the whole TKG $G = G_{R_1} \cup G_{R_2} \cup G_{R_3}$, where G_{R_i} denotes sub-KGi and $i \in \{1, 2, 3\}$.

Hyperplane Projection. We represent time as hyperplanes. Within every sub-KG $G_{R_i} = \{G_{i1}, G_{i2}, \cdots, G_{i\tau}, \cdots, G_{iT}\}$, we construct T temporal-relational hyperplanes for T time steps to segregate temporal space into T different time zones, which are represented by normal vectors $\boldsymbol{w}_{i,1}, \boldsymbol{w}_{i,2}, \cdots, \boldsymbol{w}_{i,T} \in \mathbb{R}^d$. Thus we construct $3 \times T$ temporal-relational hyperplanes in total for the whole TKG.

Considering the descriptive role of relations for entities and the diversity of both them in TKGs, for a triple (h, r, t), we apply the transformation method in TransD [9]. Each entity and relation have two representations: semantic representations $\mathbf{h}, \mathbf{t}, \mathbf{r} \in \mathbb{R}^d$ and projection representations $\mathbf{h}_p, \mathbf{t}_p, \mathbf{r}_p \in \mathbb{R}^d$, respectively. Two mapping matrices $\mathbf{M}_{rh} = \mathbf{r}_p \mathbf{h}_p^\top + \mathbf{I}^{d \times d}$ and $\mathbf{M}_{rt} = \mathbf{r}_p \mathbf{t}_p^\top + \mathbf{I}^{d \times d}$ are constructed by projection representation $\mathbf{h}_p$ and $\mathbf{t}_p$ for specific entity-relation pairs (h, r) and (t, r), respectively. Then we use $\mathbf{M}_{rh}$ and $\mathbf{M}_{rt}$ to transform entity embeddings $\mathbf{h}$ and $\mathbf{t}$ from entity space into specific relation vector space, thus we can get projected vectors $\mathbf{h}_\perp = \mathbf{M}_{rh}\mathbf{h}$ and $\mathbf{t}_\perp = \mathbf{M}_{rt}\mathbf{t}$, respectively.

To incorporate temporal-relational information, triples valid at time step j and attached with temporal property p_i are projected onto hyperplane $w_{i,j}$, where $i \in \{1, 2, 3\}, j \in \{1, 2, \cdots, T\}$. We represent $w_{i,j}$ as w_τ for simplicity, where $\tau \in [3 \times T]$. We project triple $(\mathbf{h}_\perp, r, \mathbf{t}_\perp)$ onto hyperplane w_τ as:

$$\mathbf{h}_\tau = \mathbf{h}_\perp - \left(\boldsymbol{w}_\tau^\top \mathbf{h}_\perp\right) \boldsymbol{w}_\tau, \quad \mathbf{t}_\tau = \mathbf{t}_\perp - \left(\boldsymbol{w}_\tau^\top \mathbf{t}_\perp\right) \boldsymbol{w}_\tau, \quad \mathbf{r}_\tau = \mathbf{r} - \left(\boldsymbol{w}_\tau^\top \mathbf{r}\right) \boldsymbol{w}_\tau \quad (1)$$

where $\boldsymbol{w}_\tau \in \mathbb{R}^d$ denotes the embedding of w_τ, we restrict $\|\boldsymbol{w}_\tau\|_2 = 1$, $\|\mathbf{h}_\perp\|_2 \leq 1$, $\|\mathbf{t}_\perp\|_2 \leq 1$, $\|\mathbf{h}\|_2 \leq 1$, $\|\mathbf{t}\|_2 \leq 1$, and $\|\mathbf{r}\|_2 \leq 1$. Following the expectation of translation-based models, we have $\mathbf{h}_\tau + \mathbf{r}_\tau \approx \mathbf{t}_\tau$ on hyperplane w_τ, therefore we employ the scoring function as $f_\tau(h, r, t) = \|\mathbf{h}_\tau + \mathbf{r}_\tau - \mathbf{t}_\tau\|_{11/12}$. Temporal-relational hyperplanes, along with entity and relation embeddings are learned during training by minimizing the following margin-based ranking loss:

$$\mathcal{L}_{emb} = \sum_{\tau \in 3 \times [T]} \sum_{(h,r,t) \in \mathcal{D}_\tau^+} \sum_{(\bar{h},r,\bar{t}) \in \mathcal{D}_\tau^-} \max \left(0, f_\tau((h,r,t)) - f_\tau((\bar{h},r,\bar{t})) + \gamma \right) \quad (2)$$

where γ is the margin separating positive and negative triples, $\mathcal{D}_\tau^+$ is the quadruple set valid at time step τ, and $\mathcal{D}_\tau^-$ is the corresponding negative sample set.

3.2 Evolving Modeling

According to Eq. 2, each KG snapshot is independent of the other, and triples valid at every hyperplane are learned separately. However, facts occurring on previous hyperplane usually have an effect on those on the subsequent hyperplane. Considering impressive performance of GRU on sequence modeling with few parameters, inside each relational sub-KG, GRU is adopted to preserve evolutionary dependence among hyperplanes. Taking G_{R_1} as an example, we use hyperplane $\boldsymbol{w}_{1,j}, j \in \{1, 2, \cdots, T\}$ as the input of j-th GRU unit of sub-KG G_{R_1}:

$$\boldsymbol{z}_{1,j} = GRU\left(\boldsymbol{w}_{1,j}, \boldsymbol{z}_{1,j-1}\right) \quad (3)$$

where $\boldsymbol{z}_{1,j}$ denotes the hidden state of j-th GRU unit, which carries evolving information till time j. Then $\boldsymbol{z}_{1,j}$ is used to guide the learning of next time step's hyperplane $\boldsymbol{w}_{1,j+1}$ by designing evolving loss, which transfers evolving dependency between temporal hyperplanes, that is:

$$\mathcal{L}_{evo_1} = \sum_{\tau=1}^{T-1} \|\boldsymbol{z}_{1,\tau} - \boldsymbol{w}_{1,\tau+1}\|_2 \quad (4)$$

Combined with the previous section, we sum up the embedding loss $\mathcal{L}_{emb}$ that models KG structure, and three evolving loss $\sum_{i=1}^{3} \mathcal{L}_{evo_i}$ which implement dynamic evolution of every relational sub-KG to compute the total loss as $\mathcal{L} = \mathcal{L}_{emb} + \sum_{i=1}^{3} \mathcal{L}_{evo_i}$.

3.3 Dynamic Negative Sampling

Existing TKGE models replace entities randomly, which tends to generate easily distinguishable and false negative samples. To overcome this deficiency, we propose a dynamic negative sampling mechanism to generate variational deceptive negative samples, and minimize the production of false negative samples during training. For entity and relation prediction task, we consider time-agnostic negative sampling. As shown in Fig. 1(c), given a golden quadruple (h, r, t, τ) where τ represents valid hyperplane of (h, r, t), we replace head entity h and tail entity t by other all entities, but exclude the possible generated observed triples to construct two negative sample sets $\mathcal{N}_h$ and $\mathcal{N}_t$, respectively, i.e., $\mathcal{N}_h = \{(h', r, t, \tau) \mid h', t \in \mathcal{E}, r \in \mathcal{R}, (h', r, t) \notin \mathcal{D}^+\}, \mathcal{N}_t = \{(h, r, t', \tau) \mid h, t' \in \mathcal{E}, r \in \mathcal{R}, (h, r, t') \notin \mathcal{D}^+\}$, where $\mathcal{D}^+$ denotes the time agnostic triples set. Then we compute scores of quadruples in $\mathcal{N}_h$ and $\mathcal{N}_t$, and sort them in descending order. We randomly pick up one from top-k quadruples to form a negative sample $(\bar{h}, r, t, \tau)$ and $(h, r, \bar{t}, \tau)$, respectively. The top-k quadruples with high scores are usually confusing, which facilitate adequate model training. Meanwhile, compared with selecting the top quadruple directly, randomly picking up one from top-k quadruples can avoid false negative samples to some extent. Finally, we follow the mapping property of relation-based Bernoulli distribution defined by [22] to choose between $(\bar{h}, r, t, \tau)$ and $(h, r, \bar{t}, \tau)$. Note that we do not sample negative relations since the number of relation is very small, replacing relations produces easily distinguishable negative samples, causing the model to converge quickly, which is insufficient for entity prediction. For temporal scope prediction, we directly adopt time-dependent negative sampling (TDNS) in HyTE [3].

3.4 *Expand-and-Best-Merge* Strategy (Testing phase)

Note that some triples are valid across a considerable time interval, so they may valid at multiple hyperplanes. Given a test quadruple $(h, r, t, [\tau_s, \tau_e])$, where $[\tau_s, \tau_e]$ valid at $w_\tau, \cdots, w_{\tau+k}$, HyTE only projects the triple (h, r, t) onto a single hyperplane associated with t_s during testing, i.e., w_τ, which only considers testing on a small part of the full time, and is insufficient to reflect complete test effect. For a more comprehensive test, we propose an *expand-and-best-merge* strategy. Firstly we *expand*: we project a golden triple onto all valid hyperplanes and compute ranks on each of them, which considers both temporal and relational information, e.g., given a quadruple $(h, r, t, [t_s, t_e])$, where relation r possesses temporal property p_1 and p_2 and $[\tau_s, \tau_e]$ valid at $w_\tau, \cdots, w_{\tau+k}$. Hence it can be projected into multiple hyperplanes $w_{1,\tau}, \cdots, w_{1,\tau+k}, w_{2,\tau}, \cdots, w_{2,\tau+k}$, and we can get multiple golden ranks on corresponding multiple valid hyperplanes. However, straightforwardly taking multi-ranks of one single test quadruple into evaluation causes inflated results, one test quadruple can only have one rank result. Therefore we then conduct *best-merge*: only keep the best rank among multiple ranks as the unique result and filter the rest. This strategy helps our model to realize a complete testing on all valid time intervals and maintain impartiality simultaneously.

4 Experiments

4.1 Experimental Setup

Datasets. In this work, we use the preprocessed datasets Wikidata12k and YAGO11k extracted by [3] for testing our model, where time annotations are represented in various forms, i.e., time point like $[2003\text{-}01\text{-}01, 2003\text{-}01\text{-}01]$, open time interval like $[2003, \#\#]$, and time interval like $[2003, 2005]$. We list the statistics of datasets in Table 1.

Table 1. Details of datasets

Datasets	$|\mathcal{E}|$	$|\mathcal{R}|$	Train/Valid/Test
Wikidata12k	12554	24	32.5k/ 4k/ 4k
YAGO11k	10623	10	16.4k/ 2k/ 2k

Evaluation Protocol and Baselines. We evaluate our model under three temporal link prediction tasks: *Entity Prediction*: Predicting missing head or tail entity like $(?, r, t, T)$ or $(h, r, ?, T)$; *Relation Prediction*: Predicting missing relation like $(h, ?, t, T)$; *Temporal Scope Prediction*: Predicting the most probable valid hyperplane of a triple like $(h, r, t, ?)$. For entity and relation prediction, we replace entity and relation in a golden quadruple with all possible entities and relations under *filtered* setting [1], respectively. Then we score and rank all corrupted quadruples and the golden quadruple, and find the rank of the golden quadruple. For temporal scope prediction, we project facts onto all hyperplanes to get scores on each of them and record the rank of the golden one. A particular triple may be valid at multiple hyperplanes, similarly, we select the lowest rank among all golden hyperplanes as the result [3]. We report **MR** (Mean Rank), **MRR** (Mean Reciprocal Rank), and **Hits@{1,3,10}** for all datasets.

We compare our model with several KGE and TKGE models. KGE models include TransE [1], TransD [9] and TransH [22] since they are also translation-based models, their results are taken from [3,23]. TKGE models include HyTE [3], t-TransE [10], SEDE [32], ATiSE [27], TIMEPLEX [8], TeRo [26], ToKEi [14], TeLM [25], RTFE [29], and RTGE [28]. RTFE is a framework for KGE models, we choose RTFE-TransE (**RTFE-E**) and RTFE-TransD (**RTFE-D**) for fairness due to relevance. For RTGE, we choose the best one with the same embedding setting (RTGE (d=128)) as ours. ATiSE, TIMEPLEX, TeRo, and TeLM do not distinguish head and tail predictions, follow the practice of ATiSE, we average head and tail prediction results. Additionally, we run the origin HyTE model (**HyTE-e**) for complete metrics. For a fair comparison under the *expand-and-best-merge* strategy, we propose a variant of HyTE as a baseline, denoted as **HyTE-***, which applies *expand-and-best-merge* strategy into HyTE model.

Implementation Details. Our models are implemented with TensorFlow and Pytorch. We only retain year granularity time annotation of quadruples in both datasets in our work. Following the data preprocessing in [3], we merge adjacent years with a minimum threshold of 300 triples per time step to split time annotations uniformly, resulting in 78 and 61 time steps in Wikidata12k and YAGO11k, respectively. We set batch size $b = 50k$ on both datasets for training. For all experiments, hyper-parameters are set to the best configuration reported by HyTE: embedding dimension $d = 128$, margin $\gamma = 10$, learning rate of SGD $lr = 0.0001$. We use l_1 norm in the scoring function and $k = 50$ for dynamic negative sampling.

Table 2. Entity prediction and relation prediction results on Wikidata12k. The best results are **boldfaced** and the second best results are underlined.

Datasets	Wikidata12k											
Metric	MR			MRR		H@1			H@3		H@10	
	Tail	Head	Rel	Tail	Head	Tail	Head	Rel	Tail	Head	Tail	Head
TransE	520	740	1.35	–	–	–	–	88.40	–	–	11.00	6.00
TransD	346	562	1.29	–	–	–	–	88.20	–	–	25.70	14.10
TransH	423	648	1.40	–	–	–	–	88.10	–	–	23.70	11.80
t-TransE	283	413	1.97	–	–	–	–	74.20	–	–	24.50	14.50
SEDE	158	258	1.11	–	–	–	–	97.40	–	–	59.90	31.70
RTGE	**127**	**183**	1.09	–	–	–	–	92.80	–	–	44.60	29.90
RTFE-E	–	–	1.88	36.40	14.10	20.10	5.60	73.70	47.30	13.60	**67.60**	33.40
RTFE-D	–	–	–	33.00	7.20	18.10	2.80	–	45.10	8.00	63.10	18.50
ToKEi	129	444	–	44.20	34.20	31.10	24.40	–	51.90	39.00	66.90	52.00
HyTE	179	237	1.13	–	–	–	–	92.60	–	–	41.60	25.00
HyTE-e	185	244	1.13	21.44	14.28	12.02	7.98	91.99	23.63	14.51	42.05	26.07
HyTE-*	186	244	1.13	21.43	14.26	12.05	8.01	91.97	23.63	14.41	41.73	26.14
Ours	140	254	**1.03**	**48.41**	**43.92**	**41.78**	**38.04**	**98.40**	50.83	**45.28**	62.47	**55.83**

Table 3. Entity prediction and relation prediction results on YAGO11k. The best results are **boldfaced** and the second best results are underlined.

Datasets	YAGO11k											
Metric	MR			MRR		H@1			H@3		H@10	
	Tail	Head	Rel	Tail	Head	Tail	Head	Rel	Tail	Head	Tail	Head
TransE	504	2020	1.70	–	–	–	–	78.40	–	–	4.40	1.20
TransD	138	1208	1.19	–	–	–	–	86.20	–	–	35.40	13.20
TransH	354	1808	1.53	–	–	–	–	76.10	–	–	5.80	1.50
t-TransE	292	1692	1.66	–	–	–	–	75.50	–	–	6.20	1.30
SEDE	151	745	1.20	–	–	–	–	89.00	–	–	42.00	18.20
RTGE	110	799	1.15	–	–	–	–	88.20	–	–	40.90	20.10
RTFE-E	–	–	1.43	19.80	7.60	6.30	0.40	84.10	27.00	13.10	42.50	14.60
RTFE-D	–	–	–	18.90	9.60	3.70	0.50	–	26.50	14.70	46.20	22.10
ToKEi	114	**723**	–	**50.00**	**30.70**	**40.50**	**21.80**	–	**61.80**	**35.50**	**77.90**	**47.00**
HyTE	**107**	1069	1.23	–	–	–	–	81.20	–	–	38.40	16.00
HyTE-e	110	1059	1.27	14.64	5.99	2.87	0.15	82.84	19.89	9.02	35.01	13.36
HyTE-*	110	1059	1.27	14.65	5.99	2.83	0.06	82.69	19.75	9.02	24.91	13.36
Ours	129	1532	1.08	23.42	11.52	11.17	2.22	92.76	20.81	16.38	42.66	23.35

Table 4. Entity prediction results on Wikidata12k and YAGO11k. The best results are **boldfaced** and the second best results are underlined.

Datasets	Wikidata12k					YAGO11k				
Metric	MR	MRR	H@1	H@3	H@10	MR	MRR	H@1	H@3	H@10
ATiSE	–	0.2520	14.80	28.80	46.20	–	0.1850	12.60	18.90	30.10
TIMEPLEX	–	0.3335	22.78	–	53.20	–	**0.2364**	**16.92**	–	**36.71**
TeRo	–	0.2990	19.80	32.90	50.70	–	0.1870	12.10	19.70	31.90
TeLM	–	0.3320	23.10	36.00	54.20	–	0.1910	12.90	19.40	32.10
HyTE	208	–	–	–	33.30	588	–	–	–	27.20
HyTE-e	214	0.1786	10.00	19.07	34.06	584	0.1032	1.51	14.46	24.19
HyTE-*	215	0.1785	10.03	19.02	33.94	584	0.1032	1.45	14.20	10.14
Ours	**197**	**0.4617**	**39.91**	**48.06**	**58.65**	830	0.1748	7.25	**23.69**	33.01

4.2   Results and Analysis

All baseline results (except **HyTE-e** and **HyTE-***) are taken from the reported results in respective papers. Horizontal lines indicate lacking results since there is no literature reporting the results of corresponding metrics on specific datasets.

We report the performance on entity predictions in Table 2, Table 3 and Table 4. As shown in Table 2, our method shows a large boost in performance over baselines on Wikidata12k, and achieves the best in terms of almost all metrics except Hits@{3,10} on tail prediction with a small gap. In Table 3, our model exceeds most baselines and yields the second-best on almost all evaluations on YAGO11k. For comparing with some other baselines, we average head and tail prediction results in Table 4. TRHyTE outperforms all baselines by a significant margin regarding all protocols on Wikidata12k and Hits@3 metric on YAGO11k, with the improvement of 12.82% (MRR), 16.81% (Hits@1), 12.06% (Hits@3), 4.45% (Hits@10) on Wikidata12k and 4.69% (Hits@3) on YAGO11k. Overall, our method achieves prominent performance in Wikidata12k while deficient results in YAGO11k. Since YAGO11k is much sparser than Wikidata12k dataset, we believe denser training facts in Wikidata12k better highlight the superiority of our model, whereas sparse data is inadequate for learning temporal regularity of facts and temporal properties of relations. Additionally, it's worth noting that the MR results are not satisfactory on both datasets, we claim that MR is not an appropriate metric for entity prediction where the number of entities for ranking is very large, thus MR results can be badly affected by extreme values from infrequent triples, MRR and Hits metrics better assess model performance for entity prediction, which are mainly influenced by top ranks. But for relation and temporal scope prediction where the ranking space is much smaller, MR metric is still referrible.

Table 5. Temporal scope prediction results on Wikidata12k and YAGO11k. The best results are **boldfaced** and the second best results are underlined.

Models	Wikidata12k	YAGO11k
Metric	MR	MR
HyTE	17.60	**9.88**
HyTE-e	17.96	14.38
Ours	**12.43**	10.06

We also verify the superiority of our model on relation prediction, as shown in Table 2 and Table 3, TRHyTE outperforms all competitors on all metrics. The promising performance of TRHyTE demonstrates the correctness of relational space transformation and temporal properties of relations. We also prove the validity of incorporating temporal information since static methods perform dramatically worse than temporal methods due to their inability of modeling useful temporal information.

We report MR for temporal scope prediction task, comparing with HyTE in Table 5. We run the original model (**HyTE-e**), and get similar results on Wikidata12k but 5 points higher MR on YAGO11k. Compared with **HyTE-e**, our model achieves 5 points and 4 points lower MR on Wikidata12k and YAGO11k, respectively, which demonstrates TRHyTE is capable of modeling temporal information and making temporal scope predictions.

4.3   Ablation Study and Case Study

To help explore the contribution of different components of TRHyTE, we conduct an ablation study on entity and relation prediction on Wikidata12k datasets. We use different capital letters to represent different model components, and we create variants of TRHyTE by singly adding these components from HyTE (i.e., the projection model) for comparison. **R** refers to constructing relational sub-KGs based on temporal property of relations; **G** refers to the use of GRU; **T** denotes TransD transformation; **E** denotes *expand-and-best-merge* strategy; **N** represents dynamic negative sampling. Co-occurrence of letters represents the combination of different components (e.g., +RG means that we construct relational sub-KGs and leverage GRU into the initial HyTE model), +RGTEN equals our complete model. As shown in Table 6, the *italicized* result values under each variant represent the difference in performance between the current and last variant, with the expectation of decrease on MR (-) and increase on the other metrics (+). The performance differentials present an overall pattern we expect, which illustrates that each model component plays an active role. The intact model is noticeably better than all variants. Constructing relational sub-KGs based on temporal property of relations, dynamic negative sampling mechanism, and TransD transformation contribute most to TRHyTE with prominent improvement, they consider incorporating temporal information of relations and constructing finer-grained hyperplanes for projection. Moreover, ingenious dynamic negative sampling leads to more sufficient model training. GRU also contributes model performance but is more subtle, which demonstrates that KG evolution is necessary to model, and *expand-and-best-merge* strategy guarantees a more comprehensive assessment. Overall the ablation study verifies the validity of every model component.

We conduct a case study for relation and temporal scope prediction tasks, comparing HyTE and our method of constructing relational sub-KGs (+R). For relation prediction task in Table 7, as shown in the first test quadruple with interval time[1943, 1956], HyTE wrongly gives relation <*hasWonPrize*> a higher rank than the correct relation <*worksAt*>. In contrast, since our method knows that relation <*worksAt*> has interval property while <*hasWonPrize*> has instantaneousness property, it correctly ranks <*worksAt*> ahead of <*hasWonPrize*>. Similarly, for test quadruple valid at open-interval time (example 2) and instantaneous time (example 3), our model also gives better ranks for correct answers, as it knows that relation <*isMarriedTo*>, <*isAffiliatedTo*> and <*wasBornIn*> has interval, open-interval, and instantaneousness property, respectively.

Table 6. Ablation studies on Wikidata12k. The best results are **boldfaced** and the second best results are underlined.

Datasets	Wikidata12k									
Metric	MR	MRR		H@1		H@3			H@10	
	Rel	Tail	Head	Tail	Head	Rel	Tail	Head	Tail	Head
HyTE-p	1.13	–	–	–	–	92.60	–	–	41.60	25.00
HyTE-e	1.13	21.44	14.28	12.02	7.98	91.99	23.63	14.51	42.05	26.07
+R	1.11	35.68	29.66	28.73	23.87	94.97	37.72	30.75	48.56	40.70
	−0.02	+14.24	+15.38	+16.71	+15.89	+2.98	+14.09	+16.24	+6.51	+14.63
+RG	1.11	37.21	30.01	30.33	23.53	95.02	39.49	31.31	50.38	42.57
	0.00	+1.53	+0.35	+1.60	−0.34	0.05	+1.77	+0.56	+1.82	+1.87
+RGT	1.05	41.88	35.24	<u>34.24</u>	<u>28.95</u>	97.68	44.25	36.19	56.81	47.77
	−0.06	+4.67	+5.23	+3.91	+5.42	+2.66	+4.76	+4.88	+6.43	+5.20
+RGTE	<u>1.05</u>	<u>42.09</u>	<u>35.27</u>	<u>34.24</u>	28.73	<u>97.71</u>	<u>44.74</u>	<u>36.61</u>	<u>57.08</u>	<u>48.09</u>
	0.00	+0.21	+0.03	0.00	−0.22	+0.03	+0.49	+0.42	+0.27	+0.32
+RGTEN	**1.03**	**48.41**	**43.92**	**41.78**	**38.04**	**98.40**	**50.83**	**45.28**	**62.47**	**55.83**
	−0.02	+6.32	+8.65	+7.54	+9.31	+0.69	+6.09	+8.67	+5.39	+7.74

Table 7. Case study on relation prediction. The order of prediction is in descending order. Correct one is in **bold**.

test quadruples	HyTE	+R
Doris Reynolds, ?, University of Edinburgh, [1943, 1956]	hasWonPrize, **worksAt**	**worksAt**, hasWonPrize
William Henry Young, ?, De Morgan Medal, [1917, ##]	isMarriedTo, **hasWonPrize**	**hasWonPrize**, isMarriedTo
Reg Turnbull, ?, China, [1908-02-21, 1908-02-21]	isAffiliatedTo, **wasBornIn**	**wasBornIn**, isAffiliatedTo

Table 8. Golden hyperplane rank on temporal scope prediction. Best rank is in **bold**.

Test quadruples	HyTE	+R(sub-KG1/2/3)
Lewis Price, playsFor, Brentford F.C., ?	21	**1**, 2, 27
Takaaki Kajita, hasWonPrize, Nobel Prize in Physics, ?	37	2, **1**, 15
Elizabeth F. Neufeld, wasBornIn, Paris, ?	41	30, 26, **1**

As for temporal scope prediction task in Table 8, we compare the rank of golden hyperplanes. Especially, since we construct three relational sub-KGs, we project triples into all hyperplanes inside these sub-KGs, and we compute the plausibility of the test triple on each of them. Thus we get three results for one test quadruple, representing the rank of golden hyperplanes inside every relational sub-KG, respectively. As shown in the first test quadruple with interval relation *<playsFor>* from sub-KG1, we get rank 1, 2 and 27 for golden hyperplanes inside three sub-KGs, respectively. We get the best rank from sub-KG1, which meets our expectations. A similar good phenomenon also appears in sub-KG2 and sub-KG3. The above cases prove that after clustering quadruples into three relational sub-KGs, our model can learn embeddings separately at a finer granularity, and leads to more accurate predictions.

5   Related Work

5.1   Static Knowledge Graph Embedding

Extensive researches have been done on static knowledge graph embedding. Translation-based models are the most heuristic works like TransE [1], which considers relation r as a translation from head entity h to tail entity t in the continuous vector space, i.e., $\mathbf{h} + \mathbf{r} \approx \mathbf{t}$, where $h, r, t \in \mathbb{R}^d$ are the embeddings of h, r, and t, respectively. TransE has many variants such as TransH [22] and TransD [9], etc. All of them aim at minimizing the distance between two entities translated by relation. RESCAL [17], DistMult [30], ComplEx [21], HolE [16] and SimplE [12] are matrix factorization-based or tensor decomposition-based models, which are more complicated and mathematical. There are also some exquisitely designed models which embed entities and relations into complex space, such as RotatE [19], which takes relation as a rotation from head entity to tail entity in the complex space, and QuatE [31], which extends the complex-valued space into hypercomplex by a quaternion. Though static KGE models perform satisfactorily on static KGs, they expose insufficiency on TKGs due to ignorance of useful time information.

5.2   Temporal Knowledge Graph Embedding

Recent researches have attempted to incorporate time into KGE models and have proved the effectiveness of additional time information. Series of models have emerged and differed in time modeling, which can be categorized into four branches. Time encoding-based models embed time explicitly. TA-TransE [6] and ToKEi [14] encode time directly in well-designed coding methods. Inspired by static KGE model RotatE [19], TeRo [26] regards time as a rotation from start time to current time in complex space. ChronoR [18] is another rotation-based work. HyTE [3] treats time as hyperplanes and projects facts into time-valid hyperplanes to incorporate time. Time-aware relation-based methods like t-TransE [10] and TIMEPLEX [8] treat time as a regularizer to restrain temporal sequence, regular reappearance, and time difference, etc. Since TKGs keep evolving and the current state is probably influenced by history. Evolution modeling-based approaches like RE-NET [11] and TeMP [24] leverage RNNs to capture long-term dependency of facts in TKGs. Know-Evolve [20] models the occurrence of facts as temporal point processes. CyGNet [33] learns and copies from history to generate predictions. DE-SimplE [7] and ATiSE [27] consider entity representations keep changing over time and fit changes with suitably-designed functions. Tensor decomposition-based models such as ConT [5] and TComplEx [13] consider encoding facts as tensors and apply tensor factorization operations. There are some other methods that work as a framework for turning static KGE models into temporal ones like RTFE [29], which updates parameters recursively as time progresses.

6 Conclusion

In this work, we propose TRHyTE, a temporal-relational hyperplane projection-based TKGE model. To implement a fine-grained learning, three temporal properties of relations are proposed to construct three relational sub-KGs. In each sub-KG, our model transforms entity embeddings into relation space first, and then projects entities and relations into valid hyperplanes explicitly. Considering the temporal dependency in adjacent hyperplanes, GRUs are applied to model TKG evolution. Besides, TRHyTE dynamically generates negative samples for effective training and designs an *expand-and-best-merge* strategy to realize a complete test. Through abundant experiments on real-world fact-based datasets, we demonstrate the effectiveness of TRHyTE.

Acknowledgments. This research is supported by the National Key R&D Program of China (No. 2018AAA0101900), the National Natural Science Foundation of China (Grant No. 62072323, 62102276), the Natural Science Foundation of Jiangsu Province (No. BK20191420, BK20210705, BK20211307), the Major Program of Natural Science Foundation of Educational Commission of Jiangsu Province, China (Grant No.19KJA610002, 21KJD520005), the Priority Academic Program Development of Jiangsu Higher Education Institutions, and the Collaborative Innovation Center of Novel Software Technology and Industrialization.

References

1. Bordes, A., Usunier, N., Garcia-Duran, A., Weston, J., Yakhnenko, O.: Translating embeddings for modeling multi-relational data. In: Advances in Neural Information Processing Systems 26 (2013)
2. Cho, K., et al.: Learning phrase representations using RNN encoder-decoder for statistical machine translation. In: EMNLP (2014)
3. Dasgupta, S.S., Ray, S.N., Talukdar, P.: HyTE: hyperplane-based temporally aware knowledge graph embedding. In: Proceedings of the 2018 Conference on Empirical Methods in Natural Language Processing, pp. 2001–2011 (2018)
4. Erxleben, F., Günther, M., Krötzsch, M., Mendez, J., Vrandečić, D.: Introducing Wikidata to the linked data web. In: Mika, P., et al. (eds.) ISWC 2014. LNCS, vol. 8796, pp. 50–65. Springer, Cham (2014). https://doi.org/10.1007/978-3-319-11964-9_4
5. Esteban, C., Tresp, V., Yang, Y., Baier, S., Krompaß, D.: Predicting the co-evolution of event and knowledge graphs. In: 2016 19th International Conference on Information Fusion (FUSION), pp. 98–105. IEEE (2016)
6. García-Durán, A., Dumančić, S., Niepert, M.: Learning sequence encoders for temporal knowledge graph completion. arXiv preprint arXiv:1809.03202 (2018)
7. Goel, R., Kazemi, S.M., Brubaker, M., Poupart, P.: Diachronic embedding for temporal knowledge graph completion. In: Proceedings of the AAAI Conference on Artificial Intelligence, vol. 34, pp. 3988–3995 (2020)
8. Jain, P., Rathi, S., Chakrabarti, S., et al.: Temporal knowledge base completion: new algorithms and evaluation protocols. arXiv preprint arXiv:2005.05035 (2020)

9. Ji, G., He, S., Xu, L., Liu, K., Zhao, J.: Knowledge graph embedding via dynamic mapping matrix. In: Proceedings of the 53rd Annual Meeting of the Association for Computational Linguistics and the 7th International Joint Conference on Natural Language Processing (Volume 1: Long Papers), pp. 687–696 (2015)

10. Jiang, T., Liu, T., Ge, T., Sha, L., Li, S., Chang, B., Sui, Z.: Encoding temporal information for time-aware link prediction. In: Proceedings of the 2016 Conference on Empirical Methods in Natural Language Processing, pp. 2350–2354 (2016)

11. Jin, W., et al.: Recurrent event network: Global structure inference over temporal knowledge graph (2019)

12. Kazemi, S.M., Poole, D.: Simple embedding for link prediction in knowledge graphs. arXiv preprint arXiv:1802.04868 (2018)

13. Lacroix, T., Obozinski, G., Usunier, N.: Tensor decompositions for temporal knowledge base completion. arXiv preprint arXiv:2004.04926 (2020)

14. Leblay, J., Chekol, M.W., Liu, X.: Towards temporal knowledge graph embeddings with arbitrary time precision. In: Proceedings of the 29th ACM International Conference on Information & Knowledge Management, pp. 685–694 (2020)

15. Mahdisoltani, F., Biega, J., Suchanek, F.M.: A knowledge base from multilingual wikipedias-yago3. Technical report, Telecom ParisTech (2014). http://suchanek. name/work/publications

16. Nickel, M., Rosasco, L., Poggio, T.: Holographic embeddings of knowledge graphs. In: Proceedings of the AAAI Conference on Artificial Intelligence, vol. 30 (2016)

17. Nickel, M., Tresp, V., Kriegel, H.P.: A three-way model for collective learning on multi-relational data. In: ICML (2011)

18. Sadeghian, A., Armandpour, M., Colas, A., Wang, D.Z.: ChronoR: rotation based temporal knowledge graph embedding. arXiv preprint arXiv:2103.10379 (2021)

19. Sun, Z., Deng, Z.H., Nie, J.Y., Tang, J.: RotatE: knowledge graph embedding by relational rotation in complex space. arXiv preprint arXiv:1902.10197 (2019)

20. Trivedi, R., Dai, H., Wang, Y., Song, L.: Know-evolve: deep temporal reasoning for dynamic knowledge graphs. In: International Conference on Machine Learning, pp. 3462–3471. PMLR (2017)

21. Trouillon, T., Welbl, J., Riedel, S., Gaussier, É., Bouchard, G.: Complex embeddings for simple link prediction. In: International Conference on Machine Learning, pp. 2071–2080. PMLR (2016)

22. Wang, Z., Zhang, J., Feng, J., Chen, Z.: Knowledge graph embedding by translating on hyperplanes. In: Proceedings of the AAAI Conference on Artificial Intelligence, vol. 28 (2014)

23. Wang, Z., Li, X.: Hybrid-TE: hybrid translation-based temporal knowledge graph embedding. In: 2019 IEEE 31st International Conference on Tools with Artificial Intelligence (ICTAI), pp. 1446–1451. IEEE (2019)

24. Wu, J., Cao, M., Cheung, J.C.K., Hamilton, W.L.: TeMP: temporal message passing for temporal knowledge graph completion. arXiv preprint arXiv:2010.03526 (2020)

25. Xu, C., Chen, Y.Y., Nayyeri, M., Lehmann, J.: Temporal knowledge graph completion using a linear temporal regularizer and multivector embeddings. In: Proceedings of the 2021 Conference of the North American Chapter of the Association for Computational Linguistics: Human Language Technologies, pp. 2569–2578 (2021)

26. Xu, C., Nayyeri, M., Alkhoury, F., Yazdi, H.S., Lehmann, J.: TeRo: a time-aware knowledge graph embedding via temporal rotation. arXiv preprint arXiv:2010.01029 (2020)

27. Xu, C., Nayyeri, M., Alkhoury, F., Yazdi, H., Lehmann, J.: Temporal knowledge graph completion based on time series Gaussian embedding. In: Pan, J.Z., et al. (eds.) ISWC 2020. LNCS, vol. 12506, pp. 654–671. Springer, Cham (2020). https://doi.org/10.1007/978-3-030-62419-4_37

28. Xu, Y., et al.: Time-aware graph embedding: a temporal smoothness and task-oriented approach. arXiv preprint arXiv:2007.11164 (2020)

29. Xu, Y., Song, M., Lv, X., et al.: RTFE: a recursive temporal fact embedding framework for temporal knowledge graph completion. arXiv preprint arXiv:2009.14653 (2020)

30. Yang, B., Yih, W.t., He, X., Gao, J., Deng, L.: Embedding entities and relations for learning and inference in knowledge bases. arXiv preprint arXiv:1412.6575 (2014)

31. Zhang, S., Tay, Y., Yao, L., Liu, Q.: Quaternion knowledge graph embeddings. arXiv preprint arXiv:1904.10281 (2019)

32. Zhou, Y., Peng, J., Wang, L., Zha, D., Mu, N.: SEDE: semantic evolution-based dynamic knowledge graph embedding. Aust. J. Intell. Inf. Process. Syst. **16**(4), 64–73 (2019)

33. Zhu, C., Chen, M., Fan, C., Cheng, G., Zhan, Y.: Learning from history: modeling temporal knowledge graphs with sequential copy-generation networks. arXiv preprint arXiv:2012.08492 (2020)

ExKGR: Explainable Multi-hop Reasoning for Evolving Knowledge Graph

Cheng Yan[1,2], Feng Zhao[1,2(✉)], and Hai Jin[1,2]

[1] National Engineering Research Center for Big Data Technology and System,
Services Computing Technology and System Lab, Cluster and Grid Computing Lab,
Wuhan, China
[2] School of Computer Science and Technology, Huazhong University of Science
and Technology, Wuhan, China
{yancheng,zhaof,hjin}@hust.edu.cn

Abstract. Knowledge graph reasoning is a popular approach to predict new facts in *knowledge graphs* (KGs) suffering from inherent incompleteness. Compared with the popular embedding-based approach, multi-hop reasoning approach is more interpretable. Multi-hop reasoning can be modeled as *reinforcement learning* (RL) in which the RL agent navigates in the KG. Despite high interpretability, the knowledge in real world evolves by the minute, previous approaches are based on static KG. To address the above challenges, we propose an explainable multi-hop reasoning approach (ExKGR) for practical scenario, aiming to reason the emerging entity in evolving KGs and provide evidentiary reasoning paths. Specifically, ExKGR can represent emerging entities by inductive learning of neighbors and the query. Furthermore, we restrict the RL action space of supernodes. Also, we use a dynamic reward instead of a binary reward in prior approaches. The experimental results on four benchmark datasets demonstrate that our approach significantly outperforms prior approaches.

Keywords: Knowledge reasoning · Reinforcement learning · Graph neural network

1 Introduction

Currently, an exceedingly growing number of technology companies have capitalized on the KG concept and integrated KGs into their products. For example, *Google* integrates knowledge graphs into its search engine, and knowledge graph technology is the foundation of Apple's *Siri*. Unfortunately, in most KGs, a number of facts are typically missing, making them incomplete. Therefore, we need to infer new facts from the initial KG and provide more knowledge for such products. *Knowledge graph reasoning* (KGR) is a popular approach to predict new facts from the existing facts in KGs.

Currently, embedding-based methods [3] are at the frontier of research in KGR, this method maps the KG to a corresponding multi-dimensional vector

A. Bhattacharya et al. (Eds.): DASFAA 2022, LNCS 13245, pp. 153–161, 2022.
https://doi.org/10.1007/978-3-031-00123-9_11

space. However, owing to their simplicity, these approaches tend to ignore the reasoning process. Some recent approaches have combined RL with multi-hop reasoning. Multi-hop reasoning based on RL not only achieves competitive performance but also offers substantial explainability.

Although multi-hop approach has advantages in some respects, there are still some challenges when dealing with evolving KGs. (1) First, real-world KGs evolve by the minute, and some newly added entities emerge in real-world KGs [7]. Previously proposed multi-hop reasoning approaches have only considered static KGs. (2) Furthermore, we identify the phenomenon of supernodes in real-world KGs, implying that some entities have a considerable number of neighbors. The oversized action space makes the RL policy network difficult to make decisions, and takes up a lot of memory. (3) The reward in RL-based reasoning is generally set to binary, which leads to underfitting and slow convergence.

To address the above challenges, we propose a multi-hop knowledge graph reasoning method for evolving KGs, aiming to reason new facts of the emerging entity in evolving KGs. Specifically, first we propose an encoder based on the attention mechanism to represent the emerging entity through its neighbors and the corresponding query. Furthermore, our approach is based on policy gradient RL, to avoid the action space being too large, we limit the size of the action space, which also alleviates exploration-exploitation problem in RL. Last but not least, we propose a dynamic reward to solve the reward sparsity in RL. Our main contributions are as follows:

- We tackle a realistic problem of multi-hop reasoning in a practical scenario, aiming at reasoning for the emerging entity in the evolving KG and providing explainable reasoning paths.
- To address the emerging entity which is unseen to the observed KG, we propose an encoder to represent the emerging entity by only a small number of associated triples with observed KG. We also restrict the action space and propose a dynamic reward to accelerate the convergence of the model.
- We conduct extensive experiments on four benchmark datasets. The results show our approach achieves competitive performance and provide explainable reasoning evidence.

2   Related Works

Embedding-based methods are intended to map an entity or relation to a certain vector in a high-dimensional vector space. The translation-based model is come up with for the first time by TransE [1]. ConvE [3] uses 2D convolution on the embedding to predict lost facts in the knowledge graph. Although embedding-based method is quite simple through some vector operations, it is highly efficient and accurate. However, due to its simplicity, it ignores some reasoning processes, which can result in poor performance on tasks involving complex reasoning.

Different from knowledge graph embedding, multi-hop reasoning provides more explainability. The *path-ranking algorithm* (PRA) [4] is based on random

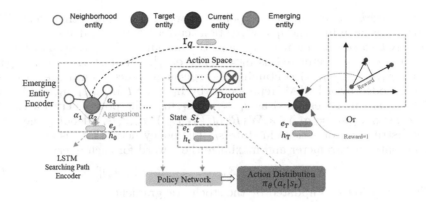

Fig. 1. Overview of *ExKGR*

walks for rule mining, each path is used as a rule to predict whether corresponding rules exist between entities. However, due to PRA paths generated from enumeration, searching the entire graph is costly. Recently the solution of integrating RL into multi-hop reasoning is gradually emerging.

Recently, due to advancements in RL, some scholars have applied RL to knowledge reasoning. DeepPath [9] was the first method to apply reinforcement learning to knowledge reasoning. DeepPath has a novel reward function that causes it to learn more and it trains an agent to reason on a KG. MINERVA [2] accomplishes query tasks based on DeepPath, it does not need pretraining and use a LSTM to encode the path. Multi-Hop [5] proposed a soft reward mechanism coupled with action dropout to reduce the impact of false negatives.

3 Methodology

3.1 Framework of ExKGR

The multi-hop reasoning problem in KGs and our proposed reinforcement learning framework is shown in Fig. 1.

Reinforcement Learning-based Multi-hop Reasoning. Formally we represent KG as $\mathcal{G} = (\mathcal{E}, \mathcal{R})$, where $\mathcal{E}$ denotes the entity set and $\mathcal{R}$ denotes the relation set. The triples in the KG are represented as $\mathcal{T} = \{(e_h, r, e_t)\} \subseteq \mathcal{E} \times \mathcal{R} \times \mathcal{E}$. The multi-hop reasoning problem can be formalized as follows: given a query $(e_s, r_q, ?)$, KGR aims to find some paths from e_s to object entity e_o based on e_s and r_q, the reasoning path can be denoted as $\{(e_s, r_1, e_1), (e_1, r_2, e_2), \ldots, (e_{n-1}, r_n, e_o)\}$.

Reinforcement learning reasoning consists of the following parts: (1) **State.** The agent makes decisions based on the current state. Each state is formally represented as $s_t = (r_q, e_t, h_t) \in \mathcal{S}$. (2) **Action.** When in state $s_t = (r_q, e_t, h_t)$, the actions the agent can take are expressed by the set of outgoing edges of the current entity e_t in $\mathcal{G}$. This action set can be formulated as $\mathcal{A}_t = \{(r, e) | (e_t, r, e) \in \mathcal{T}\}$. (3) **Reward.** If the agent successfully finds the objective entity in the KG (i.e., $e_T = e_o$), the reward is 1. If not, the reward is set to 0.

Policy Network. The aim of the policy network is to guide the agent. The current state including the query relation, current entity, and historical path is the input to the policy network, and the output is the action distribution. The action can be vectorized as $\mathbf{a}_t = [\mathbf{r}_{t+1}; \mathbf{e}_{t+1}]$. The historical search path $h_t = (e_s, r_1, e_1, \ldots, r_t, e_t)$ containing the reasoning process until the t-th step will be encoded by a LSTM, represented by $\mathbf{h}_t = LSTM(\mathbf{h}_{t-1}, \mathbf{a}_{t-1})$. At t-th step, state s_t can be encoded as $\mathbf{s}_t = [\mathbf{r_q}; \mathbf{e}_t; \mathbf{h}_t]$. The policy network can be denoted as $\pi_\theta(\mathbf{a}_t|\mathbf{s}_t) = \sigma(\mathbf{A}_t \times \mathbf{W}_1 ReLU(\mathbf{W}_2\mathbf{s}_t))$, $\pi_\theta(\mathbf{a}_t|\mathbf{s}_t)$ denotes the probability distribution of actions in state s_t. The policy network π_θ is expected to train an optimal parameter and maximize the reward for each query:

$$\mathcal{J}(\theta) = \mathbb{E}_{(e_s,r,e_o)\in\mathcal{G}}[\mathbb{E}_{a_1,\cdots,a_T\sim\pi_\theta}[R(s_T|e_s,r_q)]] \tag{1}$$

The parameter θ is updated by the stochastic gradient:

$$\nabla_\theta J(\theta) \approx \nabla_\theta J(\theta) \sum_{t=1}^{T} R(s_T|e_s,r) \log \pi_\theta(a_t|s_t) \tag{2}$$

3.2   Emerging Entities Encoder

Figure 2 illustrates the framework of emerging entities encoder, the encoder is based on the attention mechanism, which uses the Q (Query), K (Key), V (Value) approach to calculate the attention coefficients and the weighted sum. First, we linearly transform the action $(a_t = (e_j, r))$ associated with the emerging entity e_i to form the message $\mathbf{m}_{ijr} = \mathbf{W}_1[\mathbf{r}||\mathbf{W}_r\mathbf{e}_j]$, where $\mathbf{W}_r \in \mathbb{R}^{d\times d}$ is a transformation matrix of relation r mapping the entity to the relation vector space, $e_j \in \mathcal{N}_i$ denotes the neighboring entity of e_i, $\mathbf{e}_j$ denotes the embeddings of e_j. $\mathbf{W}_1 \in \mathbb{R}^{d\times 2d}$ is a linear transformation matrix and $[\cdot||\cdot]$ denotes the concatenation operator of two vectors. $\mathbf{m}_{ijr}$ will be used as Value in attention mechanism.

To construct the Query (Q) and Key (K) in attention mechanism, prior approaches tend to set Q to the neighboring entities characteristics, K and V are set to the information of the current entity. Our approach considers the knowledge graph reasoning setting, we utilize two matrix $\mathbf{W}_Q$ and $\mathbf{W}_K$ and perform linear transformation of the query relation r_q and neighboring relations $r \in \mathcal{R}_{ij}$, thus we can calculate the attention coefficient between e_i and e_j:

$$\alpha_{ijr} = \frac{\exp((\mathbf{W}_Q\mathbf{r}_q)^T(\mathbf{W}_K\mathbf{r}))}{\sum\limits_{e_j\in\mathcal{N}_i}\sum\limits_{r\in\mathcal{R}_{ij}} \exp((\mathbf{W}_Q\mathbf{r}_q)^T(\mathbf{W}_K\mathbf{r}))} \tag{3}$$

After obtaining the attention coefficient between the emerging entity and its neighbors, we can aggregate the neighboring messages $\mathbf{m}_{ijr}$ by the attention weights. To stabilize the learning process and encapsulate more hidden semantics about neighboring actions, N independent attention heads calculate corresponding aggregated messages which are concatenated, obtaining the representation of the emerging entity, $\mathbf{W}_V^n$ represents the n-th linear transformation matrix of attention mechanism:

$$e_i = ||_{n=1}^{N}(\sum_{e_j\in\mathcal{N}_i} \sum_{r\in\mathcal{R}_{ij}} \alpha_{ijr}^n\mathbf{W}_V^n\mathbf{m}_{ijr}) \tag{4}$$

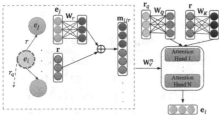
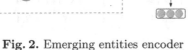
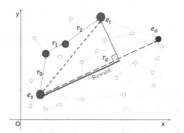

Fig. 2. Emerging entities encoder   **Fig. 3.** Dynamic reward

3.3 Dynamic Reward

The previous works suffer from reward sparsity in RL because they rely on a simple binary reward scheme. When projecting a KG in a two-dimensional coordinate system, the reasoning problem can be modeled as a process that the agent moves from the source to the target by the relation vectors as shown in Fig. 3. The green dotted line represents the query relation and the solid red line represents the reward. The reward is the projection from the red dotted line to the green dotted line. We can formalize the reasoning process as follows: $\sum_{i=0}^{T} \vec{r_i}$ represents the vector of the actual reasoning path, $\vec{e_o} - \vec{e_s}$ represents the difference vector between the inferred entity and the objective entity, and $d = \frac{(\vec{e_o} - \vec{e_s}) \cdot \sum_{i=0}^{T} \vec{r_i}}{\|\vec{e_o} - \vec{e_s}\|}$ denotes the projection from the reasoning path to the difference vector. Therefore, the reward function R in our approach is defined as follows:

$$R = \begin{cases} \sigma(d) & d \in (-\infty, \|\vec{r_q}\|) \setminus \{0\} \\ 1 & d = 0 \\ \sigma(\|\vec{r_q}\| - d) & d \in (\|\vec{r_q}\|, +\infty) \end{cases} \tag{5}$$

3.4 Action Pruning

According to the analysis of real-world knowledge graphs, supernodes having a huge number of neighbors do exist in the KG. Some entities having too many neighbors lead to oversized action space. The oversized action space makes the policy network difficult to train and only outputs almost random action distributions. Therefore, we restrict the size of supernode action space. First, we rank the entities by the number of neighbors of each entity, then select some large entities as supernodes according to the degree of the entity. We set a threshold η to restrict the action space of the supernode and calculate its top-η neighboring entities with PageRank scores. Then, the action space will be randomly dropout to further reduce the size of the action space. The dropout technique randomly masks some actions and forces the RL agent to explore diverse reasoning paths, appropriately alleviating exploration and exploitation in RL.

Table 1. Statistics of datasets

Datasets	#Entity	#Relation	#Triple	#Degree				
				Mean	Median	Min	Max	Var
UMLS	135	46	5,216	38.6	28	1	133	1057.6
Kinship	104	25	8,544	82.1	82	74	92	12.2
FB15K-237	14,505	237	272,115	19.7	14	1	1325	905.9
NELL-995	75,492	200	154,213	1.5	1	1	109	3.4

Table 2. Link prediction results (%): The Results of All Models

Model	UMLS			Kinship			FB15K-237			NELL-995		
	MRR	@1	@10	MRR	@1	@10	MRR	@1	@10	MRR	@1	@10
TransE	.201	.178	.292	.102	.075	.141	.049	.030	.068	.055	.045	.067
DisMult	.152	.124	.231	.081	.056	.123	.043	.025	.066	.057	.048	.072
RotatE	.171	.157	.269	.089	.067	.135	.045	.026	.069	.048	.037	.062
ConvE	.709	.651	.851	.581	.532	.657	.256	.197	.378	.423	.324	.593
R-GCN	.721	.671	.812	.455	.412	.509	.191	.115	.343	.584	.509	.716
MINERVA	.812	.751	.859	.734	.615	.865	.274	.192	.461	.725	.642	.821
MultiHopKG	.891	.871	.941	.812	.714	.913	.376	.292	.577	.727	.644	.822
ExKGR	.909	.882	.989	.834	.749	.968	.409	.329	.576	.732	.654	.831

4   Experiments

4.1   Setup

Dataset. Experiments are carried out on four benchmark datasets. Table 1 shows statistical characteristics (the number of entities, relations, and triples) of four datasets and the number of entity neighbors (#Degree).

Baselines and Evaluation Metrics. We compare our method with the following baselines: (1) Embedding-based methods: TransE [1], DistMult [10], RotatE [8], and ConvE [3]. (2) Graph neural network: R-GCN [6]. (3) Multi-hop reasoning: MINERVA [2] and MultiHopKG [5]. The experimental task is the link prediction for the emerging entity. The *mean reciprocal rank* (MRR) and n (Hits@n) are used to evaluate the above models.

Hyperparameters. The embedding dimension d is set to 200. A three-layer LSTM is used to encode the search path. Its hidden dimension is set to 200. Top 10% large entities are selected as supernode entities. The threshold η restricting the action space is set to 256 or 512. For emerging entities encoder, we deploy 4 attention heads. The maximum number of reasoning steps is set to 3.

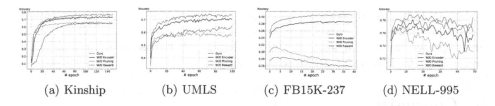

|(a) Kinship|(b) UMLS|(c) FB15K-237|(d) NELL-995|

Fig. 4. Illustration of convergence rate

Table 3. Query and reasoning path cases

Query 1 (Elon Musk, lead sorganization, ?)	**Answer 1** Tesla
Path 1 Elon Musk $\xrightarrow{ceo\_of}$ automobilemaker_tesla	
Query 2 (Oklahoma, adjoins, ?)	**Answer 2** Texas
Path 2 Oklahoma $\xrightarrow{country}$ USA $\xrightarrow{country^{-1}}$ Texas	
Query 3 (Bill Clinton, endorsed by, ?)	**Answer 3** Joe Biden
Path 3 Clinton $\xrightarrow{represents}$ USA $\xrightarrow{represents^{-1}}$ Obama $\xrightarrow{collaborates^{-1}}$ Biden	

4.2 Link Prediction Results

The link prediction results are shown in Table 2. The result shows that the embedding-based method achieves poor performances on the datasets. Embedding-based methods cannot solve emerging entities. GNN-based method R-GCN outperforms all embedding-based methods, it proves that GNN can aggregate the neighboring information. Multi-hop methods obviously achieve the greatest performance, it indicates that multi-hop reasoning may be independent of the representation of entities. Compared to multi-hop reasoning methods, the improvement of our model proves that our model is more suitable for dynamic scenarios.

4.3 Ablation Study and Analysis

To verify the effectiveness of each component, we ablate three components of our models respectively. Figure 4 shows the convergence rate of ablated models during training. W/O Reward performs worst, which indicates that binary reward restricts the RL agent to learn from the failed reasoning paths. Action pruning also has a large enhancement to the model, especially on large-scale datasets. The existence of supernodes has a negative influence on the reasoning process. Action pruning is beneficial for making meaningful decisions and speeds up the convergence rate. Emerging entities encoder has also contributed to our model.

4.4 Qualitative Analysis

To demonstrate the explainability of our model, we present some queries and corresponding reasoning paths found by our model. Table 3 shows examples

(1-hop, 2-hop, 3-hop). For example, query 1 means that "Which organization does Elon Musk lead?", we may associate Elon Musk as the CEO of Tesla, and then we can get the answer "Tesla", proving that reasoning paths are very meaningful for knowledge graph reasoning.

5   Conclusion

In this paper, we propose a multi-hop knowledge graph reasoning approach aiming to reason the new triples of emerging entities in evolving KGs. We designed an encoder based on the attention mechanism to represent the emerging entity. The phenomenon of supernodes makes the action space in RL too large, so we propose action pruning to limit the size of the action space. To allow the model to accelerate the convergence, we propose a dynamic reward.

Acknowledgment. This work was supported in part by National Key R&D Program of China under Grants No. 2018YFB1404302, National Natural Science Foundation of China under Grants No.62072203.

References

1. Bordes, A., Usunier, N., García-Durán, A., Weston, J., Yakhnenko, O.: Translating embeddings for modeling multi-relational data. In: Advances in Neural Information Processing Systems 26, pp. 2787–2795 (2013)
2. Das, R., et al.: Go for a walk and arrive at the answer: reasoning over paths in knowledge bases using reinforcement learning. In: Proceedings of 6th International Conference on Learning Representations, ICLR 2018, Vancouver, BC, Canada, April 30 - May 3, 2018, Conference Track Proceedings (2018)
3. Dettmers, T., Minervini, P., Stenetorp, P., Riedel, S.: Convolutional 2D knowledge graph embeddings. In: Proceedings of the Thirty-Second AAAI Conference on Artificial Intelligence, pp. 1811–1818 (2018)
4. Lao, N., Mitchell, T.M., Cohen, W.W.: Random walk inference and learning in a large scale knowledge base. In: Proceedings of the 2011 Conference on Empirical Methods in Natural Language Processing, pp. 529–539 (2011)
5. Lin, X.V., Socher, R., Xiong, C.: Multi-hop knowledge graph reasoning with reward shaping. In: Proceedings of the 2018 Conference on Empirical Methods in Natural Language Processing, pp. 3243–3253 (2018)
6. Schlichtkrull, M., Kipf, T.N., Bloem, P., van den Berg, R., Titov, I., Welling, M.: Modeling relational data with graph convolutional networks. In: Gangemi, A., et al. (eds.) ESWC 2018. LNCS, vol. 10843, pp. 593–607. Springer, Cham (2018). https://doi.org/10.1007/978-3-319-93417-4_38
7. Shi, B., Weninger, T.: Open-world knowledge graph completion. In: Proceedings of the Thirty-Second AAAI Conference on Artificial Intelligence, pp. 1957–1964 (2018)
8. Sun, Z., Deng, Z., Nie, J., Tang, J.: Rotate: knowledge graph embedding by relational rotation in complex space. In: Proceedings of 7th International Conference on Learning Representations, ICLR 2019, New Orleans, LA, USA, May 6–9, 2019, Conference Track Proceedings (2019)

9. Xiong, W., Hoang, T., Wang, W.Y.: Deeppath: a reinforcement learning method for knowledge graph reasoning. In: Proceedings of the 2017 Conference on Empirical Methods in Natural Language Processing, pp. 564–573 (2017)
10. Yang, B., Yih, W., He, X., Gao, J., Deng, L.: Embedding entities and relations for learning and inference in knowledge bases. In: Proceedings of 3rd International Conference on Learning Representations, ICLR 2015, San Diego, CA, USA, May 7–9, 2015, Conference Track Proceedings (2015)

Improving Core Path Reasoning for the Weakly Supervised Knowledge Base Question Answering

Nan Hu, Sheng Bi, Guilin Qi[✉], Meng Wang, Yuncheng Hua, and Shirong Shen

School of Computer Science and Engineering, Southeast University, Nanjing, China
{nanhu,bisheng,gqi,meng.wang,devinhua,ssr}@seu.edu.cn

Abstract. Core Path Reasoning (CPR) is an essential part of the knowledge base question answering (KBQA), which determines whether the answer can be found correctly and indicates the reasonableness of the path. The lack of effective supervision of the core path in weakly supervised KBQA faces great challenges in finding the correct answer through long core paths. Furthermore, even if the correct answer is found, its path might be spurious that is not semantically relevant to the question. In this paper, we focus on solving the CPR problem in weakly supervised KBQA. We introduce a CPR model that aligns questions and paths in a step-by-step reasoning manner from explicit text semantic matching and implicit knowledge bases structure matching. Additionally, we propose a two-stage learning strategy to alleviate the spurious path problem efficiently. We first find relatively correct paths and then use hard Expectation-Maximization to learn the best matching path iteratively. Extensive experiments on two popular KBQA datasets demonstrate the strong competitiveness of our model compared to previous state-of-the-art methods, especially in solving long path and spurious path problem.

Keywords: Core path reasoning · Spurious path · Knowledge base question answering · Weakly supervised

1 Introduction

Knowledge base question answering (KBQA) is an important natural language processing task, which aims to find answers to natural language questions from the knowledge base (KB). Recently, weakly supervised KBQA [1,2] has attracted more and more attention due to the high cost of annotating logical queries like SPARQL. It only needs to give question-answer pairs, which is easier than constructing complex logical queries.

Core path reasoning (CPR) is an essential part of KBQA. There are two challenges in CPR for weakly supervised KBQA: (1) **Long Path Problem**. Most weakly supervised KBQA works have shown excellent performance when the core path is short [3,4] or the KB scale is small. Several works [1,7,10] try to improve the performance of long-path KBQA. However, when the core path is long and the KB is large, neural network models are hard to converge. (2) **Spurious Path Problem**. There are many paths between topic entities of questions and answer entities. Except for the path that

© The Author(s), under exclusive license to Springer Nature Switzerland AG 2022
A. Bhattacharya et al. (Eds.): DASFAA 2022, LNCS 13245, pp. 162–170, 2022.
https://doi.org/10.1007/978-3-031-00123-9_12

exactly matches the semantics of question, the rest are spurious paths that are unrelated to the semantics of question. In that the spurious paths would not convey the meaning of the question correctly, such spurious paths will mislead the model when training. Furthermore, the model lacks interpretability even if the answer happens to be found through a spurious path. The work [4] uses a crude method like setting F1 score thresholds for pre-screening with little success. The work [6] uses prior domain knowledge to filter spurious paths in advance, which is difficult to adapt to different settings. Recent work [1] proposes a bi-directional supervision mechanism to provide the intermediate supervision signal. However, the spurious path problem is still not well solved because it does not consider the semantic relationship between the question and the core path.

To overcome the aforesaid challenges, we propose a novel multi-perspective and multi-stage CPR model. The model aligns questions and paths from an explicit perspective of text semantics and an implicit perspective of KB structure by step-by-step reasoning, which is beneficial for reasoning about long path questions. It is difficult to align questions and paths without training labels because the expression of natural language questions is different from the expression of KB relations. The explicit information from the textual perspective can distinguish coarse-grained semantic differences like "person_education" and "places_lived". The implicit information from the KB structure perspective can distinguish fine-grained semantic differences like "country_first_level_divisions" and "country_contains". In addition, considering that there is no golden logic queries to train the model to choose the correct path, we adopt the Hard-EM algorithm [9] to rely on our model itself to find the correct path during the iterative process. Based on this idea, we design two training stages with different learning goals to make the model gradually learn to filter out errors and spurious paths through iterations. In the first stage, the model filters out error paths, selecting semantically similar paths to the question. In the second stage, the model filters out spurious paths, selecting the correct path that exactly matches the question. These two training stages gradually improve the path selection ability of the model. Furthermore, we conduct experiments on two popular KBQA datasets to demonstrate the strength of our model, especially in solving long path and spurious path problems.

2   Related Work

The weakly supervised KBQA has attracted great attention because of the expensive cost of manual annotation. Previous works [3,4] used a semantic parsing method to generate logical query graphs and then scored them. However, the core path of the question they solve is short. Liang et al. [2] used reinforcement learning to generate neural programs to construct logical forms. Lan et al. [7] improved the stage query graph method to solve the long path problem by generating stage query graphs in a flexible manner. These works ignore the negative impact caused by spurious paths. Other works [1,6] tried to solve the spurious path problem, but they all had limitations. He et al. [1] proposed a bi-directional reasoning mechanism to provide intermediate supervised signals. Bhutani et al. [6] used prior domain knowledge to filter spurious paths in advance.

3 Methodology

Preliminary Our KBQA pipeline system includes three stages: Node Linking, Core Path Reasoning, and Constraint Attachment. Our work focuses on improving the CPR of the system. (1) Node Linking. We extract four types nodes from the question: Topic entity, which links the question mention to a grounded entity. Type node, which usually specifies the type of answer entity. Time node, extracted from the question using time regular expressions. Aggregate node, which is matched using a predefined superlative word list and then mapped into two functions ($argmax$ and $argmin$). (2) Core Path Reasoning. The model will find the path from topic entities to answer entities, which exactly matches the question semantics. (3) Constraint Attachment. Various constraint nodes will be used to filter answers, this idea is the same as [5].

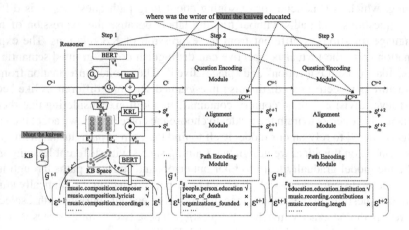

Fig. 1. The framework of our CPR model.

As shown in Fig.1, our CPR model reasoning about the path in a stepwise manner. Only the red relation in the relation set r_{ij} is correct. We will get two matching scores S_{ϕ}^t and S_m^t at each step by using a *Reasoner*. These two scores indicate the matching degree between questions and paths in terms of explicit text semantics and implicit KB structure. The *Reasoner* consists of three modules: question encoding, path encoding, and alignment module. They will be described in detail in the following sections.

3.1 Question Encoding and Path Encoding Module

The question encoding module aims to identify the specific intent of the question at each step. To accurately represent the intent of the question at each step (as shown in Fig.1), we introduce a gate mechanism that allows the model to incorporate historical information to encode specific parts of the question at each step. Specifically, the inputs of the question encoding module are historical information unit C^{t-1} and question sequence Q. The module output C^t refers to the semantic information of question focused at step t. The question encoding vector $V_q^t \in \mathbb{R}^d$ for Q is obtained by an encoder BERT [13]. An updating gate u and a forgetting gate f are used to select the information that should be kept and updated. The module output $C^t \in \mathbb{R}^d$ at the t-th step is:

$$u^t = sigmoid(W_u[V_q^t; C^{t-1}]), \tag{1}$$

$$\tilde{C}^t = tanh(W_{\tilde{C}}[u^t \odot V_q^t; V_q^t]), \tag{2}$$

$$f^t = sigmoid(W_q V_q^t + W_C C^{t-1}), \tag{3}$$

$$C^t = f^t \odot \tilde{C}^t + (1 - f^t) \odot C^{t-1}, \tag{4}$$

where $W_u, W_{\tilde{C}}, W \in \mathbb{R}^d$, are learnable weight matrices. C^0 is randomly initialized.

The path encoding module encodes each one-hop path explicitly and implicitly from the perspective of text and KB structure, respectively. Explicit textual semantic information can effectively distinguish KB relations with large semantic differences. However, it is difficult to distinguish close KB relations such as "country_first_level_divisions" and "country_contains" only by textual information. Such relations can be distinguished using the global structure information of KB. Therefore, we use the knowledge representation learning (KRL) algorithm to learn the structural information of KB to solve this problem implicitly. As shown in Fig. 1, the inputs of this module are the question-specific semantics C^t and the KB subgraph $\mathbb{G}^t$. The outputs are the textual semantic vector $V_{r_{ij}}^t$ and the KB vector E_{ei} and E_{ej}. $\mathbb{G}^t = \{e_i, r_{ij}, e_j\}$ is the one-hop neighbor subgraph of the head entity $e_i \in \xi^{t-1}$. ξ^t is the head entity set at step t, and ξ^0 is the topic entity set. r_{ij} is the KB relation between head entity e_i and tail entity e_j. In the same way as the question encoding process, we encode the token sequence r_{ij} with BERT and obtain the vector $V_{r_{ij}}^t$. In addition, the KRL algorithm TransE [14] is applied to pre-train on our KB to obtain the triple vector $\langle E_{e_i}, E_{r_{ij}}, E_{e_j} \rangle$.

3.2 Alignment Module with Two-Stage Learning Strategy

The question vector and the path vector of the two perspectives are aligned in this module. Specifically, the question-specific semantics C^t and the relation vector $V_{r_{ij}}^t$ are used to compute the textual match score $S_m^t = C^t \cdot V_{r_{ij}}^t$ in each step. The KB structure matching score is calculated from the TransE algorithm: $S_\phi^t = \phi(E_{e_i}^t, E_r^t, E_{e_j}^t)$, where $E_{e_i}, E_{e_j} \in \mathbb{R}^{d_k}$ are the head entity and tail entity vectors derived from the pre-trained TransE. $E_r^t \in \mathbb{R}^{d_k}$ is the vector of C^t in the KB vector space. It is obtained by using a transformation matrix $M_c \in \mathbb{R}^{d \times d_k}$ to transform C^t: $E_r^t = M_c C^t$.

Ideally, our model can distinguish between correct and spurious paths and assign a lower score to spurious paths. For this reason, we train the model by modeling the path matching rather than the entity distribution, which is beneficial to distinguish these paths. Then the training goal of our model is to make the correct path encoding closer to the question encoding. Inspired by the idea of Hard-EM, we utilize the model itself to find the correct path during the iterative training process, and then use the path to train the model. Due to the difficulty of finding the correct path directly, we propose a two-stage learning strategy that first lets the model find the relatively correct paths and then selects the most appropriate path. Before training, we use the depth-first search algorithm to get all paths from the topic entity to the answer entity. We keep only the paths within t-hop, and remove those paths whose F1 value of the answers to the gold answers is lower than the threshold f. The set of preserved paths is P_V.

The First Stage Learning. At this stage, the training goal is to maximize the model probability of those paths that are similar to the semantics of the question. The intuitive idea is to consider the paths with model score top-k in the path set P_V as semantically similar to the question, and the rest as semantically irrelevant paths. However, this sampling method is easy to introduce noise or exclude the correct path. We adopt a more flexible sampling approach, that is nucleus sampling [8]. We sample top-p paths, which refers to the path set P_{top-p} with the sum of probability densities greater than p. This sampling method can more flexibly and accurately select the path similar to the question semantics for training. Specifically, we construct a minimum number of path sets P_t whose probability sum greater than p:

$$P_t = \underset{P_{top-p}}{\arg\min} |P_{top-p}|, where \sum_{t \in P_{top-p}} \rho(t) > p, P_{top-p} \subset P_V, \tag{5}$$

where P_{top-p} represents all the paths set with probability sum is greater than p. $|P_{top-p}|$ denotes the number of paths in P_{top-p}. $\rho(t)$ is the probability of the path obtained from the model. Then the training objective of our model is to maximize the probability sum of the path set P_t, and we use negative log-likelihood to calculate the loss:

$$loss_{s_1} = -log \sum_{p_i \in P_t} \rho(p_i), \tag{6}$$

$$\rho(p) = \frac{(\rho_m(p \,|\, q, \mathbb{G}; \theta) + \rho_\phi(p \,|\, q, \mathbb{G}; \theta))}{2}, \tag{7}$$

where $\rho(p \,|\, q, \mathbb{G}; \theta)$ represents the probability of the path, q is a question, $\mathbb{G}$ is the subgraph obtained by querying from KB with the topic entity E^0 and θ represents our CPR model. The specific calculation is:

$$\rho_m(p \,|\, q, , \mathbb{G}; \theta) = \frac{exp(\sum_{t=0}^{|p_i|} S_m^t)}{\sum_j exp(\sum_{t=0}^{|p_j|} S_m^t)}, \tag{8}$$

$$\rho_\phi(p \,|\, q, ; \mathbb{G}; \theta) = \frac{exp(\sum_{t=0}^{|p_i|} S_\phi^t)}{\sum_j exp(\sum_{t=0}^{|p_j|} S_\phi^t)}, \tag{9}$$

where $|p_j|$ represents the number of hops of the path p_j, and p_j denotes the path in $\mathbb{G}$.

The Second Stage Learning. At this stage, the model aims to find the correct path in P_t that exactly matches the question, rather than spurious paths. We adopt Hard-EM with Uniform Prior algorithm [9] to train the model for this purpose. Specifically, this is an iterative process that uses our model to find the most probable path and then optimize it. We use the model to calculate the path p_k with the highest probability:

$$p_k = \underset{p_i \in P_t}{\arg\max} \rho_m(p_i | q, \mathbb{G}; \theta) \tag{10}$$

Then we use p_k to obtain the best path set p_{best}. For each $p_b \in p_{best}$, the name of p_b is the same as p_k. We still use negative log-likelihood to calculate the loss. Then, the loss function is: $loss_{s_2} = -log \sum_{p_b \in P_{best}} \rho(p_b)$.

4 Experiment

4.1 Experimental Setup

Datasets and Comparison Methods. WebQuestionsSP [3] (**WebQSP**) and ComplexWebQuestons [12] (**CWQ**) are two popular KBQA datasets with maximum 2-hop and 4-hop path questions, respectively. In our experimental setting, we only know the question-answer pairs but not the SPARQL queries. We compare our model with seven methods: STAGG [3], NSM [2], GrafNet [11], PullNet [10], TextRay [6], QGG [7], NSM$_h$ [1].

Implementation Details. For the CWQ dataset, the maximum step of the model and t-hop are both 4, top-p is set to 0.8. For the WebQSP dataset, maximum step and t-hop are both 2, top-p is set to 0.7. We also set a maximum sampling number of 20 for nuclear sampling. If the sampling number reaches the upper limit, sampling will be stopped even if the probability and top-p are not reached. The threshold f is set to 0.1. In addition, We added a special node and relation named $<end>$. They are used to complement paths with less than the maximum step length during the training process.

4.2 Results and Analysis

Table 1. The KBQA results of various methods. Note that NSM$_h^*$ is the result we reproduced with the same linked entities as ours because of the golden topic entities used in the original work.

Models	WebQuestionsSP		ComplexWebQuestons	
	Hit@1	F1	Hit@1	F1
STAGG	-	66.8	-	-
NSM	-	69.0	-	-
GrafNet	66.4	-	32.8	-
PullNet	68.1	-	45.9	-
TextRay	72.2	60.3	40.8	33.9
QGG	-	**74.0**	44.1	40.4
NSM$_h^*$	72.9	69.8	44.9	41.2
Our Model	**74.7**	73.0	**46.7**	**43.5**
-w/o constraints	73.7	72.1	45.3	42.5

Overall Result. As shown in Table 1, our approach shows strong competitiveness on both datasets. Compared to the SOTA method NSM_h in multi-hop KBQA without considering constraints, our model (-w/o constraints) performs better in all results, especially in the F1 score. Compared to the SOTA method QGG in complex KBQA with considering constraints, our model outperforms them in all results, except for the F1 score in WebQSP. It illustrates that our model is better at handling long path questions. However, the query graph approach, represented by QGG on the short path dataset WebQSP, would benefit from a correct selection of linked entities. Overall, the experimental results show a significant improvement in the performance of KBQA by our approach, especially in the long path complex KBQA.

Table 2. Component ablation experiment.

		WebQSP	CWQ
1	*Our model*	**72.1**	**42.5**
2	-w/o S_ϕ^t	71.6	42.1
3	-w/o S_m^t	70.5	40.3
4	-w LSTM	71.5	42.0
5	-w attention	71.9	41.8

Table 3. Learning strategy ablation experiment.

		WebQSP	CWQ
1	*Our model*	**72.1**	**42.5**
2	-w FSL	71.2	41.9
3	-w SSL	70.8	41.0
4	-w/o top-p -w top-k	71.1	41.6

Ablation Study. We conducted ablation studies to explore the contribution of each component of our model and the two-stage learning strategy. The results of the component ablation experiments for the model are shown in Table 2. We can see that both text matching (line 2) and KB structure matching (line 3) contribute to the selection of core paths. The greater contribution of text matching indicates that most KB relations can be distinguished from text semantics. Moreover, the results using LSTM instead of BERT (line 4) demonstrate the promising gains from BERT. The result (line 5) demonstrates the effectiveness of the gate mechanism by comparing a simple attention mechanism, especially in CWQ. The impact of the two-stage learning strategy on KBQA performance is shown in Table 3. Removing either the first-stage learning (FSL) strategy or the second-stage learning (SSL) strategy reduces the performance of the model, indicating a significant contribution of both. It is worth noting that using only FSL is better

Table 4. Long path and spurious path comparison experiments on two datasets. LP refers to long path and the metric is F1 (%). SP refers to spurious path and the metric is Hit@1 (%).

	WebQSP				CWQ				
	LP			SP	LP				SP
	1-hop	2-hop	Full data	Sample data	2-hop	3-hop	4-hop	Full data	Sample data
NSM_h	73.7	59.5	69.8	51.0	48.4	20.6	18.9	41.2	61.5
QGG	**78.7**	**65.0**	**73.1**	94.5	49.2	40.9	32.7	39.8	82.0
Ours	78.5	**65.0**	73.0	**96.0**	**50.1**	**44.0**	**33.5**	**43.5**	**90.5**

than using only SSL, especially on the CWQ. Since the CWQ has more spurious paths, models are difficult to converge with SSL. Besides, we experimented with using top-k sampling instead of our nucleus sampling (line 4). Using top-k sampling will cause apparent performance degradation of our model, which illustrates that top-k sampling has the drawback of introducing more noise paths.

Long Path and Spurious Path Experiments. Since no such evaluation was available previously on these two datasets, we conducted experiments to verify whether our model improves the long path and spurious path problems. We count the number of annotated core paths with different hops for a long path experimental setup. The question of missing annotations and multiple topic entities is discarded because multiple topic entities would make the number of hops difficult to measure. We finally obtain the number of 1-hop 2-hop paths as 1032, 584 on WebQSP and the number of 2-hop, 3-hop and 4-hop paths as 742, 532 and 272 on CWQ. For the spurious path experiment, we randomly selected 200 questions with F1 value of 1 from the predictions for manual evaluation. Because the core paths of a few questions are mismarked and it is reasonable to answer some questions with multiple paths. We replicated NSM_h and QGG for comparison. Their F1 values on the $full\ data$ (Table 4) were 69.8/73.1 on WebQSP and 41.2/39.8 on CWQ, respectively. The results in Table 4 show that our model is superior to the other two methods in most evaluation metrics. The advantage of our model is more obvious on CWQ because CWQ is more complex. The experimental results show that our method effectively improves the long path and spurious path problems. It is worth noting that NSM_h performs poorly on long path data in CWQ but well on the entire test set, implying that NSM_h is better at solving multiple topic entity questions.

5   Conclusion

In this paper, we focus on improving the core path reasoning problem for weakly supervised KBQA. We proposed a CPR model combining explicit text matching and implicit KB structure constraints, which can better align questions and paths. In addition, we proposed a two-stage learning strategy to solve the spurious path problem via an iterative optimization process. The experimental results showed that our model can reason the path more accurately and thus solve the spurious path and long path problem, and has a significant improvement on the performance of KBQA.

Acknowledgement. This work is supported by Natural Science Foundation of China grant (No. U21A20488).

References

1. He, G., Lan, Y., Jiang, J., Zhao, W.X., Wen, J.: Improving multi-hop knowledge base question answering by learning intermediate supervision signals. In: WSDM, pp. 553–561 (2021)
2. Liang, C., Berant, J., Le, Q.V., Forbus, K.D., Lao, N.: Neural symbolic machines: learning semantic parsers on freebase with weak supervision. In: ACL, pp. 23–33 (2017)

3. Yih, W., Richardson, M., Meek, C., Chang, M., Suh, J.: The value of semantic parse labeling for knowledge base question answering. In: ACL (2016)
4. Luo, K., Lin, F., Luo, X., Zhu, K.Q.: Knowledge base question answering via encoding of complex query graphs. In: EMNLP, pp. 2185–2194 (2018)
5. Yu, M., Yin, W., Hasan, K., Santos, C.D., Xiang, B., Zhou, B.: Improved neural relation detection for knowledge base question answering. In: ACL, pp. 571–581 (2017)
6. Bhutani, N., Zheng, X., Jagadish, H.: Learning to answer complex questions over knowledge bases with query composition. In: CIKM, pp. 739–748 (2019)
7. Lan, Y., Jiang, J.: Query graph generation for answering multi-hop complex questions from knowledge bases. In: ACL, pp. 969–974 (2020)
8. Holtzman, A., Buys, J., Forbes, M., Choi, Y.: The curious case of neural text degeneration. In: ICLR (2020)
9. Shen, T., Ott, M., Auli, M., Ranzato, M.: Mixture models for diverse machine translation: tricks of the trade. In: ICML, pp. 5719–5728 (2019)
10. Zhang, L., Winn, J., Tomioka, R.: Gaussian attention model and its application to knowledge base embedding and question answering. ArXiv: 1611.02266 (2016)
11. Zhang, Y., Dai, H., Kozareva, Z., Smola, A., Song, L.: Variational reasoning for question answering with knowledge graph. In: AAAI, pp. 6069–6076 (2018)
12. Talmor, A., Berant, J.: The web as a knowledge-base for answering complex questions. In: NAACL-HLT, pp. 641–651 (2018)
13. Devlin, J., Chang, M., Lee, K., Toutanova, K.: BERT: pre-training of deep bidirectional transformers for language understanding. In: NAACL, pp. 4171–4186 (2019)
14. Bordes, A., Usunier, N., García-Durán, A., Weston, J., Yakhnenko, O.: Translating embeddings for modeling multi-relational data. In: NIPS, pp. 2787–2795 (2013)

Counterfactual-Guided and Curiosity-Driven Multi-hop Reasoning over Knowledge Graph

Dan Shi[1,2], Anchen Li[1,2], and Bo Yang[1,2]([✉])

[1] Key Laboratory of Symbolic Computation and Knowledge Engineering of Ministry of Education, Jilin University, Changchun, China
{shidan19,liac20}@mails.jlu.edu.cn, ybo@jlu.edu.cn
[2] College of Computer Science and Technology, Jilin University, Changchun, China

Abstract. Recently, multi-hop reasoning over incomplete Knowledge Graphs (KGs) to predict missing facts has attracted widespread attention due to its desirable effectiveness and interpretability. It typically adopts the Reinforcement Learning (RL) framework and traverses over the KG to reach the target answer and find evidential paths. However, existing methods often give all reached paths equal hit rewards. Intuitively, not all paths have the same contribution to the proof of the reasoning process. Moreover, the severely sparse rewards obtained after a multi-step traversal are usually insufficient to encourage a sophisticated RL-based model to work well. In order to tackle the above two problems, we propose a novel **Co**unterfactual-guided and **Cu**riosity-driven **K**nowledge **G**raph multi-hop **R**easoning model (CoCuKGR). CoCuKGR constructs counterfactual relation reasoning tasks to estimate the semantic contribution to the query relation of each path and give each arrival path a different soft reward that can distinguish its validity. In addition, our method leverages the curiosity mechanism to generate curiosity-driven intrinsic rewards, which can not only alleviate the reward sparsity issue but also drive the agent to explore the environment more thoroughly to find more abundant paths. Experimental results show that our proposed model outperforms existing multi-hop reasoning methods significantly.

Keywords: Knowledge graph multi-hop reasoning · Counterfactuals · Curiosity mechanism

1 Introduction

Knowledge graphs (KGs), e.g., Yago, NELL and Freebase, represent numerous world knowledge structurally and have been widely adopted in many downstream

Supported by the National Key R&D Program of China under Grant Nos. 2021ZD0112501 and 2021ZD0112502; the National Natural Science Foundation of China under Grant Nos. 62172185 and 61876069; Jilin Province Key Scientific and Technological Research and Development Project under Grant Nos. 20180201067GX and 20180201044GX; and Jilin Province Natural Science Foundation under Grant No. 20200201036JC.

tasks, such as information retrieval, question answering (QA) and recommender systems. However, many knowledge graphs still suffer from serious missing, which becomes a bottleneck hindering their application capabilities.

Recently, extensive knowledge graph embedding (KGE) methods have become prevalent for reasoning on KGs. They map the entities and relations to low-dimensional dense vector space and perform link prediction to complete KGs. These methods can achieve good performance but fail to make interpretations. To solve this problem, some approaches train an agent to search over the KG to perform multi-hop reasoning using the REINFORCE algorithm. They provide not only predicted results but also paths to explain the reasoning process.

Despite the effectiveness of existing multi-hop reasoning models, there are still two problems in terms of the reward: first, most methods give hard hit reward signals to the agent, i.e., 1 for all the paths that reach the target entity and 0 for that do not reach, such as MINERVA [2], RLH [9], and so on. However, intuitively, not all paths have the same validity. For instance, suppose that for the query $(Biden, IsPresidentOf, ?)$, the following two reasoning paths are found:

$$Biden \stackrel{WorkAt}{\rightarrow} TheWhiteHouse \stackrel{LocatedIn}{\rightarrow} Washington \stackrel{IsCityOf}{\rightarrow} USA \text{ (path 1)},$$

$$Biden \stackrel{BornIn}{\rightarrow} Pennsylvania \stackrel{IsStateOf}{\rightarrow} USA \text{ (path 2)}.$$

Both paths reach the correct tail entity USA, but not all people born in Pennsylvania are the president of USA. Clearly, the semantic of path 1 better justifies the reasoning result $(Biden, IsPresidentOf, USA)$ of the query. Second, the agent gains reward only when it successfully reaches the target entity after a multi-step exploration over the incomplete KGs. The rewards are so sparse for the agent that it lacks an effective mechanism to strengthen its policy.

Several approaches have been proposed to give soft rewards. For example, DeepPath [10] employs three criteria to give rewards, including accuracy, efficiency and diversity. However, the criteria require special manual design and there is no guarantee that the paths that meet the criteria are of truly high quality. DIVINE [3] gives the same real-valued reward to all the arrival paths of each query. Therefore, giving soft rewards to arrival paths for distinguishing their validity is still a momentous but unsolved problem.

In this paper, we propose a model named CoCuKGR to enrich the reward reshaping project and further improve the reasoning performance. On the one hand, we first train a relation reasoning component that predicts the relation links based on the set of paths between entity pairs. Then, we adopt the well-trained relation reasoner for performing counterfactual reasoning to measure the semantics of different paths found by the agent and give corresponding soft rewards. On the other hand, we design intrinsic reward signals based on the prediction errors of the agent to compensate for the defect of reward sparsity.

2   Problem Formulation

A *Knowledge Graph (KG)* can be represented as $\mathcal{G} = \{(e_s, r, e_o)\} \subseteq \mathcal{E} \times \mathcal{R} \times \mathcal{E}$, where $\mathcal{E}$ and $\mathcal{R}$ denote entity set and relation set respectively. Given a KG and a query $(e_s, r, ?)$, *Multi-Hop Reasoning over Knowledge Graph* aims to predict the tail entity e_o for the query after traversing over the KG. At the same time, an evidence path $e_s \xrightarrow{r_1} e_2 \xrightarrow{r_2} \cdots e_k \xrightarrow{r_k} e_o$ is provided to prove the reasoning result (e_s, r, e_o). This corresponds to the Horn rule $r(e_s, e_o) \leftarrow r_1(e_s, e_2) \wedge \cdots \wedge r_k(e_k, e_o)$. Note that $r_i(e_i, e_j)$ is equivalent to the fact triple (e_i, r_i, e_j).

3   Methodology

The framework of CoCuKGR is illustrated in Fig. 1. We formulate multi-hop reasoning process over KGs as a Markov Decision Process (MDP) following [2]: for a given query $(e_s, r, ?)$, the agent starts from the source entity e_s and samples an outgoing edge as action according to the policy π_θ at each step. After a traversal with a pre-defined number of steps T, it finally stops at e_T and receives a hit reward if it reaches the ground-truth tail entity e_o:

$$R^H(s_T) = \mathbb{I}\{e_T = e_o\}. \tag{1}$$

We design more fine-grained rewards to guide agents to perform multi-hop reasoning by two modeling advances. First, we introduce the Intrinsic Curiosity Model to generate intrinsic reward r_t^{cu} at each step t. Second, we pack all the generated paths after T steps and construct a counterfactual reasoning task with a trained relation reasoner by removing each path from the path package to obtain the path semantic reward R^{CO}.

3.1   Path Semantic-Aware Relation Reasoner

As mentioned above, the evidence path $c_s \xrightarrow{r_1} c_2 \xrightarrow{r_2} \cdots c_k \xrightarrow{r_k} c_o$ proving the reasoning result (e_s, r, e_o) corresponds to the Horn rule $r(e_s, e_o) \leftarrow r_1(e_s, e_2) \wedge \cdots \wedge r_k(e_k, e_o)$. And if we only consider the sequential relations that have semantic information, we conclude that the path $\xrightarrow{r_1} \xrightarrow{r_2} \cdots \xrightarrow{r_k}$ proving the query relation r corresponds to the Horn rule $r \leftarrow r_1 \wedge \cdots \wedge r_k$. The causal correlation between the rule body $r_1 \wedge \cdots \wedge r_k$ and the rule head r means there is also a causal correlation between the arrival paths and the query relation. We construct a relation reasoner to model such causal correlation.

For each entity pair, we perform N traversals to obtain the paths between them. After that, we pack the paths and encode them with a semantic feature extractor, following [3]. The path representation $\boldsymbol{p}$ is obtained by summing all T relations that constitute it:

$$\boldsymbol{p} = \sum_{r_k \in p} \boldsymbol{r}_k, \tag{2}$$

where each relation r_k is mapped to an embedding vector $r_k \in \mathbb{R}^d$ pre-trained by TransE [1]. We encode the path package μ by concatenating all the embeddings of the paths in it. Next, the path package encoding $\mu \in \mathbb{R}^{nd}$ is fed into a convolutional layer and two fully-connected layers to extract its semantic features:

$$c = W_2 \text{ReLU} \left(W_1 \text{ReLU} \left(Conv \left(\mu, \omega \right) \right) \right). \tag{3}$$

We train the relation reasoner with a positive instance set Γ^+ and a negative instance set Γ^-. If an entity pair and the relation r constitute an observed fact in KG, we regard the entity pair as a positive entity pair of r. A $\gamma = (\mu, r) \in \Gamma^+$ indicates a positive instance composed of a path package μ between a positive entity pair and the relation r. On the contrary, for a negative instance, the path package μ is between a negative entity pair. For each instance, we use the dot product of the path package representation c and the relation representation r to extract the correlation, and then adopt *sigmoid* to obtain the score t_μ:

$$t_\mu = sigmoid \left(c \cdot r \right). \tag{4}$$

The objective function for a single instance is given by the cross-entropy loss:

$$\mathcal{L}_\gamma = y_\gamma \Delta \log t_\mu + (1 - y_\gamma) \Delta \log (1 - t_\mu) + \rho ||\Theta||^2, \tag{5}$$

where ρ controls the regularization strength to prevent overfitting, y_γ is 1 if $\gamma \in \Gamma^+$, otherwise is 0.

During training, we aim to minimize the total loss given by:

$$\mathcal{L} = \frac{1}{|\Gamma^+| + |\Gamma^-|} \sum_{\gamma \in (\Gamma^+ \cup \Gamma^-)} \mathcal{L}_\gamma. \tag{6}$$

3.2   Construct Counterfactuals to Give Soft Rewards

Counterfactuals describe potential consequents caused by the actions or circumstances that counter the facts and capture their causal effects. To quantify the semantic contribution of each path, we propose a counterfactual question: "What is the impact on the relation reasoning task if a particular path is not available?"

Specifically, for a certain query $(e_s, r, ?)$, we feed all the paths in the path set μ searched in KG by the policy-based agent to the relation reasoner to get its score t_μ. Then each path p is removed from the path set in turn and the changed path set is fed into the relation reasoner to get another score $t_{\mu-p}$, and the soft reward of the path p is calculated based on the score difference as follows:

$$R^{CO} = t_\mu - t_{\mu-p}. \tag{7}$$

If the path p is important to the query relation, the removing of p will lead to a degraded reasoning performance. So the result of Eq. (7) will be positive. Conversely, if p is not important or even noisy, the result will be negative.

3.3 Intrinsic Curiosity Reward

Inspired by [6], we introduce intrinsic curiosity reward signals that drive the agent to explore its environment (i.e., KGs) more thoroughly to gain more knowledge. Unlike previous work that applied curiosity on a game task where the state of the MDP is image, the state in our task is the position where the agent stays on the KG at a certain step. The state prediction occurs on the vector space rather than on the pixel space. To suit our task, we remove the inverse dynamics model in the Intrinsic Curiosity Module (ICM). We train a feedforward neural network that predicts the next state $\hat{s}_{t+1}$ based on the current state s_t and the action a_t, which is optimized by minimizing the loss function L_{CU}:

$$L_{CU} = \frac{1}{2} \left\| \hat{s}_{t+1} - s_{t+1} \right\|_2^2 . \tag{8}$$

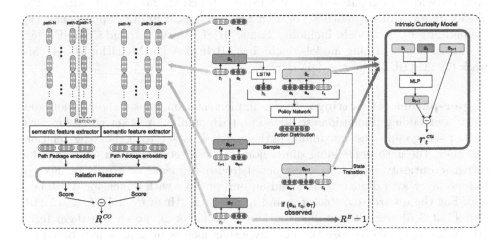

Fig. 1. The overall architecture of CoCuKGR.

The intrinsic curiosity reward signal of a single step r_t^{cu} is computed as:

$$r_t^{cu} = \frac{\eta}{2} \left\| \hat{s}_{t+1} - s_{t+1} \right\|_2^2 , \tag{9}$$

where $\eta > 0$ is a scaling factor.

3.4 Optimization and Training

The overall reward obtained for each traversal is a composition of Eqs. (1), (7) and (9) that can be written as:

$$R = R^H + \alpha R^{CO} + \sum_{t=1}^{T} r_t^{cu}, \tag{10}$$

where $\alpha > 0$ is a scalar controlling the proportion of the counterfactual reward.

We employ REINFORCE algorithm to train the policy network of the agent by maximizing the expected cumulative reward. The overall optimization problem is composed of the policy gradient loss and the ICM loss:

$$arg\min_{\theta,\theta_{CU}}\left[-\mathbb{E}_{(e_s,r,e_o)\in KG}\mathbb{E}_{p_1,p_2,\cdots,p_N\sim\pi_\theta}\left[\sum_{n=1}^{N}R_n|(e_s,r)\right]+\beta L_{CU}\right], \qquad (11)$$

where $\beta > 0$ is a scalar that weighs the importance of training the agent against learning the Intrinsic Curiosity Module.

4   Experiments

Datasets and Baselines. We evaluate the effectiveness of our proposed model on three widely used datasets: (1) WN18RR, (2) FB15K-237, and (3) NELL-995. Two categories of approaches are used for comparison in our experiments: (1) Embedding-based models, including TransE [1], DistMult [11] and ComplEX [8]. (2) Multi-hop reasoning models, including MINERVA [2], MultiHopKG [4], M-walk [7], AnyBURL [5] and DIVINE [3].

Hyper-parameter Settings. In our implementation, we set the relation and entity embedding dimensions to 100. The path number N is set to 20 for each path package and the maximum path length T is set to 3. For the relation reasoner, the path embedding dimension d is also set to 100. In addition, the intrinsic curiosity module is constructed from a sequence of a fully connected hidden layer with dimension 512 and an output layer with dimension $3 * d$, i.e., 300. For the reward scale factor α and η, we choose them from $\{1, 5, 10, 15, 20\}$ and $\{1, 2, 3, 5\}$ respectively. For the loss scale factor β, we choose them from $\{0.1, 0.2, 0.5, 1\}$. We choose the best hyperparameters by grid search based on Hits@10 on validation sets. We use Adam optimization to train the agent and the other two modules.

Table 1. The overall performance comparison results of entity prediction.

Models/Datasets	WN18RR			FB15K-237			NELL-995		
	@1	@10	MRR	@1	@10	MRR	@1	@10	MRR
TransE	28.9	56.0	35.9	24.8	45.0	36.1	51.4	75.1	45.6
DistMult	35.7	38.4	36.7	32.4	60.0	41.7	55.2	78.3	64.1
ComplEx	41.5	46.9	43.4	33.7	62.4	43.2	63.9	84.8	72.1
MINERVA	41.3	51.3	44.8	21.7	45.6	29.3	66.3	83.1	72.5
MultiHop (ConvE)	41.4	51.7	44.8	32.7	56.4	40.7	65.6	84.4	72.7
MultiHop (ComplEx)	42.5	52.6	46.1	32.9	54.4	39.3	64.4	81.6	71.2
M-walk	41.4	–	43.7	16.5	–	23.2	68.4	–	75.4
AnyBURL	43.0	52.7	46.9	23.3	48.6	⩾31.0	44.0	57.0	–
DIVINE	–	–	–	22.3	–	29.6	66.8	–	73.1
CoCuKGR	44.2	53.8	46.9	34.3	58.0	43.4	66.7	85.4	74.0

Table 2. Ablation studies results on WN18RR.

Models	Hit@1	MRR
CoCuKGR	**44.2**	**46.9**
-CU	43.4	46.0
-CO	42.0	45.4

Table 3. Performance analysis of different hyperparameter α (MAP scores).

Tasks/α	1	5	10	15	20
AthletePlaysForTeam	82.3	83.4	**83.8**	83.5	83.2
WorksFor	82.3	82.8	83.5	**83.7**	82.9

Entity Prediction Results. Table 1 shows the evaluation results on the tail entity prediction tasks. It can be seen that our framework outperforms other multi-hop reasoning baselines on WN18RR and FB15K-237 and meanwhile achieves competitive results on NELL-995. In particular, there is a significant performance improvement of our model on FB15K-237. Upon further analysis, there can be two main possible reasons. First, FB15K-237 is closer to real-world scenarios, thus there are more cases where the semantics of the paths and the reasoning relations do not match very well. We identify this difference and use it to further guide the agent, which can strengthen the pathfinding decision of the agent. Moreover, in FB15K-237, the action space is much larger. It's difficult for the baselines to search the huge action space thoroughly. Nevertheless, our curiosity module can tackle this problem to some extent.

Ablation Studies. To investigate the contribution of each proposed component to the model performance, we separately remove the intrinsic curiosity reward module (-CU) or the counterfactual relation reasoning reward module (-CO). As shown in Table 2, removing each component results in a performance drop.

Analysis of Different α. We also adjust α that weights the scale of the counterfactual soft reward on two relation prediction tasks and report in Table 3.

Case Study. In addition, to analyze the utility of the counterfactual relation reasoning, we present the paths between the entity pairs of a query relation found by CoCuKGR and their counterfactual rewards. From Table 4, we observe that the rewards are indeed higher for the paths that are more semantically relevant to the query relation in accordance with human perceptions. Using the soft rewards as confidence, we can further interpret the reasoning process in a concrete way.

Table 4. Two query cases where the counterfactual soft rewards of paths are given.

Path ID	Query relation 1: AthletePlaysForTeam	Reword
1	$AthleteLeadSportsTeam \rightarrow$ $TeamPlaysAgainstTeam^{-1}$ $\rightarrow TeamPlaysAgainstTeam$	0.0352
2	$AthleteHomeStadium \rightarrow StadiumLocatedInCity$ $\rightarrow TeamPlaysInCity^{-1}$	0.0084
Path ID	Query relation 1: WorksFor	Reword
1	$PersonLeadsOrganization \rightarrow LOOP \rightarrow LOOP$	0.0440
2	$PersonBelongsToOrganization \rightarrow$ $AgentCollaboratesWithAgent$ $\rightarrow AgentBelongsToOrganization$	0.0214
3	$AgentBelongsToOrganization \rightarrow$ $AgentActsInLocation$ $\rightarrow AgentCompetesWithAgent$	−0.0267

5   Conclusion

In this paper, we proposed a novel model called CoCuKGR to solve the problems of not distinguishing the contributions of the reasoning paths and the sparse signal of hit reward during multi-hop knowledge graph reasoning. We construct a counterfactual relation reasoning task to give different soft rewards to all the paths. Moreover, we introduce the curiosity mechanism to generate intrinsic rewards to alleviate the reward sparsity problem. Experimental results demonstrate that our approach improved the performance of the state-of-the-art multi-hop reasoning models on three benchmark KGs. In future work, we plan to employ curiosity rewards to prune the action search space of the agent.

References

1. Bordes, A., Usunier, N., Garcia-Duran, A., Weston, J., Yakhnenko, O.: Translating embeddings for modeling multi-relational data. In: Advances in Neural Information Processing Systems, vol. 26 (2013)
2. Das, R., et al.: Go for a walk and arrive at the answer: Reasoning over paths in knowledge bases using reinforcement learning. arXiv preprint arXiv:1711.05851 (2017)
3. Li, R., Cheng, X.: Divine: a generative adversarial imitation learning framework for knowledge graph reasoning. In: Proceedings of the 2019 Conference on Empirical Methods in Natural Language Processing and the 9th International Joint Conference on Natural Language Processing (EMNLP-IJCNLP), pp. 2642–2651 (2019)
4. Lin, X.V., Socher, R., Xiong, C.: Multi-hop knowledge graph reasoning with reward shaping. arXiv preprint arXiv:1808.10568 (2018)
5. Meilicke, C., Chekol, M.W., Ruffinelli, D., Stuckenschmidt, H.: Anytime bottom-up rule learning for knowledge graph completion. In: IJCAI, pp. 3137–3143 (2019)

6. Pathak, D., Agrawal, P., Efros, A.A., Darrell, T.: Curiosity-driven exploration by self-supervised prediction. In: International Conference on Machine Learning. PMLR, pp. 2778–2787 (2017)
7. Shen, Y., Chen, J., Huang, P.-S., Guo, Y., Gao, J.: M-walk: learning to walk over graphs using monte carlo tree search. arXiv preprint arXiv:1802.04394 (2018)
8. Trouillon, T.P., Bouchard, G.M.: Complex embeddings for simple link prediction, November 23 2017. US Patent App. 15/156,849
9. Wan, G., Pan, S., Gong, C., Zhou, C., Haffari, G.: Reasoning like human: hierarchical reinforcement learning for knowledge graph reasoning. In: IJCAI, pp. 1926–1932 (2020)
10. Xiong, W., Hoang, T., Wang, W.Y.: Deeppath: a reinforcement learning method for knowledge graph reasoning. arXiv preprint arXiv:1707.06690 (2017)
11. Yang, B., Yih, W., He, X., Gao, J., Deng, L.: Embedding entities and relations for learning and inference in knowledge bases. arXiv preprint arXiv:1412.6575 (2014)

Visualizable or Non-visualizable? Exploring the Visualizability of Concepts in Multi-modal Knowledge Graph

Xueyao Jiang[1], Ailisi Li[1], Jiaqing Liang[1], Bang Liu[2], Rui Xie[3], Wei Wu[3], Zhixu Li[1(✉)], and Yanghua Xiao[1,4(✉)]

[1] Shanghai Key Laboratory of Data Science, School of Computer Science, Fudan University, Shanghai, China
{xueyaojiang19,alsli19,zhixuli,shawyh}@fudan.edu.cn
[2] Mila and DIRO, Université de Montréal, Montréal, Québec, Canada
bang.liu@umontreal.ca
[3] Meituan, Shanghai, China
rui.xie@meituan.com
[4] Fudan-Aishu Cognitive Intelligence Joint Research Center, Shanghai, China

Abstract. An important task in image-based Multi-modal Knowledge Graph construction is grounding concepts to their corresponding images. However, existing research omits the intrinsic properties of different concepts. Specifically, there are some concepts that can not be characterized visually, such as *mind, texture, session cookie* and so on. In this work, we define concepts like these as non-visualizable concepts (NVC) and the others like *dog* that have clear and specific visual representations as visualizable concepts (VC). And, we propose a new task of distinguishing VCs from NVCs, which has rarely been tackled by the existing efforts. To address this problem, we propose a multi-modal classification model combining concept-related features from both texts and images. Due to the lack of enough training samples especially for NVC, we select concepts in ImageNet as the instances for VC, and also propose a webly-supervised method to get a small set of instances for NVC. Based on the small training set, we modify the basic two-step positive-unlabeled learning strategy to train the model. Extensive evaluations demonstrate that our model significantly outperforms a variety of baseline approaches.

Keywords: Visualizable concept · Multi-modal knowledge graph

1 Introduction

Nowadays, multi-modal data (mainly images) is introduced into Knowledge Graph to enrich the representation of concepts, and increasing efforts are focused on grounding entities or concepts with their corresponding images to construct image-based Multi-modal Knowledge Graph (MMKG) [6–10]. However, not all concepts in Knowledge Graphs can be characterized accurately using images, such as *mind*. Although we may associate it with images of *brain*, as shown in

A. Bhattacharya et al. (Eds.): DASFAA 2022, LNCS 13245, pp. 180–187, 2022.
https://doi.org/10.1007/978-3-031-00123-9_14

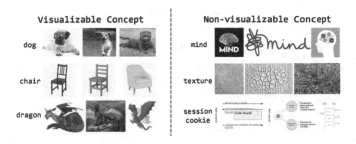

Fig. 1. Examples of visualizable concepts and non-visualizable concepts.

Fig. 1, what these images depicted are not exactly *mind* itself but something relevant with it.

Thus, we propose to classify concepts by whether they can be characterized visually. We name concepts that do not have a clear visual representation and can not be characterized visually as **Non-visualizable Concepts (NVCs)**. Oppositely, concepts that have clear and specific visual representations are called as **Visualizable Concepts (VCs)**, such as *dog*.

Most previous MMKG construction efforts focused on grounding entities or concepts with their corresponding images without carefully differentiating between VCs and NVCs. For example, TinyImage [13] (which is a MMKG built on WordNet [4]) simply eliminated all abstract concepts in the hierarchical taxonomy of WordNet while it is rough and inaccurate according to our experiments in Sect. 3. [14] analyzed the problem of non-visualizable concepts in the *person* subtree of ImageNet, but they rely on crowdsourcing to annotate the image-ability of synsets which requires a lot of manpower and material resources and is difficult to apply to large-scale MMKG. Several previous research has also addressed this problem implicitly in a learning-based way. [3] proposed to filter out non-visualizable concepts based on the visually salient score of concepts. However, they merely use webly searched images to evaluate the visually salient score, which easily suffers from the noise or bias of web data.

In this paper, we explore the visualizability of concepts by classifying VCs versus NVCs. Particularly, we design a visualizable concept classifier with multi-modal information as features, namely text and images. The classifier takes the concept description and online images of the concept (collect from the image search engine) as input and then output whether the concept is a VC or NVC. Besides, our classifier still faces the following challenges:

1) **Lack of labeled data.** We propose to automatically construct a partially annotated dataset by contrasting a symbolic knowledge base (i.e. WordNet) and an annotated image dataset organized according to this knowledge base (i.e. ImageNet [1]).

2) **Learning under PU setting.** The partially annotated dataset contains a small set of positive data and a large set of unlabeled data. Due to the severe imbalance between the number of positive data and the number of unlabeled data, we adopt the two-step PU Learning technique to tackle this problem.

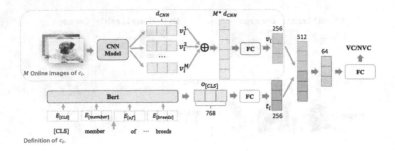

Fig. 2. Overview architecture of visualizable concept classifier.

2 Methodology

In this paper, we model this problem as a binary classification problem and propose a multi-modal visualizable concept classification model to solve it, and the Two-step PU Learning strategy is leveraged to tackle the PU setting challenge in the dataset.

2.1 Multi-modal Visualizable Concept Classifier

As shown in Fig. 2, we leverage two separate streams to get the text representation and visual representation, respectively. The representations of two modalities are further concatenated together to get the fusion of two features, which are fed into a binary classification network to classify VC and NVC.

- **Text embedding.** We use BERT to get the text embedding. The concept definition d_i is first processed into the input format of BERT as: "[CLS] d_i [SEP]" and then fed into BERT. The embedding of mark "[CLS]" is then further encoded into a 256-dim vector $t_i \in \mathbb{R}^{256}$.
- **Image embedding.** Several pretrained image classification models are leveraged to get the image representations. We input M images into the pretrained image classification model respectively, and get M feature vectors $v_i^j \in \mathbb{R}^{d_{CNN}}$ ($1 \leq j \leq M$) which is the output of the final pooling layer. d_{CNN} denotes the dimension of CNN model's average pooling layer and differs by the CNN models. These M vectors are concatenated together and fed into a fully-connected layer to get the final visual feature vector $v_i \in \mathbb{R}^{256}$.
- **Classifier.** The embeddings of text and image are then concatenated into a 512-dim vector and then fed into a binary classifier (contains a 64-dim hidden layer and an output layer) to generate the probability that the concept is a VC or not.

2.2 Training Under PU Setting

Due to the small amount of positive data and lack of negative data, we leverage Two-step PU Learning strategy to train the Visualizable Concept Classifier. As

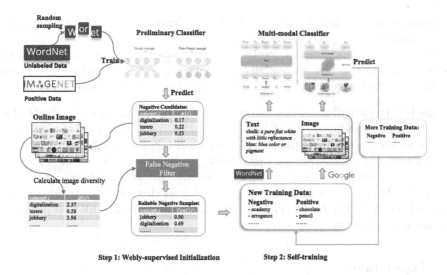

Fig. 3. Overview of the framework.

shown in Fig. 3, to complement the training dataset for binary classification, we design a Webly-supervised Initialization step that screens out a reliable negative set with the same size as that of the positive set from unlabeled data and then train the Multi-modal classifier iteratively with the self-training strategy.

2.2.1 Webly-Supervised Initialization

We propose a webly-supervised automatic method to construct a high quality negative set without manual labeling. It includes two steps: 1) Preliminary Classifier; 2) False Negative Filter.

1) Preliminary Classifier. Some unlabeled data is randomly sampled out as pseudo negatives. Together with the labeled positive data, they are used to train a preliminary classifier of VC and NVC. And then the preliminary classifier is used to predict the unlabeled data and output the confidence score of a concept c to be NVC (denoted by $p(c)$).

2) False Negative Filter. This filter is designed based on the following assumption: *Concepts with diverse online images are more likely to be non-visualizable.* We measure the image diversity of concept c by calculating the standard deviation of corresponding images' feature vectors (denoted by $D(c)$), then we recalculate the confidence score of a concept c to be NVC as $Conf(c) = softmax(1 - p(c)) + softmax(D(c))$.

2.2.2 Self-training

After getting the reliable negative data, we iteratively train our multi-modal classifier. In each iteration, we train the classifier and then use it to predict unlabeled data. Predicted data with high confidence score will be added to the training set to serve next iteration's training.

However, noise is unavoidable by adopting automated approach to sample negative data in the first step of PU Learning. Self training under such setting will face the challenge of label drift. We design two small tricks to avoid it:

- **randomly sample.** Taking sampling negative data as example, instead of using top k candidates as the pseudo labeled data, we randomly sample k candidates according to their confidence scores. For example, the concept c's confident score is 0.8, then its probability of being selected is 0.8.
- **extra visual information.** We reuse the priori knowledge that non-visualizable concepts' online images have higher diversity which can be measured by their standard deviation. The diversity score is added to predicted confidence score.

3   Experiment

3.1   Datasets and Settings

Dataset. We conduct our experiments on WordNet noun set. And we use ILSVRC dataset[1] to label positive concepts and remain the reset in WordNet unlabeled. For each concept, we collect two modalities of data. One is definition text in WordNet, the other is online images which are retrieved from Google search engine. The search query is in the form of "c d". c is the name of a concept and d is the definition from WordNet to disambiguate concepts with the same name. We manually labeled 600 concepts randomly extracted from WordNet as test set including 322 positive samples and 278 negative samples.

Experimental Settings. We apply BERT [2] to gain the text features. We conduct experiments to extract image features with three different CNN models including InceptionV3 [12], Resnet50 [5] and VGG16 [11]. During training, batch size is set to 64 and the learning rate adjustment schedule is set as [2] in each iteration while the initial learning rate of each iteration is set to 1e−3, 1e−4, 1e−6, 1e−8.

3.2   Main Results

We are the first to explicitly propose to distinguish visualizable concepts and non-visualizable concepts automatically, so there are few work that we can compare to. We compare our method with 3 baselines:

Full Set. As most previous work of the MMKG construction ignore to distinguish VC and NVC, we design a baseline approach that regards all concepts as VC.

---

[1] http://image-net.org/challenges/LSVRC/2012/browse-synsets.

Table 1. Comparison result

Model	acc	rec_{VC}	$prec_{VC}$	rec_{NVC}	$prec_{NVC}$
Full set	0.537	**1.0**	0.537	0	–
TinyImage	0.605	0.491	0.684	0.737	0.556
LEVAN	0.542	0.991	0.538	0.007	0.250
Ours	**0.828**	0.84	**0.830**	**0.810**	**0.820**

TinyImage. [13] TinyImage is a MMKG that is constructed based on WordNet hierachy. It simply regards all abstract concepts as NVCs and the others as VCs.

LEVAN. [3] LEVAN proposed to filter out NVC based on the visual salience score of concepts. Specifically, they train a binary image classifier for each concept on a dataset in which online images of the concept are labeled positive and background images are labeled negative. They regard a concept as visual salient if the well trained classifier can reach a threshold of accuracy on the validation set that has the same setting with the training set.

We compare our method with 3 baselines mentioned above. As shown in Table 1, our framework outperforms the other 3 methods in acc, $prec_{VC}$, rec_{NVC} and $prec_{NVC}$. The reason of *Full set* method has the highest recall of VC is that it regards all concepts as VC. Besides, it predicts no NVC, so the precision of NVC is not calculable. We finally use the optimal model to predict the whole unlabeled set and get 35,481 NVCs and 37,702 VCs.

3.3 Ablation Study

In this section, we provide ablation studies on Webly-supervised Initialization, multi-modal classification model and PU Learning.

Webly-Supervised Initialization. We conduct ablation experiments for both two submodules in the Webly-supervised Initialization step (step 1) by removing one of these two submodules. Besides, as for Preliminary Classifier, we test two classification models: **text based model (TM)** and **multi-modal model (MM)**. The MM is same as depicted in Sect. 2.1. TM refers to the model that removes the image embedding structure (depicted in the dotted box in Fig. 2) from MM. In conclusion, we design five experiments: (1) only FF; (2) MM w/o FF; (3) MM with FF; (4) TM w/o FF; (5) TM with FF.

To measure the quality of the datasets that outputed by above 5 combinations, we train the classifier on them and evaluate the accuracy on validation set. The experiment results are given in Table 2 (step 1). The combination of text based model (TM) and Flase Negative Filter results in a best negative set on which the multi-modal classifier is trained to achieve a highest accuracy of **0.78**. As shown, False Negative Filter brings around 7% absolute improvement.

Multi-modal Classification. For Self-training step (step 2), we also conduct experiments on TM and MM. As is shown in Table 2 (step 2), the training of

Table 2. Ablation experiments of Webly-supervised initialization

Test step	Model	False negative filter	Val accuracy
Step 1	–	✓	0.65
	MM	×	0.67
	MM	✓	0.65
	TM	×	0.71
	TM	✓	**0.78**
Step 2	+ TM	✓	0.77
	+ MM(InceptionV3)	✓	0.80
	+ MM(ResNet50)	✓	0.80
	+ MM(Vgg16)	✓	**0.83**

classifier benefits a lot from multi-modal information. The removal of multi-modal information in self-training leads to a drop of around 12% in accuracy after 4 iterations. As for influence of different CNN models, the accuracy of these three models are respectively 0.797(Inception V3), 0.804(ResNet50) and **0.828(VGG16)**, in which VGG16-based model achieves the best result.

Two-Step PU Learning. We compare the performance of our framework with classifiers directly trained on dataset constituting of randomly sampled negative data and labeled positive data. We trained the classifier 5 times and the average accuracy of these classifiers is 0.683, while our framework can reach an accuracy of **0.828**, which is approximately 7% higher.

4   Related Work

The visualization of concepts is not a new topic in computer vision. LEVAN [3] proposed to recognize "visual salient" words during constructing image dataset. In this work, the authors believed that images of a visual salient ngrams can be easily distinguished from background images and have small inter-class variances. A classifier based on SVM is trained to distinguish online images of a ngram and background images. The classifier's accuracy of VC should exceed a threshold. TinyImage [13] is a MMKG constructed based on WordNet and contains 75k noun concepts with 1,052 images per concept in average. The authors regarded all the abstract concepts as not proper to be matched with images and removed them by dropping all the hyponyms of the word "abstraction", while according to our experiments, such strategy is rough and inaccurate.

5   Conclusion

In this work, we propose a new task: *distinguishing visualizable concepts from non-visualizable concepts* and model this problem as a binary classification problem. We automatically generate a partially labeled dataset and propose a novel two-step PU learning framework to train the classifier on such dataset. Besides, multi-modal information of concepts is used to enhance the performance of the visualizable concept classifier. Extensive experimental results show that our solution achieves the state-of-the-art results compared to several baselines.

Acknowledgement. This work is supported by National Key Research and Development Project (No. 2020AAA0109302), Shanghai Science and Technology Innovation Action Plan (No. 19511120400), Shanghai Municipal Science and Technology Major Project (No. 2021SHZDZX0103) and National Natural Science Foundation of China (Grant No. 62072323).

References

1. Deng, J., Dong, W., Socher, R., Li, L.J., Li, K., Fei-Fei, L.: Imagenet: a large-scale hierarchical image database. In: 2009 IEEE Conference on Computer Vision and Pattern Recognition, pp. 248–255. IEEE (2009)
2. Devlin, J., Chang, M.W., Lee, K., Toutanova, K.: Bert: pre-training of deep bidirectional transformers for language understanding. arXiv preprint arXiv:1810.04805 (2018)
3. Divvala, S.K., Farhadi, A., Guestrin, C.: Learning everything about anything: Webly-supervised visual concept learning. In: Proceedings of the IEEE Conference on Computer Vision and Pattern Recognition, pp. 3270–3277 (2014)
4. Fellbaum, C.: Wordnet. The encyclopedia of applied linguistics (2012)
5. He, K., Zhang, X., Ren, S., Sun, J.: Deep residual learning for image recognition. In: Proceedings of the IEEE Conference on Computer Vision and Pattern Recognition, pp. 770–778 (2016)
6. Krishna, R., et al.: Visual genome: connecting language and vision using crowd-sourced dense image annotations. Int. J. Comput. Vis. **123**(1), 32–73 (2017)
7. Li, M., et al.: Gaia: a fine-grained multimedia knowledge extraction system. In: Proceedings of the 58th Annual Meeting of the Association for Computational Linguistics: System Demonstrations, pp. 77–86 (2020)
8. Mitchell, T., Fredkin, E.: Never ending language learning. In: 2014 IEEE International Conference on Big Data (Big Data), p. 1 (2014)
9. Perona, P.: Vision of a visipedia. Proc. IEEE **98**, 1526–1534 (2010)
10. Russell, B.C., Torralba, A., Murphy, K.P., Freeman, W.T.: Labelme: a database and web-based tool for image annotation. Int. J. Comput. Vis. **77**(1–3), 157–173 (2008)
11. Simonyan, K., Zisserman, A.: Very deep convolutional networks for large-scale image recognition. arXiv preprint arXiv:1409.1556 (2014)
12. Szegedy, C., Vanhoucke, V., Ioffe, S., Shlens, J., Wojna, Z.: Rethinking the inception architecture for computer vision. In: Proceedings of the IEEE Conference on Computer Vision and Pattern Recognition, pp. 2818–2826 (2016)
13. Torralba, A., Fergus, R., Freeman, W.T.: 80 million tiny images: a large data set for nonparametric object and scene recognition. IEEE Trans. Pattern Anal. Mach. Intell. **30**(11), 1958–1970 (2008)
14. Yang, K., Qinami, K., Fei-Fei, L., Deng, J., Russakovsky, O.: Towards fairer datasets: filtering and balancing the distribution of the people subtree in the imagenet hierarchy. In: Proceedings of the 2020 Conference on Fairness, Accountability, and Transparency, pp. 547–558 (2020)

Spatio-Temporal Data

JS-STDGN: A Spatial-Temporal Dynamic Graph Network Using JS-Graph for Traffic Prediction

Pengfei Li, Junhua Fang$^{(\boxtimes)}$, Pingfu Chao, Pengpeng Zhao, An Liu, and Lei Zhao

Department of Computer Science and Technology,
Soochow University, Suzhou, China
20205227067@stu.suda.edu.cn,
{jhfang,pfchao,ppzhao,anliu,zhaol}@suda.edu.cn

Abstract. Traffic prediction is a fundamental operation in real-time traffic analysis. A precise prediction of traffic condition can benefit both road users and traffic management agencies. However, since road traffic is decided by multiple static and dynamic factors, traffic prediction is still a challenging task. As the core indicator of traffic condition, many works focus on traffic speed prediction using time-series forecasting approaches. Although current methods take into account the static road topology while modelling, they fail to consider (1) the semantic closeness between road components and (2) congestion caused by upstream/downstream traffic propagation. In this paper, we introduce a Spatial-Temporal Dynamic Graph Network using JS-Graph, which considers both static road features and dynamic traffic flows when forecasting. Specifically, we first propose a data-driven 'JS-Graph' method that describes the semantic similarity between road nodes. It models the complex spatial correlations that cannot be captured by the traditional spatial adjacency graph. Secondly, we design a dynamic graph attention network that considers the traffic dynamics that happened in previous time slices when predicting the current one to capture the congestion propagation phenomena. Extensive experiments conducted on real-world datasets show that our proposed method is significantly better than baselines.

Keywords: Graph neural network · Spatial-temporal data analysis · Time series forecast

1 Introduction

In recent years, many cities have been building Intelligent Traffic Systems (ITS) to meet the increasing demand for the fast-paced lifestyle. As the foundation of ITS, an accurate monitoring/prediction of traffic condition is necessary and beneficial to most transportation applications. However, traffic prediction is still

A. Bhattacharya et al. (Eds.): DASFAA 2022, LNCS 13245, pp. 191–206, 2022.
https://doi.org/10.1007/978-3-031-00123-9_15

a challenging task due to its complex spatial-temporal correlation. On the one hand, the transportation network is complicated. The relationship between nodes in the traffic network includes not only simple geographical factors, but also social factors such as road grade, regional function and user driving behaviour. On the other hand, the traffic condition is also susceptible to various real-time events. For example, sudden weather changes or accidents can cause local traffic jam immediately, and the congestion may quickly propagate to nearby areas along the network. Therefore, modelling these spatial-temporal correlations is the key to the traffic prediction.

Recently, the success of Graph Convolutional Networks (GCN) [8] has inspired researchers to model the spatial-temporal correlation in traffic network as a graph, whose nodes represent road intersections or traffic sensors while their spatial connectivity are denoted as edges. Besides, by combining GCN with RNN [1,10,18], CNN [5,14,16], or Attention [15,19], existing works greatly improve the accuracy of traffic speed prediction over traditional statistical methods, thank to the graph model.

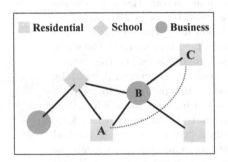

Fig. 1. Correlations between areas with similar functions in the traffic network

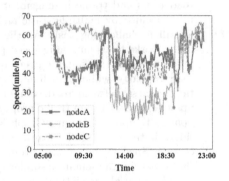

Fig. 2. Nodes that are not adjacent have similar traffic pattern.

However, current use of the spatial adjacency graph in GCN still has some shortcomings. The first disadvantage is that the graph only describes the topology of a road network from a geographical point of view. In fact, each road node is semantically abundant, and nodes with similar functions usually have identical traffic patterns. As shown in Figs. 1 and 2, although node B is geographically closer to node A compared with node C, since it is in a commercial area whereas nodes A and C are both in residential districts, the speed profiles of A and C are more similar than B, especially in the early morning and afternoon. Such semantic adjacency is not captured by any of existing spatial adjacency graph. Some work has improved the adjacency graph using adaptive matrices, which are learnable, but it loses interpretability and is easy to get over-fitting when the data is noisy.

The second disadvantage is that the current modelling of the dynamic characteristics of traffic topology is inadequate. The predefined spatial adjacency matrix and the adaptive adjacency matrix are static and not sufficient to reflect the dynamic changes on road network. Most of the existing dynamic graph modelling only considers the spatial correlation at the current moment. However, the real-world traffic conditions are propagated along the network progressively. As is shown in Fig. 3, P and Q are two sensors in a traffic network. At 9:00AM, there is a traffic jam at node P (P_1). However, Q will not get stuck until 9:10AM when the congestion finally propagates to Q_3. Previous dynamic modelling can not deal with the direct relationship between P_1 and Q_3 well, and it inspires us to consider the traffic lag when modelling dynamic spatial correlation.

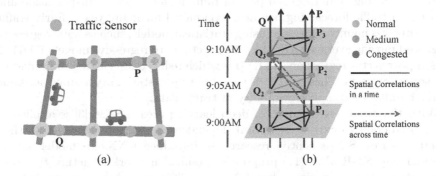

Fig. 3. Example of spatial correlations among different time. (a) Sensors in a road network. (b) Different spatial correlations. Previous work capture dynamic spatial features at every timestamp. Cross-time dynamic spatial correlation which take all nodes at all previous timestamps into account can learn the changing inter-relationship between different nodes more comprehensively.

To solve the aforementioned weaknesses, we propose a new model named Spatial-Temporal Dynamic Graph Network using JS-Graph (JS-STDGN). In our model, we first present a data driven graph construction method which captures the JS-divergence of node data distribution. Instead of using node semantic properties, like POI or road condition information, our new JS-graph is built based on node traffic history, which can better identify similar nodes with respect to their speed profiles. Secondly, we designed a dynamic graph attention network. As is shown in Fig. 3(b), our method takes the spatial network state from historical time slices into consideration to capture the cross-time dynamic spatial correlation. In addition, we follow the prediction architecture in GMAN [19] to complete the traffic speed prediction. Overall, the contributions of our work can be summarised as follows:

- We propose a new model named JS-STDGN to predict traffic speed. The model encodes both the static and dynamic traffic feature into the graph structure for better prediction accuracy.

- We propose a new graph construction method, which leverages the pairwise JS-divergence of nodes speed profiles to generate spatial adjacency graph. Compared with the traditional graph structure, it can better reflect the node association both geographically and semantically.
- We propose a new dynamic graph attention module which consider the traffic characteristics from previous time slices to better model the traffic lag.
- We conduct extensive experiments on real-world datasets and compare with multiple state-of-the-art models. Results shows that our proposed method outperforms other baselines significantly.

2   Related Work

Traffic forecasting is an essential part of building intelligent transportation and research on traffic forecasting has been conducted for many years. Early traffic forecasting was mainly addressed using statistical models, such as auto-regressive integrated moving average (ARIMA) [11], vector auto regressive model (VAR) [2] and support vector regression (SVR) [12] which reduced forecasting to individual time series forecasts. These methods were built upon stationary assumptions and cannot handle the complex variability of traffic data.

With the rapid development of deep learning area, LSTM [3] is applied to traffic prediction to extract temporal dependencies, without considering their spatial features. Subsequently, researchers introduce CNNs to model spatial dependencies. ST-ResNet [17] proposes a residual network structure to model spatial dependencies in cities, but CNNs are limited to handling regular grid structures rather than non-Euclidean graph structures dominated by unspecified road networks. To address this problem, researchers use graph convolution for modelling. Spatial correlation is captured using graph convolution in STGCN [16], while temporal correlation is captured using one-dimensional convolution on the time axis. DCRNN [10] models the dynamics of traffic flow as a diffusion process, and the network replaces the fully connected layer in the gated recursive unit (GRU) with a diffusion convolution operator. T-GCN [18] uses graph convolution to capture spatial correlation and GRU to capture temporal correlation. ASTGCN [5] introduces an attention mechanism added to GCN for prediction.

The key to graph convolutional networks is the adjacency matrix, which represents the association between nodes. Most work uses the spatial distance between sensors to construct the adjacency matrix, but graph constructed in this way ignores the complex relationships between nodes. To address this problem, Graph WaveNet [14] adds adaptive learning of the spatial adjacency matrix and extends the perceptual field along the temporal axis, which enables the original GCN to discover more hidden spatial dependencies. AGCRN [1] uses the adaptive item on DCRNN and achieves good results. MTGNN [13] adds external node features to generate an adaptive matrix through a graph learning layer. However, these methods assume a hidden relationship between any two points, which can lead to overfitting, and are not interpretable in the way they are constructed.

Besides, a major problem of graph convolutional networks is that the adjacency matrix is usually static, making it difficult to describe dynamic traffic conditions. Therefore, GMAN [19] uses a self-attention mechanism to capture spatial correlations dynamically. ASTGNN [6] proposes an improved graph convolution by multiplying a dynamic mask matrix in front of the static graph to make it dynamic, and DGCRN [9] proposes a super-network that combines the current moment velocity and the hidden state of the previous layer to obtain a dynamic graph. However, these methods only generate different adjacency matrices at each moment separately and do not consider the spatial states from past moments.

3   Preliminaries

In this section, we will introduce the definition and the problem statement.

Definition 1: Traffic Network. A traffic network is represented as a directed graph $G = (V, E, A)$, where $V = \{V_1, V_2, ..., V_N\}$ represents the set of sensors collecting traffic data, $N = |V|$ represents the number of sensors, and E refers to the set of edges, $A \in \mathbb{R}^{N \times N}$ is the spatial adjacency matrix representing the proximity or distance between nodes from V.

Definition 2: Traffic Speed Matrix. We denote the value of the speed collected by the entire traffic network G at time t as $X_t = \{x_{t,V_1}, x_{t,V_2}, ..., x_{t,V_N}\} \in \mathbb{R}^{N \times C}$,where $x_{t,V_i} \in \mathbb{R}^C$ represents the speed of traffic collected by node V_i at time t, C is the number of traffic features (e.g., volume, speed).

Problem Statement: Traffic Speed Prediction. Given a series of historical traffic speed matrices $X_{1:P} = (X_1, ..., X_P) \in \mathbb{R}^{N \times P \times C}$ over the last P time slices, the traffic network graph G and an integer Q, the traffic speed prediction aims to generate another series of traffic speed matrices $X_{P+1:P+Q} = (X_{P+1}, ..., X_{P+Q}) \in \mathbb{R}^{N \times Q \times C}$ representing the traffic speed in the next Q time slices.

Here, in our solution, as we need to construct a JS-Graph before the forecasting to achieve better prediction performance, we require additional historical speed profiles as an auxiliary input for model training. The generated graph is denoted as G_{JS}.

4   Methodology

In this section, we first present the framework of our proposed approach and introduce each component in detail.

4.1   Architecture Overview

The general architecture of our proposed model is shown in Fig. 4. We design the JS-STDGN (Spatial-Temporal Dynamic Graph Network using JS-Graph) based on the Encoder-Decoder architecture, including an encoder and a decoder. The encoder and decoder consist of the same structure. Between the encoder and

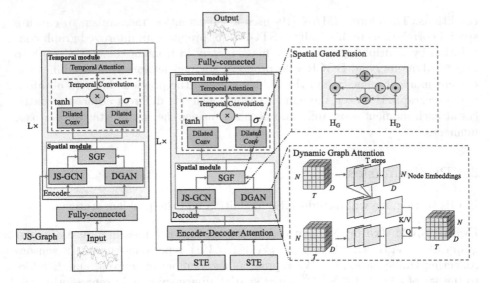

Fig. 4. Framework of JS-STDGN. JS-GCN: JS-Graph Convolution; DGAN: Dynamic Graph Attention; SGF: Spatial Gated Fusion; STE: Spatial-Temporal Embedding

decoder, we add the Encoder-Decoder Attention block to convert the encoded traffic features. In the encoder and decoder, we design the spatial module including JS-Graph convolution (JS-GCN) and dynamic graph attention (DGAN) to capture the road network's deep static features and dynamic spatial characteristics, respectively. Then the results of both are aggregated by SGF (Spatial Gated Fusion) as the representation of the spatial module. Finally, temporal module, including temporal convolution (TCN) and temporal attention, is used to extract temporal features. In addition, we use a fully-connected layer on the input for converting input $X \in \mathbb{R}^{N \times P \times C}$ dimensions to $X \in \mathbb{R}^{N \times P \times D}$, where D is the dimension of hidden state. Finally, we use a fully-connected layer on the output for prediction.

4.2 JS-Graph Convolution Network

In order to describe deeper static relationships in the spatial dimension, we further design a JS-GCN on top of the GCN. Most recent studies use the graph convolution network to model spatial dependence. GCN learns node representations by exchanging information between nodes to capture unstructured patterns hidden in the graph. Given T as P or Q, the graph convolution operation in l-th layer is defined as follows:

$$GCN\left(X^{(l)}\right) = \sigma\left(AX^{(l-1)}W^l\right) \tag{1}$$

where $X^{(l)} \in \mathbb{R}^{N \times T \times D}$, $W^{(l)} \in \mathbb{R}^{D \times D}$, σ represent node representation, weight matrix and activation function, respectively. $A \in \mathbb{R}^{N \times N}$ usually represents the interactions between nodes and is defined as follows:

$$A = \widetilde{D}^{-\frac{1}{2}} \widetilde{A} \widetilde{D}^{-\frac{1}{2}} \tag{2}$$

where $\widetilde{A}$ is the graph adjacency matrix, and $\widetilde{D}_{ij} = \sum_j \widetilde{A}_{ij}$, i, j are nodes in the graph.

The purpose of the JS-graph is to obtain node dependencies that are more representative than ordinary spatial adjacency graphs. The spatial correlation between nodes is complex. In addition to distance proximity, nodes usually have various static and invariable features, such as regional function and road grade. A regular node distance graph cannot carry such semantic properties, thus we need a new graph design to preserve these relationships.

Through the analysis of real-world traffic data, we observe that the velocity distributions of nodes with similar functions or roads of the same grade are similar. In contrast, the velocity distributions of nodes of different types differ a lot even though they are geographically close. Many metrics can be used to describe the velocity difference between different nodes, such as cosine similarity and Manhattan distance. However, when the sequence length T is set small, these methods are easily disturbed by noise, and when T is set too long, the calculated similarity value will be very close, making it indistinguishable.

So Jensen-Shannon divergence (JSD or JS divergence) is introduced to measure the similarity of nodes from a macro perspective. Compared with the above methods, JSD can better reflect the node properties and is robust to random noise. Let P_1 and P_2 represent speed distributions of two nodes. JSD can be formulated as follows:

$$JSD\left(P_1 \| P_2\right) = \frac{1}{2} KL\left(P_1 \| \frac{P_1 + P_2}{2}\right) + \frac{1}{2} KL\left(P_2 \| \frac{P_1 + P_2}{2}\right) \tag{3}$$

where $KL\left(P \| Q\right)$ can be expressed as:

$$KL(P\|Q) = \sum_{x \in X} P(x) \log \frac{1}{P(x)} + \sum_{x \in X} P(x) \log \frac{1}{Q(x)} \tag{4}$$

Therefore, based on JSD, we construct the similarity relation Graph G_{JS} between nodes, also known as JS-graph, whose matrix is A_{JS}. Algorithm 1 shows our algorithm of JS-graph construction. The general process is as follows: In line 1 and line 2, we first initialise the weight W of A to 0, and evenly divide the speed range of all nodes into n sub-intervals. Then, from line 3–7, we calculate the velocity distribution of node V_1 and V_2 denoted by P_1 and P_2, respectively. In line 6, as a smaller JSD value corresponds to a greater distribution similarity, and the range of JSD is $[0, 1]$, we use $1 - JSD(P_i, P_j)$ instead to represent the weight $W_{i,j}$. From line 8–14, we keep the top-k largest weights and set the rest to 0 to eliminate noise because the JSD is always non-zero regardless of how different two distributions are. In addition, we replace $W_{i,m}$ with $W_{i,m} - W_{i,J_{K+1}}$ for the

Algorithm 1. JS-Graph Construction

Input: Traffic speed data of all N nodes in the training set $X = \{X_{V_1}, X_{V_2}, ...X_{V_N}\}$
Output: Weighted Matrix of W JS-Graph A_{JS}

1: Let $W \in \mathbb{R}^{N \times N} = O$
2: Divide the speed distribution interval into n sub-intervals
3: **for** $i = 1, 2, ..., N$ **do**
4:    **for** $j = 1, 2, ..., N$ **do**
5:       Calculate the probability distribution P_i, P_j of X_{V_i}, X_{V_j} in each sub-interval
6:       $W_{i,j} = 1 - JSD(P_i, P_j)$
7:    **end for**
8:    Sort and select Top-K element and their index $J = \{J_1, J_2, ..., J_K, J_{K+1}, ..., J_N\}$
9:    Let $Sum = 0$
10:   **for** $m = 1, 2, ..., K$ **do**
11:      $W_{i,m} = W_{i,m} - W_{i,J_{K+1}}$
12:      $Sum+ = W_{i,m}$
13:   **end for**
14:   Let $W_{i,J_{K+1}}, W_{i,J_{K+2}}, ..., W_{i,J_N} = 0$
15:   **for** $m = 1, 2, ..., K$ **do**
16:      $W_{i,m} = W_{i,m}/Sum$
17:   **end for**
18: **end for**
19: **return** the weight W of JS-Graph A_{JS}

reason that the initial JS value is too close to show the difference in similarity, and in line 15–17, we normalise the weight of each row.

Finally, we use A_{JS} as the input graph in JS-GCN. Then, JS-GCN can be computed as follows:

$$X^{(l)} = JS\text{-}GCN\left(X^{(l-1)}\right) = \sigma\left(A_{JS}X^{(l-1)}W^l\right) \tag{5}$$

4.3 Dynamic Graph Attention Network

As the spatial correlation between roads is dynamic, designing a dynamic graph learning module is necessary. Previously, the dynamic graph leverages the spatial attention mechanism. It assigns weights to each time slice to obtain a different spatial representation for each time slice. However, such a model does not consider the state correlation between adjacent time slices.

Therefore, we propose a DGAN to solve the above problem. As shown in Fig. 5, DGAN dynamically learns the relationship between nodes at the current time, as well as taking into account past states of all nodes, to obtain more comprehensive node information. First, given T as P or Q, we split the input tensor $X \in \mathbb{R}^{N \times T \times D}$ into T timestamps. Then, to every timestamp, given the node representations X_t at time slice t, we refer to scaled dot-product attention to compute the spatial attention of node embeddings of this time slice as well as all node embeddings of previous time slices. The spatial correlation weight

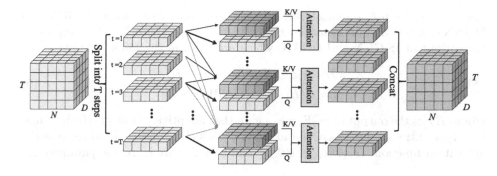

Fig. 5. DGAN structure

between node i and node j at time t is:

$$\alpha^t_{ij} = \frac{\exp\left(\left\langle f_{\alpha,1}\left([X_t]_i\right), f_{\alpha,2}\left(f_\beta\left([X_{1:t}]\right)_j\right)\right\rangle\right)/\sqrt{D}}{\sum_{v \in V} \exp\left(\left\langle f_{\alpha,1}\left([X_t]_i\right), f_{\alpha,2}\left(f_\beta\left([X_{1:t}]\right)_v\right)\right\rangle\right)/\sqrt{D}} \tag{6}$$

where $f_{\alpha,1}(x)$,$f_{\alpha,2}(x)$ are different nonlinear projections. $\langle \cdot, \cdot \rangle$ represents the dot-product operation.

Specially, we need to use a nonlinear function $f_\beta(x) = RELU\left(xW_3 + B_3\right)$ and $W_3 \in \mathbb{R}^{T \times 1}, B_3 \in \mathbb{R}^1$ to adaptively represent the state of the past moments. There are two reasons for this: firstly, it is not appropriate to directly compute the spatial correlation between the current and all historical time slices due to its extremely high complexity $O(TN^2)$; secondly, it is not necessary to bring representations of all past moments into the computation, instead, we only substitute a representation $f_\beta(X_{1:t})$ of the historical state into the computation. Before that, we mask the data after moment t to 0, and fill it into T time slices, so the complexity of computing each time slice can be reduced to $O(N^2)$.

Once the spatial attention score matrix α_t is calculated, given the input data $X_{l-1} \in \mathbb{R}^{N \times T \times D}$, DGAN in l-th layer can be computed as:

$$X^{(l)}_t = DGAN(X^{(l-1)}_t) = \sigma(\alpha_t(f_\beta(X^{(l-1)}_{1:t}))W^{(l)}) \tag{7}$$

Finally, we concatenate all the T slices, and the result is entered as a representation into the next layer.

$$X^{(l)} = Concat(X^{(l)}_1, X^{(l)}_2, ..., X^{(l)}_T) \tag{8}$$

4.4   Spatial Gated Fusion

The static and dynamic graph networks reflect the correlation between sections from different perspectives. To provide a broader range of horizons for the traffic network model, we combine the results of the static graph module with those of the dynamic graph module to improve the performance of traffic prediction.

Specifically, we use a spatial gated fusion from adaptive fusion JS-GCN and DGAN extraction of information. $H_G \in \mathbb{R}^{N \times T \times D}$ and $H_D \in \mathbb{R}^{N \times T \times D}$ represent the output of JS-GCN and DGAN, which are fused in the following way:

$$H_F = z \odot H_G + (1 - z) \odot H_D$$
$$z = \sigma (H_G W_{z1} + H_D W_{z2} + b_z) \tag{9}$$

where H_F is the output of SGF, z is a gate that adaptively controls the information flow extracted by JS-GCN and DGAN. $\odot$ is element-wise product. σ is the activation function. $W_{z1}, W_{z2} \in \mathbb{R}^{N \times N}$ and $b_z \in \mathbb{R}^D$ are learnable parameters.

4.5   Temporal Module

The traffic conditions are related to its previous observations, and this module processes the data in the temporal dimension to capture the temporal dynamics of traffic. Specifically, we use temporal gated convolution and temporal attention mechanisms to learn the local temporal feature and global temporal relation of the traffic data, respectively.

Temporal Gated Convolution. TGC proposes a method to extract hierarchical local time features in parallel by using gated convolution on the time axis. Compared with RNN, the diffusion convolution layer is more efficient in processing long time sequences.

Temporal Attention Mechanism. In order to extract the global time correlation at a deeper level, we use temporal attention to capture the large-scale time correlation of traffic data.

4.6   Other Components

Spatial-Temporal Embedding. To obtain node information better, Spatial-Temporal embedding is added to JS-STDGN as auxiliary input. Spatial embedding matrix $SE \in \mathbb{R}^{N \times D_{SE}}$ is generated by node2vec [4] and temporal embedding matrix $TE \in \mathbb{R}^{(P+Q) \times D_{TE}}$ is generated by one-hot encoding. Then they are represented as $SE \in \mathbb{R}^{N \times D}$ and $TE \in \mathbb{R}^{(P+Q) \times D}$ after two fully-connected layers. At last, we fused the spatial embedding and temporal embedding as $STE \in \mathbb{R}^{(P+Q) \times N \times D}$.

Encoder-Decoder Attention. To alleviate the error accumulation caused by long sequence prediction, Encoder-Decoder Attention is used to extract the features of coded vectors from Encoder, and transform feathers of coded vectors to generate future representation by STE of the past time and the future time.

Loss Function. In this paper, we use Mean Absolute Error (MAE) of real value and predicted value as the loss function, which can be expressed as follows:

$$loss = MAE(Output, Truth) = \frac{1}{Q} \sum_{t=P+1}^{t=P+Q} \left| Y_t - \widehat{Y}_t \right| \tag{10}$$

5   Experiments

To evaluate the performance of our proposed model, we conducted extensive experiments on two real-world datasets. The following is a detailed introduction.

Table 1. Dataset statistics.

Datasets	Nodes	Samples	Sample Rate
METR-LA	207	34272	5 min
PEMS-BAY	325	52116	5 min

5.1   Datasets

We present our results on two real-world large-scale datasets: METR-LA and PEMS-BAY, published by DCRNN. There are 207 sensors located in METR-LA and 325 in PEMS-BAY. The specification of the datasets is shown in Table 1. In the preprocessing step, we replace the missing values with 0, and apply z-score and min-max normalisation. We use 70% of the data for training, 10% for validation, and the rest for testing. According to the data distribution shown in Fig. 6, the traffic speed of BAY is mostly distributed between 60 and 70 mile/h, whereas the distribution in METR-LA is more complex and better reflects the dynamic traffic condition in urban area. Therefore, although we conduct extensive experiments on both datasets, here we mainly conduct analysis experiments on METR-LA.

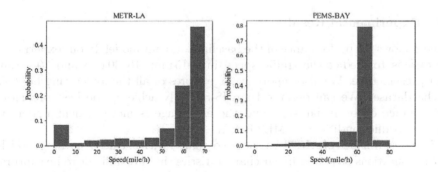

Fig. 6. Distributions of speed.

5.2   Baselines

We compare JS-STDGN with the state-of-the-art prediction models, including:

- ARIMA [11]: Auto-regressive integrated moving average model, which is a classical parametric analysis model.

- STGCN [16]: Spatial-temporal graph convolution network, which integrates graph convolution and 1D convolution.
- DCRNN [10]: Diffusion convolutional recurrent neural network, which combines GRU with two-way diffusion graph convolution.
- Graph WaveNet [14]: Graph Wavenet uses learnable adjacency matrix and uses TCN instead of 1D convolution to capture complex time correlation.
- AGCRN [1]: Adaptive graph convolution recurrent network, which adopts adaptive graph and combines GRU with adaptive graph convolution.
- GMAN [19]: Graph multi-attention network, whose spatial attention dynamically assigns weights to nodes of each time slice.

For all baselines, we experiment with their default settings. Three metrics are adopted to evaluate performance: Mean Absolute Error (MAE), Root Mean Square Error (RMSE), and Mean Absolute Percentage Error (MAPE).

5.3 Experimental Setup

Following the previous work (DCRNN, GMAN), we use $P = 12$ historical time steps (1 h) to predict the speed of the next $Q = 12$ steps (1 h). We use Adam Optimizer [7] as the optimizer, and the initial learning rate was 0.001. The number of encoder and decoder blocks L is 4, the dimension D in main blocks is 64. In terms of top-k candidate number k in JS-Graph, we conduct a series of experiments to find the best k but end up finding that the influence of k value is minor, so we chose a reasonable value as $k = 10$ because the number of each node's neighbours usually ranges from 5 to 12. In DGAN, we use multi-head attention and the number of heads H is 8.

5.4 Experimental Results

Table 2 shows the performance of the baselines and our model. In our experiment, the task is to predict the traffic speed in 0–15 min, 15–30 min and 30–60 min from present time. We also report average scores of all the forecasting horizons on the dataset. We can observe that JS-STDGN achieves the best prediction performance on both datasets, and the advantage is more evident under the complex traffic conditions of METR-LA.

The performance of ARIMA is poor compared with the deep learning models. Traffic conditions have non-linear characteristics that are difficult to be captured by traditional time series forecasting models. For deep learning models based on graph convolution, STGCN and DCRNN belong to the first class of methods using static adjacency matrices. Due to incomplete or even biased knowledge in spatial adjacency matrices, model prediction is still inaccurate. Graph-WaveNet and AGCRN are adaptive methods for learning adjacency matrices. These methods are superior to the previous ones. However, because adjacency matrices are self-learned, it lacks interpretability and is easy to produce over-fitting phenomenon. GMAN makes full use of the self-attention mechanism to simulate dynamic spatial representation. However, this method only allocates different spatial graphs in different time slices and ignores the continuous spatial features, so it has poor performance in short-term prediction. In contrast, our JS-STDGN uses the data distribution characteristics to construct a new static graph, which is better than the node adjacency graph. In addition, we simulate the dynamic spatial correlation characteristics so that JS-STDGN achieves the most advanced results on all tasks.

Table 2. Prediction accuracy comparison

Data	Method	15 min			30 min			1 h			Average		
		MAE	RMSE	MAPE	MAE	RMSE	MAPE	MAE	RMSE	MAPE	MAE	RMSE	MAPE
METR-LA	ARIMA	3.92	8.15	9.40%	5.02	10.23	12.49%	6.53	12.96	16.52%	5.16	10.45	12.80%
	STGCN	3.48	6.67	9.88%	3.84	7.45	11.20%	4.38	8.57	13.30%	4.02	7.81	11.91%
	DCRNN	2.72	5.24	7.01%	3.16	6.42	8.72%	3.70	7.72	10.91%	3.19	6.46	8.88%
	Graph WaveNet	2.64	4.89	6.71%	3.17	6.25	8.61%	3.62	7.22	10.20%	3.25	6.40	8.91%
	AGCRN	2.65	4.94	6.96%	3.23	6.57	9.09%	3.47	7.19	9.96%	3.18	6.47	8.91%
	GMAN	2.79	5.40	7.64%	3.17	6.44	9.04%	3.54	7.27	10.4%	3.26	6.60	9.36%
	JS-STDGN	**2.56**	**4.80**	**6.60%**	**2.99**	**6.01**	**8.28%**	**3.35**	**7.00**	**9.81%**	**3.06**	**6.20**	**8.63%**
PEMS-BAY	ARIMA	1.81	3.56	3.92%	2.42	4.86	5.55%	3.39	6.53	8.48%	2.54	4.98	5.98%
	STGCN	2.25	4.43	5.10%	2.51	4.95	5.72%	2.85	5.68	6.67%	2.62	5.19	6.04%
	DCRNN	1.38	2.95	2.90%	1.69	3.77	3.90%	2.04	4.58	4.71%	1.70	3.76	3.83%
	Graph WaveNet	**1.14**	**2.25**	**2.30%**	1.61	3.53	3.54%	1.93	4.32	4.49%	1.65	3.60	3.70%
	AGCRN	1.17	2.33	2.45%	1.60	3.58	3.60%	1.88	4.31	4.39%	1.64	3.73	3.70%
	GMAN	1.23	2.57	2.60%	1.62	3.66	3.59%	1.89	4.24	4.36%	1.66	3.66	3.73%
	JS-STDGN	1.17	2.31	2.34%	**1.59**	**3.50**	**3.44%**	**1.86**	**4.18**	**4.28%**	**1.62**	**3.54**	**3.59%**

5.5   Study on JS-Graph

To prove the effectiveness of our proposed JS-Graph, we select a set of nodes in METR-LA for case study. Figure 7 illustrates the distribution of different matrices of adjacency graph and JS-Graph. The rows and columns represent different

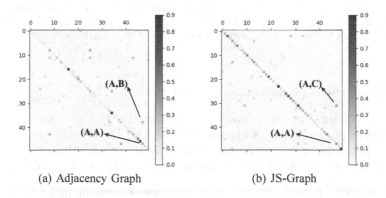

(a) Adjacency Graph   (b) JS-Graph

Fig. 7. Different graph matrices for the chosen 50 vertexes

nodes, and the colour scale bar on the right represents the correlation between nodes. Figure 7(a) shows that, node B is adjacent to node A spatially, but (b) shows that C is a node adjacent to node A in JS-graph after JSD calculation.

Figure 8 shows the velocity curves of the three nodes in a day. We can clearly see that in the same day, although A and B are spatially adjacent to each other, their traffic patterns are not related at all. Conversely, A and C are spatially disjoint, but their traffic patterns are very similar. What is shown in Fig. 9 is the distribution of velocity. We can also come to a conclusion that traffic conditions of A and B differ a lot at low speeds while A and C have high similarity in each speed interval. This demonstrates that our JS-Graph can better capture the deeper correlations between nodes.

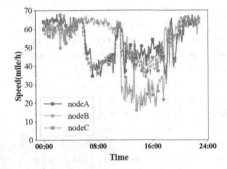

Fig. 8. Speed change in a day   **Fig. 9.** Distributions of velocity

We experiment with both graphs separately on the METR-LA dataset, and as is shown in Table 3, JS-Graph can obviously improve the performance.

Table 3. Performance comparison between JS-graph and adjacency graph

Method	15 min			30 min			1 h			Average		
	MAE	RMSE	MAPE	MAE	RMSE	MAPE	MAE	RMSE	MAPE	MAE	RMSE	MAPE
G-STDGN	2.63	4.91	6.87%	3.04	6.08	8.41%	3.38	7.05	9.86%	3.13	6.29	8.72%
JS-STDGN	**2.56**	**4.8**	**6.60%**	**2.99**	**6.01**	**8.28%**	**3.35**	**7.00**	**9.81%**	**3.06**	**6.20**	**8.63%**

5.6   Effectiveness of Each Component

The two key designs of our proposed model are the JS-GCN module for the deep inherent properties in the traffic network and the DGAN for constantly changing spatial dependencies between nodes. To investigate the effectiveness of our proposed components, we design two variants: W/O JS-GCN, W/O DGAN.

- W/O JS-GCN: In this variant, we remove the JS-GCN module.
- W/O DGAN: In this variant, we remove the DGAN module.

Except for the differences mentioned above, all other parts are identical to JS-STDGN. Figure 10 shows MAE for each prediction step of JS-STDGN and its variants. We observe that JS-STDGN is always superior to its variant, indicating the effectiveness of JS-GCN module and DGAN module in modelling complex spatial features. The introduction of the JS-GCN module can better capture the deep static characteristics of nodes. The DGAN module enables the model to obtain dynamic information to describe the dynamic topology of the road network better.

5.7   Case Study

We plot the actual and predicted values on METR-LA in Fig. 11. We can observe that: (1) JS-STDGN captures the trend of morning rush more accurately than baselines, implying that our proposed model can fit the complex traffic prediction tasks. (2) JS-STDGN is able to deal with road dynamics well and make reasonable predictions despite that the curve of ground truth changes greatly. These can be attributed to the following two reasons. First, our JS-graph is able to show the deep characteristics of nodes. Second, our DGAN module can capture dynamic spatial correlations well.

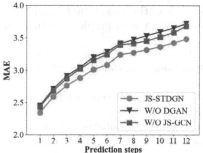

Fig. 10. MAE of each step.   Fig. 11. Visualization of predictions.

6   Conclusions

In this paper, we proposed an effective deep neural network named JS-STDGN for traffic speed prediction. In JS-STDGN, we captured traffic's in-depth features and changing features respectively from static and dynamic perspectives. Specifically, we proposed a new way to construct a traffic graph that uses the JS-divergence value between nodes to capture the non-geographical connection. In addition, we designed a new dynamic graph attention mechanism, which can dynamically change the weight between nodes without missing the information of the past time slices. Experiments on two real-world datasets showed that JS-STDGN achieved the most advanced results over state-of-the-art candidate solutions.

Acknowledgment. This work was supported by National Natural Science Foundation of China (No. 61802273, No. 62102277), Postdoctoral Science Foundation of China (No. 2020M681529), Science and Technology Plan Project of Suzhou (No. SYG202139), Natural Science Foundation of Jiangsu Province (No. BK20210703).

References

1. Bai, L., Yao, L., Li, C., Wang, X., Wang, C.: Adaptive graph convolutional recurrent network for traffic forecasting. In: NeurIPS (2020)
2. Eric Zivot, J.W.: Vector Autoregressive Models for Multivariate Time Series, pp. 385–429 (2006)
3. Fu, R., Zhang, Z., Li, L.: Using LSTM and GRU neural network methods for traffic flow prediction. In: Youth Academic Annual Conference of Chinese Association of Automation (YAC), pp. 324–328 (2016)
4. Grover, A., Leskovec, J.: node2vec: Scalable feature learning for networks. In: SIGKDD, pp. 855–864 (2016)
5. Guo, S., Lin, Y., Feng, N., Song, C., Wan, H.: Attention based spatial-temporal graph convolutional networks for traffic flow forecasting. In: AAAI, pp. 922–929 (2019)
6. Guo, S., Lin, Y., Wan, H., Li, X., Cong, G.: Learning dynamics and heterogeneity of spatial-temporal graph data for traffic forecasting. IEEE Trans. Knowl. Data Eng. (2021)
7. Kingma, D.P., Ba, J.: Adam: a method for stochastic optimization. In: ICLR (2015)
8. Kipf, T.N., Welling, M.: Semi-supervised classification with graph convolutional networks. In: ICLR (2017)
9. Li, F., Feng, J., Yan, H., Jin, G., Jin, D., Li, Y.: Dynamic graph convolutional recurrent network for traffic prediction: benchmark and solution. CoRR abs/2104.14917 (2021)
10. Li, Y., Yu, R., Shahabi, C., Liu, Y.: Diffusion convolutional recurrent neural network: Data-driven traffic forecasting. In: ICLR (2018)
11. Lippi, M., Bertini, M., Frasconi, P.: Short-term traffic flow forecasting: an experimental comparison of time-series analysis and supervised learning. IEEE Trans. Intell. Transp. Syst. **14**(2), 871–882 (2013)
12. Wu, C., Ho, J., Lee, D.: Travel-time prediction with support vector regression. IEEE Trans. Intell. Transp. Syst. **5**(4), 276–281 (2004)
13. Wu, Z., Pan, S., Long, G., Jiang, J., Chang, X., Zhang, C.: Connecting the dots: multivariate time series forecasting with graph neural networks. In: KDD. pp. 753–763 (2020)
14. Wu, Z., Pan, S., Long, G., Jiang, J., Zhang, C.: Graph wavenet for deep spatial-temporal graph modeling. In: IJCAI, pp. 1907–1913 (2019)
15. Xu, M., et al.: Spatial-temporal transformer networks for traffic flow forecasting. CoRR abs/2001.02908 (2020)
16. Yu, B., Yin, H., Zhu, Z.: Spatio-temporal graph convolutional networks: a deep learning framework for traffic forecasting. In: IJCAI, pp. 3634–3640 (2018)
17. Zhang, J., Zheng, Y., Qi, D.: Deep spatio-temporal residual networks for citywide crowd flows prediction. In: AAAI, pp. 1655–1661 (2017)
18. Zhao, L., et al.: T-GCN: a temporal graph convolutional network for traffic prediction. IEEE Trans. Intell. Transp. Syst. **21**(9), 3848–3858 (2020)
19. Zheng, C., Fan, X., Wang, C., Qi, J.: GMAN: a graph multi-attention network for traffic prediction. In: AAAI. pp. 1234–1241 (2020)

When Multitask Learning Make a Difference: Spatio-Temporal Joint Prediction for Cellular Trajectories

Yuan Xu, Jiajie Xu$^{(\boxtimes)}$, Junhua Fang, An Liu, and Lei Zhao

School of Computer Science and Technology, Soochow University, Suzhou, China
20204227016@stu.suda.edu.cn, {xujj,jhfang,anliu,zhaol}@suda.edu.cn

Abstract. Spatio-temporal joint prediction aims to simultaneously predict the next location and the corresponding switch time for a cellular trajectory. It requires to consider not only the mutual influence of spatio-temporal predicting tasks, but also the signals related to the intentions of the travel. Although multitask learning can support the joint prediction by considering both spatio-temporal signals, existing approaches neglect the effects of travel intentions and fail to model the long-term dependencies in trajectory, resulting in sub-optimal results accordingly. To solve these issues, we propose an intention-aware multitask learning method for spatio-temporal joint prediction, such that predicting travel intention is learned as an auxiliary task. Specifically, due to the implicity of travel intention, we design an effective loss function to learn meaningful intention representation, which can capture trajectory's future moving goal, so as to provide long-term information for spatio-temporal joint prediction. Furthermore, we carefully design a gating mechanism to fuse sequential and intentional information with different weights to reflect their importance in capturing current movement status. Besides, self-attention network is adopted to model the long-term dependencies of far sampling points in a dense trajectory. Finally, extensive experiments on two trajectory datasets demonstrate the superiority of our method.

Keywords: Spatio-temporal joint prediction · Multitask learning · Cellular trajectory

1 Introduction

With mobile phones becoming ubiquitous, a large number of cellular trajectories are accumulated [3,11]. Spatio-temporal joint prediction aims to simultaneously predict the next location and the corresponding switch time for a cellular trajectory, which is a multitask joint prediction process. The joint prediction is essential to decide when user data will be scheduled to which base station. It not only helps to provide high-quality communication services, but also ensures efficient resources allocation and management in mobile communications [13].

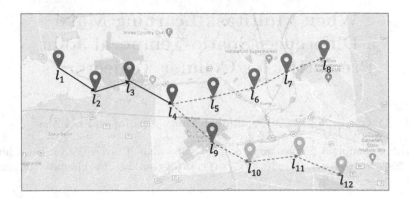

Fig. 1. The blue points denote the current trajectory. The yellow and green parts represent two possible future movement paths. Depending on different future goals, the trajectory may move to different next locations (red points). (Color figure online)

It is generally known that trajectories always carry rich spatial and temporal information. Existing trajectory prediction methods are mainly for these two tasks, i.e., location prediction and time prediction. Location prediction methods [5,8,10] model trajectory's spatial distribution by sequential models (e.g., HMM and RNNs) to predict the next location. Time prediction methods [23,24] mainly use temporal point process [1] to capture continuous time information of sequences to predict the next switch time. However, these single-task based methods cannot be directly used for another task or even predict these two tasks at the same time. Moreover, trajectory next location and time prediction tasks have strong correlation and mutual influence. Temporal information reflects trajectory's mobility patterns (e.g., underlying path and transportation mode), affecting the next location, and vice versa. Hence, spatio-temporal signals should be jointly utilized for collaborative trajectory modeling to obtain more accurate trajectory representation, so as to improve each task's accuracy.

As a technique aiming to learn meaningful information contained in multiple related tasks to improve the performance of all tasks, multitask learning (MTL) [25] is an effective method for spatio-temporal joint prediction. Although a few multitask learning methods [4,9] have considered both spatial-temporal signals and support the joint prediction, they are designed for next POI recommendation, and fail to capture dependencies in the future sequence. However, different from check-in data in POI recommendation, trajectory data has consistent travel intention, and accordingly, the state of a trajectory point is also influenced by the states of follow-up points. Considering the example in Fig. 1, it is hard to predict the next location (l_5 or l_9) for the trajectory $T = \{l_1, l_2, l_3, l_4\}$, unless we know the travel intention that can be learned from T as well. Once knowing the user is heading for the airport at l_8 as destination, we can say he is likely to follow the green path and visit l_5 next. Therefore, location switch in a cellular trajectory is subject to the travel intention, and in turn, travel intention is fulfilled by a sequence of location switches. Unfortunately, existing multitask methods cannot capture travel intention and then utilize it for improving the spatio-temporal joint prediction.

An intention-aware multitask learning method for spatio-temporal joint prediction still faces three main challenges. First, although user always has intention when traveling, the intention is implicit and modeling it from trajectory data is not straightforward. It thus calls for effective modeling of travel intention, which relies on suitable loss function to evaluate the correlations between the captured and real intentions reasonably, such that long-term information (e.g. moving direction) of each cellular trajectory can be captured. Second, the influence on next movement prediction from travel intention and sequential information is nontrivial to balance. The model should incorporate these two information into a more comprehensive representation and tune their weights to effectively capture meaningful information. Last but not least, since cellular trajectory tends to be long and dense sequence, it is supposed to capture the complex dependencies among far samplings for joint modeling of travel intention. However, this can be hardly supported by the RNN-based methods adopted in [4,9]. Therefore, an enhanced trajectory modeling method is required to capture both short-term and long-term dependencies.

To address the above issues, we propose a novel intention-aware multitask learning method called IAMT for spatio-temporal joint prediction. In the multitask framework, we take spatio-temporal joint prediction as main tasks and travel intention prediction as an auxiliary task to assist main tasks. Specifically, to obtain a meaningful intention representation, we adopt an effective distribution-aware loss function to evaluate the captured travel intention, which ensures the long-term information of the trajectory can be contained in the representation. In addition, considering that next movement status is greatly influenced by not only sequential information but also intentional information, we elaborately design a gating mechanism to effectively aggregate these two aspects of information with different weights to reflect their importance in better capturing current movement status. Finally, to ensure the long-term dependencies among far sampling points in a long and dense trajectory to be effectively modeled, we adopt a self-attention network to model the spatio-temporal trajectories. The main contributions are summarized as follows:

- We propose an intention-aware multitask learning method for spatio-temporal joint prediction, which introduces travel intention prediction as an auxiliary task to provide intentional information and improve the spatio-temporal joint prediction.
- We adopt a distribution-aware loss function to derive a meaningful intention representation, which can provide trajectory long-term information for joint prediction. Furthermore, a carefully designed gating mechanism is adopted to effectively fuse intention and sequence information with different weights to better capture meaningful knowledge for current movement status.
- We adopt self-attention to model spatio-temporal sequences to learn both short-term and long-term dependencies of dense trajectories.

- We conduct extensive experiments on two real-world datasets. The results demonstrate that our proposed method achieves better performance compared with the existing approaches.

2   Related Work

2.1   Trajectory Prediction

Existing trajectory prediction methods are mainly for two tasks: location prediction [16,19,22] and time prediction [20,23]. Recently, most methods study the prediction of the next location based on historical data. These methods model sequential information by some learning techniques like Markov Chains [14] and RNNs [8,10]. In order to add spatial and temporal contextual information, ST-RNN [10] extends RNN to model time intervals and geographical distances by time transition matrix and distance transition matrix. HST-LSTM [8] introduces spatial-temporal interval information into multiple existing gates of LSTM. Flashback [22] considers historical records with similar context by doing flashbacks on past hidden states in RNN. DeepMove [5] combines GRU with an attention mechanism to capture multi-level periodicity in historical data.

On the other hand, only a few studies [23,24] aim to predict the next arrival time. RCR [24] models the historical check-in of visitors and potential visitors to extract features and incorporates them with censored regression to make time predictions. Afterward, they further introduce a recurrent spatio-temporal point process model [23] to achieve better performance. Nevertheless, these methods assume that the location of the next check-in is pre-designated and cannot support the time prediction of trajectories without knowing the next location.

To sum up, all of the above methods neglect that trajectory prediction in most real scenarios is indeed a multitask joint prediction process, involving both the spatial and temporal prediction. These single-task methods cannot be directly used for another task or even support the spatio-temporal joint prediction.

2.2   Multitask Learning

Multitask learning (MTL) [25], as a learning paradigm in machine learning, aims to learn meaningful information contained in multiple related tasks to improve their performance. It is classified into soft or hard parameter sharing techniques according to whether hidden layers are shared among all tasks.

Recently, MTL has been successfully applied in a variety of fields, such as natural language processing [12]. Inspired by the success, a few studies adopt MTL for spatio-temporal joint prediction of next events or POIs [4,9,18]. RMTPP [4] uses RNN to model the intensity function of temporal point process to simultaneously predict the event timings and the markers. ARNPP-GAT [9] further combines an attention-based recurrent neural point process with graph attention networks to model user's short and long term preference. IRNN [18] uses two unshared RNNs to respectively model time and event sequence. However, these multitask methods ignore the effect of travel intentions contained in trajectories and cannot model trajectory's long-term dependencies. Hence, we introduce travel intention prediction as an auxiliary task to provide trajectory long-term intention information for spatio-temporal joint prediction.

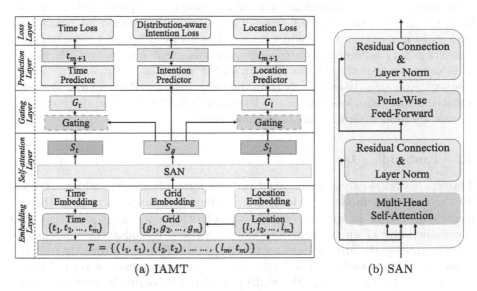

Fig. 2. (a) the framework of IAMT. The model consists of embedding layer, self-attention layer, gating layer, prediction layer, and loss layer. (b) the SAN block.

3   Problem Definition

Let $\mathbf{L} = \{l_1, l_2, \cdots, l_{n_1}\}$ denote a set of locations. A spatio-temporal point p is a tuple (l, t), where the location identification $l \in \mathbf{L}$ can refer to a base station and the positive real number $t \in \mathbf{R}^+$ presents the timestamp switching to the location l. A trajectory T is a time-ordered spatio-temporal point sequence $T = \{p_1, p_2, \cdots, p_m\}$. Besides, we represent the time interval of two consecutive spatio-temporal points as τ, i.e., $\tau_k = t_k - t_{k-1}$.

Problem Formalization. Given a set of trajectories $\mathbf{T} = \{T^1, T^2, \cdots, T^{|\mathbf{T}|}\}$, where $T^i = \{p_1^i, p_2^i, \cdots, p_m^i\}$ is the i-th trajectory, spatio-temporal joint prediction aims to predict the next spatio-temporal point p_{m+1}^i of the trajectory T^i, including the location identification l_{m+1}^i and the corresponding timestamp t_{m+1}^i derived from the time interval τ_{m+1}^i.

4   Our Model

4.1   Overview of IAMT

In Fig. 2(a), we propose an intention-aware multitask learning model called IAMT for spatio-temporal joint prediction. It is composed of five parts, i.e., embedding layer, self-attention layer, gating layer, prediction layer, and loss layer.

In the multitask framework, we take spatio-temporal joint prediction as main tasks and travel intention prediction as an auxiliary task. Specifically, embedding layer is used to embed spatial and temporal factors of the trajectory into dense vector representations. Then, we adopt self-attention layer to model both short-term and long-term dependencies to obtain sequence representation for each task. Considering the joint influence of sequence and intention information on trajectory movement, the carefully designed gating layer is used to fuse intention into trajectory representation to aggregate long-term information. Prediction layer outputs prediction results for each task. Finally, loss layer combines all the losses of three tasks to effectively train the entire network. Particularly, we design a distribution-aware loss function to learn a meaningful intention representation to provide more accurate long-term information.

4.2   Embedding Layer

Trajectory always carries rich spatial and temporal information, which is critical for the trajectory modeling. Moreover, meaningful mobility patterns of the trajectory may exist in different spatial granularity. Hence, it requires not only the embedding of spatio-temporal multi-sequence, but also the modeling of spatial multi-granularity. Based on the trajectory T, we can directly obtain a sequence of base stations $T_l = \{l_1, l_2, \cdots, l_m\}$ and a sequence of related timestamps $T_t = \{t_1, t_2, \cdots, t_m\}$. For multi-granularity modeling in spatial domain, we divide the geographical space into $\omega \times \omega$ grids to derive coarse-grained grid-based sequence $T_g = \{g_1, g_2, \cdots, g_m\}$ from T_l [21,26]. Besides, since time interval can reflect peoples' preferences on the location, we further consider the sequence of time interval between continuous spatiotemporal points, i.e., $T_\tau = \{\tau_1, \tau_2, \cdots, \tau_m\}$.

Since the length of all trajectories are unequal, we first transform each sequence into a fixed-length sequence, e.g., $T_l = \{l_1, l_2, \cdots, l_n\}$, where n is the predefined maximum length and adopt zero-padding at the end of the sequence if it does not have enough points. To represent base stations into dense vectors, we construct a randomly initialized base station embedding matrix $M_l \in \mathbf{R}^{n_1*d}$, where n_1 is the number of all base stations and d is the embedding dimension. Then we can retrieve the sequence's embedding matrix $A_l \in \mathbf{R}^{n*d}$, where $A_{l,i} = M_{l,T_{l,i}}$, that is, selecting the embedding vector of the specified row from M_l according to the identification of each base station. Similarly, we obtain the embedding matrices of time interval sequence $A_\tau \in \mathbf{R}^{n*d}$ and grid sequence $A_g \in \mathbf{R}^{n*d}$. However, unlike RNNs, self-attention ignores positional information of the sequence. Inspired by [6], we add A_τ into other two sequences' embedding matrices to solve the problem.

$$E_l = \begin{bmatrix} A_{l,1} + A_{\tau,1} \\ A_{l,2} + A_{\tau,2} \\ \cdots \\ A_{l,n} + A_{\tau,n} \end{bmatrix} \quad E_g = \begin{bmatrix} A_{g,1} + A_{\tau,1} \\ A_{g,2} + A_{\tau,2} \\ \cdots \\ A_{g,n} + A_{\tau,n} \end{bmatrix} \tag{1}$$

Besides, the final time interval input matrix is the time interval embedding matrix, i.e., $E_\tau = A_\tau$.

4.3   Self-attention Layer

After obtaining final input embeddings, we need to model sequences to capture short-term and long-term dependencies. Since cellular trajectory has dense sampling points and strong sequentiality, we adopt self-attention network to capture global dependencies of the whole sequence regardless of the distance of sampling points. Figure 2(b) illustrates the framework of self-attention network.

Multi-Head Self-Attention Network. To capture the influence between every pair of spatio-temporal points, self-attention is adopted to model sequences. We separately calculate the attention values for each input matrix with the scaled dot-product attention as follows:

$$Attention(Q_{task}, K_{task}, V_{task}) = softmax(\frac{Q_{task}K_{task}^T}{\sqrt{d}})V_{task} \qquad (2)$$

where d is the embedding dimension and the scaling factor $\frac{1}{\sqrt{d}}$ is to prevent the influence of large inner dot-product values; $task \in \{l, t, \tau\}$. Besides, Q_{task}, K_{task} and V_{task} respectively represent query matrix, key matrix, and value matrix, which are generated from the input matrix E_{task} by:

$$Q_{task} = E_{task}W^Q, K_{task} = E_{task}W^K, V_{task} = E_{task}W^V \qquad (3)$$

where $W^Q, W^K, W^V \in \mathbf{R}^{d \times d}$ denote three learnable projection matrices.

Since multi-head self-attention allows model to jointly focus on information from different representation subspaces at different locations of the squence [17], we further use multiple heads to calculate the final attention values, which can run in parallel. The process is formulated as follows:

$$H_{task} = MultiHead(Q_{task}, K_{task}, V_{task}) = Concat(head_1, \cdots, head_h)W^O$$
$$head_i = Attention(Q_{task}W_i^Q, K_{task}W_i^K, V_{task}W_i^V) \qquad (4)$$

where the projections are learnable parameter matrices $W_i^Q, W_i^K, W_i^V \in \mathbf{R}^{d \times d_h}$ and $W^O \in \mathbf{R}^{d \times d}$; h is the number of heads and $d_h = d/h$. Finally, we can obtain three sequence representation matrices H_l, H_τ and H_g.

Residual Connection and Layer Norm. Although multi-layer neural networks have been proved to better capture features, the degradation problem leads to unsatisfactory results. Residual network [7] is proposed to solve the dilemma, which spreads the low-level information to high layers through residual connections. Hence, we utilize residual connection to avoid the problem.

$$R_{task} = LayerNorm(H_{task} + E_{task}) \qquad (5)$$

where $LayerNorm$ denotes the Layer Normalization [2], which is used to normalize the inputs and make the training fast and stable.

Point-Wise Feed-Forward Network. Although self-attention can adaptively assign weights to all points in the sequence to integrate their embeddings, it is still a linear model. To learn the high-level representations of each sequence and consider the features of different latent dimensions, we adopt a point-wise feed-forward network, which is composed of two fully-connected feed-forward layers and a ReLU activation function.

$$F_{task} = ReLU(R_{task}W_{task}^1 + b_{task}^1)W_{task}^2 + b_{task}^2 \tag{6}$$

where W_{task}^1, b_{task}^1, W_{task}^2, and b_{task}^2 are learnable network parameters. Please note that, in order to better capture meaningful knowledge for each task, the point-wise feed-forward network does not share between three tasks, which are task-specific layers. Besides, we perform residual connection and layer normalization again to obtain the output of SAN.

$$S_{task} = LayerNorm(F_{task} + R_{task}) \tag{7}$$

Then, the last point's representation of the trajectory is used as the task-specific representation, that is, $S_{l,m}, S_{\tau,m}$ are the location and time sequential representations respectively. Since intention mainly provides trajectory's long-term information (e.g., future moving direction), it can be obtained from the coarse-grained grid of future goal, and thus $S_{g,m}$ is the intentional representation.

4.4   Gating Layer

Considering that the next movement status is greatly affected by not only sequential information but also intentional information, we design a gating mechanism to effectively fuse these two aspects of information to improve the prediction accuracy. By this way, the fusion representation can remain both knowledge with different proportions and our gating mechanism tends to pay more attention to the more informative features for current movement status. The gating mechanism is designed as follows,

$$\begin{aligned} G_{task,1} &= sigmoid(W_{task,1}S_{task,m} + W_{task,g,1}S_{g,m} + b_{task,1}) \\ G_{task,2} &= sigmoid(W_{task,2}S_{task,m} + W_{task,g,2}S_{g,m} + b_{task,2}) \\ G_{task} &= G_{task,1} \odot S_{task,m} + G_{task,2} \odot S_{g,m} \end{aligned} \tag{8}$$

where $W_{task,1}, W_{task,g,1}, b_{task,1}, W_{task,2}, W_{task,g,2}, b_{task,2}$ are learnable parameters, and $task \in \{l, \tau\}$. $G_{task,1}$ and $G_{task,2}$ control the way and proportion to combine sequential information with intentional information. As we expect it to integrate useful sequential and intentional information for current prediction, the gating mechanism outperforms other simple combination strategies and we further compare these strategies in experiments.

4.5   Prediction Layer

Next Location Predictor. It aims to predict the next location l_{m+1} that the user will visit. Given the final representation G_l, the probability of l_{m+1} is calculated as follows:

$$P_l = softmax(G_l W_l + b_l) \tag{9}$$

where W_l and b_l are learnable parameters.

Intention Predictor. It aims to predict the intention I of the trajectory. Similar to next location prediction layer, the probability of intention I is calculated based on the intentional representation $S_{g,m}$.

$$P_g = softmax(S_{g,m} W_g + b_g) \tag{10}$$

Switch Time Predictor. It aims to predict the switch time when the user will visit the next base station. Different from the above tasks, switch time prediction is a regression problem, which requires to predict in continuous time space. Temporal point process (TPP) [1] is an effective mathematical tool to model temporal sequential data by using the conditional intensity function $\lambda^*(t)$, which specifies the probability that next location switch occurs within a small time window $[t, t + dt]$ given the sequence T.

$$\lambda^*(t) = \frac{f^*(t)}{1 - F^*(t)} \tag{11}$$

where $f^*(t)$ calculates the probability that next location switch occurs at time t given the sequence T, and $F * (t) = \int_0^t f^*(\tau)d\tau$ is the cumulative probability distribution function of $f^*(t)$. The density function $f^*(t)$ can be calculated as:

$$f^*(t) = \lambda^*(t) \exp(-\int_{t_m}^t \lambda^*(\tau)d\tau) \tag{12}$$

Inspired by previous work [4], we calculate the condition intensity function dependent on the output of deep neural network as follows:

$$\lambda^*(t) = \exp(W_t^T G_\tau + W_\tau(t - t_m) + \lambda_0) \tag{13}$$

where W_t, W_τ, λ_0 are trainable parameters. The first term $W_t^T G_\tau$ calculates the accumulative influence among the past spatio-temporal points, the second term $W_\tau(t - t_m)$ represents the evolution of the intensity function over time, and the last term λ_0 is the base intensity. Besides, the exponential operation ensures the intensity function is a positive value. Based on the condition intensity function, we can obtain the density function $f^*(t)$.

$$f^*(t) = \exp\{W_t^T G_\tau + W_\tau(t - t_m) + \lambda_0 + \frac{1}{W_\tau} \exp(W_t^T G_\tau + \lambda_0)$$
$$- \frac{1}{W_\tau} \exp(W_t^T G_\tau + W_\tau(t - t_m) + \lambda_0)\} \tag{14}$$

Then we can compute the next switch time as $t_{m+1} = \int_{t_m}^{\infty} t f^*(t)$, which can be calculated with numerical integration technique [15].

4.6   Loss Layer

Location Loss. For next location prediction task, we apply cross entropy as our loss function, which is a multi-class logarithmic loss function frequently used corresponding to the softmax classifier. It is the negative log-likelihood of the true next location $\hat{l}_{m+1}$:

$$\mathcal{L}_l = -\hat{l}_{m+1} \log P_l \tag{15}$$

where $\hat{l}_{m+1}$ is the one-hot represented ground truth and P_l is the predicted probability distribution with respect to each base station.

Time Loss. According to the definition of TPP, the loss function for the next switch time prediction task is denoted as:

$$\mathcal{L}_\tau = -\log f^*(t_{m+1}) \tag{16}$$

where $f^*(t_{m+1})$ is the density function, which is calculated by Eq. 14.

Distribution-Aware Intention Loss. Although travel intention can be reflected by trajectory's future goal, long-term intentional information cannot be well captured if we view travel intention prediction as a simply multi-class task, because its traditional loss function treats multiple category labels as independent individuals without any correlation. However, different grids have various spatial associations, such as distance and direction consistency. It is well known that close locations can indicate a similar direction of future movement, thereby providing similar intention information. Therefore, we design a distribution-aware loss function to evaluate the correlation between captured travel intention and real intention.

$$\mathcal{L}_g = \frac{\sum_{i \in topk\_grid} D_{g_j,i} \cdot P_{g,i}}{\sum_{i \in topk\_grid} P_{g,i}} \tag{17}$$

where $topk\_grid$ denotes the grids at the top k of predicted result P_g; $D_{g_j,i}$ denotes the distance between the ground truth g_j and g_i; $j = m + m * 0.5$, m is current trajectory's length and we take its half length as the future goal, because long trajectory's destination may be unable to provide useful knowledge for current prediction; $P_{g,i}$ is the predicted probability of future goal at the ith grid. The distribution-aware loss function makes the predicted result as close to the real grid as possible in space, so as to provide similar intention.

Multi-task Loss. In this paper, we integrate all the tasks of location prediction, time prediction, and intention prediction into a multitask learning framework. The entire network is trained end-to-end by minimizing a multi-task objective, which is the weighted sum of each task's loss.

$$\mathcal{L}(\Theta) = \beta_\tau \mathcal{L}_\tau + \beta_g \mathcal{L}_g + (1 - \beta_\tau - \beta_g)\mathcal{L}_l \qquad (18)$$

where Θ are all learnable parameters in IAMT; β_τ and β_g are hyper-parameters for tuning relative influence of $\mathcal{L}_\tau$ and $\mathcal{L}_g$.

5   Experiments

5.1   Datasets

The experiments are carried out on two real-world datasets which contain cellular trajectories in Hangzhou and Xiamen. We view a base station as a location, which can provide signal to its surrounding area. If a user enters the area, the switch time and the base station identification will be recorded. A trajectory is a time-ordered spatio-temporal point sequence of a taxi within a travel order. We filter out the trajectories with less than 5 points and take each trajectory's half length as its input length. The statistics of two datasets are shown in Table 1.

Table 1. Statistics of two datasets

Datasets	# Users	# Base stations	# Records	# Trajectories
Hangzhou	10825	13902	403867	10825
Xiamen	763	7255	683972	33583

5.2   Baselines

We respectively evaluate the effectiveness of our model for two main tasks. For next location prediction, we compare our model with state-of-the-art methods, including single-task methods (STRNN, DeepMove, HST-LSTM, and Flashback) and multi-task methods (RMTPP, IRNN, ARNPP-GAT). For switch time prediction, we also compare with single-task methods (Avg, THP) and multi-task methods. Since RSTPP [23] assumes next location is prespecified, we do not compare with it because next location is unknown in advance in our tasks.

- **STRNN** [10]. It aims to predict the next location, which extends RNN by introducing distance-specific and time-specific transition matrices.
- **DeepMove** [5]. It adds a historical attention mechanism to GRU to predict the next location over lengthy and sparse trajectories.
- **HST-LSTM** [8]. It extends LSTM to consider spatial distance and time interval into the three existing gates to predict the next location.
- **Flashback** [22]. It considers historical points with similar context by doing flashbacks on past hidden states of RNN to predict the next location.
- **Avg**. It returns the average time interval of historical spatio-temporal points as the next time interval to derive the predicted next switch time.
- **THP** [6]. It couples transformer with Hawkes process to fit event sequence to predict time and use temporal encoding similar to positional encoding.

Table 2. Performance comparison results for next location prediction

Datasets	Hangzhou				Xiamen			
Method	ACC	MRR	Recall	macro-F1	ACC	MRR	Recall	macro-F1
STRNN	0.0471	0.0833	0.0254	0.0118	0.3143	0.4665	0.1836	0.1479
DeepMove	0.0480	0.0970	0.0222	0.0138	0.3375	0.4914	0.1708	0.1434
HST-LSTM	0.0476	0.0886	0.0220	0.0115	0.3156	0.4686	0.1818	0.1482
Flashback	0.0513	0.0956	0.0261	0.0203	0.3411	0.4883	0.2062	0.1737
RMTPP	0.0483	0.0961	0.0228	0.0150	0.3213	0.4784	0.1770	0.1498
IRNN	0.0559	0.0986	0.0253	0.0146	0.3422	0.4901	0.2069	0.1731
ARNPP-GAT	<u>0.0619</u>	<u>0.1077</u>	<u>0.0339</u>	<u>0.0215</u>	<u>0.3497</u>	<u>0.5186</u>	<u>0.2103</u>	<u>0.1745</u>
IAMT-TI	0.0605	0.1071	0.0328	0.0211	0.3467	0.5191	0.2097	0.1739
IAMT-I	0.0647	0.1158	0.0378	0.0227	0.3647	0.5314	0.2222	0.1865
IAMT-G	0.0684	0.1189	0.0378	0.0243	0.3707	0.5348	0.2280	0.1957
IAMT-L	0.0708	0.1253	0.0409	0.0261	0.3736	0.5409	0.2302	0.1973
IAMT	**0.0721**	**0.1302**	**0.0424**	**0.0279**	**0.3783**	**0.5458**	**0.2321**	**0.1988**

- **RMTPP** [4]. It adopts a recurrent point process to model the event sequence data and utilizes current hidden state of RNN to construct intensity function, which is used for the spatio-temporal joint prediction of event sequences.
- **IRNN** [18]. It is similar to RMTPP, but the network layers are not shared among the two tasks and adds peephole connection on LSTM.
- **ARNPP-GAT** [9]. It combines attention-based recurrent point process with GAN to learn user's short and long term preferences for next check-in inference. We remove GAN due to our datasets without user social graph.

5.3 Parameter Setup and Metrics

We implement our IAMT framework with Pytorch and use the Adam optimizer for training. We randomly split all trajectories into 70% for training, 10% for validation and 20% for testing. We set the vector dimension d as 128, batch size as 200, learning rate as 0.001, max sequence length n as 200, the number of attention head h as 8, and k as 3. β_τ and β_g are set as 0.05. ω is set as 100 and 50 in Hangzhou and Xiamen according to Fig. 4. Hyper-parameters are tuned using grid search. In detail, $h \in \{1, 2, 4, 8, 16\}$, $\beta_\tau, \beta_g \in \{0.01, 0.05, 0.1, 0.15, 0.2\}$. For location prediction, we use four widely metrics to evaluate the performance: Accuracy (ACC), Mean Reciprocal Rank (MRR), Recall, and macro-F1. For time prediction, we measure the performance of different methods by Mean Absolute Error (MAE) and Root Mean-Squared Error (RMSE). For fair comparison, we run three times of each method and take the average value as the final result.

5.4 Comparisons of Performance

Next Location Prediction Results. From Table 2, we can see that our model performs better than baselines on both datasets. In detail, IAMT is superior to all

Table 3. Performance comparison results for next switch time prediction

Datasets	Hangzhou		Xiamen	
Method	MAE	RMSE	MAE	RMSE
Avg	2.45	3.88	2.76	4.97
THP	2.36	3.22	2.51	3.51
RMTPP	2.25	3.24	2.38	3.27
IRNN	2.22	3.15	2.32	3.21
ARNPP-GAT	2.17	3.12	2.29	3.16
IAMT-LI	2.33	3.12	2.40	3.25
IAMT-I	2.11	3.10	2.25	3.09
IAMT-G	2.09	3.05	2.24	3.06
IAMT-L	2.09	3.06	2.23	3.06
IAMT	**2.08**	**3.04**	**2.21**	**3.03**

single-task methods (STRNN, DeepMove, HST-LSTM, and Flashback), because they neglect the mutual influence of spatio-temporal signals and the importance of long-term dependencies in a dense trajectory. Besides, our method performs better than multi-task methods (RMTPP, IRNN, and ARNPP-GAT), which is due to the fact that they ignore intention signal, proving the effects of intentions on spatio-temporal joint prediction. We also notice the results in Xiamen are better than those in Hangzhou. The main reason is that Hangzhou is sparser than Xiamen, which has more locations and fewer records than Xiamen. Such sparsity makes it hard for training a sequential model to learn mobility patterns.

Next Switch Time Prediction Results. As shown in Table 3, our model IAMT achieves the best performance on next switch time prediction task. Besides, Avg performs the worst since it neglects sequence and location information. Compared with the single-task method THP, our method improves the performance, because it simultaneously considers spatial information and intentional information. In addition, existing multi-task methods (RMTPP, IRNN, and ARNPP-GAT) can consider both spatio-temporal signals to support the joint prediction, but they ignore travel intention that also affects the trajectory's future movement and cannot model the long-term dependencies of sampling points in a dense trajectory, thus showing worse results than our model IAMT.

Comparison with Variants. To evaluate the effectiveness of each component in IAMT, we compare it with several variants, including:

- **IAMT-TI (IAMT-LI)** only predicts single task, and removes time (location) and intention prediction tasks.
- **IAMT-I** makes the spatio-temporal joint prediction, but removes the travel intention prediction task. Please note that it does not use gating mechanism.

- **IAMT-G** removes the gating layer, equivalent to jointly predict three tasks.
- **IAMT-L** does not use the distribution-aware loss function, but uses the cross entropy loss function similar to next location prediction task.

As Table 2 and Table 3 shown, IAMT-TI and IAMT-LI perform better than all single-task methods, which confirms long-term dependencies modeled by SAN are important in dense trajectories. However, IAMT-TI performs worse than IAMT-I, which ignores the mutual influence of spatio-temporal signals. For time prediction, IAMT-G and IAMT-L perform slightly better than IAMT-I, because intention mainly provides spatial information and has a little effect on time prediction. For location prediction, IAMT-I is worse than IAMT-G, indicating that intention can provide meaningful long-term spatial information. The comparison of IAMT-G and IAMT proves that the gating mechanism can effectively fuse intention into trajectory representation. The result of IAMT-L verifies that the distribution-aware intention loss function can learn a meaningful intention representation to provide more accurate long-term information.

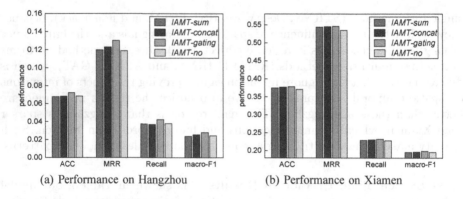

(a) Performance on Hangzhou   (b) Performance on Xiamen

Fig. 3. Effects of different aggregation strategies

Effect of Different Intention Aggregation Strategies. In order to further evaluate the performance of our designed gating mechanism, we compare it with the following strategies: (1) *IAMT-sum* sums up the two representations, i.e., $G_{task} = S_{task,m} + S_{g,m}$; (2) *IAMT-concat* concatenates two representations in the feature dimension, i.e., $G_{task} = S_{task,m} \oplus S_{g,m}$; (3) *IAMT-no* does not use the intentional representation, i.e., $G_{task} = S_{task,m}$.

The results are shown in Fig. 3. Since intention has a little effect on time prediction, we only show the effects on location prediction. *IAMT-gating* performs better than *IAMT-no* due to the aggregation of intentional information. *IAMT-sum* and *IAMT-concat* perform worse than *IAMT-gating*, because not all information is helpful for current prediction, while these methods utilize all knowledge of intention and sequence. The gating mechanism ensures that the model can capture useful information to improve the prediction accuracy.

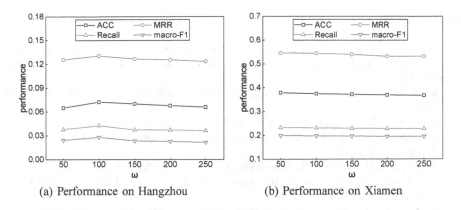

(a) Performance on Hangzhou   (b) Performance on Xiamen

Fig. 4. Effects of different grid granularity

Effect of Different Grid Granularity. In Fig. 4, we investigate the effects of grid granularity ω from $\{50, 100, 150, 200, 250\}$ on next location prediction task. When grid is divided by a proper coarse granularity, future grid can provide rough direction and long-term information. However, if ω is too small, the range of the direction reflected by future grid is too large to provide useful information; if ω is too large, it is hard for the model to accurately predict the future grid, and the provided direction may deviate from the next movement direction. According to the results, we finally set $\omega = 100$ on Hangzhou and $\omega = 50$ on Xiamen.

6   Conclusion

In this paper, we propose an intention-aware multitask learning method called IAMT for spatio-temporal joint prediction. As existing multitask methods neglect the effects of travel intentions and fail to model the long-term dependencies in trajectory, we introduce travel intention prediction as an auxiliary task to improve performance of spatio-temporal prediction tasks. Moreover, to obtain a meaningful representation of intention, we adopt a distribution-aware loss function to evaluate the captured intention. Furthermore, we carefully design a gating mechanism to incorporate intentional information into trajectory representation to utilize its useful knowledge and adopt self-attention to model the long-term dependencies of far sampling points in dense trajectories. Finally, extensive experimental results on two datasets demonstrate the effectiveness of our model.

Acknowledgement. This work was supported by National Natural Science Foundation of China projects under grant numbers (No.61872258, No.61772356, No.62072125), the major project of natural science research in universities of Jiangsu province under grant number 20KJA520005, the priority academic program development of Jiangsu higher education institutions, young scholar program of Cyrus Tang Foundation.

References

1. Aalen, O., Borgan, O., Gjessing, H.: Survival and Event History Analysis: A Process Point of View. Springer, New York (2008). https://doi.org/10.1007/978-0-387-68560-1
2. Ba, L.J., Kiros, J.R., Hinton, G.E.: Layer normalization. CoRR abs/1607.06450 (2016)
3. Blondel, V.D., Decuyper, A., Krings, G.: A survey of results on mobile phone datasets analysis. EPJ Data Sci. 4(1), 1–55 (2015). https://doi.org/10.1140/epjds/s13688-015-0046-0
4. Du, N., Dai, H., Trivedi, R., Upadhyay, U., Gomez-Rodriguez, M., Song, L.: Recurrent marked temporal point processes: embedding event history to vector. In: SIGKDD, pp. 1555–1564 (2016)
5. Feng, J., et al.: Deepmove: predicting human mobility with attentional recurrent networks. In: WWW, pp. 1459–1468 (2018)
6. Gu, Y.: Attentive neural point processes for event forecasting. In: AAAI, pp. 7592–7600 (2021)
7. He, K., Zhang, X., Ren, S., Sun, J.: Deep residual learning for image recognition. In: CVPR, pp. 770–778 (2016)
8. Kong, D., Wu, F.: HST-LSTM: a hierarchical spatial-temporal long-short term memory network for location prediction. In: IJCAI, pp. 2341–2347 (2018)
9. Liang, W., Zhang, W.: Learning social relations and spatiotemporal trajectories for next check-in inference. TNNLS (2020)
10. Liu, Q., Wu, S., Wang, L., Tan, T.: Predicting the next location: a recurrent model with spatial and temporal contexts. In: AAAI, pp. 194–200 (2016)
11. Luca, M., Barlacchi, G., Lepri, B., Pappalardo, L.: Deep learning for human mobility: a survey on data and models. CoRR abs/2012.02825 (2020)
12. Luong, M., Le, Q.V., Sutskever, I., Vinyals, O., Kaiser, L.: Multi-task sequence to sequence learning. In: ICLR (2016)
13. Lv, Z., Xu, J., Zhao, P., Liu, G., Zhao, L., Zhou, X.: Outlier trajectory detection: a trajectory analytics based approach. In: Candan, S., Chen, L., Pedersen, T.B., Chang, L., Hua, W. (eds.) DASFAA 2017. LNCS, vol. 10177, pp. 231–246. Springer, Cham (2017). https://doi.org/10.1007/978-3-319-55753-3_15
14. Rendle, S., Freudenthaler, C., Schmidt-Thieme, L.: Factorizing personalized Markov chains for next-basket recommendation. In: WWW, pp. 811–820 (2010)
15. Seiler, M.C., Seiler, F.A., et al.: Numerical recipes in c: the art of scientific computing. Risk Anal. 9(3), 415–416 (1989)
16. Sun, H., Xu, J., Zheng, K., Zhao, P., Chao, P., Zhou, X.: MFNP: a meta-optimized model for few-shot next POI recommendation. In: IJCAI, pp. 3017–3023 (2021)
17. Vaswani, A., et al.: Attention is all you need. In: NIPS, pp. 5998–6008 (2017)
18. Xiao, S., Yan, J., Yang, X., Zha, H., Chu, S.M.: Modeling the intensity function of point process via recurrent neural networks. In: AAAI, pp. 1597–1603 (2017)
19. Xu, J., Zhao, J., Zhou, R., Liu, C., Zhao, P., Zhao, L.: Predicting destinations by a deep learning based approach. TKDE 33(2), 651–666 (2021)
20. Xu, S., Zhang, R., Cheng, W., Xu, J.: MTLM: a multi-task learning model for travel time estimation. GeoInformatica (1) (2020)
21. Xue, A.Y., Zhang, R., Zheng, Y., Xie, X., Huang, J., Xu, Z.: Destination prediction by sub-trajectory synthesis and privacy protection against such prediction. In: ICDE, pp. 254–265 (2013)

22. Yang, D., Fankhauser, B., Rosso, P., Cudré-Mauroux, P.: Location prediction over sparse user mobility traces using RNNs: flashback in hidden states! In: IJCAI, pp. 2184–2190 (2020)

23. Yang, G., Cai, Y., Reddy, C.K.: Recurrent spatio-temporal point process for check-in time prediction. In: CIKM, pp. 2203–2211 (2018)

24. Yang, G., Cai, Y., Reddy, C.K.: Spatio-temporal check-in time prediction with recurrent neural network based survival analysis. In: IJCAI, pp. 2976–2983 (2018)

25. Zhang, Y., Yang, Q.: A survey on multi-task learning. CoRR abs/1707.08114 (2017)

26. Zhao, J., Xu, J., Zhou, R., Zhao, P., Liu, C., Zhu, F.: On prediction of user destination by sub-trajectory understanding: A deep learning based approach. In: CIKM, pp. 1413–1422. ACM (2018)

Efficient Retrieval of Top-k Weighted Spatial Triangles

Ryosuke Taniguchi, Daichi Amagata$^{(\boxtimes)}$, and Takahiro Hara

Osaka University, Osaka, Japan
{taniguchi.ryosuke,amagata.daichi,hara}@ist.osaka-u.ac.jp

Abstract. Due to the proliferation of location-based services and IoT devices, a lot of spatial points are being generated. Spatial data analysis is well known to be an important task. As spatial data analysis tools, graphs consisting of spatial points, where each point has edges to its nearby points and the weight of each edge is the distance between the corresponding points, have been receiving much attention. We focus on triangles (one of the simplest sub-graph patterns) in such graphs and address the problem of retrieving the top-k weighted spatial triangles. This problem has important real-life applications, e.g., group search, urban planning, and co-location pattern mining. However, this problem is computationally challenging, because the number of triangles in a graph is generally huge and enumerating all of them is not feasible. To solve this challenge, we propose an efficient algorithm that returns the exact result. Our experimental results on real datasets show the efficiency of our algorithm.

Keywords: Spatial points · Weighted graph · Top-k retrieval

1 Introduction

Due to the proliferation of location-based services and IoT devices, a lot of spatial (or geo-location) points are being generated nowadays. Analyzing such spatial points yields useful observations. Many spatial point processing techniques [1–4, 9] and systems [10, 13, 15] have therefore been devised. Recently, as spatial point analysis tools, graph-based approaches have been receiving attention [6, 14, 16].

Given a set P of spatial points and a distance threshold r, a spatial neighbor graph of P consists of a set of vertices that correspond to points in P and a set of edges where an edge is created between two points iff the distance between them is not larger than r and the weight of this edge is the distance. Graph-based structures provide intuitive relationships between spatial points, so techniques that mine some patterns (i.e., sub-graphs) from spatial neighbor graphs are often required. Triangles are particularly considered in graph contexts, because triangle is one of the simplest yet important primitive sub-graph patterns (e.g., clique) having many applications [8, 11]. For example, spatial triangles can be utilized in group search [7], co-location pattern mining [16], and urban planning

A. Bhattacharya et al. (Eds.): DASFAA 2022, LNCS 13245, pp. 224–231, 2022.
https://doi.org/10.1007/978-3-031-00123-9_17

[6]. Note that the number of triangles in a spatial neighbor graph is generally huge. Enumerating all of them is therefore not feasible, and the output size should be controllable (by a user-specified parameter k) [8]. In spatial databases, given a subset of points in P (e.g., that form triangles), the cohesiveness of the subset is a factor in measuring its importance [16].

Motivated by the above applications and observations, this paper addresses the problem of retrieving the top-k weighted spatial triangles. The weight of the triangle formed by points p_x, p_y, and p_z is defined as $dist(p_x, p_y) + dist(p_y, p_z) + dist(p_x, p_{zy})$, where $dist(\cdot, \cdot)$ measures the Euclidean distance between two points, which takes into account the cohesiveness. Then, given P and the output size k, this problem retrieves k spatial triangles with the minimum weight among all triangles in the spatial neighbor graph of P. This problem is computationally challenging, as seen below. A straightforward solution for this problem is to enumerate all triangles and then output k triangles with the minimum weight. The number of triangles in the spatial neighbor graph is $O(\binom{n}{2})$, where $n = |P|$, so this solution is not feasible. To alleviate this computational cost, we can use DHL [8], which is a heuristic algorithm and was proposed originally for graph databases. DHL assumes that edges are sorted by weights, and it greedily accesses the edges in this order, so as to avoid enumerating triangles with large weights. However, to employ DHL, we face substantial time incurred by building a spatial neighbor graph of P and sorting a large amount of edges.

To solve the above issues, we propose an efficient algorithm that returns the exact answer. We find an observation that a subset of the spatial neighbor graph, which usually contains the top-k weighted triangles, can be built offline. Besides, from this partial graph, for each point $p \in P$, we can enumerate a triangle having p with a small weight in $O(1)$ time offline. These n triangles provide a tight threshold for the top-k result, which helps filter unnecessary points and triangles, resulting in improvement of online computation. Thanks to these observations, our algorithm does not need to correctly build the spatial graph and sort all edges. To summarize, our main contributions are as follows:

- We address the problem of retrieving the top-k weighted spatial triangles. To our knowledge, we are the first to tackle this problem in spatial databases.
- We propose a simple, efficient, and exact solution for this problem.
- We conduct experiments on real datasets, and the results show that our solution for static data is up to *three orders of magnitude faster* than a baseline algorithm.

2   Preliminary

Let P be a set of spatial (or geo-location) points in a Euclidean space. A spatial point $p \in P$ has 2-dimensional coordinates $\in \mathbb{R}^2$. We use $dist(p, p')$ to denote the Euclidean distance between p and p'. We assume that P is memory resident.

Given a distance threshold r, where r is a tolerable distance between points to regard them as being located close to each other, we can build a spatial neighbor graph of P defined below:

Definition 1 (SPATIAL NEIGHBOR GRAPH). *Given a set P of points and a distance threshold r, the spatial neighbor graph of P is an undirected graph consisting of a set of vertices that correspond to the points in P and a set of edges where an edge is created between p_i and p_j iff $dist(p_i, p_j) \leq r$. The edge between p_i and p_j is represented as $e_{i,j}$ and has a weight $w(e_{i,j})$ where $w(e_{i,j}) = dist(p_i, p_j)$.*

In the spatial neighbor graph, there are triangles consisting of three points fully connected to each other. We define their weight:

Definition 2 (WEIGHT OF A TRIANGLE). *Given a triangle $\triangle_{x,y,z}$ consisting of three points p_x, p_y, and p_z, the weight of this triangle, $w(\triangle_{x,y,z})$, is:*

$$w(\triangle_{x,y,z}) = dist(p_x, p_y) + dist(p_y, p_z) + dist(p_x, p_z). \tag{1}$$

Then, our problem in Sect. 3 is defined as follows:

Definition 3 (TOP-k WEIGHTED TRIANGLE RETRIEVAL PROBLEM). *Given a set P of points, an output size k, and a distance threshold r, this problem is to retrieve at most k triangles in the spatial neighbor graph of P with the minimum weight[1].*

3   Our Solution

Main Idea. To efficiently retrieve k triangles with the minimum weight, it is desirable to prune points that do not contribute to the top-k result. Assume that triangle $\triangle_{x,y,z}$ is included in the top-k result. From Eq. (1) and Definition 3, it is intuitively seen that, for p_x, edges $e_{x,y}$ and $e_{x,z}$ would be (two of) the t nearest neighbors (t-NNs) of p_x, where t is a small constant. This suggests that the top-k triangles can be retrieved from the t-NN graph and that correct building of the spatial neighbor graph of P is not necessary.

This idea brings an important advantage: the spatial neighbor graph of P needs to be built online (since it depends on r), whereas the t-NN graph of P can be built *offline* (for $t = O(1)$). Furthermore, if we have the t-NN graph of P, for each $p \in P$, we can enumerate a promising triangle having p, i.e., the triangle formed by p and its 2 nearest neighbors, in the same offline step. Even if these triangles are not included in the top-k result, they usually have small weights, yielding a tight threshold for online computation in practice. This threshold helps prune unnecessary points (and thus triangles), so the above ideas improve the efficiency of online computation.

Our algorithm is designed based on the above ideas and consists of a one-time offline computation and online computation. In the next subsections, we present how to prepare these triangles and how to compute the exact top-k result in detail.

---

[1] When r is too small, the spatial neighbor graph of P can be very sparse and there may be less than k triangles in the graph. In this case, this problem is easily solved, thus we assume that r is reasonably specified and there are many triangles in the graph.

Offline Processing. The objectives of this offline processing are to (i) build a
B-NN graph of P, where $B \geq 3$ is a batch size, and (ii) enumerate triangles with
small weights. The batch size B is tuned empirically. We use $p.E$ to denote the
set of edges held by a point $p \in P$.

Given P and B, for each $p_x \in P$, we compute the B-NNs of p_x in $P \backslash \{p_x\}$ by
using a kd-tree [5]. The B-NNs are maintained in $p.E$ and sorted in ascending
order of weight (i.e., distance). Moreover, for each $p_x \in P$, we compute the
triangle $\triangle_{x,y,z}$, where p_y and p_z are respectively the NN and 2-NN of p_x. This
triangle is maintained in T, so T has at most n triangles (we remove duplicated
triangles). Last, we sort the triangles in T in ascending order of weight.

REMARK. The kd-tree of P is built in $O(n \log n)$ time. For a fixed B (i.e., $B =
O(1)$), the B-NNs of $p_x \in P$ are retrieved in $O(Bn^{1-1/d}) = O(\sqrt{n})$ time [12].
We can therefore build the B-NN graph in $O(n^{1.5})$ time. Last, sorting triangles
in T needs $O(n \log n)$ time. Our offline algorithm hence needs $O(n^{1.5})$ time.

Building the spatial neighbor graph of P incurs $O(n(\sqrt{n} + s_{avg}))$ time, where
s_{avg} is the average number of edges held by each point. Compared with this,
our offline algorithm is cheaper, and it is general to any k and r. We exploit
the B-NN graph of P and the set T of triangles to efficiently retrieve the top-k
weighted spatial triangles.

Online Processing. To efficiently retrieve the top-k weighted spatial triangles,
we consider *edge access order*. Let τ be an intermediate threshold of the top-k
result (i.e., the weight of the intermediate top k-th triangle). From τ and triangle
inequality, for any edges, we can obtain a weight θ that has to be satisfied to form
the top-k weighted spatial triangles. That is, any triangles that have edges with
weights larger than θ do not have to be enumerated. We exploit this observation
along with the triangles in T and the B-NN graph obtained offline.

Let P_{cand} be the set of points that may form top-k triangles, and $P_{cand} = P$
at initialization. Our online algorithm has the following steps:

1. We first initialize the top-k result R and the threshold τ from the n triangles obtained offline in DETERMINE-THRESHOLD(P_{cand}, r). Then, from τ, we
   compute a threshold θ for edges. As seen later, any edges with weights larger
   than θ cannot form top-k triangles.
2. (If necessary, we update the B-NN graph by increasing B.) In REDUCE-
   CANDIDATES(P_{cand}, i, θ), we remove points with no edges satisfying θ any
   more from P_{cand}.
3. For each point in P_{cand}, we additionally enumerate triangles that could be in
   the top-k result and update R if necessary.
4. We repeat steps 2 and 3 until we have $P_{cand} = \varnothing$, and then R is returned.

Below, we detail steps 1, 2, and 3.

- *Step 1.* Recall that T is a sorted set of triangles obtained offline. Each triangle in T is formed by a point p, its NN, and 2-NN. (We remove all triangles in T that have edges with weights larger than r.) In DETERMINE-
  THRESHOLD(P_{cand}, r), we initialize R by the first k triangles in T, and τ is

the weight of the k-th triangle. Let $\triangle_{x,y,z}$ be the k-th triangle. We set the threshold θ for edges as follows:

$$\theta = \tau - \max\{dist(p_x, p_y), dist(p_y, p_z), dist(p_x, p_z)\}. \tag{2}$$

This is used in the next step.

- *Step 2.* We next filter unnecessary points in P_{cand} by using θ. Let p_{x_j} be the j-th NN of p_x. Consider the i-th iteration of REDUCE-CANDIDATES(P_{cand}, i, θ). For $p_x \in P_{cand}$, if $w(e_{x,x_{i+2}}) > \theta$, triangles including $e_{x,x_{i+2}}$ can be ignored. (Recall that NN and 2-NN were considered in the offline processing.)

Proposition 1. *For a point $p_x \in P_{cand}$, if $w(e_{x,x_{i+2}}) > \theta$, any triangles that have $e_{x,x_{i+2}}$ cannot be the top-k weighted spatial triangles.*

Proof. Consider a triangle $\triangle_{x,x_{i+2},y}$. We have $w(e_{x,x_{i+2}}) \le w(e_{x,y}) + w(e_{x_i,y})$ from triangle inequality. Equation (2) shows that θ is the sum of the weights of two edges of the (intermediate) top k-th triangle. Therefore, if $w(e_{x,x_{i+2}}) > \theta$, the weights of any triangles that have $e_{x,x_{i+2}}$ are larger than τ. $\square$

From this observation, we see that, if $w(e_{x,x_{i+2}}) > \theta$, all unseen triangles having p_x do not have to be enumerated and p_x can be safely removed from P_{cand}. REDUCE-CANDIDATES(P_{cand}, i, θ) does this point removal. (For a point $p_x \in P_{cand}$, if we do not have $e_{x,x_{i+2}}$, we update the B-NN graph by increasing B before REDUCE-CANDIDATES(P_{cand}, i, θ).)

The triangles enumerated offline practically have small weights, as they are based on NN and 2-NN. Therefore, τ and θ are tight even when i is small, and we can effectively reduce the size of P_{cand} in early iterations.

- *Step 3.* After filtering unnecessary points in the above step, we enumerate triangles that may become the top-k result in ENUMERATE-TRIANGLES(P_{cand}, r, i). Consider the i-th iteration of this step. For each $p_x \in P_{cand}$, we enumerate triangles formed by p_x, $p_{x_{i+2}}$, and p_{x_j}, where $j \in [1, ..., i+1]$, while updating the top-k result R, τ, and θ.

W.r.t. p_{x_j}, we access it in order of $p_{x_1}, ..., p_{x_{i+1}}$. Then, it is important to notice that $w(e_{x,x_j}) + w(e_{x,x_{i+2}})$ monotonically increases. When we have $w(e_{x,x_j}) + w(e_{x,x_{i+2}}) \ge \tau$, we see that triangles with these edges cannot be the top-k result, thus we can stop enumerating triangles without losing correctness.

Analysis. We analyze the theoretical performance of our online algorithm. Step 1 needs $O(1)$ time, since T is sorted offline. Consider the i-th iteration of step 2, and let n_i be the size of P_{cand} in this iteration. (Notice that n_i is affected by k.) In step 2, for each $p_x \in P_{cand}$, we check $e_{x,x_{i+2}}$. It hence needs $O(n_i)$ time[2]. Next, consider the i-th iteration of step 3. Let n'_i be the size of P_{cand} in this iteration. Notice that $n'_i \le n_i$, since step 2 reduces the size of P_{cand}. In step 3, for each $p_x \in P_{cand}$, we enumerate triangles formed by p_x, $p_{x_{i+2}}$, and p_{x_j}. Although we

---

[2] When we need to update the B-NN graph, we need $O(n_i\sqrt{n})$ additional time.

can early terminate this enumeration, its worst number is $i + 1$. That is, we need $O(i)$ time for p_x, thus step 3 requires $O(i \cdot n_i')$ time. Consequently, our online algorithm needs $O(\sum_{i=1}^{I}(n_i + i \cdot n_i'))$ time, where I is the number of iterations of step 3.

4   Experiment

This section introduces our experimental results. All experiments were conducted on a machine with 3.6 GHz Intel Core i9-9900K CPU and 128 GB RAM. In addition, all algorithms tested were single threaded and compiled by g++ 9.3.0 with -O3 flag.

We compared it with DHL [8], which can compute the exact answer from the spatial neighbor graph of P. For DHL, we used the original implementation[3].

Dataset. We used two real datasets, CaStreet[4] and Places[5]. CaStreet consists of the minimum bounding rectangles of road segments in the U.S.A. We used bottom-left and upper-right points, and its cardinality is 4,499,454. Places consists of the geo-locations of public places in the U.S.A, and its cardinality is 9,356,750.

Parameter. We set $n = 1,000,000$ (via random sampling) and $k = 100$ by default. In all experiments, $r = 0.01$ (and it did not affect the performance of our algorithm). We set $B = 10$.

Impact of k. Next, we investigate the impact of the result size k, and Fig. 1 shows our experimental result. Our algorithm is significantly faster than DHL. For example, when $k = 100$, our algorithm is 2021 and 1465 times faster than DHL on CaStreet and Places, respectively. DHL suffers from the overhead cost incurred by dealing with the spatial neighbor graph (while we do not have this drawback.)

It can be also observed that the tendency of our algorithm is different between CaStreet and Places. We found that Places is denser than CaStreet. Due to this feature, compared with CaStreet case, our algorithm needed to enumerate more triangles and update the top-k result more frequently on Places. However, it still returns the result in 1.13 [sec] even when $k = 1000$. Recall that it needed 0.16 [sec] when $k = 100$. We therefore see that our algorithm scales linearly to k.

Impact of n. Figure 2 studies the scalability of our algorithm to the cardinality of dataset n. Our algorithm has a linear scalability to n, while DHL is superlinear w.r.t. n. This clarifies the advantage of our algorithm. When we used all points of CaStreet and Places, our algorithm is 2807 and 6193 times faster than DHL on CaStreet and Places, respectively.

---

[3] https://github.com/raunakkmr/Retrieving-top-weighted-triangles-in-graphs.
[4] http://chorochronos.datastories.org/?q=node/59.
[5] https://archive.org/details/2011-08-SimpleGeo-CC0-Public-Spaces.

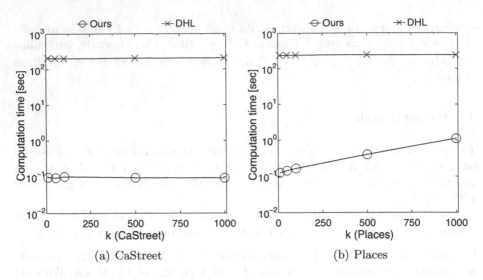

(a) CaStreet

(b) Places

Fig. 1. Impact of k

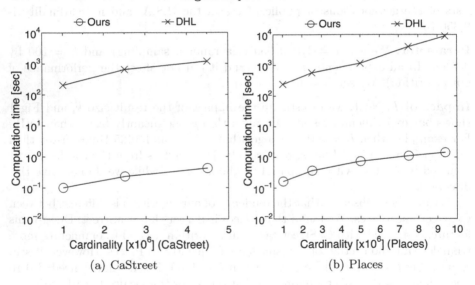

(a) CaStreet

(b) Places

Fig. 2. Impact of cardinality of dataset

5 Conclusion

The number of location-based services is increasing, and a lot of spatial points are being generated nowadays. This fact strengths the importance of analyzing spatial points, and much efforts have been made to devise techniques for spatial point analysis.

As a spatial point analysis tool, we proposed the problem of retrieving the top-k weighted spatial triangles. Because the number of triangles in a set of spatial points can be huge, simply enumerating triangles is time-consuming. To avoid this issue, we proposed an efficient algorithm that returns the exact answer. We conducted experiments on real datasets, and the results demonstrate the efficiencies of our solution.

Acknowledgements. This research is partially supported by JSPS Grant-in-Aid for Scientific Research (A) Grant Number 18H04095, JST CREST Grant Number J181401085, and JST PRESTO Grant Number JPMJCR21F2.

References

1. Amagata, D., Hara, T.: Monitoring maxrs in spatial data streams. In: EDBT, pp. 317–328 (2016)
2. Amagata, D., Hara, T.: A general framework for maxrs and maxcrs monitoring in spatial data streams. ACM Trans. Spatial Algorithms Syst. (TSAS) **3**(1), 1–34 (2017)
3. Amagata, D., Hara, T.: Identifying the most interactive object in spatial databases. In: ICDE, pp. 1286–1297 (2019)
4. Amagata, D., Tsuruoka, S., Arai, Y., Hara, T.: Feat-sksj: fast and exact algorithm for top-k spatial-keyword similarity join. In: SIGSPATIAL, pp. 15–24 (2021)
5. Bentley, J.L.: Multidimensional binary search trees used for associative searching. Commun. ACM **18**(9), 509–517 (1975)
6. Fang, Y., Li, Y., Cheng, R., Mamoulis, N., Cong, G.: Evaluating pattern matching queries for spatial databases. VLDB J. **28**(5), 649–673 (2019). https://doi.org/10.1007/s00778-019-00550-3
7. Fang, Y., et al.: On spatial-aware community search. IEEE Trans. Knowl. Data Eng. **31**(4), 783–798 (2018)
8. Kumar, R., Liu, P., Charikar, M., Benson, A.R.: Retrieving top weighted triangles in graphs. In: WSDM, pp. 295–303 (2020)
9. Nishio, S., Amagata, D., Hara, T.: Lamps: location-aware moving top-k pub/sub. IEEE Trans. Knowl. Data Eng. **34**(1), 352–364 (2022)
10. Pandey, V., Kipf, A., Neumann, T., Kemper, A.: How good are modern spatial analytics systems? PVLDB **11**(11), 1661–1673 (2018)
11. Park, H.M., Myaeng, S.H., Kang, U.: PTE: enumerating trillion triangles on distributed systems. In: KDD, pp. 1115–1124 (2016)
12. Toth, C.D., O'Rourke, J., Goodman, J.E.: Handbook of Discrete and Computational Geometry. CRC Press, Boca Raton (2017)
13. Tsuruoka, S., Amagata, D., Nishio, S., Hara, T.: Distributed spatial-keyword KNN monitoring for location-aware pub/sub. In: SIGSPATIAL, pp. 111–114 (2020)
14. Wang, Y., Yu, S., Dhulipala, L., Gu, Y., Shun, J.: Geograph: a framework for graph processing on geometric data. ACM SIGOPS Oper. Syst. Rev. **55**(1), 38–46 (2021)
15. Yu, J., Sarwat, M.: Geosparkviz: a cluster computing system for visualizing massive-scale geospatial data. VLDB J. **30**(2), 237–258 (2021)
16. Zhang, C., Zhang, Y., Zhang, W., Qin, L., Yang, J.: Efficient maximal spatial clique enumeration. In: ICDE, pp. 878–889 (2019)

DIOT: Detecting Implicit Obstacles from Trajectories

Yifan Lei[1,2], Qiang Huang[1(✉)], Mohan Kankanhalli[1], and Anthony Tung[1]

[1] School of Computing, National University of Singapore, Singapore, Singapore
{leiyifan,huangq,mohan,atung}@comp.nus.edu.sg
[2] Tencent Inc., Shenzhen, China

Abstract. In this paper, we study a new data mining problem of obstacle detection. Intuitively, given two kinds of trajectories, i.e., reference and query trajectories, the obstacle is a region such that most query trajectories bypass this region, whereas the reference trajectories go through as usual. We introduce a density-based definition for the obstacle within a new normalized Dynamic Time Warping distance and the density functions tailored for the sub-trajectories to estimate the density variations. With this definition, we introduce a novel framework DIOT to detect implicit obstacles. Experimental results show that DIOT can capture the nature of obstacles yet detect the obstacles efficiently and effectively.

Keywords: Obstacle detection · Trajectory · Dynamic Time Warping · Kernel density estimation · Nearest Neighbor Search

1 Introduction

With the prevalence of location devices, many trajectory data have been used for data analytics. A trajectory is often represented as a sequence of geo-locations of moving objects such as cars, vessels, and anonymous persons. In this paper, we study a new data mining problem of obstacle detection based on trajectory data. Given two kinds of trajectories, i.e., reference and query trajectories, the obstacle is a region such that most query trajectories bypass this region; in contrast, the reference trajectories go through as usual. For ease of illustration, we use $\mathcal{T}$ to denote the reference trajectories and $\mathcal{Q}$ to indicate the query trajectories.

Example 1. Obstacles are ubiquitous. Figure 1 shows a real-life example of the obstacle. We plot the vessel trajectories in May 2017 and August 2017 in Figs. 1(a) and (b), respectively. According to the official document of Singaporean Notices to Mariners in June 2017,[1] there is a temporary exclusion zone for operations on a sunken vessel *Thorco Cloud* from March to June 2017, and a red polygon shows its geo-location. From Fig. 1, most trajectories in May 2017 bypass the red polygon, whereas the trajectories in August 2017 can go through this zone as usual. Thus, this zone can be regarded as an obstacle.

---

[1] https://www.mpa.gov.sg/web/wcm/connect/www/b10f0a7b-09fe-4642-bc30-0282ff8b48f4/notmarijun17.pdf?MOD=AJPERES.

© The Author(s), under exclusive license to Springer Nature Switzerland AG 2022
A. Bhattacharya et al. (Eds.): DASFAA 2022, LNCS 13245, pp. 232–240, 2022.
https://doi.org/10.1007/978-3-031-00123-9_18

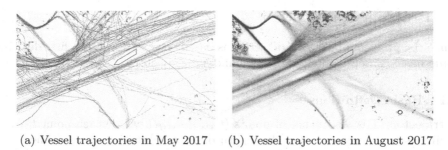

(a) Vessel trajectories in May 2017   (b) Vessel trajectories in August 2017

Fig. 1. An example of obstacle

Obstacle detection arises naturally in many real-life scenarios, such as urban planning and transportation analysis.

Scenario 1: Urban Planning. The government can leverage the trajectories from different kinds of anonymous people to detect obstacles for urban planning. For example, suppose there are two kinds of trajectories (e.g., youngsters' trajectories $\mathcal{T}$ and elderlies' trajectories $\mathcal{Q}$) passing through a housing estate. The government can detect implicit obstacles (e.g., steep slope with stairs) for elderlies based on the density variations of $\mathcal{Q}$ compared with $\mathcal{T}$ and redesign the housing estates to make elderly easier to move through.

Scenario 2: Transportation Analysis. Suppose there is a highway road with two lanes. $\mathcal{T}$ denotes a set of car trajectories from suburb to downtown; $\mathcal{Q}$ is a set of car trajectories vice versa. In the morning, the lane of $\mathcal{Q}$ is an obstacle because most people live in the suburbs, and they need to drive to downtown to work. Thus, the lane of $\mathcal{T}$ has much higher density than that of $\mathcal{Q}$. Similarly, the lane of $\mathcal{T}$ is an obstacle in the afternoon. Based on such inferences, the traffic management department can change this road as a tidal road.

One might wonder why we do not consider 2D histograms or road networks. The 2D histograms satisfy Scenario 1, but they cannot carry directional information, which does not satisfy Scenario 2. The road networks are suitable for Scenario 2. Nonetheless, they cannot model the randomly moving data, which is not suitable for Scenario 1. Moreover, Compared to the long trajectories with variable sizes, obstacles are often small regions. Thus, we partition trajectories into fixed-size sub-trajectories and use them as the primary data representation.

In this paper, we first formalize the definition of the obstacle based on the density variation of sub-trajectories. To accurately characterize the obstacles, we design a new normalized Dynamic Time Warping (nDTW) distance and develop the density functions to estimate the density variation of sub-trajectories and their succeeds. Then, we propose a novel framework DIOT to Detect Implicit Obstacles from Trajectories. Given a collection of reference sub-trajectories, for a set of query sub-trajectories, the insight of DIOT is to recursively identify the reference sub-trajectory whose density variation in query sub-trajectories is significantly larger than that in reference sub-trajectories. Extensive experiments on two real-life data sets demonstrate the superior performance of DIOT.

2   Problem Formulation

Since the problem of obstacle detection is based on trajectories, we first describe some basic concepts about trajectory and sub-trajectory.

2.1   Basic Definitions

A trajectory T is a sequence of points $(t_{(1)}, \cdots, t_{(l)})$, where each point $t_{(i)}$ is a d-dimensional vector and l is the length of T. A sub-trajectory is defined as a consecutive sub-sequence of a trajectory, i.e., $t = T[i : j]$ where $1 \leq i < j \leq l$. In this paper, we assume each point $t_{(i)}$ represents as a 2-dimensional geo-location (latitude, longitude). DIOT can be easily extended to support obstacle detection for d-dimensional points for any $d \geq 3$.

Given a sliding window w and a step s $(s < w)$, we *partition* a trajectory T into a set of sub-trajectories, i.e., $P(T) = \{t_1, t_2, t_3, \cdots\}$, where $t_1 = T[1 : w]$, $t_2 = T[1+s : w+s]$, $t_3 = T[1+2s : w+2s]$, etc. Furthermore, given a set of sub-trajectories $\{t_1, t_2, t_3, \cdots\}$ which are partitioned by sequential order from the same T, t_2 is the *succeed* of t_1 and t_3 is the succeed of t_2, i.e., $t_2 = succ(t_1)$ and $t_3 = succ(t_2)$. Hereafter, we use $P(\mathcal{T}) = \{P(T) \mid T \in \mathcal{T}\}$ to be the reference sub-trajectories and $P(\mathcal{Q}) = \{P(Q) \mid Q \in \mathcal{Q}\}$ to represent the query sub-trajectories.

2.2   Distance Function

As is well known, Dynamic Time Warping (DTW) [6] is one of the most robust and widely used distance functions for the trajectory and time-series data [3]. Given any two sub-trajectories t and q of the same length w, we can compute $DTW(t,q)$ in $O(w^2)$ time via dynamic programming. However, it might not be sufficient to use DTW as the distance function of sub-trajectories for obstacle detection. For example, the sub-trajectory pair (t_1, t_2) in Fig. 2 shows the same pattern as (t_3, t_4). They should have the same density. Nevertheless, $DTW(t_1, t_2) < DTW(t_3, t_4)$. Thus, we propose a *normalized DTW (nDTW)* as the distance

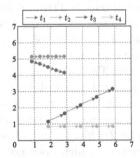

Fig. 2. An example of sub-trajectories for DTW

function of sub-trajectories for density estimation. Compared with DTW, we use the length of sub-trajectories for normalization.

Definition 1 (nDTW). *Given any two sub-trajectories* $t = (t_{(1)}, \cdots, t_{(w)})$ *and* $q = (q_{(1)}, \cdots, q_{(w)})$, *the* $nDTW(t,q)$ *is computed as follows:*

$$nDTW(t,q) = \frac{DTW(t,q)}{\sqrt{\sum_{i=1}^{w-1} \|t_{(i)} - t_{(i+1)}\|} \sqrt{\sum_{j=1}^{w-1} \|q_{(j)} - q_{(j+1)}\|}}.$$

According to Definition 1, $nDTW(t_1, t_2) = nDTW(t_3, t_4)$. Thus, by utilizing $nDTW(t, q)$ as the distance function, the pairs of sub-trajectories with the same pattern have the same density for obstacle detection.

2.3   Density Function

Density of Sub-Trajectories. Given a set of sub-trajectories $P(\mathcal{T})$, for any sub-trajectory t, one often adopts the popular Gaussian KDE [8] to estimate its density, i.e., $\hat{f}_{P(\mathcal{T})}(t) = \frac{1}{|P(\mathcal{T})|} \sum_{t_i \in P(\mathcal{T})} \exp(-\frac{nDTW(t,t_i)^2}{2\sigma^2})$, where σ determines the bandwidth of the Gaussian kernel. Note that for those $t_i \in P(\mathcal{T})$ that are far from t, their contributions to $\hat{f}_{P(\mathcal{T})}(t)$ can be neglected. Thus, we only consider k *Nearest Neighbors* (kNNs) of t, i.e., $N_{\mathcal{T}}(t)$. Moreover, if some sub-trajectories $t_i^* \in N_{\mathcal{T}}(t)$ are from the same trajectory, $\hat{f}_{P(\mathcal{T})}(t)$ might be high but the actual density is low. To remedy this issue, we add an extra condition to $N_{\mathcal{T}}(t)$ such that they come from *distinct* trajectories. Let $\tilde{N}_{\mathcal{T}}(t)$ be the k nearest sub-trajectories of t from k distinct trajectories of $\mathcal{T}$. We use $\tilde{N}_{\mathcal{T}}(t)$ instead of $P(\mathcal{T})$ to estimate the density of t, i.e., $\hat{f}_{\tilde{N}_{\mathcal{T}}}(t) = \frac{1}{|\mathcal{T}|} \sum_{t_i^* \in \tilde{N}_{\mathcal{T}}(t)} \exp(-\frac{nDTW(t,t_i^*)^2}{2\sigma^2})$.

Density of Succeed Sub-trajectories. To evaluate the density variation of t, we need to estimate $succ(t)$'s density, i.e., $\hat{f}_{\tilde{N}_{\mathcal{T}},succ}(t) = \frac{1}{|\mathcal{T}|} \sum_{t_i^* \in \tilde{N}_{\mathcal{T}}(t)} \exp(-\frac{\Delta_i^2}{2\sigma^2})$, where $\Delta_i = \max\{nDTW(t,t_i^*), nDTW(succ(t),succ(t_i^*))\}$. Compared to $\hat{f}_{\tilde{N}_{\mathcal{T}}}(t)$, we only use the succeeds from the same $\tilde{N}_{\mathcal{T}}(t)$ to estimate $\hat{f}_{\tilde{N}_{\mathcal{T}},succ}(t)$ because we aim to evaluate the density variation of t, so we only consider the density of $succ(t)$ from the same direction of t. Moreover, to evaluate $\hat{f}_{\tilde{N}_{\mathcal{T}},succ}(t)$ precisely, we use Δ_i to add penalty if any $succ(t_i^*)$ is no longer close to $succ(t)$.

2.4   Obstacle Detection

Finally, we follow the Association Rule [1] and DBSCAN [4] and adopt the standard z-test of hypothesis testing to define the obstacle.

Definition 2 (Obstacle). *Given two thresholds τ ($\tau > 0$) and δ ($\delta > 0$), obstacles are detected from two kinds of sub-trajectories $P(\mathcal{T})$ and $P(\mathcal{Q})$ (Relativity). An obstacle is a set of last points from a subset of $P(\mathcal{T})$ such that for a subset of close query sub-trajectories $q \in P(\mathcal{Q})$, each $t \in N_{\mathcal{T}}(q)$ should satisfy:*

- (Significance) *The density variation score of t is significant, i.e.,*

$$score(t) = \frac{\hat{p}_1 - \hat{p}_2}{\sqrt{\hat{p}(1-\hat{p})(\frac{1}{\hat{f}_{\tilde{N}_{\mathcal{T}}}(t)} - \frac{1}{\hat{f}_{\tilde{N}_{\mathcal{Q}}}(t)})}} > \tau, \qquad (1)$$

where $\hat{p}_1 = \frac{\hat{f}_{\tilde{N}_{\mathcal{T}},succ}(t)}{\hat{f}_{\tilde{N}_{\mathcal{T}}}(t)}$, $\hat{p}_2 = \frac{\hat{f}_{\tilde{N}_{\mathcal{Q}},succ}(t)}{\hat{f}_{\tilde{N}_{\mathcal{Q}}}(t)}$, *and* $\hat{p} = \frac{\hat{f}_{\tilde{N}_{\mathcal{T}}}(t)\cdot\hat{p}_1 + \hat{f}_{\tilde{N}_{\mathcal{Q}}}(t)\cdot\hat{p}_2}{\hat{f}_{\tilde{N}_{\mathcal{T}}}(t) + \hat{f}_{\tilde{N}_{\mathcal{Q}}}(t)}$.

- (Support) $\tilde{N}_{\mathcal{T}}(t)$ *and* $\tilde{N}_{\mathcal{Q}}(t)$ *are close to t, i.e.,*

$$\hat{f}_{\tilde{N}_{\mathcal{T}}}(t) > \delta \text{ and } \hat{f}_{\tilde{N}_{\mathcal{Q}}}(t) > \delta. \qquad (2)$$

In Definition 2, inspired by Association Rule, we use the ratio $\hat{p}_1$ and $\hat{p}_2$ to denote the density variation of t in $P(\mathcal{T})$ and $P(\mathcal{Q})$, respectively. We then adopt the one-sided z-test to compute the significance of the density variation (Inequality 1). We use Inequality 2 to keep the closeness between t and its nearest sub-trajectories. Additionally, we follow the definition of DBSCAN such that: (1) the query sub-trajectories are close to each other; otherwise, the obstacle can be divided into multiple regions; (2) we use the last points of the selected $t \in P(\mathcal{T})$ to construct an obstacle so that it can be of arbitrary shape.

According to Definition 2, the obstacles are usually different depending on $P(\mathcal{Q})$. Thus, we formalize the obstacle detection as an online query problem:

Definition 3 (Obstacle Detection). *Given a set of reference sub-trajectories $P(\mathcal{T})$ and two thresholds τ ($\tau > 0$) and δ ($\delta > 0$), the problem of obstacle detection is to construct a data structure which, for a collection of query sub-trajectories $P(\mathcal{Q})$, finds all implicit obstacles as defined in Definition 2.*

3   DIOT

3.1   The Basic Framework

The basic DIOT framework consists of two phases: pre-processing phase and query phase, which are described as follows.

In the pre-processing phase, we first partition the reference trajectories $\mathcal{T}$ into $P(\mathcal{T})$. Then, we build an HNSW graph $G_{\mathcal{T}}$ for $P(\mathcal{T})$. We choose HNSW [5] because (1) it is very efficient for Nearest Neighbor Search (NNS) [2]; (2) $G_{\mathcal{T}}$ directly stores $N_{\mathcal{T}}(t)$ for each $t \in P(\mathcal{T})$. To determine $\tilde{N}_{\mathcal{T}}(t)$, we only need to check whether the new candidate comes from distinct trajectories.

In the query phase, given a set of query trajectories $\mathcal{Q}$, we first partition $\mathcal{Q}$ into $P(\mathcal{Q})$ and build an HNSW graph $G_{\mathcal{Q}}$ for $P(\mathcal{Q})$. After indexing $\mathcal{Q}$, we initialize an empty set $\mathcal{S}$ to store obstacles and use a bitmap to flag each $q \in P(\mathcal{Q})$ checked or not. Then, we find the candidate sub-trajectories $\mathcal{C}$ using the Depth First Search (DFS) method for each $q \in P(\mathcal{Q})$. We construct an obstacle O by the last points of $\mathcal{C}$ and add O to $\mathcal{S}$ if $|\mathcal{C}| > 0$. We return $\mathcal{S}$ as the answer.

We then illustrate the DFS method to find the candidate sub-trajectories $\mathcal{C}$. For each $q \in P(\mathcal{Q})$, we first find its $N_{\mathcal{T}}(q)$ from $G_{\mathcal{T}}$. For each $t \in N_{\mathcal{T}}(q)$, we compute $\hat{f}_{\tilde{N}_{\mathcal{T}}}(t)$, $\hat{f}_{\tilde{N}_{\mathcal{T}},succ}(t)$, $\hat{f}_{\tilde{N}_{\mathcal{Q}}}(t)$, $\hat{f}_{\tilde{N}_{\mathcal{Q}},succ}(t)$ and validate whether both Inequalities 1 and 2 are satisfied or not. We add t to $\mathcal{C}$ if both satisfied. After checking all $t \in N_{\mathcal{T}}(q)$, if $\mathcal{C}$ is not empty, i.e., $|\mathcal{C}| > 0$, which means there may exist an obstacle, we continue to find the candidate sub-trajectories from the close query sub-trajectories of q, i.e., $N_{\mathcal{Q}}(q)$, until no further new candidate can be found or all $N_{\mathcal{Q}}(q)$'s have been checked.

3.2   Optimizations

The basic obstacle detection framework can perform well on many data sets. We now develop four insightful strategies for further optimization.

Pre-compute $\tilde{N}_T(t)$. For each $t \in N_T(q)$, we need to conduct distinct k-NNS twice to find $\tilde{N}_T(t)$ from G_T and $\tilde{N}_Q(t)$ from G_Q, respectively. Notice that the operation to find $\tilde{N}_T(t)$ is independent of Q. Thus, we determine all $\tilde{N}_T(t)$'s in the pre-processing phase. Although the complexity remains the same, this strategy can speedup the query phase as it is a very frequent operation.

Build a Bitmap of $P(T)$. During the query phase, some $t \in P(T)$ may be checked multiple times. For example, suppose $q_1, q_2 \in P(Q)$ and they are close to each other. If $t \in N_T(q_1)$, it is very likely that $t \in N_T(q_2)$. As such, we need to check t twice. To avoid redundant computations, we also build a bitmap of $P(T)$ to flag each $t \in P(T)$ checked or not.

Skip the Close $q_i \in N_Q(q)$. As we call the DFS method recursively, we do not need to consider all $q_i \in N_Q(q)$ as many $t \in N_T(q_i)$'s have been checked in previous recursions. Let ϵ be a small distance threshold. Before the recursion of q_i, we first check its closeness to q. If $nDTW(q_i, q) < \epsilon$, as $N_T(q_i)$ are almost identical to $N_T(q)$, we set $flag[q_i] = true$ and skip this recursion.

Skip the Close $t \in N_T(q)$. Similar to the motivation of skipping the close $q_i \in N_Q(q)$, we do not need to check all $t \in N_T(q)$. Specifically, for each $t \in N_T(q)$, if there exists a $t_i \in N_T(t)$ that has been checked and $nDTW(t, t_i) < \epsilon$, we directly follow the same operation of t_i to t to avoid duplicated computations.

4   Experiments

We study the performance of DIOT for obstacle detection. We use two real-life data sets Vessel and Taxi for validation, which are described as follows.

Vessel is a collection of GPS records of the vessels near Singapore Strait during May to September 2017. We find three sunken vessels, i.e., *Harita Berlian, Thorco Cloud, and Cai Jun 3*, with available operating geo-location area. Thus, we select the trajectories in non-operating time as references and the trajectories around the operating region in the operating time as queries.

Taxi [7] is a set of trajectories of 14,579 taxis in Singapore. We study the effect of Electronic Road Pricing (ERP) gantries, which is an electronic system of road pricing in Singapore. We select the taxi trajectories with free state as references and those in *morning peak hour* and *afternoon peak hour* when the ERP is working as queries. We suppose taxi drivers do not pass through the ERP gantries when the taxi is free. Thus, we use the location of working ERP gantries as the ground truths.

We use interpolation to align the trajectories. Due to the different nature of Vessel and Taxi, we use 600 and 30 s as the interval of interpolation, respectively. We set $w = 6$ and $s = 1$ for partitioning and use $k = 8$ for kNNs.

4.1   Quantitative Analysis

We first conduct the quantitative analysis of DIOT. We report the highest F1-scores of DIOT from a set of $\delta \in \{0.5, 1.0, \cdots, 4.0\}$ and $\tau \in \{1.282, 1.645, 1.960, 2.326, 2.576\}$ using grid search. The results are depicted in Table 1.

Table 1. The results of quantitative analysis, where Time refers to the query time.

Query set	DIOT without optimization				DIOT with optimization			
	Precision	Recall	F1-score	Time	Precision	Recall	F1-score	Time
Harita Berlian	100.0	100.0	100.0	209.16	100.0	100.0	100.0	109.87
Thorco Cloud	50.0	100.0	66.7	18.46	100.0	100.0	100.0	12.31
Cai Jun 3	25.0	100.0	40.0	15.06	20.0	100.0	33.3	7.23
Morning ERP	51.3	88.2	64.9	18.14	50.0	82.4	62.2	4.67
Afternoon ERP	41.7	68.0	51.7	25.47	47.6	52.0	49.7	7.32

For Vessel, since each query has only one ground truth obstacle, DIOT can detect all of them with 100% recall. As the obstacle pattern of Harita Berlian is obvious, its F1-score is uniformly higher than that of Thorco Cloud and Cai Jun 3. DIOT has a slightly lower F1-score for Cai Jun 3 because its operating area is not at the centre of the vessel route. For Taxi, more than 50% and 40% detected obstacles fit the ERP gantries for Morning and Afternoon ERP queries, respectively. These results validate our assumption that taxi drivers tend to avoid ERP gantries when their taxis are free. Table 1 also shows that DIOT with optimization is 2~4 times faster than that without optimization under the similar accuracy, which confirms the effect of the optimization strategies.

4.2 Case Studies

To validate the actual performance of DIOT, we conduct case studies for some typical obstacles found from Vessel and Taxi.

Vessel: Thorco Cloud. Figure 3(a) shows the obstacle discussed in Sect. 1. The orange polygon represents the operating area (actual obstacle). One can regard the convex hull formed by the red circles as the returned obstacle. During the operating time, the vessels (blue curves) have clear pattern to avoid the operating area, while during the non-operating time, the vessels (green curves) move freely. This discrepancy is successfully captured by DIOT, and the location of the detected obstacle region (red circles) fits the operating area.

Taxi: Morning ERP. Figure 3(b) depicts the typical obstacles caused by ERP gantries. The orange stars are the location of ERP gantries. One can find explicit correlations between the returned obstacles and the ERP gantries. For example, as shown in Rectangle A, the star represents the ERP gantry in the Bt Timah Expressway street whose operating time is 7:30–9:00 am weekdays. The query trajectories (blue curves) have significantly less tendency to go towards the ERP gantry. Moreover, some obstacles that are not directly associated to the ERP gantries might be caused by the ERP as well. For instance, the detected obstacle in Rectangle B is directly towards to the Central Express Street in Singapore that ends with some ERP gantries. Thus, their correlation might be even higher than the precision and recall values shown in Table 1.

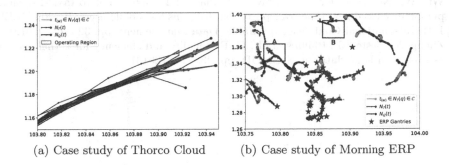

(a) Case study of Thorco Cloud   (b) Case study of Morning ERP

Fig. 3. Case study (Color figure online)

5   Conclusions

In this paper, we study a new data mining problem of obstacle detection that has applications in many scenarios. We focus on the trajectory data and introduce a density-based definition for the obstacle. Moreover, we introduce a novel framework DIOT for obstacle detection and develop four insightful strategies for optimization. Experimental results on two real-life data sets demonstrate that DIOT enjoys superior performance yet captures the essence of obstacles.

Acknowledgements. This research is supported by the National Research Foundation, Singapore under its Strategic Capability Research Centres Funding Initiative. Any opinions, findings and conclusions or recommendations expressed in this material are those of the author(s) and do not reflect the views of National Research Foundation, Singapore.

References

1. Agrawal, R., Imieliński, T., Swami, A.: Mining association rules between sets of items in large databases. In: SIGMOD, pp. 207–216 (1993)
2. Aumüller, M., Bernhardsson, E., Faithfull, A.: Ann-benchmarks: a benchmarking tool for approximate nearest neighbor algorithms. Inf. Syst. **87**, 101374 (2020)
3. Ding, H., Trajcevski, G., Scheuermann, P., Wang, X., Keogh, E.: Querying and mining of time series data: experimental comparison of representations and distance measures. PVLDB **1**(2), 1542–1552 (2008)
4. Ester, M., Kriegel, H.P., Sander, J., Xu, X., et al.: A density-based algorithm for discovering clusters in large spatial databases with noise. In: KDD, vol. 96, pp. 226–231 (1996)
5. Malkov, Y.A., Yashunin, D.A.: Efficient and robust approximate nearest neighbor search using hierarchical navigable small world graphs. TPAMI (2018)

6. Myers, C.S., Rabiner, L.R.: A comparative study of several dynamic time-warping algorithms for connected-word recognition. Bell Syst. Tech. J. **60**(7), 1389–1409 (1981)
7. Wu, W., Ng, W.S., Krishnaswamy, S., Sinha, A.: To taxi or not to taxi?-Enabling personalised and real-time transportation decisions for mobile users. In: 2012 IEEE 13th International Conference on Mobile Data Management, pp. 320–323 (2012)
8. Zheng, Y., Jestes, J., Phillips, J.M., Li, F.: Quality and efficiency for kernel density estimates in large data. In: SIGMOD, pp. 433–444 (2013)

Exploring Sub-skeleton Trajectories for Interpretable Recognition of Sign Language

Joachim Gudmundsson, Martin P. Seybold, and John Pfeifer[✉]

University of Sydney, Sydney, Australia
johnapfeifer@yahoo.com

Abstract. Recent advances in tracking sensors and pose estimation software enable smart systems to use trajectories of skeleton joint locations for supervised learning. We study the problem of accurately recognizing sign language words, which is key to narrowing the communication gap between hard and non-hard of hearing people.

Our method explores a geometric feature space that we call 'sub-skeleton' aspects of movement. We assess similarity of feature space trajectories using natural, speed invariant distance measures, which enables clear and insightful nearest neighbor classification. The simplicity and scalability of our basic method allows for immediate application in different data domains with little to no parameter tuning.

We demonstrate the effectiveness of our basic method, and a boosted variation, with experiments on data from different application domains and tracking technologies. Surprisingly, our simple methods improve sign recognition over recent, state-of-the-art approaches.

1 Introduction

The problem of automatically and accurately identifying the meaning of human body movement has gained research interest due to advances in motion capture systems, artificial intelligence algorithms, and powerful hardware. Motion capture systems such as Microsoft's Kinect or the Leap Motion controller have been used to capture motion of human actors for recently released benchmark data sets (e.g. KinTrans [8], LM [5], NTU RGB+D [13]).

This work studies the problem of recognizing patterns in human sign language. Human sign languages are systems of communication that use manual movement patterns of arms and hands as well as non-manual elements, such as head, cheek and mouth posture, to convey meaning. We focus on data sets that consist of extracted, single word labeled inputs along with the sequences

---

See https://arxiv.org/abs/2202.01390 for the full version of this paper.

Supplementary Information The online version contains supplementary material available at https://doi.org/10.1007/978-3-031-00123-9_19.

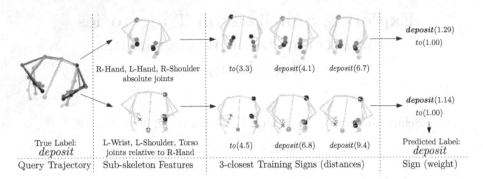

Fig. 1. Example of the proposed kNN-m classification ($k = 3$) of a query trajectory based on speed-invariant distances of one *absolute* (top) and one *relative* (bottom) sub-skeleton movement pattern.

of skeletal joint locations of the actors. Our goals are to attain high classification accuracy with acceptable learning and query latency on diverse and evolving data sources, requiring little to no parameter tuning. In contrast to deep learning models that assume static environments, we are particularly interested in simple, interpretable methods that, in turn, provide insight for curation of evolving, publicly available catalogs of sign language.

Contribution and Paper Structure. The main idea of our method is to map human body skeletal movement into a feature space that captures absolute and relative movement of sets of joints (see Fig. 1). Based on the high-dimensional feature trajectories, we apply simple classifiers that use geometric similarity measures (see Sect. 2). The main technical difficulty is determining the most discriminative way to transform spatial joint locations into high-dimensional trajectories. Motivated by sign language, we effectively navigate the vast space of absolute and relative movement patterns that subsets of joints describe, which we call 'sub-skeleton' features.

We propose a general, novel method that *automatically mines* a set of maximal discriminating sub-skeleton features, irrespective of the present skeleton formats (see Sect. 3). Our main contributions are as follows:

– Our mining method discovers sub-skeleton aspects in the data sets that are highly discriminating. Since all feature trajectories directly relate to the input data, the merits of each classification result can be *interpreted* and visualized naturally in terms of a geometric similarity measure (e.g. Fig. 2).
– Simple Nearest Neighbor and Ensemble Boosted Classification achieve improved accuracy on sign-language benchmarks of diverse tracking technologies. Competitive accuracy on Human Action benchmarks show that our method generalizes to other recognition problems. Particularly noteworthy are our high accuracy results on *very small* training sets (see Sect. 4).

– To the best of our knowledge, we are the first to assess sub-skeleton features with trajectory similarity measures (e.g. the Fréchet distance) for sign language and human action classification problems.

Furthermore, our publicly available[1] implementation achieves average query latency below 100 ms on standard, non-GPU, desktop hardware.

2 Setup and Problem Definition

A skeleton G is an undirected connected graph where each vertex represents a joint. In our setting G models a part of a human body skeleton where vertices are adjacent if their joints have a rigid connection (e.g. a bone). As input for a word signed by an actor, we are given a *sequence* S of n frames $\langle G_1, \ldots, G_n \rangle$, and each G_i holds the 3D location of the joints at time step i. Frame frequency is typically constant but depends heavily on the capture technology. Sequences have varying duration, for example, between 0.2 and 30 s (cf. Table 1). Every sequence is labeled with one class and the actor's body movement is performed once, or a small number of times, per sign class.

The input data $\mathbb{D}$ is a set of sequences $\{S_1, \ldots, S_\ell\}$ and is partitioned into two sets: the training set $\mathbb{D}_r$ and the testing set $\mathbb{D}_t$. Our aim is to classify each sequence in the test set, using only sequences from the training set, as accurately and efficiently as possible.

3 Mining Sub-skeleton Features

Sign language conveys meaning based on the movement patterns of skeletal joints. Our goal is to mine joint movement and determine which combination of joints best discriminates classes. We first describe three feature mining concepts that are important in our setting, and then present our mining algorithm.

Absolute and Relative Joints. The movement of a joint can be viewed in an absolute space or relative space. We say that a joint is absolute if its position is described in a fixed coordinate system. For example, the left index fingertip joint moving through space is an absolute joint. A joint whose movement is described in terms of its position relative to another joint whose spatial location serves as the center of a moving coordinate system is called a relative joint, e.g., the joints of the right arm relative to the neck joint.

Feature Space. Consider an underlying feature space F that contains all possible absolute and relative joint subset combinations derived from skeleton G. Note that F contains $|G|^2$ *singletons*, i.e., $|G|$ absolute and $|G|^2 - |G|$ relative joints. Any *feature $f \in F$* can be derived as the union of $|f|$ singletons, regardless of f being an absolute or relative sub-skeleton. Thus, the size of F is *exponential* in the number of joints $|G|$ of the skeleton.

---

[1] https://github.com/japfeifer/frechet-queries.

Feature Trajectory. Any given feature $f \in F$ maps input skeletal movement to feature trajectories, i.e. f maps $\langle G_1, \ldots, G_n \rangle$ to $\langle p_1, \ldots, p_n \rangle$ and each point p_i has dimensionality $3|f|$, where $|f|$ denotes the number of joints in the set.

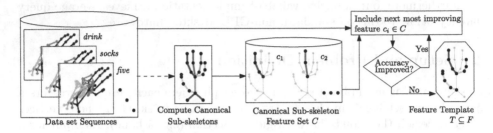

Fig. 2. Automated mining from sub-skeleton features produces a highly discriminative feature template. A feature set C of canonical sub-skeletons is determined from skeletal joints from G. Then, a greedy algorithm constructs a feature template T by testing and adding the next best feature $c_i \in C$, until the accuracy no longer improves.

Greedy Feature Template Mining Algorithm. Since the feature space is huge, we desire an algorithm that quickly chooses a small number of features from F, and achieves high accuracy on a classifier that is simple and interpretable. Thus, we now describe our mining algorithm that *greedily* selects a small number of discriminating features $\{f_1, \ldots, f_l\}$ from F.

The set of *canonical sub-skeletons* $C \subseteq F$ form the basis of our greedy exploration. In the case where $|G|$ is not too large, C contains all singletons. In the case where $|G|$ is very large, we derive a smaller set C by merging similar singletons together into a single set (e.g., four singleton joints of the index finger are merged into a single 'index finger' set). The singleton groups are computed by identifying all chains of degree 2 vertices in G (e.g., right leg, left arm), and for the case of deriving relative canonical sub-skeletons we use central joints (e.g., neck or torso) as reference joints. Figure 2 shows an example of merging similar hand singletons together to reduce the size of C.

The mining algorithm computes a discriminating *feature template* $T \subseteq F$. There are up to $1 + |G|$ features in T: one feature that contains the union of one or more *absolute* singletons, and $|G|$ *relative* features, each of which contain the union of one or more singletons that have the *same* reference joint. We denote with $l(T)$ the number of features from F that are defined by T, so $l(T) \leq 1 + |G|$.

The algorithm constructs T by iteratively performing *adapted union* operations, denoted $\tilde{\cup}$. In this context, a given relative joint $c \in C$ is added to the feature $f \in T$ that has the same reference joint, and a given absolute joint $c \in C$ is added to the feature $f \in T$ that contains the absolute joints. For example, say T contains two features f_1 and f_2, where $f_1 \in T$ is relative to the right elbow and $f_2 \in T$ contains absolute joints. If a joint c is relative to the right elbow, then it will be added to f_1. If a joint c is an absolute joint, then it will be added to f_2. If a joint c is relative to the neck, then it will be added to an initially empty feature f_3.

We use standard classifiers to determine which elements of C to add to the feature template. The simplest is a Nearest Neighbor method that finds the closest label in $\mathbb{D}_r$ based on a trajectory distance measure. To employ underlying classifiers during our greedy exploration, we initially partition the training set $\mathbb{D}_r$ randomly with a 1:2-split into $\mathbb{D}'_r$ and $\mathbb{D}'_t$ and proceed as follows:

1. Start with the empty set $T = \emptyset$.
2. Compute $\forall c \in C \setminus T$ the classification accuracy of $T \cup c$ on $\mathbb{D}'_r$ and $\mathbb{D}'_t$.
3. If one such best c improves over the last iteration,
   then add c to T and GOTO 2.
4. Return T.

The resulting set T, which contains a subset of canonical sub-skeletons, describes a small number of absolute and relative features by means of the aforementioned adapted union operation. Hence we have an equal number of generated feature trajectories for each input sequence. Clearly, the algorithm can be executed for different classifiers and trajectory similarity measures to select the best feature set found.

Since this deterministic greedy exploration is based on very simple classifiers, with very few or no parameters, the 'overfitting' problem of complex models is avoided. Nevertheless, the obtained classification accuracy is already comparatively high (see Sect. 4).

Note that the construction of the feature template from $\mathbb{D}'_r$ and $\mathbb{D}'_t$ is only to accommodate a fair comparison of the final accuracy against other recognition methods. For the purpose of pure pattern discovery, one may well use the whole data set $\mathbb{D}$ in the selection process.

4   Experimental Setup and Results

We implemented our methods and ran experiments on five skeleton-based data sets of actors that perform American Sign Language and Human Actions, each captured with different tracking technologies. Through experiments, we investigate if (i) sub-skeleton feature mining discovers highly discriminative movement patterns, (ii) classification accuracy improves on state-of-the art methods for the benchmarks and on small training sets, and (iii) classification queries are answered quickly. Experiments ran on a standard desktop computer (3.6GHz Core i7-7700 CPU, non-GPU) with MATLAB R2020a software on Windows 10, except for those using competitor ST-GCN [15], which ran on a high-performance computing environment with an NVIDIA V100 GPU and up to 32 CPUs.

Table 1. Test data sets, showing type (Sign Language or Human Action), number of skeletal joints, sequences, classes and subjects; Mean±SD of frames per sequence, frames per second in Hz, and mean sequences per class.

Benchmark	Type	Jts.	Seq	Class	Sub	Frames/Seq	F./s	Seq/Cl.
KinTrans	SL-Body	10	5,166	73	n/a	40 ± 13	30	71
LM	SL-Hands	54	17,312	60	25	71 ± 21	30	289
NTU60	HA-Body	25	44,887	60	40	78 ± 34	30	748
UCF	HA-Body	15	1,280	16	16	66 ± 34	30	80
MHAD	HA-Body	15	660	11	12	$3,602 \pm 2,510$	480	60

Table 2. Classification accuracy results comparing our methods against others. For sign language data sets, we achieve the best results by a large margin on low training information, our simple kNN-m classifier is often better than others (which use complex neural-networks), and our DM-m method performs best in all tests. Although our focus is sign language recognition, our human action data set results show that our methods generalize well and achieve high accuracies. Accuracies in *italics* were reported in previous work.

Benchmark	Paper	Method	Acc.
KinTrans 2 T/C	[15]	ST-GCN	57.6
	Ours	2NN-m (DTW)	79.4
		DM-m (CF)	75.0
		DM-m (DTW)	**80.1**
KinTrans 3 T/C	[15]	ST-GCN	76.7
	Ours	2NN-m (DTW)	84.5
		DM-m (CF)	84.9
		DM-m (DTW)	**90.2**
KinTrans 10% T/C	[15]	ST-GCN	96.7
	Ours	2NN-m (DTW)	96.4
		DM-m (CF)	96.2
		DM-m (DTW)	**98.4**
KinTrans 20% T/C	[15]	ST-GCN	99.2
	Ours	2NN-m (DTW)	99.3
		DM-m (CF)	96.7
		DM-m (DTW)	**99.5**
LM 5-fold XSub	[5]	Kine.-LSTM	*91.1*
	Ours	1NN-s (CF)	80.6
		3NN-m (CF)	71.3
		DM-m (DTW)	**94.3**

(a) Sign language data sets.

Benchmark	Paper	Method	Acc.
NTU60 XSub	[15]	ST-GCN	**78.7**
	Ours	DM-m (DTW)	70.8
NTU60 XView	[15]	ST-GCN	**86.8**
	Ours	DM-m (DTW)	82.8
UCF 4-fold	[3]	Log. Reg.	*95.9*
	[11]	SVM	*97.1*
	[12]	dHMM	*97.7*
	[14]	SVM	*97.9*
	[16]	kNN+Vote	*98.5*
	[7]	SVM	*98.8*
	[2]	DP+kNN	*99.2*
	[15]	ST-GCN	**99.7**
	Ours	1NN-s (DTW)	96.9
		DM-m (DTW)	99.5
MHAD XSub	[9]	k-SVM	*80.0*
	[15]	ST-GCN	89.8
	[10]	SVM	*95.4*
	[6]	CNN	*98.4*
	[1]	MKL-SCM	*100*
	Ours	1NN-s (CF)	94.9
		DM-m (DTW)	**100**

(b) Human Action data sets.

Feature Template Mining. The KinTrans feature mining took 47 min for a specific normalization combination, which we consider reasonable on a standard desktop computer. As a guide, the respective ST-GCN training in Table 2a took more than 11 h on the high-performance GPU computing environment.

Classification Accuracy. We compare the accuracy of our simple classifiers against various state-of-the-art skeleton-based recognition approaches using five diverse data sets from Table 1. To investigate the effectiveness of our method on sign language data sets, we compare our accuracy results against: (i) publicly available code of the recent ST-GCN [15] method (that exclusively supports body skeletons) for our KinTrans data set, and (ii) LM data set accuracy results of [5] (see Table 2a).

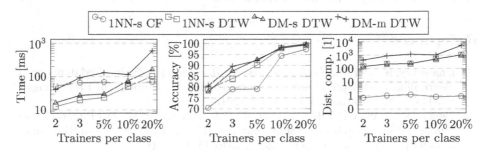

Fig. 3. Query latency of our classifiers for various training sets sizes using the KinTrans data set. Shown are avg. time per query (left), overall classification acc. (middle), and avg. num. of trajectory distance computations per query (right). (Color figure online)

In the lowest training information experiment (2 trainers per class) we have almost 40% accuracy improvement over ST-GCN. As more training information is introduced, we still outperform ST-GCN, even with our simple kNN-s classifier.

The experiments show that the accuracy performance of our simple methods generalizes well over different data domains (sign language/human action), training set sizes (small/large), and skeleton capture formats (coarse body/detailed hands).

Query Time. To investigate the query latency of our distance based classifiers, we run experiments on training sets of different sizes from the KinTrans data set with each of our methods. The testing sets always contain the remainder of the whole data set and the DM methods compute the full number of distance columns. For the metric distance measure CF, we use the k Nearest-Neighbor search structure from [4]. The results in Fig. 3 show average wall-clock time per query, overall classification accuracy, and average number of necessary distance computations per query for our 1NN-s (blue) and DM methods (black). All 1NN-s and DM-s classifiers show an average query time under 200 ms.

5 Conclusion

Our work on skeleton-based sign language recognition introduced the sub-skeleton feature space and studied it using speed-invariant similarity measures. Our method automatically discovers absolute and relative movement patterns, which enables highly accurate Nearest Neighbor classification, with acceptable latency, on training data of varying domains, skeleton formats, and sizes. Our distance based classifiers are interpretable and train on basic computing hardware, which make them particularly interesting for data sets that change frequently.

Acknowledgements. The authors acknowledge the technical assistance provided by the Sydney Informatics Hub, a Core Research Facility of the University of Sydney. This work was supported under the Australian Research Council Discovery Projects funding scheme (project number DP180102870).

References

1. Chaudhry, R., Ofli, F., Kurillo, G., Bajcsy, R., Vidal, R.: Bio-inspired dynamic 3D discriminative skeletal features for human action recognition. In: Proceedings of IEEE CVPR Workshops, pp. 471–478 (2013)
2. Devanne, M., Wannous, H., Berretti, S., Pala, P., Daoudi, M., Del Bimbo, A.: 3-D human action recognition by shape analysis of motion trajectories on Riemannian manifold. IEEE Trans. Cybern. **45**(7), 1340–1352 (2014)
3. Ellis, C., Masood, S., Tappen, M., LaViola, J., Sukthankar, R.: Exploring the trade-off between accuracy and observational latency in action recognition. Int. J. Comput. Vis. **101**(3), 420–436 (2013)
4. Gudmundsson, J., Horton, M., Pfeifer, J., Seybold, M.P.: A practical index structure supporting Fréchet proximity queries among trajectories. ACM Trans. Spat. Alg. Syst. **7**(3), 1–33 (2021). https://doi.org/10.1145/3460121
5. Hernandez, V., Suzuki, T., Venture, G.: Convolutional and recurrent neural network for human activity recognition: application on American sign language. PLoS ONE **15**(2), 1–12 (2020)
6. Ijjina, E.P., Mohan, C.K.: Human action recognition based on mocap information using convolution neural networks. In: 2014 13th International Conference on Machine Learning and Applications, pp. 159–164. IEEE (2014)
7. Kerola, T., Inoue, N., Shinoda, K.: Spectral graph skeletons for 3D action recognition. In: Cremers, D., Reid, I., Saito, H., Yang, M.-H. (eds.) ACCV 2014. LNCS, vol. 9006, pp. 417–432. Springer, Cham (2015). https://doi.org/10.1007/978-3-319-16817-3_27
8. linedanceAI: The KinTrans Project (2020). https://www.kintrans.com
9. Ofli, F., Chaudhry, R., Kurillo, G., Vidal, R., Bajcsy, R.: Berkeley MHAD: a comprehensive multimodal human action database. In: 2013 IEEE Workshop on Applications of Computer Vision, pp. 53–60. IEEE (2013)
10. Ofli, F., Chaudhry, R., Kurillo, G., Vidal, R., Bajcsy, R.: Sequence of the most informative joints (SMIJ): a new representation for human skeletal action recognition. JVCIR **25**(1), 24–38 (2014)
11. Ohn-Bar, E., Trivedi, M.: Joint angles similarities and HOG2 for action recognition. In: Proceedings of IEEE CVPR Workshops, pp. 465–470 (2013)

12. Presti, L.L., La Cascia, M., Sclaroff, S., Camps, O.: Hankelet-based dynamical systems modeling for 3D action recognition. IVCJ **44**, 29–43 (2015)
13. Shahroudy, A., Liu, J., Ng, T.T., Wang, G.: NTU RGB+D: a large scale dataset for 3D human activity analysis. In: Proceedings of IEEE CVPR, pp. 1010–1019 (2016)
14. Slama, R., Wannous, H., Daoudi, M., Srivastava, A.: Accurate 3D action recognition using learning on the Grassmann manifold. Pattern Recogn. **48**(2), 556–567 (2015)
15. Yan, S., Xiong, Y., Lin, D.: Spatial temporal graph convolutional networks for skeleton-based action recognition. In: Proceedings of 32nd AAAI, pp. 7444–7452 (2018)
16. Zanfir, M., Leordeanu, M., Sminchisescu, C.: The moving pose: an efficient 3D kinematics descriptor for low-latency action recognition and detection. In: Proceedings of IEEE ICCV, pp. 2752–2759 (2013)

Significant Engagement Community Search on Temporal Networks

Yifei Zhang, Longlong Lin, Pingpeng Yuan$^{(\boxtimes)}$, and Hai Jin

National Engineering Research Center for Big Data Technology and System, Services Computing Technology and System Lab, Cluster and Grid Computing Lab, School of Computer Science and Technology, Huazhong University of Science and Technology, Wuhan 430074, China
{yfzhangsz,longlonglin,ppyuan,hjin}@hust.edu.cn

Abstract. Community search, retrieving the cohesive subgraph which contains the user-specified query vertex, has been widely touched over the past decades. The existing studies on community search mainly focus on static networks. However, real-world networks, such as scientific cooperation networks and communication networks, usually are temporal networks whose each edge is associated with timestamps. Therefore, the previous methods do not work when handling temporal networks. Inspired by this, we study the problem of identifying the significant engagement community to which the user-specified query belongs. Specifically, given an integer k and a query vertex u, then we search for the subgraph $\mathcal{H}$ which satisfies (i) $u \in \mathcal{H}$; (ii) the de-temporal graph of $\mathcal{H}$ is a connected k-core; (iii) In $\mathcal{H}$ that u has the maximum engagement level. To address our problem, we first develop a top-down greedy peeling algorithm named $TDGP$, which iteratively removes the vertices with the maximum temporal degree. To further boost the efficiency, we then design a bottom-up local search algorithm named $BULS$ with several powerful pruning strategies. Lastly, we empirically show the superiority of our proposed solutions on six real-world temporal graphs.

Keywords: Temporal networks · Community search · k-core

1 Introduction

There are numerous community structures presented in real-world networks. Therefore, mining communities is an important tool for analyzing network structure and organization. Generally, there are two main research directions on community mining: (1) community detection identifies all communities by some predefined criteria [2,5,16]. However, it has intractable computational bottleneck and is not customized for user-specified query vertices. (2) community search aims to identify the community containing the user-specified query vertices [4,19], which is more efficient and personalized. Besides, community search

---

Y. Zhang and L. Lin—Contribute equally to this work.

© The Author(s), under exclusive license to Springer Nature Switzerland AG 2022
A. Bhattacharya et al. (Eds.): DASFAA 2022, LNCS 13245, pp. 250–258, 2022.
https://doi.org/10.1007/978-3-031-00123-9_20

can also be applied to numerous high-impact applications, including friend recommendation, link-spam detection, and drug discovery.

The relationships of real-world networks vary over time. For instance, a researcher collaborates with others on a project or a paper at some time. Persons call their friends from time to time. Such time-related connections among entities can be naturally modeled as temporal graphs [11,18], in which each edge is attached a timestamp to indicate when the connections occur. In such networks, an entity actively engages in a community via frequent connections with other entities at different periods while others may incur occasional relationships. Moreover, the entity has different engagement levels in different communities. It is more useful and challenging to study the engagement level of the entity in a community and identify the target community with the highest engagement level from all communities. Motivated by this, we introduce a new problem of identifying the significant engagement community to which a specified vertex belongs. In a nutshell, our contributions are reported as follows:

i) We propose a novel community search model named *SECS*, which comprehensively considers the structural cohesiveness of the community and the engagement level of the query vertex.

ii) To solve our problem, we propose a top-down greedy peeling algorithm *TDGP* and a more efficient bottom-up local search algorithms *BULS*.

iii) We conduct extensive experiments on six real-world temporal graphs, which reveal that our solutions perform well in efficiency and effectiveness.

2   Related Work

Community is a general concept appears in physics, computational biology, and computer science, and so on. Notable methods include modularity optimization [16], spectral analysis [5] and cohesive subgraph discovering [2]. Community search has recently been proposed for semi-supervised learning task that can recover the community in which the query vertex is located [7]. Namely, they aim at identifying the specified communities containing the given query vertices [14, 19]. There are more complete researches, such as community search on keyword-based graphs [15], location-based social networks [6], multi-valued graphs [10], and heterogeneous information networks [8]. However, they ignore the temporal properties of networks that frequently appear in applications.

Temporal networks can model the complex networks in a fine-grained manner, in which each interaction between vertices occurs at a specific time. Many models on temporal networks have been investigated [18]. Until recently, some work have been done on community mining over massive temporal networks [9,11,12,17]. For example, Lin et al. [13] introduced the diversified lasting cohesive subgraphs on temporal graphs. Li et al. [11] studied the problem of identifying maximum persistent communities from temporal networks. Unfortunately, they cannot tell the differences how the vertices participate in those subgraphs.

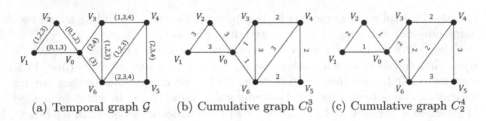

(a) Temporal graph $\mathcal{G}$ (b) Cumulative graph C_0^3 (c) Cumulative graph C_2^4

Fig. 1. Example of temporal graph $\mathcal{G}$ and its cumulative graphs

3   Significant Engagement Community Search

Here, we consider an undirected temporal graph $\mathcal{G}(\mathcal{V}, \mathcal{E}, \mathcal{T})$, in which $\mathcal{V}$ is the set of vertices inside the $\mathcal{G}$, $\mathcal{E} = \{(u, v, t) | u, v \in \mathcal{V}\}$ is the set of temporal edges and $\mathcal{T} = \{t | (u, v, t) \in \mathcal{E}\}$ is the timestamps set of $\mathcal{G}$. Additionally, we define $\mathcal{H} = (\mathcal{V}_\mathcal{H}, \mathcal{E}_\mathcal{H}, \mathcal{T}_\mathcal{H})$ as the temporal subgraph of $\mathcal{G}$ when $\mathcal{V}_\mathcal{H} \subseteq \mathcal{V}$, $\mathcal{E}_\mathcal{H} \subseteq \mathcal{E}$, and $\mathcal{T}_\mathcal{H} \subseteq \mathcal{T}$. $\mathcal{G}$'s de-temporal graph is $G(V, E)$, which meets the conditions that: $V = \mathcal{V}$ and $E = \{(u, v) | \exists (u, v, t) \in \mathcal{E}\}$. Namely, the de-temporal graph G is a static graph that ignores the temporal information carried on the edges. Similarly, we denote $H(V_H, E_H)$ as a subgraph of G when there it satisfies $V_H \subseteq V$ and $E_H \subseteq E$. To help formalize our problem, we put out several definitions as follows.

Definition 1 (Edge Occurrences). *Edge occurrence is a measure to demonstrate how many times the connections between two vertices occur within an interval of time. We first define the following function to indicate whether an edge exists:*

$$\pi(u, v, t) = \begin{cases} 0 & (u, v, t) \notin \mathcal{E} \\ 1 & (u, v, t) \in \mathcal{E} \end{cases} \tag{1}$$

So, the edge occurrences of (u, v) over time interval $[t_s, t_e]$ is defined as:

$$o_{(u,v)}(t_s, t_e) = \sum_{i=t_s}^{t_e} \pi(u, v, i) \tag{2}$$

Definition 2 (Cumulative Graph). *The cumulative graph of temporal graph $\mathcal{G}$ for time interval $[t_s, t_e]$ is a weighted graph $C_{t_s}^{t_e}(\mathbb{V}_{t_s}^{t_e}, \mathbb{E}_{t_s}^{t_e}, w_{t_s}^{t_e})$, in which the $\mathbb{V}_{t_s}^{t_e} = \{u | (u, v, t) \in \mathcal{E}, t \in [t_s, t_e]\}$, $\mathbb{E}_{t_s}^{t_e} = \{(u, v) | (u, v, t) \in \mathcal{E}, t \in [t_s, t_e]\}$, and $w_{t_s}^{t_e}(u, v) = o_{(u,v)}(t_s, t_e)$. Let $C_\mathcal{H}$ be the cumulative graph of $\mathcal{H}$ when the time interval is $[min(\mathcal{T}_\mathcal{H}), max(\mathcal{T}_\mathcal{H})]$. Besides, we have $\mathbb{N}_{u,C_\mathcal{H}} = \{v | (u, v, t) \in \mathcal{E}_\mathcal{H}\}$ and $\mathbb{D}_{u,C_\mathcal{H}} = |\{(u, v) | (u, v, t) \in \mathcal{E}_\mathcal{H}\}|$.*

Definition 3 (Temporal Degree). *The temporal degree of the vertex u w.r.t. $[t_s, t_e]$ and temporal graph $\mathcal{G}$ is defined as following:*

$$d_{u,\mathcal{G}}(t_s, t_e) = \sum_{i=t_s}^{t_e} \pi(u, v, i) = \sum_{v \in \mathcal{G}} w_{t_s}^{t_e}(u, v) \tag{3}$$

So, temporal degree of u in $\mathcal{G}$ is $d_{u,\mathcal{G}} = d_{u,\mathcal{G}}(min(\mathcal{T}_\mathcal{G}), max(\mathcal{T}_\mathcal{G}))$.

Definition 4 (Engagement Level). *For the temporal subgraph* $\mathcal{H}$, *engagement level of vertex* u *in* $\mathcal{H}$ *is the impact on* $\mathcal{H}$ *which* u *achieves. It is defined as:*

$$Eng_u(\mathcal{H}) = \frac{d_{u,\mathcal{H}}}{\sum_{v \in \mathcal{H}} d_{v,\mathcal{H}}} \tag{4}$$

Definition 5 (k-core [1]). *For a de-temporal graph* G, H *is a subgraph of* G. *We say* H *is a* k-core *in* G *if* $|\{v|(u,v) \in H\}| \geq k$ *for any vertex* $u \in H$ *holds.*

Example 1. Figure 1(a) shows a temporal graph $\mathcal{G}$ in which there are 7 vertices with 27 temporal edges, Fig. 1(b) and (c) are the cumulative graphs of $\mathcal{G}$ with time interval $[0, 3]$ and $[2, 4]$ respectively. There is a temporal subgraph $\mathcal{H}$ that $\mathcal{V}_{\mathcal{H}} = \{V_0, V_1, V_2\}$, $\mathcal{T}_{\mathcal{H}} = \mathcal{T}_{\mathcal{G}}$, and $\mathcal{E}_{\mathcal{H}} = \{(u,v,t)|(u,v,t) \in \mathcal{E}_{\mathcal{G}}, u, v \in \mathcal{V}_{\mathcal{H}}, t \in \mathcal{T}_{\mathcal{H}}\}$, we have $Eng_{V_0}(\mathcal{H}) = \frac{6}{18} = \frac{1}{3}$. Meanwhile, we can observe that $\mathcal{H}$ is a 2-core;

Our Problem (Significant Engagement Community Search: *SECS*). Given a temporal graph $\mathcal{G}$, a query vertex u, and a parameter k, our goal is to find a temporal subgraph $\mathcal{H}$ which meets: i) $u \in \mathcal{V}_{\mathcal{H}}$; ii) the de-temporal graph of $\mathcal{H}$ is a connected k-core; iii) $Eng_u(\mathcal{H}) \geq Eng_u(\mathcal{H}')$ for all temporal subgraph $\mathcal{H}'$. For simplicity, we call $\mathcal{H}$ is a significant engagement community (*SEC* for short) of u.

4   The Top-Down Greedy Peeling Algorithm

In this section, we introduce our proposed top-down greedy peeling algorithm (*TDGP*), which is shown in Algorithm 1. The first thing we need to do is generating the cumulative graphs from the temporal graph $\mathcal{G}$ (Line 1). Since there are $|T|$ timestamps, we can get in total $(1 + |T|)|T|/2$ time intervals. Each time interval corresponds to a cumulative graph. It should be noticed that we only consider the time intervals of which u has edges occur on its two ends. Though we cannot make sure that with this pruning strategy whether some k-core structures are ruined, however in this way we can pay attention to the time interval that u has action instead of the whole time interval of the temporal graph $\mathcal{G}$. Considering the cohesive constraint for *SEC*, for each cumulative graph $C_{\mathcal{H}}$, we need to maintain the de-temporal graph $\mathcal{H}$ as a k-core and check whether it contains the query vertex u (Line 3). While $C_{\mathcal{H}}$ meeting all these requirements, we try to reduce its extent to maximize $Eng_u(\mathcal{H})$. For this part, since there is no direct correlation between $d_{u,\mathcal{H}}$ and $Eng_u(\mathcal{H})$, we delete the vertices with the maximum temporal degree greedily in order to maximize $Eng_u(\mathcal{H})$, until it can not satisfy the conditions for *SEC* mentioned above. After all of these, we can finally get the community *SEC* in which u has the maximum engagement level. In this algorithm, we need to deal with in total $(1 + |T|)|T|/2$ amount of cumulative graphs. Here we take m to represent $|T|$ and n to represent the scale of the graph, the time complexity for the algorithm is $O(nm^2)$. We use a mitosis and BFS way to consider the time interval, there are at most $1 + T$ amount of cumulative graphs exists at one time, the whole space complexity is $O(mn)$.

Algorithm 1. Top-Down Greedy Peeling Algorithm

Input: temporal graph $\mathcal{G}$, query vertex u, integer k
Output: significant engagement community $\mathcal{SEC}$
1: $\mathcal{C} \leftarrow$ compute all the cumulative subgraphs of $\mathcal{G}$
2: **for** each $C_{\mathcal{H}} \in \mathcal{C}$ contains u **do**
3:    **while** $C_{\mathcal{H}}$ is a k-core contains u **do**
4:       select a vertex v ($v \neq u$) with the maximum temporal degree
5:       $C_{\mathcal{H}} \leftarrow C_{\mathcal{H}} - v$
6: $\mathcal{SEC} \leftarrow argmax_{C_{\mathcal{H}}} Eng_u(\mathcal{H})$
7: **return** $\mathcal{SEC}$

Algorithm 2. Framework of Candidate Generation Algorithm

Input: cumulative graph $C_{\mathcal{H}}$, query vertex u, integer k
Output: alternative subgraph $\mathcal{AS}$
1: $\mathcal{AS} \leftarrow \emptyset; Q \leftarrow \emptyset$
2: Q.push(u); $\mathcal{AS}$.push(u)
3: **while** $Q \neq \emptyset$ **do**
4:    $s \leftarrow Q$.pop()
5:    **for** each $v \in \mathbb{N}_{s,C_{\mathcal{H}}}$ **do**
6:       **if** v meets the requirements for candidates **then**
7:          Q.push(v); $\mathcal{AS}$.push(v)
8: **return** $\mathcal{AS}$

---

5   The Bottom-Up Local Search Algorithm

In this section, we develop a bottom-up local search algorithm (*BULS*). The core concept of this local search method is to generate an alternative subgraph $\mathcal{AS}$ from the query vertex u, then we can use the greedy peeling algorithm on $\mathcal{AS}$ instead of the whole cumulative graph $C_{\mathcal{H}}$ to receive our results. The framework of candidate generation algorithm is shown in Algorithm 2. For the naive candidate generation algorithm, the expanding strategy to make judgement in Line 6 is choosing the vertices with degrees no less than k to be included into our $\mathcal{AS}$. Besides, we develop the advanced candidate generation algorithm additionally using the relationships among the temporal degrees and vertex engagement. Specifically, we let the query vertex u and its neighbors with degree no less than k in $C_{\mathcal{H}}$ to form a private community N. For vertex $v \in N$, if its degree in $N : \mathbb{D}_{v,N} \geq k$, we do not need to further extend from it. Meanwhile, for vertex v that $v \in N$ and $\mathbb{D}_{v,N} < k$, we choose it to start to expand from, we put it into a queue K. Then at each time we pop a vertex x from K and use a queue Q to separately handle its neighbors, here we apply two different expanding strategies.

Non-reference Strategy. Assume we start to expand the alternative graph from x, and we get a vertex m from Q. When we consider a vertex n from the neighbor set of m, we regard the vertices in a line from x to n as a whole. The increment for $d_{u,\mathcal{AS}}$ is $w_{C_{\mathcal{H}}}(u,x)$ and increment for $\sum_{v \in \mathcal{AS}} d_{v,\mathcal{AS}}$ is at least the sum of edge weights one way to connect from u to n, here we take $ac(m)$ to represent the sum from u to m. We do not want to let the $Eng_u(\mathcal{AS})$ decrease in each step, we have: if $n \notin \mathcal{AS}$, $\mathbb{D}_{n,\mathcal{AS}} \geq k$, and $\dfrac{d_{u,\mathcal{AS}}+w_{C_{\mathcal{H}}}(u,x)}{\sum_{v \in \mathcal{AS}} d_{v,\mathcal{AS}}+w_{C_{\mathcal{H}}}(m,n)+ac(m)} >$ $\dfrac{d_{u,\mathcal{AS}}}{\sum_{v \in \mathcal{AS}} d_{v,\mathcal{AS}}}$, we execute Q.push(n) and $\mathcal{AS}$.push(n). Since this is a greedy expanding strategy, to simplify the analysis and operation we set $\mathcal{AS}$ to be a fixed one (which exists before dealing the vertices in K) in practical calculations.

Reference Strategy. Considering that we use a top-down way to deal with various cumulative graphs for $\mathcal{H}$, there are some non-terminal results for $\mathcal{SEC}$ in this progress. Let *bestresult* represent the engagement level for u in current $\mathcal{SEC}$. *bestresult* can be used as the threshold to judge whether to include more vertices. Specifically, we only consider the increment for a single vertex at each time, here is the strategy: if $n \notin \mathcal{AS}$, $\dfrac{d_{u,\mathcal{AS}}}{\sum_{v \in \mathcal{AS}} d_{v,\mathcal{AS}}+w_{C_{\mathcal{H}}}(m,n)} > bestresult$, we execute Q.push(n) and $\mathcal{AS}$.push(n). It should be noticed that the $\mathcal{AS}$ here is not a fixed one, and we take all the vertices in N into the $\mathcal{AS}$ before expanding.

Here we formally introduce our local search algorithm *BULS*. When dealing with the first cumulative graph $C_{\mathcal{H}}$, we use the naive candidate generation algorithm to generate the alternative graph. Besides, for the cumulative graphs in following steps, we use the advanced candidate generation algorithm with reference strategy to expand, the rest of process is the same with *TDGP*. We also develop another local search algorithm *BULS+*. The main difference is that we use the advanced candidate generation algorithm with non-reference strategy to deal with the first cumulative graph $C_{\mathcal{H}}$. With it we can get the $\mathcal{AS}$ with smaller size. Additionally, since this algorithm might miss the results in some cases, we will turn to use the naive candidate generation algorithm when we find there is no such k-core containing u after the expanding for the alternative subgraph. The time complexity of the algorithm is still $O(nm^2)$. The space complexity is unchanged as $O(mn)$.

6   Experimental Evaluation

We evaluate our solutions on real-world temporal networks[1] with different types and sizes (Table 1), including social (Facebook, Twitter, Wiki), email (Enron, Lkml), and scientific collaboration (DBLP) networks. Besides the *TDGP*, *BULS*, and *BULS+* for our *SECS*. We choose *TopkDBSOL* [3] and *CST* [4] as competitors. *TopKDBSOL* is the online algorithm to find the top-k density bursting subgraphs. *CST* refers to the algorithm to handle the problem of community search with threshold constraint. We use engagement level (EL) and temporal density (TD) [3] to evaluate the quality of the results. Temporal density is the metric that measures the denseness of the community. To be more reliable, we randomly select 100 vertices as query vertices and report the average running time and quality.

---

[1] http://snap.stanford.edu/, http://konect.cc/, http://www.sociopatterns.org/.

Table 1. Datasets statistics. TS is the time scale of the timestamp.

Dataset	$n = |V|$	$m = |\mathcal{E}|$	$\bar{m} = |E|$	$|\mathcal{T}|$	TS
Facebook	45,813	461,179	183,412	223	Week
Twitter	304,198	464,653	452,202	7	Day
Wiki	1,094,018	3,224,054	2,787,967	77	Month
Enron	86,978	697,956	297,456	177	Week
Lkml	26,885	328,092	159,996	98	Month
DBLP	1,729,816	12,007,380	8,546,306	72	Year

Table 2. Running time of different algorithms with $k = 2$ (second)

	DBLP	Lkml	Enron	Facebook	Twitter	Wiki
CST	109.04	1.07	2.22	1.55	3.74	30.71
TopkDBSOL	1,707	2,178	1,920	39	26,872	20,217
TDGP	279.38	25.49	71.77	70.60	5.45	30.35
BULS	111.90	5.36	12.98	18.42	6.28	25.55
BULS+	36.27	5.04	8.55	10.49	4.36	16.48

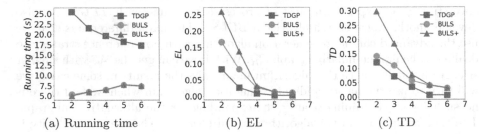

(a) Running time   (b) EL   (c) TD

Fig. 2. Comparison of different algorithms with various k in Lkml

Table 2 shows the running time of different algorithms with $k = 2$. For our problem, the *BULS+* has the least running time. To test how the parameter k affects the results, we vary k from 2 to 6 (Fig. 2). We can observe that the *BULS+* still has the best performance, it costs much less running time than *TDGP* while getting the results with the best quality. The values of *EL* and *TD* are reducing when k grows bigger, for that the communities have more vertices. Due to the similar reason, the *bestresult* is smaller and the $\mathcal{AS}$ has a larger size, so the running time of *BULS* and *BULS+* has an upward trend at the beginning.

7 Conclusion

In this paper, we first introduce the definition of engagement level, and then raise a novel problem called significant engagement community search. To tackle this problem, we develop a global algorithm called *TDGP*. To further improve the efficiency, we then devise a local search algorithm called *BULS* and its enhanced version *BULS+*. Finally, we evaluate our solutions on six real-world temporal graphs and the results show the superiority of our solutions.

Acknowledgements. The research is supported by the National Key Research and Development Program of China (No. 2018YFB1402802), NSFC (Nos. 62072205 and 61932004).

References

1. Batagelj, V., Zaversnik, M.: An o(m) algorithm for cores decomposition of networks. CoRR cs.DS/0310049 (2003)
2. Chang, L., Qin, L.: Cohesive subgraph computation over large sparse graphs. In: Proceedings of ICDE, pp. 2068–2071. IEEE (2019)
3. Chu, L., Zhang, Y., Yang, Y., Wang, L., Pei, J.: Online density bursting subgraph detection from temporal graphs. PVLDB **12**(13), 2353–2365 (2019)
4. Cui, W., Xiao, Y., Wang, H., Wang, W.: Local search of communities in large graphs. In: Proceedings of SIGMOD, pp. 991–1002. ACM (2014)
5. Donetti, L., Munoz, M.A.: Detecting network communities: a new systematic and efficient algorithm. J. Stat. Mech. Theory Exp. **2004**(10), P10012 (2004)
6. Fang, Y., Cheng, R., Li, X., Luo, S., Hu, J.: Effective community search over large spatial graphs. Proc. VLDB Endow. **10**(6), 709–720 (2017)
7. Fang, Y., et al.: A survey of community search over big graphs. VLDB J. **29**(1), 353–392 (2020)
8. Fang, Y., Yang, Y., Zhang, W., Lin, X., Cao, X.: Effective and efficient community search over large heterogeneous information networks. Proc. VLDB Endow. **13**(6), 854–867 (2020)
9. Galimberti, E., Barrat, A., Bonchi, F., Cattuto, C., Gullo, F.: Mining (maximal) span-cores from temporal networks. In: Proceedings of CIKM, pp. 107–116. ACM (2018)
10. Li, R., et al.: Skyline community search in multi-valued networks. In: Proceedings of SIGMOD (2018)
11. Li, R., Su, J., Qin, L., Yu, J.X., Dai, Q.: Persistent community search in temporal networks. In: Proceedings of ICDE, pp. 797–808. IEEE Computer Society (2018)
12. Lin, L., Yuan, P., Li, R.H., Wang, J., Liu, L., Jin, H.: Mining stable quasi-cliques on temporal networks. IEEE Trans. Syst. Man Cybern. Syst. 1–15 (2021)
13. Lin, L., Yuan, P., Li, R., Jin, H.: Mining diversified top-r lasting cohesive subgraphs on temporal networks. IEEE Trans. Big Data 1 (2021)
14. Liu, Q., Zhao, M., Huang, X., Xu, J., Gao, Y.: Truss-based community search over large directed graphs. In: Proceedings of SIGMOD, pp. 2183–2197 (2020)
15. Liu, Q., Zhu, Y., Zhao, M., Huang, X., Xu, J., Gao, Y.: VAC: vertex-centric attributed community search. In: Proceedings of ICDE, pp. 937–948 (2020)
16. Newman, M.E.: Fast algorithm for detecting community structure in networks. Phys. Rev. E **69**(6), 066133 (2004)

17. Qin, H., Li, R., Yuan, Y., Wang, G., Yang, W., Qin, L.: Periodic communities mining in temporal networks: Concepts and algorithms. IEEE Trans. Knowl. Data Eng. 1 (2020)
18. Rozenshtein, P., Gionis, A.: Mining temporal networks. In: Proceedings of KDD, pp. 3225–3226 (2019)
19. Sozio, M., Gionis, A.: The community-search problem and how to plan a successful cocktail party. In: Proceedings of KDD, pp. 939–948 (2010)

Influence Computation for Indoor Spatial Objects

Yue Li[1], Guojie Ma[1(✉)], Shiyu Yang[2], Liping Wang[1], and Jiujing Zhang[2]

[1] East China Normal University, Shanghai, China
51194501052@stu.ecnu.edu.cn, gjma@fem.ecnu.edu.cn,
lipingwang@sei.ecnu.edu.cn
[2] Guangzhou University, Guangzhou, China
syyang@gzhu.edu.cn, jiujingzhang@e.gzhu.edu.cn

Abstract. Studies have shown that people spend more than 85% of their time in indoor spaces. Providing varies location-based services (LBS) for indoor space is of great demand and has drew attentions from both industry and academic. The influence computation for spatial objects is one of the important applications in LBS and has broad applications in indoor facility location and indoor marketing. The influence query has been studied extensively in outdoor spaces. However, due to the fundamental difference between indoor and outdoor space, the outdoor techniques can not be applied for indoor space. In this paper, we propose the first indoor influence computation algorithm IRV to efficiently process indoor influence query. The proposed algorithm is based on the state-of-art indoor index structure VIP-Tree and several pruning rules are also presented to reduce the computation cost. The experiment results on both real and synthetic data sets show our proposed method outperforms the baseline algorithm.

Keywords: Indoor space · Reverse k nearest neighbor · Influence computation

1 Introduction

Studies have shown that people spend more than 85% of their time in indoor spaces [6], such as supermarkets, hospitals, apartments, office buildings and etc. Providing varies location-based services (LBS) for indoor space is of great demand and has drew attentions from both industry and academic. The influence computation for spatial objects is one of the important applications in LBS. According to the definition in [12], a facility f is said to influence a user u if f is one of the k closest facilities of the user u. This is because the users usually prefer to visit/use near by facilities. The influence set of a facility f is the set of users influenced by f. The influence set, also called reverse k nearest neighbors (RkNN), which has been extensively studied in the past. In this paper, we use influence query and RkNN query to refer the same problem.

The influence computation has broad applications in LBS, especially for the indoor space. Considering the example show in Fig. 1, there are 7 users and 7

A. Bhattacharya et al. (Eds.): DASFAA 2022, LNCS 13245, pp. 259–267, 2022.
https://doi.org/10.1007/978-3-031-00123-9_21

advertising boards in a shopping mall. Suppose users are usually drawn to close advertising boards, the users in the influence set are the potential customers and may be influenced by the advertising board.

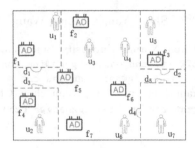

Fig. 1. RkNN example

Although, the influence computation has been well studied for outdoor space, the existing solutions for outdoor space can not be applied to indoor space. First, the indoor space is characterized by entities such as doors, rooms, and corridors. These entities greatly restrict the movement of objects in the indoor space. Therefore, outdoor spatial distance metrics such as Euclidean distance and road network distances can not be used to measure indoor distances. Second, the topology of indoor space is usually complicated. For an office building, there are multiple floors connected by lifts or stairs. The complicated topology of indoor space provide challenges of data modelling and query processing.

In this paper, we formally define the influence query (also called RkNN query) for indoor space, and propose efficient query processing algorithm namely IRV based on the VIPTree index. We also present effective pruning rules to reduce the computation cost. The experiment results on real and synthetic data sets shows the performance of our proposed method.

2 Related Work

2.1 Outdoor Techniques

RkNN queries have been extensively studied in outdoor space [5,9–11,13–15]. The existing outdoor techniques use a pruning and verification framework. The two most notable pruning method are half-space pruning (TPL [10], Finch [11], InfZone [5]) and regions-based pruning (Six-regions [9], SLICE [14,15]).

2.2 Indoor Techniques

Indoor Data Modelling. Many model are proposed for indoor space, such as CityGML [1], AB graph [7], Distance matrix [7], and IP-Tree [8]. The state-of-art indoor index IP-Tree is an indexing structure based on a D2D graph.

The basic idea of IP-Tree is to combine adjacent indoor partitions (e.g., rooms, hallways, stairs) to form leaf nodes and then iteratively combine adjacent leaf nodes until all nodes are combined into a single root node. Due to its excellent index structure, many spatial queries in IP-Tree are efficient, such as shortest distance query and kNN query. Our algorithm is also based on IP-Tree.

3   Preliminaries

3.1   Problem Definition

In indoor space, object motion would be greatly restricted by entities such as doors, rooms and hallways. Therefore the traditional spatial network distance and Euclidean distance cannot be used to measure the distance in indoor space and we define Indoor Distance. We use $dist(p, q)$ to represent the Euclidean distance between point p and point q in this paper.

Definition 1. *Indoor Distance.* *Given two points p and q in indoor space, the **Indoor Distance** between p and q is the length of the shortest path(not directly across obstacles such as walls) between p and q. We record the **Indoor Distance** of p and q as $distID(p, q)$ in this paper.*

Definition 2. *Indoor RkNN Query.* *Consider a set of facilities F and a set of users U in indoor space. Given a query point $q \in F$ and k, **Indoor RkNN Query** returns every user $u \in U$ for which q is one of its k-closest facilities using Indoor Distance rather than Euclidean distance when calculating distance.*

Definition 3. *Prune State.* *We define the result of pruning a region by a facility f as **Prune State**. If all points in an area (node or room) are closer to facility f than the query point q, then the **Prune State** of this area is **Pruned** for f. If all points in an area (node or room) are closer to the query point q than facility f, then the **Prune State** of this area is **Result** for f. If the **Prune State** is neither **Pruned** nor **Result**, then the **Prune State** is **Candidate**.*

3.2   Observation

The easiest way of solving *RkNN* query is to calculate *kNN* for all user points but it costs a lot. We observe some conclusions based on geometric properties that help reduce the query time of *RkNN*.

Observation 1. If the facility point f and the query point q are not in the node N, and for any access door d_i in node N meets that $distID(f, d_i) < distID(q, d_i)$. Then facility f meets that $distID(f, p) < distID(q, p)$ for any point p in node N and the *Prune State* of node N is *Pruned* for facility f.

Proof. Suppose there is a point p in node N that satisfies $distID(f, p) >= distID(q, p)$. Because q is not in node N, the shortest path from q to p must pass through one access door d_i of of N. So $distID(q, p) =$

$distID(q, d_i) + distID(d_i, p) <= distID(f, p) <= distID(f, d_i) + distID(d_i, p)$ and $distID(q, d_i) <= distID(f, d_i)$ which contradicts the assumption. Therefore, $distID(f, p) < distID(q, p)$ for any point p in node N. The proofs of the latter two observations are similar and we will no longer give proofs.

Observation 2. If the facility point f and the query point q are not in the node N, and for any access door d_i in node N meets that $distID(f, d_i) > distID(q, d_i)$. Then facility f meets that $distID(f, p) > distID(q, p)$ for any point p in N and the *Prune State* of node N is *Result* for facility f.

Observation 3. If the facility point f is in the node N and the query point q is not in the node N, and for any access door d_i in node N meets that $distID(f, d_i) < distID(q, d_i)$. Then for any point p in node N, facility f meets that $distID(f, p) < distID(q, p)$ and the *Prune State* of node N is *Pruned* for facility f.

4   IRV Algorithm

4.1   Solution Overview

Our algorithm uses a pruning and verification framework based on IP-Tree which supports shortest distance/path query in indoor space. In the pruning phase, the algorithm prunes nodes first and then prunes rooms in candidate nodes using facilities. We can divide the whole indoor space into three parts: *Pruned area, Result area, Candidate area.* In the verification phase, users that lie in the *Candidate area* are identified and they are then verified as RkNN if its kNN contains the query point q. While users that lie in the *Result area* must be RkNN of the query point. *Baseline Algorithm* performs the *verification phase* on all user points to determine whether the query point is their kNN.

4.2   Pruning Algorithm

Our pruning algorithm contains two parts: pruning nodes and pruning rooms. In the pruning nodes phase, we traverse the IP-tree from top to bottom and prune nodes. In the pruning rooms phase, we pruned rooms of the candidate nodes obtained in the pruning nodes phase.

Algorithm 1. $Prune(O, q, f)$

Input: Object O(node N or room R), k, facility point f
Output: *Prune State:Pruned, Result, or Candidate*
1: **if** for each access_door ad of node N meets that $distID(q, ad) > distID(f, ad)$ &&
   q is not in O **then**
2:     return *Pruned*;
3: **else if** for each access_door ad of node N meets that $distID(q, ad) < distID(f, ad)$
   && f is not in O **then**
4:     return *Result*;
5: **else**
6:     return *Candidate*;

Prune Algorithm. According to the observations put forward above, we can apply them to the *Prune Algorithm* to prune nodes and rooms. We use the position relationship between points and object and the distance between points and access doors to get *Prune State*. *Prune Algorithm* returns *Prune State: Pruned, Result, Candidate*. *Prune State* is *Pruned*, indicating that object O is pruned by facility f. *Prune State* is *Result*, indicating that all points in object O is farther to facility f. When *Prune State* is *Candidate*, we need to further pruning the node or verifying user points located in object O.

PruneNodes Algorithm. In the pruning nodes phase, we divide nodes into *Candidate node*, *Result node* and *Pruned node*. *Candidate nodes* are nodes where we don't know whether the point located in it is RkNN of the query point. *Result node* is the node that all points are RkNN of the query point in which. *Pruned node* is the node that all points are not RkNN of the query point in which. We traverse the IP-tree from top to bottom and prune nodes. Each time we traverse to a node, we use *Prune Algorithm* to get *Prune state* of all facilities. If the number of facilities whose *Prune state* is *Pruned* is greater than or equal to k, node N currently traversed is a *Prune node*. If the number of facilities whose *Prune state* is *Result* is greater than or equal to the number of facilities minus k, node N currently traversed is a *Result node*. If node N is neither a *Pruned node* nor a *Result node*, it is a *Candidate node* and we use *PruneRooms Algorithm* to further pruning rooms located in it.

Algorithm 2. PruneNodes

Input: *IPTree, int k, users, facilities, query point q*
Output: *candidate_nodes, result_nodes*
1: Heap $h \leftarrow \{root\_id\}$
2: **while** $!h.empty()$ **do**
3:   de-heap node N from h
4:   **for** facility f in facilities **do**
5:     $PruneState = Prune(N, q, f)$;
6:     **if** $PruneState == Pruned$ && $pruneCount >= k$ **then**
7:       break; // this node is pruned
8:     **if** $PruneState == Result$ && $resultCount >= facility\_num - k$ **then**
9:       result_nodes.push(N); break; // this node is a result node
10:   **if** node N is neither a pruned node nor a result node **then**
11:     **if** N is a leaf node **then**
12:       candidate_nodes.push(N); $PruneRooms(N)$;
13:     **else**
14:       push all children of N into h;

PruneRooms Algorithm. For the candidate nodes obtained in the pruning nodes step, we prune the rooms located in these nodes. Similar to the *PruneNodes Algorithm*, we divide rooms in the pruning nodes into *Candidate room*, *Result room* and *Pruned room*. If the number of facilities whose *Prune state* is *Pruned* is greater than or equal to k, room r is a *Pruned room*. If the number of facilities whose *Prune state* is *Result* is greater than or equal to the number of facilities minus k, room r currently traversed is a *Result room*. If room r is neither a *Pruned room* nor a *Result room*, it is a *Candidate room*. Due to limited space, we omit the corresponding pseudocode.

4.3 Verification Algorithm

In the verification phase, for *Result node* and *Result room*, we only need add users located in they to the result set. For users located in *Candidate room*, we use *IsRkNN* algorithm to determine whether they are the RkNN of the query point. Compared to computing RkNN, *IsRkNN* algorithm employs a simple optimization. Due to limited space, we will not introduce it here.

5 Experimental

5.1 Experimental Settings

We compare our algorithm *IRV* with *Baseline* which does not perform pruning and uses *IsRkNN* for all user points. The two algorithms are implemented in C++. The experiments are run on a 64-bit PC with Intel Xeon 2.40 GHz dual CPU and 128 GB memory running Redhat Linux.

Table 1 shows the detailed settings, and the default values are shown in bold. User points and facility points are generated randomly. We use one synthetic dataset (denoted as Syn) which consists of 338 rooms spread over 10 levels and three real datasets: Melbourne Central [2] (denoted as MC), Menzies building [3] (denoted as Men) and Clayton Campus [4] (denoted as CL). Melbourne Central is a major shopping centre in Melbourne which consists of 297 rooms spread over 7 levels. Menzies building is the tallest building at Clayton campus of Monash University consisting of 14 levels and 1306 rooms. The Clayton dataset corresponds to 71 buildings in Clayton campus of Monash University.

Table 1. Experimental settings.

Parameter	Range
Number of facilities	50, 100, **200**, 500, 1000
Number of users	200, 500, 1000, **2000**, 5000, 10000, 20000
The value of k	1, 2, 3, 4, **5**, 6, 7, 8, 9, 10

5.2 Experiment Results

Effect of Number of Users. In Fig. 2, we study the effect of the number of users in different datasets for *IRV* and *Baseline*, respectively. As we expected, *IRV* is up to two orders of magnitude more efficient than *Baseline*. And from Fig. 2 we can see that query time of *IRV* and *Baseline* increases as the number of users increases. This is because that the number of users is more, the number of candidate users is more, and the number of calls of *IsRkNN algorithm* is more.

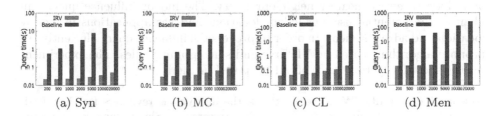

| (a) Syn | (b) MC | (c) CL | (d) Men |

Fig. 2. Query time varies with the number of users.

Effect of Number of Facilities. In Fig. 3, we study the effect of the number of facilities. From Fig. 3 we can see that query time of *Baseline* decrease. This is because that the execution times of *IsRkNN algorithm* decrease as the number of facilities increases. And we can see that query time of *IRV* increases. This is because that the number of calls of *Prune Algorithm* increases as the number of facilities increases in the pruning phase.

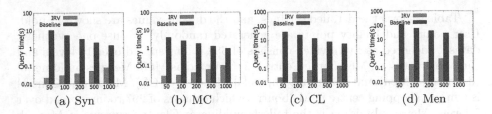

Fig. 3. Query time varies with the number of facilities.

Effect of k. In Fig. 4, we study the effect of the value of k. From Fig. 4 we can see that the query time of *Baseline* increases as the value of k increases. And we can see that *IRV* performs much better than *Baseline* no matter how k changes.

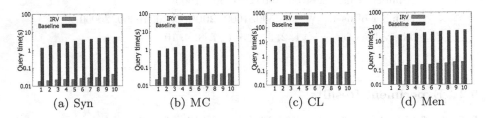

Fig. 4. Query time varies with k value.

6 Conclusion

In this paper, we propose an new indoor spatial query namely influence query or indoor reverse k nearest neighbour query. The indoor influence query can identify the potential users of indoor locations and has applications in facility location and advertisement. In order to process the influence query efficiently, we propose algorithms based on VIPTree and some pruning rules. The experiment results show the performance of our proposed method.

Acknowledgements. We sincerely thank the anonymous reviewers for their feedback which helped improve our work. The research of Shiyu Yang is supported by NSFC61802127 and GuangDong Basic and Applied Basic Research Foundation 2019B1515120048.

References

1. http://www.citygml.org/
2. http://www.melbournecentral.com.au/
3. http://lostoncampus.com.au/15641
4. https://www.monash.edu/pubs/maps/3-Claytoncolour.pdf

5. Cheema, M.A., Lin, X., Zhang, W., Zhang, Y.: Influence zone: efficiently processing reverse k nearest neighbors queries. In: 2011 IEEE 27th International Conference on Data Engineering, pp. 577–588. IEEE (2011)
6. Jenkins, P.L., Phillips, T.J., Mulberg, E.J., Hui, S.P.: Activity patterns of Californians: use of and proximity to indoor pollutant sources. Atmos. Environ. Part A. Gen. Top. **26**(12), 2141–2148 (1992)
7. Lu, H., Cao, X., Jensen, C.S.: A foundation for efficient indoor distance-aware query processing. In: 2012 IEEE 28th International Conference on Data Engineering, pp. 438–449. IEEE (2012)
8. Shao, Z., Cheema, M.A., Taniar, D., Lu, H.: VIP-tree: an effective index for indoor spatial queries. Proc. VLDB Endow. **10**(4), 325–336 (2016)
9. Stanoi, I., Agrawal, D., El Abbadi, A.: Reverse nearest neighbor queries for dynamic databases. In: ACM SIGMOD Workshop on Research Issues in Data Mining and Knowledge Discovery, vol. 20. Citeseer (2000)
10. Tao, Y., Papadias, D., Lian, X.: Reverse KNN search in arbitrary dimensionality. In: Proceedings of the Very Large Data Bases Conference (VLDB), Toronto (2004)
11. Wu, W., Yang, F., Chan, C.Y., Tan, K.L.: Finch: evaluating reverse k-nearest-neighbor queries on location data. Proc. VLDB Endow. **1**(1), 1056–1067 (2008)
12. Yang, S., Cheema, M.A., Lin, X.: Impact set: computing influence using query logs. Comput. J. **58**(11), 2928–2943 (2015)
13. Yang, S., Cheema, M.A., Lin, X., Wang, W.: Reverse k nearest neighbors query processing: experiments and analysis. Proc. VLDB Endow. **8**(5), 605–616 (2015)
14. Yang, S., Cheema, M.A., Lin, X., Zhang, Y.: Slice: reviving regions-based pruning for reverse k nearest neighbors queries. In: 2014 IEEE 30th International Conference on Data Engineering, pp. 760–771. IEEE (2014)
15. Yang, S., Cheema, M.A., Lin, X., Zhang, Y., Zhang, W.: Reverse k nearest neighbors queries and spatial reverse top-k queries. VLDB J. **26**(2), 151–176 (2017)

A Localization System for GPS-free Navigation Scenarios

Jiazhi Ni[1], Xin Zhang[1], Beihong Jin[2,3]($\boxtimes$), Fusang Zhang[2,3], Xin Li[1],
Qiang Huang[1], Pengsen Wang[1], Xiang Li[1], Ning Xiao[1], Youchen Wang[1],
and Chang Liu[1]

[1] Localization Technology Department, Tencent Inc, Beijing, China
[2] State Key Laboratory of Computer Sciences, Institute of Software,
Chinese Academy of Sciences, Beijing, China
Beihong@iscas.ac.cn
[3] University of Chinese Academy of Sciences, Beijing, China

Abstract. Localization is crucial to mobile navigation, which mainly depends on satellite navigation systems such as GPS. Unfortunately, sometimes GPS localization might be missing due to hardware failure or the urban canyon effect. Network localization is a promising way to improve localization experience for users under these conditions. However, there has not been any report on the large-scale applications of network localization methods to GPS-free outdoor scenarios. In this paper, we describe the challenges and build a novel navigation fingerprint based localization system, which fuses road network information and Wi-Fi signals scanned outdoors. With large amount of navigation trajectories as supervised information, we train a ranking model to obtain a reasonable metric for localization. The localization system has been deployed in Tencent Map, improving navigation accuracy in GPS-free outdoor scenarios for 1.6 million users.

Keywords: Mobile navigation · Network localization · Road network

1 Introduction

In recent years, due to the popularity of mobile phones and the development of navigation technology, mobile navigation plays a more important role in people's daily life, including driving, walking, cycling and other navigation scenarios. Modern navigation technology can form the road network information of the entire city by means of various surveying and mapping methods. With the help of mobile navigation, anyone can go anywhere at anytime. It calculates shortest distances to destinations and reduces travel costs.

In general, Global Positioning System (GPS) is widely used as the preferable localization method during mobile navigation. However, according to the statistics from Tencent Map, nearly 67% of negative user feedback are related to weak or no GPS signals. This hinders the user experience of Map apps a lot. Table 1

A. Bhattacharya et al. (Eds.): DASFAA 2022, LNCS 13245, pp. 268–273, 2022.
https://doi.org/10.1007/978-3-031-00123-9_22

Table 1. Statistics of user feedback on navigation problems from Tencent Map

Feedback category	Count	Ratio
Weak or no GPS signal	1976	69.7%
Low GPS accuracy	624	22%
Cannot locate	27	0.9%
Locate too slow	25	0.9%
Wrong direction	103	3.6%
Inaccurate speed	79	2.8%

shows an example. We collect 2834 user feedback in a month from Tencent Map which are related to navigation problems. As shown in Table 1, these feedback are classified into six categories, and 1976 of the feedback are related to GPS-free navigation scenarios which we attribute to the failure of GPS module or the urban canyon effect. GPS hardware failure is more likely to happen in low-end mobile phones which cost less than 400 dollars and account for 70% of Tencent Map users in 2019. Additionally, the urban canyon will cause GPS signal loss for all mobile phones.

In recent years, network localization using various signal data have attracted intensive research interests, proposing different methods such as Wi-Fi based methods [1,2], Bluetooth based methods [3–5], UWB based methods [6–8], etc. Among these techniques, Wi-Fi based localization methods have been intensively studied and applied to indoor localization because of low cost and wide deployment of Wi-Fi devices. Unfortunately, few studies exist on the Wi-Fi based localization methods in GPS-free outdoor scenarios. We note that in GPS-free scenarios, there are two reasons why the traditional Wi-Fi based methods cannot work: 1) road network information is not fully exploited in the process of fingerprint construction. 2) the traditional fingerprint matching model cannot effectively identify the differences of Wi-Fi signals in indoor and outdoor navigation scenarios.

In the paper, we build a novel navigation fingerprint based localization system, providing single-point localization for outdoor users using Tencent Map. The system takes the road network information into consideration and designs an effective fingerprint matching model to get localization results with high accuracy.

To the best of our knowledge, we are the first to mine Wi-Fi signals to provide the outdoor localization for urban GPS-free scenarios. Although this is only a preliminary work, it offers a practical way to fuse road network information and Wi-Fi signals scanned outdoors, and provides high-accuracy localization results for navigation applications.

2   System Overview

From the perspective of Tencent Map users, the closer the distance between localization point given by system and the user's real location is, the higher the corresponding rating of the localization system's accuracy is. Thus, the localization system first extracts the GPS information from historical trajectories and then calculates the distance between the fingerprint and GPS location at every timestamp, which are used to measure the localization system's accuracy at that time, the distances are finally transforms into localization ratings, denoted as Y, as shown in Eq. (1).

$$Y = \begin{bmatrix} y_{lng_1,lat_1} & \cdots & y_{lng_1,lat_k} \\ \vdots & \ddots & \vdots \\ y_{lng_k,lat_1} & \cdots & y_{lng_k,lat_k} \end{bmatrix}_{k_{lng} \times k_{lat}} \tag{1}$$

where k_{lng} is the number of fingerprints divided along the longitude direction, similarly, k_{lat} along the latitude direction. (lng_i, lat_j) represents the geographical coordinates of the lower right corner of fingerprint grid $FP_{(i,j)}$. y_{lng_i,lat_j} represents the ratings of localization in fingerprint grid $FP_{(i,j)}$.

In addition, we introduce the road network information [9,10], including link length, link type, link level, link connectivity, etc., to construct the road topology matrix R with dimension $m \times n_R$, where m represents the number of samples and n_R represents the number of road topology features. We also collect the Wi-Fi signals scanned outdoors [11,12], mainly utilize these Wi-Fi signals' statistics (such as min/max/std, etc.) of Received Signal Strength Indicator(RSSI) and occurrence frequency to construct the signal energy matrix S with dimension $m \times n_S$, where n_S represents the number of signal energy features.

In essence, fingerprint localization is a ranking task, which sorts all candidate fingerprints. Thus, our goal is to select the optimal matching fingerprint as the final localization result based on road topology matrix R and signal energy matrix S and using localization ratings matrix Y as supervised information, illuminated in Eq. (2).

$$FP_{optimal} = \arg\min_{idx} \; Loss(Y, Rank(FP_{idx}, [R|S])) \tag{2}$$

where $Rank(.)$ is the function to compute the sorted list of candidate fingerprints FP_{idx} according to feature matrix $[R|S]$, and $Loss(.)$ is the loss function to measure the error between ground truth Y and the sorted candidate fingerprint list.

We propose a localization method based on Wi-Fi fingerprints for GPS-free navigation scenarios, which is a variant of fingerprint localization [13–15] and build a navigation fingerprint based localization system. Our localization system consists of four modules, which are shown in Fig. 1.

- Data Preprocessing Module: This module first preprocesses the road network data, Wi-Fi signal records scanned outdoors and historical trajectories records, where the GPS signals in trajectories are used as the supervised information to generate the localization rating matrix Y.
- Model Construction Module: This module devises the specialized aggregation network as well as an integrated loss functions to achieve the fingerprint

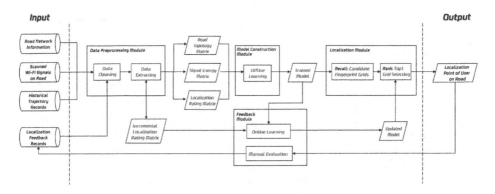

Fig. 1. System architecture

matching. After being offline trained using road topology matrix R, signal
energy matrix S as input, the module outputs the trained ranking model.

- Localization Module: This module utilizes the trained model to predict the
  rating for recalled fingerprints and provides the final localization point of
  the outdoor user. Specifically, there are two stages, i.e., the recall stage and
  the ranking stage. At the recall stage, we employ the customized rules (e.g.
  directly select relevant fingerprints according to the scanned Wi-Fi list or
  draw a range based on the localization result from cells of base stations as
  reference) to recall the candidate set of fingerprints. At the ranking stage,
  we employ the trained ranking model to predict the ratings of fingerprints in
  candidate set and select the optimal matching fingerprint as the final result.
- Feedback Module: This module periodically collects the wrong localization
  cases during navigation, and incrementally fine-tunes the model after the
  manual evaluation. Specifically, we first obtain the ground truth of the wrong
  cases by manual evaluation and get their ratings through preprocessing module, and then employ online learning to update the parameters in our model,
  and finally use the updated model to provide localizaition services.

3   System Deployment

Our navigation fingerprint based localization system has already provided services via Tencent Map application (see in Fig. 2) for millions of users.

In GPS-free navigation scenarios, the Wi-Fi signals scanned outdoors are
sent to servers of Tencent Map which call the proposed system to match the
most likely road link. Our system can provide single-point localization before
or during driving for users using the app on mobile phones. Specifically, the
localization process of Tencent Map can seamlessly switch between our system
and GPS positioning if the latter is appropriate.

For model evolution, we regularly update the fingerprint database with the
latest Wi-Fi signal data, road network data and historical trajectories, and feed
data into the model. Furthermore, we also improve user experience through

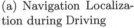

(a) Navigation Localization during Driving   (b) Navigation Localization before Driving   (c) Navigation Localization Feedback

Fig. 2. GUIs of navigation localization in Tencent Map app

online training model based on the feedback. The fingerprint database and models are stored in servers and users can get the efficient services without updating local application.

References

1. Roy, P., Chowdhury, C.: A survey of machine learning techniques for indoor localization and navigation systems. J. Intell. Robot. Syst. **101**(3), 1–34 (2021)
2. Mendoza-Silva, G., et al.: Environment-aware regression for indoor localization based on WiFi fingerprinting. IEEE Sensors J. (2021)
3. Chen, J., et al.: A data-driven inertial navigation/Bluetooth fusion algorithm for indoor localization. IEEE Sensors J. (2021)
4. HajiAkhondi-Meybodi, Z., et al.: Bluetooth low energy and CNN-based angle of arrival localization in presence of Rayleigh fading. In: ICASSP 2021–2021 IEEE International Conference on Acoustics, Speech and Signal Processing (ICASSP). IEEE (2021)
5. Xu, S., et al.: Bluetooth, floor-plan, and MEMS assisted wide-area audio indoor localization system: apply to smartphones. IEEE Trans. Ind. Electron. (2021)
6. Djosic, S., et al.: Fingerprinting-assisted UWB-based localization technique for complex indoor environments. Expert Syst. Appl. **167**, 114188 (2021)
7. Zhao, M., et al.: ULoc: low-power, scalable and cm-accurate UWB-tag localization and tracking for indoor applications. Proc. ACM Inter. Mob. Wearable Ubiquitous Technol. **5**(3), 1–31 (2021)
8. Wang, C., et al.: A high-accuracy indoor localization system and applications based on tightly coupled UWB/INS/Floor map integration. IEEE Sensors J. **21**(16), 18166–18177 (2021)
9. Win, M.Z., Shen, Y., Dai, W.: A theoretical foundation of network localization and navigation. Proc. IEEE, **106**(7), 1136–1165 (2018)

10. Li, D., Lei, Y., Li, X., Zhang, H.: Deep learning for fingerprint localization in indoor and outdoor environments. ISPRS Int. J. Geo-Inf. **9**(4), 267 (2020)
11. Singh, N., Choe, S., Punmiya, R.: Machine learning based indoor localization using Wi-Fi RSSI fingerprints: an overview. IEEE Access (2021)
12. Wu, C.-Y., et al.: Effects of road network structure on the performance of urban traffic systems. Phys. A: Stat. Mech. Appl. **563**, 125361 (2021)
13. Morelli, A.B., Cunha, A.L.: Measuring urban road network vulnerability to extreme events: an application for urban floods. Trans. Res. Part D Transp. Environ. **93**, 102770 (2021)
14. Niu, K., et al.: Understanding WiFi signal frequency features for position-independent gesture sensing. IEEE Trans. Mob. Comput. (2021)
15. Olivares, E., et al.: Applications of information channels to physics-informed neural networks for WiFi signal propagation simulation at the edge of the industrial internet of things. Neurocomputing **454**, 405–416 (2021)

Systems

HEM: A Hardware-Aware Event Matching Algorithm for Content-Based Pub/Sub Systems

Wanghua Shi and Shiyou Qian[✉]

Shanghai Jiao Tong University, Shanghai, China
{s-whua,qshiyou}@sjtu.edu.cn

Abstract. Content-based publish/subscribe (CPS) systems are widely used in many fields to achieve selective data distribution. Event matching is a key component in the CPS system. Many efficient algorithms have been proposed to improve matching performance. However, most of the existing work seldom considers the hardware characteristics, resulting in performance degradation due to a large number of repetitive operations, such as comparison, addition and assignment. In this paper, we propose a Hardware-aware Event Matching algorithm called HEM. The basic idea behind HEM is that we perform as many bit OR operations as possible during the matching process, which is most efficient for the hardware. In addition, we build a performance analysis model that quantifies the trade-off between memory consumption and performance improvement. We conducted extensive experiments to evaluate the performance of HEM. On average, HEM reduces matching time by up to 86.8% compared with the counterparts.

Keywords: Event matching · Hardware-aware · Bitset OR operation

1 Introduction

As a flexible communication paradigm, content-based publish/subscribe (CPS) systems are widely applied in manifold applications, such as stock trading system [1,16], recommendation system [7], social message dissemination system [2] and monitoring system [10]. There are three basic roles in CPS systems to achieve fine-grained data distribution: subscriber, publisher and broker. Subscribers commit subscriptions to brokers according to their interests and focuses. Publishers produce events and send them to brokers. For each incoming event, brokers execute event matching algorithms to search for matching subscriptions and forward the event to the subscribers to which those matches belong.

Apparently, matching algorithms play an irreplaceable role in CPS systems. With the growth of event arrival rate, subscription number and the size of event and subscription, it is inevitable for event matching algorithm to be a performance bottleneck of the entire CPS system. The urge to decrease the end-to-end

A. Bhattacharya et al. (Eds.): DASFAA 2022, LNCS 13245, pp. 277–292, 2022.
https://doi.org/10.1007/978-3-031-00123-9_23

data distribution latency and improve the throughput in CPS systems in the past decades has driven algorithm design in event matching.

Diverse algorithms have been proposed to boost event matching. According to the first search targets (matching or unmatching subscriptions), there are forward algorithms (such as TAMA [20] and OpIndex [19]) and backward algorithms (such as REIN [13] and GEM [6]). Each matching algorithm has its own design idea. For example, TAMA [20] designs an index table to locate all the predicates matching a given event value and maintains a counter for each subscription. REIN [13] locates two unmatching cell lists for each event value and traverses the cells to mark unmatching subscriptions. From the hardware perspective, TAMA and OpIndex do numerous addition operations one by one while REIN and GEM perform arduous traversal and marking operations bit by bit, thereby consuming much time to do repetitive operations. In effect, it is more efficient for computers to do in-flash bitwise operations because the overhead of data movement between caches and memory is mitigated [8]. In addition, as the memory price is falling and large capacity of memory becomes more common nowadays, cache mechanism becomes a potential method to trade space for time.

Motivated by the above discussion, we propose a novel hardware-aware event matching algorithm (HEM) based on bitset OR operations. HEM adopts a backward method to obtain matching results, similar to REIN [13] and GEM [6]. The basic idea is to fully utilize the hardware characteristic of doing a set of bitwise OR operations in the parallel way. Specifically, HEM uses a subscription pre-mark cache (SPC) mechanism to replace repeated traversal and marking operations with efficient bitwise OR operations in the matching process, following the concept of trading space for time. We build a theoretical model to quantify the trade-off between performance improvement and memory consumption.

Extensive experiments are conducted to evaluate the performance of HEM based on synthetic and real-world stock dataset. First of all, verification experiments validate the conclusion of the theoretical analysis: doubling the cache size halves the marking time. Secondly, compared with four counterparts, namely REIN [13], Ada-REIN [15], OpIndex [19] and TAMA [20], metric experiments show that HEM reduces matching time by 86.8%, 86.3%, 82.1% and 45.3% on average. The main contributions of our work are as follows:

- We propose a hardware-aware event matching algorithm called HEM which aims to reduce the time of repeated operations in the matching process.
- We propose a subscription pre-mark cache method to optimize the matching efficiency of HEM by trading space for time.
- We build a theoretical performance analysis model that quantifies the trade-off between memory consumption and performance improvement.
- We evaluate the performance of HEM through extensive experiments based on synthetic and real-world dataset.

2   Related Work

In the past few decades, improving matching performance has been one of the hot topics in the CPS system. Copious efficient event matching algorithms have been proposed, such as REIN [13], Ada-REIN [15], Comat [5], TAMA [20], OpIndex [19], H-Tree [14], MO-Tree [4], GEM [6], GSEC [18], PS-Tree [9].

Classification of Matching Algorithms. Matching algorithms can be classified from different perspectives. For instance, the work [11] classifies event matching algorithms according to whether the underlying data structure is subscription-grouping (such as H-Tree [14] and MO-Tree [4]) or predicate-grouping (such as REIN [13] and TAMA [20]). The work [21] reviews matching algorithms from three aspects: single-thread algorithms (such as Ada-REIN [15] and OpIndex [19]), parallel algorithms (such as PhSIH [11] and CCM [17]), and algorithms for elasticity (such as GSEC [18] and CAPS [12]).

Moreover, the work [6] divides matching mechanisms into single dimensional based and all dimensional based. The work [13] evaluates algorithms by whether it is forward matching or backward matching. Based on the search strategies, the work [3] regards the algorithms as filtering-based matching and counting-based matching. The work [4] provides a new perspective of whether the matching algorithm supports event matching and subscription matching.

Furthermore, event matching algorithms can also be distinguished from exact matching (such as REIN [13] and PS-Tree [9]) or approximate matching (such as Ada-REIN [15] and TAMA [20]), multi-algorithm composition matching (such as Comat [5]) or single algorithm matching.

Analysis of Operations in Matching Algorithms. In this paper, we assess a matching algorithm by whether it has a bottleneck on repeated operations. For example, REIN [13] is a backward matching algorithm indexing predicates. In the matching process, REIN mainly performs cell traversal and excessive repeated bit-marking operations. Each unmatching subscription is marked ψ'_S times for each event where ψ'_S is the number of unsatisfied predicates. REIN performs well with high matching probability of subscriptions because the workload of marking unmatches is small. GEM [6] is an analogous backward algorithm and has a similar problem. It designs a cache method to boost the removal operations, which is consistent with the idea of alleviating repeated operations.

TAMA [20] is a forward and counting-based matching algorithm. The core idea is to obtain the satisfied predicates rapidly and increase the counters of the corresponding subscriptions. TAMA performs well with low matching probability because the workload of counting satisfied predicates is small. However, all the satisfied predicates of both matching and unmatching subscriptions are counted. As a result, counting operations is a performance bottleneck. Differently, OpIndex [19] starts from the pivot attribute to search, which filters abundant unmatching subscriptions if the event does not contain the pivot attribute. If subscriptions including a set of certain interval predicates defined on different attributes are indexed together, we can replace multiple plus one operations with one direct addition operation. GSEC [18] constructs data structure in this way.

Distinct from the above algorithms suffering from repeatedly executing operations (such as assignment operations in REIN and addition operations in

TAMA) one by one, HEM stores subscription states in caches and mainly performs efficient bitwise OR operations during the matching process.

3   Problem Definition

For brevity, in model design and analysis, we regard the value domain of each attribute as $[0, 1]$. Let d be the number of attributes in the content space.

Definition 1. *An event $E(e_1, e_2, ..., e_{\psi_E})$ consists of ψ_E attribute-value pairs, which is a data point in the space. $e_i(a_j, v)$ means the ith value v of E is defined on attribute a_j. Event size ψ_E is the number of nonempty attributes in E. Generally, ψ_E is much smaller than d in a high-dimensional space.*

Definition 2. *A predicate $p(a_j, [l, h])$ is an interval defined on attribute a_j, which includes a low value l and a high value h in the closed form. The width w of $p(a_j, [l, h])$ is $h - l$. $p(a_j, [l, h])$ matches an event value $e(a_i, v)$ only if $a_j = a_i \wedge l \le v \le h$.*

Definition 3. *A subscription $S(p_1, p_2, ..., p_{\psi_S})$ is composed by ψ_S predicates defined on distinct attributes. Usually ψ_S is much smaller than ψ_E. S matches E if each predicate of S matches the value of E on the same attribute.*

Definition 4. *Given an event and a set of subscriptions, event matching algorithm searches all the matches of the event from the set.*

4   Design

4.1   Overview

It is challenging to design matching algorithm to adapt to a wide range of application requirements. First, the number and size of subscriptions may vary widely. This change should not cause large fluctuations in the matching time. Second, the dimension of the content space may be between tens to tens of thousands. Both low-dimensional and high-dimensional situations should be adopted. Third, multiple event types should be supported and the size of the event should not seriously affect the matching performance. Fourth, the insertion and deletion of subscriptions should bring little overhead. Fifth, the skewed distribution of attributes, event values and predicate values should not cause the major performance loss. Sixth, the width of the predicates in subscriptions should not have much impact on the matching time. The first three points are regarded as hard parameters and the last two ones are soft parameters.

When the number of subscriptions is large, for most algorithms, matching an event requires lots of repeated operations, such as comparison, addition or bit marking. From a hardware point of view, bit OR operations are more efficient than the operations performed repeatedly by most existing matching algorithms. Therefore, replacing other types of operations with bit OR operations is a feasible solution to improve matching performance. Taking this idea into consideration, we design a hardware-aware data structure to index subscriptions.

Checking Time   Marking Time 6.79ms, 93.56%   Comparing Time
0.422ms, 5.82%   0.045ms, 0.62%

Fig. 1. The matching time distribution of REIN

4.2 Data Structure of HEM

The data structure of HEM consists of a three-level index layer and a collection of bitsets for each attribute. The first level of indexing is based on attributes. The second level is indexed by the low value end (LVE) and high value end (HVE) of the predicate. The third level is constructed by dividing the value domain of the attribute into c cells. Each cell maps to a bucket that stores the low or high value of the predicate and the corresponding subscription ID. Each LVE or HVE is associated with a collection of bitsets that are used to pre-mark certain subscriptions as mismatches. When inserting subscription S into the structure of HEM, the low/high value of each predicate in S is mapped to the cell responsible for the value at the LVE/HVE of the attribute respectively. The predicate value (low or high) and the subscription ID are stored as a pair in the cell. An example of the structure is shown in Fig. 2.

Similar to GEM [6] and REIN [13], HEM uses a backward matching method, first searching for unmatching subscriptions to obtain matches indirectly. Given the partitioned cells with an attribute, if an event value v falls into the cell c_j, all predicates with low values larger than v will definitely not match the event at LVE. Specifically, unsatisfied predicates are stored in the cells from c_{j+1} to the last cell. Similarly, at HVE, all predicates with high values less than v stored in the cells from c_1 to c_{j-1} should also be marked as unmatching. At LVE and HVE, the event should compare with the pairs in cell c_j one by one to determine unmatching subscriptions.

When matching event, REIN [13] iterates through each cell that contains unmatching subscriptions, and marks each mismatch in the bitset. These repeated marking operations account for most of the matching time, as shown in Fig. 1. HEM avoids marking mismatches as much as possible by a caching mechanism. Multiple cells are allocated to one group and each group has a bitset to record whether each subscription is in the group. HEM pre-marks the subscriptions in each group as mismatches in the corresponding bitset. This optimization is called the subscription pre-mark cache (SPC) method. In this way, HEM avoids costly traversal and marking operations, and instead uses bitwise OR operations that are more efficient for hardware.

Subscription Pre-mark Cache (SPC) Method. Let g be the number of groups and c be the number of cells at LVE or HVE. In particular, each group g_i contains a list of continuous $i * \frac{c}{g}$ cells. Since the number of bitsets at LVE or HVE is equal to the number of groups, it is reasonable to let bitset B_i record the subscriptions in group g_i that contains the first/last $i * \frac{c}{g}$ cells for HVE/LVE. In this way, the coverage lengths of the bitsets constitute an arithmetic sequence with a common difference of $\frac{c}{g}$ cells.

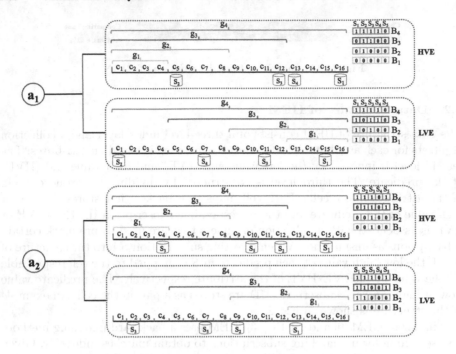

Fig. 2. HEM data structure

Table 1. Sample subscriptions

ID	a_1	a_2	ID	a_1	a_2	ID	a_1	a_2
S_1	[0.9, 0.95]	[0.8, 0.9]	S_2	[0.0, 0.3]	[0.5, 0.7]	S_3	[0.63, 0.69]	[0.1, 0.2]
S_4	[0.38, 0.76]	–	S_5	–	[0.4, 0.57]	–		

Figure 2 shows an example structure of HEM which builds indexes on two attributes a_1 and a_2. The value domain of each attribute is divided into sixteen cells assigned to four groups. Specifically, at the LVE of each attribute, group g_1 includes cells from c_{13} to c_{16}, g_2 from c_9 to c_{16} and so on. In contrast, the way of grouping cells at the HVE starts from c_1. The length of the bitset associated with each group is equal to the number of subscriptions n. Each bitset is used to mark the subscriptions stored in the cells belonging to the corresponding group.

Insertion Algorithm. Inserting a subscription S into HEM has two steps. Firstly, for each predicate in S, according to the attribute and low/high value, the predicate value and the subscription ID are stored as a pair in the cell that covers the value. Secondly, according to the cell grouping scheme, the subscription is marked in one or more bitsets. Algorithm 1 shows the pseudo code of insertion.

Figure 2 also shows the state of the data structure indexing the five sample subscriptions listed in Table 1. For example, when inserting S_1, for its first pred-

---

Algorithm 1: Insertion Algorithm

Input: Subscription S

1 **for** *each predicate* $p(a_i, [l, h])$ *in* S **do**
2   Insert a pair $(l, S.ID)$ into the cell covering l and mark S in the corresponding bitsets of the groups that cover the cell at LVE;
3   Insert a pair $(h, S.ID)$ into the cell covering h and mark S in the corresponding bitsets of the groups that cover the cell at HVE;

---

icate, $\lfloor 0.9 * 16 + 1 \rfloor = 15$ and $\lfloor 0.95 * 16 + 1 \rfloor = 16$, so S_1 is mapped to cell c_{15} and c_{16} at the LVE and HVE of a_1 respectively. Predicate values are omitted for brevity. Notice that c_{15} is in all the four groups at the LVE, so S_1 is marked in the four bitsets. At the HVE, c_{16} is only contained by group g_4, so S_1 is marked in the corresponding bitset B_4. The second predicate of S_1 is processed similarly.

4.3   Matching Procedure of HEM

HEM uses a bitset B to record the matching results. Each unmarked bit represents a matching subscription. Algorithm 2 gives the matching procedure of HEM, which can be divided into six steps. The first four steps are to process each attribute-value pair of the event. Step 1 performs comparisons in the cell into which the event value falls at LVE and HVE respectively and marks the unmatching subscriptions in B (lines 3–4). Step 2 selects the largest-size group that does not contain the cell into which the event value falls. Note that at LVE/HVE, when the cell ID is larger or equal than $\frac{c(g-1)}{g}$ /smaller or equal than $\frac{c}{g}$, no such group is available (line 5). Step 3 performs bit OR operations between B and the bitset of the selected group if available at LVE and HVE respectively for each nonempty attribute in the event (line 6). Step 4 marks the unmatching subscriptions in B that are stored in the cells not covered by the selected group (line 7). Step 5 does a series of bit OR operations between B and the bitset of the largest-size group for each null attribute of the event (line 8). When an event does not contain an attribute, all predicates defined on the attribute are not satisfied. Step 6 checks the unmarked bits in B to obtain the matching results (line 9).

Let event $E = \{(e_1(a_1, 0.64), e_2(a_2, 0.32)\}$ as an example. Based on the data structure shown in Fig. 2, the first attribute-value pair e_1 of E falls into the cell $\lfloor 0.64 * 16 + 1 \rfloor = 11$ on a_1. The low/high values of the predicates stored in cell c_{11} need to be compared with e_1 one by one at LVE/HVE respectively. Since the low value 0.63 of S_3 in c_{11} is smaller than the event value 0.64, it is not marked as unmatching in B. Next, $B_1(10000)$ and $B_2(01000)$ are selected to do bit OR operations with B since g_1 and g_2 are the largest-size group that does not cover c_{11} at the LVE and HVE of a_1 respectively. As a result, $B = (11000)$. Subsequently, cell c_{12} at LVE and cells c_9 and c_{10} at HVE should be traversed since they store unmatching subscriptions and are not covered by the selected groups. In this case, there are no subscriptions to be marked as unmatching in

---

Algorithm 2: Matching Algorithm

Input: Event E
Output: Matching results B

1 Initialize a zero bitset B whose length is the number of subscriptions;
2 **for** *each attribute-value pair* (a_j, v) *in Event E* **do**
3   Find the cell c_l at LVE and c_h at HVE that the attribute value falls into;
4   Compare v with the predicate values in c_l and c_h, and marks the IDs of subscriptions as unmatching in B;
5   Select the largest-size group g_l and g_h not covering c_l and g_h at LVE and HVE respectively;
6   Do bit OR operations between B and the bitsets of g_l and g_h to obtain all unmatching subscriptions of the groups if the group is not null;
7   Mark the IDs of subscriptions as unmatching for the rest unmatching cells not covered by g_l and g_h in B at LVE and HVE respectively;
8 Do $(d - \psi_E)$ times of bit OR operations between B and the bitset of the largest-size group to obtain the mismatches defined on null attributes of E;
9 Check the unmarked bits in B to output matching results.

B. In REIN [13], cells from c_{12} to c_{16} at LVE and from c_1 to c_{10} at HVE need to be traversed one by one, but in HEM only three cells need to be traversed. The second attribute-value pair e_2 in E is processed similarly. At the comparing step, S_5 stored in cell c_7 at the LVE of a_2 is marked as unmatching in $B(11001)$. $B_2(11000)$ and $B_1(00100)$ at the LVE and HVE respectively of a_2 do bit OR operations with B. At the final checking step, $B(11101)$ has one unmarked bit, meaning that S_4 is the match of E.

HEM reduces the traversal and marking operations in the matching process by setting up a set of caches that pre-mark certain subscriptions as unmatching. Therefore, HEM mainly does bit OR operations when matching events. Generally, compared to marking each unmatching subscription by traversing each cell, it is more efficient to perform a bit OR operation to collectively mark the unmatching subscriptions stored in multiple cells. When the selected largest-size group covers all the cells that need to be traversed, the marking time of HEM can be optimized to be close to zero. For each attribute, the total number of cells to be traversed at LVE and HVE is a constant, namely $\frac{c}{g} - 1$.

5 Theoretical Analysis

5.1 Complexity Analysis

Time Complexity of Insertion Algorithm. For a subscription S with size ψ_S, it needs to insert $2\psi_S$ pairs into the corresponding cells, which takes $O(\psi_S)$. In addition, marking the subscription in one or more bitsets has a cost at most $O(g\psi_S)$. Therefore, the time complexity of insertion is $O(g\psi_S)$.

Time Complexity of Matching Algorithm. The matching procedure of HEM can be divided into six steps to analyze. The comparison step (Lines 3–4) checks pairs in two cells with average cost $O(\frac{n\psi_S}{dc})$. The cost of two bit OR operations (Lines 5–6) is $O(n)$. The marking step (Line 7) traverses $\frac{c}{g} - 1$ cells for both value ends, so the cost is $O(\frac{n\psi_S}{dc} * (\frac{c}{g} - 1)) = O(\frac{n\psi_S}{dg})$. Given the event size ψ_E, these three steps have a cost $O(\psi_E n(1 + \frac{\psi_S}{dg}))$ since $g \leq c$. The time to process null attributes (Line 8) is $O((d-\psi_E)n)$. The final check step (Line 9) takes $O(n)$. Therefore, the time complexity of the matching algorithm is $O(dn + \frac{n\psi_E\psi_S}{dg})$.

Space Complexity. Given the dimensionality d of the content space and the number of cells c divided on each attribute, the total number of cells is dc. The total number of bitsets is $2dg$. For n subscriptions with size ψ_S, the total number of predicates is $n\psi_S$. Thus, the space complexity of HEM is $O(dc + dng + n\psi_S)$.

5.2 Performance Analysis

In this section, we build a analysis model for HEM to quantify the relationship between performance improvement and memory consumption. To explore the improvement on marking time, we take the HVE of one attribute as a breakthrough since the LVE and HVE of each attribute have similar characteristics.

Lemma 1. *Given the number of subscriptions n with size ψ_S, the marking time of HEM is proportional to $0.5n\psi_S$ without the SPC method.*

Proof. The marking time of HEM is proportional to the times of marking subscriptions one by one, which is computed as:

$$n\psi_S \int_0^1 (x - 0)dx = 0.5n\psi_S \tag{1}$$

where dx can be seen as a probability and x is the possible value of an event. The interval range to be traversed is $[0, x]$. □

Lemma 2. *Given n, ψ_S and g, the marking time of HEM is halved by doubling g with the SPC method.*

Proof. The adoption of the SPC method limits the number of traversing cells to $\frac{c}{g} - 1$. Hence, with the SPC method, the number of marked subscriptions is computed as:

$$n\psi_S \sum_{i=0}^{g-1} \int_{\frac{i}{g}}^{\frac{i+1}{g}} (x - \frac{i}{g})dx$$

$$= n\psi_S \int_0^1 xdx - \frac{n}{g}\psi_S \sum_{i=0}^{g-1} i \int_{\frac{i}{g}}^{\frac{i+1}{g}} 1dx \tag{2}$$

$$= 0.5n\psi_S - \frac{(g-1)n\psi_S}{2g} = \frac{n\psi_S}{2g}$$

Table 2. Parameter settings in the experiments

Name	Description	Experimental values
$\mathcal{R}$	The value domain of attribute	$[1, 1M]$
α	Parameter of Zipf	**0**, 1–5
d	Number of attributes	**20**, 30, 100, 300–900
n	Number of subscriptions	0.3M, **1M**, 3M–9M
ψ_E	Event size	**20**, 30–80
ψ_S	Subscription size	5, **10**, 15–30
w	Predicate width	0.1, 0.2, **0.3**–0.9

where $x - \frac{i}{g}$ is the length of interval to be marked one by one for any event value $x \in [\frac{i}{g}, \frac{i+1}{g}], i \in [0, g-1]$. Since the marking time is proportional to the times of marking subscriptions one by one, doubling g halves the marking time.   □

Theorem 1. *Given* n, ψ_E, ψ_S, d *and* g, *when* $\psi_E = d, g > 1$ *and the predicate values are uniformly distributed in the value domain* $[0, 1]$, *the improvement ratio of the marking time of HEM is* $1 - \frac{1}{g}$ *with the SPC method.*

Proof. Based on Lemma 1 and 2, the improvement ratio of the marking tasks of HEM is $1 - \frac{n\psi_S}{2g * 0.5n\psi_S} = 1 - \frac{1}{g}$.   □

Theorem 1 means that 50% or 96.875% of marking operations are avoided when g is set to 2 or 32 respectively, presenting an inverse proportional relationship. The ratio does not hold when $g = 1$ because it assumes that the unique bitset on LVE or HVE covers all cells. Nevertheless, we can configure that the unique bitset on LVE or HVE covers only half of the cells when $g = 1$. Thus, the improvement of $g = 1$ is equivalent to that of $g = 2$ when $\psi_E = d$.

6   Experiments

6.1   Setup

Workloads. The event data comes from a real-world stock dataset with 50 attributes after being cleaned up. The dataset was collected from the Chinese stock market on June 8, 2018. The subscription data is generated based on the stock dataset. For high-dimensional testing, we synthesize event data. Table 2 lists the parameter settings where the default values are marked in bold.

Baselines. We compare HEM with four event matching algorithms reviewed in Sect. 2, namely REIN [13], Ada-REIN [15], TAMA [20] and OpIndex [19]. Based on REIN, Ada-REIN ignores some marking tasks on attributes to reduce matching time, resulting in some false positive matches. REIN and Ada-REIN originally only support single-type event matching ($\psi_E = d$). We re-implemented them to match events with multiple types, namely $\psi_E \ll d$. Besides, they are set with the same c as HEM (1,000 by default). The false positive rate of Ada-REIN is set to 0.05. TAMA counts the matching predicates for each subscription which exist only in ψ_E attributes. The discretization level of TAMA is set to 13. We re-implemented TAMA to achieve exact matching with negligible overhead. OpIndex classifies subscriptions by their pivot attributes and only pivot attributes of events are processed.

Testbed. All the algorithms are implemented in C++ language and compiled by g++ 9.3.0 with -O3 optimization enabled on Ubuntu 20.04 system. All the experiments are conducted on an AMD 3.7 GHz machine with 64 GB RAM.

Metrics. We evaluate the performance of the five algorithms in terms of three metrics: matching time, insertion time and memory consumption. The matching time is measured from the beginning of matching an event to the end of obtaining the whole matching result. 500 events are processed to calculate the average matching time in each experiment. Insertion time refers to the time of inserting a subscription into the data structure. Memory consumption refers to the total memory used by the underlying data structure of the matching algorithm after inserting a subscription dataset.

6.2   Verification Experiments

We design a benchmark experiment to verify the performance analysis model in Sect. 5.2 and investigate the trade-off between matching time and memory usage. In this experiment, the parameters are set to the default values.

Figure 3 presents the marking time of HEM with different number of groups g from 1 to 512. Starting from $g = 2$, the marking time of HEM halves for each time g is doubled. For example, the marking time is 0.60 ms and 0.31 ms for $g = 16$ and $g = 32$ respectively. In this experiment, we set $\psi_E = d$, so the group covering all the cells is not used. Consequently, the marking time for $g - 1$ is approximately equal to that for $g = 2$. The average marking time without the SPC method is 6.79 ms, smaller than twice of the marking time when $g = 2$. This is because the groups are statically divided and the predicates are not absolutely evenly distributed. When $g = 32$, the performance improvement ratio of HEM is $1 - \frac{0.31}{6.79} \approx 95.4\%$, which is close to the theoretical value 96.875% based on Theorem 1. Overall, the ratio of marking time in the total matching time decreases from 93.7% to 2.5% when g increases from 1 to 512, indicating that the bottleneck operation (Fig. 1) has been alleviated. In summary, the SPC optimization method is effective and Theorem 1 is validated.

Figure 4 depicts the matching time and memory usage of HEM varying g from 1 to 512. The matching time dwindles exponentially and the memory usage grows exponentially with the exponential increase of g. When the SPC optimization method is not enabled, the matching time of HEM is 7.56 ms and the memory

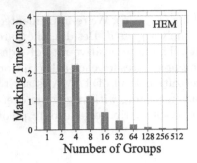

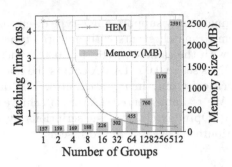

Fig. 3. Marking time of HEM with different g

Fig. 4. Matching time and memory usage of HEM with different g

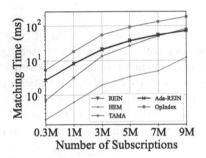

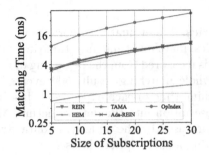

Fig. 5. Effect of n

Fig. 6. Effect of ψ_S

consumption is 152 MB. We finally set $g = 32$ to do the metric experiments since the matching time of HEM drops by 89.7% and the memory usage is nearly doubled, which makes a good trade-off.

6.3 Metric Experiments

The performance of the event matching algorithm is affected by many parameters. We change their settings to observe their effects in the experiments.

Number of Subscriptions n. The number of subscriptions is a core parameter to measure the workload, which has a vital impact on the matching time. As shown in Fig. 5, all algorithms have higher matching time as n increases. HEM performs best in all situations. Compared with REIN, TAMA, Ada-REIN and OpIndex, HEM reduces the matching time by 90.1%, 83.5%, 90.0% and 95.9% respectively on average. When n grows from 3M to 7M, the predicates become more densely distributed in the cells and the optimization space becomes larger. Hence, the matching time of HEM increases slower than before. When n is 9M, the matching time of HEM grows more quickly because the bit OR operations cost a lot. The matching time of TAMA increases faster than REIN and Ada-REIN because the workload of counting satisfied predicates is time-consuming.

OpIndex performs worst because predicates are evenly distributed in attributes and all the attributes are elected as pivot attributes.

Subscription Size ψ_S. To measure the effect of ψ_S, we set $d = \psi_E = 30, w = 0.7$ and vary ψ_S from 5 to 30. From Fig. 6 we can see that the matching time of the five algorithms increases linearly with ψ_S. Therefore, ψ_S is more related to the real workload compared to n. The performance of Ada-REIN is almost the same as REIN. This is attributable to the low false positive rate, low matching probability and the uniform distribution of predicates. Compared with REIN, TAMA, Ada-REIN and OpIndex, HEM reduces the matching time by 83.7%, 82.6%, 83.1% and 95.2% respectively average.

Event Size ψ_E. In the event size experiment, we set $d = 80$ and vary ψ_E from 30 to 80. Generally, the event size is proportional to the matching time of the forward matching algorithms (TAMA and OpIndex) and inversely proportional to the matching time of the backward matching algorithms (REIN and Ada-REIN), as shown in Fig. 7. Nevertheless, HEM, as a backward matching algorithm, has a slowly increasing matching time with ψ_E. This is because both comparing and marking operations are avoided and only one bit OR operation is needed to process each null attribute. On average, the matching time of HEM is reduced by 88.5%, 86.5%, 86.3% and 50.3% compared with OpIndex, REIN, Ada-REIN and TAMA, respectively.

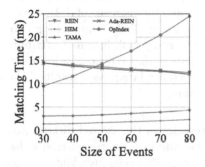

Fig. 7. Effect of ψ_E

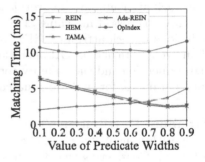

Fig. 8. Effect of w

Predicate Width w. The matching probability of subscriptions is relevant to ψ_S and w. Figure 8 presents how w affects the matching time by varying w from 0.1 to 0.9. ψ_S is set to 5 to ensure a nonzero matching probability in the experiment. This setting makes the predicates sparsely distributed in cells. As a forward algorithm, TAMA takes a longer time with the increasing of w. The performance of OpIndex is not affected by w. As backward algorithms, REIN and Ada-REIN run faster with w. However, when $w = 0.9$, the predicates are dense and the comparing time of REIN and Ada-REIN becomes unnegligible so

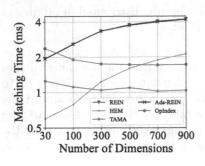

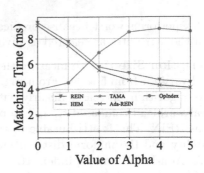

Fig. 9. Effect of d Fig. 10. Effect of α

their matching time increase a little. HEM is nearly immune to w and exhibits a steady performance because the number of cells to be marked is limited to $\frac{c}{g} - 1$ for each nonempty attribute. On average, in comparison with REIN, TAMA, Ada-REIN and OpIndex, HEM reduces the matching time by 88.0%, 85.0%, 87.3% and 95.9% respectively.

Number of Attributes d. To simulate sparse workloads, we set $w = 0.5$ and vary d from 30 to 900. Figure 9 indicates that all algorithms behave monotonically. TAMA performs best after d is up to 300. The two forward matching algorithms show a similar trend with d because they process each event value rather than each attribute and the workload decreases with the increase of d. However, the three backward matching algorithms have to mark all the unmatching subscriptions in each attribute. HEM replaces the marking task in a null attribute of an event with one bit OR operation. Unfortunately, that still costs a lot under high dimension and the memory consumption becomes large. As a result, d is a hard parameter for backward matching algorithms.

Attribute Distribution. There are two categories of the skewed distribution of algorithm input. One is the value distribution and the other is attribute distribution. Considering that the matching time basically keeps invariant with an uneven value distribution of subscriptions and events, we only give the experiment results under skewed attribute distribution of both subscriptions and events in Fig. 10, where $d = 50, \psi_S = 5$ and $w = 0.5$. In Zipf distribution, a larger α means a more serious skewed distribution and a more heavy workload. HEM and TAMA overcomes the skewed problem well while the other three counterparts fluctuate greatly with α. Ada-REIN skips from 1 attribute and about 26 k predicates to 40 attributes and about 5M predicates when α varies from 1 to 5. Thus its matching time is smaller than that of REIN.

6.4   Maintainability

Two experiments are conducted to test the maintenance cost of HEM. Figure 11 reveals that the average insertion time of HEM increases by about 42.6%

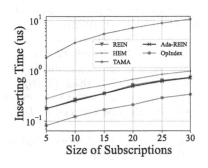

Fig. 11. Insertion time ψ_S

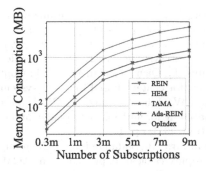

Fig. 12. Memory usage(MB) by n

compared to REIN because HEM needs to pre-mark a new subscription in one or more bitsets for each attribute on which the predicate is defined. TAMA inserts the subscription ID into a set of cells for each predicate, thereby resulting in the highest insertion time. The deleting time of HEM is very close to that of REIN because the time to find the deleted pairs in cells accounts a lot. The experiment results are omitted. Figure 12 shows the memory usage of the five algorithms with different n. All the curves rise logarithmically. The memory usage of HEM is about twice of REIN's and 64.3% of TAMA's.

7 Conclusion

When processing a large number of subscriptions, most matching algorithms need to repeatedly perform a lot of operations, which becomes a performance bottleneck. In this paper, we propose a hardware-aware event matching algorithm termed HEM, aiming to execute efficient operations in the matching process. The experiment results reveal the superiority of HEM to its counterparts. HEM shows excellent performance in terms of matching time and stability under various conditions. In the future, we plan to design a state reduction method to optimize the bit OR operations of HEM under high dimension, and extend HEM to multi levels to accommodate input with multiple hotspots.

Acknowledgments. This work was supported by the National Key Research and Development Program of China (2019YFB1704400), the National Natural Science Foundation of China (61772334, 61702151), and the Special Fund for Scientific Instruments of the National Natural Science Foundation of China (61827810).

References

1. Barazzutti, R., Heinze, T., et al.: Elastic scaling of a high-throughput content-based publish/subscribe engine. In: ICDCS, pp. 567–576. IEEE (2014)
2. Chen, L., Shang, S.: Top-k term publish/subscribe for geo-textual data streams. VLDB J. **29**(5), 1101–1128 (2020)

3. Ding, T., Qian, S.: SCSL: optimizing matching algorithms to improve real-time for content-based pub/sub systems. In: IPDPS, pp. 148–157. IEEE (2020)
4. Ding, T., et al.: MO-Tree: an efficient forwarding engine for spatiotemporal-aware pub/sub systems. IEEE Trans. Parallel Distrib. Syst. **32**(4), 855–866 (2021)
5. Ding, T., Qian, S., Zhu, W., et al.: Comat: an effective composite matching framework for content-based pub/sub systems. In: ISPA, pp. 236–243. IEEE (2020)
6. Fan, W., Liu, Y., Tang, B.: GEM: an analytic geometrical approach to fast event matching for multi-dimensional content-based publish/subscribe services. In: IEEE INFOCOM, pp. 1–9 (2016)
7. Fontoura, M., Sadanandan, S., Shanmugasundaram, J., et al.: Efficiently evaluating complex Boolean expressions. In: ACM SIGMOD, pp. 3–14 (2010)
8. Gao, C., Xin, X., et al.: ParaBit: processing parallel bitwise operations in NAND flash memory based SSDs. In: IEEE/ACM MICRO-54, pp. 59–70 (2021)
9. Ji, S.: Ps-tree-based efficient Boolean expression matching for high-dimensional and dense workloads. Proc. VLDB Endow. **12**(3), 251–264 (2018)
10. Ji, S., Jacobsen, H.A.: A-tree: a dynamic data structure for efficiently indexing arbitrary boolean expressions. In: ACM SIGMOD, pp. 817–829 (2021)
11. Liao, Z., Qian, S., Cao, J., et al.: PhSIH: a lightweight parallelization of event matching in content-based pub/sub systems. In: ICPP, pp. 1–10 (2019)
12. Ma, X., Wang, Y., Pei, X., Xu, F.: A cloud-assisted publish/subscribe service for time-critical dissemination of bulk content. Concurr. Comput. Pract. Exp. **29**(8), e4047 (2017)
13. Qian, S., Cao, J., Zhu, Y., Li, M.: REIN: a fast event matching approach for content-based publish/subscribe systems. In: IEEE INFOCOM, pp. 2058–2066 (2014)
14. Qian, S., Cao, J., Zhu, Y., Li, M., Wang, J.: H-tree: an efficient index structure for event matching in content-based publish/subscribe systems. IEEE Trans. Parallel Distrib. Syst. **26**(6), 1622–1632 (2015)
15. Qian, S., Mao, W., Cao, J., Mouël, F.L., Li, M.: Adjusting matching algorithm to adapt to workload fluctuations in content-based publish/subscribe systems. In: IEEE INFOCOM, pp. 1936–1944 (2019)
16. Sadoghi, M., Labrecque, M., Singh, H., Shum, W., Jacobsen, H.A.: Efficient event processing through reconfigurable hardware for algorithmic trading. Proc. VLDB Endow. **3**(1–2), 1525–1528 (2010)
17. Shah, M.A., Kulkarni, D.: Multi-GPU approach for development of parallel and scalable pub-sub system. In: Iyer, B., Nalbalwar, S.L., Pathak, N.P. (eds.) Computing, Communication and Signal Processing. AISC, vol. 810, pp. 471–478. Springer, Singapore (2019). https://doi.org/10.1007/978-981-13-1513-8_49
18. Wang, Y.: A general scalable and elastic content-based publish/subscribe service. IEEE Trans. Parallel Distrib. Syst. **26**(8), 2100–2113 (2014)
19. Zhang, D., Chan, C.Y., Tan, K.L.: An efficient publish/subscribe index for e-commerce databases. Proc. VLDB Endow. **7**(8), 613–624 (2014)
20. Zhao, Y., Wu, J.: Towards approximate event processing in a large-scale content-based network. In: IEEE ICDCS, pp. 790–799 (2011)
21. Zhu, W., et al.: Lap: a latency-aware parallelism framework for content-based publish/subscribe systems. Concurr. Comput. Pract. Exp. e6640 (2021)

RotorcRaft: Scalable Follower-Driven Raft on RDMA

Xuecheng Qi, Huiqi Hu$^{(\boxtimes)}$, Xing Wei, and Aoying Zhou

School of Data Science and Engineering, East China Normal University,
Shanghai, China
{xcqi,simba_wei}@stu.ecnu.edu.cn, {hqhu,ayzhou}@dase.ecnu.edu.cn

Abstract. State machine replication plays a fundamental role in meeting both the scalability and the fault-tolerance requirement in cloud services. However, the single-point leader is easy to become a bottleneck of scalability because it needs to handle all read and write requests and independently replicate logs in order for all followers. Moreover, machine resources are shared through cloud services where the scale-up of the leader is very expensive. In this paper, we propose a variant of Raft protocol using RDMA named RotorcRaft to significantly offload burden from the leader to followers to relieve the single-point bottleneck. First, RotorcRaft assigns a follower-driven log replication mechanism that exploits hybrid RDMA primitives to relieve part of the burden of leader to followers in log replication. Then, RotorcRaft proposes a quorum follower read that enables followers to handle read requests without the involvement of the leader. Experimental results demonstrate that RotorcRaft has excellent scalability and up to 1.4x higher throughput with 84% latency compared against the state-of-the-art work.

Keywords: Follower-driven · RDMA · Raft

1 Introduction

In cloud services, State Machine Replication (SMR) is a key infrastructure that ensures both the scalability and the fault-tolerance requirement. SMR enables the reliability of services by ensuring logs are recorded in the same order by the majority of servers before they are applied to the state machine. Specifically, Raft is a leader-based state machine consensus protocol that is widely used in commercial systems like etcd [1], TiDB [2] and PolarFS [3], because of its simplicity of implementation and strong consistent replication.

However, the single-point of leader can easily become a bottleneck of system scalability. Figure 1 shows the leader node bottlenecks for a classic Raft: (1) the leader needs to independently replicate logs to all followers and ensure the order of records. (2) the leader takes on heavy workloads that needs to process all the read and write requests from clients. Adding node in the consensus protocol can increase the fault tolerance of the system, but accordingly hurts the performance.

© The Author(s), under exclusive license to Springer Nature Switzerland AG 2022
A. Bhattacharya et al. (Eds.): DASFAA 2022, LNCS 13245, pp. 293–308, 2022.
https://doi.org/10.1007/978-3-031-00123-9_24

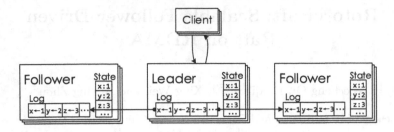

Fig. 1. Architecture of Raft

Some work [3,4] has exploited high-speed network cards that support Remote Direct Memory Access (RDMA) to boost the performance of consensus protocols. However, they do not address the underlying problems of leader-based protocols: the consumption of the leader's CPU and network is too heavy, while the resources of followers are left idle. Not only does the leader need to replicate the logs to all nodes, but the leader also plays a key role in follower read. The follower needs to get the latest commit index (ReadIndex) from the leader. And, before sending the commit index to the follower, the leader requires a quorum confirmation of leadership by communicating with majority nodes, which requires significant CPU and network overhead.

In this paper, we propose a variant of Raft protocol named RotorcRaft, a scalable follower-driven Raft on RDMA that significantly offloads burden from the leader to followers to relieve the single-point leader bottleneck. We are aimed to answer two main questions: (1) *Can we partially offload the overhead of the leader to followers in log replication?* (2) *Can we fully offload the overhead of the leader to followers without the involvement of the leader in follower read?* Main contributions are summarized ad follows:

- To answer the first question, RotorcRaft designs a follower-driven log replication mechanism that exploits hybrid RDMA primitives to offload tasks on the critical path in log replication from the leader to followers, effectively reducing the CPU and network overhead of the leader.
- To answer the second question, RotorcRaft proposes a quorum follower read scheme that enables followers to directly handle read requests by leveraging majority vote among followers and remote direct ReadIndex read in a follower-driven manner without the involvement of the leader.
- We implement RotorcRaft with 100 Gpb/s RMDA InfiniBand NICs and use Intel Optane DCPMM device to store logs. Experimental results demonstrate that RotorcRaft has excellent scalability and up to 1.4x higher throughput with 84% latency compared against the state-of-the-art work.

The remainder of the paper is organized as follows: Sect. 2 describes the preliminary of RDMA network and Intel Optane DCPMM, then reviews the related works. Section 3 describes the overview design of RotorcRaft. Section 4 demonstrates the follower-driven log replication mechanism. Section 5 introduces the quorum follower read scheme. Section 6 discusses communication complexity between Raft and RotorcRaft. In Sect. 7, we present the results of our performance evaluation. Section 8 concludes the paper.

2 Preliminary

2.1 RDMA Network

RDMA network provides high bandwidth and low round-trip latency accessing to the memory of the remote machine. This is achieved by using zero-copy networking bypassing the OS kernel and remote CPU. RDMA supports Memory verbs, Message verbs and RDMA multicast:

Memory verbs, also known as **one-sided verbs** (i.e., *read* and *write*), require no involvement of the remote CPU, namely, these verbs entirely bypass the remote CPU. Client can *write* or *read* directly to the memory of server that the server is unaware of client's operations.

Message verbs, also called as **two-sided verbs** (i.e., *send* and *receive*), provide user-level two-sided message passing that involves remote machine's CPU, before a *send* operation, a pre-posted *receive* operation should be specified by the remote machine.

RDMA multicast [5], supports one-to-multiple RDMA communication in one multicast group. It is only supported in Unreliable Datagram (UD) transport type, but corrupted or out-of-sequence packets are silently dropped [6].

2.2 Intel Optane DCPMM

Intel Optane DCPMM is the first commercial Non-Volatile Memory (NVM) hardware module that promises byte-addressability, persistence and high density. It can be configured in two modes, Memory mode and AppDirect mode:

Memory mode, simply uses Optane DCPMM to extend the capacity of DRAM without persistence. The conventional DRAM that is transparent to memory controller serves as a "**L4 cache**" between CPU cache and Optane DCPMM. CPU and operating system directly consider it as main memory with larger capacity, usage of Optane DCPMM is the same as DRAM.

AppDirect mode, uses Optane DCPMM as an individual persistent memory device that provides persistence and DRAM is not used as a cache. Softwares can access Optane DCPMM with load and store instructions. Because of CPU reordering, extra flush instruction (e.g., *clushopt* or *clwb*) and fence instruction (e.g., *sfence* or *mfence*) are needed to persist data from CPU cache to the persistent memory after store instructions. Alternatively, non-temporal instruction (i.e., *ntstore*) can bypass CPU cache and directly write to memory.

2.3   Related Works

RDMA and NVM Backend Systems. AsymNVM [7] proposes a generic framework for asymmetric disaggregated persistent memories using RDMA networks. It implements the fundamental primitives for building recoverable persistent data structures and NVM space can be shared by multiple servers. Clover [8] is a disaggregated key-value store to separate the location of data and metadata to improve scalability. It proposes a hybrid disaggregation model (MS+DN) that locates metadata at metadata server using two-sided RDMA verbs, stores data remotely in data node using one-sided RDMA verbs. FileMR [9] proposes an abstraction combining NVM regions and files that allows direct remote access to an NVM-backed file through RDMA. These works take full advantages of the characteristics of RDMA and NVM, but none of them introduces the RDMA multicast to build a scalable, high-available system.

Primary-Backup Replication. Some works [10–13] exploit RDMA and NVM for primary-backup replication. FaRM [10] is a distributed computing platform that uses RDMA *write* to transfer logs to the NVM of the backups. Query Fresh [11] employs an append redo-only storage with parallel log replay to avoid conventional dual-copy design and utilizes RDMA *write with imm* for log shipping. Erfan et al. [12] design a replication protocol to minimize the CPU processing by respectively sending log records and data items into NVM of backups using one-sided RDMA *write*. Mojim [13] is a replication framework that uses two-sided RDMA verbs to replicate all data to the NVM of backups. Primary-backup replication usually maintains backup nodes with weakly consistent copies of data while quorum-based replication guarantees strong consistency. Due to the difference, these works that optimize log replication for primary-backup replication cannot be directly applied to the quorum-based replication.

State Machine Replication on High-speed Network. DARE [4] redesigns and speeds up Raft protocol by leveraging the advantage the RDMA verbs. PolarFS [3] proposes ParallelRaft built on RDMA network that enables out-of-order concurrent log replication by recording the write block address modified by the previous N log entries in look behind buffer. HovercRaft [14] enhances the scalability of Raft in log replicaiton by offloading leader duties to the programmable switch and adopting load balancing replies. However, DARE [4] and ParallelRaft [3] have exploited RDMA primitives to speed up Raft but do not focusing on solving the single-point leader bottleneck. HovercRaft [14] requires specialized programmable switch hardware that is not practical at present. RotorcRaft assigns the follower-driven log replication and quorum follower read mechanism that significantly offloads the burden of leader to followers so that relieves the single-point bottleneck.

Implementations of Follower Read. The original implementation for follower read [15] is that followers forward read requests sent by the client to the leader to process. Alternatively, etcd [1] implemented the ReadIndex mechanism proposed in [15] that the leader only sends the last commit index to the follower after the confirmation heartbeat of leadership. TiDB [2] further introduces a

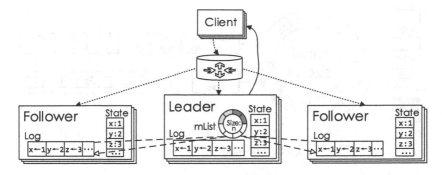

Fig. 2. Architecture of RotorcRaft

LeaseRead scheme that allows follower read in a lease interval but under the ideal assumption that CPU clocks of most servers are accurate. Although the ReadIndex and LeaseRead methods effectively improve the ability of follower read, both of them require the participation of the leader. To solve the bottleneck of single-point leader, RotorcRaft proposes the quorum follower read mechanism that significantly offloads duties of the leader to followers.

3 RotorcRaft Overview

Figure 2 illustrates the architecture of a 3-node RotorcRaft cluster, connected by an RDMA switch. RotorcRaft maintains a mList which stores metadata of log records at the leader-side. mList is implemented by a lock-free circular array with fixed size N, which is shown in Fig. 2. The space of mList is pre-allocated and can be reused to avoid the overhead of garbage collection. More specific details are in Sect. 4.1. RotorcRaft is a follower-driven Raft that log replication is issued by followers and supports quorum follower read. We have summarized the two main modules as follows:

Follower-Driven Log Replication: In RotorcRaft, log replication has been divided into three phases: request phase, pull phase, response phase. We take full advantage of the hybrid RDMA primitives in different phases and significantly reduce the overhead on the leader. Considering that there will be slow followers in Raft, mList is well-tuned for slow followers to chase after the latest log records. In RotorcRaft, log chase is initiated by the slow follower node as well, rather than the leader patching the logs directly.

Quorum Follower Read: Without the involvement of the leader, we design a mechanism of quorum follower read that can be done entirely among followers. We use RDMA read to complete the process of confirming the leadership and ReadIndex acquisition to fully utilize the CPU and network resource of followers. In RotorcRaft, the quorum follower read mechanism significantly offloads the burden from the leader to followers to serve follower read requests.

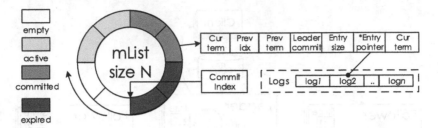

Fig. 3. The structure of mList

4   Follower-Driven Log Replication

4.1   The Structure of mList

The mList plays a key role in the follower-driven log replication. Because Rotor-cRaft separates replication from ordering logs of the leader, mList is maintained on the leader-side to decide the order of log records. Figure 3 illustrates the structure of mList, mList is implemented by a lock-free circular array with fixed size N. The space of mList is pre-allocated and can be reused to avoid the overhead of garbage collection. In mList, entries have four kinds of status, *empty, active, committed,* and *expired. empty* means the entry is unoccupied and available; *active* means a log is being processed and the entry is not available; *committed* means the log has been confirmed by the majority and has been committed, but the entry is not available now. *expired* means that the committed log is expired and the entry is available.

The log metadata in Raft is fixed size and the size of entry in mList is the same as it. Among them are *current_term, prev_log_idx, prev_log_term, leader_commit, entry_size* and an *entry_pointer* to point the log entry in the NVM. Upon receiving the request, the leader immediately generates metadata of the request and adds it to its mList at the position of (*commit_index* % N), then changes the entry status as active. Because Raft does not tolerate log holes so logs cannot be committed out of order. If the leader finds that the status of entry in the ((*commit_index - 1*) % N) position is active, i.e., the last log has not been committed, the leader will wait for the last log to be committed before processing the current log. Meanwhile, followers directly read the remote ((*commit_index* - 1) % N)$_{th}$ entry in mList using RDMA read. To check whether the read metadata is complete, we use the same method as Cell [16], which redundantly adds the *cur_term* field at the end of the metadata and validate it after read. After the majority of nodes commit the log, the status of the entry will be set to *committed*. When there is no more free entries, the *committed* entry is set to *expired* status and can be reused. We do not adopt the strategy of immediate recycling *committed* entries for the consideration of slow followers in the cluster. The design of mList not only serves log replication effectively but is also friendly for slow follower nodes to chase logs, detailed in Sect. 4.3.

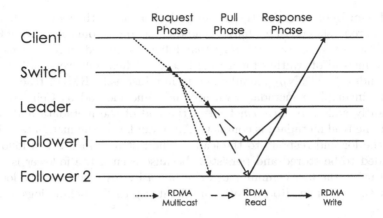

Fig. 4. Mechanism of follower-driven log replication

4.2   Mechanism of Follower-Driven Log Replication

In the native Raft protocol [15], it depends on a strong leader in charge of ordering and replicating client requests across all followers, while actively processing all the write/read requests, which incurs a major impediment to the scalability of the system. It's challenging to offload the burden from the leader to followers that utilize the CPU and network resources of followers in log replication. Fortunately, emerging RDMA-capable NIC is promising to achieve this goal.

To solve the issue, we propose a follower-driven log replication mechanism that relieves the burden of the leader. The key point of the follower-driven mechanism is to separate replication from ordering in leader and enable a follower-driven manner that the log replication is completed by followers. As Fig. 4 shows, we have decomposed the entire log replication into three phases (1) Request Phase. (2) Pull Phase. (3) Response Phase. Each phase adopts different RDMA primitives as the best consideration.

Request Phase: We adopt RDMA multicast to convert leader-to-multipoint interactions into point-to-point interactions. Rather than targeting a specific server, clients directly send requests in a multicast group containing the leader and the followers. Without the leader individually replicating to each of the follower nodes, all nodes in the group receive client requests that contain the record entries that need to be modified via RDMA multicast. Obviously, RDMA multicast does not guarantee log order. In RotorcRaft, the leader decides the order of client requests. Upon receiving the request, the leader immediately generates metadata of the request and adds it to its mList while followers insert the request into a set of unordered requests. The mList is a lock-free circular array with fixed size N to buffer metadata of log records for followers to read in Pull Phase. In RDMA networks, multicast only supports Unreliable Datagram (UD) transport type. But even with unreliable transports (UD), packets are never lost due to buffer overflow [17]. Therefore, in the common case, with no packet loss occurs, the leader is only in charge of ordering requests but not data replication.

Pull Phase: Upon receiving the request, followers insert the request into a set of unordered requests and immediately read the metadata from mList using RDMA read. We adopt an instant strategy that followers immediately read metadata remotely instead of waiting for a time interval. Then, followers pull the ((last commit index + 1) % N)$_{th}$ metadata entry in mList using RMDA read. To avoid accessing incomplete metadata, we use the same method as Cell [16], which redundantly adds the *cur_term* field at the end of the metadata and validate whether the read metadata is complete. After validating the metadata, followers persist the log and respond to the leader. The last read location of followers is not needed to be stored and persisted, because even if the follower is crashed and restarted, the last commit index can be easily restored from the logs. Due to the design of mList, RotorcRaft can tolerate up to N-1 behind logs from slow followers.

Response Phase: After completing log persistence in pull phase, followers will send responses to the leader via RDMA write. When receiving responses from the majority, the leader commits the log and sets the corresponding entry of mList to *committed* status. After that, the leader responds to the client. After the log is committed, the leader updates the pointer of the corresponding entry in mList to point to the address of the log record in NVM.

4.3   Log Chase

Although RDMA network is stable in common cases, it is inevitable that appears sporadic follower is slower than the leader. In Raft, the leader should always require multiple rounds of network communication to determine which logs are missing and explicitly send missing logs to the slow follower. In RotorcRaft, we support the log chase in a follower-driven manner with the assistance of mList as well. RotorcRaft can tolerate up to N-1 logs behind. To achieve this goal, mList adopts a lazy garbage collection policy. When the log is committed by the majority of replicas, the metadata is obsolete but mList will not reclaim the space of the entry at once.

When the latest log position in mList is $((commit\_index_{leader} - 1) \% N)$, the slow node still goes to read the $((commit\_index_{slow\_node} - 1) \% N)$ position of mList. If this gap is less than N - 1, the slow node is able to read the metadata of missing logs on its own and fill the logs. If it even loses the requests sent by the client, it can read the log records via RDMA read according to the corresponding address stored in the entry of mList. Users can determine the level of tolerating slow nodes by setting the size of N. If the gap is greater than N - 1, we recommend using snapshot for log chase.

4.4   Log Replication RPC

The native Raft has a major bottleneck of scalability because of the single-point leader. To solve the bottleneck, we take advantage of hybrid RDMA primitives and have crafted follower-driven RPCs for log replication and follower read to

---

Algorithm 1: LogReplicationRPC

---

1 **Function** f_join_multicast():
2     $qp \leftarrow$ initialize a Queue Pair;
3     $channel \leftarrow$ initialize a Completion Channel;
4   //Join in the multicast group
5   rdma_join_multicast($qp, channel$);
6   //Verify if successfully joined the multicast group
7     $ret \leftarrow$ get_cm_event($channel$, RDMA_CM_EVENT_MULTICAST_JOIN);
8   **if** $ret == ERROR$ **then**
9   //Leave the multicast group
10   rdma_leave_multicast($channel$);

11 **Function** f_multi_send():
12     $qp \leftarrow$ acquire a Queue Pair;
13     $ah \leftarrow$ create an Address Handle for UD QP;
14   //send requests in multicast group
15   ibv_post_send(ah,qp);

16 **Function** f_meta_read():
17     $qp \leftarrow$ acquire a Queue Pair;
18     $last\_buf\_id \leftarrow$ The last buffer id in mList;
19     $latest\_buf\_id = (latest\_buf\_id + 1)) \bmod N$
20   //Read the metadata of latest log record in mList on leader-side
21   ibv_post_read(qp, REMOTE_ADDR + $latest\_buf\_id$ * BUF_SIZE);

---

realize the design of RotorcRaft. We explain the implementation of RPCs of RotorcRaft in detail. Here are two main RPCs: LogReplicationRPC for log replication and FollowerReadRPC (detail in Sect. 5.2) for follower read.

In LogReplicationRPC, there has $f\_join\_multicast()$, $f\_multi\_send()$, $f\_meta\_read()$ functions. As algorithm 1 shows, the node (leader/follower/client) first initializes a queue pair "qp" for communication and a completion channel "channel" which is used for monitoring events (line 2–3). Then, the node registers the qp into the same multicast group (line 3–5). Next, verify if the node is successfully joined the multicast group, if not, leave the node from the multicast group (line 6–10). In $f\_multi\_send()$ function, the client creates an address handle for UD QP and calls $ibv\_post\_send()$ to send requests in multicast group (line 12–15). In the $f\_meta\_read()$ function, the follower first gets the latest buffer id of the mList, that is, the $last\_buf\_id$ (stored locally) +1 and then mod the size of mList (line 17–19). After that, the follower directly reads the metadata of the latest log record in mList on the leader side (line 20–21). Finally, the follower persists the log record and replies to the leader.

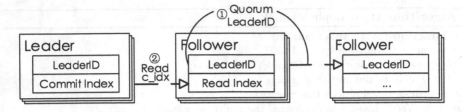

Fig. 5. Quorum follower read

5 Quorum Follower Read

5.1 Mechanism of Quorum Follower Read

We propose a mechanism of quorum follower read, which means that follower read is conducted by majority vote among followers without the involvement of the leader. The mechanism can effectively enhance the scalability of the system and improve the performance of the follower reads. In the original Raft, follower read needs to be forwarded for the leader to process: (1) The leader first needs to confirm whether it has the current leadership, so it sends a heartbeat packet that contains the latest commit index to all followers. After receiving the positive responses of the majority nodes, the leader confirms its leadership. (2) Then the leader sends the confirmed commit index to the follower. After that, the follower records the commit index as ReadIndex and waits until the state machine applies to the ReadIndex, and then processes the client's read request.

Obviously, the above method needs to consume the CPU and network resources of the leader and is not a "real" follower read. With the advantage of RDMA read, we craft a mechanism of quorum follower read that almost needs no involvement of the leader. The key to this design is the majority vote among followers and the ingenious use of RDMA read. As Fig. 5 illustrates, the processes of quorum follower read are shown as follows:

(1) **Preparation.** Before processing a quorum follower read, some preparatory work needs to be done in advance. On the follower-side, the LeaderID field in DRAM stores the ID of the current leader node. Each node should put the 64-bit field in the first 64-bit address of the RDMA registered memory region that can be read via RDMA read directly. On the leader-side, in addition to the LeaderID field, the Commit Index filed is placed in a fixed memory address as well. Both fields are modified using CAS atomic operations so that the 8B data can never be partially written.

(2) **Majority vote.** We have adopted a follower-driven process to determine the current leadership. Frist, the follower which processes read requests reads the leaderID field of all followers directly via RDMA read, and puts them in a list (including its own). Then, the follower confirms the current leadership by quorum. In the right case, the leadership should just be owned by its current leader. Otherwise, the follower will abort the follower read requests and establish connection with the new leader.

Algorithm 2: FollowerReadRPC

---

1 **Function** f_quorum_leadership():
2     $qp[n-1] \leftarrow$ acquire Queue Pair array;
3     $leader[n-1] \leftarrow$ acquire Leader ID array;
4   **for** $i = 0$ *to FOLLOWER_NUM* **do**
5         $qp[i] \leftarrow$ get the i_{th} QP;
6         $leader[i] \leftarrow$ ibv_post_read($qp[i]$, FOLLOWERi_LEADER_ADDR);
7   //Quorum confirmation of the leadership among followers
8     $leader\_id \leftarrow$ = quorum_leader(*$leader$);
9 **Function** f_get_readindex():
10     $qp \leftarrow$ acquire a Queue Pair;
11   //Directly read read_index from the leader
12     $read\_index \leftarrow$ ibv_post_read(qp, LEADER_READINDEX_ADDR);

---

(3) **Read ReadIndex.** In the common case, after the successful confirmation of leadership, the follower then gets the latest Commit Index from the leader. It exploits RDMA read to read the remote Commit Index on the leader-side and saves it locally as ReadIndex. Finally, the follower waits until its state machine applies to the ReadIndex, and processes the client's read request.

5.2 Follower Read RPC

FollowerReadRPC has *f_quorum_leadership()* and *f_get_readindex()* functions. As Algorithm 2 shows, to serve the read requests from the client, the follower first acquires the queue pair array for communicating with all other followers and the leader id array to store *leader_id* read from majority followers (including itself) (line 2–3). Then, the follower directly reads the *leader_id* using RDMA read from all other followers and store them in the leader id array (line 4–6). Next, the follower uses quorum confirmation of leadership among followers to determine the current leader (line 7–8). After that, the confirmation of the leadership is completed. In *f_get_readindex()* function, after the leader has been confirmed, the follower directly reads the latest commit index from the remote leader node via RDMA read and stores it locally as read index (line 10–12). Finally, the follower waits until the state machine applies to the read index and then processes read requests from the client.

6 Communication Complexity

Table 1 summarizes the communication complexity at the leader and followers in log replication and follower read for a cluster with N nodes (1 leader and N-1 followers). In the case of Raft, the leader replicates the log record to N 1 followers and replies to the client, while followers send N-1 responses to the leader

Table 1. Comparison of network complexity in Raft and RotorcRaft

Request/System	Raft		RotorcRaft	
	Leader	Followers	Leader	Followers
LogReplication	(N-1)+1	N-1	1	2*(N-1)
FollowerRead	(N-1)+1	2	0	N-2+1+1

in log replication. In follower read, the leader needs to send heartbeat for leadership confirmation to N-1 followers, and send the latest commit index to the follower while the follower sends a ReadIndex request to the leader and replies to the client. In the case of RotorcRaft, N-1 number of followers directly read the metadata in leader and send N-1 replies to the leader in replication while the leader just needs to reply to the client. In follower read, the follower directly reads the LeaderID index among N-1 followers (excluding itself) and ReadIIndex from the leader, and replies to the client while the leader does nothing. In summary, RotorcRaft successfully offloads burden from the leader to followers in log replication and follower read that significantly improves system scalability.

7   Evaluation

7.1   Experimental Setup

In this section, we analyze the overall performances of RotorcRaft, and the benefits from each optimization. Experimental setup and the benchmark used in this evaluation are given below.

Hardware Platform: Our experiments run on a cluster of 9 machines, one for the leader and the other 8 for followers. Each machine is equipped with a 2-socket Intel(R) Xeon(R) Gold 6240M CPU @ 2.60GHz processor (36 cores in total), 192 GB of DRAM memory and 1.5TB (6 × 256 GB) Intel DCPMM NVM device of installed with CentOS 7.5. All nodes and connected with Mellanox ConnectX-4 EDR 100Gb/sec Infiniband cards.

Comparison Target: We use DARE [4] which is the implementation of Raft on RDMA to denote the Raft-R. We further compare RotorcRaft with ParallelRaft [3], the state-of-art design of Raft on RDMA that supports out-of-order concurrent log commit. We have implemented these three works on RDMA and NVM, and all three implementations execute requests in batches. The default number of replicas is 5.

Workloads: We adopt YCSB [18] workloads to evaluate three methods. The default datasize is 10 GB. The distrbution of our workloads are uniform, the keys are chosen uniformly at random. The size of each record is about 64B.

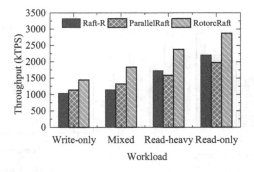

Fig. 6. Overview throughput under different workloads

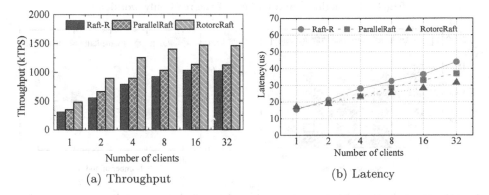

(a) Throughput   (b) Latency

Fig. 7. Throughput and latency unnder the write-only workload

7.2 Overview Performance

We first measure the throughput under varying workloads: write-only (100% write), mixed (50% read, 50% write), read-heavy (95% read, 5% write) and read-only (100% read) workloads. All methods have five replicas with 16 clients.

The overview throughput is shown in Fig 6. We can observe that the throughput of RotorcRaft outperforms ParallelRaft and Raft-R under varying workloads. In general, the throughput of RotorcRaft is 1.28x than ParallelRaft and 1.40x than Raft-R under write-only workload, 1.45x than ParallelRaft and 1.3x than Raft-R under the read-only workload. This result demonstrates that RotorcRaft successfully offloads the burden from the leader to followers that improves parallelism in log replication and supports follower read without the involvement of the leader.

7.3 Log Replication Performance

Figure 7 illustrates the system throughput and latency over the increasing number of clients under the write-only workload. The throughput of all three implementations rises sharply as the number of clients increases from 0 to 16, and

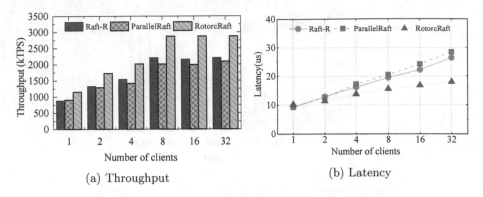

(a) Throughput

(b) Latency

Fig. 8. Throughput and latency of the read-only workload

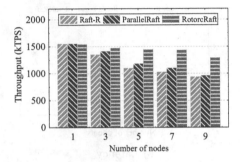

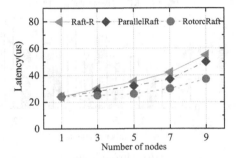

Fig. 9. Throughput under varying number of nodes

Fig. 10. Latency under varying number of nodes

remains stable after the number of clients exceeds 16. Because Raft observes strict serialization and the leader needs to independently replicate logs to all follower machines and ensure the order of records, Raft-R has the worst throughput and highest latency. Although ParallelRaft can replicate in parallel so it has better performance than Raft-R, but suffers high overhead in commit checking and status maintenance. RotorcRaft not only utilizes the resources of the followers to achieve maximum parallelism in log replication but also enables followers to replicate only the metadata of logs from the leader that effectively reduces network traffic. As a result, RotorcRaft has the best performance and the lowest latency in replication, 1.4x than ParallelRaft with only 84% of its latency.

7.4   Follow Read Performance

Figure 8 shows the system throughput and latency over the increasing number of clients under the read-only workload. The trend of the results is similar to that in Fig. 7. Due to the overhead of consensus for read index from the single-point leader, the throughput of Raft is limited and less than RotorcRaft. It is worth noting that ParallelRaft is the worst over the three implementations in follow

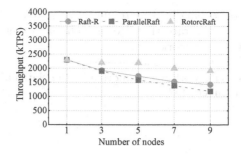

Fig. 11. Throughput under varying number of nodes

read performance because it needs to wait for the log hole to be filled before serving the requests. We observe that RotorcRaft outperforms the other two implementations, since the quorum follower read effectively relives the bottleneck of the single point of the leader.

7.5 Scalability

We further investigate the scalability of the three implementations. We measure the throughput and latency with varying numbers of replicas with 16 clients under the write-only and read-heavy workload. The result in Fig. 9 shows that the performance of Raft-R and ParallelRaft decreases significantly when the number of replicas changes from 3 to 5. For the reason that the single-point leader needs to independently replicate logs to all followers, so the latency is obviously increased as Fig. 10 shows. Furthermore, we can observe that the performance of RotorcRaft remains stable with a little decline with the increasing number of replicas. Figure 11 shows the throughput with varying numbers of replicas under the read-heavy workload. We use the leader to handle only write requests and random followers serve read requests. The throughput of Raft-R and ParallelRaft decreases sharply when the number of replicas increases. Because they both require the help of the leader to process the ReadIndex requests, which requires significant network overhead. The performance of RotorcRaft remains stable with a little decline with the increasing number of replicas as well. Therefore, RotorcRaft performs excellent good scalability both in log replication and follower read.

8 Conclusion

In this paper, we propose RotorcRaft to significantly offload the burden from the leader to followers to relieve the single-point bottleneck. We solve the issue by two effective mechanisms, i.e., follower-driven log replication and quorum follower read. Follower-driven log replication significantly relieves part of the burden of leader in log replication by exploiting hybrid RDMA primitives. Quorum follower read enables followers to handle read requests without the involvement of the leader. Evaluations demonstrate that RotorcRaft has excellent scalability and outperforms state-of-the-art work.

Acknowledgements. This work was supported by the National Science Foundation of China under grant number 61977025.

References

1. ECTD. https://etcd.io/
2. Tidb. https://pingcap.com/
3. Cao, W., Liu, Z., Wang, P.: PolarFS: an ultra-low latency and failure resilient distributed file system for shared storage cloud database. PVLDB **11**(12), 1849–1862 (2018)
4. Poke, M., Hoefler, T.: DARE: high-performance state machine replication on RDMA networks. In: HPDC, pp. 107–118 (2015)
5. Mellanox docs. https://www.mellanox.com/related-docs/prod_software/RDMA_Aware_Programming_user_manual.pdf
6. Kalia, A., Kaminsky, M., Andersen, D.G.: Design guidelines for high performance RDMA systems. In: ATC, pp. 437–450 (2016)
7. Ma, T., Zhang, M., Chen, K.: Asymnvm: an efficient framework for implementing persistent data structures on asymmetric NVM architecture. In: ASPLOS, pp. 757–773 (2020)
8. Tsai, S.-Y., Shan, Y., Zhang, Y.: Disaggregating persistent memory and controlling them remotely: an exploration of passive disaggregated key-value stores. In: ATC, pp. 33–48 (2020)
9. Yang, J., Izraelevitz, J., Swanson, S.: Filemr: rethinking RDMA networking for scalable persistent memory. In: NSDI, pp. 111–125 (2020)
10. Dragojević, A., Narayanan, D., Castro, M., Hodson, O.: Farm: fast remote memory. In: NSDI, pp. 401–414 (2014)
11. Wang, T., Johnson, R., Pandis, I.: Query fresh: log shipping on steroids. PVLDB **11**(4), 406–419 (2017)
12. Zamanian, E., Xiangyao, Yu., Stonebraker, M.: Rethinking database high availability with RDMA networks. PVLDB **12**(11), 1637–1650 (2019)
13. Zhang, Y., Yang, J., Memaripour, A.: Mojim: a reliable and highly-available nonvolatile memory system. In: ASPLOS, pp. 3–18 (2015)
14. Kogias, M., Bugnion, E.: Hovercraft: achieving scalability and fault-tolerance for microsecond-scale datacenter services. In: EuroSys, pp. 25:1–25:17 (2020)
15. Ongaro, D., Ousterhout, J.K.: In search of an understandable consensus algorithm. In: ATC, pp. 305–319 (2014)
16. Mitchell, C., Montgomery, K., Nelson, L.: Balancing CPU and network in the cell distributed b-tree store. In: ATC, pp. 451–464 (2016)
17. Kalia, A., Kaminsky, M., Andersen, D.G.: Using RDMA efficiently for key-value services. In: SIGCOMM, vol. 44, pp. 295–306. ACM (2014)
18. Cooper, B.F., Silberstein, A.: Benchmarking cloud serving systems with YCSB. In: SOCC, pp. 143–154. ACM (2010)

Efficient Matrix Computation for SGD-Based Algorithms on Apache Spark

Baokun Han[1,2], Zihao Chen[1,2], Chen Xu[1,2(✉)], and Aoying Zhou[1,2]

[1] East China Normal University, Shanghai, China
{bkhan,zhchen}@stu.ecnu.edu.cn
[2] Shanghai Engineering Research Center of Big Data Management, Shanghai, China
{cxu,ayzhou}@dase.ecnu.edu.cn

Abstract. With the increasing of matrix size in large-scale data analysis, a series of Spark-based distributed matrix computation systems have emerged. Typically, these systems split a matrix into matrix blocks and save these matrix blocks into a RDD. To implement matrix operations, these systems manipulate the matrices by applying coarse-grained RDD operations. That is, these systems load the entire RDD to get a part of matrix blocks. Hence, it may cause the redundant IO when running SGD-based algorithms, since SGD only samples a min-batch data. Moreover, these systems typically employ a hash scheme to partition matrix blocks, which is oblivious to the sampling semantics. In this work, we propose a *sampling-aware* data loading which uses *fine-grained* RDD operation to reduce the partitions without sampled data, so as to decrease the redundant IO. Moreover, we exploit a *semantic-based* partition scheme, which gathers sampled blocks into the same partitions, to further reduce the number of accessed partitions. We modify SystemDS to implement *Emacs*, efficient matrix computation for SGD-based algorithms on Apache Spark. Our experimental results show that Emacs outperforms existing Spark-based matrix computation systems by 37%.

Keywords: Matrix computation · Redundancy IO reduction · Distributed system

1 Introduction

Matrix computation are widely employed in various application areas including machine learning, data mining, and statistical science. In those areas, a typical application is to solve optimization problems, which involves using stochastic gradient descent (SGD). SGD performs in an iterative way till convergence. In each iteration, it samples a group of data items from the input dataset to form batch data items, and calculate the gradient on the batch data items. Then, this gradient is adopted to update the objective function.

In big data era, a single machine is not able to provide sufficient computation and storage resources to process large-scale matrices. There are varying types of systems, e.g., SystemDS [2], MLlib [8], ScaLAPACK [10], SciDB [4] and MADlib [7], supporting distributed matrix computation. Based on a comparative evaluation [11], SystemDS offers perhaps the best balance between physical data independence and performance at scale among the aforementioned systems. In particular, SystemDS encapsulates the interfaces for distributed matrix computation based on Spark [15]. To exploit Spark, matrix computation systems (MCS) split the matrices into small blocks and store these blocks into RDDs, and implement matrix operation via RDD operations.

Nonetheless, existing Spark-based MCSs do not fully explore the property of sampling in SGD. In general, SystemDS and MLlib implement matrix operation by coarse-grained RDD operations [15]. These operations blindly access all elements in RDDs. If an RDD is not entirely in memory, accessing the whole RDD would incur a high disk IO. However, in each iteration, SGD only needs to access the sampled batch data rather than the entire dataset. Consequently, the implementation via coarse-grained RDD operations may lead to redundant IO. Moreover, the semantics of SGD specifies sampling data in either a row-based or column-based way. Yet, Spark-based MCSs typically employ a hash scheme to partition matrix blocks, which is oblivious to the sampling semantics. In particular, the hash partition scheme scatters the blocks of the same row or column into different partitions, so that the sampling still accesses numbers of partitions, and results in redundant IO. Although DMac [13] and MatFast [14] utilize row-oriented and column-oriented partitions to accelerate matrix computation, they still rely on coarse-grained RDD operations, alleviating the benefits.

In this paper, we exploit the property of sampling in SGD to perform efficient matrix computation for SGD-based algorithms on Spark. To reduce redundant IO, we load data in a *sampling-aware* manner. That is, instead of coarse-grained RDD operations, we generate the runtime code using *fine-grained* RDD operations which depends on the partition indexes of sampled data to access partitions. In this way, the fine-grained loading reduces the redundant IO. Furthermore, according to the sampling semantics, we explore a *semantic-based* partition scheme that gathers sampled blocks into the same partitions, to reduce the number of accessed partitions. We implement a prototype system Emacs on top of SystemDS. Our experimental results show Emacs outperforms SystemDS and MatFast by 37% and 25%, respectively.

In the rest of our paper, we highlight the motivation in Sect. 2. Our paper makes the following contributions.

- We propose to generate the runtime code via *fine-grained loading* operations in Sect. 3, in order to avoid the redundant IO to load the partitions without any sampled data.
- We exploit a *semantic-based partition scheme* in Sect. 4 to gather sampled data, and consequently reduce IO.
- We discuss the implementation of Emacs in Sect. 5, and our experimental results in Sect. 6 demonstrate it outperforms state-of-the-art solutions.

In addition, we introduce related work in Sect. 7, and summarize our work in Sect. 8.

2   Motivation

In this section, we first illustrate the redundant IO in SGD algorithm to motivate the sampling-aware data loading. Then, we discuss the impact of partition scheme on the amount of IO to motivate the sampling-aware partition scheme.

2.1   Motivation for Sampling-Aware Data Loading

Coarse-Grained Loading. The existing Spark-based MCSs implement matrix operations with coarse-grained operations, leading to coarse-grained loading when memory is insufficient. In specific, existing Spark-based MCSs typically split a matrix into small matrix blocks with fixed size, and store the blocks as an RDD persisted in memory or on disk. Also, Spark-based MCSs use RDD operations (e.g., `map`, `join`, `combineByKey`) to implement matrix operations (e.g., matrix multiplication, element-wise addition, sampling). Nonetheless, those operations are coarse-grained, since they require to access all RDD partitions.

```
1  X = read("...")
2  while (loop_condition) {
3      S = sample(X)   # sample multiple rows
4      ....
5  }
```

Listing 1.1. User program of SGD in SystemDS

```
1  val X = sc.hadoopFile("...")
2             .persist(DISK_AND_MEMORY)
3  while (loop_condition) {
4      val rowIds = random_row_indexes()
5      val S = X.mapToPair(block => select_rows(block, rowIds))
6             .combineByKey(mergeBlocks)
7      ...
8  }
```

Listing 1.2. Runtime execution code of SGD in SystemDS

Example 1. To demonstrate the procedure of data loading with coarse-grained operations, we take an example of the SGD algorithm running on SystemDS. As shown in Listing 1.1, in each iteration, the SGD algorithm randomly samples multiple rows from the input matrix X to form a matrix S, i.e., a batch of data, and calculates gradient on S. Subsequently, the gradient is adopted to update the weights of objective function. According to this program, SystemDS will generate the runtime execution code, as shown in Listing 1.2. First, SystemDS employs a `mapToPair` operation to take the sampled rows from the original matrix. Then, SystemDS uses a `combineByKey` operation to merge the sampled rows to a small matrix. Both the `mapToPair` and `combineByKey` are coarse-grained RDD operations, i.e., accessing all partitions.

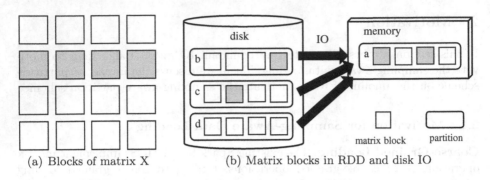

(a) Blocks of matrix X   (b) Matrix blocks in RDD and disk IO

Fig. 1. The distributed matrix representation and disk IO

Redundant IO. Given that the coarse-grained RDD operations access all partitions, they are not suitable for sampling operations. Especially when memory is insufficient, the coarse-grained operations have to load the RDD partitions persisted on disk into memory. However, in each iteration, a sampling operation only samples a part of matrix rows. This leads to redundant IO of loading data from disk to memory.

Example 2. Following Example 1, we illustrate the redundant IO incurred by coarse-grained operations. As shown in Fig. 1(a), we suppose the input matrix X is split into sixteen matrix blocks of four rows and four columns, stored as an RDD. The RDD has four partitions. Partition a is in memory, and partitions b, c, and d are spilled to disk. Via coarse-grained operations, SystemDS loads all partitions into memory to get the sampled rows, including partition d. However, the SGD algorithm only needs the second row to calculate gradient, while partition d does not contain the blocks of the second row. Hence, it is redundant to load partition d from disk.

The SGD algorithm only needs a part of matrix in each iteration, yet the coarse-grained operations access all RDD elements. This conflict leads to redundant IO when lacking memory. The observation motivates us to propose sampling-aware data loading to reduce the redundant IO, which only loads the partitions of sampled blocks (e.g., partitions a, b, and c in Example 2).

2.2   Motivation for Sampling-Aware Data Partition

Partition Hitting. Our sampling-aware data loading strategy is to load the partitions that contain sampled blocks, so the number of loaded partitions has a key impact on the amount of IO. To simplified presentation, we refer to the partition contains the matrix blocks sampled by the sampling operation as a hit partition. If a hit partition is stored on disk, then we have to load this partition to get this matrix block. Hence, more hit partitions means higher disk IO costs.

The hash partition scheme is widely used in Spark-based MCS in default, since it achieves good load balance [2]. However, the hash scheme is not suitable for SGD algorithm. In specific, the SGD algorithm samples a number of rows in each iteration, while the hash scheme scatters the blocks belonging to the same row into different partitions. As a result, the number of hit partitions may be close to the number of all partitions. Overall, the hash scheme weakens the effect of our sampling-aware data loading strategy on reducing redundant IO.

Example 3. Following Example 2, we take an example to explain the redundant IO caused by hash scheme. Since the hit partitions b and c are on disk, there is inevitably disk IO when loading sampled blocks. However, there are not sampled blocks in partitions b and c, which are redundant to load as well.

The shortage of the hash scheme is, the matrix blocks of the sampled rows are scattered into numbers of partitions, preventing our sampling-aware data loading strategy to effectively reduce redundant IO. This motivates us to propose the sampling-aware data partition scheme which decreases the number of hit partitions to reduce redundant IO further.

3 Sampling-Aware Data Loading

In this section, we propose the sampling-aware data loading strategy which exploits fine-grained operations to reduce redundant IO.

3.1 Amount of Redundant IO

In this section, we quantify the amount of redundant IO. In the process of iterative calculation, since the coarse-grained execution plan loads all partitions on disk, the amount of IO of the coarse-grained execution plan is: $IO_{all} = D - M$, where M is the storage memory, and D is the size of dataset. However, Spark provides some partition-wise RDD operations that are able to load data at partition granularity (e.g., `mapPartition`, `mapPartitionWithIndex`). Hence, ideally, we only load the hit partitions to obtain sampled blocks. To simplified presentation, we define partition hit rate, as follows.

Definition 1. *Partition hit rate is the ratio of the hit partition number on disk to the total partition number on disk.*

Hence, the ideal amount of IO is: $IO_{ideal} = P(D - M)$, where P is the partition hit rate, as defined in Definition 1. Therefore, the amount of redundant IO is:

$$IO_{redundant} = (1 - P)(D - M). \qquad (1)$$

We illustrate the runtime execution plan of Example 2 in Fig. 2(a). In this execution plan, the `mapToPair` and `combineByKey` operations are coarse-grained, so all partitions on disk are loaded into memory (e.g., partitions b, c, and d). In this example, we have $D = 4$, $M = 1$, and thus $IO_{all} = 3$. However, since partition d is not hit, $P = \frac{2}{3}$. Hence, we have $IO_{ideal} = 2$ and $IO_{redundant} = 1$.

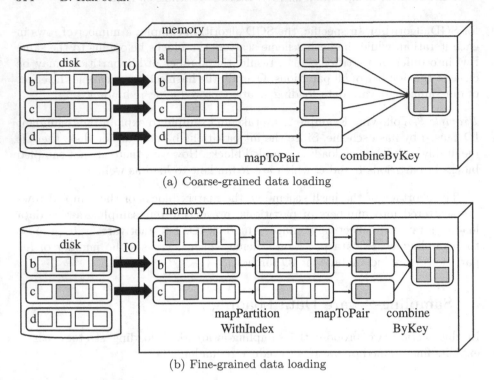

(a) Coarse-grained data loading

(b) Fine-grained data loading

Fig. 2. Execution plans of sampling the second row

3.2   Fine-Grained Data Loading

In order to reduce redundant IO, we exploits fine-grained operations on data loading. In specific, we employ the partition-wise RDD operations in Spark to load hit partitions only.

As described in Listing 1.3, the fine-grained data loading consists two steps. First, we calculate the indexes of all hit partitions based on the row column numbers of sampled data. Second, we exploit partition-wise RDD operation `mapPartitionWithIndex` to remove the partitions that do not need to be loaded. Consequently, the filtered RDD is smaller than the original RDD. We use the filtered RDD to proceed the upcoming operations and do not modify the upcoming operations, i.e., the `mapToPair` operation and `combineByKey` operation.

To demonstrate the redundant IO reduced by fine-grained data loading, we illustrate the execution plan and the running status of the second step of fine-grained data loading in Fig. 2(b). In the first step, we get that the indexes of hit partitions are a, b, and c. Therefore, in the second step, the `mapPartitionWithIndex` operation keeps partitions a, b, and c, but removes partition d. After the processing of `mapPartitionWithIndex`, the partitions a, b and c are kept, and the following operations (i.e., `mapToPair` and `combineByKey`) continue to perform the sampling operation, so the computation result are not changed.

The partition d is removed and does not participate in the remaining calculation, so it is not loaded into memory. Hence, the fine-grained data loading reduces the redundant IO of loading partition d. As shown in Fig. 2(b), by using the fine-grained operations on data loading, we only load 2 partitions. In comparison to the coarse-grained data loading in Fig. 2(a), we reduce the redundant IO.

```
1  val X = sc.hadoopFile("...")
2             .persist(DISK_AND_MEMORY)
3  while (loop_condition) {
4      val rowIds = random_row_indexes()
5      val pIds = get_partitions(rowIds)
6      val S = X.mapPartitionWithIndex((pId,blocks)->{
7                  if (pIds.contains(pId)) {
8                      return blocks
9                  } else {
10                     return nothing
11                 }
12             }).mapToPair(block => select_rows(block, rowIds))
13             .combineByKey(mergeBlocks)
14     ...
15 }
```

Listing 1.3. Runtime execution code by fine-grained data loading

4   Sampling-Aware Data Partition

According to Eq. (1), the effectiveness of fine-grained data loading is strongly influenced by the partition hit rate. The partition hit rate varies by partition scheme, so we discuss the partition scheme in this section. In particular, we first discuss why the fine-grained data loading fails when using hash scheme, and then introduce the semantic-based scheme which is more suitable for SGD algorithm.

4.1   Hash Partition

The hash partition scheme is widely adopted in MCSs, e g , SystemDS [?] and ScaLaPack [10]. The reason is that, by using the hash partition scheme, MCSs achieve good load balance, and avoid skew issue. However, for SGD, the hash scheme leads to high partition hit rate, and leads to a failure of the fine-grained data loading. As shown in Fig. 2(b), via the hash partition scheme, only one partition (i.e., partition d) is not hit. That means we can only reduce redundant IO of at most one partition. As for other partitions in disk (i.e., partitions b and c), there are also redundant IO in them, but the fine-grained data loading can not reduce it.

Next, we evaluate the partition hit rate of the hash scheme. We suppose that Spark-based MCS splits one matrix into m rows and n columns of blocks, and stores them in p partitions. If we sample k rows to calculate the gradient, then there are kn sampled blocks randomly distributed in p partitions. If we randomly sample one block, then the probability for one partition not contain the sampled

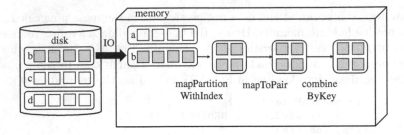

Fig. 3. Execution plans using the semantic-based partition scheme

block is $1 - \frac{1}{p}$. If we randomly sample kn block, then the probability for one partition not contain the sampled block is $(1 - \frac{1}{p})^{kn}$. Hence, the partition hit rate is:

$$P_{hash} = 1 - (1 - \frac{1}{p})^{kn}. \tag{2}$$

According to Eq. 2, P_{hash} increases with kn. Typically, P_{hash} is close to 1. For example, if $p = 1000$, $m = 1000$, $n = 100$, and $k = 100$, then $P_{hash} = 0.999$. That is, even though we sampled only 10% rows, almost all partitions are hit. So the fine-grained data loading have to load almost all partitions and can not reduce redundant IO to accelerate the execution. In order to make the fine-grained data loading work, we should switch to another partition scheme to get the partition hit rate close to 0.

4.2 Semantic-Based Partition

To reduce the partition hit rate, we propose the semantic-based partition scheme. It is suitable for the sampling operation because it distributes the sampled blocks into smallest number of partitions. The semantic-based partition scheme distributes the matrix blocks according to the semantic of sampling operation and the organization of input data. If the sampling operation samples rows as data items, the semantic-based partition scheme tries its best to distribute matrix blocks of one row into one partition. If the sampling operation samples columns as data items, the semantic-based partition scheme tries its best to distribute matrix blocks of one column into one partition.

Figure 3 illustrates an example of semantic-based partition scheme with the fine-grained data loading. We suppose the input data in organized by row, so the semantic-based partition scheme tries it best to distribute the matrix blocks of each row into one partition. Since the SGD algorithm samples the second row to calculate gradient, only partition b contains the sampled matrix blocks, so only partition b is loaded. Hence, using semantic-based partition scheme and fine-grained data loading, the redundant IO of partition c and d are reduced. Compared to hash partition scheme in Fig. 2(b), the semantic-based partition scheme reduces more redundant IO.

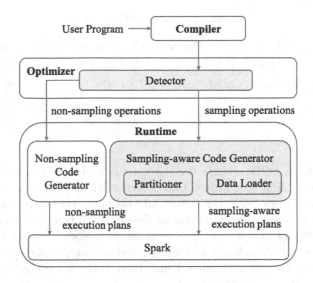

Fig. 4. System architecture of Emacs

Similar to hash scheme, the partition hit rate of semantic-based partition scheme is:

$$P_{semantic} = 1 - (1 - \frac{1}{p})^k. \tag{3}$$

Clearly, $k \leq kn$, so $P_{semantic} \leq P_{hash}$. Using the same example values of hash scheme, i.e., $p = 1000$ and $k = 100$, then $P_{semantic} = 0.095$. This means only 9.5% partitions are hit, so the fine-grained data loading can reduce 90.5% redundant IO to accelerate the execution. The partition hit rate of semantic-based partition scheme is less than the one of hash scheme. Hence, semantic-based partition scheme achieves better performance for SGD-based algorithms using fine-grained data loading.

5   System Implementation

Recently, a variety of Spark-based matrix computations systems have been proposed, e.g., SystemDS [2], MLlib [8]. To use MLlib, users must manually tune partitioning, caching behavior, and select from a set of matrix data structures supporting inconsistent operators [11]. Nonetheless, SystemDS automatically compiles ML algorithms into efficient execution plans on top of Spark. It is convenient to write and debug ML algorithms using the declarative DSL of SystemDS. SystemDS automatically generates the optimal execution plan based on memory estimates, data, and cluster characteristics. Therefore, we implemented the sampling-aware data loading strategy and semantic-based partition scheme on top of SystemDS, due to its high performance and high availability.

As shown in Fig. 4, the system architecture of Emacs consists of a *detector* and a *sampling-aware code generator*. First, given a DAG compiled from a user

Table 1. Statistics of datasets.

Dataset	# of Rows	# of Columns	# of non-zero values	Footprint
criteo1	19584198	121083	753527640	11.3 G
criteo2	97920991	121083	3762193624	56.6 G
criteo3	156673586	121083	6018213853	90.5 G
criteo4	195841983	121083	7637804468	113.2 G

program involving SGD, we use the detector to recursively traverse the entire DAG to find sampling operations. The detector delivers the sampling operation to our sampling-aware code generator, and delivers other non-sampling operations to non-sampling code generator of SystemDS. Then, for a sampling operation, the sampling-aware code generator employs a *data loader* to implement the sampling-aware data loading strategy, and employs a *partitioner* to implement the semantic-based partition scheme. Finally, the sampling-aware execution plans are submitted to Spark cluster.

6   Experimental Studies

In this section we discuss the experimental studies on the efficiency of Emacs, and compare its performance against the state-of-the-art solutions.

6.1   Experimental Setting

Cluster Setup. We conduct experiments on a cluster of 4 nodes. Each node has one 24-core Intel Xeon CPU, 32 GB memory, one 2 TB HDD and 1 Gbps Ethernet. We deploy Spark 2.4.5 on the cluster, with one node as the Spark master and the other three nodes as Spark workers.

Workloads. We choose linear regression and logistic regression based on SGD algorithm for our experiments, since they are common in enterprise analytics, as well as social and physical sciences [11].

Datasets. We evaluate the algorithms above on the real-word dataset *criteo*[1]. We generated four sampled datasets of different sizes from the first day of the *criteo* dataset, named as criteo1, criteo2, criteo3 and criteo4 from smallest to largest. Table 1 provides the numbers of rows, columns, and non-zero values, and the memory footprint of datasets. Since the storage memory is 45 G, only criteo1 can be held in memory, whereas the other datasets are spilled to disk.

6.2   Efficiency of Fine-Grained Data Loading

In this section, we evaluate the performance of the sampling-aware data loading strategy, and demonstrate the effect of partition hit rate on execution time.

---

[1] https://labs.criteo.com/2013/12/download-terabyte-click-logs-2/.

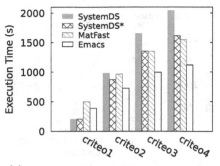

(a) Execution time of linear regression

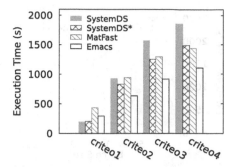

(b) Execution time of logistic regression

Fig. 5. Efficiency of sampling-aware data loading strategy and semantic-based partition scheme of datasets of different sizes

Execution Time. To evaluate the impact of data loading strategies on the performance, we ran SystemDS [2] and SystemDS* on datasets of different sizes. Here, SystemDS uses the coarse-grained data loading. We replace the coarse-grained data loading in SystemDS with fine-grained data loading which is denoted as SystemDS*. Figure 5 illustrates the overall execution time on datasets criteo1, criteo2, criteo3, and criteo4. Here, the batch size is 50. As shown in Fig. 5, when dataset size is less than the cluster memory (e.g., criteo1), the overall execution time of SystemDS and SystemDS* are same, because all data is stored in memory and no disk IO could be reduced. However, when the dataset size is larger than the storage memory (e.g., criteo2, criteo3, and criteo4), the redundant IO occurs. Via the fine-grained data loading, SystemDS* mitigates this issue, so as to reduce the execution time by 16.8% on average compared to SystemDS. Specifically, SystemDS* reduces the execution time of linear regression by 18.3% compared to SystemDS on criteo3, as shown in Fig. 5(a). SystemDS* reduces the execution time of logistic regression by 19.8% compared to SystemDS on criteo4, as shown in Fig. 5(b).

Partition Hit Rate. To evaluate the affect of partition hit rate on IO, we conducted the performance analysis of the coarse-grained and fine-grained data loading with different partition hit rates. Here, we adjust partition hit rates by changing the batch size. In specific, the partition hit rate increases with the batch size. Figures 6(a) and 6(b) illustrate the overall execution time of coarse-grained and fine-grained data loading on different datasets and batch sizes. On criteo1 dataset, there is no redundant IO, so the execution time of two kinds of data loading are the same. On other datasets except criteo1, as the batch size decreases, the effect of fine-grained loading becomes more significant. As shown in Fig. 6(a), when batch size is 10, fine-grained data loading achieves a speedup of 38.9% over coarse-grained data loading on criteo4 of linear regression. However, when the batch size is 200, fine-grained data loading degenerates into coarse-grained data loading. As shown in Fig. 6(b), when the batch size is 200,

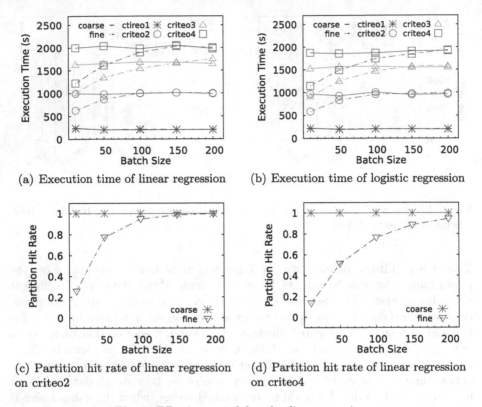

(a) Execution time of linear regression   (b) Execution time of logistic regression

(c) Partition hit rate of linear regression   (d) Partition hit rate of linear regression
on criteo2   on criteo4

Fig. 6. Effectiveness of data loading strategies

fine-grained data loading is 0.7% slower than coarse-grained data loading on criteo4 of logistic regression.

To further study the poor performance on large batch size, we evaluate the partition hit rates of coarse-grained and fine-grained data loading with different batch sizes on criteo2 and criteo4, as depicted in Figs. 6(c) and 6(d). For any batch size, since coarse-grained data loading always loads all partitions, its partition hit rate is 1.0. Therefore the execution time of coarse-grained data loading are similar on any batch size. As shown in Figs. 6(c), on criteo2, when batch size is 10, the partition hit rate of fine-grained data loading is 0.257. Fine-grained data loading loads less partitions than coarse-grained data loading, therefore its execution time is shorter than coarse-grained data loading. However, the partition hit rate of fine-grained data loading is 0.998 when batch size is 200, which is close to 1.0 of coarse-grained data loading. Figures 6(d) shows the similar result on criteo4. With the batch size growing from 10 to 200, the partition hit rate of fine-grained data loading on criteo4 grows from 0.135 to 0.945. Certainly, the hash scheme leads to a high partition hit rate and therefore causes a failure of the sampling-aware data loading strategy.

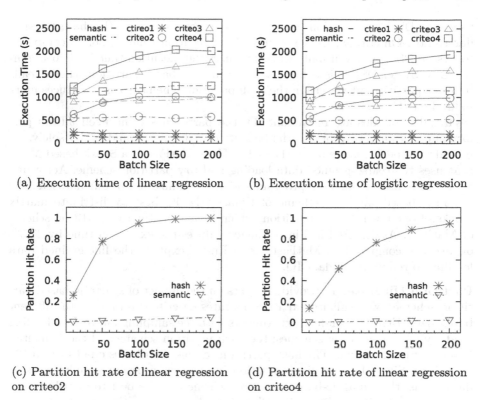

(a) Execution time of linear regression   (b) Execution time of logistic regression

(c) Partition hit rate of linear regression   (d) Partition hit rate of linear regression
on criteo2   on criteo4

Fig. 7. Effectiveness of partition schemes

In general, by using the sampling-aware data loading strategy, fine-grained data loading improves the performance of SGD algorithm when batch size is low. However, when batch size is high, the hash scheme leads to a high partition hit rate close to 1, and thus prevent the sampling-aware data loading strategy from reducing redundant IO. Hence, we propose the semantic-based partition scheme to reduce the partition hit rate.

6.3   Efficiency of Semantic-Bases Partition Scheme

In this section, we evaluate the performance of semantic-based partition scheme. In addition, to show the insights of the performance, we demonstrate the effect of partition scheme on partition hit rate and execution time.

Execution Time. To demonstrate the impact of partition schemes on IO, we further conducted the performance analysis of Emacs, which uses the sampling-aware data loading strategy and semantic-based partition scheme. Figure 5 illustrates the execution time of Emacs. Specifically, as shown in Fig. 5(a), on criteo4, Emacs reduces the execution time of linear regression by 30.8% compared to SystemDS*, and eventually Emacs reduces the execution time by 45.2% compared to SystemDS. The execution time of logistic regression in Fig. 5(b) shows the similar trends. On criteo3, Emacs reduces the execution time by 26.8% compared to SystemDS*, and eventually Emacs reduces the execution time by 41.6% compared to SystemDS. This is because, the semantic-based partition scheme leads to lower partition hit rate than the hash partition scheme, so as to reduces more redundant IO.

In addition, despite redundant IO, the partition scheme itself affects performance as well. To evaluate the benefits of reducing redundant IO solely, we provide the performance of MatFast [14]. Here, MatFast is a Spark-based MCS, that uses the coarse-grained data loading and row partition scheme. According to the semantic of our user program and the organization of our datasets, the semantic-based partition scheme of Emacs tries its best to distribute matrix blocks of one row into one partition, which is same as the row partition scheme of MatFast. As depicted in Fig. 5, Emacs reduces the execution time by 27.3% on average compared to MatFast, since Emacs exploits the fine-grained data loading to reduce redundant IO.

Partition Hit Rate. To provide insights on the impact of semantic-based partition schemes, we evaluate partition hit rates on criteo4 dataset with different batch sizes. Here, we conduct experiments with the sampling-aware data loading strategy of two partition schemes: the hash partition scheme and the semantic-based partition scheme. The hash partition scheme distributes one row into 122 partitions. Since the sampling operation in our user program samples rows as data items, the semantic-based partition scheme tries its best to distribute one row into one partition. Figure 7(a) illustrates the execution time of linear regression. As the batch size grows from 10 to 200, the execution time on criteo4 of the hash scheme and the semantic-based partition scheme grows 64.1% and 12.3%, respectively. Especially, when batch size is 200, the semantic-based partition scheme achieves a speedup of 37.6% over the hash partition scheme on criteo4. Figure 7(b) shows the similar trends of logistic regression. When batch size is 200, the semantic-based partition scheme achieves a speedup of 40.9% over the hash partition scheme on criteo4 of logisitc regression. Overall, the execution time of the hash partition scheme grows significantly along with batch size, while that of the semantic-based partition scheme does not.

To demonstrate why the semantic-based partition scheme outperforms the hash partition scheme, especially when batch size is 200, we illustrate their partition hit rates of different batch sizes on criteo2 and criteo4 in Figs. 7(c) and 7(d). As the batch size grows, the partition hit rate of the semantic-based partition scheme grows slower than the rate of the hash scheme. Specifically, as shown in Figs. 7(c), when batch size is 200, on criteo2, the partition hit rates of the hash partition scheme and the semantic-based partition scheme are 0.998 and 0.046, respectively. As shown in Figs. 7(c), on criteo4, the partition hit rates of the hash partition scheme and the semantic-based partition scheme are 0.945 and 0.023, respectively. If one row is in fewer partition, the partition hit rate is certainly lower. The semantic-based partition scheme distributes one row into fewer partitions than the hash partition scheme, so its partition hit rate is lower than that of the hash partition scheme.

7   Related Work

This section discusses the existing work related to the matrix computation systems in two aspects: redundant IO reduction and partition scheme optimization.

Redundant IO Reduction. The reduction of redundant IO has been widely studied in matrix computation systems. For optimization at the operator implementation level, the replication based matrix multiplication method replicates the input matrices several times, therefore leads to redundant IO. SystemDS [2] proposes the cross product based matrix multiplication method and the broadcast based matrix multiplication method, and DistMe [6] proposes the CuboidMM method, to reduce the redundant IO. For optimization at the execution plan level, SPORES [12], SPOOF [1] and MatFast [14] use fused operators to eliminate unnecessary materialization of intermediates, or unnecessary scans, and to exploit sparsity across entire chains of operations, so as to reduce redundant IO. SystemML [3] exploits the common subexpression elimination method to eliminate the redundancy of computing common subexpression several times. However, when lacking memory, their execution plan for SGD-based algorithms formed by coarse-grained RDD operations leads to redundant disk IO. Unlike them, we propose the sampling-aware data loading strategy to reduce redundant IO by using fine-grained RDD operations.

Partition Scheme Optimization. There are works focusing on the optimization of partition scheme. DMac [13] and MatFast [14] exploit the matrix dependencies and cost model to generate efficient execution plans, so as to reduce communication overhead. However, they do not consider the relationship between the semantic of sampling operations and partition scheme, whereas we exploit the semantic-based partition scheme to reduce redundant IO in SGD-based algorithms. In the area of graph processing, in order to obtain batter performance, some graph partitioning strategies exploit the structure of graph instead of naive random hash partitioner to partition the graph. WASP [5] dynamically adjust partitions regarding to the frequency of active edges of the existing query workload. Makoto Onizuka et.al. [9] proposed a clustering-based algorithm to partition the graph. However, the characteristics of matrix computations are different from graph computations. Our semantic-based partition scheme exploits the characteristics of the SGD algorithm in matrix computations.

8   Conclusion

In this paper, we propose a new system called Emacs, an efficient matrix computation system for SGD-based algorithms on Apache Spark. In Emacs, first, we exploit the sampling-aware data loading strategy and semantic-based partition scheme to reduce redundant IO when memory is insufficient, so as to reduce the overall execution time. The sampling-aware data loading strategy reduces the redundant IO by using the fine-grained operation to remove unwanted partitions. Second, we exploit the semantic-based partition scheme to enhance the effectiveness of the sampling-aware data loading strategy. Our experiments illustrate that the sampling-aware data loading strategy with the semantic-based partition scheme method significantly outperforms the existing methods, in terms of performance. We have implemented Emacs on top of SystemDS. However, the sampling-aware data loading strategy and the semantic-based partition scheme are also fit for other Spark-based distributed matrix computation systems.

Acknowledgments. This work was supported by the National Natural Science Foundation of China (No. 61902128), Shanghai Sailing Program (No. 19YF1414200).

References

1. Boehm, M., et al.: On optimizing operator fusion plans for large-scale machine learning in systemml. In: PVLDB, pp. 1755–1768 (2018)
2. Boehm, M., et al.: Systemds: a declarative machine learning system for the end-to-end data science lifecycle. In: CIDR (2020)
3. Böhm, M., et al.: Systemml's optimizer: plan generation for large-scale machine learning programs. IEEE Data Eng. Bull. **37**(3), 52–62 (2014)
4. Brown, P.G.: Overview of SciDB: large scale array storage, processing and analysis. In: SIGMOD, pp. 963–968 (2010)
5. Davoudian, A., et al.: A workload-adaptive streaming partitioner for distributed graph stores. Data Sci. Eng. **6**(2), 163–179 (2021)
6. Han, D., et al.: Distme: a fast and elastic distributed matrix computation engine using gpus. In: SIGMOD, pp. 759–774 (2019)
7. Hellerstein, J.M., et al.: The madlib analytics library or MAD skills, the SQL. In: PVLDB, pp. 1700–1711 (2012)
8. Meng, X., et al.: Mllib: machine learning in apache spark. JMLR, 34:1–34:7 (2016)
9. Onizuka, M., et al.: Graph partitioning for distributed graph processing. Data Sci. Eng. **2**(1), 94–105 (2017)
10. ScaLAPACK: http://www.netlib.org/scalapack/
11. Thomas, A., Kumar, A.: A comparative evaluation of systems for scalable linear algebra-based analytics. Proc. VLDB Endowment **11**(13), 2168–2182 (2018)
12. Wang, Y.R., et al.: SPORES: sum-product optimization via relational equality saturation for large scale linear algebra. PVLDB, 1919–1932 (2020)
13. Yu, L., et al.: Exploiting matrix dependency for efficient distributed matrix computation. In: SIGMOD, pp. 93–105 (2015)
14. Yu, Y., et al.: In-memory distributed matrix computation processing and optimization. In: ICDE, pp. 1047–1058 (2017)
15. Zaharia, M., et al.: Resilient distributed datasets: a fault-tolerant abstraction for in-memory cluster computing. In: NSDI, pp. 15–28 (2012)

Parallel Pivoted Subgraph Filtering with Partial Coding Trees on GPU

Yang Wang[iD], Yu Gu, and Chuanwen Li[✉][iD]

School of Computer Science and Engineering, Northeastern University,
Shenyang, China
lichuanwen@mail.neu.edu.cn

Abstract. The pivoted subgraph isomorphism problem is a special subgraph isomorphism problem that focuses on the pivoted nodes rather than the entire subgraphs. The key challenge in adapting existing techniques to the pivoted problem is eliminating their redundant intermediate results. In this paper, we propose a GPU-based pivoted subgraph isomorphism filtering technique, where information of each node is encoded into a series of codes. When performing a pivot subgraph search, the candidate nodes satisfying the coding requirements are collected parallelly on GPU while others are filtered away. Then the final result can be effectively retrieved by a verification process on the filtered nodes. As demonstrated by the experimental results, our method dramatically reduces the processing time of the pivoted subgraph isomorphism problem. Compared to the state-of-the-art GPU-friendly subgraph matching method GpSM which also focuses on filtering effect, the algorithm's execution time is halved, confirming that our approach can effectively process pivoted subgraph isomorphism queries.

Keywords: Pivoted subgraph isomorphism · Subgraph isomorphism · Parallel acceleration · Coding tree · GPU

1 Introduction

With the growing importance of graphs, the management of graph data has become a focus of research. Pivoted subgraph isomorphism [1] is commonly used in a variety of real-world applications, such as protein network function prediction [3], frequent subgraph mining [4,5], etc. [9]. For a given query graph and a node in the query graph as the pivot, the pivoted subgraph isomorphism locates the nodes in the data graph that correspond to the pivot, and each of these nodes must satisfy the requirement that there is at least one subgraph that matches the query graph and contains this node. An example is illustrated in Fig. 1. The figure contains a query graph Q with the pivot u_1 and a data graph G with the candidate pivots v_1, v_5. In the data graph, only v_1 is a valid match for the pivot u_1.

In this paper, the pivot query processing takes into account both the structure and label feature of each node. A parallel adjacency matrix processing method

© The Author(s), under exclusive license to Springer Nature Switzerland AG 2022
A. Bhattacharya et al. (Eds.): DASFAA 2022, LNCS 13245, pp. 325–332, 2022.
https://doi.org/10.1007/978-3-031-00123-9_26

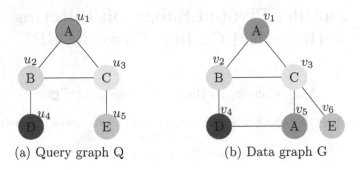

(a) Query graph Q   (b) Data graph G

Fig. 1. (a) A query graph of pivoted subgraph isomorphism, u_1 is the pivot of the query graph; (b) A data graph of pivoted subgraph isomorphism.

is developed to improve coding performance. We construct an adjacency matrix for each node in the data graph and utilize it to generate a series of smaller matrices to illustrate each incomplete coding tree. All algorithms are designed to run in parallel on a GPU.

2   Partial Coding Tree

This paper proposes a GPU-based pivoted subgraph isomorphism algorithm. By transforming the neighbor structure of graph nodes into a tree structure, the eigenvalues of the adjacency matrix of the tree structure are obtained, and then the node coding is performed. Then we connect the candidate nodes after encoding and filtering [6], and finally get the solution of the pivoted subgraph isomorphism.

Definition 1. *(Pivoted Subgraph isomorphism) Given a query graph $Q = \{V_Q, E_Q, L_Q\}$ with an pivot $u_p \in V_Q$ and a data graph $G = \{V_G, E_G, L_G\}$, the pivot $v_p \in V_G$, there exists an injective function $f : V_Q \rightarrow V_G$, and the pivoted subgraph isomorphism needs to meet the following three conditions: (1) $\forall u, v \in V_Q, f(u), f(v) \in V_G$ and $L_Q(u) = L_G(f(u)), L_Q(v) = L_G(f(v))$, (2) $\forall (u, v) \in E_Q, (f(u), f(v)) \in E_G$, (3) $v_p = f(u_p)$.*

Definition 2. *(Partial Label Tree) Each label is classified into a label bucket by a certain form(hashing, dividing .etc.), all the label buckets with the i-th bit of 1 in the binary form of the label bucket serial number are classified into the i-th group, and there are $\lceil log_2^n \rceil$ groups in total. The neighbor nodes of the label type in the same label bucket group will form a tree together with the root node, which is called a partial label tree (PLT).*

Theorem 1. *Given a tree T with n vertices and a tree P with m vertices, their adjacency matrices are denoted as A and B respectively. For a matrix A, the* eigenvalues are $\lambda_1 \geq \lambda_2 \geq \cdots \geq \lambda_n$. For a matrix B, the eigenvalues are $\beta_1 \geq \beta_2 \geq \cdots \geq \beta_m$. If P is an induced subtree of T, then $\lambda_{n-m+i} \leq \beta_i \leq \lambda_i, (i = 1, \cdots, m)$ [10].

Theorem 2. *Given two trees P and T, we can generate multiple PLT to obtain the PLT sequence according to Definition 2, denoted as $\{p_1, p_2 \cdots, p_n\}, \{t_1, t_2, \cdots, t_n\}$. If all p_i in the sequence are induced subtrees of t_i, then P must be an induced subtree of T.*

Proof. Due to the addition of label, if all p_i are induced subtrees of t_i in the sequence, then all p_i are still induced subtrees of t_i when the labels are removed. Thus P is a induced subtree of T.

Taking query graph node u_1 and data graph node v_1 as examples, we first generate a three-layer tree (without considering labels), as shown in Fig. 2. For the 5 kinds of labels included in Fig. 1, we can divide them into [ACE],[BC][DE] according to whether the binary numbers of the labels are 1 or not, thereby dividing one unlabeled tree into three PLTs of $\_\_1PLT, \_1\_PLT, 1\_\_PLT$, as shown in Fig. 3.

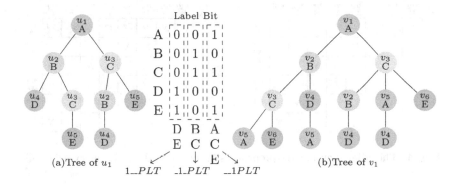

Fig. 2. 3-layer trees generated by query graph node u_1 and data graph node v_1.

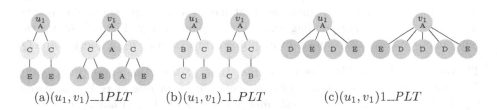

Fig. 3. Three labeled trees generated by nodes u_1 in graph Q and v_1 in graph G.

3 Partial Adjacency Matrix

Although combining the neighbor label features of the node can significantly improve the filtering effect, we need to split a tree into multiple PLTs, and the cost of generating multiple PLTs is too expensive. But the final eigenvalues required are all calculated through the adjacency matrix, so we consider directly dividing the original adjacency matrix into several small adjacency matrices.

Definition 3. *Given a tree with m nodes, the upper triangular matrix of the $m \times m$ adjacency matrix can be used to indicate the connection relationship between the nodes, and the continuous bottom rows with all zeros are deleted, which is recorded as a partial upper triangular matrix (PUTM).*

When we cut off a node in the tree, its parent node should be connected to its child node to form a tree structure. First, we can pre-process the original adjacency matrix so that the nodes that may be connected are pre-connected. Then we will choose to keep or delete according to the situation. Figure 4 shows the PUTM after preprocessing.

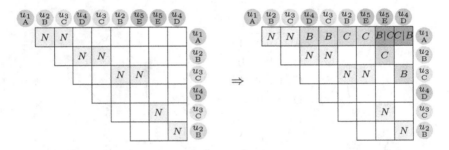

Fig. 4. The adjacency matrix preprocessing of the node u_1 in query graph Q

After preprocessing the original PUTM, we also need to know the position of the label nodes that meet the conditions in the original adjacency matrix in different label trees. We judge whether all nodes meet the conditions of the corresponding label bucket in parallel, and store them in the array as 0 or 1. By calculating the prefix sum, the nodes that meet the corresponding label bucket are stored in the corresponding subscript mapping array. Figure 5 shows the generation process of the subscript mapping arrays of node u_1 in query graph Q.

Theorem 3. *In PUTM, given that the value of the element in (m, n) is 0, (1) assuming that the column with the value of N in row m is $k...k_m$, where $1 \leq k...k_m \leq m$, then there is only one position in $(k..k_m, n)$ having a value of N; (2) assuming the column whose value is not N and not 0 in row m is $k...k_m$, then there is only one position in $(k..k_m, n)$ having a value of N.*

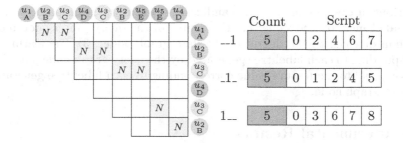

Fig. 5. The three label subscript mapping arrays generated by nodes u_1 in graph Q.

Proof. Since N indicates a connection between nodes, the upper triangular matrix we discussed only retains the top-down connection of the tree. If the above (1) and (2) exist, two-parent nodes are connected to a child node. There is a contradiction with the definition of a tree, so the above two situations do not exist.

Finally we constructs multiple PUTMs by mapping arrays and original PUTM. We calculate all the combinations of the elements in the subscript mapping array, and allocate the corresponding number of threads for parallel processing. When its value in PUMT is 0 or meets the label bucket conditions, it is stored in 0. Otherwise, it is stored in 1. Figure 6 shows that the original PUTM of node u_1 is divided into three PUTMs by the subscript mapping arrays.

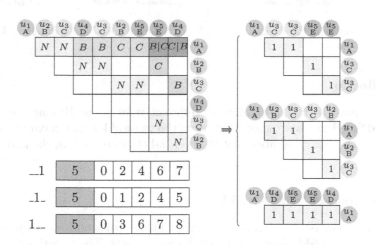

Fig. 6. Three labeled adjacency matrices generated by nodes u_1 in graph Q.

For the multi-label tree of each node in the query graph, we can encode vertex as $MCT = \{label, degree, seq = \{a_1, a_2, b_1, b_2, ...\}\}$. When the vertex code of the node in the data graph and the vertex code of the query graph meet the following conditions, it can be stored in the candidate set of the query graph node: (1) the node label must be the same; (2) the degree of the query graph node must be less than or equal to the degree of the data graph node; (3) The feature value corresponding to each label tree generated by the query graph node must be less than or equal to the feature value corresponding to each label tree generated by the data graph node.

4   Experimental Results

This section analyzes the performance of the method proposed in this paper through experiments. In the experiment, the GPU-based subgraph isomorphism algorithm GpSM is compared with the algorithm in this paper.

All codes are implemented in C++ with CUDA Toolkit 10.6 and run on an NVIDIA RTX4000 GPU.

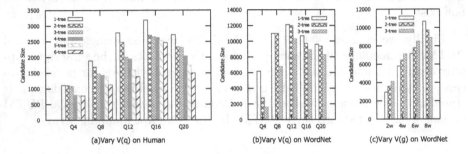

Fig. 7. Filtering ability of candidate nodes of different trees under different datasets.

4.1   Effect of K

Figures 7 show the filtering ability of different trees in the Human dataset and the WordNet. The results show that when the value of k is the maximum number of digits in the binary number that the node label can represent, the performance is optimal.

4.2   Comparison with GpSM

This paper compares the performance of the pivoted subgraph isomorphism algorithm and the GpSM algorithm through experiments. The experimental results for different query graph sizes and data graph sizes are given.

Filtering Time vs. Query Vertices and Data Vertices. Figure 8 shows that the filtering time of this algorithm is less than GpSM and is relatively stable, whether it is the change of query graph nodes or the change of data graph nodes.

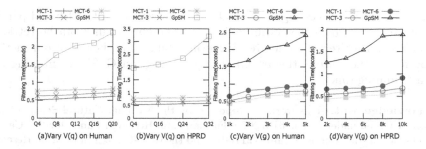

Fig. 8. Algorithm filtering time under different datasets.

Execution Time vs. Query Vertices and Data Vertices. Figure 9 shows that the execution time of the GpSM algorithm increases linearly. In contrast, the execution time of the pivoted subgraph isomorphism algorithm is relatively stable first and then tends to increase linearly. Still, the overall time is less than that of the GpSM algorithm.

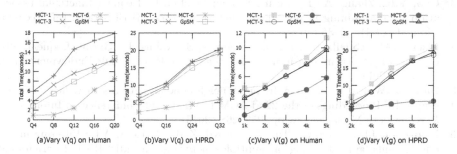

Fig. 9. Algorithm total time under different datasets.

5 Related Work

In recent years, with the rapid development of GPU, more researchers try to convert serial computation on CPU into parallel computation [8] on GPU. For example, the GRASS [2] proposed by Vincenzo Bonnici et al. also uses GPUs to parallel filter out candidate nodes of the current query node that are not satisfied. GpSM [7] is proposed by Ha-Nguyen Tran et al., which is based on the filtering and joining framework. It filters candidate nodes of query nodes in parallel by generating spanning trees.

6 Conclusion

This paper proposes a novel multi-coding tree-based filtering method, which converts the neighbor structure and neighbor node labels into a series of codes for each node. In the processing of the codes, we devise several GPU-based parallel algorithms to enhance the coding efficiency. The final results can be retrieved by verifying the filtered candidates. Experimental results show that our method outperforms the state-of-the-art GPU-based algorithm GpSM in filtering and overall running time.

Acknowledgement. This work is supported by the National Natural Science Foundation of China under Grant 61872071, Fundamental Research Funds for the Central Universities of China under Grant N2116010 and the CCF-Huawei Innovation Research Plan.

References

1. Abdelhamid, E., Abdelaziz, I., Khayyat, Z., Kalnis, P.: Pivoted subgraph isomorphism: The optimist, the pessimist and the realist. In: EDBT, pp. 361–372 (2019)
2. Bonnici, V., Giugno, R., Bombieri, N.: An efficient implementation of a subgraph isomorphism algorithm for GPUs. In: 2018 IEEE International Conference on Bioinformatics and Biomedicine (BIBM), pp. 2674–2681. IEEE (2018)
3. Cho, Y.R., Zhang, A.: Predicting protein function by frequent functional association pattern mining in protein interaction networks. IEEE Trans. Inf. Technol. Biomed. **14**(1), 30–36 (2009)
4. Elseidy, M., Abdelhamid, E., Skiadopoulos, S., Kalnis, P.: Grami: frequent subgraph and pattern mining in a single large graph. Proc. VLDB Endowment **7**(7), 517–528 (2014)
5. Han, J., Wen, J.R.: Mining frequent neighborhood patterns in a large labeled graph. In: Proceedings of the 22nd ACM International Conference on Information & Knowledge Management, pp. 259–268 (2013)
6. Moorman, J.D., Chen, Q., Tu, T.K., Boyd, Z.M., Bertozzi, A.L.: Filtering methods for subgraph matching on multiplex networks. In: 2018 IEEE International Conference on Big Data (Big Data), pp. 3980–3985. IEEE (2018)
7. Tran, H.-N., Kim, J., He, B.: Fast subgraph matching on large graphs using graphics processors. In: Renz, M., Shahabi, C., Zhou, X., Cheema, M.A. (eds.) DASFAA 2015. LNCS, vol. 9049, pp. 299–315. Springer, Cham (2015). https://doi.org/10.1007/978-3-319-18120-2_18
8. Wang, X., et al.: Efficient subgraph matching on large RDF graphs using mapreduce. Data Sci. Eng. **4**(1), 24–43 (2019)
9. Wu, Y., Zhao, J., Sun, R., Chen, C., Wang, X.: Efficient personalized influential community search in large networks. Data Sci. Eng. **6**(3), 310–322 (2021)
10. Zou, L., Chen, L., Yu, J.X., Lu, Y.: A novel spectral coding in a large graph database. In: Proceedings of the 11th International Conference on Extending Database Technology: Advances in Database Technology, pp. 181–192 (2008)

TxChain: Scaling Sharded Decentralized Ledger via Chained Transaction Sequences

Zheng Xu[1,2], Rui Jiang[1,2], Peng Zhang[1,2], Tun Lu[1,2(✉)], and Ning Gu[1,2]

[1] School of Computer Science, Fudan University, Shanghai, China
`{zxu17,jiangr20,zhangpeng_,lutun,ninggu}@fudan.edu.cn`
[2] Shanghai Key Laboratory of Data Science, Fudan University, Shanghai, China

Abstract. The blockchain has become the most prevalent distributed ledger (DL). Sharding has emerged as a major solution to the scalability bottleneck of DLs. From the underlying data structure of existing sharding schemes, although the Directed Acyclic Graph (DAG)-based topology improves the scalability of DLs compared to chained blocks, the security and reliability of consensus mechanisms in DAG-based DLs have not been verified. Moreover, these schemes suffer from high communication overhead when scaling out. To address these issues, we propose a sharded DL named TxChain, which adopts a novel data structure manipulated by the unit of transaction and constituted by chained transaction sequences of each account. TxChain optimistically processes concurrent transactions and ensures the consistency of all shards via transaction sequence conversion (TSC)-based consensus mechanism. Shards maintain the full replica of TxChain and execute transactions by trustworthy validators, which reduce the frequency of communication with other shards. We theoretically prove the consistency of shards maintained by TSC and demonstrate TxChain's throughput scales with low latency through extensive experiments.

1 Introduction

Blockchain makes the distributed ledgers (DLs) evolve into a irreversible and decentralized data maintenance technology. Each node in the DL serially processes generated transactions, however, the performance of a single node has become the throughput bottleneck of the DL. Sharding has become a feasible and proven horizontal scaling approach which assigns nodes to multiple shards to handle a portion of transactions and enables parallelization of the computation and storage in DLs for high throughput. Maintainers of shards process transactions in their local shard and update the state of the entire system.

The underlying data of existing sharded DLs is in the structure of chained blocks or the Directed Acyclic Graph (DAG). Although DAG-based data structure can improve the scalability of DLs, it is difficult and unstable to achieve the final consistency. Moreover, mainstream sharded DLs such as RapidChain [8],

A. Bhattacharya et al. (Eds.): DASFAA 2022, LNCS 13245, pp. 333–340, 2022.
https://doi.org/10.1007/978-3-031-00123-9_27

OmniLedger [4] and Elastico [5] use the unspent transaction outputs (UTXO)-based transaction model which has weak programmability, high computational complexity and large storage redundancy. The Byzantine fault-tolerant (BFT) consensus mechanism in these DLs may cause high communication costs when the system scales up. These above problems indicate that existing sharded DLs still have a lot of room for improvement in terms of the data structure and parallel processing performance.

Based on the above problems and challenges, we propose a novel DL named TxChain. To achieve high scalability, TxChain processes each transaction unit parallelly by selected validators based on sharding technology. Validators execute intra-shard transactions instantly in the local shard and broadcast them to remote shards. All shards store a full replica of TxChain and collaboratively maintain their states by synchronizing transactions in accounts' TxSEQs. To ensure the consistency of all shards, three kinds of transaction dependencies and a transaction sequence conversion (TSC)-based consensus mechanism are proposed to determine the ordinal relationship between transactions. Based on this consensus mechanism, TxSEQs can have the same transaction order among different shards in accordance with the original dependency. Moreover, to optimize the latency of processing transactions, honest and trustworthy validators are selected according to their stake ratio and are protected by the trusted execution environment (TEE), which reduce the communication overhead of TxChain compared to BFT-based DLs. Besides, the modification of each shard can be completely and consistently recorded in each replica, thus a transaction can be confirmed in a low time delay without querying the state of other shards repeatedly. We summarize our main contributions as follows:

- We propose a novel sharded DL named TxChain which is parallelly manipulated by the unit of transaction and constituted by accounts' transaction sequences.
- We propose a consensus mechanism based on the transaction sequence conversion (TSC) to maintain transaction dependencies and consistency among shards. TxChain's high scalability with low confirmation latency has been demonstrated via extensive experiments.

2   System Overview and Problem Definition

2.1   System Model

Nodes are assigned to different shards $S = \{S_1, S_2, \cdots, S_n\}$. Validators are selected according to their stake ratio in the local shard (the validator in shard S_1 is represented by V_1). Nodes with higher stakes are more likely to be selected. The selected validator verifies transactions, consents on the order of them, broadcasts them to remote shards and maintains the consistency of TxChain on behalf of nodes in the shard. All validator are protected by the TEE, and have no intention to be evil because of the stake-based selection. Then combining with the settings and assumptions in prior sharding-based blockchains [1,8], all validators can be regard as honest and trustworthy. TxChain adopts the structure of Merkle Patricia tree (MPT), but its leaf node is one account composed of chained transaction sequences. Figure 1 shows the data structure of TxChain. Intra-shard transactions are executed instantly in the local replica, and then are broadcast to remote shards to synchronize the account status. When transactions are received by a remote shard, the validator first verifies the signature of each transaction to ensure its validity, and then orders and executes each transaction.

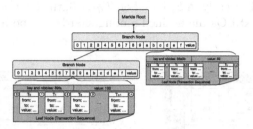

Fig. 1. Data structure of TxChain.

2.2 Transaction Model

TxChain adopts a state-based account whose state is cumulative results of related transactions. The transaction is defined as $\langle from, to, value, s, p, ts, fee \rangle_\sigma$ where $from$ is the sender of the transaction T, to is the recipient of T, $value$ is the transaction amount, s is the number of the shard generating T, p is the position of T in the TxSEQ of the generating account, ts is the timestamp, fee is the transaction fee, and σ is the signature of T.

In addition to executing local transactions, the validator V also needs to execute the received remote transaction T by inserting them into the TxSEQ of the specified account. Before executing T, the validator V needs to determine its relative position in the TxSEQ. V has to determine the ordinal relationship between transactions, which is determined by their dependencies. It is worth noting that only transactions in the same account constitute a dependency. All shards cannot have the exact same time in the decentralized environment, which means that it is impossible to directly determine the orders of transactions from different shards. Therefore, in order to determine the position of T in the TxSEQ and maintain consistency of its order in all shards, this paper proposes transaction dependencies. T_a and T_b are from shards S_i and S_j respectively and are executed on the same account, and their dependencies are defined as follows:

Definition 1. Dependencies. *T_a and T_b satisfy the causality relationship "$\succ$" iff (1) $T_a.ts < T_b.ts$ when $S_i = S_j$, or (2) T_a has executed in S_j before T_b generates when $S_i \neq S_j$; T_a and T_b satisfy the concurrency relationship "$\|$" iff neither $T_a \succ T_b$ nor $T_b \succ T_a$; T_a and T_b satisfy the conflict relationship "$\otimes$" iff $T_a \| T_b$ and $T_a.p = T_b.p$.*

2.3   Problem Definition

Transactions are executed instantly in the local shard and then are broadcast to remote shards for synchronization. After the remote shard receives the transaction T, the shard's state may change and it cannot execute T directly in the current view according to the position specified by $T.p$. As shown in Fig. 2, V_1 attempts to append T to T_2 in the TxSEQ of account "0x3889fa", i.e., $T.p = 3$. V_1 then sends T after S_2. However, V_2 has concurrently executed the transaction T_2' at the position 2 of account "0x3889fa" because $T_1 \succ T_2' \succ T_2$, which changes $T_2.p$ from 2 to 3. Thus, T should be ordered at position 4 which is after $T_2.p$ in S_2. If T is executed directly in S_2 based on $T.p = 3$, the transaction dependencies are violated and TxSEQs maintained among shards will be inconsistent.

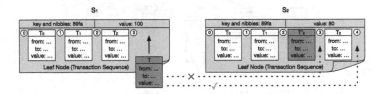

Fig. 2. Inconsistency example in TxChain.

TxChain supports the parallel execution of transactions in all shards. However, shards' state may change after transactions reach remote shards, resulting in violations of transaction dependencies and inconsistent shard states. Therefore, how to design a consensus mechanism which executes transactions in obedience to their dependencies and maintains the consistency of all shards becomes the key problem faced by TxChain.

3   Consensus Mechanism in TxChain

3.1   Prerequisites of Transaction Execution

The consensus in TxChain is to order and execute transactions based on transaction dependencies and ensures the consistency of these orders among shards. In order to maintain the transaction dependencies, TxChain adopts the vector-based timestamp [2] to determine the execution order of transactions. Each shard in TxChain has a collection of accounts $\mathcal{L} = \{L_1, L_2, \cdots, L_m\}, L_j \in \mathcal{L}$. In addition to the local state of the accounts, each shard needs to record the state of other shards. Each shard's state is composed of all the account states in it. We use the execution vector $EV_{S_i}^{L_j}$ to denote the state of L_j in S_i. Each element of EV represents the state of the same account in different shards and starts with 0, i.e., $EV_{S_i}^{L_j}[S_i] = 0, S_i \in \mathcal{S}$. After the shard S_j executes a transaction from S_i, L_j's execution vector $EV_{S_j}^{L_j}[S_i] = EV_{S_j}^{L_j}[S_i] + 1$. The timestamp of a transaction

T is $T.ts$ which represents the EV when it is generated. When remote shards receive T, they maintain the dependencies and execution order of transactions according to $T.ts$. Therefore, based on the timestamp of transactions, we present how to determine whether a transaction can be instantly executed. Meanwhile, The consistency constraints of TxChain should make transactions match their causality relationship and let transactions with higher transaction fee be ordered in the front.

Definition 2. Transaction Execution. *A transaction T can be executed instantly in the shard iff (1) T is from this local shard, or (2) T is from the remote shard and transactions which is causally before T have been executed.*

3.2 Transaction Sequence Conversion Algorithm

To solve the problem in Sect. 2.3, according to the consistency constraints of TxChain, we design the consensus mechanism based on the transaction sequence conversion (TSC) algorithm. T_a and T_b are two concurrent transactions for the same account L in S_i and S_j respectively, and are executed instantly in their local shards. When T_b arrives at S_i, the execution vector of L in S_i is $EV_{S_i}^L$. The ordering position of T_b in S_i shifts because the execution of T_a, so T_b cannot be executed directly at the $T_b.p$ in S_i. Therefore, TSC refers to the idea of address space transformation (AST) [3,7] to convert the account's TxSEQ to a specific state, and then execute transactions according to Definition 2.

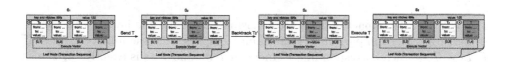

Fig. 3. Transaction sequence conversion.

We still use the example in Sect. 2.3 for illustrating the process of TSC, which has been presented in Fig. 3. Before V_1 attempts to execute T in the account $L = $ "$0x3889fa$", the execution vector of L is $EV_{S_1}^L = [0,3]$. When T arrives at S_2, $EV_{S_2}^L = [0,4]$. TSC traces S_2 back to the state when T is generated, i.e., $EV_{S_2}^L = T.ts = [0,3]$. In other words, TSC sets the impact of T_2' to be invisible. After T is executed in S_2, the $EV_{S_2}^L$ is updated to $[1,4]$. Finally, TSC restores the impact of T_2'. Similarly, S_1 follows the above steps to synchronize T_2'.

Based on the above content, TSC is presented as Algorithm 1. The transaction T is generated from account L of the remote shard S_j and it will be executed in L of S_i whose execution vector is $EV_{S_i}^L$. The first step of TSC is to verify the signature of T for its integrity. Next, TSC traces the TxSEQ of L back to the state when T is generated, i.e., adjusts $EV_{S_i}^L$ to equal $T.ts$. The process of backtrack determines which transactions have been executed when $EV_{S_i}^L = T.ts$. For one transaction T' in the TxSEQ of L maintained by S_i, if

---
Algorithm 1. The Transaction Sequence Conversion Algorithm

Input: Transaction sequence $TxSEQ$, transaction T;
1: **if** Verify (T) **then**
2:    Backtrack $(TxSEQ, T.ts)$
3:    Execute T in $TxSEQ$
4:    $EV_{S_i}^L[S_j] = EV_{S_i}^L[S_j] + 1$
5:    Backtrack $\left(TxSEQ, EV_{S_i}^L\right)$
6: **else**
7:    Abort T
8: **end if**

Algorithm 2. The Sequencing Algorithm

Input: Sequencing interval (p_x, p_y), transaction T
Output: Ordering position $Site$
1: $Site \leftarrow null$
2: **for** $T'.p$ between (p_x, p_y) **do**
3:    **if** $T \parallel T'$ **then**
4:       **if** $TOrder(T) < TOrder(T')$ && $Site = null$ **then**
5:          $Site \leftarrow T'.p$
6:       **end if**
7:    **else if** $T \succ T'$ **then**
8:       $Site \leftarrow T'.p$
9:       break
10:   **end if**
11:   $T' \leftarrow T'.next$
12: **end for**
13: **if** $Site = null$ **then**
14:    **return** $T'.p$
15: **else**
16:    **return** $Site$
17: **end if**

$T.ts \geq T'.ts$, the impact of T' is set as visible, otherwise the impact of T' is set as invisible. The comparison of timestamp complies with the following rules. If $T'.ts = T.ts$, each corresponding element in $T'.ts$ and $T.ts$ is exactly the same. If $T'.ts < T.ts$, each corresponding element in $T'.ts$ is not greater than $T.ts$, and the sum of the elements in $T'.ts$ is less than $T.ts$. If $T'.ts > T.ts$, $T'.ts$ contains at least one corresponding element greater than $T.ts$.

After TSC has traced the shard to the state when T is executed, the ordering position of T in the TxSEQ has been determined, which is the interval between two positions. However, due to the backtrack process, there may be several invisible transactions in that interval. Therefore, it is also necessary to compare the relationship between T and these invisible transactions to finally confirm the ordering position of T. In order to compare the relationship between T and these invisible transactions, we can also use $TOrder$ [6] in addition to transaction dependencies defined in Sect. 2.2.

Definition 3. TOrder. *Given two transactions* T_a *and* T_b, $TOrder(T_a) <$ $TOrder(T_b)$ *iff (1)* $SUM(T_a.ts) < SUM(T_b.ts)$, *or (2)* $T_a.fee > T_b.fee$ *when* $SUM(T_a.ts) = SUM(T_b.ts)$, *or (3)* $i < j$, *when* $SUM(T_a.ts) = SUM(T_b.ts)$ *and* $T_a.fee = T_b.fee$. *Meanwhile,* $SUM(T.ts) = \sum_{i=1}^{n} T.ts[S_i]$.

We have known T will be inserted between two positions p_x and p_y. Then we traverse each transaction T' in the sequencing interval (p_x, p_y). If T' and T are concurrent and $TOrder(T) < TOrder(T')$, $T'.p$ becomes the ordering position of T. If T causally precedes T', $T'.p$ is the ordering position of T and the traversal process ends. The process of ordering transactions in (p_x, p_y) can be described as Algorithm 2. Then T will be executed according to its ordering position.

4 Performance Evaluation

4.1 Experimental Setup

The prototype of TxChain is implemented in Golang, and its nodes spread across 16 machines. Every machine is equipped with the Intel Xeon E5128 CPU, 32 GB of RAM, 10 Gbps network link and Ubuntu 16.04 LTS operating system. In order to emulate the realistic network environment, we limit the bandwidth between nodes to 20Mbps and artificially insert a delay of 100 ms to communication links.

4.2 Throughput Scalability and Transaction Latency of TxChain

We evaluate the throughput scalability and transaction latency of TxChain in terms of the number of shards in TxChain $|S|$.

As shown in Fig. 4(a), the throughput of TxChain is improved when $|S|$ increases. When we double $|S|$, the throughput of TxChain can be increased by 1.59 to 1.89 times. Compared to RapidChain [8] which achieves 1.57 to 1.70 times when doubling its network size, TxChain can achieve higher scalability. TxChain have better scalability especially when $|S|$ is small because each shard maintains the full replica of TxChain and TSC has greater computational overhead when the amount data in TxChain expands.

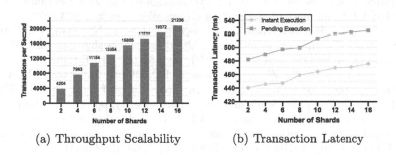

(a) Throughput Scalability   (b) Transaction Latency

Fig. 4. Performance of TxChain.

Figure 4(b) plots the latency of processing a transaction which can be executed instantly or not respectively for different $|S|$. Transactions which cannot be executed instantly is pending for a ordering position by TSC. We take the execution of a pending transaction as the example and find that the latency increases slightly from 482.26 milliseconds (ms) to 525.31 ms when $|\mathcal{S}|$ varies from 2 to 16. With the same $|\mathcal{S}|$, a pending transaction has roughly 10% more latency than the transaction can be executed instantly. These results show that the time delay is mainly due to TxSEQs conversion in TSC.

5   Conclusion

In this paper, we propose TxChain which is a novel DL based on sharding and constituted by accounts' transaction sequences. TxChain enable shards to update the global state via the unit of transaction in parallel, which greatly scales its throughput. The transaction sequence conversion (TSC)-based consensus mechanism traces accounts' transaction sequences to the state when transactions can be ordered correctly for maintaining the shards' consistency. We implement TxChain and demonstrate its high scalability and low transaction latency.

Acknowledgements. This work was supported by the Science and Technology Commission of Shanghai Municipality under Grant No. 21511101501 and the National Natural Science Foundation of China (NSFC) under Grant No. 61932007.

References

1. Dang, H., Dinh, T.T.A., Loghin, D., Chang, E.C., Lin, Q., Ooi, B.C.: Towards scaling blockchain systems via sharding. In: Proceedings of the 2019 International Conference on Management of Data, pp. 123–140 (2019)
2. Ellis, C.A., Gibbs, S.J.: Concurrency control in groupware systems. In: Proceedings of the 1989 ACM SIGMOD International Conference on Management of Data, pp. 399–407 (1989)
3. Gu, N., Yang, J., Zhang, Q.: Consistency maintenance based on the mark & retrace technique in groupware systems. In: Proceedings of the 2005 International ACM SIGGROUP Conference on Supporting Group Work, pp. 264–273 (2005)
4. Kokoris-Kogias, E., Jovanovic, P., Gasser, L., Gailly, N., Syta, E., Ford, B.: Omniledger: a secure, scale-out, decentralized ledger via sharding. In: 2018 IEEE Symposium on Security and Privacy (SP), pp. 583–598. IEEE (2018)
5. Luu, L., Narayanan, V., Zheng, C., Baweja, K., Gilbert, S., Saxena, P.: A secure sharding protocol for open blockchains. In: Proceedings of the 2016 ACM SIGSAC Conference on Computer and Communications Security, pp. 17–30 (2016)
6. Sun, C., Jia, X., Zhang, Y., Yang, Y., Chen, D.: Achieving convergence, causality preservation, and intention preservation in real-time cooperative editing systems. ACM Trans. Comput.-Hum. Inter. (TOCHI) 5(1), 63–108 (1998)
7. Yang, J., Wang, H., Gu, N., Liu, Y., Wang, C., Zhang, Q.: Lock-free consistency control for web 2.0 applications. In: Proceedings of the 17th international conference on World Wide Web, pp. 725–734 (2008)
8. Zamani, M., Movahedi, M., Raykova, M.: Rapidchain: scaling blockchain via full sharding. In: Proceedings of the 2018 ACM SIGSAC Conference on Computer and Communications Security, pp. 931–948 (2018)

Zebra: An Efficient, RDMA-Enabled Distributed Persistent Memory File System

Jingyu Wang[1], Shengan Zheng[2(✉)], Ziyi Lin[3], Yuting Chen[1], and Linpeng Huang[1(✉)]

[1] Shanghai Jiao Tong University, Shanghai, China
{wjy114,chenyt,lphuang}@sjtu.edu.cn
[2] MoE Key Lab of Artificial Intelligence, AI Institute, Shanghai Jiao Tong University, Shanghai, China
shengan@sjtu.edu.cn
[3] Alibaba Group, Hangzhou, China
cengfeng.lzy@alibaba-inc.com

Abstract. Distributed file systems (DFSs) play important roles in datacenters. Recent advances in persistent memory (PM) and remote direct memory access (RDMA) technologies provide opportunities in enhancing distributed file systems. However, state-of-the-art distributed PM file systems (DPMFSs) still suffer from *a duplication problem* and *a fixed transmission problem*, leading to high network latency and low transmission throughput. To tackle these two problems, we propose *Zebra*, an efficient RDMA-enabled distributed PM file system—Zebra uses *a replication group design* for alleviating the heavy replication overhead, and leverages *a novel transmission protocol* for adaptively transmitting file replications among nodes, eliminating the fixed transmission problem. We implement Zebra and evaluate its performance against state-of-the-art distributed file systems on an Intel Optane DC PM platform. The evaluation results show that Zebra outperforms CephFS, GlusterFS, and NFS by 4.38×, 5.61×, and 2.71× on average in throughput, respectively.

Keywords: RDMA · Adaptive transmission · Persistent memory · Distributed file system

1 Introduction

Nowadays, distributed file system plays an increasingly important role in datacenters. A distributed file system is a client/server-based application that allows clients to access and process remote files as they do with the local ones. Many efforts have been spent on leveraging persistent memory (PM) and Remote Direct Memory Access (RDMA) technologies to enhance the distributed file systems. Persistent memories are of large capacities and near-DRAM performance,

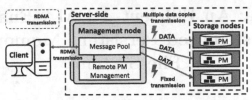

(a) **Architecture of a typical DPMFS.** It adopts a fixed transmission mode, suffering from heavy replication overhead between the management and the storage nodes.

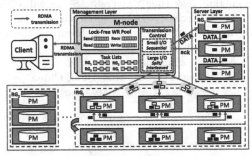

(b) **Architecture of Zebra.** We let each replication group contain three D-nodes and each file has three file replications.

Fig. 1. A comparison of a typical DPMFS and Zebra.

while RDMA supports direct accesses of PMs. PM and RDMA supplement each other, enabling efficient access to remote files.

Several RDMA-enabled distributed PM file systems (DPMFS), such as Octopus [5], Orion [12] and Assise [2] achieve high performance through coupling PM and RDMA features. A DPMFS, as Fig. 1(a) shows, is composed of a management node and many storage nodes. The management node is responsible for receiving clients' requests and/or managing file replications. The storage nodes are organized into a chain structure or a star topology, on which file replications are stored. Persistent memories are deployed on the central and storage nodes, as they are of large capacities and near-DRAM accessing speeds, and RDMA is employed, allowing users to directly access PMs and efficiently synchronize file replications [6]. Conventional distributed file systems adopt RDMA by substituting their communication modules with RDMA libraries.

The problem of file replication can be cast into a problem of file transmission: file replications are transmitted from the management node to the storage nodes, or vice versa, such that clients can access files close to them. Meanwhile, the performance of existing DPMFSs suffers from two transmission problems.

Problem 1: Duplication Problem. File replications need to be intensively transmitted. In a chain-structured DPMFS [4,9], the management node is followed by a chain of storage nodes—a file is transmitted along the chain and much effort needs to be spent on synchronizing file replications so as to avoid inconsistency. Comparatively, in a DPMFS with a star topology, file replications are only transmitted between the central node and the storage nodes [4, 8, 11], whilst the loads are not balanced, as the central node suffers from heavy replication overhead.

Problem 2: Fixed Transmission Problem. A DPMFS, or even a traditional DFS, usually adopts a fixed transmission strategy. Such a DPMFS does not adapt to different granularity, and thus suffers from a mismatch between the file replications and the transmission block sizes. A DPMFS needs to adopt a much more flexible transmission strategy in small/large file transmission scenarios (in which files of small/large sizes are transmitted), binding files to different transmission modes.

To tackle the duplication and the fixed transmission problems, we present *Zebra*, an efficient, RDMA-enabled distributed PM file system. The key idea is to (1) use *a replication group design* for alleviating the heavy replication overhead, and (2) design *an adaptive RDMA-based transmission protocol* that adaptively transmits file replications among nodes, solving the fixed transmission problem. In addition, to support multithreaded data transmission, Zebra leverages a lock-free work request (WR) pool and a conflict resolution mechanism to accelerate transmission. This paper makes the following contributions:

- *Design:* Zebra uses a replication group design that alleviates the heavy replication overhead between the central node and storage nodes.
- *Protocol:* Zebra provides an adaptive RDMA transmission protocol for distributed PM file systems, significantly improving the transmission throughput.
- *Implementation and evaluation:* We implement Zebra and evaluate its performance against state-of-the-art distributed file systems on an Intel Optane DC PM platform. The evaluation results clearly show the efficiency of Zebra in small/large file transmission scenarios.

2   The Zebra System

2.1   Design

Zebra is an efficient DPMFS that uses an adaptive transmission mode to speed up accesses to PMs. As Fig. 1(b) shows, Zebra consists of a management layer containing management nodes (*M-nodes*) which are responsible for coordinating resources, and a server layer containing a set of storage nodes (*D-nodes*):

Group of D-nodes. Zebra provides a replication group (RG) mechanism that divides D-nodes into groups. The D-nodes in a group work as backups for each other. Each RG contains a master D-node that coordinates the D-nodes, guarantees their orderliness, and feedback to the M-node.

M-nodes. Zebra decouples its metadata and data to improve the scalability and performance. File I/O requests are received, processed, and broadcasted by the M-node.

Zebra also pre-allocates memories for work requests (WR) during the registration phase, reducing the overhead of memory allocation during RDMA transmissions. As Fig. 2 shows, Zebra uses a *lock-free WR pool* for storing WRs, which is produced for each transfer.

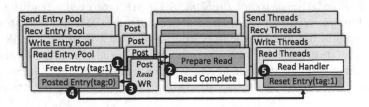

Fig. 2. Workflow in a lock-free work request pool.

2.2 An Adaptive Replication Transmission Protocol

All of the nodes in Zebra system are connected through an RDMA network. During file transmission, Zebra first establishes a *critical transmission process* that transmits data and transmission control information from the M-node to one or more D-nodes; Zebra then asynchronously transmits data among D-nodes, exchanging file replications.

Let $\{d_1, d_2, \ldots\}$ be a set of data packets to be transmitted. Let $g(D_i)$ be the D-node D_i in a replication group g. Zebra flexibly chooses the transmission modes on the basis of file transmission scenarios, adapting to I/O sizes:

1) When transmitting a small number of files and/or files of small sizes, Zebra uses a *sequential transmission* mode due to the sufficiency of the transmission bandwidth. The file replications can be transmitted directly to the D-nodes using a chain replication style, where (1) the M-node M picks up a D-node, say D_i, as a master D-node and writes the data to it; (2) D_i receives the transmission control information and writes data to the other n nodes in the RG g. D_{i-2} (the D-node next to last) sends an ACK message to the M-node for completing the writing activity.

2) Zebra uses a novel, *split/interleaved transmission* mode in large I/O scenarios (when a large number of files and/or files of large sizes are transmitted). This mode reduces the transmission overhead from the M-node to the D-nodes. First, Zebra takes a split transmission process, $M \rightarrow g(D_1) = \{d_1, d_2, ..., d_k\}$, $M \rightarrow g(D_2) = \{d_{k+1}, d_{k+2}, ..., d_j\}, \ldots$ in which the data is split and transmitted from the M-node M to the D-nodes. The successive interleaved transmission are $g(D_1) \leftrightarrow g(D_2) \leftrightarrow \cdots \leftrightarrow g(D_n)$, indicating that the file replications are interleaved between the D-nodes in the replication group, rather than between the M-node and the replication group.

For example, the upper bound of the RDMA throughput (T_R) is approximately 12 GB/s in Mellanox ConnectX-5, whereas the bandwidth of PM (T_{PM}) is approximately 35 GB/s in read and 20 GB/s in write. In the case, even though the maximal bandwidth of RDMA is reached, the bandwidth of PM is not fully leveraged. The split/interleaved mode expands the transmission bandwidth, reducing the latency to $L_s = \frac{S_d/m}{T_R}$, where S_d is the total size of data in transmission and the data is split into m parts. Here we let the number of splits be $m = \frac{T_{PM}}{T_R}$, where T_{PM} is the maximal bandwidth of PM. The latency of transmitting unsplit data/files is m times of that of taking the *split/interleaved transmission* mode, i.e., $L_{us} = \frac{S_d}{T_R}$. The latency increases exponentially after the bandwidth reaches a bottleneck.

Load Balance. The M-node manages all space allocations in Zebra. It balances the storage usage and the throughput by sampling the realtime load across the RGs: the M-node gathers the operations; a load-balance controller records the total capacity and realtime usage of each RG, and computes the throughput of each node according to the number of ongoing RDMA operations.

Let g be a replication group and C_g be the percentage of its available capacity. Let $Task$ be the set of unfinished tasks and $Task_g$ be the set of unfinished tasks w.r.t. g. Let *availability* of g be $availability_g = \alpha \times C_g + \beta \times \frac{|Task|}{|Task_g|} \times 100\%$, where α and β are two real values defined by human engineers. The RG with a smaller availability (i.e., with less running tasks) is assigned a higher possibility for storing file replications. An RG with the largest C is chosen when several RGs are of the same availability values. To avoid frequent calculations, Zebra selects RG candidates with the highest availability values and updates them periodically. This reduces the load-balancing overhead.

2.3   Multithreaded RDMA Transmission

Multithreaded transmission are frequent in DFSs. Traditionally, lock operations need to be performed in resource pools to guarantee the availability of resources, while it incurs excessive lock contention among multiple threads. Zebra takes a lock-free mechanism that uses kernel atomic verbs to manage WR resources in the WR pools. When Zebra performs a transmission, a daemon thread calls `atomic_dec_and_test` to search for an available WR entry.

In order to avoid conflicts caused by linear detection among multi-threading, we adopt a decentralized query strategy to speed up usable WR acquisition process. Threads process work requests scattered within the WR pool. Every time a thread obtains an available WR, it linearly searches and points to the next available WR in advance. When a thread detects that the current WR has been occupied, it searches for the next available WR. If a thread detects that the previously available WR is now occupied, and then performs a location hash. This ensures a high level of parallelism.

3   Evaluation

We evaluated the performance of Zebra against other state-of-the-art distributed file systems using a set of benchmarks. The evaluation is designed to answer the following research questions (Fig. 3):

- RQ1. How effective is Zebra in transmitting files of different sizes?
- RQ2. How effective is Zebra in processing multithreaded workloads, compared to the other distributed file systems?
- RQ3. Is Zebra scalable for real-world applications?

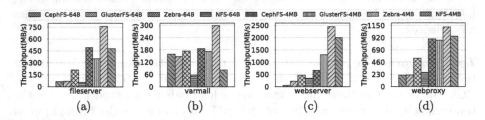

Fig. 3. Systems' throughput w.r.t. filebench with different workloads.

3.1 Setup

We conduct the evaluation on an Intel Optane DC PM platform. Each node is equipped with six 256 GB Intel Optane DCPMMs that are configured in the App-Direct interleaved mode. The nodes communicate with each other through a Mellanox ConnectX-5 RNIC with the Infiniband mode. The workload threads are pinned to local NUMA nodes. Zebra is deployed on a cluster, on which every two D-nodes are organized into a replication group.

We compare Zebra against four file systems: CephFS [11], GlusterFS [7], and NFS [3]. We use *Filebench* [10], to evaluate the adaptive transmission design in large/small file transmission scenarios. We use *fileserver*, a workload in *Filebench*, to emulate the multithreaded scenario. Furthermore, we evaluate the overall performance of Zebra using four *Filebench* workloads. In addition, we evaluate Zebra's scalability with Redis [1], which is a popular persistent key-value store.

3.2 Sensitivity to I/O Size

We use four filebench workloads (fileserver, varmail, webproxy, and webserver) to evaluate Zebra. For small I/O sizes (Fig. 4(a)(c) where the I/O size is smaller than 64 KB), Zebra adopts a sequential transmission mode. This reduces the overhead of one-to-many transmission, increasing the throughput. Figure 4(b)(d) corresponds to the scenario in which the I/O sizes are larger than 256 MB. The threshold of switching the transmission modes depends on the I/O size, which is also determined by the peak throughput of the other systems. This ensures that the use of efficient split/interleaved transmission can alleviate the performance degradation caused by throughput bottlenecks. In the case of large data transmission scenarios (i.e., 1 or 2 GB), the performance of CephFS, GlusterFS, and NFS decreases by 21%, 27% and 58% on average, respectively. The performance of Zebra decreases by only 2.3%.

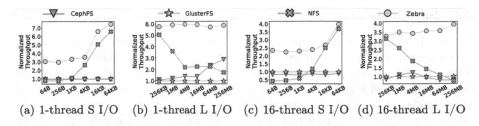

(a) 1-thread S I/O   (b) 1-thread L I/O   (c) 16-thread S I/O   (d) 16-thread L I/O

Fig. 4. Performance for various I/O sizes and threads. It is normalized w.r.t. the throughput of GlusterFS.

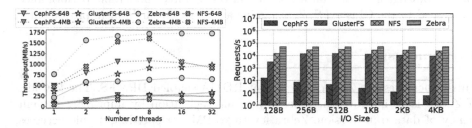

Fig. 5. Concurrency performance.   **Fig. 6.** Performance of running Redis.

In a small I/O size scenario, Zebra outperforms CephFS, GlusterFS, and NFS transmission protocols by 3.7×, 3.4×, and 2.9×, respectively. In a large I/O size scenario, Zebra outperforms CephFS, GlusterFS, and NFS by 4.7×, 6.2×, and 2.3×, respectively.

3.3 Concurrency

Figure 5 shows the performance of the systems running with multi threads. We set the I/O size to 64 B and 4 MB and run fileserver with the thread numbers ranging from 1 to 32. Zebra adopts a two-tier concurrency mechanism, including a lock-free WR pool and an adaptive transmission strategy that improves its scalability.

In CephFS, each replica applies the update locally and subsequently sends an acknowledgment to the primary node. This results in heavy network transmission on the critical path. GlusterFS performs the worst because of its heavy data management overhead. When NFS reaches the maximal number of threads, any subsequent requests need to wait, resulting in timeouts. As NFS is a TCP-based network file system, a large number of threads with unbalanced loads can lead to give up of file transmission. Zebra allows data to be transmitted in a multithreaded manner and achieves the highest performance among the file systems in the multithreaded transmission scenario. Zebra outperforms CephFS, GlusterFS, and NFS on average by 2.1×, 2.3×, and 1.8×, respectively.

3.4 Scalability

We use Redis [1], a popular persistent key-value store, to further evaluate the performance of Zebra in real world applications. We choose AOF mechanism in Redis to achieve data persistence, which synchronizes data after every modification. As shown in Fig. 6, Zebra achieved the optimum performance on the basis of optimization in small data transmission. The underlying storage consistency protocol of CephFS, makes it vulnerable to small data accesses. The performance of CephFS drops 98.2% from 64 B to 4096 B. Zebra and GlusterFS use block storage, which stores a set of data in chunks. Block storage provides high throughput against small data I/O and also guarantees preferable performance. Compared to the GlusterFS, Zebra gains a 17–21× throughput improvement.

4 Conclusion

This paper presents Zebra, an efficient RDMA-enabled DPMFS. The key idea of Zebra is to improve the efficiency of file replication by improving the efficiency of data transmission. Zebra leverages PMs for speeding up file accesses, uses an adaptive RDMA protocol for reducing network latency and improving transmission throughput. It uses the method of dividing large data into blocks and complete data reliability backup. The evaluation results show that Zebra outperforms other state-of-the-art systems, such as CephFS, GlusterFS, NFS, and Octopus in transmitting data streams and synchronizing file replications.

Acknowledgements. This research is supported by National Natural Science Foundation of China (Grant No. 62032004), Shanghai Municipal Science and Technology Major Project (No. 2021SHZDZX0102) and Natural Science Foundation of Shanghai (No. 21ZR1433600).

References

1. Redis (2018). https://redis.io/
2. Anderson, T.E., Canini, M., Kim, J., et al.: Assise: performance and availability via NVM colocation in a distributed file system. CoRR (2020)
3. Callaghan, B., Lingutla-Raj, T., Chiu, A., et al.: NFS over rdma. In: Network-I/O Convergence: Experience, Lessons, Implications (2003)
4. Davies, A., Orsaria, A.: Scale out with glusterfs. Linux J. (2013)
5. Lu, Y., Shu, J., Chen, Y., et al.: Octopus: an RDMA-enabled distributed persistent memory file system. In: ATC (2017)
6. Nielsen, L.H., Schlie, B., Lucani, D.E.: Towards an optimized cloud replication protocol. In: SmartCloud (2018)
7. Noronha, R., Panda, D.K.: IMCA: a high performance caching front-end for glusterfs on infiniband. In: ICPP (2008)
8. Shan, Y., Tsai, S.Y., Zhang, Y.: Distributed shared persistent memory. In: SOCC (2017)
9. Shvachko, K., Kuang, H., Radia, S., Chansler, R.: The hadoop distributed file system. In: MSST (2010)

10. Tarasov, V., Zadok, E., Shepler, S.: Filebench: a flexible framework for file system benchmarking. Usenix Mag. (2016)
11. Weil, S.A., Brandt, S.A., et al.: Ceph: a scalable, high-performance distributed file system. In: OSDI (2006)
12. Yang, J., Izraelevitz, J., Swanson, S.: Orion: a distributed file system for non-volatile main memory and RDMA-capable networks. In: FAST (2019)

Data Security

ADAPT: Adversarial Domain Adaptation with Purifier Training for Cross-Domain Credit Risk Forecasting

Guanxiong Zeng[1], Jianfeng Chi[1], Rui Ma[1], Jinghua Feng[1(✉)],
Xiang Ao[2,3,4(✉)], and Hao Yang[1]

[1] Alibaba Group, Hangzhou, China
{moshi.zgx,bianfu.cjf,qingyi.mr,jinghua.fengjh,
youhiroshi.yangh}@alibaba-inc.com
[2] Key Lab of Intelligent Information Processing of Chinese Academy of Sciences
(CAS), Institute of Computing Technology, CAS, Beijing 100190, China
aoxiang@ict.ac.cn
[3] University of Chinese Academy of Sciences, Beijing 100049, China
[4] Institute of Intelligent Computing Technology, Suzhou, CAS, Suzhou, China

Abstract. Recent research on transfer learning reveals that adversarial domain adaptation effectively narrows the difference between the source and the target domain distributions, and realizes better transfer of the source domain knowledge. However, how to overcome the *intra/inter-domain imbalance problems* in domain adaptation, e.g. observed in cross-domain credit risk forecasting, is under-explored. The intra-domain imbalance problem results from the extremely limited throngs, e.g., defaulters, in both source and target domain. Meanwhile, the disparity in sample size across different domains leads to suboptimal transferability, which is known as the inter-domain imbalance problem. In this paper, we propose an unsupervised purifier training based transfer learning approach named **ADAPT** (**A**dversarial **D**omain **A**daptation with **P**urifier **T**raining) to resolve the intra/inter-domain imbalance problems in domain adaptation. We also extend our ADAPT method to the multi-source domain adaptation via weighted source integration. We investigate the effectiveness of our method on a real-world industrial dataset on cross-domain credit risk forecasting containing 1.33 million users. Experimental results exhibit that the proposed method significantly outperforms the state-of-the-art methods. Visualization of the results further witnesses the interpretability of our method.

Keywords: Multi-source domain adaptation · Class-imbalance · Purifier training · Credit risk forecasting

1 Introduction

The outbreak of COVID-19 has almost devastated the global economy, and it was reported that large number of businesses have problems with financial

G. Zeng, J. Chi—Contributed equally.

difficulties[1]. This recession, on the other hand, burgeons on the online financial services, which aims to provide inclusive financial services to businesses to help them tide over the difficulties. For example, Alibaba Group, a Chinese e-commerce corporation, has offered multiple financial services to its registered users (businesses) in 1688.com and alibaba.com, etc.

Based on the previous experience on online financial service, overdue and fraud are inevitable, and they are in general the culprits of asset loss of a platform [2,8–11]. It is conceivable that the financial services provided on new platforms will also confront the same risks. Even worse, few available data and ground truth in new platforms make its financial risk management more challenged on perceiving the high-risk customers in advance and reducing the risk exposure of the platform.

Domain Adaptation (DA) aims to transfer the representation of a source domain with wealth data and information to a target domain with few or no labeled data. Recent years have witnessed successive successful applications of domain adaptation in large-scale datasets and various research tasks, e.g., image recognition [21,23] and fraud detection [26]. It thus seems to be a plausible solution that applying DA techniques to adapt useful knowledge from mature financial service platforms to a brand-new or barely performing financial service platform. However, some special challenges raised by the class-imbalance problem of credit risk forecasting render conventional DA techniques ineffective in this scenario, and we summarize the challenges as follows.

Intra-domain Class Imbalance. Although the risk of defaulters is very high, the number of defaulters is much smaller than that of benign users. The average number of benign users (negative) is about more than a hundred times than that of defaulters (positive) per month [2,8]. Therefore, it suffers from a severe class-imbalance problem [5] in both source and target domain.

Inter-domain Sample Imbalance. Recall that an ideal adaptation is to transfer more knowledge about the defaulters of the source domain to unearth the defaulters in the target domain. That is, we attempt to mainly transfer the knowledge of the minor samples that might be easily ignored by the model. Hence, how to identify valid samples of the source domain and alleviate suboptimal even negative transfer [15,22,23] caused by the sample imbalance across the domains is the second challenge.

To remedy these special challenges in cross-domain credit risk forecasting, we propose a novel approach coined ADAPT (*Adversarial Domain Adaptation with Purifier Training*) in this paper. ADAPT is an unsupervised purifier training based transfer learning approach resolve the intra/inter-domain imbalance problems in domain adaptation simultaneously. First, a self-adaptive re-weighted approach is devised to overcome the intra-domain class-imbalance problem by estimating the uncertainty of the source samples. Second, we cherry-pick the source samples based on the target domain classification through the purifier training to remedy the issues brought by inter-domain sample imbalance.

---

[1] https://www.pwc.com/us/en/library/covid-19.html.

Finally, we extend the ADAPT method to the multi-source domain adaptation via weighted source integration. This can make up for the problem of insufficient or even no labels in the target domain and is conducive to the cold start of financial service on a new platform.

The main contributions of this paper are summarized as follows:

- To the best of our knowledge, we are the very first attempt to simultaneously overcome the intra/inter-domain imbalance problem in domain adaptation, which is observed in cross-domain financial credit risk forecasting and is ubiquitous in many other scenarios.
- We propose ADAPT, an unsupervised transfer learning approach, which adopts self-adaptive re-weighted loss to resolve the class-imbalance problem within domains and a purifier training that cherry-picks the samples to perform better inter-domain knowledge transfer. We also extend ADAPT to a multi-source domain adaptation version to facilitate the generalization of our model.
- Experiments on four real-world datasets demonstrate the effectiveness of proposed ADAPT, which can outperform other recent DA learning methods. Furthermore, our model provides good interpretability.

The remainder of this paper is organized as follows. Section 2 surveys the related researches in the literature. Section 3 introduces problem statement in our paper. Section 4 details the proposed ADAPT method. Section 5 demonstrates the experiment settings and main results. Section 6 concludes the paper.

2   Related Work

2.1   Domain Adaptation

Domain Adaptation (DA) is a frequently used technique once applied in many research fields to improve the performance of models. From the perspective of the method, it is mainly divided into two categories. One is to use a loss function to minimize the differences between the source domain and the target domain, such as L1, L2 or cosine similarity, KL or JS divergence, Maximum Mean Discrepancy (MMD) [3]. The role of these loss functions is to narrow the distance between the source domain and the target domain in the latent space. Adversarial training is another popular method of transferring domain information. For example, Adversarial Discriminative Domain Adaptation (ADDA) [17] combines discriminative modeling, multi-head feature extraction, and the GAN loss. Adversarial Domain Adaptation with Domain Mixup (DM-ADA) [21] guarantees domain invariance in a more continuous latent space and guides the domain discriminator to determine samples' difference relative to the source and target domains.

Considering the number of source domains, DA can be divided into single-source domain adaptation and multi-source domain adaptation. Most single-source domain adaptation methods can be extended to multi-source domain adaptation. MDDA [23] separately pre-trains feature extractor and classifier for each domain and matches the features between source and target by Wasserstein distance. Then it combines the different predictions from each source classifier by a weighting strategy based on the discrepancy between each source and target.

2.2 Credit Risk Forecasting

Credit risk forecasting (CRF) is the core of financial services management, and relevant research has been made as early as 1968 [1]. However, at that time, researchers mainly employ some linear models and artificial strategies to do simple population risk stratification. Financial institutions will prevent some potential high-risk customers from opening relevant financial services through the stratification of the population. Later, random forest methods are proposed to assess the potential risk and obtain better performance than traditional financial institutions [14].

A lot of recent works have introduced deep neural networks to identify specific risk groups, e.g., heterogeneous information network based defaulter detection [25], meta-learning based potential defaulters mining [2,19], cross-domain fraud detection via domain adaptation [26], and multi-task based credit risk forecasting [8]. In fact, all the above methods suffer from serious class imbalance, because the number of high risk users is far less than that of benign users. Some methods alleviate these problems through simple undersampling [8,25], while others introduce relatively complex loss functions [26].

2.3 Class-Imbalance

Main methods to solve the class-imbalance problem can be divided into two categories: data-based method and algorithm-based method. The data-based method is to undersample the majority classes or oversample the minority classes. For example, CGAN [18] uses generative adversarial networks to augment the training dataset. For class-imbalance financial risk prediction, [13] propose an adversarial data augmentation method to generate high-risk samples to improve the effect. A typical undersampling method is the random majority undersampling (RUS), which may lose some important information. Recently, [12] propose an imbalanced learning approach named PC-GNN to overcome the problem in the GNN based methods, which uses label-balanced sampler to pick nodes and edges and neighborhood sampler to choose neighbors with a learnable distance function. In this way, they can oversample the minority classes and undersample the majority classes.

One of the typical algorithm-based methods is the cost-sensitive learning method by assigning relatively high costs to minority classes, thus overcoming the class-imbalance problem. And cost-sensitive learning methods don't change the distribution of training data, and it's easier to apply. For example, [24] propose a cost-sensitive hierarchical classification for imbalance classes, which constructs a cost-sensitive factor to balance the majority and minority classes based on the probability of each hierarchical subtask.

3   Business Setting and Problem Statement

3.1   Business Setting

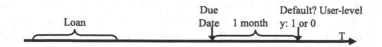

Fig. 1. Timeline of y in a billing cycle in the online credit loan services.

In this work, our business settings are based on online credit loan services provided to the users by the Alibaba Group. According to the general definition in online finance service platforms, we assign a label $y \in \{0, 1\}$ on each user to indicate whether he/she is a defaulter or not. Hence, the cross-domain credit risk forecasting task is to predict the default probability of the user. For instance, a billing cycle is shown in Fig. 1 [2,8]. If a user borrows money from a financial platform by credit, there will be a due date according to the lending service agreement. If the user fails to repay the loan one month after the due date, she/he would be regarded as a one-month delinquent defaulter (c.f. y in the figure). Because more than half of the above-mentioned users may be overdue for a longer period of time, ultimately causing economic losses. Considering the brand new platform, we also hope it have its credit rating based on the existing data of the platform, which can help existing users enjoy the new service of consumer loans and improve capital turnover rate, and better overcome the fund difficulties. However, there is no financial performance, except the profile information and business behavior for the users in the new platform. The lack of financial performance challenges the performance of traditional models.

3.2   Problem Statement

Why Domain Adaptation. We want to use these data which produce on existing platforms with rich information to better predict the probability for a user to be a defaulter in the new platform. And, we also need to find the correlation between the platforms, allowing us to have more prior knowledge in the operation of financial products on the new platform.

However, the characteristics of most platforms are not completely different and users in different platforms also have their own uniqueness. Fortunately, the nature of the features are similar, such as member level, purchasing behavior, etc. So, we adopt popular adversarial training method to transfer domain information [17]. The Unsupervised Domain Adaptation (UDA) scenario is considered here because that the target domain is a brand new platform and it has almost no labels. To summarise, in this work, the goal is to predict the probability of the user to be a defaulter through Unsupervised Domain Adaptation or other traditional methods.

Data Composition. There are k labeled source domains $\mathcal{D}_S^1, \ldots, \mathcal{D}_S^k$ and one target domain $\mathcal{D}_T$ without financial performance. For the ith source domain, $\mathcal{D}_S^i$ is the observed users with corresponding labels in our ith financial platforms. For simplicity, we set $\mathcal{X}_S^i \in \mathbb{R}^n$ as feature space in the ith source domain $\mathcal{D}_S^i$ and $\mathcal{Y}_S^i = \{0, 1\}$ as label space, respectively. Let $y = 0$ indicates benign user and $y = 1$ indicates the defaulter. Then, samples in source domain are labeled 0 or 1, which form source dataset $\mathcal{D}_S^i = \{\mathcal{X}_S^i, \mathcal{Y}_S^i\}$. Similarly, the samples $\mathcal{X}_T$ in target domain form the target dataset $\mathcal{D}_T = \{\mathcal{X}_T\}$ which is unlabeled due to the lack of corresponding financial performance. In practice, we use the labeled samples $\mathcal{D}_T^v = \{\mathcal{X}_T^v, \mathcal{Y}_T^v\}$ as the validation set to evaluate the effect of different methods. The samples in validation set are also labeled 0 or 1. These very few positive samples (assigning as 1) are those users who have been penalized for such as business violation/fraud.

Definition of Cross-domain Credit Risk Forecasting. Given the source dataset $\mathcal{D}_S$, $\mathcal{D}_T$ as the training set and the labeled samples $\mathcal{D}_T^v$ as the validation set. Our purpose is to build a model $\mathcal{F} : \{\mathcal{X}_S, \mathcal{X}_T\} \rightarrow \hat{y}$ through different methods, where $\hat{y}$ is the default probability of the users in the target domain. The transfer task aims to improve the model performance on the unlabeled $\mathcal{D}_T$ with the help of $\mathcal{D}_S$.

4   The Proposed Model

4.1   The Model

In this section, we present the proposed ADAPT (Adversarial Domain Adaptation with Purifier Training), its overall architecture is shown as Fig. 2. Figure 2(a) is the core model architecture of the ADAPT, which is mainly divided into two network structures. The left is an embedding layer, and the right is a Re-weighted Adversarial Discriminative Domain Adaptation (Re-weighted ADDA) Layer. We extend the ADAPT method to a multi-source scenario in Fig. 2(b). The input is the data of all domains, including $1 \sim k$ source domain and one target domain, and the output is the inference of the target domain data integrated by all ADAPT modules.

We use the subscript $(\cdot)_S$ or $(\cdot)_T$ to distinguish the source and target respectively and the superscript $(\cdot)^i, i \in \{1, ..., k\}$ to distinguish different source domains.

Embedding Layer
In our scenario, almost all data are tabular statistical features. We collect users' data under the premise of complying with security and privacy policies. We utilize feature extraction F^{pre} to deal with these original inputs. The encoded

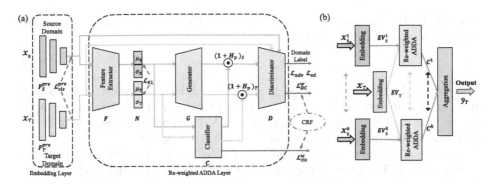

Fig. 2. (a) Model architecture of the ADAPT module. **(b)** Extending ADAPT method to MSDA.

representations are denoted as follows with the same dimensions in all source and target domains:

$$EV_S^i = (F_S^{pre})^i(\mathcal{X}_S^i) \quad (i \in \{1, ..., k\})$$
$$EV_T = F_T^{pre}(\mathcal{X}_T) \tag{1}$$

All feature extraction layers of the target and source domains are independent. The non-shared feature extractor network can obtain the discriminant feature representation and accurate classifier of each source domain. When aggregating information from multiple source domains downstream of the network, the final target prediction will be better improved [23]. In the ith source domain, the classification loss function is defined as:

$$\mathcal{L}_{cls}^{pre} = - \mathop{\mathbb{E}}_{x_S^i \sim \mathcal{P}_S^i} \frac{1}{C_\#} \sum_{c=1}^{C_\#} (y_s^i \log((C_S^{pre})^i(EV_S^i))) \tag{2}$$

where $C_\#$ is the number of the class, y_s^i is the ground truth of the corresponding pre-training task, $(C_S^{pre})^i$ is the corresponding classifier. It is worth noting that the target domain is unlabeled. We use the user level classification of the corresponding platform as the domain pre-training task.

Re-weighted ADDA Layer
The Re-weighted ADDA Layer can be divided into five parts: the encoder F, the gaussian normalization layer N, the classifier C, the generator G, and the discriminator D. The core of the Re-weighted ADDA Layer is Adversarial Discriminative Domain Adaptation (ADDA) [17]. Details are described in the following.

1) The first part is the common encoder F^i for each pair (EV_S^i, EV_T). The embedding after the encoder is $(F^i(EV_S^i), F^i(EV_T))$.

2) The embedding pair $(F^{ii}(EV_S^i), F^{ii}(EV_T))$ are mapped to (μ_S^i, σ_S^i) and (μ_T^i, σ_T^i) respectively. And they are regularized by a standard gaussian priori over the latent distribution as conventional VAE in the gaussian normalization

layer N [7,21], the aim of which is to narrow the Kullback-Leibler divergence between a posterior and a priori.

$$\mathcal{L}_{KL} = D_{KL}(\mathcal{N}(\mu, \sigma) \| \mathcal{N}(0, \boldsymbol{I})) \tag{3}$$

where μ and σ are the encoded mean and standard deviation of the source and target samples. In this way, the output of gaussian normalization layer N^i follows the standard Gaussian distribution $\mathcal{N}(0, \boldsymbol{I})$, which is conducive to the matching of the tabular characteristics in the latent space.

3) The classifier $\boldsymbol{C}^i$ is optimized with classification loss defined on the ith source domain. The input of the classifier $\boldsymbol{C}^i$ is $[\mu_S^i, \sigma_S^i]$ for each sample, where $[\cdot]$ denotes concatenation. In this paper, the distribution of the sample in the source domain is extremely imbalanced. We add a self-adaptive re-weighted method to the classification loss. We use the weighted entropy to measure the uncertainty of samples. Sample (user) with low entropy can be regarded as a well-classified sample. Otherwise, it is a poorly classified sample [20]. Thus, we could increase the weight for samples that are poorly classified sample. We add the self-adaptive weight $\mathcal{H}_p$ into the classification loss to better overcome the intra-domain imbalance problem.

$$\mathcal{H}_p = -\frac{1}{C_\#} \sum_{c=1}^{C_\#} p_c \cdot \log(p_c) \cdot \alpha_c \tag{4}$$

where p_c is the probability of predicting a sample to class c, and α_c is the hard weight to corresponding class c. It would pay more attention to the class with higher hard weight α_c during training. We adopt the conditional entropy as the indicator to weight the classification loss and the classification loss is extended as:

$$\mathcal{L}_{cls}^w = -\mathop{\mathbb{E}}_{x_S^i \sim \mathcal{P}_S^i} \frac{1}{C_\#} \sum_{c=1}^{C_\#} ((1 + \mathcal{H}_p) y_s^i \log(\boldsymbol{C}^i(\boldsymbol{F}^i(\boldsymbol{EV}_S^i)))) \tag{5}$$

The form of the $\mathcal{L}_{cls}$ is the same as that of Eq. 2. It is worth noting that there is no classification error for the target domain here. We only do inference on the samples of the target domain to get the performance on labeled samples $\mathcal{D}_T^v$. These labels are very few, we only use them for evaluation and do not put them directly into our model in this paper. And the loss is also used to train the encoder $\boldsymbol{F}^i$.

4) In order to strengthen the diversity of generator $\boldsymbol{G}$, we add Gaussian noise to its input. We also add the one-hot object class label l_{cls}^i into the input to the generation of the auxiliary samples [21]. In the source domain, the l_{cls}^i is the one-hot code corresponding to the sample label. As for the target domain, the default one-hot is $[0, 1]$. After that, generator $\boldsymbol{G}^i$ generates the auxiliary generated sample (representations) $\boldsymbol{EVG}$ as below:

$$\begin{aligned} \boldsymbol{EVG}_S^i &= \boldsymbol{G}^i([\mu_S^i, \sigma_S^i, z, l_{cls}^i]) \\ \boldsymbol{EVG}_T^i &= \boldsymbol{G}^i([\mu_T, \sigma_T, z, [0, 1]]) \end{aligned} \tag{6}$$

where z is the noise vector randomly sampled from standard Gaussian distribution.

5) The Discriminator D^i is mainly used to distinguish the auxiliary generated sample EVG_S^i, EVG_T^i produced by G^i from the original sample EV_S^i, EV_T. We restrict not only the domain invariability to the source domain and the target domain but also the generation of the different classes of samples. The min-max optimization objective on different domains are defined as follows [21]:

$$\min_{F^i,G^i} \max_{D^i} ((\mathcal{L}_{adv})_S + (\mathcal{L}_{adv})_T) \tag{7}$$

$$(\mathcal{L}_{adv})_S = \mathop{\mathbb{E}}_{x_S^i \sim \mathcal{P}_S^i} \log(D^i(EV_S^i)) + \log(1 - D^i(EVG_S^i)) \tag{8}$$

$$(\mathcal{L}_{adv})_T = \mathop{\mathbb{E}}_{x_T \sim \mathcal{P}_T} \log(1 - D^i(EVG_T^i)) \tag{9}$$

To ensure that features of the different classes in the source domains are highly differentiated in the latent space, we have added a classification loss as the Eq. 5, in which the self-adaptive weight $\mathcal{H}_p$ is to learn the hard samples between different domains for better overcoming the inter-domain imbalance problem. We also add a Euclidean Distance in the discriminator, which takes the class information into account and measures the class discrepancy in the source domains.

$$\mathcal{L}_{DC}^w = - \mathop{\mathbb{E}}_{x_S^i \sim \mathcal{P}_S^i} \frac{1}{C_\#} \sum_{c=1}^{C_\#} ((1 + \mathcal{H}_p)y_s^i \log(DC^i(EVG_S^i))) \tag{10}$$

$$\mathcal{L}_{ed} = \frac{d(EVG_S^i, EVG_T^i)}{d(EVG_S^i|c=1, EVG_S^i|c=0)} \tag{11}$$

$$\mathcal{L} = \mathcal{L}_{DC}^w + \lambda((\mathcal{L}_{adv})_S + (\mathcal{L}_{adv})_T + \mathcal{L}_{ed}) \tag{12}$$

where DC^i represents the classifier in discriminator D^i, $d(\cdot)$ represents the average Euclidean distance between two different embeddings, λ is a trade-off parameter.

4.2   Multi-source Adversarial Domain Adaptation

In this section, we extend ADAPT method to the multi-source domain adaptation via weighted source integration. The overall framework of multi-source adversarial domain adaptation is shown in Fig. 2(b). The input is the all domains data, including source domain $\mathcal{X}_S^1, ..., \mathcal{X}_S^k$ and the only target domain $\mathcal{X}_T$. In the testing phase, the output is the inference of the target domain data integrated by all classifiers. So, the final prediction of the target domain is:

$$\hat{y}_T = \sum_{i=1}^{k} w_i C^i(N^i(F^i(EV_T))) \tag{13}$$

The discriminator D^i can distinguish whether the sample is generated by G^i, and can also distinguish which domain the sample comes from. So, we can obtain the probability that the sample x_T comes from the ith source domain: $p_S^i = D^i(F^i(EV_T))$. Then we can get the weight w_i:

$$w_i = \frac{\exp(p_S^i)}{\sum_{i=1}^{k} \exp(p_S^i)} \tag{14}$$

4.3   The Training Method

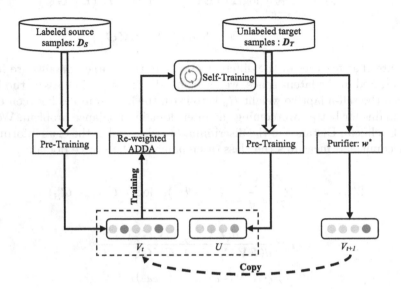

Fig. 3. The flow chart of the ADAPT method.

This section presents the details of the ADAPT method in the training and testing phase. As shown in Fig. 3, the first step is pre-training the feature extractor from both the labeled source domain $\mathcal{D}_S$ with the CRF task and the unlabeled target domain $\mathcal{D}_T$ with additional labeled task, and its goal is to achieve the optimal performance of the classifier on the evaluation task. Then, we can get dataset V_0 the initial source domain and fixed dataset U for the target domain. And the self-training in the block diagram is a parallel internal iterative loop process. Then, we can get the domain loss and the classification loss. The above loss would be fed back to guide the purifier w^*, which aims to select the samples in the source domain which are more similar to the samples in the target domain.

The details of building purifier module w^* for dataset $\mathcal{D}_S$ are as follows. In the tth step, we can cherry-pick dataset V_{t+1} for the source domain based on the dataset V_t and the discriminator D^i which can distinguish whether the sample comes from the source domain or the target domain. In the self-training phase, we will select the top $\eta\%$ samples from $\mathcal{D}_S$ based on the distance between the sample in the source domain and all the target domain.

$$\text{dis}(x_S^i, \mathcal{X}_T) = ||D^i(F^i(EV_S^i)) - \mathop{\mathbb{E}}_{x_T \sim \mathcal{P}_T} (D^i(F^i(EVG_T^i)))||^2 \quad (15)$$

$$w^*(\mathcal{X}_S^i) = \{x_S^i | x_S^i \in \mathcal{X}_S^i \wedge (\text{dis}(x_S^i, \mathcal{X}_T) - \text{dis}(x_S^i, \mathcal{X}_T)_{\eta\%}) < 0\} \quad (16)$$

where $\text{dis}(x_S^i, \mathcal{X}_T)_{\eta\%}$ is the distance from $\eta\%$ after all distances are sorted in ascending order within the ith source domain. To speed up the process of parameter search, multiple thresholds η with different models are trained at the same time. We would cherry-pick the best dataset defined as V_{t+1} and the corresponding ADAPT module. Finally, the model with the best evaluation results on classification loss and domain loss will be retained.

5   Experiments

5.1   Dataset

We collect a real-world dataset from an online E-Commerce consumer lending service provided by the Alibaba Group which services both personal and enterprise users. Specifically, in this paper, there are four platforms, SP1–3 denotes three different source domains consisting of domestic and foreign trade, respectively, and the TP denotes the brand new platform (target domain) that contains few financial service yet. Our finally task is to make use of the information of the SP1–3 as much as possible to predict credit risks of users in the TP.

Table 1. The statistics of feature sets used in the CRF task.

Description	Example	Dimension
Complaint rate	0.0, 0.1, 0.2	1
Refund rate	0.0, 0.1, 0.2	1
Registration time	1 m, 3 m, 4 m	1
Log.sales	1, 2, 3	1
Member type	0 A, 1 A, 2 A	6
...	...	...

The features we used in the experiments are all tabular expert characteristics collecting from different platforms, and their sketched descriptions are detailed in Table 1. It is worth noting that the interval between the training set and the test set should not be less than one month because the data of the next month is required when defining the labels of the CRF task. The sample in both source and target domain for the training is from 2019/07 to 2019/12 and for testing that is from 2020/02 to 2020/03, chronologically. The data distribution of different domains is described in Table 2.

Table 2. The statistical information of dataset in different domains.

Domain	#Total	#Pos	Pos_rate	Dimension	Transaction cycle
SP1	725,000	890	0.12%	389	9
SP2	513,000	1,109	0.22%	177	5
SP3	389,000	550	0.14%	113	45
TP	115,000	81	0.07%	215	30

5.2 Baselines and Compared Methods

The following describes the tree-based methods that are representative in the industry and the latest state-of-the-art methods.

(1) Target-only based method.

- **FDIF**: Feedback Isolation Forest (FDIF) is a tree-based semi-supervised anomaly detection approach [16]. It reduces the effort of the model by incorporating feedback on whether they are interested in investigating anomalies.
- **RUS**: It is the random majority undersampling (RUS) method, which is widely used in general class-imbalance problem [4,8]. In this spaper, we use GBDT as the base classifier in the RUS method, which is a competitive supervised gradient boosting approach that has been widely used in industry [8].
- **TRUST**: Trainable undersampling with self training (TRUST) is a semi-supervised meta-learning based approach which can effectively use the unlabeled data by self-training to overcome the class-imbalance problem [2].

(2) Multi-source based method.

- **HEN**: Hierarchical Explainable Network (HEN) can model users' behavior sequences and improve the performance of fraud detection [26]. The method of feature extraction and pre-training is the same as in this paper. All source domains are mixed together for domain adaptation.
- **MDDA**$_{\mathcal{H}_p}$: Multi-source distilling domain adaptation (MDDA) investigates the different similarities of the source samples to the target ones and the distance between multiple sources and the target [23]. This method cannot be directly reused in our experiments due to the lack of consideration for class-imbalance. So, we strengthen the MDDA by adding self-adaptive weight $\mathcal{H}_p$ and get **MDDA**$_{\mathcal{H}_p}$.
- **ADAPT**$_{\backslash \mathcal{H}_p}$, **ADAPT**$_{\backslash PT}$: Two submodels of ADAPT by removing self-adaptive weight $\mathcal{H}_p$ and purifier training, respectively.
- **ADAPT**: Our proposed full method. We investigate some problems in cross-domain credit risk forecasting and propose an unsupervised transfer learning and purifier training based approach.

5.3 Implementation Details

To demonstrate the superiority of our method, the AUC is adopted to measure the performance of different experiments in this paper, which is widely used in the class-imbalance problem. AUC metric is a common evaluation index to measure the quality of binary classification models. The higher AUC indicates the better performance of the method.

We implement the methods on the Pytorch platform and choose Adam [6] as the optimizer. We set the batch size to 256, the learning rate to 0.005, and the trade-off parameter $\lambda = 0.3$, the thresholds $\eta = 50$ by grid-searching. The number of trees in FDIF and GBDT is 300.

5.4   Main Results in the CRF Task

We compared performances with different target-only based methods and recent state-of-the-art domain adaptation-based methods to demonstrate the effectiveness of our method. Table 3 demonstrates the performances of all compared methods. The main findings are summarized as follows.

Table 3. Comparisons of different methods in various tasks.

Task	FDIF	RUS	TRUST	HEN	MDDA$_{\mathcal{H}_p}$	ADAPT
Target-only	0.700	0.710	0.793	–	–	–
Single-source best	–	–	–	0.853	0.873	**0.883**
Double-sources best	–	–	–	0.867	0.885	**0.910**
All sources	–	–	–	0.879	0.894	**0.915**

We first consider a series of target-only based methods. As can be seen from the first line in Table 3, the AUC of the ADAPT (0.915) improves 30.71% compared with the tree-based methods FDIF. The improvements are 28.87% and 15.38% compared with the undersamping based methods RUS and TRUST respectively. The improvement is shrinking step by step, which shows that it is very important to deal with the class imbalance problem in this paper. The last column shows that the performance of the ADAPT method increases as the number of source domains increases. The final result improves about 3.62%. Comparing the results of HEN and ADAPT, it is found that mixing all domains directly with training is not conducive to the combination of information in multiple source domains. The improvement shows that the ADAPT can better integrate information from different source domains and select appropriate source domain samples. The AUC can continue improve by 2.01% via adding the purifier training instead of distilling samples only in MDDA$_{Hp}$. This shows that purifier training can better cherry-pick the samples in the source domain that are more similar to the target domain. In general, the results of the ADAPT are optimal regardless of the average performance or overcoming the class-imbalanced domain adaptation problem under different situations.

5.5   Ablation Test

As shown in Table 4, when removing any component of the ADAPT, all metrics deteriorate to some extent. Specifically, self-adaptive weight $\mathcal{H}_p$ is more conducive, since the AUC of ADAPT$_{\backslash \mathcal{H}_p}$ is reduced by 10.11%. Furthermore, the AUC of ADAPT$_{\backslash PT}$ which removing the purifier training reduces by 2.01%. It exhibits that the purifier training can better cherry-pick the useful knowledge in the source domain.

Table 4. Ablation study of the ADAPT methods in combination of different source domains in the TP.

Method	SP1	SP2	SP3	SP1&2	SP1&3	SP2&3	SP1&2&3
ADAPT$_{\backslash \mathcal{H}_p}$	0.809	0.743	0.716	0.813	0.821	0.771	0.831
ADAPT$_{\backslash PT}$	0.866	0.800	0.750	0.873	0.890	0.828	0.897
ADAPT	**0.883**	0.830	0.769	0.890	**0.910**	0.844	**0.915**

Considering different data sources, the SP1 has the best effect on the TP domain. The results in the source domain combination of the SP1 and SP2 are not much better than that from the SP1 alone. We speculate that the SP1 and SP2 may have a lot of information that is redundant as they are domestic trade. On the contrary, the same analysis can show that the combination of the SP1 and SP3 can provide better information supplement because the AUC is increased by 3.06% and 18.33% respectively compared with the SP1 and SP3 alone. The same results can also be obtained from the combination of SP2 or SP3. The result of the combination of all source domains is the best, with a maximum increase of 27.79% over the single source domain. Besides, we can draw a conclusion that self-adaptive weight $\mathcal{H}_p$ plays a more important role than purifier training in our tasks. The above experimental results also show that the selected source domains should not have too much redundant information. Single domain with poor effect might provide important additional information when combining multi-source domains.

5.6   Result Visualization

In order to verify the effectiveness of the purifier training, we analyze the distribution of the samples in the SP1. To make the visualization more intuitive, we adopt the t-SNE method on the high-level features. In Fig. 4(a), the black lower triangles represent the samples in the SP1 which are more similar to the target domain TP. The location of the black lower triangle is where the source domain and target domain are highly similar. And in Fig. 4(b), the black lower triangles represent the positive samples. From Fig. 4(a) and Fig. 4(b), we can conclude that the few positive samples play a more important role during the domain adaption and the behaviors of the potential defaulters are relatively similar in

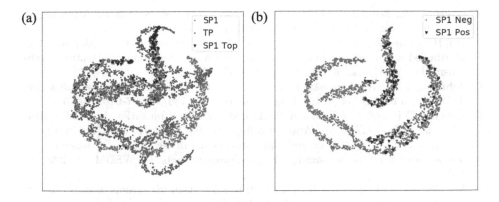

Fig. 4. (a) The t-SNE visualization of the samples in the SP1 and TP domains after the ADAPT method. **(b)** The t-SNE visualization of the samples in the SP1 domain after the ADAPT method.

different platforms. It also shows that our method can accomplish knowledge transfer focus on these limited but important positive samples.

6 Conclusion

In this paper, we investigated main challenges in the cross-domain credit risk forecasting task, namely intra-domain class-imbalance and inter-domain sample imbalance problems. We proposed a novel method named **ADAPT** to solve these special challenges. Furthermore, our model provided a general framework, in which unsupervised multi-source adversarial domain adaptation methods can be integrated. Various experiments on real-world industrial datasets were performed to evaluate the performance of our method. Experimental results in all of the experimental datasets demonstrate the superiority of the proposed method under different settings. Meanwhile, the effectiveness of the ADAPT method is demonstrated by series of visualizations.

Acknowledgment. The research work supported by Alibaba Group through Alibaba Innovative Research Program and the National Natural Science Foundation of China under Grant (No.61976204, 92046003, U1811461). Xiang Ao is also supported by the Project of Youth Innovation Promotion Association CAS, Beijing Nova Program Z201100006820062.

References

1. Altman, E.I.: Financial ratios, discriminant analysis and the prediction of corporate bankruptcy. J. Financ. **23**(4), 589–609 (1968)
2. Chi, J., et al.: Learning to undersampling for class imbalanced credit risk forecasting. In: ICDM, pp. 72–81 (2020)

3. Gretton, A., Borgwardt, K.M., Rasch, M.J., Schölkopf, B., Smola, A.: A kernel two-sample test. J. Mach. Learn. Res. **13**(1), 723–773 (2012)
4. Hu, B., Zhang, Z., Shi, C., Zhou, J., Li, X., Qi, Y.: Cash-out user detection based on attributed heterogeneous information network with a hierarchical attention mechanism. In: AAAI, vol. 33, no. 01, pp. 946–953 (2019)
5. Johnson, J.M., Khoshgoftaar, T.M.: Survey on deep learning with class imbalance. J. Big Data **6**(1), 1–54 (2019). https://doi.org/10.1186/s40537-019-0192-5
6. Kingma, D.P., Ba, J.: Adam: a method for stochastic optimization. In: ICLR (2015)
7. Kingma, D.P., Welling, M.: Auto-encoding variational Bayes. In: ICLR (2013)
8. Liang, T., et al.: Credit risk and limits forecasting in e-commerce consumer lending service via multi-view-aware mixture-of-experts nets. In: WSDM, pp. 229–237 (2021)
9. Lin, W., et al.: Online credit payment fraud detection via structure-aware hierarchical recurrent neural network. In: IJCAI (2021)
10. Liu, C., Sun, L., Ao, X., Feng, J., He, Q., Yang, H.: Intention-aware heterogeneous graph attention networks for fraud transactions detection. In: KDD, pp. 3280–3288 (2021)
11. Liu, C., et al.: Fraud transactions detection via behavior tree with local intention calibration. In: KDD, pp. 3035–3043 (2020)
12. Liu, Y., et al.: Pick and choose: a GNN-based imbalanced learning approach for fraud detection. In: WWW, pp. 3168–3177 (2021)
13. Liu, Y., Ao, X., Zhong, Q., Feng, J., Tang, J., He, Q.: Alike and unlike: resolving class imbalance problem in financial credit risk assessment. In: CIKM, pp. 2125–2128 (2020)
14. Malekipirbazari, M., Aksakalli, V.: Risk assessment in social lending via random forests. Expert Syst. Appl. **42**(10), 4621–4631 (2015)
15. Shen, J., Qu, Y., Zhang, W., Yu, Y.: Wasserstein distance guided representation learning for domain adaptation. In: AAAI (2018)
16. Siddiqui, M.A., Fern, A., Dietterich, T.G., Wright, R., Theriault, A., Archer, D.W.: Feedback-guided anomaly discovery via online optimization. In: KDD, pp. 2200–2209 (2018)
17. Tzeng, E., Hoffman, J., Saenko, K., Darrell, T.: Adversarial discriminative domain adaptation. In: CVPR, pp. 7167–7176 (2017)
18. Wang, C., Yu, Z., Zheng, H., Wang, N., Zheng, B.: CGAN-plankton: towards large-scale imbalanced class generation and fine-grained classification. In: ICIP, pp. 855–859 (2017)
19. Wang, D., et al.: A semi-supervised graph attentive network for financial fraud detection. In: ICDM, pp. 598–607 (2019)
20. Wang, S., Zhang, L.: Self-adaptive re-weighted adversarial domain adaptation. In: IJCAI (2020)
21. Xu, M., et al.: Adversarial domain adaptation with domain mixup. In: AAAI, vol. 34, no. 4, pp. 6502–6509 (2020)
22. Zhao, H., Zhang, S., Wu, G., Moura, J.M., Costeira, J.P., Gordon, G.J.: Adversarial multiple source domain adaptation. In: NeurIPS (2018)
23. Zhao, S., et al.: Multi-source distilling domain adaptation. In: AAAI, vol. 34, no. 7, pp. 12975–12983 (2020)
24. Zheng, W., Zhao, H.: Cost-sensitive hierarchical classification for imbalance classes. Appl. Intell. **50**(8), 2328–2338 (2020). https://doi.org/10.1007/s10489-019-01624-z

25. Zhong, Q., et al.: Financial defaulter detection on online credit payment via multi-view attributed heterogeneous information network. In: WWW, pp. 785–795 (2020)
26. Zhu, Y., et al.: Modeling users' behavior sequences with hierarchical explainable network for cross-domain fraud detection. In: WWW, pp. 928–938 (2020)

Poisoning Attacks on Fair Machine Learning

Minh-Hao Van[1] , Wei Du[1] , Xintao Wu[1(✉)] , and Aidong Lu[2]

[1] University of Arkansas, Fayetteville, AR 72701, USA
{haovan,wd005,xintaowu}@uark.edu
[2] University of North Carolina at Charlotte, Charlotte, NC 28223, USA
Aidong.Lu@uncc.edu

Abstract. Both fair machine learning and adversarial learning have been extensively studied. However, attacking fair machine learning models has received less attention. In this paper, we present a framework that seeks to effectively generate poisoning samples to attack both model accuracy and algorithmic fairness. Our attacking framework can target fair machine learning models trained with a variety of group based fairness notions such as demographic parity and equalized odds. We develop three online attacks, adversarial sampling, adversarial labeling, and adversarial feature modification. All three attacks effectively and efficiently produce poisoning samples via sampling, labeling, or modifying a fraction of training data in order to reduce the test accuracy. Our framework enables attackers to flexibly adjust the attack's focus on prediction accuracy or fairness and accurately quantify the impact of each candidate point to both accuracy loss and fairness violation, thus producing effective poisoning samples. Experiments on two real datasets demonstrate the effectiveness and efficiency of our framework.

Keywords: Poisoning attacks · Algorithmic fairness · Adversarial machine learning.

1 Introduction

Both fair machine learning and adversarial machine learning have received increasing attention in past years. Fair machine learning (FML) aims to learn a function for a target variable using input features, while ensuring the predicted value be fair with respect to some sensitive attributes based on given fairness criterion. FML models can be categorized into pre-processing, in-processing, and post-processing (see a survey [13]). Adversarial machine learning focuses on vulnerabilities in machine learning models and has been extensively studied from perspectives of attack settings and defense strategies (see surveys [4,21]).

There have been a few works on attacking FML models very recently. Solans et al. [18] developed a gradient-based poisoning attack to increase demographic disparities among different groups. Mehrabi et al. [14] also focused on demographic disparity and presented anchoring attack and influence attack. Chang

A. Bhattacharya et al. (Eds.): DASFAA 2022, LNCS 13245, pp. 370–386, 2022.
https://doi.org/10.1007/978-3-031-00123-9_30

et al. [5] focused on attacking FML models with equalized odds. To tackle the challenge of intractable constrained optimization, they developed approximate algorithms for generating poisoning samples. However, how to effectively generate poisoning samples to attack algorithmic fairness still remains challenging due to its difficulty of quantifying impact of each poisoning sample to accuracy loss or fairness violation in the trained FML model.

In this paper, we present a poisoning sample based framework (PFML) for attacking fair machine learning models. The framework enables attackers to adjust their attack's focus on either decreasing prediction accuracy or increasing fairness violation in the trained FML model. Our framework supports a variety of group based fairness notions such as demographic parity and equalized odds. We present three training-time attacks, adversarial sampling, adversarial labeling, and adversarial feature modification. All of these attacks leave the test data unchanged and instead perturb the training data to affect the learned FML model. In adversarial sampling, the attacker is restricted to select a subset of samples from a candidate attack dataset that has the same underlying distribution of the clean data. Adversarial labeling and adversarial feature modification can further flip the labels or modify features of selected samples. All three developed attacking methods are online attacks, which are more efficient than those offline poisoning attacks. Our framework enables attackers to flexibly adjust the attack's focus on prediction accuracy or fairness and accurately quantify the impact of each candidate point to both accuracy loss and fairness violation, thus producing effective poisoning samples. Experiments on two real datasets demonstrate the effectiveness and efficiency of our framework.

2   Background

2.1   Fair Machine Learning

Consider a binary classification task $f_\theta : \mathcal{X} \to \mathcal{Y}$ from an input $x \in \mathcal{X}$ to an output $y \in \mathcal{Y}$. Let $l : \Theta \times \mathcal{X} \times \mathcal{Y} \to \mathbb{R}_+$ be a loss function, $\mathcal{D}$ be the training set and each $(x, y) \in \mathcal{D}$ be a data point. The classification model minimizes, $\mathcal{L}(\theta, \mathcal{D}) = \sum_{(x,y) \in \mathcal{D}} l(\theta; x, y)$, the cumulative loss of the model over the training data set $\mathcal{D}$, to obtain the optimal parameters. Without loss of generality, we assume $\mathcal{X}$ contains one binary sensitive feature $S \in \{0, 1\}$. FML aims to train a model such that its predictions are fair with respect to S based on a given fairness notion, e.g., disparate impact, equal opportunity and equalized odds.

Definition 1. *A binary classifier f_θ is δ-fair under a fairness notion Δ if $\Delta(\theta, \mathcal{D}) \leq \delta$, where $\Delta(\theta, \mathcal{D})$ is referred as the empirical fairness gap of the model and δ is a user-specified threshold. The model satisfies exact fairness when $\delta = 0$.*

Definition 2. *We denote demographic parity and equalized odds as Δ_{DP} and Δ_{EO}, respectively. They are defined as:*

$$\Delta_{DP}(\theta, \mathcal{D}) := |\Pr(f_\theta(X) = 1 \, S = 1) - \Pr(f_\theta(X) = 1 \, S = 0)| \tag{1}$$

$$\Delta_{EO}(\theta, \mathcal{D}) := \max_{y \in \{0,1\}} |Pr[f_\theta(X) \neq y | S = 0, Y = y] - Pr[f_\theta(X) \neq y | S = 1, Y = y]| \tag{2}$$

Algorithm 1. Online Learning for Generating Poisoning Data

Require: $\mathcal{D}_c$, $n = |\mathcal{D}_c|$, feasible poisoning set $\mathcal{F}(\mathcal{D}_k)$, number of poisoning data ϵn, learning rate η.

Ensure: Poisoning dataset $\mathcal{D}_p$.
1: Initialize $\theta^0 \in \Theta$, $\mathcal{D}_p \leftarrow Null$
2: **for** $t = 1 : \epsilon n$
3:   $(x^t, y^t) \leftarrow argmax_{(x,y) \in \mathcal{F}(\mathcal{D}_k)} [l(\theta^{t-1}; x, y)$
4:   $\mathcal{D}_p \leftarrow \mathcal{D}_p \cup \{(x^t, y^t)\}$, $\mathcal{F}(\mathcal{D}_k) \leftarrow \mathcal{F}(\mathcal{D}_k) - \{(x, y)\}$
5:   $\theta^t \leftarrow \theta^{t-1} - \eta \frac{\nabla \mathcal{L}(\theta^{t-1}; \mathcal{D}_c \cup \mathcal{D}_p)}{n+t}$
6: **end for**

Demographic parity requires that the predicted labels are independent of the protected attribute. Equalized odds [9] requires the protected feature S and predicted outcome $\hat{Y}$ are conditionally independent given the true label Y. Equalized opportunity is a weaker notion of equalized odds and requires non-discrimination only within the advantaged outcome group. Our framework naturally covers equalized opportunity. The FML model achieves δ-fairness empirically by minimizing the model's empirical accuracy loss under the fairness constraint:

$$\hat{\theta} = \arg \min_{\theta \in \Theta} \frac{1}{|\mathcal{D}|} \mathcal{L}(\theta; \mathcal{D}) \text{ s.t. } C(\theta, \mathcal{D}) = \Delta(\theta, \mathcal{D}) - \delta \leq 0 \tag{3}$$

2.2   Data Poisoning Attack

Data poisoning attacks [2,3,15] seek to increase the misclassification rate for test data by perturbing the training data to affect the learned model. The perturbation can generally include inserting, modifying or deleting points from the training data so that the trained classification model can change its decision boundaries and thus yields an adversarial output. The modification can be done by either directly modifying the labels of the training data or manipulating the input features depending on the adversary's capabilities. In this study, we assume that an attacker can access to the training data during the data preparation process and have the knowledge of the structure and fairness constraint of the classification model. We focus on three data poisoning attacks, adversarial sampling, adversarial labeling, and adversarial feature modification, against group-based FML models. In all three attacks, the adversary can select the feature vector of the poisoning data from an attack dataset $\mathcal{D}_k$, which is sampled from the same underlying distribution of the clean dataset $\mathcal{D}_c$, and can control sampling, labeling, or modifying for a fraction of training data in order to reduce the test accuracy.

Algorithm 1 shows the general online gradient descent algorithm for generating poisoning samples. The input parameter n denotes the size of the clean set $\mathcal{D}_c$, ϵ is the fraction of the size of generated poisoning data over the clean data in the training data set, $\mathcal{F}(\mathcal{D}_k)$ is feasible poisoning set. Specifically, $\mathcal{F}(\mathcal{D}_k)$ is the same as $\mathcal{D}_k$ for adversarial sampling. A fraction of data points $(x, y) \in \mathcal{D}_k$

are changed to $(x, 1 - y)$ for adversarial labeling, and to $(\tilde{x}, y)$ for adversarial feature modification where $\tilde{x}$ is a modified version of feature vector x. In line 1, it first initializes the model with $\theta^0 \in \Theta$. Using the feasible set of poisoning points, the algorithm generates ϵn poisoning data points iteratively. In line 3, it selects a data point with the highest impact on the loss function with respect to θ^{t-1}. In line 4, it adds the generated data point to $\mathcal{D}_p$. In line 5, the model parameters θ are updated to minimize the loss function based on the selected data point (x^t, y^t).

3 Data Poisoning Attack on FML

3.1 Problem Formulation

The attacker's goal is to find a poisoning dataset that maximizes the linear combination of the accuracy loss and the model's violation from the fairness constraint. The fairness constraint is defined as $C(\theta, \mathcal{D}) = \Delta(\theta, \mathcal{D}) - \delta \leq 0$. We formulate the data poisoning attack on algorithmic fairness as a bi-level optimization problem:

$$\max_{\mathcal{D}_p} \mathbb{E}_{(x,y)}[\alpha \cdot l(\hat{\theta}; x, y) + (1 - \alpha) \cdot \gamma \cdot l_f(\hat{\theta}; x, y)]$$

$$\text{where} \quad \hat{\theta} = \arg\min_{\theta \in \Theta} \frac{\mathcal{L}(\theta; \mathcal{D}_c \cup \mathcal{D}_p)}{|\mathcal{D}_c \cup \mathcal{D}_p|} \tag{4}$$

$$\text{s.t.} \quad C(\theta, \mathcal{D}_c \cup \mathcal{D}_p) = \Delta(\theta, \mathcal{D}_c \cup \mathcal{D}_p) - \delta \leq 0$$

where $\alpha \in [0, 1]$ is a hyperparameter that controls the balance of the attack's focus on accuracy and fairness, $l(\hat{\theta}; x, y)$ is the prediction accuracy loss of the sample (x, y), $l_f(\hat{\theta}; x, y)$ is the fairness loss, and γ is a hyperparameter to have l_c and l_f at the same scale.

We can solve Eq. 4 by optimizing user and attacker's objectives separately. Intuitively, the user (inner optimization) minimizes the classification loss subject to fairness constraint. The attacker (outer optimization) tries to maximize the joint loss $\mathbb{E}_{(x,y)}[\alpha \cdot l(\hat{\theta}; x, y) + (1 - \alpha) \cdot \gamma \cdot l_f(\hat{\theta}; x, y)]$ by creating a poisoning set $\mathcal{D}_p$ based on $\hat{\theta}$ obtained by the user to degrade the performance of classifier either from accuracy or fairness aspect. For example, if the value of α approaches to 1, then the attacker tends to degrade more on the accuracy of the model. Note that the loss expectation is taken over the underlying distribution of the clean data. Inspired by [5], we also approximate the loss function in the outer optimization via the loss on the clean training data and the poisoning data and have

$$\max_{\mathcal{D}_p}[\alpha \cdot \mathcal{L}(\hat{\theta}; \mathcal{D}_c \cup \mathcal{D}_p) + (1 - \alpha) \cdot \gamma \cdot \Delta(\hat{\theta}; \mathcal{D}_c \cup \mathcal{D}_p)] \tag{5}$$

The accuracy loss $\mathcal{L}$ and fairness loss Δ may be at different scales due to the use of different loss functions and data distribution. Figure 1 shows the curves of accuracy loss and fairness loss of equalized odds when we increase the generated

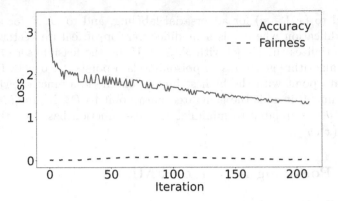

Fig. 1. Accuracy and fairness loss (in terms of equalized odds) with different iterations of PFML-AS ($\alpha = 0.8$) on COMPAS.

poisoning samples from 1 to 211 on COMPAS dataset (see experiment section for the detailed experimental setting). This shows the importance of introducing hyperparameter γ to have accuracy loss and fairness loss at the same scale.

The user tries to achieve optimal and fair $\hat{\theta}$ under the poisoning set $\mathcal{D}_p$. As the constrained optimization is intractable, we further transform the inner optimization to its dual form as the following:

$$\hat{\theta} = \min_{\theta \in \Theta} \left(\frac{1}{n+t} \mathcal{L}(\theta; \mathcal{D}_c \cup \mathcal{D}_p) + \lambda \Delta(\theta, \mathcal{D}_c \cup \mathcal{D}_p) \right) \tag{6}$$

where λ is the Lagrange multiplier and t is the current size of poisoning samples $\mathcal{D}_p$. By Eq. 5 and Eq. 6, we effectively capture the contribution of each poisoning point (x, y) to both accuracy loss and fairness gap.

3.2   Convex Relaxation of Fairness Constraint

The dual optimization problem in Eq. 6 involves the calculation of $\Delta(\theta, \mathcal{D}_c \cup \mathcal{D}_p)$ over the current $\mathcal{D}_c \cup \mathcal{D}_p$. However, fairness notions such as demographic parity and equalized odds are non-convex. We adopt simplifications proposed by [22] for demographic parity and [7] for equalized odds to reach convex relaxations of fairness constraints. Demographic parity can be approximated by the decision boundary fairness. The decision boundary fairness over $\mathcal{D}_c \cup \mathcal{D}_p$ is defined as the covariance between the sensitive attribute and the signed distance from the non-sensitive attribute vector to the decision boundary. It can be written as:

$$C(\theta, \mathcal{D}_c \cup \mathcal{D}_p) = \frac{1}{n+t} \sum_{i=1}^{n+t} (s_i - \bar{s}) d_{\theta(x_i)} \tag{7}$$

where t is the size of the current poisoning samples $\mathcal{D}_p$, s_i is the value of the sensitive attribute of the sample x_i, $d_{\theta(x_i)} = \theta^T x_i$ is the distance to the decision boundary of the classifier f_θ, $\bar{s}$ is the mean value of the sensitive attribute over $\mathcal{D}_c \cup \mathcal{D}_p$. We require that $|C(\theta, \mathcal{D}_c \cup \mathcal{D}_p)| \leq \tau$ to achieve fairness.

We adopt the fairness definition for equalized odds by balancing the risk among two sensitive groups. Let the linear loss be $\mathcal{L}_l$ (e.g., $\mathcal{L}_l = 0.5(1 - f_\theta(x))$ for SVM model) and denote $\mathcal{D} = \mathcal{D}_c \cup \mathcal{D}_p$. We can write down the convex relaxation for the fairness gap of equalized odds as the following:

$$C(\theta, \mathcal{D}) = \frac{1}{2} \sum_{y=0,1} |R^{y,s=0}(\theta, \mathcal{D}) - R^{y,s=1}(\theta, \mathcal{D})| \qquad (8)$$

where $R^{y,s}(\theta, \mathcal{D}) = \frac{1}{n^{y,s}} \sum_{(x,y) \in \mathcal{D}_{y,s}} \mathcal{L}_l(x, y; \theta)$. $\mathcal{D}_{y,s}$ is the dataset of points with group s and label y and $n^{y,s}$ is the size of $\mathcal{D}_{y,s}$. Similar to the approximation of equalized odds, we can use $C(\theta, \mathcal{D}) = |R^{y=1,s=0}(\theta, \mathcal{D}) - R^{y=1,s=1}(\theta, \mathcal{D})|$ for the convex relaxation of equalized opportunity.

Algorithm 2. Poisoning Attack on Fair Machine Learning (PFML)

Require: $\mathcal{D}_c$, $n = |\mathcal{D}_c|$, feasible poisoning set $\mathcal{F}(\mathcal{D}_k)$, number of poisoning data cn, penalty parameter (Lagranger multiplier) λ, learning rate η, scaling factor γ, balance ratio α, fairness notion Δ.

Ensure: Poisoning dataset $\mathcal{D}_p$.

1: Initialize $\theta^0 \in \Theta$

2: **for** $i = 1 : I$

3:     $\theta^i \leftarrow \theta^{i-1} - \eta \left(\frac{\nabla \mathcal{L}(\theta^{i-1}; \mathcal{D}_c)}{n} + \nabla \left[\lambda \Delta \left(\theta^{i-1}, \mathcal{D}_c \right) \right] \right)$

4: **end for**

5: $\theta^0 \leftarrow \theta^I$, $\mathcal{D}_p \leftarrow Null$

6: **for** $t = 1 : \epsilon n$

7:     $(x^t, y^t) \leftarrow argmax_{(x,y) \in \mathcal{F}(\mathcal{D}_k)} [\alpha \cdot l(\theta^{t-1}; x, y) +$

8:                  $(1 - \alpha) \cdot \gamma \cdot \Delta \left(\theta^{t-1}, \mathcal{D}_c \cup \mathcal{D}_p \cup \{(x,y)\} \right)]$

9:     $\mathcal{D}_p \leftarrow \mathcal{D}_p \cup \{(x^t, y^t)\}$, $\mathcal{F}(\mathcal{D}_k) \leftarrow \mathcal{F}(\mathcal{D}_k) - \{(x,y)\}$

10:    $\theta^t \leftarrow \theta^{t-1} - \eta \left(\frac{\nabla \mathcal{L}(\theta^{t-1}; \mathcal{D}_c \cup \mathcal{D}_p)}{n+t} + \nabla \left[\lambda \Delta \left(\theta^{t-1}, \mathcal{D}_c \cup \mathcal{D}_p \right) \right] \right)$

11: **end for**

3.3 Attack Algorithm

Algorithm 2 shows pseudo code of our poisoning attack framework on fair machine learning (PFML). Our three algorithms are denoted as PFML-AS for adversarial sampling, PFML-AF for adversarial flipping, and PFML-AM for adversarial feature modification. In each algorithm, we can adjust the attack's focus on prediction accuracy or fairness by choosing different α values. For example, when 1 (0), the attack's focus is purely on accuracy (fairness) and when 0.5, the focus is on the combination of fairness and accuracy. In line 2–4, we first train FML model on the clean data $\mathcal{D}_c$ and use the fitted parameter θ^I to start

Table 1. Test accuracy and fairness gap of fair reduction [1] and post-processing [9] with **equalized odds** under PFML and baselines (COMPAS).

Method	Accuracy					Fairness				
	Fair reduction (δ)				Post (δ)	Fair reduction (δ)				Post (δ)
	0.12	0.1	0.07	0.05	0	0.12	0.1	0.07	0.05	0
Benign	0.950	0.949	0.949	0.948	0.877	0.108	0.103	0.086	0.082	0.095
RS	0.936	0.930	0.919	0.912	0.839	0.101	0.105	0.104	0.103	0.081
LF	0.935	0.931	0.919	0.911	0.839	0.062	0.066	0.072	0.080	0.109
HE	0.915	0.908	0.899	0.891	0.829	0.076	0.082	0.100	0.109	0.131
INFL	0.850	0.848	0.845	0.841	0.653	0.089	0.081	0.078	0.081	0.054
KKT	0.890	0.891	0.891	0.886	0.701	0.136	0.137	0.137	0.142	0.096
min-max	0.891	0.887	0.878	0.874	0.678	0.096	0.125	0.089	0.075	0.082
AS	0.830	0.824	0.816	0.810	0.740	0.051	0.069	0.111	0.143	0.156
AF	0.823	0.817	0.808	0.803	0.740	0.046	0.059	0.100	0.130	0.136
PFML-, α										
AS, 0	0.853	0.847	0.833	0.802	0.753	0.126	0.148	0.164	0.185	0.190
AS, 0.2	0.843	0.837	0.820	0.792	0.728	0.112	0.124	0.138	0.127	0.188
AS, 0.5	0.824	0.820	0.814	0.809	0.705	0.110	0.118	0.130	0.142	0.143
AS, 0.8	0.820	0.816	0.809	0.800	0.715	0.101	0.105	0.116	0.120	0.099
AS, 1.0	0.811	0.807	0.800	0.796	0.724	0.083	0.071	0.061	0.061	0.074
AF, 0	0.847	0.841	0.832	0.805	0.752	0.120	0.144	0.172	0.184	0.193
AF, 0.2	0.843	0.838	0.817	0.792	0.728	0.107	0.117	0.125	0.126	0.186
AF, 0.5	0.818	0.814	0.808	0.804	0.711	0.101	0.110	0.126	0.139	0.136
AF, 0.8	0.804	0.797	0.791	0.786	0.714	0.093	0.090	0.097	0.107	0.090
AF, 1.0	0.803	0.797	0.794	0.788	0.722	0.088	0.068	0.043	0.039	0.097
AM, 0	0.908	0.906	0.904	0.897	0.764	0.195	0.200	0.215	0.207	0.198
AM, 0.2	0.811	0.805	0.798	0.794	0.731	0.102	0.086	0.076	0.076	0.153
AM, 0.5	0.793	0.788	0.780	0.775	0.688	0.079	0.059	0.071	0.080	0.124
AM, 0.8	0.791	0.789	0.782	0.773	0.696	0.082	0.055	0.053	0.077	0.096
AM, 1.0	0.828	0.823	0.817	0.813	0.696	0.063	0.045	0.055	0.073	0.076

generating poisoning samples. We then execute the loop of line 6–9 to iteratively generate ϵn poisoning samples. In line 7, when generating the data point (x^t, y^t) with highest impact on a weighted sum of the accuracy loss and the fairness violation with respect to θ^{t-1}, we add both the previously generated data points in $\mathcal{D}_p$ and the data point (x^t, y^t) to $\mathcal{D}_c$. As a result, we can measure the incremental contribution of that data point to the fairness gap $\Delta\left(\theta^{t-1}, \mathcal{D}_c \cup \mathcal{D}_p \cup \{(x, y)\}\right)$. Note that in this step, the accuracy loss can be simply calculated over each point $(x, y) \in \mathcal{F}(\mathcal{D}_k)$ as the accuracy loss of existing data points from $\mathcal{D}_c \cup \mathcal{D}_p$ is unchanged. In line 8, we add the chosen poisoning point (x^t, y^t) to $\mathcal{D}_p$ and also remove it from the feasible poisoning set. In line 9, when updating the model parameters θ, we minimize the penalized loss function over $\mathcal{D}_c$ and $\mathcal{D}_p$. We see the execution time is mostly spent on line 6–11. In fact, line 9 and line 10 only involve one time operation. The time complexity of line 8 is $\mathcal{O}(m)$, where m is

Table 2. Test accuracy and fairness gap of fair reduction and post-processing with **demographic parity** (COMPAS).

Method	Accuracy					Fairness				
	Fair reduction (δ)				Post (δ)	Fair reduction (δ)				Post (δ)
	0.12	0.1	0.07	0.05	0	0.12	0.1	0.07	0.05	0
Benign	0.887	0.867	0.803	0.768	0.859	0.175	0.169	0.107	0.095	0.046
RS	0.882	0.839	0.813	0.767	0.867	0.187	0.155	0.130	0.076	0.023
LF	0.890	0.852	0.814	0.775	0.868	0.194	0.166	0.138	0.099	0.021
HE	0.901	0.859	0.808	0.766	0.840	0.205	0.181	0.135	0.098	0.036
INFL	0.879	0.855	0.774	0.748	0.784	0.200	0.186	0.097	0.108	0.015
KKT	0.884	0.875	0.788	0.768	0.817	0.221	0.214	0.127	0.136	0.016
min-max	0.870	0.870	0.843	0.818	0.810	0.201	0.204	0.182	0.167	0.036
PFML-, α										
AS, 0	0.853	0.829	0.771	0.750	0.824	0.195	0.168	0.109	0.099	0.041
AS, 0.2	0.847	0.819	0.766	0.736	0.798	0.189	0.171	0.106	0.100	0.039
AS, 0.5	0.844	0.812	0.763	0.731	0.795	0.182	0.167	0.101	0.092	0.038
AS, 0.8	0.845	0.811	0.758	0.731	0.791	0.175	0.166	0.096	0.094	0.036
AS, 1.0	0.829	0.816	0.757	0.722	0.790	0.171	0.151	0.083	0.075	0.032
AF, 0	0.848	0.822	0.786	0.761	0.822	0.192	0.185	0.098	0.080	0.057
AF, 0.2	0.841	0.805	0.766	0.742	0.806	0.188	0.163	0.095	0.086	0.056
AF, 0.5	0.842	0.809	0.762	0.733	0.801	0.174	0.136	0.087	0.086	0.036
AF, 0.8	0.838	0.803	0.755	0.729	0.798	0.167	0.134	0.086	0.079	0.027
AF, 1.0	0.831	0.808	0.752	0.721	0.793	0.160	0.132	0.082	0.069	0.032
AM, 0	0.883	0.853	0.816	0.791	0.833	0.246	0.219	0.183	0.159	0.031
AM, 0.2	0.840	0.820	0.762	0.730	0.814	0.218	0.208	0.138	0.128	0.038
AM, 0.5	0.838	0.802	0.757	0.733	0.793	0.212	0.170	0.120	0.114	0.030
AM, 0.8	0.826	0.800	0.758	0.720	0.787	0.193	0.147	0.115	0.065	0.031
AM, 1.0	0.853	0.805	0.767	0.726	0.815	0.184	0.138	0.105	0.060	0.029

the size of feasible poisoning set $\mathcal{F}(\mathcal{D}_k)$. Therefore, the time complexity of the loop from line 6–11 is $\mathcal{O}(\epsilon nm)$. In practice, the size of $\mathcal{F}(\mathcal{D}_k)$ is fixed, and we can simplify time complexity as $\mathcal{O}(cn)$.

Remarks. Chang et al. [5] presented an online gradient descent algorithm that generates poisoning data points for fair machine learning model with equalized odds. As the fairness gap is not an additive function of the training data points, they used $\mathcal{D}_c \cup \{(x^t, y^t)^{\epsilon n}\}$ (denoted as $\mathcal{D}_t$) to measure the contribution of that data point to the fairness gap where $\mathcal{D}_t$ is equivalent to adding ϵn copies of (x^t, y^t) to $\mathcal{D}_c$. The weighted loss function used for selecting poisoning samples is shown as $\left[\epsilon \cdot l\left(\theta^{t-1}; x, y\right) + \lambda \cdot \Delta\left(\theta^{t-1}, \mathcal{D}_t\right)\right]$. The algorithm then updates the model parameters θ via the gradient descent, i.e., $\theta^t \leftarrow \theta^{t-1} - \eta\left(\frac{\nabla \mathcal{L}\left(\theta^{t-1}; \mathcal{D}_c\right)}{n} + \nabla\left[\epsilon \cdot l\left(\theta^{t-1}; x^t, y^t\right) + \lambda \cdot \Delta\left(\theta^{t-1}, \mathcal{D}_t\right)\right]\right)$. However, both the use of $\mathcal{D}_c \cup \{(x, y)^{\epsilon n}\}$ to quantify the (x^t, y^t)'s contribution to the fairness gap and the use of parameters (ϵ and λ) to define the weighted loss are heuristic, thus hard to produce effective poisoning samples on algorithmic fairness.

Moreover, different from [5] that covers only a single fairness notion (i.e., equalized odds) and two attacks (adversarial sampling and adversarial label flipping), our paper presents a general framework with algorithms for three group based fairness notions and a new important adversarial feature modification attack. In our evaluation, we compare our methods with [5] and three new state-of-the-art baselines (influence attack, KKT, and min-max attack) from [10].

4   Experiments

Datasets. We conduct our experiments on COMPAS [11] and Adult [8] which are two benchmark datasets for FML community. COMPAS is a collection of personal information such as criminal history, demographics, jail and prison time. Adult is also a collection of individual's information including gender, race, martial status, and so forth. The task for COMPAS is a binary classification to predict whether the individual will be re-offended based on personal information, while the task for Adult dataset is to predict if an individual's annual income will be over \$50k based on his personal information. We use race (only black/white) as the sensitive attribute for COMPAS and gender as sensitive attribute for Adult. After preprocessing, COMPAS has 5278 data points and 11 features, while Adult has 48842 data points and 14 features. For each dataset, we first train a SVM model on the entire dataset. For the 60% data with the smallest loss, we randomly split them into clean dataset $\mathcal{D}_c$, attack candidate dataset $\mathcal{D}_k$, and test dataset $\mathcal{D}_{test}$ with ratio 4:1:1. The rest 40% data is treated as hard examples and added into $\mathcal{D}_k$. For COMPAS, $\mathcal{D}_c$ contains 2111 samples and $\mathcal{D}_{test}$ has 528 samples. $\mathcal{D}_k$ has 2639 samples including 2112 hard examples. For adversarial label flipping, we randomly flip the label of 15% data from $\mathcal{D}_k$ to build the feasible poisoning candidate set $\mathcal{F}(\mathcal{D}_k)$. For adversarial feature modification, we randomly flip one binary feature of each data point from $\mathcal{D}_k$ and include them to $\mathcal{F}(\mathcal{D}_k)$. Following the similar pre-processing strategy, $\mathcal{D}_c$, $\mathcal{D}_{test}$, and $\mathcal{D}_k$ of Adult contain 15385 samples, 6594 samples, and 26863 samples, respectively. Due to space limit, we report detailed results of COMPAS in the majority of this experiment section and only show the summarized results of Adult in Fig. 2 at the end of this experiment section.

Baselines. We consider the following baselines: (a) Random Sampling (RS): attacker randomly selects data samples from $\mathcal{D}_k$; (b) Label Flipping (LF): attacker randomly selects data samples from $\mathcal{D}_k$ and flips their labels; (c) Hard Examples (HE): attacker randomly selects data samples from hard examples set; (d) influence attack (INFL), (e) KKT attack, (f) min-max attack, (g) adversarial sampling (AS), and (h) adversarial flipping (AF). Attacks (d)–(f) are stronger data poisoning attacks breaking data sanitization defenses and all control both the label y and input features x of the poisoned points [10]. Attacks (g) and (h) are designed for attacking FML from [5]. As attacks (g) and (h) are only designed for equalized odds, we exclude them from baselines when reporting comparisons based on demographic parity.

Fair Classification Models. We use SVM as the classification model and choose fair reduction [1] and post-processing [9] as FML under attack. Post-processing adjusts an unconstrained trained model to remove discrimination based on fairness notions such as demographic parity and equalized odds. After adjustment, the unconstrained model behaves like a randomized classifier that assigns each data point a probability conditional on its protected attribute and predicted label. These probabilities are calculated by a linear program to minimize the expected loss. Fair reduction is an advanced in-processing approach that reduces fair classification to a sequence of cost-sensitive classification and achieves better accuracy-fairness tradeoff than previous FML models. Hence, we do not report results from other in-processing FML models.

Hyperparameters. In our default setting, we choose the number of pretrain steps with $\mathcal{D}_c$ as 2000, learning rate lr as 0.001, penalty parameter λ as 5, and ϵ as 0.1. The scaling factor γ is calculated as the ratio of accuracy loss and fairness loss over $\mathcal{D}_c$. **Metrics.** We run our attacks, PFML-AS, PFML-AF and PFML-AM, each with five α values, and baseline attacks to generate the poisoning data $\mathcal{D}_p$ and then train fair reduction (with four δ values as fairness threshold) and post-processing models with $\mathcal{D}_c \cup \mathcal{D}_p$. Finally we run the trained FML models on the test data $\mathcal{D}_{test}$ and report the test accuracy and fairness gap. For each experiment, we report the average value of five runs. Due to space limit, we skip reporting their standard deviation and instead we summarize comparisons based on t-test.

Reproducibility. All datasets, source code and setting details are released in GitHub with https://github.com/minhhao97vn/PFML for reproducibility.

4.1   Evaluation of PFML with Equalized Odds

Table 1 shows the comparison of our PFML attacks under different α with other baseline models in terms of both accuracy and fairness on two FML models (fair reduction and post-processing) trained with equalized odds under different fairness threshold values of δ. In each cell of Table 1, we report the average value of five runs. Due to space limit, we skip reporting their standard deviation and instead we summarize comparisons based on t-test at the end of this subsection.

First, the accuracy of FML model under all three PFML attacks (PFML-AS, PFML-AF and PFML-AM) is significantly lower than the benign case. For each fixed δ, both the accuracy value and fairness gap of FML under PFML attacks decrease when we increase α. Recall that larger α indicates that PFML attacks more on accuracy and smaller α indicates more attack's focus on fairness. Note that larger fairness gap caused by smaller α indicates higher model unfairness. Taking PFML-AS as an example, the accuracy of fair reduction with $\delta = 0.12$ is 0.853 and the fairness gap is 0.126 when $\alpha = 0$; the accuracy is 0.811 and the fairness gap is 0.083 when $\alpha = 1$. This result demonstrates that controlling α is flexible and effective for attackers to tune attack target on either prediction accuracy or fairness. Second, PFML-AF and PFML-AM outperform PFML-AS in terms of attacking performance from both accuracy and fairness perspectives, which shows modifying input features or flipping labels is more powerful than adversarial sampling. Third, compared to RS, LF and HE, our PFML attacks can reduce more accuracy or incur more unfairness with the same δ for both fair reduction and post-processing. Taking PFML-AS with $\delta = 0.12$ and $\alpha = 0$ as an example, the accuracy is 0.811, which is 0.125, 0.124, and 0.104 lower than that of RS, LF and HE, respectively. Compared to previous FML attacks (AS, AF) [5] and sanitization attacks (INFL, KKT, min-max) [10], our PFML attacks achieve better attack performance in terms of accuracy (fairness) with large (small) α values, which is consistent with our expectation. In particular, the accuracy of PFML-AS with $\alpha = 1$ and $\delta = 0.12$ is 0.811, which is 0.020 lower than AS. If we choose smaller α, the attack performance of PFML on accuracy under performs [5], which is consistent with our expectation.

We also notice, for each fixed α, the accuracy of fair reduction under our PFML attacks decreases when we decrease δ. The fairness gap of fair reduction under PFML-AS (PFML-AF) attack increases when we decrease δ, which indicates the fair reduction model is less robust or more vulnerable when stricter fairness constraint is enforced. However, the fairness gap of fair reduction under PFML-AM attack instead decreases along the decrease of δ. Theoretical analysis is needed to understand the robustness of fair reduction approach with equalized odds under different poisoning attacks.

4.2   Evaluation of PFML with Demographic Parity

Table 2 shows the comparison results of adversarial fair machine learning with demographic parity. Note that we do not compare with online FML attacks (AS, AF) as they do not support demographic parity. Generally we see similar patterns as equalized odds shown in Table 2. For each fixed δ, both the accuracy and fairness gap of FML models under all three PFML attacks decrease when we increase α. This is because smaller α means more attack's focus on fairness. Compared to RS, LF and HE, our PFML attacks can reduce more model accuracy of FML with the same δ for both fair reduction and post-processing. Compared to INFL, KKT and min-max attacks, PFML attacks achieve better attack performance in terms of accuracy drop (fairness gap) of FML models when we set large (small) α values.

For each fixed α, the accuracy of fair reduction under PFML attacks decreases when we decrease δ. This pattern is similar as equalized odds. However, the fairness gap of fair reduction has a clear decreasing trend when δ decreases, which is different from equalized odds. This result actually indicates the fair reduction model with stricter fairness requirement (small δ) is less vulnerable under poisoning attacks.

4.3   Sensitivity Analysis of Hyperparameters

In this section, we evaluate the sensitivity of PFML attacks under different hyperparameters. Table 3 shows the accuracy, fairness gap and execution time for COMPAS with equalized odds when we change the size of poisoning samples

ϵ against fair reduction. In all experiments, we fix $\delta = 0.07$ and $\alpha = 0.8$. In general, with increasing ϵ, the accuracy of fair reduction drops while its fairness gap increases when fair reduction is under each of our PMFL attacks. Note that larger ϵ corresponds to injecting more poisoning data points into the training data, thus causing more accuracy drop and unfairness of the trained FML model.

Table 3. Effects of ratio ϵ for fairness reduction with equalized odds (COMPAS).

Dataset		$\epsilon = 0.025$	$\epsilon = 0.05$	$\epsilon = 0.1$	$\epsilon = 0.15$
Accuracy	INFL	0.891	0.857	0.845	0.820
	KKT	0.912	0.899	0.891	0.884
	min-max	0.918	0.902	0.878	0.850
	PFML-AS	0.882	0.821	0.809	0.799
	PFML-AF	0.867	0.824	0.791	0.794
	PFML-AM	0.891	0.839	0.782	0.777
Fairness Gap	INFL	0.068	0.054	0.078	0.063
	KKT	0.086	0.102	0.137	0.158
	min-max	0.083	0.108	0.089	0.215
	PFML-AS	0.086	0.082	0.116	0.134
	PFML-AF	0.089	0.081	0.097	0.142
	PFML-AM	0.089	0.092	0.077	0.107
Exec. Time (s)	INFL	497.1	915.7	1569.1	2009.5
	KKT	1633.6	2903.3	5503.3	8400.0
	min-max	337.9	597.1	1137.6	1714.6
	PFML-AS	4.8	6.1	8.8	12.1
	PFML-AF	5.1	6.7	9.9	13.9
	PFML-AM	6.5	10.9	16.5	22.5

We also compare with baseline attack models (INFL, KKT, and min-max). We can see with the same ϵ our PFML attacks can degrade the model accuracy more than the baselines, and cause similar or higher level of model unfairness than the baselines in most scenarios. We also report the execution time in Table 3 and we can see the execution time of our PFML attacks increases linearly with increasing ϵ, which is consistent with our time complexity analysis in Algorithm 2. Compared to the baseline models, our PFML attacks takes two or three orders of magnitude less time to generate poisoning samples than baselines.

Table 4. Effects of penalty parameter λ on PFML-AS for post-processing with equalized odds (COMPAS).

	$\lambda = 1$	$\lambda = 5$	$\lambda = 10$	$\lambda = 15$	$\lambda = 50$	$\lambda = 150$
Accuracy	0.709	0.715	0.718	0.720	0.726	0.723
Fairness Gap	0.094	0.099	0.122	0.118	0.125	0.156

Table 4 shows the accuracy and fairness gap with equalized odds when we use PFML-AS ($\alpha = 0.8$) to attack post-processing FML [9] under different λ values. The post-processing approach has strict fairness constraint $\delta = 0$. We can see with larger λ, the PFML attack focuses more on attacking fairness, which leads to larger fairness gap and smaller accuracy drop of the FML model.

4.4 Significance Testing

For each experiment, we have run our methods and other baseline models five times as shown in all our tables. We apply independent two-sample t-test to compare each of three PFML models (with a given fairness notion and α) with each of baseline models in terms of accuracy reduction and fairness respectively. The t-test results show our PFML attacks significantly outperform baselines from both accuracy and fairness perspectives. Due to space limit, we only report our summarized results here. All p-values except three are less than 0.01 and the left three are still less than 0.1, which demonstrates statistical significance of our PFML methods over baselines.

4.5 Summarized Results of Adult Dataset

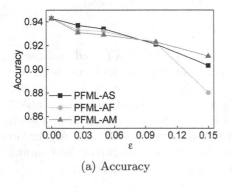

(a) Accuracy

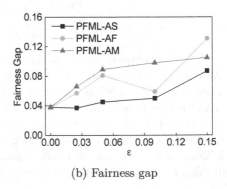

(b) Fairness gap

Fig. 2. Effects of ratio ϵ for fairness reduction with equalized odds (Adult).

We also report our summarized results on Adult with the setting of $\delta = 0.1$, $\alpha = 0.8$, and varied ϵ values[1]. Figures 2a and 2b plot the curves of accuracy and fairness gap for each of three PFML attacks with the increasing ϵ. The accuracy of fair reduction under PFML-AS, PFML-AF and PFML-AM attacks decreases when we increase ϵ. This pattern is consistent with our observation on COMPAS. As we analyzed previously, larger ϵ indicates stronger attack as more poisoning data are injected during the model training thus cause more performance degradation. Similarly, the fairness gap under PFML-AS, PFML-AF and PFML-AM generally increases with increasing ϵ. In terms of execution time, our PFML methods take from 187.4 s to 1399.2 s with increasing ϵ, which is significantly less than the baseline models (e.g., two orders of magnitude faster than min-max attack).

5 Related Work

The bulk of recent research on adversarial machine learning has focused on test-time attacks where the attacker perturbs the test data to obtain a desired classification. Train-time attacks leave the test data unchanged, and instead perturb the training data to affect the learned model. Data poisoning attacks are among the most common train-time attack methods in adversarial learning.

Barreno et al. [2] first proposed poisoning attacks which modify the training dataset to potentially change the decision boundaries of the targeted model. The modification can be done by either direct modifying the labels of the training data or manipulating the input features depending on the adversary's capabilities. Biggio et al. [3] developed an approach of crafting poisoning samples using gradient ascent. Shortly speaking, the method identifies the inputs corresponding to local maxima in the test error of the classification model. Mei et al. [15] developed a method that finds an optimal change to the training data when the targeted learning model is trained using a convex optimization loss and its input domain is continuous. Recent approaches include optimization-based methods [10] (e.g., influence, KKT, and min-max), poisoning Generative Adversarial Net (pGAN) model [16], and class-oriented poisoning attacks against neural network models [23]. The influence attack is a gradient-based attack that iteratively modifies each attack sample to increase the test loss, the KKT attack selects poisoned samples to achieve pre-defined decoy parameters, and the min-max attack efficiently solves for the poisoned samples that maximize train loss as a proxy for test loss. All three attacks control both the label and input features of the poisoned points. The structure of pGAN includes a generator, a discriminator, and an additional target classifier. The pGAN model generates poisoning data points to fool the model and degrade the prediction accuracy. Defense methods [10,19] typically require additional information, e.g., a labeled set of outliers or a clean set, and apply supervised classification to separate outliers from normal samples.

---

[1] We conduct experiments on Adult in other settings as COMPAS and observe similar patterns. We skip them due to space limit.

There have been a few works on attacking fair machine learning models very recently [5,14,17,18]. Solans et al. [18] introduced an optimization framework for poisoning attacks against algorithmic fairness and developed a gradient-based poisoning attack to increase classification disparities among different groups. Mehrabi et al. [14] also focused on attacking FML models trained with fairness constraint of demographic disparity. They developed anchoring attack and influence attack and focused on demographic disparity. Chang et al. [5] formulated the adversarial FML as a bi-level optimization and focused on attacking FML models trained with equalized odds. To tackle the challenges of the non-convex loss functions and the non-additive function of equalized odds, they further developed two approximate algorithms. Roh et al. [17] developed a GAN-based model that tries to achieve accuracy, fairness and robustness against adversary attacks.

6   Conclusions and Future Work

In this paper, we present a poisoning sample based framework that can attack model accuracy and algorithmic fairness. Our attacking framework can target fair machine learning models trained with a variety of group based fairness notions such as demographic parity and equalized odds. Our framework enables attackers to flexibly adjust the attack's focus on prediction accuracy or fairness and accurately quantify the impact of each candidate point to both accuracy loss and fairness violation, thus producing effective poisoning samples. We developed three online attacks, adversarial sampling, adversarial labeling, and adversarial feature modification. All three attacks effectively and efficiently produce poisoning samples via sampling, labeling, or modifying a fraction of training data in order to reduce the test accuracy. The three attacks studied in this paper are special cases of gradient-based attacks and belong to indiscriminate attacks. In our future work, we will extend our approach to other attacks, e.g., the targeted attacks that seek to cause errors on specific test examples. We will also investigate robust defense approaches against attacks on fair machine learning models, e.g., by applying multi-gradient algorithms for multi-objective optimization [6,12] and robust learning [20].

Acknowledgement. This work was supported in part by NSF grants 1564250, 1937010 and 1946391.

References

1. Agarwal, A., Beygelzimer, A., Dudík, M., Langford, J., Wallach, H.: A reductions approach to fair classification. In: International Conference on Machine Learning. pp. 60–69. PMLR (2018)
2. Barreno, M., Nelson, B., Sears, R., Joseph, A.D., Tygar, J.D.: Can machine learning be secure? In: Proceedings of the 2006 ACM Symposium on Information, computer and communications security. pp. 16–25 (2006)

3. Biggio, B., Nelson, B., Laskov, P.: Poisoning attacks against support vector machines. In: Proceedings of the 29th International Conference on Machine Learning, ICML 2012, Edinburgh, Scotland, UK, June 26 - July 1, 2012. icml.cc / Omnipress (2012)
4. Chakraborty, A., Alam, M., Dey, V., Chattopadhyay, A., Mukhopadhyay, D.: Adversarial attacks and defences: A survey. arXiv preprint arXiv:1810.00069 (2018)
5. Chang, H., Nguyen, T.D., Murakonda, S.K., Kazemi, E., Shokri, R.: On adversarial bias and the robustness of fair machine learning. arXiv preprint arXiv:2006.08669 (2020)
6. Désidéri, J.A.: Multiple-gradient descent algorithm (mgda) for multiobjective optimization. Comptes Rendus Mathematique 350(5–6), 313–318 (2012)
7. Donini, M., Oneto, L., Ben-David, S., Shawe-Taylor, J., Pontil, M.: Empirical risk minimization under fairness constraints. In: Bengio, S., Wallach, H.M., Larochelle, H., Grauman, K., Cesa-Bianchi, N., Garnett, R. (eds.) Advances in Neural Information Processing Systems 31: Annual Conference on Neural Information Processing Systems 2018, NeurIPS 2018, December 3–8, 2018, Montréal, Canada. pp. 2796–2806 (2018), https://proceedings.neurips.cc/paper/2018/hash/83cdcec08fbf90370fcf53bdd56604ff-Abstract.html
8. Dua, D., Graf, C.: Adult dataset. https://archive.ics.uci.edu/ml/datasets/adult (1994)
9. Hardt, M., Price, E., Srebro, N.: Equality of opportunity in supervised learning. In: Advances in neural information processing systems. pp. 3315–3323 (2016)
10. Koh, P.W., Steinhardt, J., Liang, P.: Stronger data poisoning attacks break data sanitization defenses. arXiv preprint arXiv:1811.00741 (2018)
11. Larson, J., Mattu, S., Kirchner, L., Angwin, J.: Compas dataset. https://github.com/propublica/compas-analysis (2017)
12. Liu, S., Vicente, L.N.: The stochastic multi-gradient algorithm for multi-objective optimization and its application to supervised machine learning. arXiv preprint arXiv:1907.04472 (2019)
13. Mehrabi, N., Morstatter, F., Saxena, N., Lerman, K., Galstyan, A.: A survey on bias and fairness in machine learning. arXiv preprint arXiv:1908.09635 (2019)
14. Mehrabi, N., Naveed, M., Morstatter, F., Galstyan, A.: Exacerbating algorithmic bias through fairness attacks. CoRR abs/2012.08723 (2020), https://arxiv.org/abs/2012.08723
15. Mei, S., Zhu, X.: Using machine teaching to identify optimal training-set attacks on machine learners. In: Proceedings of the AAAI Conference on Artificial Intelligence. vol. 29 (2015)
16. Muñoz-González, L., Pfitzner, B., Russo, M., Carnerero-Cano, J., Lupu, E.C.: Poisoning attacks with generative adversarial nets. arXiv preprint arXiv:1906.07773 (2019)
17. Roh, Y., Lee, K., Whang, S., Suh, C.: Fr-train: A mutual information-based approach to fair and robust training. In: International Conference on Machine Learning. pp. 8147–8157. PMLR (2020)
18. Solans, D., Biggio, B., Castillo, C.: Poisoning attacks on algorithmic fairness. arXiv preprint arXiv:2004.07401 (2020)
19. Steinhardt, J., Koh, P.W., Liang, P.: Certified defenses for data poisoning attacks. arXiv preprint arXiv:1706.03691 (2017)
20. Taskesen, B., Nguyen, V.A., Kuhn, D., Blanchet, J.: A distributionally robust approach to fair classification. arXiv preprint arXiv:2007.09530 (2020)

21. Yuan, X., He, P., Zhu, Q., Li, X.: Adversarial examples: Attacks and defenses for deep learning. IEEE transactions on neural networks and learning systems **30**(9), 2805–2824 (2019)
22. Zafar, M.B., Valera, I., Rogriguez, M.G., Gummadi, K.P.: Fairness constraints: Mechanisms for fair classification. In: AISTATS (2017)
23. Zhao, B., Lao, Y.: Class-oriented poisoning attack. arXiv preprint arXiv:2008.00047 (2020)

Bi-Level Selection via Meta Gradient for Graph-Based Fraud Detection

Linfeng Dong[1,2], Yang Liu[1,2], Xiang Ao[1,2,3(✉)], Jianfeng Chi[4], Jinghua Feng[4], Hao Yang[4], and Qing He[1,2,5]

[1] Key Lab of Intelligent Information Processing of Chinese Academy of Sciences (CAS), Institute of Computing Technology CAS, Beijing 100190, China
{donglf19s,liuyang17z,aoxiang,heqing}@ict.ac.cn
[2] University of Chinese Academy of Sciences, Beijing 100049, China
[3] Institute of Intelligent Computing Technology, Suzhou, CAS, Suzhou, China
[4] Alibaba Group, Hangzhou, China
{bianfu.cjf,jinghua.fengjh,youhiroshi.yangh}@alibaba-inc.com
[5] Henan Institutes of Advanced Technology, Zhengzhou University, Zhengzhou 450052, China

Abstract. Graph Neural Networks (GNNs) have achieved remarkable successes by utilizing rich interactions in network data. When applied to fraud detection tasks, the scarcity and concealment of fraudsters bring two challenges: class imbalance and label noise. In addition to overfitting problem, they will compromise model performance through the message-passing mechanism of GNNs. For a fraudster in a neighborhood dominated by benign users, its learned representation will be distorted in the aggregation process. Noises will propagate through the topology structure as well. In this paper, we propose a **Bi-Level Selection (BLS)** algorithm to enhance GNNs under imbalanced and noisy scenarios observed from fraud detection. BLS learns to select instance-level and neighborhood-level valuable nodes via meta gradient of the loss on an unbiased clean validation set. By emphasizing BLS-selected nodes in the model training process, bias towards majority class (benign) and label noises will be suppressed. BLS can be applied on most GNNs with slight modifications. Experimental results on two real-world datasets demonstrate that BLS can significantly improve GNNs performance on graph-based fraud detection.

Keywords: Graph Neural Network · Fraud detection · Imbalanced learning

1 Introduction

Recently, Graph Neural Networks (GNNs) are widely used in fraud detection tasks [2,6]. These approaches build an end-to-end learning paradigm. First, each node is encoded into a representation by aggregating and transforming the information of its neighbors, namely the *representation learning phase*. Then, the

A. Bhattacharya et al. (Eds.): DASFAA 2022, LNCS 13245, pp. 387–394, 2022.
https://doi.org/10.1007/978-3-031-00123-9_31

learned representation is passed to a classifier to identify the fraudsters from the benign users, namely the *classification phase*.

Despite the remarkable success existing GNN-based methods achieved, the severe class imbalance and noisy label are still vital problems in fraud detection. Due to the contingency of fraudulent activities, the number of positive (fraud) samples is far less than the number of negative (benign) samples in fraud detection tasks. Meanwhile, the concealment of fraudulent activity leads to noisy label problem. Real-world users labeled as benign could either be benign or potentially fraudulent. As a result, the negative instances in the training set may consist of noisy labels.

Based on these observations, we emphasize two key challenges of GNN-based fraud detection as follows:

Neighborhood-Level Imbalance and Noise: In the representation learning phase, due to the propagation mechanism on topology, excess benign neighbors will dominate the network structure and dilute the feature of fraudsters, resulting in inaccurate embeddings of fraudulent nodes.

Instance-Level Imbalance and Noise: In the classification phase, the majority class will dominate the training loss during the gradient descent step, leading to a biased decision boundary. Undiscovered fraudsters (noise) will contribute wrong gradient direction, thus polluting the learned classification boundary.

To tackle the bi-level imbalanced and noisy problems, we propose **Bi-Level Selection (BLS)**, a lightweight algorithm for GNN-based fraud detection that learns to select valuable nodes on instance level and neighborhood level through a meta-learning paradigm (Fig. 1). BLS first forms an small unbiased and clean meta validation set by picking nodes from training set with high assortativity (the ratio of 1-hop neighbors that share the same label as itself). Then BLS uses the meta set to guide the training process. It selects valuable nodes according to their potential impact on the meta gradient of validation loss. It follows such assumptions: a better selection of valuable training nodes will improve the model performance and reduce the validation loss.

We integrate BLS with three GNN frameworks: GCN, GAT, and Graph-SAGE. Experiments on two real-world fraud detection datasets demonstrate that our algorithm can effectively improve the performance of GNN under imbalanced and noisy settings. BLS enhanced GNNs also outperform state-of-the-art.

Our contributions can be summarized as follows:

- We propose BLS, a meta gradient based algorithm to address the imbalanced and noisy label problem in graph-based fraud detection. In both representation learning and classification phase, BLS adopts a unified meta-learning paradigm to select instance-level and neighborhood-level valuable nodes.
- Compared to existing methods, BLS is the first work that considers the impact of class imbalance and noisy label on the message-passing mechanism of GNNs.
- The proposed BLS algorithm has high portability that can be applied on any GNN framework. By applying BLS on widely-used GNNs, we achieved

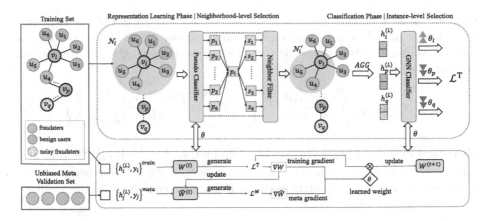

Fig. 1. Illustration of the BLS's workflow.

significant improvement compared to base models and state-of-the-art on two real-world datasets.

2 Methodology

In this section, we propose **Bi-Level Selection** (BLS), a lightweight meta-gradient-based method that can fit into general GNN structures. BLS addresses the class imbalance and noisy label problems in fraud detection with the following two key strategies: (1) select instance-level valuable nodes in classification phase by assigning weights θ_i, detailed in Sect. 2.1; (2) select neighborhood-level valuable nodes in representation learning phase by a filtered neighborhood $\mathcal{N}_i'$, detailed in Sect. 2.2.

2.1 Instance-level Node Selection

To select valuable nodes on instance level, we learn a weight θ_i for each node v_i in the training set by a meta-learning mechanism as soft selection. We denote the training set as $\{v_i, y_i\}_{i=1}^{N}$, the unbiased meta validation set as $\{v_j, y_j\}_{j=1}^{M}$, where $M \ll N$. The prediction of GNN is denoted as $F(h_i^{(L)}, W_f)$, where W_f is the parameters of GNN classifier. The cross-entropy loss function is denoted as $l(\cdot, \cdot)$. Searching for optimal GNN parameters W_f^* and optimal weights θ^* is a nested loops of optimization. To reduce computation cost, following the analysis of [9], we compute θ by one-step gradient approximation. At each iteration step t, the optimizer updates W_f from current parameter $W_f^{(t)}$ with step size α and uniform weights $\theta_i = \frac{1}{n}$ according to training loss on a mini-batch $\{v_i, y_i\}_{i=1}^{n}$:

$$\hat{W}_f^{(t)}(\theta) = W_f^{(t)} - \alpha \cdot \sum_{i=1}^{n} \theta_i \cdot \frac{\partial}{\partial W_f} l(y_i, F(h_i^{(L)}, W_f^{(t)})) \tag{1}$$

Then we slightly perturb the weights θ to evaluate the impact of each training node on the model performance on the meta validation set. We search for the optimal weight θ^* to minimize the meta loss by taking a single gradient step on the meta-validation set:

$$\theta_i \propto -\beta \cdot \frac{1}{M} \sum_{j=1}^{M} \left(\frac{\partial l_j(\hat{W}_f)}{\partial \hat{W}_f} \Big|_{\hat{W}_f = \hat{W}_f^{(t)}} \right)^{\top} \left(\frac{\partial l_i(W_f)}{\partial W_f} \Big|_{W_f = W_f^{(t)}} \right) \quad (2)$$

where $l_i(W_f)$ is $l(y_i, F(h_i^{(L)}, W_f))$, and $l_j(\hat{W}_f)$ is $l(y_j, F(h_j^{(L)}, \hat{W}_f))$. We take $\theta_i = \max(\theta_i, \frac{\theta_{th}}{n})$ and then batch-normalize θ_i, where $\theta_{th} \in [0,1]$ is a hyperparameter representing the minimal weight threshold. Then we can compute the weighted cross-entropy (CE) loss $\mathcal{L}_{GNN}$:

$$\mathcal{L}_{GNN} = -\sum_{i=1}^{N} \theta_i \cdot (y_i \log p_i + (1 - y_i) \log(1 - p_i)) \quad (3)$$

2.2   Neighborhood-level Node Selection

To select valuable nodes on neighborhood-level, BLS forms a subset $\mathcal{N}_i' \subseteq \mathcal{N}_i$ for each center node, where $\mathcal{N}_i$ is the original neighborhood. We append a pseudo classifier before the aggregation each GNN layer to inference pseudo labels. Then we filter the neighborhood according to the pseudo label affinity scores. The ℓ-th layer pseudo classifier $G^{(\ell)}$ parameterized by $W_g^{(\ell)}$ takes the representation $h_i^{(\ell-1)}$ of node v_i from the previous layer as input and generates a pseudo label. Then, the pseudo label affinity score between center node v_i and its neighbor $v_j \in \mathcal{N}_i$ is computed by the L-1 distance function:

$$\hat{p}_i^{(\ell)} = G^{(\ell)}(h_i^{(\ell-1)}, W_g^{(\ell)}) \quad (4)$$

$$S_{ij}^{(\ell)} = 1 - \|\hat{p}_i^{(\ell)} - \hat{p}_j^{(\ell)}\|_1 \quad (5)$$

We sort the neighbors by $S_{ij}^{(\ell)}$ in descending order and select top-k neighbors to form the filtered neighborhood $\mathcal{N}_i^{(\ell)}$, $k = \lceil \rho \cdot |\mathcal{N}_i| \rceil$. With the filtered neighborhood $\mathcal{N}_i'^{(\ell)}$ of center node v_i, we apply neighbor aggregation on v_i:

$$h_i^{(\ell)} = \sigma(W^{(\ell)}(h_i^{(\ell-1)} \oplus AGG(\{h_j^{(\ell-1)} | v_j \in \mathcal{N}_i'\}))) \quad (6)$$

The quality of filtered neighborhood highly depends on the accuracy of predicted pseudo labels. Therefore, we adopt a layer-wise direct supervised weighted loss $\mathcal{L}_{PSE}^{(\ell)}$ similar to Eq. (3). The overall loss function can be formulated as the combination of layer-wise pseudo classifier loss and GNN loss:

$$\mathcal{L} = \mathcal{L}_{GNN} + \sum_{\ell=1}^{L-1} \mathcal{L}_{PSE}^{(\ell)} \quad (7)$$

3   Experiments

In this section, we evaluate BLS-enhanced GNNs on two graph-based fraud detection datasets. Specifically, we aim to answer the following research questions: (RQ1.) How much improvement does BLS bring to the base models under imbalanced and noisy circumstances? (RQ2.) Does BLS outperform other imbalanced learning methods on fraud detection tasks? (RQ3.) How do the key components of BLS contribute to the overall fraud detection performance?

3.1   Experimental Setup

Datasets.   We adopt two real-world graph-based fraud detection datasets YelpChi [8] and Amazon [7], collected from online platforms Yelp.com and Amazon.com. Reviews (node) with less than 20% helpful vote are considered as fraudulent nodes. The statistics of the two datasets are shown in Table 1, where $|N|$, $|E|$, $|R|$ stand for number of nodes, edges and edge types. PR is the ratio of positive nodes (fraudsters).

Table 1. Statistics of two graph-based fraud detection datasets.

Dataset	$|N|$	$|E|$	$|R|$	PR
YelpChi	45,954	3,846,979	3	14.5%
Amazon	11,944	4,398,392	3	6.9%

Baselines and Evaluation Metrics. BLS is a lightweight method that can be applied to various existing GNN architectures. We select three widely-used GNNs (GCN, GraphSAGE and GAT) and their multi-relational extensions (GCN_M, GAT_M and $GraphSAGE_M$) as base models. We also compare BLS-enhanced GNNs with the state-of-the-art graph-based fraud detection methods: Graph-Consis [6], CARE-GNN [2] and PC-GNN [5]. We adopt two widely used metrics AUC score and G-Mean [5] for evaluation.

Experimental Settings. We set node embedding dimension d and hidden layer dimension as 64, L as 2, learning rate of Adam optimizer lr as 0.01, training epochs as 1000, batch size as 1024 for YelpChi dataset and 256 for Amazon dataset. For BLS, we set the preserving proportion ρ to 0.5, the minimal weight threshold ω_{th} to 0.01. The train/valid/test ratio are 40%, 20%, 40%. We use 266 (5%) nodes in the YelpChi training set and 107 (9%) nodes in the Amazon training set as meta validation. We conduct 10 runs on two datasets with all the compared models and report the average value with standard deviation of the performance metrics.

3.2   Overall Evaluation (RQ1)

We evaluate the performance of all compared methods on the graph-based fraud detection task with two datasets. According to the main results shown in Table 2, we have the following observations: (1) By incorporating BLS, all three base models gain significant improvements in terms of AUC and G-Mean. (2) Compared to the baselines, BLS-enhanced GAT_M achieves the best performance on both datasets. By selection under meta guidance, BLS can generate more credible layer-wise pseudo label affinity scores. Thus, the filtering process based on affinity scores is able to reduce the noise in the neighborhood of fraudulent nodes. Then we evaluate the performance of BLS-enhanced GNNs on against noisy-label circumstances. We randomly choose 0% to 50% fraudulent nodes and flip their labels. By label flipping, we simulate unidentified fraudsters which are labeled as benign. Figure 2 shows the change on AUC scores respecting noisy label ratio. We can observe that BLS-enhanced GNNs always achieve a higher AUC score than the corresponding base models, proving the robustness of BLS toward label noises.

Table 2. Performance comparison on two graph-based fraud detection datasets.

Dataset	YelpChi		Amazon	
Methods	AUC	G-Mean	AUC	G-Mean
GCN	59.02±1.08	55.61±2.96	79.83±1.38	73.38±4.29
GraphSAGE	58.46±3.03	46.90±5.82	81.13±2.93	75.17±5.28
GAT	64.18±1.84	59.53±4.37	88.48±1.36	85.69±4.72
GraphConsis	69.83±3.42	58.57±3.85	87.41±3.34	76.77±4.86
CARE-GNN	78.44±0.69	70.13±2.17	93.14±0.74	85.63±0.71
PC-GNN	79.87±0.14	71.60±1.30	95.86±0.14	90.30±0.44
GCN_M	74.62±1.38	68.72±1.92	92.93±2.04	83.22±3.85
$GraphSAGE_M$	77.12±2.56	69.15±3.96	93.63±3.17	85.92±4.20
GAT_M	81.73±1.48	75.33±3.52	93.71±1.06	85.82±3.65
BLS+GCN_M	83.28±0.86	76.31±2.74	94.42±1.55	87.41±0.78
BLS+$GraphSAGE_M$	86.50±0.78	80.02±2.93	94.71±1.33	87.44±2.02
BLS+GAT_M	**89.26±1.04**	**81.82±3.02**	**95.93±0.73**	**90.72±1.64**

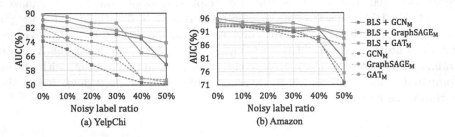

Fig. 2. Performance of BLS-enhanced GNNs w.r.t noisy label.

3.3 Comparison with Imbalanced Learning Methods (RQ2)

To further observe the effectiveness of meta selection strategy of BLS, we use two imbalanced learning methods Focal Loss [4] and CB Loss [1] to replace the node-level selection on GAT. We can observe that BLS achieves highest scores on both datasets, demonstrating that BLS is able to filter unidentified fraudsters with meta knowledge and prevent over-fitting on the majority class (Table 3).

Table 3. Performance comparison of BLS with other imbalanced learning methods

Dataset	YelpChi		Amazon	
Strategy	AUC	G-Mean	AUC	G-Mean
Focal Loss	84.21	77.62	93.84	87.79
CB Loss	86.28	79.09	93.96	88.15
BLS	**89.26**	**81.82**	**95.93**	**90.72**

3.4 Ablation Study (RQ3)

In this subsection, we explore how the two key components in BLS, i.e., instance-level selection and neighborhood-level selection, improve GNN models. We take BLS-enhanced GAT for demonstration, $BLS_{\backslash N}$ removes neighborhood-level selection, $BLS_{\backslash I}$ removes instance-level selection, $BLS_{\backslash NI}$ removes both strategies. As Fig. 3 illustrated, the complete model achieves the best performance on all metrics, removing each component will cause performance dropping except G-mean on Amazon, proving that both components are effective for graph-based fraud detection tasks.

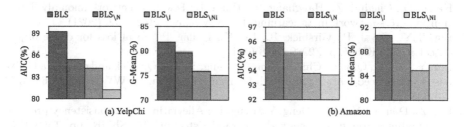

Fig. 3. Ablation study of two key components of BLS on YelpChi and Amazon.

4 Related Work

Graph-based fraud detection focus on analyzing the interactions and connectivity patterns to identify fraudulent activities. GraphConsis [6] and CARE-GNN [2]

are GNN-based anti-spam model, tackle the inconsistency problems and camouflages in fraud detection. PC-GNN [5] is designed for imbalanced supervised learning on graphs. It incorporates a two-step resampling method to reconstruct label-balanced sub-graphs. Meta-GDN [3] uses deviation loss and cross-network meta-learning algorithm for network anomaly detection tasks. Unlike our problem setting, Meta-GDN treats negative nodes as unlabeled nodes and requires extra auxiliary networks.

5 Conclusion

In this work, we propose a lightweight algorithm named Bi-Level Selection (BLS) that can be incorporated into general GNN architectures to handle the class imbalance and noisy label problems usually appeared in graph-based fraud detection tasks. BLS selects valuable nodes on two levels guided by meta gradient of validation loss. Experiments on two benchmark graph-based fraud detection datasets demonstrate the effectiveness of our algorithm.

Acknowledgement. The research work is supported by the National Natural Science Foundation of China under Grant (No.61976204, 92046003, U1811461). Xiang Ao is also supported by the Project of Youth Innovation Promotion Association CAS, Beijing Nova Program Z201100006820062.

References

1. Cui, Y., Jia, M., Lin, T.Y., Song, Y., Belongie, S.: Class-balanced loss based on effective number of samples. In: CVPR, pp. 9268–9277 (2019)
2. Dou, Y., Liu, Z., Sun, L., Deng, Y., Peng, H., Yu, P.S.: Enhancing graph neural network-based fraud detectors against camouflaged fraudsters. In: CIKM, pp. 315–324 (2020)
3. Kaize, D., Qinghai, Z., Hanghang, T., Huan, L.: Few-shot network anomaly detection via cross-network meta-learning. In: WWW, pp. 2448–2456 (2021)
4. Lin, T.Y., Goyal, P., Girshick, R., He, K., Dollár, P.: Focal loss for dense object detection. In: ICCV, pp. 2980–2988 (2017)
5. Liu, Y., Ao, X., Qin, Z., Chi, J., Feng, J., Yang, H., He, Q.: Pick and choose: a GNN-based imbalanced learning approach for fraud detection. In: WWW, pp. 3168–3177 (2021)
6. Liu, Z., Dou, Y., Yu, P.S., Deng, Y., Peng, H.: Alleviating the inconsistency problem of applying graph neural network to fraud detection. In: SIGIR, pp. 1569–1572 (2020)
7. McAuley, J.J., Leskovec, J.: From amateurs to connoisseurs: modeling the evolution of user expertise through online reviews. In: WWW, pp. 897–908 (2013)
8. Rayana, S., Akoglu, L.: Collective opinion spam detection: bridging review networks and metadata. In: KDD, pp. 985–994 (2015)
9. Ren, M., Zeng, W., Yang, B., Urtasun, R.: Learning to reweight examples for robust deep learning. In: ICML, pp. 4334–4343 (2018)

Contrastive Learning for Insider Threat Detection

M. S. Vinay[1] (iD), Shuhan Yuan[2] (iD), and Xintao Wu[1]([⊠]) (iD)

[1] University of Arkansas, Fayetteville, AR 72701, USA
{vmadanbh,xintaowu}@uark.edu
[2] Utah State University, Logan, UT 84322, USA
Shuhan.Yuan@usu.edu

Abstract. Insider threat detection techniques typically employ supervised learning models for detecting malicious insiders by using insider activity audit data. In many situations, the number of detected malicious insiders is extremely limited. To address this issue, we present a contrastive learning-based insider threat detection framework, CLDet, and empirically evaluate its efficacy in detecting malicious sessions that contain malicious activities from insiders. We evaluate our framework along with state-of-the-art baselines on two unbalanced benchmark datasets. Our framework exhibits relatively superior performance on these unbalanced datasets in effectively detecting malicious sessions.

Keywords: Insider threat detection · Contrastive learning · Cyber-security

1 Introduction

Insider threat refers to the threat arising form the organizational insiders who can be employees, contractors or business partners etc. These insiders usually have an authorization to access organizational resources such as systems, data and network etc. A popular approach to detect malicious insiders is by analyzing the insider activities recorded in the audit data [14] and applying supervised learning models. Usually, the insider audit data is unbalanced because only a few malicious insiders are detected. Hence, applying supervised learning models on such unbalanced datasets can result in poor detection accuracy. To address this limitation, we present a framework, CLDet, to detect malicious sessions (containing malicious activities from insiders) by using contrastive learning.

Our CLDet framework has two components: self-supervised pre-training and supervised fine tuning. Specifically, the self-supervised pre-training component generates encodings for user activity sessions by utilizing contrastive learning whereas the supervised fine tuning component classifies a session as malicious or normal by using these encodings. Contrastive learning requires data augmentations for generating augmented versions of an original data point and ensures

A. Bhattacharya et al. (Eds.): DASFAA 2022, LNCS 13245, pp. 395–403, 2022.
https://doi.org/10.1007/978-3-031-00123-9_32

that these augmented versions have close proximity with each other when compared to the augmented versions of the other data points. Since each user activity session can be modelled as a sentence and each activity as a word of this sentence [14], we adapt sentence based data augmentations from the Natural Language Processing (NLP) domain [10] in our framework. We conduct an empirical evaluation study of our framework and evaluation results demonstrate noticeable performance improvement over state-of-the-art baselines.

2   Related Work

Insider Threat Detection. Traditional insider threat detection models employ handcrafted features extracted from user activity log data to detect insider threats. Yuan et al. [11] argued that utilizing hand crafted features for detecting insider threats can be tedious and time consuming and hence proposed to utilize deep learning model to automatically learn the features. Specifically they employed a LSTM model to extract encoded features from user activities and then detected malicious insiders through a Convolutional Neural Network (CNN). Similarly, Lin et al. [5] used unsupervised Deep Belief Network to extract features from user activity data and applied one-class Support Vector Machine (SVM) to detect the malicious insiders. Lu et al. [6] modeled the user activity log information as a sequence and extract user specific features through a trained LSTM model. Yuan et al. [13] combined a RNN with temporal point process to utilize both intra- and inter-session time information. The closely related work is [14] where the authors proposed a few-shot learning based framework to specifically addresses the data imbalance issue in insider threat detection. The developed framework applies the word-to-vector model for generating encoded features from user activity data and then uses a trained BERT language model to refine the encoded features. We refer readers to a survey [12] for other related works.

Contrastive Learning. Contrastive learning has been extensively studied in the literature for image and NLP domains. Jaiswal et al. [3] presented a comprehensive survey on contrastive learning techniques for both image and NLP domains. Marrakchi et al. [7] effectively utilized contrastive learning on unbalanced medical image datasets to detect skin diseases and diabetic retinopathy. The developed algorithm utilizes a supervised pre-training component, which is designed by employing a Residual Network, and generates image representations. These generated image representations are further fed as input to a fine tuning component which is designed by using a single linear layer. In our framework, we utilize some of data augmentation concepts presented in [10] and [9]. Wu et al. [10] presented a contrastive learning based framework for analyzing text similarity. Their framework employs sentence based augmentation techniques for self-supervised pre-training. Wang et al. [9] presented a new contrastive loss function for the image domain.

3 Framework

User activities are modeled through activity sessions. Specifically, each session consists of multiple user activities. Let S_k denote the kth activity session of a user. Here, $S_k = \{e_{k_1}, e_{k_2}, \ldots, e_{k_T}\}$, where $e_{k_i}(1 \leq i \leq T)$ is the ith user activity. Let $D = \{S_i, y_i\}_{i=1}^m$ denote the insider threat dataset where m denotes the number of sessions, y_i is the label of S_i. Here, $y_i = 1$ and $y_i = 0$ denote that S_i is malicious and normal session respectively. The two main components of our CLDet framework are self-supervised pre-training and supervised fine tuning. The pre-training component is responsible for generating session encodings and the fine tuning component, using these session encodings as input classifies a given input session as a malicious or normal session.

3.1 Self-supervised Pre-training Component

3.1.1 Encoder and Projection Head

Each activity in the session is represented through trained word-to-vector model. Let $\mathbf{x}_{k_i} \in \mathbb{R}^d$ denote the word-to-vector model representation of activity e_{k_i}, where d denotes the number of representation dimensions. Each activity of an input session is converted to its corresponding word-to-vector representation and it is fed as an input to a specially designed *Encoder*. We choose Recurrent Neural Network (RNN) to design our encoder. The encoder is responsible for generating the session encoding $\mathbf{x}_k \in \mathbb{R}^d$ of session S_k. Finally, a projection head will project $\mathbf{x}_k$ to a new space representation $\mathbf{z}_k \in \mathbb{R}^d$. The projection head is only used in the training of the self-supervised component. After this training, the projection head will be discarded and only the encoded session representation will be used as an input to the supervised fine tuning component.

The encoder consists of a RNN and a linear layer. The RNN consists of two hidden layers denoted as $H^{(1)}$ and $H^{(2)}$ respectively. The first hidden layer $H^{(1)}$ is represented as $\mathbf{h}_{k_t}^{(1)} = tanh(W_1^1 \mathbf{x}_{k_t} + \mathbf{b}_1^1 + W_2^1 \mathbf{h}_{k_{t-1}}^{(1)} + \mathbf{b}_2^1)$ where $1 \leq t \leq T$, W_1^1 and W_2^1 are $(d \times d)$ weight matrices, $\mathbf{b}_1^1 \subset \mathbb{R}^d$ and $\mathbf{b}_2^1 \in \mathbb{R}^d$ are the bias vectors, and $\mathbf{h}_{k_t}^{(1)}$ denotes the encoded output of $H^{(1)}$ for the input $\mathbf{x}_{k_t}$. The second hidden layer $H^{(2)}$ is similarly represented as $\mathbf{h}_{k_t}^{(2)} = tanh(W_1^2 \mathbf{h}_{k_t}^{(1)} + \mathbf{b}_1^2 + W_2^2 \mathbf{h}_{k_{t-1}}^{(2)} + \mathbf{b}_2^2)$ where W_1^2 and W_2^2 are $(d \times d)$ weight matrices, $\mathbf{b}_1^2 \in \mathbb{R}^d$ and $\mathbf{b}_2^2 \in \mathbb{R}^d$ are the bias vectors, and $\mathbf{h}_{k_t}^{(2)}$ denotes the encoded output of $H^{(2)}$ for the input $\mathbf{h}_{k_t}^{(1)}$. Finally $\{\mathbf{h}_{k_i}^{(2)}\}_{i=1}^T$ is flattened to denote the output of RNN as $\mathbf{v}_k \in \mathbb{R}^{Td}$, which is then fed to the linear layer $L^{(1)}$ to obtain the session encoding $\mathbf{x}_k$. This linear layer is represented as $\mathbf{x}_k = A_1 \mathbf{v}_k + \mathbf{b}_1$ where A_1 is a $(d \times Td)$ weight matrix and $\mathbf{b}_1 \in \mathbb{R}^d$ is a bias vector. The projection head is denoted as $L^{(2)}$ and is represented as $\mathbf{z}_k = A_2 \mathbf{x}_k + \mathbf{b}_2$ where A_2 is a $(d \times d)$ weight matrix and $\mathbf{b}_2 \in \mathbb{R}^d$ is a bias vector.

3.1.2   Contrastive Loss

A contrastive learning loss function is used for a contrastive prediction task, i.e., predicting positive augmentation pairs. We adapt the $SimCLR$ contrastive loss function [10] in our framework and augment each batch of sessions. Let $B_s = \{S_1, S_2, ..., S_N\}$ denote a batch of sessions. Each $S_k \in B_s$ is subjected to data augmentation and two augmented sessions denoted as S_k^1 and S_k^2 are generated. Let $B_s^a = \{S_1^1, S_1^2, S_2^1, S_2^2, ..., S_N^1, S_N^2\}$ denote a batch of augmented sessions. The augmented sessions (S_k^1, S_k^2) form a positive sample pair and all the remaining sessions in B_s^a are considered as the negative samples. Let $\mathbf{z}_k^1$ and $\mathbf{z}_k^2$ denote the projection head representations of the augmented sessions S_k^1 and S_k^2 respectively. The loss function for the positive pair $(\mathbf{z}_k^1, \mathbf{z}_k^2)$ is represented as

$$l(\mathbf{z}_k^1, \mathbf{z}_k^2) = -log \frac{exp(cos(\mathbf{z}_k^1, \mathbf{z}_k^2)/\alpha)}{exp(cos(\mathbf{z}_k^1, \mathbf{z}_k^2)/\alpha) + \sum_{i=1}^{N} \mathbf{1}_{[i \neq k]} \sum_{j=1}^{2} exp(cos(\mathbf{z}_k^1, \mathbf{z}_i^j)/\alpha)} \quad (1)$$

Here, $cos()$ denotes the cosine similarity function, $\mathbf{1}_{[i \neq k]}$ denotes an indicator variable, and α denotes the tunable temperature parameter. This pair loss function is not symmetric, because $l(\mathbf{z}_k^1, \mathbf{z}_k^2) \neq l(\mathbf{z}_k^2, \mathbf{z}_k^1)$. For the batch of augmented sessions B_s^a, we can easily see there are N positive pairs. The contrastive loss function for B_s^a, which is defined as the sum of all positive pairs' loss in the batch, is represented as $CL(B_s^a) = \sum_{i=1}^{N} l(\mathbf{z}_i^1, \mathbf{z}_i^2) + l(\mathbf{z}_i^2, \mathbf{z}_i^1)$.

For session S_k, we adapt three basic NLP based sentence augmentation techniques [10]: 1) Activity Replacement (Rpl), we generate the augmented session S_k^1 (S_k^2) by randomly replacing g_1 (g_2) number of activities with a set of token activities; 2) Activity Reordering (Rod), we generate the augmented session S_k^1 (S_k^2) by randomly selecting a sub-sequence with length g_1 (g_2) and shuffling all activities in the chosen sub-sequence while keeping all other activities unchanged; 3) Activity Deletion (Del), g_1 and g_2 number of activities in S_k are deleted to generate the augmented sessions S_k^1 and S_k^2 respectively. We will investigate the effectiveness of other complex data augmentation techniques in our future work.

3.2   Supervised Fine Tuning Component

The supervised fine tuning component has two layers denoted as $L^{(3)}$ and $L^{(4)}$. The first layer $L^{(3)}$ is represented as $\mathbf{m}_k = A_3 \mathbf{x}_k + \mathbf{b}_3$. Here, A_3 is a $(d \times d)$ weight matrix, $\mathbf{b}_3 \in \mathbb{R}^d$ is a bias vector, and $\mathbf{m}_k \in \mathbb{R}^d$ denotes the output encoding of $L^{(3)}$. The output layer $L^{(4)}$ is represented as $\mathbf{o}_k = Softmax(A_4 \mathbf{m}_k + \mathbf{b}_4)$. Here, A_4 is $(2 \times d)$ weight matrix, $\mathbf{b}_4 \in \mathbb{R}^2$ is a bias vector, and $\mathbf{o}_k$ denotes the output of the supervised fine-tuning component. We use the Softmax activation function in the output layer. The supervised fine tuning component is trained by using the cross entropy loss function.

4   Experiments

4.1   Experimental Setup

4.1.1   Datasets

The empirical evaluation study of our proposed framework is conducted on two datasets: CERT Insider Threat [2] and UMD-Wikipedia [4]. In CERT, each user activities are chronologically recorded over 516 days. To perform our empirical analysis, we split the dataset into training and test sets using chronological ordering. Specifically, the user activities recorded until the first 460 days and between 461 to 516 days are used in the training and test sets respectively. Additionally, we further split the training set for training the pre-training and fine tuning components, wherein, the user activities recorded until the first 400 days and between 401 to 460 days are used for training the pre-training and fine tuning components respectively. For the supervised fine-tuning component, four scenarios are utilized in the training phase. Each scenario involves different number of malicious sessions. Specifically, 5, 8, 10 and 15 malicious sessions are utilized in the training phase. The UMD-Wikipedia dataset is relatively more balanced than CERT dataset. Since our framework is specifically designed to effectively operate on unbalanced datasets, we only use a limited number of malicious sessions for training the supervised fine tuning component. The training set is split between pre-training and fine tuning components, wherein, 4436 and 50 normal sessions are used for training the pre-training and fine tuning components respectively and similarly, 3577 and 50 malicious sessions are used for training the pre-training and fine tuning components respectively. Again, we use four scenarios in the training phase of the supervised fine-tuning component. Specifically, 5, 15, 30 and 50 malicious sessions are utilized in the training phase. We show the detailed settings in Table 1.

Table 1. Training and test sets

Dataset	Partition		# of Malicious Sessions	# of Normal Sessions
CERT	Training set	Pre Train	23	1,217,608
		Fine Tune	15	50
	Test set		10	1000
UMD-Wikipedia	Training set	Pre-Train	3577	4436
		Fine Tune	50	50
	Test set		1000	1000

4.1.2 Training Details

The activity features are extracted through a trained word-to-vector model. Specifically, the word-to-vector model is trained through the skip-gram approach with the minimum word frequency parameter as 1. The hyper-parameters g_1 and g_2 employed in our data augmentation techniques control the amount of distortion caused by augmenting the original session. For activity replacement and deletion-based data augmentation techniques we set $g_1 = 1$ and $g_2 = 1$, and for activity reordering based data augmentation technique we set $g_1 = 3$ and $g_2 = 3$. We set the number of dimensions of activity and session encodings d as 5 and the temperature parameter α in the contrastive loss function as 1. Four metrics are utilized to quantify the performance of our framework: Precision, Recall, F_1 and FPR.

4.1.3 Baselines

We compare our CLDet framework with three baselines: Few-Shot [14], Deep-Log [1] and BEM [8]. Few-Shot has a similar design as our framework, wherein, it has both self-supervised pre-training and supervised fine tuning components. Specifically, the self-supervised pre-training component is used for generating session encodings and these encodings are utilized to detect malicious sessions through the supervised fine tuning component. The session encodings are generated through the BERT language model. The self-supervised pre-training component is trained by using the Mask Language Modeling (MLM) loss function. We train both self-supervised pre-training and supervised fine tuning components by using the same settings shown in Table 1. BEM employs LSTM to model user activity sessions. Specifically, it considers the past user activities and predicts the probabilities of future activities through LSTM. If the predicted probability of an activity in the session is low, then that session is flagged as a malicious session. The LSTM model employs a single hidden layer and the model training is performed by using cross entropy loss. We train this baseline by using the same training set which we have used for training the fine-tuning component of our framework. Deep-Log differs from BEM in two ways: (1) It employs two hidden layers in its LSTM model. (2) It predicts the probabilities of the top-K future activities, if some activity in the session is not in the list of predicted top-K activities, then that session is flagged as a malicious session. Deep-Log training is performed by using cross entropy loss. We use the same training settings which was used for BEM to train this baseline.

Table 2. Performance of our framework and baselines under different scenarios. The higher the better for Precision, Recall, and F1. The lower the better for FPR. The cells with—indicate the extreme scenario where all sessions are predicted as normal. Best values are bold highlighted. M denotes the number of malicious sessions.

Models	Scenario	CERT					UMD-Wikipedia				
		M	Precision	Recall	F1	FPR	M	Precision	Recall	F1	FPR
Deep-Log	1	5	—	—	—	—	5	—	—	—	—
	2	8	—	—	—	—	15	—	—	—	—
	3	10	0.7600	0.8125	0.7294	0.4500	30	—	—	—	—
	4	15	**1.0000**	0.5875	0.7394	**0.0000**	50	0.6765	0.9200	0.7797	0.4400
BEM	1	5	—	—	—	—	5	—	—	—	—
	2	8	0.5000	0.3125	0.3846	**0.0000**	15	—	—	—	—
	3	10	0.6724	0.5165	0.4971	0.4500	30	0.5000	0.1500	0.2307	**0.0000**
	4	15	0.7500	0.8100	0.7179	0.5000	50	0.6282	**0.9800**	0.7656	0.5800
Few-Shot	1	5	—	—	—	—	5	—	—	—	—
	2	8	0.3666	0.1125	0.1709	0.1861	15	—	—	—	—
	3	10	0.5833	0.1875	0.2832	0.1361	30	0.4286	0.1200	0.1875	0.1600
	4	15	0.4000	0.4125	0.3709	0.5111	50	0.4894	0.9200	0.6389	0.9600
CLDet(Rpl)	1	5	**0.9444**	0.5875	0.6195	**0.0000**	5	0.6234	**0.9600**	0.7559	0.5800
	2	8	0.9158	0.9026	0.9070	0.0812	15	**0.8718**	0.6800	**0.7640**	**0.1000**
	3	10	0.9111	0.9210	0.9117	0.0812	30	**0.8750**	0.7000	0.7778	0.1000
	4	15	0.9236	0.9333	0.9243	0.0715	50	0.8039	0.8200	**0.8119**	0.2000
CLDet(Del)	1	5	—	—	—	—	5	**0.6935**	0.8600	0.7679	**0.3800**
	2	8	**0.9444**	0.6500	0.7013	**0.0000**	15	0.7551	**0.7400**	0.7175	0.2400
	3	10	**1.0000**	0.7250	0.7806	**0.0000**	30	0.7636	**0.8400**	0.8000	0.2600
	4	15	**1.0000**	0.9000	0.9150	**0.0000**	50	**0.8222**	0.7400	0.7789	**0.1600**
CLDet(Rod)	1	5	0.9206	**1.0000**	**0.9584**	0.0800	5	0.6667	**0.9600**	**0.7860**	0.4800
	2	8	**0.9444**	**1.0000**	**0.9706**	**0.0000**	15	0.7826	0.7200	0.7500	0.2000
	3	10	**1.0000**	**1.0000**	**1.0000**	**0.0000**	30	0.7778	**0.8400**	**0.8077**	0.2400
	4	15	**1.0000**	**1.0000**	**1.0000**	**0.0000**	50	0.8039	0.8200	**0.8119**	0.2000

4.2 Experimental Results

We consider the three versions of our CLDet framework based on the specific data augmentation technique employed for pre-training. The performance of the three versions of our framework and the baselines w.r.t four different scenarios is shown in Table 2. Our CLDet framework consistently shows better overall performance than the baselines in all the considered scenarios and datasets. The main reason for this performance is that the self-supervised pre-training component by utilizing contrastive learning generates favorable encoding for each session and by using these favorable encodings as inputs, the supervised fine-tuning component can effectively separate the malicious and normal sessions. We would point out that we purposely introduce the first scenario where the number of malicious sessions used in the training is only 5 for both CERT and UMD-Wikipedia datasets. Under this extreme setting, all baselines completely fail (all sessions in the test data are predicted as normal). On the contrary, our framework can still achieve reasonable performance except the version of using the activity deletion on CERT. There is a no clear winner among the three data augmentation techniques used in our framework when all the scenarios and datasets are considered. However, all the three data augmentation techniques can be considered as quite effective in achieving the main goal of our framework.

Table 3. Ablation analysis results. M denotes the number of malicious sessions.

Dataset	Scenario	M	Precision	Recall	F1	FPR
CERT	1	5	—	—	—	—
	2	8	0.2531	0.5000	0.3361	0.5000
	3	10	0.4423	0.3026	0.3594	0.0384
	4	15	0.4706	1.0000	0.6400	1.0000
UMD-Wikipedia	1	5	—	—	—	—
	2	15	—	—	—	—
	3	30	0.9294	0.4210	0.5795	0.0320
	4	50	0.6487	0.9750	0.7791	0.5280

Ablation Analysis. We conduct one ablation study by removing the self supervised pre-training component from our framework and only utilizing the supervised fine-tuning component. The supervised fine-tuning component consists of only linear layers and cannot model sequence data. To resolve this limitation for our ablation study, we suitably format the input data and layer $L^{(3)}$ of the fine tuning component. Consider the word-to-vector representations of the activities belonging to the session $S_k = \{e_{k_1}, e_{k_2}, ..., e_{k_T}\}$ which are denoted as $\{\mathbf{x}_{k_1}, \mathbf{x}_{k_2}, ..., \mathbf{x}_{k_T}\}$. We flatten this sequence $\{\mathbf{x}_{k_t}\}_{t=1}^{T}$ into a vector and feed this flattened vector as an input to layer $L^{(3)}$ of the fine-tuning component. Table 3 shows the results of this ablation study. For both datasets, the supervised fine tuning component when used in isolation for detecting malicious sessions, underperforms against our framework in all the four scenarios. Clearly, this ablation study demonstrates that self-supervised pre-training component is crucial for our framework to achieve good performance.

5   Conclusion

We presented a contrastive learning-based framework to detect malicious insiders. Our framework is specifically designed to operate on unbalanced datasets. Our framework has self-supervised pre-training and supervised fine tuning components. The former is responsible for generating user session encodings. These session encodings are generated through the aid of contrastive learning and are then used by the supervised fine tuning component to detect malicious sessions. We presented an empirical study and results demonstrated our framework's better effectiveness than the baselines.

Acknowledgement. This work was supported in part by NSF grants 1564250, 1937010 and 2103829.

References

1. Du, M., Li, F., Zheng, G., Srikumar, V.: Deeplog: anomaly detection and diagnosis from system logs through deep learning. In: Proceedings of the 2017 ACM SIGSAC Conference on Computer and Communications Security, CCS 2017, Dallas, TX, USA, October 30 - November 03, 2017. pp. 1285–1298. ACM (2017)
2. Glasser, J., Lindauer, B.: Bridging the gap: a pragmatic approach to generating insider threat data. In: 2013 IEEE Symposium on Security and Privacy Workshops, San Francisco, CA, USA, May 23–24, 2013. pp. 98–104. IEEE Computer Society (2013)
3. Jaiswal, A., Babu, A.R., Zadeh, M.Z., Banerjee, D., Makedon, F.: A survey on contrastive self-supervised learning. CoRR arXiv:2011.00362 (2020)
4. Kumar, S., Spezzano, F., Subrahmanian, V.: Vews: a wikipedia vandal early warning system. In: Proceedings of the 21th ACM SIGKDD International Conference on Knowledge Discovery and Data Mining, p. 607–616. KDD 2015 (2015)
5. Lin, L., Zhong, S., Jia, C., Chen, K.: Insider threat detection based on deep belief network feature representation. In: 2017 International Conference on Green Informatics (ICGI), pp. 54–59 (2017)
6. Lu, J., Wong, R.K.: Insider threat detection with long short-term memory. In: Proceedings of the Australasian Computer Science Week Multi-conference. New York, NY, USA (2019)
7. Marrakchi, Y., Makansi, O., Brox, T.: Fighting class imbalance with contrastive learning. In: de Bruijne, M., Cattin, P.C., Cotin, S., Padoy, N., Speidel, S., Zheng, Y., Essert, C. (eds.) MICCAI 2021. LNCS, vol. 12903, pp. 466–476. Springer, Cham (2021). https://doi.org/10.1007/978-3-030-87199-4_44
8. Tuor, A., Baerwolf, R., Knowles, N., Hutchinson, B., Nichols, N., Jasper, R.: Recurrent neural network language models for open vocabulary event-level cyber anomaly detection. CoRR arXiv:1712.00557 (2017)
9. Wang, X., Qi, G.: Contrastive learning with stronger augmentations. CoRR arXiv:2104.07713 (2021)
10. Wu, Z., Wang, S., Gu, J., Khabsa, M., Sun, F., Ma, H.: CLEAR: contrastive learning for sentence representation. CoRR arXiv:2012.15466 (2020)
11. Yuan, F., Cao, Y., Shang, Y., Liu, Y., Tan, J., Fang, B.: Insider threat detection with deep neural network. In: Shi, Y., Fu, H., Tian, Y., Krzhizhanovokaya, V.V., Lees, M.H., Dongarra, J., Sloot, P.M.A. (eds.) ICCS 2018. LNCS, vol. 10860, pp. 43–54. Springer, Cham (2018). https://doi.org/10.1007/978-3-319-93098-7_4
12. Yuan, S., Wu, X.: Deep learning for insider threat detection: Review, challenges and opportunities. Comput. Secur. **104**, 102221 (2021). https://doi.org/10.1016/j.cose.2021.102221
13. Yuan, S., Zheng, P., Wu, X., Li, Q.: Insider threat detection via hierarchical neural temporal point processes. In: 2019 IEEE International Conference on Big Data (Big Data), pp. 1343–1350 (2019)
14. Yuan, S., Zheng, P., Wu, X., Tong, H.: Few-shot insider threat detection. In: CIKM 2020: The 29th ACM International Conference on Information and Knowledge Management, Virtual Event, Ireland, October 19–23, 2020. pp. 2289–2292. ACM (2020)

Metadata Privacy Preservation for Blockchain-Based Healthcare Systems

Lixin Liu[1,2], Xinyu Li[3], Man Ho Au[3], Zhuoya Fan[1], and Xiaofeng Meng[1(✉)]

[1] School of Information, Renmin University of China, Beijing, China
{lixinliu,fanzhuoya,xfmeng}@ruc.edu.cn
[2] School of Information Engineering, Inner Mongolia University of Science and Technology, Baotou, China
[3] Department of Computer Science, The University of Hong Kong, Hong Kong, China
allenau@cs.hku.hk

Abstract. Blockchain-based healthcare systems provide a patient-centric and accountable way to manage and share electronic health records. Thanks to its unique features, the blockchain is employed to record the metadata and to carry out access control. Nevertheless, the transparent nature of blockchain also poses a new challenge to these systems. We identify that the metadata stored on the blockchain leaks the relationship between doctors and patients. Based on this relationship, the adversary can launch linkage attacks to infer patients' information. Hence, it is necessary to protect metadata privacy. However, strong privacy protection may reduce accountability. Striking a balance between accountability and privacy preservation is a major challenge for the blockchain and its applications. In this paper, we first elaborate on the reasons why the metadata could leak the privacy of patients in blockchain-based healthcare systems. After that, we propose privacy-preserving and accountable protocols to deal with this problem for two different healthcare scenarios: the single doctor case and the group consultation case. Finally, the theoretical analysis demonstrates the practicality of our protocols.

Keywords: Privacy preservation · Accountability · Healthcare system

1 Introduction

Blockchain-based Healthcare systems have shown great benefits in the management and sharing of the Electronic Health Records (EHRs) for patients and medical institutions. They provide a technical opportunity to establish a patient-centric and accountable EHRs management and sharing framework [1,5–7,9]. Typically, they employ the blockchain to record the metadata and define auditable access policies in the smart contract, and only store encrypted EHRs in

Supported by organization x.

an off-chain storage (e.g. cloud server) [2,6,9]. In this way, the patients can utilize the metadata for accountability and define the access policies by themselves after the EHRs are uploaded and the metadata is recorded. However, less attention has been paid to security and privacy issues of the phase when the EHRs are uploaded and the metadata is recorded. The previous work proposes a protocol to solve the security issues when the EHRs are uploaded to the cloud server [5]. However, it can not protect the privacy derived from the metadata (**i.e. metadata privacy**). More specifically, adversaries may get the relationship between doctors and patients from the metadata. Based on this relationship, adversaries can make linkage attacks to infer patients' private information by linking the metadata with other databases. For example, if adversaries know that the doctor is a specialist in a sensitive disease (such as HIV), they can infer that the patient may suffer from that disease.

The root cause of the problem is privacy issues of the blockchain [7]. **First, the metadata stored on the blockchain may be utilized to make linkage attacks because of the transaction privacy issues**. The metadata on the blockchain usually includes the unique identifier of the patient and the unique identifier of the doctor in plaintext form [6,9]. Hence, the relationship between them is exposed to all participants, including adversaries. **Second, the metadata stored on the blockchain may also be employed to make linkage attacks because of the identity privacy issues**. The doctor is usually delegated to upload EHRs and to record the metadata [5]. That is, the doctor sends the transaction to the blockchain as the sender. The transaction address of the sender is pseudonymity. Therefore, the adversary still can get the relationship between the doctor and the patient. However, existing preserving-privacy protocols of the blockchain cannot directly solve these problems [9].

In this paper, we mainly focus on the metadata privacy preservation in Ethereum-based healthcare systems, because Ethereum is popularly used in healthcare system [4,5,8]. We propose different protocols for two application cases. More specifically, our contributions are summarized as follows.

- We introduce the metadata privacy breach in the blockchain-based healthcare systems, and illustrate why it may cause linkage attacks. To the best of our knowledge, we are the first ones that point out the privacy breach issue of the metadata in the blockchain-based healthcare systems.
- We propose privacy-preserving and accountable protocols for different medical scenarios. First, we propose a succinct protocol for the popular single doctor case. Then, we design an efficient protocol for the complicated group consultation case. Specially, our solutions are compatible with the Ethereum-based healthcare systems and can be directly applied to those systems.
- Security and privacy analysis illustrate that our scheme is secure in the rational adversary model.

The remainders of the paper are organized as follows. In Sect. 2, we present the formulation of the problem, adversaries' models and security requirements. In Sect. 3, we elaborate the details of the system and the analysis of the protocols. Finally, we conclude our work in Sect. 4.

2   Problem Formulation

The system architecture is shown in Fig. 1, which consists of six different entities: a hospital administrator, patients, doctors, a cloud server, an auditor and a blockchain system. When a patient visits a doctor for a diagnosis, the process can be divided into five phases.

- The initialization phase: the hospital administrator initiates the system.
- The appointment phase: the patient first needs to provide necessary information and registers with the hospital administrator, and then consults with the hospital administrator to set up a diagnosis key for encrypting the EHRs.
- The delegation phase: the patient should delegate the doctor to outsource EHRs, because the data generator (i.e. the doctor) and the data owner (i.e. the patient) are different.
- The storage phase: the doctor diagnoses for the patient and generates corresponding EHRs. Then, the doctor outsources the EHRs to the cloud server and records the metadata to the blockchain. Finally, the cloud server performs the authentication of the doctor according to the delegation information and the metadata.
- The audit phase: the auditor can perform the audit.

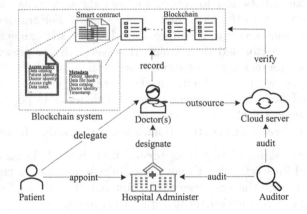

Fig. 1. System architecture

In practical medical applications, there are also cases where multiple doctors diagnose one patient together (known as the group consultation case). Every doctor of the group gives the corresponding diagnosis result and generates the EHRs based on the previous doctors' EHRs successively. In this situation, the patient needs to delegate the doctors to generate EHRs, and every doctor should be responsible for the EHRs that she or he generates.

We consider the adversary model from the external adversaries and internal adversaries.

- The external adversaries are the external entities and the nodes in the blockchain system. The external adversaries mainly target two types of attacks. First, external attackers may analyze blockchain transactions to get the identity privacy and the transaction privacy. Second, an external attackers may perform guessing attacks to get the diagnosis key in order to learn the EHRs.
- The internal adversaries include semi-honest doctors and the rational cloud server [3]. Specially, the rational model means the cloud server may deviate from the agreed principles to get more benefit. The internal adversaries mainly carry out four aspects of attacks. First, the delegated doctors may try to cover up the target EHRs by themselves in the medical malpractice. Second, the delegated doctors may incentivize the cloud server to collude with them to cover up the targeted EHRs. Third, the cloud server may deviate from the prescribed scheme to modify or even delete the EHRs to save the costs. Finally, the undelegated doctor may pretend to be the delegated doctor to forge EHRs in order to frame the delegated doctor.

According to the adversaries and the adversary model, the healthcare system should satisfy the following security and privacy requirements.

- Data confidentiality. The EHRs cannot be learned by other entities except the delegated doctor(s) and the patient.
- Resistance to linkage attacks. The relationship between the delegated doctor and the patient cannot be used to link with other databases by any adversary.
- Resistance to modification or forgery attacks. Only the delegated doctors can generate and outsource the EHRs. And any modification or forgery can be detected after the EHRs are outsourced.
- Resistance to collusion attacks. The delegated doctors cannot collude with the cloud server to cover up the EHRs.
- Accountability. When abnormal issues occur, the auditor can perform trace to find the adversary.

3   The Proposed Scheme

In this section, we first present the overview of our protocols, and then describe the design details together with the security and privacy analysis.

3.1   Overview

Considering that the transaction address of sender is derived from the public key in Ethereum, we adopt the public key as the identifier of the doctor in the metadata. Hence, the transaction address of doctor and the identifier of the doctor can be protected simultaneously. Based on this, we propose different protocols for the single doctor case and the group consultation case. In the single doctor case, we design a secure protocol to generate a blinded key pair for the doctor to resist the linkage attacks. Meanwhile, the corresponding private key can only be obtained by the delegated doctor for accountability. In the group consultation case, we combine the group signature and the same key pair instead. The group signature is used for accountability, while the same key pair is used for anonymity. Besides, secure key exchange protocol is employed to get the key for encryption, and the short signature is employed for secure delegation.

3.2 Construction of Our Scheme

We introduce our protocols in details for the single doctor case and the group consultation case, respectively. The notations used are listed in Table 1.

Table 1. Notations

Notations	Descriptions	Notations	Descriptions
HA	Hospital administrator	(PK_{ha}, SK_{ha})	Key pair of HA: $PK_{ha} = g_2^{SK_{ha}}$
P_i	Patient i	(PK_{pi}, SK_{pi})	Key pair of P_i: $PK_{pi} = g_2^{SK_{pi}}$
D_i	Doctor i	(PK_{di}, SK_{di})	Key pair of D_i: $PK_{di} = g_2^{PK_{di}}$
M_i	EHRs generate by D_i	KEY_{pi}	Diagnosis key of P_i
CS	Cloud server	Au	Auditor
$E()$	Symmetric encryption algorithm	$Addr_{CS}$	Ethereum address of CS
$Para_{pub}$	Public parameters	$Addr_{di}$	Ethereum address of D_i

Construction for the Single Doctor Case. P_i consults D_i for diagnosis. D_i diagnoses and generates EHRs M_i for P_i. We illustrate the communication diagram demonstrating the phases of the protocols for the single doctor case in Fig. 2.

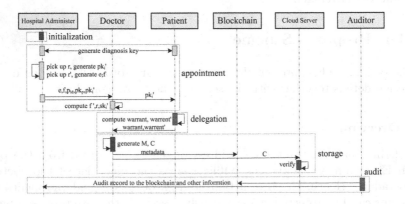

Fig. 2. Communication diagram for the single doctor case

We propose a protocol for generation of blinded key pair. Then we combine it with key exchange protocol, BLS short signature to archive our security and privacy goals. In order to resist linkage attacks, HA randomly chooses $r \in Z_p$ to generate the blinded public key for D_i: $PK'_{di} = PK_{di}{}^r$. Then, HA transfers r to D_i and makes sure that only D_i can compute the corresponding blinded private key SK'_{di}. Based on this, the Au and HA can simply release the random r to trace the real-world identity of the delegated doctor D_i in audit phase. Besides, in order to securely get the diagnosis key KEY_{pi}, HA and P_i perform the key exchange protocol to get KEY_{pi}, and HA securely transfers KEY_{pi} to D_i in the appointment phase. In addition, in order to securely generate the warrant, P_i performs short signature for the delegation to the blinded public key of D_i.

Construction for the Group Consultation Case. When P_i attends a group consultation, HA designates a group of doctors denoted as $\{D_1, D_2, ..., D_n\}$ for P_i. Every D_j ($j \in 1, 2, ...n$) has public key and private key $PK_{di} = g_2{}^{SK_{di}}$. The doctors generate the EHRs for P_i together, and every doctor generates the different parts of the EHRs one by one (denoted as $\{M_1, M_2, ..., M_n\}$, respectively. We illustrate the communication diagram demonstrating the phases of the protocols for the group consultation case in Fig. 3.

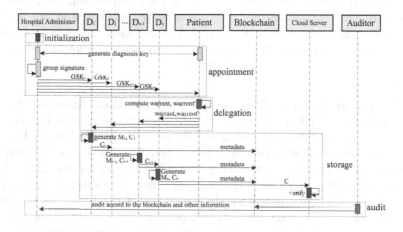

Fig. 3. Communication diagram for the group consultation case

Different from the single doctor case, we mainly adopt group signature to archive our security and privacy goals besides key exchange protocol and BLS short signature. HA acts as the group manager and sets up the group signature. The doctors act as the group members. Then, HA designates the group of doctors for P_i. At the same time, HA generates and distributes the same key pair (PK_{gt}, SK_{gt}) to the group members. Hence, P_i only needs to delegate for the group by performing the short signature, instead of delegating the doctors one by one in the delegation phase. Correspondingly, Every D_j $(j \in 1, 2, ...n)$ performs signature algorithm of the group signature, and records group signature as the metadata to the blockchain. Therefore, Au asks HA to open the group signature to identify the malicious doctor in the audit phase.

3.3 Security and Privacy Analysis

We analyze how protocols are resilient to the following security and privacy attacks.

The Confidentiality of the Patients' EHRs Can Be Ensured in the Proposed Scheme. The EHRs is encrypted by the diagnosis key. The diagnosis key is generated by the key exchange protocol and transmitted by the public key encryption. The key exchange protocol is secure under the CDH assumption. The public key encryption makes only the delegated doctor can get the diagnosis key. Hence, the EHRs cannot be learned by any adversary.

The Proposed Scheme Can Resist Linkage Attacks. We adopt the blinded key pair to protect the identities of the doctors in the single doctor case, which is secure under the DDH assumption[1]. While in the group consultation case, each doctor signs the transaction by using the same private key, and the metadata only includes the group information instead of the doctor identifies. Hence, the linkage attacks are resisted both in single doctor case and group consultation case.

The Proposed Scheme Can Resist Modification Attacks, Forgery Attacks and Collusion Attacks. When there are no collusion attacks, the doctors cannot perform modification and forgery attacks. On the one hand, the semi-trusted doctor outsources new EHRs to the cloud server, as well as records the metadata to the blockchain. Then, the cloud server covers up the target EHRs or stores the EHRs as a new one, and records metadata to the transaction. The transaction guarantees the timeliness of the EHRs, so the auditor can still trace the adversary when the dispute occurs. On the other hand, the doctor may not record the metadata or record the incorrect metadata to the transaction. In this case, the cloud server can discover the misbehaving doctor and do not store the EHRs. If the rational cloud server colludes with the doctor to replace the existing EHRs, the cloud server and the doctor have to fork the blockchain to record the new metadata.

---

[1] Specifically, we require DDH holds in group G_2 equipped with a bilinear map. Sometimes this assumption is referred to as the symmetric external Diffie-Hellman assumption.

The Goal of Accountability Can Be Achieved in the Proposed Scheme. The auditor can trace the real-world identity of a malicious attacker with the help of the hospital administrator when disputes occur. The hospital administrator charges for the generation the blinded key pair in the single doctor case, and also manages the group signature in the group consultation case. Hence, the auditor can trace the adversary with the help of the hospital administrator.

4   Conclusion and Future Works

In this paper, we first introduced the metadata privacy and linkage attacks in the blockchain-based healthcare systems. Further, we elaborated on why the metadata privacy and linkage attacks might breach the patients' privacy. However, this issue received relatively few attention in previous works. With this observation in mind, we proposed different metadata privacy preservation protocols for different medical scenarios. Specially, the proposed scheme is compatible with the existing Ethereum-based healthcare systems, and facilitates the use of blockchain for healthcare system without compromising privacy. In the future, we plan to evaluate the performance of the protocols with experiments, including communication costs, computation costs and gas price of a transaction. We also plan to optimise the protocols.

Acknowledgements. This work was supported in part by the NSFC grants (61941121, 62172423, and 91846204). And part of this work was done when the first author was visiting the Hong Kong Polytechnic University.

References

1. Aguiar, E.J.D., Faiçal, B.S., Krishnamachari, B., Ueyama, J.: A survey of blockchain-based strategies for healthcare. ACM Comput. Surv. **53**(2), 1–27 (2020)
2. Amofa, S., et al.: A blockchain-based architecture framework for secure sharing of personal health data. In: 20th IEEE International Conference on e-Health Networking, Applications and Services, Healthcom 2018, Ostrava, Czech Republic, September, pp. 1–6 (2018)
3. Armknecht, F., Bohli, J., Karame, G.O., Liu, Z., Reuter, C.A.: Outsourced proofs of retrievability. In: Proceedings of the 2014 ACM SIGSAC Conference on Computer and Communications Security, Scottsdale, AZ, USA, 3–7 November 2014, pp. 831–843 (2014)
4. Azaria, A., Ekblaw, A., Vieira, T., Lippman, A.: MedRec: using blockchain for medical data access and permission management. In: International Conference on Open & Big Data (2016)
5. Cao, S., Zhang, G., Liu, P., Zhang, X., Neri, F.: Cloud-assisted secure ehealth systems for tamper-proofing EHR via blockchain. Inf. Sci. **485**, 427–440 (2019)
6. Dubovitskaya, A., Xu, Z., Ryu, S., Schumacher, M., Wang, F.: Secure and trustable electronic medical records sharing using blockchain. In: AMIA 2017, American Medical Informatics Association Annual Symposium, Washington, DC, USA, 4–8 November 2017 (2017)

7. Huang, H., Zhu, P., Xiao, F., Sun, X., Huang, Q.: A blockchain-based scheme for privacy-preserving and secure sharing of medical data. Comput. Secur. **99**, 102010 (2020)
8. Kuo, T., Rojas, H.Z., Ohno-Machado, L.: Comparison of blockchain platforms: a systematic review and healthcare examples. J. Am. Med. Inform. Assoc. **26**(5), 462–478 (2019)
9. Shi, S., He, D., Li, L., Kumar, N., Khan, M.K., Choo, K.R.: Applications of blockchain in ensuring the security and privacy of electronic health record systems: a survey. Comput. Secur. **97**, 1–6 (2020)

Blockchain-Based Encrypted Image Storage and Search in Cloud Computing

Yingying Li[1], Jianfeng Ma[1(✉)], Yinbin Miao[1], Ximeng Liu[2], and Qi Jiang[1]

[1] School of Cyber Engineering, Xidian University, Xi'an, China
lyylyingying@163.com, jfma@mail.xidian.edu.cn,
{ybmiao,jiangqixdu}@xidian.edu.cn
[2] College of Mathematics and Computer Science, Fuzhou University, Fuzhou, China
snbnix@gmail.com

Abstract. In the traditional cloud computing system, existing ranked image search mechanisms will leak the similarity of search results, even cannot prevent malicious cloud servers from tampering with or forging results. Therefore, with the help of blockchain technology, we propose an encrypted image storage and search system. A threshold encryption method based on the matrix similarity transformation theory is designed to protect the similarity of search results. Besides, a hybrid storage structure based on blockchain and cloud server is designed to ensure the correctness and completeness of search results. Finally, we implement the prototype in Python and Solidity, and conduct performance evaluations on local Ethereum client. The theoretical analysis and performance evaluations using the real-world dataset demonstrate that our proposed system is secure and feasible.

Keywords: Images storage and search · Blockchain · Smart contract · Cloud computing · Verifiable

1 Introduction

Although the cloud computing paradigm allows resource-limited data owners to enjoy flexible and convenient image storage and search services, the security and privacy concerns still impede its widely deployments in practice due to semi-honest cloud servers. Encryption is an alternative solution to guarantee data security, but it makes the search over encrypted images impossible. Moreover, the ranked similarity image search [6–8,11] based on the secure k-Nearest Neighbor (kNN) algorithm [10] may leak the similarity of search results as the cloud server can obtain the search results by ranking the calculated inner products. If the cloud server infers the content of queried image through some background knowledge, it can further infer the content of search result with the highest similarity, thereby knowing the content of outsourced image. The consequences of data leakage would be too ghastly to contemplate. Therefore, we need to design a similarity search method that does not reveal the similarity of search results.

In addition, traditional schemes [3,6,7,11] store the encrypted images and indexes in the cloud server, and enable the cloud server to return relevant results. However, the cloud server may be malicious, which executes a fraction of search operations and returns some tampered or forged results. To this end, scheme [8] employs the Merkle hash tree to verify the correctness of search results. With the transparency and immutability of blockchain, there are also some schemes [1,4,5] using blockchain and smart contract technologies, in which the hash values of encrypted images are stored in the blockchain and the smart contract or data user checks the correctness of search results. However, these schemes still cannot guarantee the completeness of search results.

In this paper, with the help of blockchain, smart contract and cloud computing technologies, we design an encrypted images storage and search system, where the data owner stores encrypted images in the cloud server and encrypted indexes in the blockchain. When the data user issues a query, the smart contract searches indexes and the cloud server searches images. Specifically, the main contributions of our work are as follows:

- First, we build a hybrid storage structure based on cloud server and blockchain, where the cloud server stores encrypted images and the blockchain stores hash values and encrypted indexes by designing a STORAGE contract.
- Second, we design a threshold encryption method of encrypting image indexes and queries, which not only realizes similarity search but also protects the privacy of search results.
- Third, we guarantee the correctness and completeness of search results by designing a SEARCH contract. Meanwhile, we allow the data owner to update (*i.e.*, add, delete, modify) images by designing an UPDATE contract.

2   Related Work

Data Storage and Search with Blockchain. The transparency and immutability of blockchain attract attention in the field of searchable encryption. Hu *et al.* [5] constructed a decentralized privacy-preserving search scheme by replacing the central server with a smart contract, in which data owner can receive correct search results. Cai *et al.* [1] utilized the smart contract and fair protocol to avoid service peers returning partial or incorrect results and clients slandering the service peers. Guo *et al.* [4] proposed a verifiable and dynamic searchable encryption scheme with forward security based on blockchain. However, these schemes realize the encrypted text search rather than image search.

Secure Encrypted Image Search. There are many schemes aimed at achieving high accuracy, efficiency and security. Li *et al.* [8] employed the pre-trained Convolutional Neural Network (CNN) model to obtain the high search accuracy. In addition, scheme [3] using Homomorphic Encryption (HE) algorithm and scheme [9] using secure multi-party computing to encrypt feature vectors achieved the same accuracy as plaintext search. However, they incurred high computational costs. To end this, schemes *et al.* [6,11] built an efficient clustering index tree encrypted with the secure kNN algorithm. Moreover, scheme [7] encrypted feature vectors with the secure kNN based on Learning with Errors (LWE-kNN) method, which further improved the data security.

3   Proposed System

Problem Formulation. Suppose a certain **Data Owner** (DO, *e.g.*, Alice) extracts n feature vectors $\{\vec{f_1}, \vec{f_2}, \cdots, \vec{f_n}\}$ from n images $\mathcal{M} = \{m_1, m_2, \cdots, m_n\}$, where $\vec{f_{ID}} = (f_{ID,1}, f_{ID,2}, \cdots, f_{ID,d})$ is a d-dimension vector, $ID \in [1, n]$. Then, Alice encrypts each $\vec{f_{ID}}$ as I_{ID} and each m_{ID} as c_{ID}. Finally, Alice outsources all encrypted vectors $\mathcal{I} = \{I_1, I_2, \cdots, I_n\}$ to blockchain and encrypted images $\mathcal{C} = \{c_1, c_2, \cdots, c_n\}$ to **Cloud Server** (CS). Now, consider an authorized **Data User** (DU, *e.g.*, Bob) with query image m_q, Bob extracts the feature vector $\vec{q} = (q_1, q_2, \cdots, q_d)$ from m_q, then encrypts $\vec{q}$ as $\mathcal{T}_q$ and sends $\mathcal{T}_q$ to blockchain. During querying process, $\mathcal{T}_q$, $\mathcal{I}$ and $\mathcal{C}$ should not be revealed. We denote search results by $\mathcal{R}$. Then, the problem we research can be defined: $Search(\mathcal{T}_q, \mathcal{I}, \mathcal{C}) \rightarrow \mathcal{R}$. We emphasize that the correctness, completeness and privacy of $\mathcal{R}$ should be ensured and protected.

System Constructions. Our system consists of four phases: initialization, storage, retrieval and update.

Phase 1: Initialization. First, DO (or DU) generates the image encryption key k_m and the search key $SK = \{M_1, M_2, M_1^{-1}, M_2^{-1}, \pi\}$ when he/she joins the system. Here, $M_1, M_2 \in \mathbb{R}^{(d+5)\times(d+5)}$ are two random matrices, M_1^{-1}, M_2^{-1} are corresponding inverse matrices, $\pi : \mathbb{R}^{d+5} \rightarrow \mathbb{R}^{d+5}$ is a random permutation, d is the dimension of extracted feature vector. Then, DO deploys the STORAGE, SEARCH and UPDATE contracts.

Phase 2: Storage. First, DO encrypts $\mathcal{M}$ with the symmetric encryption algorithm (*i.e.*, AES) and uploads the ciphertexts $\mathcal{C} = \{\langle 1, c_1 \rangle, \langle 2, c_2 \rangle, \cdots, \langle n, c_n \rangle\}$ to CS. Then, for the feature vector $\vec{f_{ID}} = (f_{ID,1}, f_{ID,2}, \cdots, f_{ID,d})$ of image m_{ID}, DO generates the encrypted index I_{ID} by using Algorithm 1. All indexes are set as the index set $\mathcal{I} = \{I_1, I_2, \cdots, I_n\}$. Finally, DO calls STORAGE contract

---

Algorithm 1: *GenIndex* (for the image m_{ID})

Input: $\vec{f_{ID}} = (f_{ID,1}, f_{ID,2}, \cdots, f_{ID,d})$, SK, c_{ID}.
Output: I_{ID}.

1 Select two random numbers α and r_f;
2 Extend $\vec{f_{ID}}$ to $\vec{f'_{ID}} = (\alpha \vec{f_{ID}}, \alpha, -\alpha \sum_{i=1}^{d} f_{ID,i}^2, \alpha, 0, r_f)$;
3 Compute $\vec{f''_{ID}} = \pi(\vec{f'_{ID}})$;
4 Transform $\vec{f''_{ID}}$ to a diagonal matrix F_{ID} with $\text{diag}(F_{ID}) = \vec{f''_{ID}}$;
5 Generate a random $(d+5) \times (d+5)$ lower triangular matrix $S_{f,ID}$ with the diagonal entries fixed as 1;
6 Compute $C_{ID} = M_1 \cdot S_{f,ID} \cdot F_{ID} \cdot M_2$;
7 Compute $h_{ID} = h(ID||c_{ID})$;
8 Set $\langle ID, h_{ID}, C_{ID} \rangle$ as the index I_{ID} of m_{ID}.

---

functions (see to Algorithm 2) to store all indexes on the blockchain. Since a matrix cannot be sent to the smart contract through a transaction, DO first sends the ID and h_{ID}, and then sends the C_{ID} row by row. The indexes of n images are uploaded one by one.

Algorithm 2: Contracts

```
// STORAGE contract                          // UPDATE contract
1 Mapping ID to structure Index;             // Function
  // Function StIDhash(ID, h_ID):                InsertIDhash(ID', h'_ID):
1 Index[ID].ID = ID;                         1 STORAGE.StIDhash(ID', h'_ID);
2 Index[ID].h_ID = h_ID;                      // Function
  // Function StCID(ID, i, C_ID[i]):             InsertCID(ID', i, C'_ID[i]):
1 for each row i do                          1 STORAGE.StCID(ID', i, C'_ID[i]);
2 |   Index[ID].C_ID[i] = C_ID[i];            // Function
  end                                            ModifyIDhash(ID, h''_ID):
                                             1 STORAGE.StIDhash(ID, h''_ID);
  // SEARCH contract                          // Function
  // Function UploadT(T_q):                      ModifyCID(ID, i, C''_ID[i]):
3 for each row i do                          1 STORAGE.StCID(ID, i, C''_ID[i]);
4 |   Store T_q[i];                           // Function DeleteIDhash(ID):
  end                                        1 delete Index[ID].ID;
  // Function Compute():                     2 delete Index[ID].h_ID;
1 for each I_ID do                            // Function DeleteCID(ID, i):
2 |   Compute S_ID,q = Tr(C_ID · T_q);       1 delete Index[ID].C_ID[i];
3 |   if S_ID,q ≥ 0 then
4 |   |   Add ⟨ID, h_ID⟩ to the proof list
  |   |   PL;
5 |   |   return PL;
  |   end
  end
```

Algorithm 3: GenQuery

Input: $\vec{q} = (q_1, q_2, \cdots, q_d)$, SK.
Output: T_q.

1 Select two random numbers β and r_q, and define a threshold θ;
2 Extend $\vec{q}$ to $\vec{q}' = (2\beta\vec{q}, -\beta\sum_{i=1}^{d} q_i^2, \beta, \beta\theta^2, r_q, 0)$;
3 Permute $\vec{q}'$ with the random permutation π to obtain $\vec{q}'' = \pi(\vec{q}')$;
4 Transform $\vec{q}''$ to a diagonal matrix Q with $\text{diag}(Q) = \vec{q}''$;
5 Generate a random $(d+5) \times (d+5)$ lower triangular matrix S_q with the diagonal entries fixed as 1;
6 Compute $T_q = M_2^{-1} \cdot Q \cdot S_q \cdot M_1^{-1}$;
7 Set T_q as the query of m_q.

Phase 3: Retrieval. First, the authorized DU extracts the feature vector $\vec{q} = (q_1, q_2, \cdots, q_d)$ from the queried image m_q. Then, DU generates a query T_q with Algorithm 3. Next, DU calls the SEARCH contract function $UploadT(T_q[i])$ to upload T_q and calls $Compute()$ to retrieve $\mathcal{I}$ (see to Algorithm 2). The proof list PL including ID and h_{ID} will be returned. To obtain the corresponding search result $\mathcal{R}$, DU submits IDs to CS. Finally, DU calculates the hash value of encrypted image in $\mathcal{R}$ one by one and verifies whether it is equal to the returned h_{ID}. If they are equal, DU accepts and decrypts $\mathcal{R}$ with symmetric encryption algorithm (*i.e.*, AES), otherwise rejects.

Phase 4: Update. 1) Insert m'_{ID}. DO first encrypts m'_{ID} to c'_{ID} and sends c'_{ID} to CS. Then, DO submits a transaction to insert ID' and h'_{ID} by calling the **UPDATE** contract function $InsertIDhash(ID', h'_{ID})$ (see to Algorithm 2). Finally, DO inserts C'_{ID} by calling $InsertCID(ID', i, C'_{ID}[i])$. *2) Modify* m_{ID} *to* m''_{ID}. DO first encrypts m''_{ID} to c''_{ID} and sends c''_{ID} to CS. Then, DO submits a transaction to modify h_{ID} to h''_{ID} by calling the **UPDATE** contract function $ModifyIDhash(ID, h''_{ID})$. Finally, DO modifies C_{ID} to C''_{ID} by calling $ModifyCID(ID, i, C''_{ID}[i])$. *3) Delete* m_{ID}. DO first sends ID to CS. Then, DO submits a transaction to delete ID and h_{ID} by calling the **UPDATE** contract function $DeleteIDhash(ID)$, where **delete** in Algorithm 2 is the library function of Solidity. Finally, DO deletes C_{ID} by calling $DeleteCID(ID, i)$.

4 Theoretical Analysis

Privacy Preservation. Under the known-plaintext attack model, the adversary $\mathcal{A}$ (*i.e.*, CS) has some queries and corresponding ciphertexts. To prove that our system is secure against known-plaintext attack, we first define the experiment $Exp_\mathcal{A}$: $\mathcal{A}$ selects two query vectors $\vec{q_0}$ and $\vec{q_1}$ to the challenger $\mathcal{B}$. Then, $\mathcal{B}$ chooses a uniform bit $b \in \{0, 1\}$, computes the ciphertext $\mathcal{T}$ of $\vec{q_b}$ and returns $\mathcal{T}$ to $\mathcal{A}$. After that, $\mathcal{B}$ sends some query vectors and the corresponding ciphertexts to $\mathcal{A}$. Finally, $\mathcal{A}$ outputs a bit b'. The output of experiment $Exp_\mathcal{A}$ is 1 if $b = b'$, and 0 otherwise. We say that the privacy of index and query is secure against known-plaintext attack if for any polynomial-time adversary $\mathcal{A}$, there is a negligible function *negl* such that the probability $|Pr(Exp_\mathcal{A} = 1) - \frac{1}{2}| \leq negl$.

Theorem 1. *Our system can ensure the confidentiality of index and query under the known-plaintext attack.*

Proof. The main idea of the proof of this theorem is as follows[1]. For $\vec{q_0}$ and $\vec{q_1}$, according to Algorithm 3, $\vec{q_0}$ is encrypted as $\mathcal{T}_0 = M_2^{-1} \cdot Q_0 \cdot S_0 \cdot M_1^{-1}$, where S_0 is a random matrix. According to $Exp_\mathcal{A}$ and Algorithm 1, we know that M_2^{-1} and M_1^{-1} are fixed, β, θ, r_0 and S_0 are one-time random numbers. Thus, the entries in $\mathcal{T}_0$ look random to $\mathcal{A}$. This means that, for any query vector $\vec{q}$ and its corresponding ciphertext, $\mathcal{A}$ cannot distinguish which vector is actually encrypted. Thus, $\mathcal{A}$ can only output b' by randomly guessing. As a result, we have $|Pr(Exp_\mathcal{A} = 1) - \frac{1}{2}| \leq negl$. For index vectors, the proof is almost the same as the one for query vector and is omitted.

Correctness. We prove the correctness of threshold encryption. For a square matrix X, the trace $\mathsf{Tr}(X)$ denotes the sum of diagonal elements of X. Multiplying X by the same size of matrix A and its inverse matrix A^{-1} is called

---

[1] The readers can refer to the reference [12].

similarity transformation of X. According to linear algebra, the trace of a square matrix remains unchanged under similarity transformation. Thus, we have $\mathsf{Tr}(X) = \mathsf{Tr}(AXA^{-1})$. Furthermore, we have the following theorem.

Theorem 2. $S_{ID,q} \geq 0$ *if and only if* $||\vec{f_{ID}} - \vec{q}||^2 \leq \theta^2$.

Proof. Since $S_{ID,q} = \mathsf{Tr}(C_{ID} \cdot T_q)$, according to the law of matrix multiplication, the diagonal of $C_{ID} \cdot T_q$ is the same as that of $F_{ID} \cdot Q$. Thus we have $\mathsf{Tr}(C_{ID} \cdot T_q) = \vec{f_{ID}} \cdot \vec{q}^{\top}$. Since the square of Euclidean distance $||\vec{f_{ID}} - \vec{q}||^2 = \sum_{i=1}^{d} f_{ID,i}^2 + \sum_{i=1}^{d} q_i^2 - 2\vec{f_{ID}} \cdot \vec{q}^{\top}$, we have $\vec{f_{ID}} \cdot \vec{q}^{\top} = \alpha\beta(2\vec{f_{ID}} \cdot \vec{q}^{\top} + \theta^2 - \sum_{i=1}^{d} f_{ID,i}^2 - \sum_{i=1}^{d} q_i^2) = \alpha\beta(\theta^2 - ||\vec{f_{ID}} - \vec{q}||^2)$. Since α and β are positive random numbers, $S_{ID,q} \geq 0$ if and only if $||\vec{f_{ID}} - \vec{q}||^2 \leq \theta^2$. As a result, the search results can be found by judging whether $S_{ID,q} \geq 0$.

Completeness. In traditional schemes, the correctness and completeness of search results depend on whether CS performs the search operation honestly. However, the malicious CS may tamper with or forge results, resulting in DU cannot enjoy the reliable services. In our system, the SEARCH contract processes the query request and returns the similar results to DU. The consensus property of blockchain ensures that the deployed smart contracts can be executed correctly. Therefore, the search results obtained by DU are correct and complete.

5   Performance Evaluations

We evaluate the performance on the real-world dataset Caltech256 [2]. For threshold encryption (denoted by TE), we compare its costs of the encryption and search operations with the secure kNN encryption (denoted by kNN)[2] and the homomorphic encryption (denoted by HE)[3] technologies. We implement them using Python and construct the Ethereum smart contracts using Solidity 0.4.21. The experimental environment is deployed on a PC with Ubuntu 18.04 LTS operating system. The smart contracts are compiled using Remix and tested with the Ethereum blockchain using a local simulated network Geth. The consensus protocol used in Geth is POA (Proof of Authority).

Threshold Encryption. We give the costs of encrypting and searching data in Table 1. Although TE has more search cost than kNN, its security is higher. HE has higher security, but its cost is so expensive that it is impractical. Therefore, TE is a better choice than kNN and HE in terms of security and efficiency.

Smart Contracts. The gas costs of deploying STORAGE, SEARCH and UPDATE contracts are 284711, 534050 and 338372, respectively. When DO stores indexes on blockchain, the time and gas costs are displayed in Table 2. In order to avoid

---

[2] Splitting index and query vectors with a random Boolean vector, and then encrypting them with two random matrices and their inverse matrices.

[3] Using Paillier homomorphic algorithm to encrypt index and query vectors.

Table 1. The time costs of encrypting and searching data

Operations	Methods	d	1000	2000	3000	4000	5000	6000	7000	8000	9000	10000
Encrypt index	kNN(s)	32	0.255	0.357	0.501	0.621	0.743	0.803	0.91	1.038	1.157	1.726
		64	0.271	0.438	0.673	0.777	0.883	0.935	1.021	1.453	1.57	1.788
		128	0.326	0.476	0.688	0.831	0.98	1.127	1.331	1.71	1.841	2.044
	HE(h)	32	1.031	2.009	2.953	3.925	5.026	5.923	7.076	7.883	8.795	8.895
		64	1.955	3.960	5.811	8.109	9.794	11.991	13.692	16.302	18.618	20.785
		128	4.118	8.046	12.346	16.167	21.401	25.389	29.237	33.673	39.459	44.889
	TE(s)	32	0.361	0.504	0.749	1.042	1.321	1.455	1.877	2.39	2.796	2.92
		64	0.407	0.841	1.049	1.128	1.57	1.856	2.087	2.436	3.349	4.755
		128	0.326	0.476	0.688	0.831	0.98	1.127	1.331	1.71	1.841	2.044
Search	kNN(s)	32	0.009	0.014	0.017	0.022	0.028	0.033	0.04	0.045	0.048	0.053
		64	0.017	0.02	0.033	0.032	0.039	0.047	0.054	0.063	0.066	0.071
		128	0.019	0.034	0.051	0.057	0.071	0.091	0.097	0.121	0.149	0.18
	HE(min)	32	2.023	3.961	5.882	8.174	10.529	12.063	14.214	15.789	17.676	19.944
		64	3.407	7.049	10.597	14.186	18.002	21.292	27.424	29.088	34.593	38.055
		128	6.843	13.520	20.257	27.182	36.317	42.030	49.097	57.020	69.193	79.226
	TE(s)	32	0.014	0.028	0.044	0.096	0.146	0.149	0.151	0.199	0.16	0.247
		64	0.024	0.049	0.071	0.095	0.147	0.176	0.178	0.241	0.272	0.41
		128	0.096	0.118	0.214	0.279	0.98	0.323	0.383	0.42	0.494	0.554

Table 2. The time and gas costs of storage and search

Operations	d	100	200	300	400	500	600	700	800	900	1000
Index storage time (h)	32	0.03	0.07	0.11	0.15	0.21	0.23	0.28	0.32	0.37	0.43
	64	0.07	0.17	0.28	0.39	0.52	0.67	0.79	0.91	1.08	1.26
	128	0.27	0.64	1	1.65	2.09	2.36	2.66	3.20	5.35	7.75
Index storage gas ($\times 10^9$)	32	2.22	4.44	6.66	8.88	11.11	13.33	15.54	17.76	19.98	22.21
	64	7.80	16.10	24.14	32.19	40.24	48.30	56.33	64.39	72.45	80.45
	128	30.58	61.12	91.67	122.21	152.79	183.31	213.84	244.42	275.02	305.53
Search time (s)	32	3.31	7.09	8.67	7.89	10.06	14.73	20.54	25.4	22.95	25.46
	64	9.53	17.79	33.09	39.55	59.11	65.43	87.15	103.91	109.9	142.45
	128	24.52	50.04	78.02	110.03	126.96	116.64	192.33	226.73	239.31	255.56

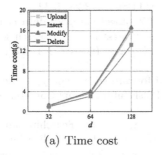

(a) Time cost

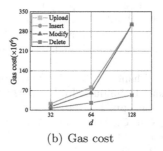

(b) Gas cost

Fig. 1. The time and gas costs of uploading a query and update.

exceeding the gas limit of block, uploading a row vector maybe require two or more transactions. When the number of transactions increases, the time and gas costs also increase. As for uploading a query, see to Fig. 1, when the d increases from 32 to 128, the time and gas costs increase by 4 times because the size of $\mathcal{T}_q$ increases by 4 times. In the search phase, compared with the search time in Table 1, the search time in Table 2 is longer because the Solidity takes longer to calculate the matrix multiplication than Python. When DO update indexes, as shown in Fig. 1, the cost of insertion is close to that of modification, as the insertion and modification functions in Algorithm 2 both invoke the STORAGE contract functions.

6   Conclusion

In this paper, we propose a threshold encryption method, which returns all similar images within a given threshold. Further, we propose a dynamic and verifiable encrypted image storage and search system based on blockchain and cloud computing, which not only verifies the correctness and completeness of search results but also allows the image owner to update the images and indexes. The final theoretical analysis and experimental analysis demonstrate the feasibility of our proposed system. Although the security has been improved in this paper, search efficiency still needs to be improved. In future, we will continue to study vector encryption methods that improve security without sacrificing efficiency.

Acknowledgements. This work was supported by the National Natural Science Foundation of China (No. 62072361), the Fundamental Research Funds for the Central Universities (No. JB211505), Henan Key Laboratory of Network Cryptography Technology, State Key Laboratory of Mathematical Engineering and Advanced Computing (No. LNCT2020-A06).

References

1. Cai, C., Weng, J., Yuan, X., Wang, C.: Enabling reliable keyword search in encrypted decentralized storage with fairness. IEEE Trans. Dependable Secure Comput., 1 (2018). https://doi.org/10.1109/TDSC.2018.2877332
2. Griffin, G., Holub, A., Perona, P.: Caltech-256 object category dataset (2007)
3. Guo, C., Su, S., Choo, K.R., Tang, X.: A fast nearest neighbor search scheme over outsourced encrypted medical images. IEEE Trans. Industr. Inf. **17**(1), 514–523 (2021)
4. Guo, Y., Zhang, C., Jia, X.: Verifiable and forward-secure encrypted search using blockchain techniques. In: ICC 2020–2020 IEEE International Conference on Communications, pp. 1–7 (2020)
5. Hu, S., Cai, C., Wang, Q., Wang, C., Luo, X., Ren, K.: Searching an encrypted cloud meets blockchain: a decentralized, reliable and fair realization. In: IEEE INFOCOM 2018 - IEEE Conference on Computer Communications, pp. 792–800 (2018)

6. Li, X., Xue, Q., Chuah, M.: Casheirs: cloud assisted scalable hierarchical encrypted based image retrieval system. In: Proceedings of the IEEE Conference on Computer Communications, pp. 1–9. IEEE (2017)
7. Li, Y., Ma, J., Miao, Y., Wang, Y., Liu, X., Choo, K.R.: Similarity search for encrypted images in secure cloud computing. IEEE Trans. Cloud Comput., (2020). https://doi.org/10.1109/TCC.2020.2989923
8. Li, Y., Ma, J., Miao, Y., Liu, L., Liu, X., Choo, K.K.R.: Secure and verifiable multikey image search in cloud-assisted edge computing. IEEE Trans. Industr. Inf. **17**(8), 5348–5359 (2020)
9. Shen, M., Cheng, G., Zhu, L., Du, X., Hu, J.: Content-based multi-source encrypted image retrieval in clouds with privacy preservation. Futur. Gener. Comput. Syst. **109**, 621–632 (2020)
10. Wong, W.K., Cheung, D.W., Kao, B., Mamoulis, N.: Secure KNN computation on encrypted databases. In: Proceedings of the 2009 ACM SIGMOD International Conference on Management of Data, pp. 139–152 (2009)
11. Yuan, J., Yu, S., Guo, L.: Seisa: secure and efficient encrypted image search with access control. In: 2015 IEEE Conference on Computer Communications (INFOCOM), pp. 2083–2091. IEEE (2015)
12. Zhou, K., Ren, J.: Passbio: privacy-preserving user-centric biometric authentication. IEEE Trans. Inf. Forensics Secur. **13**(12), 3050–3063 (2018)

Applications of Algorithms

Improving Information Cascade Modeling by Social Topology and Dual Role User Dependency

Baichuan Liu⬛, Deqing Yang(✉)⬛, Yuchen Shi, and Yueyi Wang

School of Data Science, Fudan University, Shanghai 200433, China
{bcliu20,yangdeqing}@fudan.edu.cn
{ycshi21,yueyiwang21}@m.fudan.edu.cn

Abstract. In the last decade, information diffusion (also known as information cascade) on social networks has been massively investigated due to its application values in many fields. In recent years, many sequential models including those models based on recurrent neural networks have been broadly employed to predict information cascade. However, the user dependencies in a cascade sequence captured by sequential models are generally unidirectional and inconsistent with diffusion trees. For example, the true trigger of a successor may be a non-immediate predecessor rather than the immediate predecessor in the sequence. To capture user dependencies more sufficiently which are crucial to precise cascade modeling, we propose a non-sequential information cascade model named as **TAN-DRUD** (**T**opology-aware **A**ttention **N**etworks with **D**ual **R**ole **U**ser **D**ependency). TAN-DRUD obtains satisfactory performance on information cascade modeling through capturing the dual role user dependencies of information sender and receiver, which is inspired by the classic communication theory. Furthermore, TAN-DRUD incorporates social topology into two-level attention networks for enhanced information diffusion prediction. Our extensive experiments on three cascade datasets demonstrate that our model is not only superior to the state-of-the-art cascade models, but also capable of exploiting topology information and inferring diffusion trees.

Keywords: Information cascade · Information diffusion · User dependency · Social networks · Diffusion tree

1 Introduction

With the development of online social networks, various information spreads more quickly and broadly on web. The diffusion of a piece of information generally forms a cascade among different users in the network, which is often observed as a sequence of *activated users* who have disseminated the information. The

This work was supported by Shanghai Science and Technology Innovation Action Plan No. 21511100401.

A. Bhattacharya et al. (Eds.): DASFAA 2022, LNCS 13245, pp. 425–440, 2022.
https://doi.org/10.1007/978-3-031-00123-9_35

precise prediction of information diffusion (a.k.a. information cascade) has been proven crucial to some valuable applications, including market forecast [17], community detection [1], etc.

Inspired by the success of deep neural networks in computer vision, recommender systems, etc., more and more researchers also employed DNNs to model information cascade. In recent years, some deep models based on recurrent neural networks (RNNs) [9,25], including the RNNs coupled with attention mechanism [21], have been proposed to predict information diffusion, since an information cascade is often modeled as a sequence.

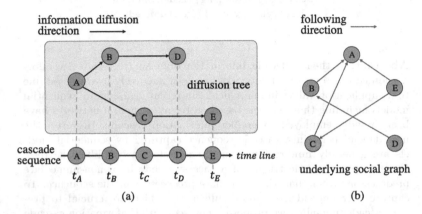

Fig. 1. An example of an information cascade sequence, diffusion tree and social graph.

Despite the effects of these RNN-based models, their sequential assumption may model the user dependencies inconsistent with the diffusion tree. We take the example shown in Fig. 1 to elaborate it. Given the following information cascade $c = \{(A, t_A), (B, t_B), (C, t_C), (D, t_D), (E, t_E)\}$ where user $A \sim E$ are sorted according to their activation (disseminating information) time. User A is the cascade source while user E is last one who disseminates this piece of information. There is a real diffusion tree behind this cascade sequence which does not strictly follow the sequential assumption. Suppose a set of chronological diffusion behaviors is denoted as $\{(A, B, t_B), (A, C, t_C), (B, D, t_D), (C, E, t_E)\}$ where a triplet (A, B, t_B) means A spreads information to B at time t_B. Based on this diffusion behavior set, a real diffusion tree can be reconstructed as shown in the yellow rectangle of Fig. 1(a). The diffusion tree indicates that D is influenced by B rather than C to be activated. The RNN-based models fed only with c would model D to be influenced by C more than B, since they assume one object in a sequence depends on an immediate predecessor rather than a non-immediate predecessors. Such modeled user dependency is inconsistent with the diffusion tree, but could be alleviated if the underlying social graph of these five users (as depicted in Fig. 1(b)) is provided to guide information cascade modeling. It shows that the social topology plays an important role in precise cascade modeling.

Although some sequential models [3,13,14,20,22,26] have taken social topology into account when modeling cascade process, they are still not satisfactory due to their inherent sequential assumption. More recently, a non-sequential model with hierarchical attention networks [24] has demonstrated good performance through capturing non-sequential dependencies among the users in a cascade, but this model still neglects the underlying social graph.

In addition, the user dependencies captured by most of previous diffusion models [21,24,26] are only unidirectional. They suppose a successor is only influenced by a predecessor during information diffusion, whereas the opposite dependency is rarely considered. According to *Laswell's '5W' Communication Model* [10]: Who (sender) says What in Which channel to Whom (receiver) with What effect, we believe that a user's role in the process of an information cascade is not just a single role of sender or receiver, but a *dual role of both sender and receiver*. In other words, information diffusion depends not only on how each user in a cascade behaves as a sender of his/her successors, but also on how each user behaves as a receiver of his/her predecessors. For example, a Twitter user is easy to be influenced by his/her followees who often disseminate appealing information to him/her. Meanwhile, a users may also be influenced by his/her followers when he/she decides to disseminate information, i.e., he/she would consider what kind of information the followers are more likely to receive. Accordingly, capturing such dual role user dependencies instead of single role (unidirectional) user dependencies is helpful to precise cascade prediction. Unfortunately, such intuition was overlooked by previous cascade models.

Inspired by above intuitions, in this paper we propose a non-sequential deep model of information cascade, named as **TAN-DRUD** (**T**opology-aware **A**ttention **N**etworks with **D**ual **R**ole **U**ser **D**ependency). TAN-DRUD is built with two-level attention networks to learn optimal cascade representations, which are crucial to predict information diffusion precisely. At first, the *user-level attention* network is used to learn the *dependency-aware representation* of a user that serves as the basis of a cascade's representation. Specifically, each user is first represented by two separate embeddings corresponding to his/her dual role of information sender and receiver, respectively. In order to exploit the social topology's indicative effects on information cascade modeling, we employ Node2Vec [6] to learn *topological embeddings* upon the social graph, which are used to adjust the attention values. Through our empirical studies, we have verified that Node2Vec is more effective than GNN-based graph embedding model [29] in our model framework. With the topology-adjusted attentions, the user dependencies among a cascade are encoded into *dependency-aware user representations* dynamically and sufficiently. Then, the *cascade-level attention* network is fed with the combination of dependency-aware user representations, topological embeddings and time decay information, to learn the cascade's representation. Since the cascade-level attentions can be regarded as a historical activated user's probability of activating the next user, the diffusion trees can be inferred based on our model. Our contributions in this paper are summarized as follows:

1. We propose a non-sequential cascade model TAN-DRUD with two-level attention networks, which demonstrates satisfactory performance through capturing dual role user dependencies in information cascades. What's more, TAN-DRUD's performance is enhanced with the aid of social topology.
2. In the user-level attention network of TAN-DRUD, we particularly design a sender attention module and a receiver attention module to learn two separate embeddings for a user in a cascade, which encode dual role user dependencies sufficiently. Such manipulation's rationale is inspired by the classic communication model, and has been proven more effective than modeling user dependencies in terms of single role.
3. We conducted extensive experiments on three real cascade datasets, which not only justify our model's advantage over the state-of-the-art diffusion models, but also demonstrate our model's capability of inferring diffusion tree.

The rest of this paper is organized as below. In Sect. 2, we introduce some research works related to information cascade modeling. Next, we formalize the problem addressed in this paper and introduce our proposed model in detail in Sect. 3. In Sect. 4, we display our experiment results based on which we provide further analysis. At last, we conclude our work in Sect. 5.

2   Related Work

Information cascade models in deep learning domain can be divided into two types: diffusion path based models and topological-based diffusion models.

2.1   Diffusion Path Based Methods

DeepCas [13] is the first end-to-end deep learning model for information diffusion prediction. It uses the same way as DeepWalk [15] to sample node sequences from cascade graphs, then bidirectional gated recurrent units (GRU) [4] and attention mechanism are used to process node sequences and get cascade representations for prediction. DCGT [12] is an extended model of DeepCas which incorporates the content of information to predict cascades. [2] is an RNN-based model fed with diffusion sub-paths, in which a non-parametric time attenuation factor is applied on the last hidden state of all sub-paths, to represent the self-excitation mechanism [30] in information diffusion. DeepDiffuse [9] utilizes timestamps and cascade information to focus on specific nodes for prediction. [25] employs self-attention and CNNs to alleviate RNN's disadvantage of long-term dependency. Unlike above sequential models, HiDAN [24] is a non-RNN hierarchical attention model which captures dependency-aware user representations in a cascade, resulting in precise prediction.

2.2   Topological-Based Diffusion Model

TopoLSTM [20] utilizes directed acyclic (social) graph to augment node embedding learning in a cascade. [27] studies the influence of interactions between users' social roles during information diffusion. A role-aware information diffusion model is proposed to combine social role information and diffusion sequences information. [3] builds a shared presentation layer to discover the complementary information of sequence representations and topology representations, which can be applied to micro and macro tasks. FOREST [26] employs reinforcement learning on social graph modeling to solve multi-scale tasks for information diffusion prediction. [14] uses multi self-attention and social graph information to discovery the long-term dependency, which is crucial for diffusion prediction. Inf-vae [16] learns topological embeddings through a VAE framework as encoders and decoders, then an co-attentive fusion network is used to capture complex correlations between topological and temporal embeddings to model their joint effects. [23] discovers relations between information diffusion and social network through an identity-specific latent embeddings. DyHGCN [28] implements a dynamic heterogeneous graph convolutional network to capture users' dynamic preferences between global social graph and temporal cascade graph.

3   Methodology

3.1   Problem Definition

Suppose an observed cascade sequence consists of i users along with their timestamps of information dissemination. This sequence is denoted as

$$c_i = \{(u_1, t_1), (u_2, t_2), \ldots, (u_i, t_i)\}$$

where $u_j (1 \leq j \leq i)$ is a user and t_j is u_j's timestamp. All users are sorted chronologically, i.e., $t_1 < t_2 < \ldots < t_i$. Moreover, an underlying social graph $\mathcal{G} = (\mathcal{V}, \mathcal{E})$ including all users may be obtained, where each user corresponds to a node in set $\mathcal{V}$ and $|\mathcal{V}| = N$. Then, the prediction task of our model is formalized as: given c_i, the model should predict the next user to be activated, denoted as u_{i+1}, through computing the conditional probability $p(u_{i+1}|c_i)$.

3.2   Model Framework

The framework of our model is depicted in Fig. 2. In the user-level attention network, each user is first represented by a *sender embedding* and a *receiver embedding* corresponding to his/her dual role. Then, a user's *receiver-role representation* is learned as the attentive sum of his/her predecessors' sender embeddings. Likewise, a user's *sender-role representation* is learned as the attentive sum of his/her successors' receiver embeddings. Furthermore, we utilize the social *topological embeddings* learned by Node2Vec to adjust the attention values. Next, a forget gate mechanism is used to aggregate a user's dual role representation into the user's *dependency-aware representation*. In cascade-level attention network, dependency-aware user representations, social topological embedding, and time information are combined to generate a cascade's representation for diffusion prediction.

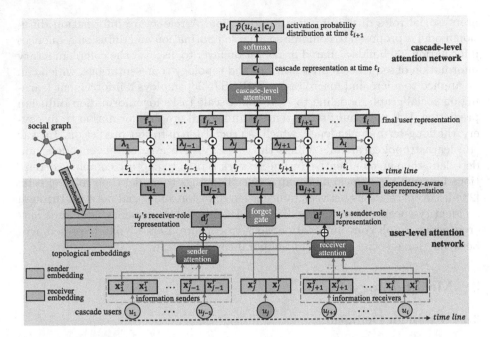

Fig. 2. The framework of our proposed TAN-DRUD consisting of two-level attention networks. In the user-level attention network, each user in a cascade is first represented by a sender embedding and a receiver embedding. With these two embeddings and the topological embeddings, the user's dependency-aware representation is learned. In the cascade-level attention network, a cascade's representation is generated based on dependency-aware user representations, topological embeddings and time decay, with which the next activated user is predicted.

3.3 Embedding Preparation

Sender and Receiver Embeddings. Two separate embeddings are first initialized to represent a given user. We use $\mathbf{X}^s, \mathbf{X}^r \in \mathbb{R}^{N \times d}$ to represent sender embedding matrix and receiver embedding matrix respectively. Each row of the two matrices represents a user. N is the total number of users, and d is embedding dimension (size). All embeddings in $\mathbf{X}^s, \mathbf{X}^r$ are initialized in random.

Social Topological Embeddings. Since $\mathcal{G}$ is a homogeneous network, the models towards heterogeneous networks such as Metapath2Vec [5] are not suitable for our scenario. In addition, we do not prefer the semi-supervised graph neural networks (GNNs), e.g., GCN, because such graph embedding models were primarily designed to aggregate node features which are subject to specific downstream tasks. Through our empirical study, we finally adopted Node2Vec to learn topological embeddings.

3.4   Two-Level Attention Networks

User-Level Attention Network. Given u_j in a cascade $\{(u_1, t_1), \ldots, (u_i, t_i)\}$, we first design a *sender attention module* to learn u_j's receiver-role representation $\mathbf{d}_j^r$ which is computed based on the sender embeddings of $\{u_1, \ldots, u_{j-1}\}$. We symmetrically learn u_j's sender-role representation $\mathbf{d}_j^s$, which is computed based on the receiver embeddings of $\{u_{j+1}, \ldots, u_i\}$ in *receiver attention module* .

Formally, suppose $u_k (1 \leq k \leq j-1)$ is a predecessor of u_j, u_k's sender attention to u_j is α_{kj}^s and computed as

$$\alpha'^s_{kj} = \frac{\exp\left(\langle \mathbf{W}_s^o \mathbf{x}_k^s, \mathbf{W}_r^c \mathbf{x}_j^r \rangle\right)}{\sum_{l=1}^{j-1} \exp\left(\langle \mathbf{W}_s^o \mathbf{x}_l^s, \mathbf{W}_r^c \mathbf{x}_j^r \rangle\right)}, \tag{1}$$

$$\mathbf{E} = \frac{\mathbf{G}_c \mathbf{G}_c^\top}{\|\mathbf{G}_c\| \times \|\mathbf{G}_c^\top\|}, \tag{2}$$

$$\alpha_{kj}^s = \frac{\exp\left(\alpha'^s_{kj} e_{kj}\right)}{\sum_{l=1}^{j-1} \exp\left(\alpha'^s_{lj} e_{lj}\right)}. \tag{3}$$

In Eq. 1, $\mathbf{x}_k^s$ is u_k's sender embedding obtained from $\mathbf{X}^s$ and $\mathbf{x}_j^r$ is u_j's receiver embedding obtained from $\mathbf{X}^r$. $\mathbf{W}_s^o, \mathbf{W}_r^c \in \mathbb{R}^{d \times d}$ are transformation matrices and $<,>$ is inner product. According to Eq. 1, the attention α'^s_{kj} captures u_j's dependency on u_k in terms of the information receiver of u_k. In Eq. 2, $\mathbf{G}_c \in \mathbb{R}^{i \times d_g}$ is the matrix consisting of the topological embeddings of $\{u_1, \ldots, u_{i-1}, u_i\}$. Thus, $\mathbf{E}$ quantifies the social similarities between $u_1 \sim u_i$. Two social similar users are more likely to influence each other during information diffusion. Hence we utilize $\mathbf{E}$'s element e_{kj}, e_{lj} to adjust α'^s_{kj} into α_{kj}^s as Eq. 3. Then u_j's receiver-role representation $\mathbf{d}_j^r$ is computed as follows,

$$\mathbf{d}'^r_j = \sum_{k=1}^{j-1} \alpha_{kj}^s \mathbf{x}_k^s, \quad \mathbf{d}_j^r = \mathbf{d}'^r_j + \mathbf{x}_j^r. \tag{4}$$

Similarly, u_j's sender-role representation $\mathbf{d}_j^s$ is computed as

$$\alpha'^r_{kj} = \frac{\exp\left(\langle \mathbf{W}_r^o \mathbf{x}_k^r, \mathbf{W}_s^c \mathbf{x}_j^s \rangle\right)}{\sum_{l=j+1}^{i} \exp\left(\langle \mathbf{W}_r^o \mathbf{x}_l^r, \mathbf{W}_s^c \mathbf{x}_j^s \rangle\right)}, \tag{5}$$

$$\alpha_{kj}^r = \frac{\exp\left(\alpha'^r_{kj} e_{kj}\right)}{\sum_{l=j+1}^{i} \exp\left(\alpha'^r_{lj} e_{lj}\right)}, \tag{6}$$

$$\mathbf{d}'^s_j = \sum_{k=j+1}^{i} \alpha_{kj}^r \mathbf{x}_k^r, \quad \mathbf{d}_j^s = \mathbf{d}'^s_j + \mathbf{x}_j^s. \tag{7}$$

Next, we need to aggregate u_j's sender-role representation and receiver-role representation into one dependency-aware representation $\mathbf{u}_j$. We use a forget gating mechanism [20] to implement this operation as follows, since it can wisely decide how much input information should be reserved or forgotten to compose the output.

$$\mathbf{m} = \sigma(\mathbf{W}_m^s \mathbf{d}_j^s + \mathbf{W}_m^r \mathbf{d}_j^r + \mathbf{b}_m), \tag{8}$$
$$\mathbf{n} = \sigma(\mathbf{W}_n^s \mathbf{d}_j^s + \mathbf{W}_n^r \mathbf{d}_j^r + \mathbf{b}_n), \tag{9}$$
$$\mathbf{u}_j = (\mathbf{1} - \mathbf{m}) \odot \mathbf{d}_j^s + (\mathbf{1} - \mathbf{n}) \odot \mathbf{d}_j^r \tag{10}$$

where σ is Sigmoid function and $\odot$ is element-wise product. $\mathbf{1} \in \mathbb{R}^d$ is a unit vector. $\mathbf{W}_{m/n} \in \mathbb{R}^{d \times d}$ and $\mathbf{b}_{m/n} \in \mathbb{R}^d$ in above equations are trainable parameters. $\mathbf{d}_j^s$ and $\mathbf{d}_j^r$ generated by the user level attention mechanism are input into this gated model to obtain the user-level representation of the user. Specifically, vector $\mathbf{m}, \mathbf{n} \in \mathbb{R}^d$ are used to control how much information we should forget for each type of inputs.

Cascade-Level Attention Network. Three kinds of information are used in this step: dependency-aware user representation, topological information and time decay. We first map all topological embeddings in $\mathbf{G}_c$ into the embeddings of d dimensions as follows:

$$\mathbf{G}_{new} = \tanh(\mathbf{W}_g \mathbf{G}_c + \mathbf{b}_g) \tag{11}$$

where $\mathbf{G}_{new} \in \mathbb{R}^{i \times d}$ is transformed topological embedding matrix. $\mathbf{W}_g \in \mathbb{R}^{d \times d_g}$ and $\mathbf{b}_g \in \mathbb{R}^d$ are trainable parameters.

To consider time decay, we first set a unit of time interval as $\Delta^t = T_{\max}/T$, where T_{max} is the max time interval observed from all cascades, and T is the number of time intervals. Then, given u_j's time decay interval $\Delta t_j = t_i - t_j$, we get its time decay vector $\mathbf{t}_j \in \mathbb{R}^T$ (one-hot vector), in which only the n-th element is 1 when $n = int(\Delta t_j / \Delta^t)$. $int(\cdot)$ is rounding up operation. We also map $\mathbf{t}_j$ into an embedding with the same size as user representations by:

$$\boldsymbol{\lambda}_j = \sigma(\mathbf{W}_t \mathbf{t}_j + \mathbf{b}_t) \tag{12}$$

where $\mathbf{W}_t \in \mathbb{R}^{d \times T}$ and $\mathbf{b}_t \in \mathbb{R}^d$ are trainable parameters.

Then, u_j's final representations is computed as

$$\mathbf{f}_j = \boldsymbol{\lambda}_j \odot (\mathbf{g}_j + \mathbf{u}_j) \tag{13}$$

where $\mathbf{g}_j$ is u_j's topological embedding obtained from $\mathbf{G}_{new}$ in Eq. 11.

Next, u_j's time-aware influence to the next activated user can be quantified by the following attention:

$$\beta_j = \frac{\exp(\langle \mathbf{w}, \mathbf{f}_j \rangle)}{\sum_{k=1}^{i} \exp(\langle \mathbf{w}, \mathbf{f}_k \rangle)} \tag{14}$$

where $\mathbf{w} \in \mathbb{R}^d$ is a trainable embedding.

At last, for a cascade observed until time t_i, i.e., c_i, its representation $\mathbf{c}_i$ is computed based on the adjusted representations of the users in c_i. Thus we get

$$\mathbf{c}_i = \sum_{j=1}^{i} \beta_j \mathbf{f}_j. \tag{15}$$

3.5 Prediction and Optimization

Given c_i, the activation probability distribution of all users at time t_{i+1} is denoted as $p_i \in \mathbb{R}^N$, and computed as:

$$p_i = \text{softmax}(W_c c_i + b_c) \tag{16}$$

where $W_c \in \mathbb{N}^{d \times T}$ and $b_c \in \mathbb{R}^N$ are trainable parameters.

Given the training set containing M cascade sequences in which the m-th cascade is denoted as c^m and its length is n_m, TAN-DRUD's learning objective is to minimize the following log-likelihood loss:

$$\mathcal{L} = -\frac{1}{M} \sum_{m=1}^{M} \sum_{i=1}^{n_m-1} \log \hat{p}(u_{i+1} \mid c_i^m) + \lambda L_2 \tag{17}$$

where u_{i+1} is the truly activated user at time t_{i+1} given c^m, and $\hat{p}(u_{i+1} \mid c_i^m)$ is fetched from the p_i computed according to Eq. 16. λ is the controlling parameter of L2 regularization. We use stochastic gradient descent and Adam algorithm for optimization.

4 Experiments

In this section, we try to answer the following research questions through our empirical studies.

RQ1: Is our TAN-DRUD more effective and efficient than the state-of-the-art diffusion models?

RQ2: Are the two separate embeddings corresponding to a user's dual role helpful for enhanced prediction performance?

RQ3: Is the incorporated social topology helpful for prediction performance?

RQ4: Is the graph embedding model sensitive to TAN-DRUD's final performance?

RQ5: Can the diffusion trees be recovered approximately by our model?

4.1 Datasets and Baselines

We conducted experiments upon the following three datasets often used in information diffusion prediction to evaluate our model.

Twitter [7]: As the most prevalent social media, tweet spreading among Twitter users is a representative kind of information diffusion in social networks. This dataset contains the tweets with URLs posted in Oct, 2010. The tweets with the same URL are regarded as an information cascade, thus their publishers are the activated users in this cascade.

Douban [31]: It is a review sharing website for various resources including book, movie, music, etc., where users seek their favorite resources based on others' reviews. In this dataset, each book is regarded as a piece of information. A user is regarded as being activated and joining the cascade of a book if he/she has read this book. Similar to Twitter, each user in Douban can also follow others.

MemeTracker [11]: It collects massive news stories and blogs from online websites and tracks popular quotes and phrases as memes. We treat each meme as a piece of information and each website URL as an activated user. Thus, social topology does not exist in this dataset, which is used to evaluate the models without topological information.

The detailed statistics of the datasets are shown in Table 1. We divided all cascades in each datasets into training set, validation set and test set, according to the ratio of 8:1:1. TAN-DRUD's source codes and our experiment samples on https://github.com/JUNJIN0126/TAN-DRUD.

Table 1. Statistics of the three used datasets.

Dataset	Twitter	Douban	Meme
User number	12,627	23,123	4,709
Cascade number	3,442	10,602	12,661
Average cascade length	32.60	27.14	16.24
Social link number	309,631	348,280	–

We compared TAN-DRUD with the following cascade models, including sequential models and non-sequential models to verify our TAN-DRUD's advantages.

DeepDiffuse [9]: It employs embedding technique on timestamps and incorporates cascade information to focus on specific nodes for prediction.

Bi-LSTM [8]: The dual role user dependencies can also be regarded as bidirectional user dependencies in a sequence. So Bi-LSTM consisting of forward LSTM and backward LSTM can be used to model cascade sequences, and then predict the next activated user.

TopoLSTM [20]: It is an LSTM-based model incorporated with diffusion topology, i.e., directed acyclic graph for diffusion prediction.

SNIDSA [22]: It uses attention networks to extract user dependencies from social graph, then adopts a gate mechanism to combine user information and sequential information.

FOREST [26]: It employs reinforcement learning framework fed with social graphs to solve multi-scale tasks for information diffusion.

HiDAN [24]: It is a non-sequential model built with hierarchical attention networks. Compared with TAN-DRUD, it does not establish two separate embeddings for users, and omits social graphs.

In addition, we propose a variant **AN-DRUD** of TAN-DRUD for ablation study, in which social topological embeddings are absent.

4.2   Experiment Settings

We introduce some important settings of our experiments as follows.

Evaluation Metrics. The next infected user prediction can be regarded as a ranking problem based on users' potential probabilities of spreading the information. Thus, we adopted *Precision on top-K ranking* (P@K) and *Reciprocal Rank* (RR) as our evaluation metrics, since they are popular to evaluate sequential ranking [21,24]. Specifically, given a test sample, its P@K = 100% if the true next activated user u_{i+1} is in its top-K list ranked according to $\hat{p}$, otherwise P@K = 0.

General Settings. We ran the experiments on a workstation with GPU of GeForce GTX 1080 Ti. For the baseline diffusion models and graph embedding models, we directly used their public source codes, and tuned their hyperparameters in terms of optimal prediction performance. The topological embedding size (d_g) was set to 128.

TAN-DRUD's Settings. TAN-DRUD's hyper parameters includes learning rate, topological embedding size, user dual role embedding size, time interval number and so on. We set learning rate to 0.001, $\lambda = 1e-5$, and also used Dropout with the keep probability of 0.8 to enhance our model's generalization capability. Due to space limitation, we only display the results of tuning user dual role embedding dimension d and time interval number T.

In general, the embedding dimension (size) in deep learning models is set empirically. User information in cascades will not be captured sufficiently if user dual-role embedding size is too small, while overfitting and high training time consumption may happen if it is too large. Therefore, we set d to different values to investigate its influence on our model's performance. According to the results in Table 2, we set $d = 64$ for our model in the subsequent comparison experiments.

Table 2. TAN-DRUD's prediction performance (score %) with dual role embedding sizes (d).

d	RR	P@10	P@50	P@100
16	14.31	24.33	43.56	53.50
32	15.53	26.30	45.12	54.58
64	**16.62**	**28.13**	**45.61**	**55.43**
128	15.82	27.53	45.17	54.62

The time interval number T is used to generate the time decay vector in Eq. 12. Small T results in coarse-grained representations of time decay between different users in a cascade. Thus the temporal features can not contribute to precise cascade modeling. By contrast, large T leads to the time decay vector of large size, resulting in high training time consumption. According to the results in in Table 3, we set $T = 50$ for our model in the subsequent comparison experiments.

Table 3. TAN-DRUD's prediction performance (score %) with different time interval numbers (T).

T	RR	P@10	P@50	P@100
1	15.95	27.12	45.53	**55.91**
10	15.84	27.26	45.06	55.04
50	**16.62**	**28.13**	45.61	55.43
100	16.32	27.52	**46.04**	55.36

Table 4. Prediction performance (score %) of all compared models for the three datasets. The best performance scores among all compared models are indicated in bold. The performance scores of leading baseline are underlined.

Model	Dataset											
	Twitter				Douban				Meme			
	RR	P@10	P50	P100	RR	P@10	P@50	P@100	RR	P@10	P@50	P@100
DeepDiffuse	2.21	4.45	14.35	21.61	3.23	9.02	14.93	19.13	6.48	13.45	30.10	41.31
Bi-LSTM	7.12	13.41	26.71	36.06	7.95	15.97	29.89	37.41	12.32	24.73	46.27	56.33
Topo-LSTM	4.56	10.17	21.37	29.29	3.87	8.24	16.61	23.09	–	–	–	–
SNIDSA	–	23.37	35.46	43.39	–	11.81	21.91	28.37	–	–	–	–
FOREST	<u>17.49</u>	<u>24.63</u>	<u>37.73</u>	<u>46.20</u>	8.19	13.58	23.47	29.69	<u>16.76</u>	28.49	45.85	55.19
HiDAN	12.99	22.45	35.51	43.01	<u>8.78</u>	<u>17.40</u>	<u>32.37</u>	<u>40.49</u>	15.31	<u>29.03</u>	<u>50.01</u>	<u>60.07</u>
AN-DRUD	13.54	23.28	36.90	45.28	8.91	17.72	32.73	41.01	16.32	**29.48**	**51.09**	**61.33**
TAN-DRUD	16.62	28.13	45.61	55.43	**9.41**	**18.21**	**34.26**	**42.02**	–	–	–	–
Improv. rate%	−4.97	14.21	20.89	19.98	7.18	4.66	5.84	3.78	−2.63	1.56	2.16	2.10

4.3 Results and Analysis

In this subsection, we display the results of our comparison experiments, based on which we further provide some insights.

Efficacy Performance Comparison. To answer RQ1, we first exhibit all models' mean performance scores (averaged over 5 runnings for each model) in Table 4[1], where Topo-LSTM, SNIDASA and TAN-DRUD's scores in Meme are absent since they need social topology. The best performance scores among all compared models are indicated in bold. The performance scores of leading baseline are underlined, which has the best performance among all baseline models. We also provide the improvement rate of our model (TAN-DRUD in Twitter and Douban, AN-DRUD in Meme) w.r.t. the leading baseline in the table's bottom row.

Based on the results, we propose the following analysis.

1. TAN-DRUD outperforms all baselines remarkably in Douban, especially in P@K. In Meme where social graphs are missing, AN-DRUD also has the best

---

[1] For SNIDSA, we directly cited the results in its original paper where RR is missing.

performance except for RR compared with FOREST. Please note that FOR-EST predicts the number of potential users in the cascade at first, and then explores the subsequent users. With a special module of *cascade simulation for macroscopic prediction*, FOREST is more capable of capturing some true next activated users on first place. As a result, FOREST has the highest *RR* in Twitter and Meme. However, FOREST's lower P@k implies that the rest true next activated users in it have lower ranks than that of our model.

2. AN-DRUD's superiority over HiDAN shows that, using two separate embeddings to capture dual role user dependencies is more effective than capturing single role (unidirectional) user dependencies (to answer RQ2). AN-DRUD also outperforms Bi-LSTM, justifying that our designed mechanism is better than Bi-LSTM to capture bi-directional user dependencies in a cascade sequence. This conclusion is confirmed by the comparison results of both RNN-based models and non-sequential models, justifying that a user joins a cascade as a dual role of information sender and receiver rather than a single role.

3. TAN-DRUD's superiority over AN-DRUD justifies social topology's positive effects on diffusion prediction (to answer RQ3). Moreover, TAN-DRUD's superiority in Twitter is more prominent than Douban. It is because that Twitter is a more typical social platform where most information spreads along social links. Hence, Twitter's social topology plays a more important role in information diffusion.

Efficiency Performance Comparisons. Then, we only compared TAN-DRUD's efficiency with the baselines which are also fed with social topology, since topology computation is generally the most time-consuming step of these models. For average time consumption of one epoch (including training and test), Topo-LSTM takes 4,915s, FOREST takes 1,800s while TAN-DRUD takes 23.7s. Even added with the time consumption of Node2Vec (695.2s), TAN-DRUD's time cost is much lower than Topo-LSTM and FOREST. The reason of our model's higher efficiency is two-fold: 1. TAN-DRUD is a non-sequential model in which matrix computation can be parallelized. 2. We used Node2Vec to generate topological embeddings as pre-training, which avoids topology computation in every epoch.

Influence of Graph Embedding Model. To answer RQ4, we further tested different graph embedding models' influence to TAN-DRUD's performance. In our experiments, we selected five graph embedding models suitable for homogeneous graphs, i.e., SDNE [19], LINE [18], DeepWalk [15], Node2Vec [6] and SCE [29] to learn the topological embeddings fed into TAN-DRUD. The corresponding performance scores upon Twitter are shown in Table 5.

On various social networks, many users may interact with or are influenced by remote users rather than their direct neighbors. Thus, the models which can capture high-order connections would be more competent to our scenario. Unlike SDNE and LINE that only model first and second order connections,

DeepWalk and Node2Vec can capture high-order connections, resulting in better performance. Moreover, Node2Vec can sample more high-order neighbors than DeepWalk with breadth-first and depth-first sampling, thus it has the best performance. SCE is a new unsupervised graph embedding model based on GNN, which uses a contrastive objective based on sparsest cut problem. It may be not suitable for this relatively sparse Twitter's social graph, thus showing unsatisfactory performance compared with other graph embedding models.

Table 5. TAN-DRUD's prediction performance (score %) upon Twitter with different graph embedding models.

Model	RR	P@10	P@50	P@100
SDNE	16.27	26.68	41.96	51.58
LINE	15.50	26.71	43.66	53.12
DeepWalk	16.55	27.84	**46.06**	55.25
Node2Vec	**16.62**	**28.13**	45.60	**55.43**
SCE	13.59	24.20	40.54	50.75

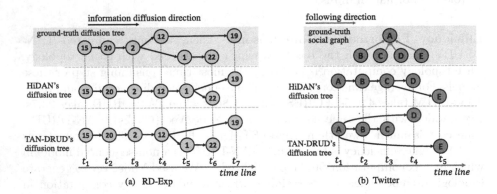

Fig. 3. Case study of diffusion tree inference.

Diffusion Tree Inference. In fact, the cascade-level attentions of TAN-DRUD and HiDAN quantify the probabilities of different historical activated users triggering the next user. Specifically, given a user u_{i+1} activated at time t_{i+1}, we assume that the user in c_i who has the highest attention value (β_j in Eq. 14) triggers u_{i+1}, i.e., is u_{i+1}'s parent node in the diffusion tree. Accordingly, the diffusion tree of an observed cascade can be inferred. We illustrate two case studies to answer RQ5. The first case was extracted from the dataset RD-Exp [21] including social networks. This cascade sample's real diffusion tree was provided by [24]. So we can compare the diffusion trees inferred by TAN-DRUD and HiDAN with the ground truth, as shown in Fig. 3(a). It shows that, TAN-DRUD's inferred diffusion tree is more approximate to the upper real diffusion tree than HiDAN's tree. In addition, Fig. 3(b) only displays the real social graph of a cascade's users in Twitter, since real diffusion trees do not exist in Twitter dataset. It also shows that TAN-DRUD's diffusion tree is more approximate to the real social graph, implying that TAN-DRUD's attentions are more precise than HiDAN's attentions.

5   Conclusion

In this paper, we propose a non-sequential information cascade model TAN-DRUD built with two-level attention networks. TAN-DRUD obtain satisfactory performance through capturing dual role user dependencies in a cascade sequence, and incorporating social topology into cascade modeling. Our extensive experiments on three real datasets demonstrate TAN-DRUD's higher efficacy and efficiency over the state-of-the-art diffusion models, and also prove that diffusion tree can be inferred approximately by our model.

References

1. Barbieri, N., Bonchi, F., Manco, G.: Cascade-based community detection. In: Proceedings of WSDM, pp. 33–42 (2013)
2. Cao, Q., Shen, H., Cen, K., Ouyang, W., Cheng, X.: Deephawkes: bridging the gap between prediction and understanding of information cascades. In: Proceedings of CIKM, pp. 1149–1158 (2017)
3. Chen, X., Zhang, K., Zhou, F., Trajcevski, G., Zhong, T., Zhang, F.: Information cascades modeling via deep multi-task learning. In: Proceedings of SIGIR, pp. 885–888 (2019)
4. Cho, K., Van Merrienboer, B., Bahdanau, D., Bengio, Y.: On the properties of neural machine translation: encoder-decoder approaches. Computer Science (2014)
5. Dong, Y., Chawla, N.V., Swami, A.: metapath2vec: scalable representation learning for heterogeneous networks. In: Proceedings of KDD (2017)
6. Grover, A., Leskovec, J.: node2vec: Scalable feature learning for networks. In: Proceedings of KDD, pp. 855–864 (2016)
7. Hodas, N.O., Lerman, K.: The simple rules of social contagion. Sci. Rep. 4, 4343 (2014)
8. Huang, Z., Xu, W., Yu, K.: Bidirectional lstm-crf models for sequence tagging. arXiv preprint arXiv:1508.01991 (2015)
9. Islam, M.R., Muthiah, S., Adhikari, B., Prakash, B.A., Ramakrishnan, N.: Deepdiffuse: predicting the 'who' and 'when' in cascades. In: Proceedings of ICDM, pp. 1055–1060. IEEE (2018)
10. Lasswell, H.D.: The structure and function of communication in society. Commun. Ideas 37(1), 136–139 (1948)
11. Leskovec, J., Backstrom, L., Kleinberg, J.: Meme-tracking and the dynamics of the news cycle. In: Proceedings of KDD, pp. 497–506 (2009)

12. Li, C., Guo, X., Mei, Q.: Joint modeling of text and networks for cascade prediction. In: Proceedings of the International AAAI Conference on Web and Social Media, vol. 12 (2018)
13. Li, C., Ma, J., Guo, X., Mei, Q.: Deepcas: an end-to-end predictor of information cascades. In: Proceedings of the 26th International Conference on World Wide Web, pp. 577–586 (2017)
14. Liu, C., Wang, W., Jiao, P., Chen, X., Sun, Y.: Cascade modeling with multihead self-attention. In: Proceedings of IJCNN, pp. 1–8. IEEE (2020)
15. Perozzi, B., Al-Rfou, R., Skiena, S.: Deepwalk: online learning of social representations. In: Proceedings of KDD, pp. 701–710 (2014)
16. Sankar, A., Zhang, X., Krishnan, A., Han, J.: Inf-vae: a variational autoencoder framework to integrate homophily and influence in diffusion prediction. In: Proceedings of the 13th International Conference on Web Search and Data Mining, pp. 510–518 (2020)
17. Shen, H.W., Wang, D., Song, C., Barabási, A.L.: Modeling and predicting popularity dynamics via reinforced poisson processes. arXiv preprint arXiv:1401.0778 (2014)
18. Tang, J., Qu, M., Wang, M., Zhang, M., Yan, J., Mei, Q.: Line: large-scale information network embedding. In: Proceedings of WSDM, pp. 1067–1077 (2015)
19. Wang, D., Cui, P., Zhu, W.: Structural deep network embedding. In: Proceedings of KDD, pp. 1225–1234 (2016)
20. Wang, J., Zheng, V.W., Liu, Z., Chang, K.C.C.: Topological recurrent neural network for diffusion prediction. In: Proceedings of ICDM, pp. 475–484. IEEE (2017)
21. Wang, Y., Shen, H., Liu, S., Gao, J., Cheng, X.: Cascade dynamics modeling with attention-based recurrent neural network. In: Proceedings of IJCAI, pp. 2985–2991 (2017)
22. Wang, Z., Chen, C., Li, W.: A sequential neural information diffusion model with structure attention. In: Proceedings of CIKM, pp. 1795–1798 (2018)
23. Wang, Z., Chen, C., Li, W.: Joint learning of user representation with diffusion sequence and network structure. IEEE Trans. Knowl. Data Eng. (2020)
24. Wang, Z., Li, W.: Hierarchical diffusion attention network. In: Proceedings of IJCAI, pp. 3828–3834 (2019)
25. Yang, C., Sun, M., Liu, H., Han, S., Liu, Z., Luan, H.: Neural diffusion model for microscopic cascade prediction. arXiv preprint arXiv:1812.08933 (2018)
26. Yang, C., Tang, J., Sun, M., Cui, G., Liu, Z.: Multi-scale information diffusion prediction with reinforced recurrent networks. In: Proceedings of IJCAI, pp. 4033–4039 (2019)
27. Yang, Y., et al.: Rain: social role-aware information diffusion. In: Proceedings of the AAAI Conference on Artificial Intelligence, vol. 29 (2015)
28. Yuan, C., Li, J., Zhou, W., Lu, Y., Zhang, X., Hu, S.: Dyhgcn: a dynamic heterogeneous graph convolutional network to learn users' dynamic preferences for information diffusion prediction. arXiv preprint arXiv:2006.05169 (2020)
29. Zhang, S., Huang, Z., Zhou, H., Zhou, Z.: Sce: scalable network embedding from sparsest cut. In: Proceedings of the 26th ACM SIGKDD International Conference on Knowledge Discovery & Data Mining pp. 257–265 (2020)
30. Zhao, Q., Erdogdu, M.A., He, H.Y., Rajaraman, A., Leskovec, J.: Seismic: a self-exciting point process model for predicting tweet popularity. In: Proceedings of KDD, pp. 1513–1522 (2015)
31. Zhong, E., Fan, W., Wang, J., Xiao, L., Li, Y.: Comsoc: adaptive transfer of user behaviors over composite social network. In: Proceedings of KDD, pp. 696–704 (2012)

Discovering Bursting Patterns over Streaming Graphs

Qianzhen Zhang, Deke Guo[✉], and Xiang Zhao[✉]

Science and Technology on Information Systems Engineering Laboratory,
National University of Defense Technology, Changsha, China
{dekeguo,xiangzhao}@nudt.edu.cn

Abstract. A streaming graph is a constantly growing sequence of directed edges, which provides a promising way to detect valuable information in real time. A bursting pattern in a streaming graph represents some interaction behavior which is characterized by a sudden increase in terms of arrival rate followed by a sudden decrease. Mining bursting pattern is essential to early warning of abnormal or notable event. While Bursting pattern discovery enjoys many interesting real-life applications, existing research on frequent pattern mining fails to consider the bursting features in graphs, and hence, may not suffice to provide a satisfactory solution. In this paper, we are the first to address the continuous bursting pattern discovering problem over the streaming graph. We present an auxiliary data structure called BPD for detecting the burst patterns in real time with a limited memory usage. BPD first converts each subgraph into a sequence, and then map it into corresponding tracks based on hash functions to count its frequency in a fixed time window for finding the bursting pattern. Extensive experiments also confirm that our approach generate high-quality results compared to baseline method.

1 Introduction

A streaming graph G is an unbounded sequence of items that arrive at a high speed, and each item indicates an edge between two nodes. Together these items form a large dynamic graph. Typical examples of streaming graphs include social media streams and computer network traffic data. Streaming graph analysis is gaining importance in various fields due to the natural dynamicity in many real graph applications. Various types of queries over streaming graphs have been investigated, such as subgraph match [4,9,10], frequent pattern mining [2,11], and triangle counting [6]. However, discovering bursting patterns in real-world streaming graphs remains an unsolved problem.

A *Burst pattern* is a subgraph that is characterized by a sudden increase in terms of arrival rate followed by a sudden decrease. The arrival rate of a subgraph refers to the number of matching results via isomorphism [7] in a fixed time window. Bursting pattern often indicates the happening of abnormal or notable events. We next use an example of monitoring the happening of abnormal business shifting to illustrate its basic idea.

© The Author(s), under exclusive license to Springer Nature Switzerland AG 2022
A. Bhattacharya et al. (Eds.): DASFAA 2022, LNCS 13245, pp. 441–458, 2022.
https://doi.org/10.1007/978-3-031-00123-9_36

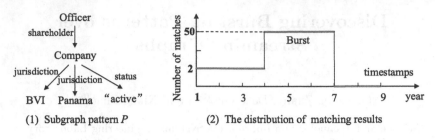

(1) Subgraph pattern P (2) The distribution of matching results

Fig. 1. Business shifting pattern and its burstiness

Example 1. Consider trading networks, e.g., Panama[1], which is a real-word network, where each vertex is an offshore entity (e.g., company, country, jurisdiction) and each edge represents corresponding relationship (e.g., establish, close) between two entities. In such a streaming graph, a bursting pattern is helpful to unveiling the burst of some financial activities among companies. Figure 1 shows a business shifting pattern P that we have mined from Panama (see Fig. 1(1)) and its corresponding matching results during different time intervals (see Fig. 1(2)). It states that the statuses of lots of companies become active with jurisdiction areas of those companies changed from "BVI" (British Virgin Islands) to "Panama" during [4, 7]. Interestingly, P corresponds to a fact that when BVI cracks down the bearer shares, many companies moved bearer share clients to Panama during time interval [4, 7]. Based on P, it is easy to find the companies that are breaking the law of BVI through subgraph matching calculations.

Specifically, given a streaming graph G and an integer k, continuous bursting pattern discovering problem is to find the k-edge subgraph patterns that consist of a sudden increase and a sudden decrease in terms of arrival rate in the graph.

Challenges. In practice, the large scale and high dynamicity of streaming graph make it both memory and time consuming to discovering bursting patterns accurately. It is a natural choice to resort to efficiently compute approximations with limited memory. In the literature, there are solutions to solve another related problem: frequent subgraph pattern mining problem in a streaming graph [2,11]. The main idea is to maintain a uniform sample of subgraphs via *reservoir sampling* [16], which in turn allows to ensure the uniformity of the sample when an edge insertion occurs and then estimate the frequency of different patterns.

This process can be extended to support continuous bursting pattern discovering: estimate the frequency of each k-edge pattern P at each time window based on the sampling and then verify whether the frequency of P is characterized by a sudden increase in terms of arrival rate followed by a sudden decrease. Since the estimation accuracy depends on the sample size, the algorithm needs to maintain a large number of k-edge subgraphs for mining the bursting patterns accurately, which is memory consuming. What's more, the algorithm needs to conduct expensive subgraph matching calculations for these sampled subgraphs to estimate the frequency of each subgraph pattern after

---

[1] https://offshoreleaks.icij.org/pages/database.

all updates have occurred at current timestamp, which is time consuming. In this light, advanced techniques are desiderated to discovery bursting patterns efficiently.

Our Solution. Based on the above discussion, existing frequent subgraph pattern mining approach over the streaming graph is not suitable for mining bursting patterns. Our paper aims for a new way to solve the problem. Our main idea is as follows: instead of using the sampling techniques to maintain the k-edge subgraphs, we propose to design an auxiliary data structure called BPD to accurately detect bursting patterns in real time. We use d buckets, each k-edge subgraph will be mapped into one cell of the buckets by hash functions $h_1(\cdot), \cdots, h_d(\cdot)$ to count the frequency directly. In this way, we can avoid storing any k-edge subgraph in the mining process.

Contributions. In this paper, we make the following contributions: 1) We are the first to propose the problem of continuous discovering the bursting patterns over real streaming graph. 2) We propose the BPD for counting the frequency of each k-edge subgraph pattern with accuracy and efficiency guarantee under limited memory. 3) We design a new graph invariant that map each subgraph to its sequence space representation in the BPD for deriving high efficiency. 4) We propose an edge sampling strategy to speed up the subgraph pattern mining process. 5) Extensive experiments confirm that our method outperforms the baseline solution in terms of efficiency, memory size and estimation accuracy.

2 Preliminaries

A streaming graph G is a constantly time evolving sequence of items $\{e_1, e_2, e_3, \cdots, e_n\}$, where each item $e_i = (v_{i_1}, v_{i_2}, t(e_i))$ indicates a directed edge from vertices v_{i_1} to v_{i_2} arriving at time $t(e_i)$ and the subscripts of the vertices are vertex IDs. This sequence continuously arrives from data sources like routers or monitors with high speed. It should be noted that the throughput of the streaming graph keeps varying. There may be multiple (or none) edges arriving at each time point. For simplicity of presentation, we only consider vertex labelled graphs and ignore edge labels, although handling the more general case is not more complicated. A streaming graph G is given in Fig. 2.

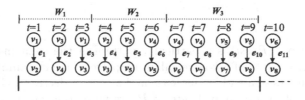

Fig. 2. Streaming graph

Definition 1 (Snapshot graph). *A snapshot graph at timestamp t, denoted as G_t, is a graph induced by all the edges in G that have been observed up to and including time t.*

A subgraph $S_k = (V_S, E_S)$ is referred to as a k-edge subgraph if it is induced by k edges in G_t. For any $t \geq 0$, at time $t + 1$ we receive an edge insertion e and add it into G_t to obtain G_{t+1}. For each newly inserted edge e in G_{t+1}, we use the notation $E_k(e)$ to denote the set of k-edge subgraphs that contain e in G_{t+1}.

Definition 2 (Subgraph isomorphism). *Two subgraphs S'_k and S''_k are isomorphic if there exists a bijection $f: V'_S \to V''_S$ such that 1) $\forall v \in V'_S$, $L(v) = L(f(v))$, and 2) $\forall (v_i, v_j) \in E'_S$, $(f(v_i), f(v_j)) \in E''_S$.*

Let $\mathcal{C}$ be a set of k-edge subgraphs that have isomorphism relation. We call the generic graph $P = (V_P, E_P, L)$ that is isomorphic to all the members of $\mathcal{C}$ the k-edge pattern of $\mathcal{C}$, where V_P is a set of vertices in P, E_P is a set of directed edges with size k, L is a function that assigns a label for each vertex in V_P. Note that, P can be obtained by deleting the IDs (resp. timestamps) of the vertices (resp. edges) of any k-edge subgraph in $\mathcal{C}$.

Given a newly inserted edge e in G_t and a k-edge subgraph pattern P, we use $\eta(e, P)$ to denote the number of k-edge subgraphs in $E_k(e)$ that are isomorphic to P. In this way, the frequency of P at timestamp t, denoted by $fre(t, P)$, can be represented as the sum of $\eta(e, P)$ for each edge e with $t(e) = t$ in G_t.

Table 1. Notations

Notations	Description
G/G_t	The temporal graph/The snapshot graph of G at time t
S_k/P	A k-edge subgraph/A k-edge subgraph pattern
$E_k(e)$	The k-edge subgraphs that contain each newly inserted edge e
$fre(t/W, P)$	The frequency of P at time t/in window W
$W/\sigma/\mathcal{B}/\mathcal{L}$	Window size/Burst ration/Burst threshold/Burst width
$FT(P)$	The frequencies set of P at recent $(\mathcal{L} + 2) \cdot W$ timestamps
$\mathcal{S}/M$	The set of sampled subgraphs/The size of $\mathcal{S}$

Burst Detection. *Burst,* is a particular pattern of the changing behavior in terms of the arrival rate of a k-edge subgraph pattern in a streaming graph, and the pattern consists of a sudden increase and a sudden decrease. Given a k-edge pattern P, to obtain the arrival rate of P, we need to calculate the frequency of P in a fixed window. In specific, we divide the streaming graph into time-based fixed-width windows, i.e., $W_1, \cdots, W_n$, from current timestamp t, each of which has size W. The frequency of P in window W_m, denoted by $fre(W_m, P)$, is the sum of $fre(t_l, P)$ for each timestamp $t_l \in W_m$. A sudden increase means in two adjacent windows, the frequency of P in the second window is no less than σ times of that in the first window. Similarly, a sudden decrease is that the frequency of P in the second window is no more than $\frac{1}{\sigma}$ of that in the first window. We do not consider infrequent bursting patterns as bursts, for they are not useful in most applications, so the frequency of a burst pattern should exceed a burst threshold $\mathcal{B}$. In practice, a burst occurs over a short period of time. Therefore, we set a limitation $\mathcal{L}$ for the width of a burst, namely, the number of windows that the burst lasts.

Definition 3 (Burst pattern). *Given a snapshot graph G_t and a k-edge subgraph pattern P. P is a bursting pattern if there exists four windows W_i, W_{i+1}, W_j, W_{j+1} from t such that 1) $fre(W_{i+1}, P) \geq \sigma \cdot fre(W_i, P) \wedge fre(W_{j+1}, P) \leq \frac{1}{\sigma} \cdot fre(W_j, P) \wedge j > i$; 2) $fre(W_m, P) \geq \mathcal{B}$, $\forall m \in \{i+1, \cdots, j\} \wedge j - i \leq \mathcal{L}$.*

Problem Statement. Given a streaming graph G, and parameters $\mathcal{W}$, $\mathcal{B}$ and $\mathcal{L}$, bursting patterns discovery computes the set of k-edge subgraph patterns that consists of a sudden increase and a sudden decrease in terms of arrival rate.

Frequently used variables are summarized in Table 1.

3   The Baseline Solution

In the literature, the state-of-the-art algorithm proposed in [2] resorts to the sampling framework, aiming to estimate the frequency of a k-edge pattern by maintaining a uniform sample when an edge update occurs. To obtain a reasonable baseline, in this section, we extend the algorithm proposed in [2], and design an sample-and-verify algorithm to compute the bursting patterns by estimating the frequency of a k-edge pattern P in each window from current timestamp t. According to Definition 3, we need at most $\mathcal{L} + 2$ windows from t to verify whether P is a bursting pattern. As a result, we need to maintain $fre(t_l, P)$ where $t_l \in (t - (\mathcal{L} + 2) \cdot \mathcal{W}, t]$ to estimate the frequency of P in each window.

The Sample-and-Verify Algorithm. We briefly introduce the sample-and-verify algorithm (Algorithm 1). We use a set $PatternSet$ to store the generated k-edge patterns and their frequencies at recent $(\mathcal{L}+2) \cdot \mathcal{W}$ timestamps from time t. Each item in the $PatternSet$ is a triple $(P, \widehat{fre}(t, P), \mathsf{FT}(P))$, where P is a k-edge pattern, $\widehat{fre}(t, P)$ is an estimation of $fre(t, P)$ and $\mathsf{FT}(P)$ is a queue with limited size $(\mathcal{L} + 2) \cdot \mathcal{W}$ that is used to store the frequencies set of P. Initially, it calls initializeFre to initialize the $PatternSet$ (Line 1). That is, initializeFre sets $\widehat{fre}(t, P) \leftarrow 0$ for each pattern P in the $PatternSet$. Then, it updates the $PatternSet$ by calling estFrequency (Line 2). After that, if $t - (\mathcal{L} + 2) \cdot \mathcal{W} \geq 0$, for each pattern P in the $PatternSet$, it estimates the frequency of P at each time window based on $\mathsf{FT}(P_i)$ to verify whether P satisfies the bursting feature (Line 3–5). Finally, it returns all bursting patterns at timestamp t (Line 6).

Algorithm 1: findBP

Input : G_t is the snapshot graph at time t; E_t is the set of edge insertions at time t;
           $k, \mathcal{W}, \mathcal{B}, \mathcal{L}, \sigma, M$, are the parameters.
Output : The set of bursting patterns.

1   $PatternSet \leftarrow$ initializeFre($PatternSet$);
2   $BurstSet \leftarrow \emptyset$, $PatternSet \leftarrow$ estFrequency($E_t, G_t, M, PatternSet$);
3   **if** $t - (\mathcal{L} + 2) \cdot \mathcal{W} \geq 0$ **then**
4   **foreach** $(P, \mathsf{FT}(P))$ in the $PatternSet$ **do**
5   **if** BurstCheck(FT(P)) = **true then** $BurstSet \leftarrow BurstSet \cup \{P\}$;

6   **return** $BurstSet$;

    Function estFrequency(E_t, G_t, M)
1     $N_t \leftarrow 0$, $b \leftarrow 0$, $\mathcal{S} \leftarrow \emptyset$;
2   **foreach** edge insertion e in E_t **do**
3       $E_k(e) \leftarrow$ findSubgraph(e, G_t);
4   **foreach** subgraph S_k in $E_k(e)$ **do**
5         $N_t \leftarrow N_t + 1$;
6   ReservoirSampling($\mathcal{S}, M, N_t, S_k$);

7   calculate the k-edge patterns from $\mathcal{S}$;
8   **if** $t - (\mathcal{L} + 2) \cdot \mathcal{W} < 0$ **then** $b \leftarrow t$;
9   **else** $b \leftarrow (\mathcal{L} + 2) \cdot \mathcal{W}$;
10   **foreach** k-edge pattern P_i **do**
11   **if** P is in the $PatternSet$ **then** $\widehat{fre}(t, P) \leftarrow \frac{fre^{\mathcal{S}}(t, P)}{M} \cdot N_t$;
12   **if** P is not in the $PatternSet$ **then** add $b - 1$ zeros to FT(P), insert
        $(P, \widehat{fre}(t, P), \mathsf{FT}(P))$ into the $PatternSet$;

13   **foreach** k-edge pattern P in the $PatternSet$ **do**
14   add $\widehat{fre}(t, P_i)$ to FT(P);

15   **return** $PatternSet$;

Function estFrequency. estFrequency maintains a uniform sample $\mathcal{S}$ of fixed size M of k-edge subgraphs based on the *standard reservoir sampling*. Let N_t be the number of k-edge subgraphs at time t that is initialized as 0 (Line 1). Whenever an edge insertion e occurs at timestamp t, estFrequency first calls findSubgraph (Omitted) to calculates $E_k(e)$ (Line 3). In detail, findSubgraph explores a candidate subgraph space in a tree shape in G_t, each node representing a candidate subgraph, where a child node is grown with one-edge extension from its parent node. The intention is to find all possible subgraphs with size k grown from e. To avoid duplicate enumeration of a subgraph, findSubgraph checks whether two subgraphs are composed of the same edges at each level in the tree space.

Then, for each k-edge subgraph S_k in $E_k(e)$, estFrequency sets $N_t \leftarrow N_t + 1$ and checks whether $|\mathcal{S}| < M$; if so, estFrequency adds S_k into the sample $\mathcal{S}$ directly. Otherwise, if $|\mathcal{S}| = M$, estFrequency removes a randomly selected subgraph in $\mathcal{S}$ and inserts the new one S_k with probability M/N_t (Lines 2–6). After dealing with all edge insertions, estFrequency partitions the set of subgraphs in $\mathcal{S}$ into T_k equivalence classes based on subgraph isomorphism, denoted by $\mathcal{C}_1, \cdots, \mathcal{C}_{T_k}$, and calculate the k-edge subgraph pattern P of each equivalence class $\mathcal{C}_i$ ($i \in [1, T_k]$) (Line 7). The frequency of P in $\mathcal{S}$ at timestamp t, denoted by $fre^{\mathcal{S}}(t, P)$, is the number of subgraphs in corresponding equivalence class. As proofed in [2], $|\frac{fre^{\mathcal{S}}(t,P)}{|\mathcal{S}|} - \frac{fre(t,P)}{N_t}| \leq \frac{\epsilon}{2}$ holds with probability at least $1 - \delta$ if we set $M = log(1/\delta) \cdot (4 + \epsilon)/\epsilon^2$ where $0 < \epsilon, \delta < 1$ are user defined constants. We denote $\widehat{fre}(t,P) = \frac{fre^S(t,P)}{M} \cdot N_t$ as an (ϵ, δ)-approximation to $fre(t,P)$. After that, estFrequency updates the $PatternSet$. Let b be the number of elements in each $FT(\cdot)$ at timestamp t (Lines 8–9). There are two possible cases: (1) if P is in the $PatternSet$, estFrequency sets $\widehat{fre}(t,P) \leftarrow \frac{fre^S(t,P)}{M} \cdot N_t$ (Line 11); (2) if P is not in the $PatternSet$, estFrequency adds $b-1$ zeros to $FT(P)$ and inserts $(P, \widehat{fre}(t,P), FT(P))$ into the $PatternSet$ (Line 12). Finally, estFrequency adds $\widehat{fre}(t,P)$ to $FT(P)$ for each pattern P in the $PatternSet$ (Lines 13–14).

Complexity Analysis. There are four main steps in algorithm findBP. (1) In the k-edge subgraphs enumeration process, given an edge insertion e in G_t, let n be the number of vertices of the subgraph extended from e with radius k. findSubgraph takes $O(2^{n^2})$ to find the k-edge subgraphs that contain e. (2) For each k-edge subgraph, estFrequency takes $O(1)$ to add each new produced subgraph into the reservoir. (3) Let ϵ and be the average unit time to verify whether two k-edge subgraphs are isomorphic. estFrequency takes $O((M^3-M)\cdot\epsilon)$ to partition the subgraphs in S into T_k equivalence classes. (4) Let D be the number of patterns in the $PatternSet$. estFrequency takes $O(T_k \cdot D \cdot \epsilon)$ to update the $Patternset$ and takes $O(1)$ to verify whether a pattern is a bursting pattern.

4   A New Approach

In this section, we first analyze the drawbacks of the baseline solution, and then introduce our proposed approximate data structure called BPD to significantly reduce the memory and computational cost in quest of bursting patterns.

4.1   Problem Analysis

Why Costly? Algorithm findBP is not scalable enough to handle large streaming graphs with high speed due to the following three drawbacks:

- *Drawback 1: Large Memory Cost.* Recall that findBP needs to maintain $log(1/\delta) \cdot (4 + \epsilon)/\epsilon^2$ k-edge subgraphs if the throughput of the streaming graph is huge at time t. As a result, if we use the parameters setting in [2], i.e., $\delta = 0.1$, $\epsilon = 0.01$, there are more than 10^4 k-edge subgraphs to store at time t, which consumes a large amount of memory.
- *Drawback 2: Repeated Subgraph Matching.* To update the $PatternSet$ at each timestamp, estFrequency first partitions the set of subgraphs in sample S and calculates the k-edge subgraph patterns based on subgraph isomorphism. Then, estFrequency needs to re-execute subgraph isomorphism calculation for each pattern to check whether it exists in the $PatternSet$, which can be detrimental.

Our Idea. We devise a new algorithm for bursting pattern discovery, which can overcome the drawbacks introduced above. In the new algorithm, we propose an auxiliary data structure called BPD to calculate the frequency of a pattern

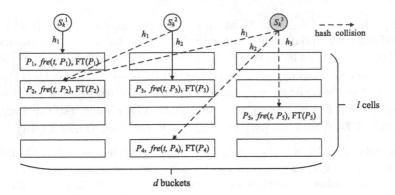

Fig. 3. Data structure of BPD

at each timestamp. Specially, for each new produced k-edge subgraph, we use a hash function to map it into a fixed position in the BPD. In this way, we can count the frequency of the pattern directly without storing any subgraphs, and thus avoiding the repeated subgraph isomorphism calculation in the sample.

BPD Structure (Fig. 3). BPD consists of d buckets, each of which consists of l cells. Let $H_i[j]$ be the jth cell in the i bucket. Each cell has three fields: a pattern P_i, $fre(t, P_i)$ and the frequencies set $\mathsf{FT}(P_i)$ (See Algorithm 1). The d buckets are associated with d pairwise independent hash functions $h_1(\cdot), \cdots, h_d(\cdot)$, respectively. Each k-edge subgraph S_k will be mapped to the fixed cell in a bucket based on the hash functions. That is, for each hash function h_i ($i \in [1, d]$), we map S_k into $H_i[\cdot]$ based on h_i; if S_k is not isomorphic to the pattern in $H_i[\cdot]$, we map S_k into $H_{i+1}[\cdot]$ based on h_{i+1}. It is worth noting that the number of buckets determines the maximal number of hash collisions for storing a new pattern. Therefore, we recommend using enough buckets to achieve higher accuracy.

4.2   The Progressive Algorithm Framework

The new algorithm findBP$^+$ is shown in Algorithm 2, which follows the same framework of Algorithm 1 with different frequency estimation process. It first calls initializeFre to initialize the auxiliary data structure BPD (Line 1). Then it calls the new procedure updateBPD to estimate the frequencies set $\mathsf{FT}(P)$ for each pattern P in the BPD, which is described as follows (Line 2).

Function updateBPD. Initially, updateBPD calculates the constant b and the subgraphs set $E_k(e)$ for each edge insertion $e \in E_t$ (Lines 1–5). Recall that b is the number of the elements in each $\mathsf{FT}(\cdot)$ at timestamp t. Then, for each new produced k-edge subgraph S_k, updateBPD hashes S_k into d mapping buckets according to two cases:

  Case 1: S_k is isomorphic to the pattern P in $H_i[h_i(S_k)]$. updateBPD just increments the frequency of P at time t by 1 (Lines 7–8).

Algorithm 2: findBP$^+$

Input   : G_t is the snapshot graph at time t; E_t is the set of edge insertions at time t;
             $k, \mathcal{W}, \mathcal{B}, \mathcal{L}, \sigma$ are the parameters.
Output : The set of bursting patterns.

1 $BurstSet \leftarrow \emptyset$, BPD $\leftarrow$ initializeFre(BPD);
2 BPD $\leftarrow$ updateBPD(E_t, G_t, BPD);
3 **if** $t - (\mathcal{L} + 2) \cdot \mathcal{W} \geq 0$ **then**
4   **foreach** cell $H_i[j]$ in the BPD **do**
5         $\lfloor$ **if** $H_i[j] \neq \emptyset$ **and** BurstCheck(FT(P)) = **true then** $BurstSet \leftarrow BurstSet \cup \{P\}$;
6   **return** $BurstSet$;

Function updateBPD(E_t, G_t, BPD)
1      $b \leftarrow 0$;
2   **if** $t - (\mathcal{L} + 2) \cdot \mathcal{W} < 0$ **then** $b \leftarrow t$;
3   **else** $b \leftarrow (\mathcal{L} + 2) \cdot \mathcal{W}$;
4   **foreach** edge insertion e in E_t **do**
5         $E_k(e) \leftarrow$ findSubgraph(e, G_t);
6   **foreach** subgraph S_k in $E_k(e)$ **do**
7   **foreach** $i \in [1, d]$ **do**
8   **if** S_k is isomorphic to $H_i[h_i(S_k)]$ **then** $fre(t, P) \leftarrow fre(t, P) + 1$, **break**;
9   **if** S_k is not isomorphic to $H_i[h_i(S_k)]$ **and** $H_{i+1}[h_{i+1}(S_k)]$ is empty **then**
10   calculate the pattern P of S_k, add $b - 1$ zeros to FT(P);
11   insert $(P, 1, \mathsf{FT}(P))$ into $H_{i+1}[h_{i+1}(S_k)]$;

12   **foreach** cell $H_i[h_i(S_k)]$ in BPD **do**
13        $\lfloor$ **if** $H_i[h_i(S_k)]$ is not empty **then** add $fre(t, P)$ into FT(P);
14   **return** BPD;

Case 2: S_k is not isomorphic to the pattern P in $H_i[h_i(S_k)]$ and the cell $H_{i+1}[h_{i+1}(S_k)]$ is empty. updateBPD first calculates the pattern of S_k by deleting its vertex IDs and edge timestamps, and adds $b - 1$ zeros to FT(P) (Lines 9–10). Then updateBPD inserts $(P, 1, \mathsf{FT}(P))$ into $H_{i+1}[h_{i+1}(S_k)]$ (Line 11).

updateBPD next adds $fre(t, P)$ into FT(P) for each nonempty bucket in the BPD and returns the updated BPD (Lines 12–14).

Example 2. Figure 3 shows an running example of the hash process. In this example, subgraph S_k^1 is hashed into cell $H_1(h_1(S_k^1))$ directly since S_k^1 is isomorphic to P_1. When considering subgraph S_k^2, we first hash it into $H_1(h_1(S_k^2))$. Since S_k^2 is not isomorphic to P_2, we then hash it into $H_2(h_2(S_k^2))$. Note that, we cannot detect the pattern of subgraph S_k^3, since S_k^3 is not isomorphic to the pattern in any cell, and none of the cell is empty.

Algorithm Analysis. Compared to Algorithm 1, Algorithm 2 needs not to store the sampled k-edge subgraphs since it uses the hash functions to map each k-edge subgraph into the fixed cell in the BPD. This significantly reduces the memory consumption and avoids the repeated subgraph matching calculations. Note that, users can tune the parameter d to make a trade off between accuracy and speed depending on the application requirements. As shown in our experiments, the recall rate increases as d becomes larger. However, a larger value of d will slow down its efficiency because we have to check $d - 1$ more buckets for each new produced k-edge subgraph. In other words, increasing d means higher accuracy but will lower speed.

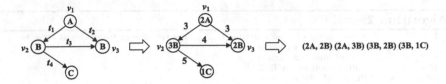

Fig. 4. Sequence representation of a 4-edge subgraph

4.3   Mapping Subgraphs to Sequences

To realize the algorithm framework findBP$^+$ in Algorithm 2, we still need to solve the following issue: how to map a k-edge subgraph to sequence in the hashing process.

Let $m: S_k \rightarrow Seq_k$ be a function to map graph S_k to its sequence space representation Seq_k. The goal in this conversion procedure is to map the subgraph into a string representation such that: if k-edge subgraph S'_k is isomorphic to subgraph S''_k, then $m(S'_k) = m(S''_k)$. This condition can be satisfies by using *graph invariants*.

Definition 4 (Graph invariant). *A graph invariant is a function m, such that $m(S'_k) = m(S''_k)$, whenever S'_k and S''_k are isomorphic graphs.*

There are several possible graph invariants [8,17], but most of them impose a lexicographic order among the subgraphs, which is clearly as complex as graph isomorphism, and thus expensive to compute. In this paper, we generate a degree sequence as our graph invariant that can achieve a higher efficiency. Specially, we map a subgraph S_k to a sequence in the following manner. First, we push the degree and lable of a vertex into together as its new label. Let $l(v)$ denote the new label of vertex v. Extending the same procedure, for each edge $e = (v_i, v_j, t(e))$, we label $l(e) = (l(v_i), l(v_j))$. Now, we consider the edge order of a subgraph. We assign each single-edge pattern a weight in the streaming graph, which is equal to the order of the occurrence of the pattern. Then, each edge can also be assigned a weight according to corresponding single-edge pattern. Let $w(e)$ denote the weight of edge e. Specifically, if $w(e_i) < w(e_j)$, then $e_i < e_j$. Else, if $w(e_i) = w(e_j)$, $e_i < e_j$ if $l(e_i) < l(e_j)$, i.e., the vertex degrees of e_i is lexicographically smaller (ties are broken arbitrarily). Finally, the mapping $m(S_k)$ of a subgraph S_k containing edges $\{e_1, \cdots, e_n\}$ where $e_i < e_{i+1}$, is "$l(e_1)l(e_2)\cdots l(e_n)$."

Example 3. Figure 4 shows the sequence representation of a 4-edge subgraph. We can see that the lable of vertex v_1 is changed from A to 2A since we add the degree of v_1 into its label. What's more, we can also find that edge $(v_1, v_2, t_1) < (v_1, v_3, t_2)$ since (2A, 2B) is lexicographically smaller than (2A, 3B).

4.4   Optimization: Edge Sampling

In Algorithm 2, we still need to call expensive procedure findSubgraph to find all k-edge subgraphs for each newly inserted edge e to calculate the frequency of each pattern, which is too time consuming. Therefore, we propose a sampling algorithm: For each edge isertion e, we randomly sample it and compute $match(e, P)$ with fixed probability p. Here, $match(e, P)$ denotes the number of subgraphs in $E_k(e)$ that match the pattern P in the BPD. Then, we require an unbiased estimator $\widetilde{fre}(t, P)$ of $fre(t, P)$ by adding up $match(e, P)$ for each sampled edge e, i.e., $\widetilde{fre}(t, P) = \frac{1}{p} \sum_{e \in \widehat{E_t}} match(e, P)$, where $\widehat{E_t}$ is the set of sampled edges. Next, we analyze the estimate $\widetilde{fre}(t, P)$ theoretically.

Theorem 1. $\widetilde{fre}(t, P)$ *is an unbiased estimator for* $fre(t, P)$ *at time* t, *i.e., the expected value* $\mathbb{E}[\widetilde{fre}(t, P)]$ *of* $\widetilde{fre}(t, P)$ *is* $fre(t, P)$.

Proof. We consider the edge insertions in E_t are indexed by $[1, m]$ and use an indicator $I(i)$ to denote whether the i-th edge e_i is sampled. Here, $I(i) = 1$ if $e_i \in \widehat{E_t}$ and 0 otherwise. Then, we have

$$\widetilde{fre}(t, P) = \frac{1}{p} \sum_{e \in \widehat{E_t}} match(e, P) = \frac{1}{p} \sum_{i=1}^{m} I(i) \times match(e_i, P). \tag{1}$$

Next, based on Eq. (3) and the fact that $\mathbb{E}[I(i)] = p$, we have

$$\mathbb{E}[\widetilde{fre}(t, P)] = \frac{1}{p} \sum_{i=1}^{m} \mathbb{E}[I(i)] \times match(e_i, P) = fre(t, P). \tag{2}$$

and conclude the proof.

Theorem 2. *The variance* $Val[\widetilde{fre}(t, P)]$ *of* $\widetilde{fre}(t, P)$ *returned by the sampling method is at most* $\frac{1-p}{p} \times fre^2(t, P)$.

Proof. According to Eq. (1) we have

$$Val[\widetilde{fre}(t, P)] = \sum_{i,j=1}^{m} \frac{match(e_i, P)}{p} \times \frac{match(e_j, P)}{p} \times Cov(I(i), I(j)). \tag{3}$$

Since the indicators $I(i)$ and $I(j)$ are independent if $i \neq j$, we have $Cov(I(i), I(j)) = 0$ for any $i \neq j$. In addition, $Cov(I(i), I(i)) = Val[I(i)] = p - p^2$. Based on the above results, we have

$$Val[\widetilde{fre}(t, P)] = \sum_{i=1}^{m} \frac{match^2(e_i, P)}{p^2} \times (p - p^2) = \frac{1-p}{p} \sum_{i=1}^{m} match^2(e_i, P)$$

$$\leq \frac{1-p}{p} \times (\sum_{i=1}^{m} match(e_i, P))^2 = \frac{1-p}{p} \times fre^2(t, P), \tag{4}$$

and conclude the proof.

Theorem 3. $Pr[|\widetilde{fre}(t, P) - fre(t, P)| < \alpha \times fre(t, P)] > 1 - \beta$ *for parameters* $0 < \alpha, \beta < 1$.

Proof. By applying the **two-sided Chernoff bounds**, we have $Pr[|\widetilde{fre}(t,P) - fre(t,P)| \geq \alpha \times fre(t,P)] \leq \frac{Val[\widetilde{fre}(t,P)]}{\alpha^2 \times fre^2(t,P)}$. By substituting $Val[\widetilde{fre}(t,P)]$ with $\frac{1-p}{p^2} \times fre^2(t,P)$, then we have $Pr[|\widetilde{fre}(t,P) - fre(t,P)| < \alpha \times fre(t,P)] > 1 - \beta$, when $p = \frac{1}{1+\beta\alpha^2}$.

Algorithm Analysis. Using the edge sampling can efficiently improve the speed of Algorithm 2 since we need not calculate the k-edge subgraphs for each edge insertion. However, edge sampling strategy will lower accuracy of Algorithm 2 since we only get an unbiased estimator for $fre(t,P)$. Therefore, users can tune the edge sampling probability p to make a trade off between accuracy and speed. In our experiments, we find that findBP$^+$-S is much faster than findBP$^+$ and still has a higher accuracy than findBP for $p = 0.1$ with limited memory.

5 Experiments

In this section, we report and analyze experimental results. All the algorithms were implemented in C++, run on a PC with an Intel i7 3.50GHz CPU and 32GB memory. In all experiments, we use BOB Hash[2] to implement the hash functions. Every quantitative test was repeated for 5 times, and the average is reported.

Datasets. We use three real-life datasets:

- *Enron*[3] is an email communication network of 86K entities (e.g., ranks of employees), 297 K edges (e.g., email), with timestamps corresponding to communication data.
- *Citation*[4] is a citation network of 4.3 M entities (e.g., papers, authors, publication venues), 21.7 M edges (e.g., citation, published at), and 273 labels (e.g., key-words, research domain), with timestamps corresponding to publication date.
- *Panama* (See footnote 1) contains in total 839K offshore entities (e.g., companies, countries, jurisdiction), 3.6 M relationships (e.g., establish, close) and 433 labels covering offshore entities and financial activities including 12 K active days.

Algorithms. We implement and compare three algorithms:

- findBP: Our baseline method for mining bursting patterns;
- findBP$^+$: Our advanced algorithm that uses the auxiliary data structure BPD;
- findBP$^+$-S: findBP$^+$ equipped with the proposed edge sampling optimization.

Metrics. We use the following four metrics:

---

[2] http://burtleburtle.net/bob/hash/evahash.html.
[3] http://konect.uni-koblenz.de/networks/.
[4] https://aminer.org/citation.

- Recall Rate (RR): The ratio of the number of correctly reported to the number of true instances.
- Precision Rate (PR): The ratio of the number of correctly reported to the number of reported instances.
- F1 Score: $\frac{2 \times RR \times PR}{RR + PR}$. It is calculated from the precision and recall of the test, and it is also a measure of a test's accuracy.
- Throughput: Kilo insertions handled per second (KIPS).

Parameter Settings. We measure the effects of some key parameters, namely, the number of hash functions d, the number of cells in a bucket l, the burst threshold $\mathcal{B}$, and the ratio between two adjoin windows for sudden increase or sudden decrease detection σ.

In specific, we vary d from 2 to 8 with a default 6 and very l from 4 to 32 with a default 16. $\mathcal{B}$ could be set by domain scientists based on domain knowledge and is selected from 20 to 160 with a default 80. σ is selected from 2 to 8 with a default 4. In addition, we fix the subgraph size $k = 4$ and fix the edge sampling probability $p = 0.1$ (refer to the Optimization). Without otherwise specified, when varying a certain parameter, the values of the other parameters are set to their default values.

5.1   Experiments on Different Datasets

In this section, we evaluate findBP$^+$'s performance with F1 score and KIPS on three real-life datasets using bounded-size memory. To construct the ground truth dataset, we identify the total bursting patterns using algorithm findBP by replacing the subgraphs reservoir with all k-edge subgraphs at each time t. Note that, we need to store the entire streaming graph to work. Therefore, we reserve space for storing all edges in each dataset. Each edge has 2 vertex IDs, 2 vertex labels and one timestamp, each of which occupies 8 bytes. As the edges are organized as a linked list, an additional pointer is needed by each edge. Therefore 48 bytes are needed for each edge in the streaming graph. To this end, we fix the total memory size of *Enron*, *Panama* and *Citation* to 40 MB, 220 MB and 1 GB, respectively.

Figure 5(1)–(2) show the F1 score and KIPS of findBP$^+$ and its competitors on three datastes with default parameters. Similar results can also be observed under the other parameter settings. From Fig. 5(1), we can see that the F1 score of findBP$^+$ is much higher than all other competitors and the F1 score of findBP$^+$-S is also higher than findBP. For example, on *Enron*, the F1 score achieves 100% for findBP$^+$, and is smaller than 90% for the baseline findBP. From Fig. 5(2), we find that the insertion throughput of findBP$^+$-S is always higher than that of other algorithms and the throughput of findBP$^+$ is also higher than findBP. In specific, findBP$^+$-S outperforms findBP$^+$ by up to 5 times on *Citation* and findBP$^+$ outperforms findBP by up to 3 times on *Panama*. The performance of findBP$^+$ in three datasets are only slightly different, and the trends are very similar. The results show the robustness of findBP$^+$, so in the following experiments, we only use *Panama* dataset.

Analysis. The experiment results show that findBP$^+$ and its optimized version greatly outperform the baseline solution. The main reason is that findBP needs to store enough subgraphs to guarantee the accuracy, which will cause low performance when the memory is limited and also cause redundant subgraph matching calculations. In contrast, first, findBP$^+$ does not store any subgraph, which is less affected by the memory. Second, findBP$^+$ uses the proposed auxiliary data structure BPD to count the frequency of each pattern in the BPD exactly, which can avoid redundant calculations. What's more, findBP$^+$-S can further improve the efficiency since it needs not call expensive procedure findSubgraph for each edge insertion and can also achieve an unbiased estimator for $fre(t, P)$.

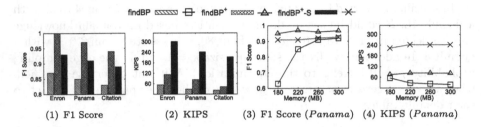

| (1) F1 Score | (2) KIPS | (3) F1 Score (*Panama*) | (4) KIPS (*Panama*) |

Fig. 5. Experimental results - I

5.2   Experiments on Varying Memory

In this section, we evaluate the accuracy and speed of findBP$^+$ and it competitors with varying memory size on *Panama*. We vary the memory size from 180MB to 300MB. The curves of F1 score and KIPS for all the algorithms are shown in Fig. 5(3)–(4), respectively. From Fig. 5(3), we can see the increase of memory size can increase the F1 score of findBP and has little effect on that of findBP$^+$ and findBP$^+$-S. We also find that findBP begins to have F1 score larger than 90% only when the memory is larger than 260MB. On the other hand, findBP$^+$ and findBP$^+$-S get same accuracy with only 180MB. In other words, our algorithms achieve competitive performance with much less space. From Fig. 5(4), we can see that the increase of memory size can decrease the throughput of findBP and has little effect on that of findBP$^+$ and findBP$^+$-S. What's more, findBP$^+$-S is much faster than other algorithms.

Analysis. When the memory size is small, findBP achieves lower accuracy since it has no enough space to store the sampled subgraphs for estimating the frequency of each pattern exactly. However, the throughput of findBP is higher because we need less time to partition the set of subgraphs in S into T_k equivalence classes. For findBP$^+$ and findBP$^+$-S, they use the auxiliary data structure BPD without storing any sampled k-edge subgraph. Since the BPD only stores the k-edge patterns and their frequency sets, we can store it into memory directly. As a result, findBP$^+$ and findBP$^+$-S is less affected by the memory size, which indicates that our algorithms works well with very limited memory.

5.3   Experiments on Varying Parameters

In this section, we evaluate the RR, PR and KIPS of findBP$^+$ and findBP$^+$-S with varying parameters on *Panama* using bounded-size memory, i.e., 220 MB. Note that, when varying a parameter, we keep other parameters as default. The results on the other datasets are consistent.

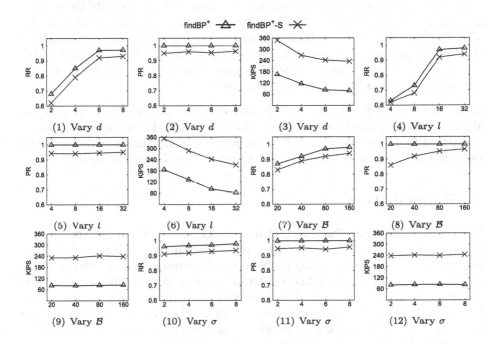

Fig. 6. Experimental results - II

Effect of d **(Fig. 6 (1)–(3)).** In this experiment, we vary d from 2 to 8. Especially, we observe that the increase of d can increase the recall rate and decrease the of throughput of findBP$^+$ and findBP$^+$-S. The reason could be that for a larger d, potential bursting patterns has more opportunities to be stored into the BPD and the recall rate of findBP$^+$ and findBP$^+$-S will be increase. However, the throughput of findBP$^+$ and findBP$^+$-S will be decrease since they have to check $d - 1$ more buckets for each edge insertion. Therefore, users can adjust d to strike a good trade off between accuracy and speed. Furthermore, the precision rate of findBP$^+$ is 100% since it can count the frequency of each pattern in the BPD exactly.

Effect of l **(Fig. 6 (4)–(6)).** The experimental results show that the increase of l can increase the recall rate and decrease the of throughput of findBP$^+$ and findBP$^+$-S. This is because when l increases, there are more tracks in the BPD and we can detect more patterns simultaneously. However, resulting in more subgraph matching calculations since we need to count the frequency of the pattern in each track of the BPD at each timestamp. The precision rate of findBP$^+$ and findBP$^+$-S is insensitive to l since l does not affect the frequency of the k-edge patterns in the BPD.

Effect of $\mathcal{B}$ **(Fig. 6 (7)–(9)).** Our experimental results show that the increase of $\mathcal{B}$ can increase the recall rate of findBP$^+$ and findBP$^+$-S. This is because for a smaller $\mathcal{B}$, the ground truth could be very large and we can only detect fixed number of patterns in the BPD. Therefore, resulting in a lower recall rate. We also find that the increase of $\mathcal{B}$ can increase the precision rate of findBP$^+$-S since more false positives can be filtered safely due to burstness constraint. The throughput of findBP$^+$ and findBP$^+$-S is insensitive to $\mathcal{B}$ since $\mathcal{B}$ does not affect the number of the k-edge patterns in the BPD.

Effect of σ **(Fig. 6 (10)–(12)).** Our experimental results show that our algorithms perform well even when the ratio is very high. As the ratio σ varies, the RR, PR and KIPS of findBP$^+$ and findBP$^+$-S are stable, which indicates that the performances of findBP$^+$ and findBP$^+$-S are insensitive to σ.

6   Related Work

Frequent Subgraph Pattern Mining in Dynamic Graphs. Our work is related to the studies on frequent subgraph pattern mining. Aslay et al. [2] studied the frequent pattern mining problem in a streaming scenario and proposed a sampling-based method to find the latest frequent pattern when edge updates occur on the graph. Ray et al. [13] considered the frequent pattern mining problem in a single graph with continuous updates. Their approach, however, is a heuristic applicable only to incremental streams and comes without any provable guarantee. Abdelhamid et al. [1] proposed an exact algorithm for frequent pattern mining which borrows from the literature on incremental pattern mining. The algorithm keeps track of "fringe" subgraph patterns, which have frequency close to the frequency threshold. Borgwardt et al. [3] looked at the problem of finding dynamic patterns in graphs, i.e., patters over a graph time series, where persistence in time is the key property that makes the pattern interesting.

Bursting Subgraph Mining in Temporal Networks. There is a number of studies for mining dense bursting subgraphs in temporal networks [5,12,15]. Qin et al. [12] defined a bursting subgraph as a dense subgraph such that each vertex in the subgraph satisfies the degree constraint in a period of time. Chu et al. [5] defined a bursting subgraph as a dense subgraph that accumulates its density at the fastest speed during a time interval. Rozenshtein et al. [14] studied the problem of mining dense subgraphs at different time intervals and they define the subgraph that is densest in multiple time interval as bursting subgraph [15]. Compared to them, our work adopts a different definition of burstiness and considers a subgraph pattern that is characterized by a sudden increase in terms of arrival rate followed by a sudden decrease.

7 Conclusion

Real-time burst detection is important in many applications. In this work, we tackle the novel problem of discovering bursting patterns continuously in a streaming graph. We propose an auxiliary data structure called BPD for counting the frequency of a pattern without storing any sampled subgraph, which is fast, memory efficient, and accurate. We further design a new graph invariant that map each subgraph to its sequence space and explore an optimization strategies by using edge sampling to speed up the pattern mining process. Experimental results show that our algorithms can achieve high accuracy and efficiency with limited memory usage in real-time bursting pattern detection.

Acknowledgement. This work is partially supported by National key research and development program under Grant Nos. 2018YFB1800203, National Natural Science Foundation of China under Grant No. U19B2024, National Natural Science Foundation of China under Grant No.61872446 and Natural Science Foundation of Hunan Province under Grant No. 2019JJ20024.

References

1. Abdelhamid, E., Canim, M., Sadoghi, M., Bhattacharjee, B., Chang, Y., Kalnis, P.: Incremental frequent subgraph mining on large evolving graphs. IEEE Trans. Knowl. Data Eng. **29**(12), 2710–2723 (2017)
2. Aslay, Ç., Nasir, M.A.U., Morales, G.D.F., Gionis, A.: Mining frequent patterns in evolving graphs. In: Proceedings of the 27th ACM International Conference on Information and Knowledge Management, CIKM 2018, Torino, Italy, 22–26 October, 2018. pp. 923–932 (2018)
3. Borgwardt, K.M., Kriegel, H., Wackersreuther, P.: Pattern mining in frequent dynamic subgraphs. In: Proceedings of the 6th IEEE International Conference on Data Mining, 18–22 December 2006, Hong Kong, China, pp. 818–822 (2006)
4. Choudhury, S., Holder, L.B., Jr., G.C., Agarwal, K., Feo, J.: A selectivity based approach to continuous pattern detection in streaming graphs. In: Proceedings of the 18th International Conference on Extending Database Technology, EDBT 2015, Brussels, Belgium, 23–27 March, 2015
5. Chu, L., Zhang, Y., Yang, Y., Wang, L., Pei, J.: Online density bursting subgraph detection from temporal graphs. Proc. VLDB Endow. **12**(13), 2353–2365 (2019)
6. Gou, X., Zou, L.: Sliding window-based approximate triangle counting over streaming graphs with duplicate edges. In: SIGMOD '21: International Conference on Management of Data, Virtual Event, China, June 20–25, 2021, pp. 645–657 (2021)
7. Kim, J., Shin, H., Han, W., Hong, S., Chafi, H.: Taming subgraph isomorphism for RDF query processing. Proc. VLDB Endow. **8**(11), 1238–1249 (2015)
8. Kuramochi, M., Karypis, G.: Frequent subgraph discovery. In: Proceedings of the 2001 IEEE International Conference on Data Mining, 29 November - 2 December 2001, San Jose, California, USA, pp. 313–320 (2001)
9. Li, Y., Zou, L., Özsu, M.T., Zhao, D.: Time constrained continuous subgraph search over streaming graphs. In: 35th IEEE International Conference on Data Engineering, ICDE 2019, Macao, China, 8–11 April, 2019, pp. 1082–1093 (2019)

10. Min, S., Park, S.G., Park, K., Giammarresi, D., Italiano, G.F., Han, W.: Symmetric continuous subgraph matching with bidirectional dynamic programming. Proc. VLDB Endow. **14**(8), 1298–1310 (2021)
11. Nasir, M.A.U., Aslay, Ç., Morales, G.D.F., Riondato, M.: Tiptap: approximate mining of frequent k-subgraph patterns in evolving graphs. ACM Trans. Knowl. Discov. Data **15**(3), 48:1–48:35 (2021)
12. Qin, H., Li, R., Wang, G., Qin, L., Yuan, Y., Zhang, Z.: Mining bursting communities in temporal graphs. CoRR (2019)
13. Ray, A., Holder, L., Choudhury, S.: Frequent subgraph discovery in large attributed streaming graphs. In: Proceedings of the 3rd International Workshop on Big Data, Streams and Heterogeneous Source Mining: Algorithms, Systems, Programming Models and Applications, BigMine 2014, New York City, USA, 24 August, 2014, vol. 36, pp. 166–181
14. Rozenshtein, P., Bonchi, F., Gionis, A., Sozio, M., Tatti, N.: Finding events in temporal networks: segmentation meets densest subgraph discovery. Knowl. Inf. Syst. **62**(4), 1611–1639 (2019). https://doi.org/10.1007/s10115-019-01403-9
15. Rozenshtein, P., Tatti, N., Gionis, A.: Finding dynamic dense subgraphs. ACM Trans. Knowl. Discov. Data **11**(3), 27:1–27:30 (2017)
16. Vitter, J.S.: Random sampling with a reservoir. ACM Trans. Math. Softw. **11**(1), 37–57 (1985)
17. Yan, X., Han, J.: gspan: Graph-based substructure pattern mining. In: Proceedings of the 2002 IEEE International Conference on Data Mining, 9–12 December 2002, Maebashi City, Japan, pp. 721–724 (2002)

Mining Negative Sequential Rules from Negative Sequential Patterns

Chuanhou Sun[1], Xiaoqi Jiang[1], Xiangjun Dong[1(✉)], Tiantian Xu[1],
Long Zhao[1], Zhao Li[1], and Yuhai Zhao[2]

[1] Department of Computer Science and Technology, Qilu University of Technology
(Shandong Academy of Sciences), Jinan, China
`d-xj@163.com, zhaolong@qlu.edu.cn, liz@sdas.org`
[2] Northeastern University, Shenyang, China
`zhaoyuhai@mail.neu.edu.cn`

Abstract. As an important tool for behavior informatics, negative
sequential rules (NSRs, e.g., $<ab> \Rightarrow \neg<cd>$) are sometimes much more
informative than positive sequential rules (PSRs, e.g., $<ab> \Rightarrow <cd>$),
as they can provide valuable decision-making information from both negative and positive sides. Very limited NSR mining algorithms are available now and most of them discover NSRs only from positive sequential
patterns (PSPs, e.g., $<abcd>$) rather than from negative sequential patterns (NSPs, e.g., $<a\neg bc\neg d>$), which may result in a loss of important
information. However, discovering NSRs (e.g., $<a\neg b> \Rightarrow <c\neg d>$) from
NSPs is much more difficult than mining NSRs from PSPs because NSPs
do not satisfy the downward closure property. In addition, it is very difficult to find which kind of NSRs should be generated. This paper proposes
a novel algorithm named nspRule to address all these difficulties. The
experiment results on real-life and synthetic datasets show that nspRule
can mine NSRs correctly and efficiently w.r.t. several aspects including
rule number, runtime, memory usage and scalability.

Keywords: Negative sequential rule · Negative sequential pattern ·
Positive sequential rule · Positive sequential pattern

1 Introduction

Behavior permeates all aspects of our lives and how to understand a behavior is
a crucial issue in providing a competitive advantage to support decision-making.
Sequential pattern mining (SPM), which discovers frequent sub-sequences as
patterns in a sequential database, is an important tool for behavior informatics
with broad applications, such as the analysis of students' learning behaviors,

Partly supported by the National Natural Science Foundation of China (62076143,
61806105, 61906104) and the Natural Science Foundation of the Shandong Province
(ZR2019BF018).
C. Sun and X. Jiang—Contributed equally to this work.

customer purchase behaviours, continuous natural disasters, ordered outlying patterns, a series of disease treatments, and so on. Many SPM algorithms, such as GSP, SPAM, SPADE, FreeSpan and PrefixSpan have been proposed. However, these algorithms only consider the support (frequency) of sub-sequences, which is not sufficient to make predictions and recommendations [6]. For example, given a pattern <*Mon-breakfast, Tue-breakfast, Wed-breakfast, Thu-breakfast, good*>(*sup* = 7%), where '*Mon-breakfast*',...,'*Thu-breakfast*' means the student has breakfast from Monday to Thursday and 'good' means the student's academic grade is good. Only using the support of this pattern is not sufficient to predict that students' breakfast behaviour affects their grades, as we don't know the probability of students attaining a 'good' grade under the condition of having breakfast regularly.

To solve this problem, sequential rule mining (SRM) has been proposed. Sequential rules are often expressed in the form $X \Rightarrow Y$, i.e., if X (the antecedent) occurs in a sequential pattern then Y (the consequent) also occurs in that sequential pattern following X by satisfying two thresholds: support (*sup*) and confidence (*conf*). In the above example, a sequential rule <*Mon-breakfast, Tue-breakfast, Wed-breakfast, Thu-breakfast*> $\Rightarrow$ <*good*>(*sup* = 7%, *conf* = 67%) means that the probability of this rule accounts for 7% in a sequential database, and the probability of the student attaining a 'good' grade under the condition of having breakfast regularly accounts for 67%. According to this rule, we can predict that students who have breakfast regularly attain better academic grades.

To date, SRM mining has been applied in many domains such as behavior analysis, drought management, stock market analysis, house allocation e-learning and e-commerce, and many algorithms have been proposed [2,6,7,12,13]. However, these algorithms only focus on occurring (positive) behaviors (OBs), without taking non-occurring (negative) behaviors (NOBs) into consideration. For example, two items are rarely purchased in the same basket, or one item is rarely bought after another. In order to explore the negative relations between basket items, negative sequential rules (NSRs) have been proposed[15–17]. NSRs can reflect the negative relationships between patterns, which can thus provide more comprehensive information than the previous sequential rules which only consider positive relationships (also called positive sequential rules (PSRs)) in many applications. For example, PSR mining algorithms can only mine rules like <*ab*> $\Rightarrow$ <*cd*>, while NSR mining algorithms can mine rules like <*ab*> $\Rightarrow$ ¬<*cd*> or ¬<*ab*> $\Rightarrow$ <*cd*>.

Only a few NSR mining algorithms are available and most of them discover NSRs only from positive sequential patterns (PSPs, e.g., <*abcd*>)) [16,17], rather than from negative sequential patterns (NSPs, e.g., <*a¬bc ¬d*>) [15], which may result in a lot of important information being missing. The reason is that NSPs consider positive and negative relationships between items (itemsets) [1,8,11] and mining NSRs from such NSPs can provide more comprehensive information than mining NSRs from PSPs only. For example, mining NSRs from a PSP <*abcd*> can generate NSRs like <*ab*> $\Rightarrow$ <*cd*>, <*ab*> $\Rightarrow$ ¬<*cd*>,

¬*<ab>* ⇒ *<cd>* and ¬*<ab>* ⇒ ¬*<cd>*. While based on *<abcd>*, we may obtain the corresponding NSPs like *<a¬bcd>*, *<ab¬cd>*, *<abc¬d>*, *<¬ab¬cd>*, *<¬abc¬d>* and *<a¬bc¬d>*, and from these NSPs we may obtain NSRs like *<a¬b>* ⇒ *<cd>*, *<ab>* ⇒ *<¬cd>*, *<ab>* ⇒ *<c¬d>*, *<¬ab>* ⇒ *<¬cd>*, *<¬ab>* ⇒ *<c¬d>*, *<a¬b>* ⇒ *<c¬d>*, etc. Clearly, these NSRs obtained from NSPs can provide more comprehensive information for decision-making.

Unfortunately, very few methods are available to discover NSRs from NSPs. The only one we can find is e-HUNSR [15], which uses the utility-confidence framework and doesn't involve the problems this paper discusses. In fact, NSP mining and NSR mining are still at an early stage and many problems need to be solved [1]. Although a few NSR mining algorithms can discover NSRs from PSPs [16,17], they cannot be used to discover NSRs from NSPs. This is because mining NSRs from NSPs is much more difficult than from PSPs, particularly due to the following three intrinsic challenges.

Challenge 1) Because NSPs do not satisfy the downward closure property [1], the sub-sequences (e.g., *<a¬b>* or *<c¬d>*) of a frequent NSP (e.g., *<a¬bc¬d>*) may not be frequent and their support may not be included in the frequent NSP set, which would make it difficult to calculate the support and the confidence of the corresponding rule (e.g., *<a¬b>* ⇒ *<c¬d>*).

Challenge 2) For the same reason as Challenge (1), the support of an NSP (e.g., *<a¬bc¬d>*) may be greater than the support of its sub-sequence (e.g., *<a¬bc>*). This may result in the confidence of an NSR being greater than one, which is not concordant with the traditional support-confidence framework. This may cause a lot of confusion to users in selecting an appropriate minimum confidence.

Challenge 3) How do we know which kind of rules should be mined? In existing NSR mining algorithms, at least four kinds of rules *<ab>* ⇒ *<cd>*, *<ab>* ⇒ ¬*<cd>*, ¬*<ab>* ⇒ *<cd>* and ¬*<ab>* ⇒ ¬*<cd>* have been generated from PSP *<abcd>*. For NSR mining from NSPs, can we still follow the same procedure? For NSP *<a¬bc¬d>*, should we also generate rules in the form of *<a¬b>* ⇒ *<c¬d>*, *<a¬b>* ⇒ ¬*<c¬d>*, ¬*<a¬b>* ⇒ *<c¬d>* or ¬*<a¬b>* ⇒ ¬*<c¬d>*, etc.? As we can see, there are many negative antecedents or consequents of rules including negative items. Clearly, these kinds of rules are very confusing to explain and should not be mined. So, which kind of rules should be mined?

To address the afore mentioned problem and its challenges, a novel algorithm named nspRule is proposed to mine NSRs from NSPs correctly and efficiently. The contributions of this paper are as follows.

1) We address the problem of how to calculate the support and confidence of NSRs when the sub-sequences of a frequent NSP are not frequent.
2) We address the problem that the confidence of NSRs may be greater than one by normalizing the confidence and letting the value of the confidence meet the traditional support-confident framework.
3) We analyse which kind of rules should be generated by applying a correlation measure to mining suitable rules.

4) Finally, we propose a new algorithm nspRule to mine NSRs. The experiment results on real-life and synthetic datasets show that nspRule can mine NSRs correctly and efficiently w.r.t. various aspects including rule number, runtime, memory usage and scalability.

The rest of this paper is organized as follows. The related work is discussed in Sect. 2. Section 3 introduces the preliminaries. Section 4 details the nspRule algorithm. Section 5 discusses the experimental results. Section 6 presents the conclusions and future work.

2   Related Work

In this section, some related research on NSP mining, SRM and NSR mining is briefly reviewed.

2.1   NSP Mining

Several algorithms have been proposed to mine NSPs. Early algorithms such as NegGSP [19], PNSP [10], GA-based algorithm [18] have lower time efficiency because they calculate the support of negative sequential candidates (NSCs) by re-scanning the database. Later algorithms such as e-NSP [1], e-RNSP [4], F-NSP+ [3], NegI-NSP [11], VM-NSP [14], Topk-NSP [5], NegPSpan [9], sc-NSP [8] have better time efficiency because they calculate the support of NSCs by fast methods, without re-scanning the database. By converting the problem of negative containment to positive containment, e-NSP algorithm can quickly calculate the support of NSCs only using the corresponding PSPs' information. Based on e-NSP, e-RNSP can capture more comprehensive NSPs with repetition properties in a sequence. F-NSP+ uses a novel data structure bitmap to store the PSP information and then obtain the support of NSC only by bitwise operations, which is much faster than e-NSP. NegI-NSP and VM-NSP loosen the constraints in e-NSP and can obtain more valuable NSPs. Topk-NSP can mine the top-k useful NSPs, without setting minimum support. NegPSpan restrains the maxgap of two adjacent elements and extracts NSPs by a PrefixSpan depth-first method, which is more time-efficient than e-NSP for mining long sequences. sc-NSP is more efficient in dense datasets than NegPSpan because it utilizes an improved PrefixSpan algorithm of a bitmap storage structure and a bitwise-based support calculation method to detect NSPs.

2.2   Sequential Rule Mining

SRM has wide applications in several domains to make predictions and many algorithms have been proposed [2,6,7,12,13]. The work in [13] uses a dynamic bit vector data structure and adopts a prefix tree in the mining process to mine non-redundant sequential rules. The TWINCLE algorithm adds time, cost, consistency, and length constraints to address low-cost mining problem [2]. Viger et al.

proposed the CMRules algorithm to discover rules common to many sequences [6]. In this algorithm, the antecedents and consequents of the rule are both unordered, which are called partially-ordered sequential rules. The RuleGrowth algorithm uses a pattern growth method to find partially-ordered sequential rules with high efficiency [7]. Furthermore, to fit with their research, Setiawan et al. introduced time_lapse concept to partially-ordered sequential rule algorithm for solving the practicability and reliability issue [12].

2.3 NSR Mining

To date, only a few algorithms are available for mining NSR [15–17]. Zhao et al. presented a method to find event-oriented negative sequential rules, where the right side of the rule is a single event. This method only generates the four forms of rules [16], namely $\neg X \Rightarrow \neg Y$, $X \Rightarrow \neg Y$, $\neg X \Rightarrow Y$ and $X \Rightarrow Y$. They also presented a new notion of impact-oriented NSR, where the left side of the rule is PSPs or NSPs and the right side of the rule is a positive or negative target outcome and they proposed an SpamNeg algorithm based on the SPAM algorithm to mine such rules [17]. Both these algorithms only mine NSRs from PSPs and only generate four forms of rules, such as $\neg X \Rightarrow \neg Y$, $X \Rightarrow \neg Y$, $\neg X \Rightarrow Y$ and $X \Rightarrow Y$. However, they cannot generate rules like $<ab> \Rightarrow <\neg cd>$, $<ab> \Rightarrow <c\neg d>$, $<a\neg b> \Rightarrow <cd>$, $<\neg ab> \Rightarrow <cd>$ and so on, which may result in important information being missing. Zhang et al. proposes a method [15] which can mine high utility negative sequential rules with the aforementioned formats. The method generates HUNSRCs from HUNSPs, then uses the utility-confidence framework to evaluate the usefulness of rules rather than the traditional support–confidence framework, and finally, acquires HUNSRs from HUNSPs.

3 Preliminaries

In this section, some important concepts of PSPs and NSPs are introduced.

3.1 Positive Sequential Patterns

Let $I = \{x_1, x_2, ..., x_n\}$ be a set of items. An itemset is a subset of I. A sequence is an ordered list of itemsets. A sequence s is denoted by $<s_1 s_2...s_l>$, where $s_j \subseteq I(1 \leq j \leq l)$. s_j is also called an element of the sequence and is denoted as $(x_1 x_2...x_m)$, where x_k is an item, $x_k \in I(1 \leq k \leq m)$. For simplicity, the brackets are omitted if an element only has one item, i.e. element (x) can be expressed as x.

Definition 1. *Length of sequence. The length of sequence s is the total number of items in all elements in s, expressed as length(s). If length(s) = m, this means that s is a m-length sequence.*

For example, a sequence $s = <(ab)cd>$ is a 4-length sequence, i.e., length$(s) = 4$.

Definition 2. *Size of sequence. The size of sequence s is the total number of elements in s, expressed as $size(s)$. If $size(s) = m$, it means that s is a m-size sequence.*

For example, a sequence $s = <(ab)cd>$ is a 3-size sequence, i.e., $size(s) = 3$.

Definition 3. *Sub-sequence. There are two sequences $s_\alpha = <\alpha_1\alpha_2...\alpha_i>$ and $s_\beta = <\beta_1\beta_2...\beta_k>$, if there exists $1 \le j_1 < j_2 < ... < j_i \le k$ such that $\alpha_1 \subseteq \beta_{j1}, \alpha_2 \subseteq \beta_{j2}, ..., \alpha_i \subseteq \beta_{ji}$, we call sequence $s_\alpha = <\alpha_1\alpha_2...\alpha_i>$ a sub-sequence of sequence $s_\beta = <\beta_1\beta_2...\beta_k>$, expressed as $s_\alpha \subseteq s_\beta$ and s_β is called a super-sequence of s_α.*

For example, $s_1 = <(ab)cd>$ is a super-sequence of $s_2 = <(ab)>$.

Definition 4. *Support of sequence. The number of tuples in sequential database D is expressed as $|D|$, where the tuples are $<sid$ (sequence $- ID$), ds (data sequence$)>$. The set of tuples containing sequence s is expressed as $<s>$. The support of s is the number of tuples that are contained in $<s>$, expressed as $sup(s)$, i.e., $sup(s) = |\{<s>\}| = |\{<sid, ds>, <sid, ds> \in D \wedge (s \subseteq ds)\}|$. min_sup is a minimum support threshold which is predefined by users. If $sup(s) \ge min\_sup$, sequence s is called a frequent sequential pattern. By contrast, if $sup(s) < min\_sup$, s is called infrequent.*

3.2 Negative Sequential Patterns

In real applications, the number of NSCs and NSPs may be very huge if no constraints are added, and many of them are meaningless. So many existing NSP mining methods introduce some constraints similar to e-NSP. This paper also uses the same constraints as e-NSP, so we introduce them and some important definitions in e-NSP as follows.

A negative item/element is a non-occurring item/element and positive item/element is an occurring item/element. A negative sequence includes at least one negative item/element. For instance, a sequence $s_1 = <abcF>$ is a positive sequence; $s_2 = <ab\neg cF>$ is a negative sequence because it contains a negative item $\neg c$.

Definition 5. *Positive part. The positive part of a negative element $(\neg ab)$ is (ab), expressed as $p((\neg ab))$. That is, $p((\neg ab)) = (ab)$, while the positive part of positive element (ab) is (ab) itself, i.e., $p((ab)) = (ab)$.*

Constraint 1: *Frequency constraint.* For simplicity, this paper only focuses on the negative sequences ns whose positive partners are frequent, i.e., $sup(p(ns)) \ge min\_sup$.

Constraint 2: *Format constraint.* Continuous negative elements in an NSC are not allowed. For example, $<\neg(ab)\neg cd>$ is not allowed.

Constraint 3: *Negative element constraint.* The smallest negative unit in an NSC is an element. If an element consists of more than one item, either all or none of the items is allowed to be negative. For example, $<(a\neg b)cd>$ is not allowed, because in element $(a\neg b)$, only one item is negative while another one is positive.

Definition 6. *Maximum positive sub-sequence. A negative sequence ns = <a¬bb¬a(ijF)>, <ab(ijF)> is the sub-sequence which includes all positive elements <a¬bb¬a(ijF)>. We denote <ab(ijF)> as a maximum positive sub-sequence, expressed as MPS(ns).*

Definition 7. *1-neg-length maximum sub-sequences. For a negative sequence ns, its sub-sequence that includes MPS(ns) and one negative element e are called a 1-neg-length maximum sub-sequence, expressed as $1 - negMS_{ns}$. The sub-sequence set of $1 - negMS_{ns}$ is called $1 - negMSS_{ns}$.*

Definition 8. *Negative containment. Given a data sequence ds and a negative sequence ns, ds contains ns if and only if two conditions hold: (1)MPS(ns) ⊆ ds; and (2) ∀1 − negMS ∈ 1 − negMSS_{ns}, p(1 − negMS) ∉ ds.*

For example, given $ds = <a(bc)d(cde)>$, 1) if $ns = <a¬dd¬d>$, $1 - negMSS_{ns} = \{<a¬dd>, <ad¬d>\}$, then ds does not contain ns because $p(<a¬dd>) = <add> ⊆ ds$; 2) if $ns' = <a¬bb¬a(cde)>$, $1 - negMSS'_{ns} = \{<a¬bb(cde)>, <ab¬a(cde)>\}$, then ds contains ns' because $MPS(ns') = <ab(cde)> ⊆ ds ∧ p(<a¬bb(cde)>) ∉ ds ∧ p(<ab¬a(cde)>) ∉ ds$.

This definition is consistent with set theory [1]. Through this set theory, the negative containment can be converted to a positive containment, so e-NSP can calculate the support of NSCs directly by matching the corresponding PSPs without re-scanning the database.

Definition 9. *Negative Sequential Patterns. A negative sequence ns is a negative sequential pattern (NSP) if sup(ns) ≥ min_sup.*

4   The nspRule Algorithm

In this section, we first briefly introduce the steps in the e-NSP algorithm because nspRule is built on e-NSP. Then the steps and the corresponding pseudocode of the nspRule are given.

4.1   Review of e-NSP Algorithm

The e-NSP algorithm can efficiently mine positive and negative sequential patterns (PNSPs). Each step of this algorithm is as follows:

Step 1. All PSPs are discovered from the sequential database using a PSP mining algorithm. In this paper, we use the Spam algorithm to mine PSPs.

Step 2. An *NSC − generation* method based on the three constraints is used to generate NSCs from the above PSPs.

The *NSC − generation* method is as follows: For a k-size PSP, its NSCs are generated by changing any m non-contiguous elements to their negative elements, $m = 1, 2, ..., \lceil k/2 \rceil$, where $\lceil k/2 \rceil$ is a minimum integer that is not less than $k/2$. For example, when $m = 1$, the NSCs of a PSP $<ab(cd)>$ is $<¬ab(cd)>$, $<a¬b(cd)>$ and $<ab¬(cd)>$; when $m = 2$, the NSC is only $<¬ab¬(cd)>$.

Step 3. Convert the NSCs into PSPs according to the Definition 8.

Step 4. Calculate the support of NSCs using the following equations.

Given a k-size and j-neg-size negative sequence ns, for $\forall 1 - negMS_i \in 1 - negMSS_{ns}(1 \leq i \leq j)$, the support of ns in sequential database D is:

$$sup(ns) = sup(MPS(ns)) - \mid \cup_{i=1}^{j}\{p(1 - negMS_i)\} \mid \qquad (1)$$

Specially, if there is only one negative element in sequential database $|D|$, the support of $<\neg a>$ can be calculated by Eq. (2):

$$sup(<\neg a>) = |D| - sup(<a>) \qquad (2)$$

Step 5. Output PNSPs.

4.2   The Steps of the nspRule Algorithm

The main steps of this algorithm are as follows:

Step 1. Get all PNSPs and record their support using the e-NSP algorithm.

Step 2. Generate the sequential rule candidate (SRC).

A sequential rule *candidate* (SRC)-generation method is proposed to generate SRCs, i.e., $X \Rightarrow Y$, from the above PNSPs.

The *SRC-generation* method is as follows: For a PNSP $= <e_1e_2e_3...e_k>(k \geq 2)$, its SRCs are generated by dividing PNSP into two parts, i.e., $\forall i \in \{2...k\}$, the antecedent $X = <e_1e_2...e_{i-1}>$ and the consequent $Y = <e_i...e_k>$, where $X, Y \neq \emptyset$ and $X \bowtie Y$=PNSP. For example, $<a\neg bc>$ is an NSP, and it can generate SRCs $<a> \Rightarrow <\neg bc>$ and $<a\neg b> \Rightarrow <c>$.

Step 3. Calculate the support of SRCs.

As mentioned in challenge 1 in Sect. 1, the antecedents and consequents of NSRs may be not frequent and their support may not be included in the frequent NSP set, which means the support and confidence of the corresponding NSR cannot be calculated. To address this problem, we eliminate the SRCs in which the antecedents and consequents of the rule is not frequent before calculating the support of SRCs.

Definition 10. *Support of rule. For a given sequential database D and a given sequential rule $X \Rightarrow Y$, the number of tuples in sequential database D is expressed as $|D|$, where tuples is $<sid$ (sequence $- ID$), ds (data sequence)$>$. The number of tuples containing $X \bowtie Y$ in D is expressed as $sup(X \bowtie Y)$. The support of rule $X \bowtie Y$ in D, denoted as $sup(X \Rightarrow Y)$. The formula to calculate $sup(X \Rightarrow Y)$ is as follows:*

$$sup(X \Rightarrow Y) = sup(X \bowtie Y)/|D| \qquad (3)$$

We only consider SRCs in which the antecedents and consequents are both frequent and calculate the support of SRCs using Eq. (3).

Step 4. Choose which kind of rules should be generated.

As discussed in challenge 3 in Sect. 1, one of the issues to be considered is to identify which kind of rules should be generated in mining NSRs from NSPs. For example, for an NSP $<a\neg bc\neg d>$, should we generate rules in the form of $<a\neg b> \Rightarrow <c\neg d>$, $<a\neg b> \Rightarrow \neg<c\neg d>$, $\neg<a\neg b> \Rightarrow <c\neg d>$ or $\neg<a\neg b> \Rightarrow \neg<c\neg d>$, etc.? Could all of these be used to make decisions? For NSR $<a\neg b> \Rightarrow \neg<c\neg d>$, the negative consequent ($\neg<c\neg d>$) also includes negative items ($\neg d$). It is very confusing to explain the meaning of this kind of rule. So, this kind of rule should not be generated. In order to avoid generating them, we take into consideration the rule's correlation which is widely used in association rules.

Definition 11. *Correlation of rule. The correlation of a positive and negative sequential rule $X \Rightarrow Y$ is used to determine whether the two events are related, denoted as $corr(X \Rightarrow Y)$. If $corr(X \Rightarrow Y) > 1$, this means that the two events are positively related; if $corr(X \Rightarrow Y) < 1$, this means that the two events are negatively related; if $corr(X \Rightarrow Y) = 1$, this means that the two events are independent. The formula to calculate $corr(X \Rightarrow Y)$ is as follows:*

$$corr(X \Rightarrow Y) = sup(X \bowtie Y)/(sup(X)sup(Y)) \tag{4}$$

We calculate the correlation of SRCs and only choose the positively related SRCs, i.e., the $corr(X \Rightarrow Y)$ is greater than 1.

Step 5. Calculate the confidence of SRCs.

Definition 12. *confidence of rule. A sequential rule $X \Rightarrow Y$ has a measure of its strength called confidence (denoted as $conf(X \Rightarrow Y)$). The formula to calculate $conf(X \Rightarrow Y)$ is as follows:*

$$conf(X \Rightarrow Y) = sup(X \bowtie Y)/sup(X) \tag{5}$$

We calculate the confidence of SRCs using Eq. (5) and put those SRCs which satisfy the min_conf (minimum confidence given by users) threshold into the SRC set.

Step 6. Normalize the confidence of SRC.

As discussed in challenge 2 in Sect. 1, the confidence of NSRs may be greater than one, which is not concordant with the traditional support-confidence framework so this may cause a lot of confusion for users when selecting the appropriate minimum confidence. To address this problem, we propose a method to normalize the value of confidence to fall into [0,1] and we propose a new measure named normal_confidence to express the new confidence rule.

Definition 13. *normal_confidence of rule. normal_confidence is the normalization of the confidence rule, denoted as $nor\_conf(X \Rightarrow Y)$. The values of $nor\_conf(X \Rightarrow Y)$ is fall in [0,1]. Suppose the maximum confidence of the rule in the rule set is max_confidence. The formula to calculate $nor\_conf(X \Rightarrow Y)$ is as follows:*

$$nor\_conf(X \Rightarrow Y) = conf(X \Rightarrow Y)/max\_confidence \tag{6}$$

We calculate the *normal_confidence* of SRCs and eliminate the SRCs which do not satisfy the *min_nor_conf* (minimum_normal_confidence given by users) threshold from the SRC set.

Step 7. Output the rule from the SRC set as positive and negative sequential rules (PNSRs).

Definition 14. *Positive and negative sequential rule. A sequential rule $X \Rightarrow Y$ is a positive and negative sequential rule if $sup(X \Rightarrow Y) \geq min\_sup$, $nor\_conf(X \Rightarrow Y) \geq min\_nor\_conf$ and $corr(X \Rightarrow Y) > 1$.*

4.3   Algorithm Pseudocode

The pseudocode of the nspRule algorithm is shown in Algorithm 1. This algorithm first mines all PSPs and records their support using the Spam algorithm (line 1). Then it mines NSPs and records their support using the e-NSP algorithm (line 2) and puts all PNSPs into PNSPset (line 3). If the size of pnsp is greater than 1, it generates SRCs using the *CSR-generation* method (line 7–8). Then, it judges whether X and Y are both frequent. If X and Y are frequent, it calculates the $sup(X \Rightarrow Y)$ using Eq. (3). If $sup(X \Rightarrow Y) \geq min\_sup$, it calculates the $corr(X \Rightarrow Y)$ using Eq. (4). If $corr(X \Rightarrow Y) > 1$, it calculates the $conf(X \Rightarrow Y)$ using Eq. (5) (line 9–15). Then, it finds the maximum confidence of SRCs and gives it to a new parameter *max_confidence* which is defined in line 4 (line 16–18). If $conf(X \Rightarrow Y) \geq min\_conf$, it adds SRC $X \Rightarrow Y$ to SRCset (line 19–21). Then, it calculates the $nor\_conf(X \Rightarrow Y)$ using Eq. (6) and eliminates the SRC which does not satisfy the *min_nor_conf* from SRCset (line 26–31). Finally, it returns the SRCset as PNSRs (line 32).

4.4   Analysis of the Time Complexity

We provide a brief analysis of the time complexity of nspRule. First, the time complexity of converting a sequence database into a transaction database is linear with respect to the number of sequences and their sizes. The time complexity of the e-NSP algorithm is more difficult to establish. It depends on which sequential pattern mining algorithm is used. In this paper, we used the Spam algorithm to mine PSPs. The search strategy of the Spam algorithm combines a vertical bitmap representation of the database with efficient support counting, so this algorithm is especially efficient. The e-NSP algorithm is run in two phases: mining PSPs and converting PSPs into NSPs through set theory. Therefore, the first phase of e-NSP is the most costly, so it is acceptable to ignore the second phase when estimating the time complexity. So, the time complexity of e-NSP can be expressed by the Spam algorithm. The Spam algorithm uses a depth-first search strategy that integrates a depth-first traversal of the search space, so the complexity of this algorithm is $O(d^{2n})$ where d is the number of different items and n is the number of transactions in the database. After obtaining all PNSPs by using the e-NSP algorithm, nspRule splits each PNSP into SRCs, then it checks if each SRC satisfies the minimum threshold. In the best case

---

Algorithm 1. nspRule

Input: Sequential database D, min_sup, min_conf, min_nor_conf;
Output: PNSRs;
1: mine all PSPs and their support(sup) by Spam algorithm;
2: mine all NSPs and their sup by e-NSP algorithm;
3: $PNSPset \leftarrow (pnsp, sup)$;
4: $SRCset \leftarrow \emptyset$;
5: $max\_confidence \leftarrow 0$;
6: **for** each pnsp in PNSPset **do**
7:   **if** size(pnsp)>1 **then**
8:     use *CSR-generation* method to generate SRCs($X \Rightarrow Y$);
9:     **if** X and Y are both frequent **then**
10:       calculate the $sup(X \Rightarrow Y)$ by Equation (3);
11:       **if** $sup(X \Rightarrow Y) \geq min\_sup$ **then**
12:         calculate the $corr(X \Rightarrow Y)$ by Equation (4);
13:         **if** $corr(X \Rightarrow Y) > 1$ **then**
14:           calculate the $conf(X \Rightarrow Y)$ by Equation (5);
15:         **end if**
16:         **if** $conf(X \Rightarrow Y) > max\_confience$ **then**
17:           max_confience=$conf(X \Rightarrow Y)$;
18:         **end if**
19:         **if** $conf(X \Rightarrow Y) \geq min\_conf$ **then**
20:           $SRCset \leftarrow SRCset \cup \{X \Rightarrow Y\}$;
21:         **end if**
22:       **end if**
23:     **end if**
24:   **end if**
25: **end for**
26: **for** each SRC $X \Rightarrow Y$ in SRCset **do**
27:   calculate the $nor\_conf(X \Rightarrow Y)$ by Equation (6);
28:   **if** $nor\_conf(X \Rightarrow Y) < min\_nor\_conf$ **then**
29:     eliminate $X \Rightarrow Y$ from SRCset;
30:   **end if**
31: **end for**
32: **return** $SRCset$;

---

and worst case, there are respectively $PNSP_{(size>1)}.count \times |S|$ SRCs will be generated, where $PNSP_{(size>1)}.count$ means the total number of PNSPs whose size is greater than 1 and $|S|$ means the size of each PNSP. These generation and pruning steps are done in linear time. Thus, the time complexity of the last step is linear with respect to the number of PNSPs and their size.

5   Experiment with the nspRule Algorithm

We conduct experiments on one real-life (Dataset 1) and three synthetic datasets (Datasets 2–4) to assess the influence of min_sup and min_nor_conf on the performance of the nspRule algorithm. We also assess the scalability of the nspRule algorithm using a different number of sequences contained in the sequential database.

Our proposed algorithm is the first study to mine NSRs form NSPs so there are no baseline algorithms with which to compared. E-HUNSR is not suitable to compare with nspRule because it uses the utility-confidence framework, which is very different with the support-confidence framework used in nspRule. The nearest algorithm that can be compared is the SpamNeg algorithm [16], which mines both PSRs and NARs from PSPs. It mines PSRs from frequent PSPs and mines NSRs from infrequent PSPs. In order to make a fair comparison, we unify the input of the two algorithms, i.e., we change SpamNeg to mine NSRs also from frequent patterns. In our experiments, we compare our proposed algorithm with the SpamNeg algorithm with respect to candidates, the final derived rules, runtime and memory usage. The algorithms are implemented in Eclipse, running on Windows 7 PC with 8 GB memory, Intel (R) Core (TM) i7-6700 CPU 3.40 GHz. All the programs are written in Java.

Dataset 1 (DS1): BMSWebView1 (Gazelle) (KDD CUP 2000).
Dataset 2 (DS2): C5_T6_S6_I4_DB3k_N100.
Dataset 3 (DS3): C10_T8_S12_I10_DB2k_N100.
Dataset 4 (DS4): C6_T4_S4_I4_DB10k_N100.

5.1   Experiment to Assess the Influence of min_sup

In the first experiment, different min_sup values and a fixed min_nor_conf value are used to assess the influence of min_sup on datasets DS1 to DS4.

We compare our proposed algorithm with an NSR mining algorithm, named SpamNeg [16], with respect to candidates, final derived rules, runtime and memory usage. In the SpamNeg algorithm, there is no min_nor_conf measure, so we use min_conf while mining NSRs. On DS1-DS4, the min_conf and min_nor_conf are set to 0.1. For acquiring sufficient rules to observe the difference of two algorithms, there are different ranges of min_sup with different datasets because of their inherent attributes. Figure 1 (a) shows the different number of candidates and the final derived rules generated by nspRule and the SpamNeg algorithm respectively with different min_sup values and (b)(c) shows the comparison of nspRule and the SpamNeg algorithm in terms of runtime and memory. From Fig. 1(a), we can observe that when the min_sup increases to a certain value, the number of NSRs mined by the two algorithms is close, and with the decrease of min_sup, the number increases gradually. We can also see that the number of PNSRs mined by the nspRule algorithm is greater than that of the SpamNeg algorithm. The reason is that our algorithm mines PNSRs from both PSPs and NSPs, while SpamNeg only mines PNSRs from PSPs. The number of NSPs is much larger than that of PSPs. From Fig. 1(b), we can see that the runtime of nspRule and the SpamNeg algorithm is close. We can also see that with a decrease of the min_sup, the runtime of the nspRule and SpamNeg algorithm increases. Also, with the decrease of the min_sup, the memory usage of nspRule and SpamNeg algorithm increases, as shown in Fig. 1(c). Figure 1(b)

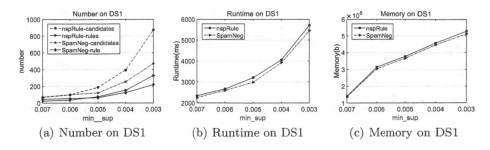

Fig. 1. Influence of *min_sup* on DS1

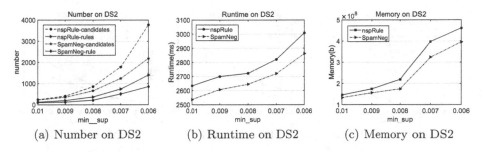

Fig. 2. Influence of *min_sup* on DS2

and (c) show that our algorithm costs more time and memory than the Spam-Neg algorithm. The reason for this is that our algorithm first mines NSPs from PSPs, then it mines PNSRs from these PSPs and NSPs, while SpamNeg mines PNSRs directly from these mined PSPs. From Figs. 2, 3 and 4, we can see that a similar trend occurs for both synthetic and real-life datasets from DS1 to DS4.

5.2 Experiment to Assess the Influence of min_nor_conf

In the second experiment, our proposed algorithm compares the SpamNeg algorithm with different min_nor_conf (min_conf) values and a fixed min_sup value to assess the influence of min_nor_conf (min_conf) on DS1 to DS4.

On DS1, the nspRule algorithm and SpamNeg algorithm are run with a fixed $min\_sup = 0.004$ and min_nor_conf $(min\_conf) = 0.1,0.12,...,0.18$. On DS2, the two algorithms run with a fixed $min\_sup = 0.007$ and min_nor_conf (min_conf) $= 0.1,0.12,...,0.18$. On DS3, the two algorithms run with a fixed $min\_sup = 0.06$ and min_nor_conf $(min\_conf) = 0.1, 0.2,...,0.5$. On DS4, the two algorithms run with a fixed $min\_sup = 0.006$ and min_nor_conf $(min\_\ conf)$. The reason why we set a different min_nor_conf (min_conf) on DS3 is that the data in DS3 is concentrated and is not sensitive to min_nor_conf (min_conf). If we set min_nor_conf (min_conf) from 0.12 to 0.18, the number of PNSRs will change a little. Figure 5(a) shows that with an increase of the min_nor_conf (min_conf), the number of NSRs in both algorithms decreases gradually. The rules in which the confidence is less than min_nor_conf (min_conf) are pruned, so with an

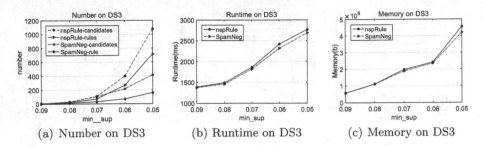

(a) Number on DS3   (b) Runtime on DS3   (c) Memory on DS3

Fig. 3. Influence of min_sup on DS3

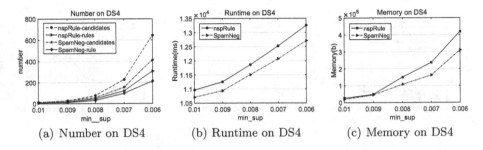

(a) Number on DS4   (b) Runtime on DS4   (c) Memory on DS4

Fig. 4. Influence of min_sup on DS4

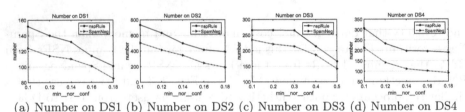

(a) Number on DS1 (b) Number on DS2 (c) Number on DS3 (d) Number on DS4

Fig. 5. Influence of min_nor_conf on DS1–DS4

increase of the min_nor_conf (min_conf) threshold, more rules will be eliminated. We can see that a similar trend occurs for DS2, DS3 and DS4, as shown in Fig. 5 (b)(c)(d). The mining results in Fig. 5 clearly show that the number of PNSRs mined by nspRule algorithm is greater than that of the SpamNeg algorithm.

5.3   Experiment to Assess the Influence of $|S|$

A third experiment is performed to assess the scalability of the nspRule algorithm with respect to a different scale of sequence databases. In this experiment, memory and runtime are analyzed to assess the performance of the nspRule algorithm on both the real-life and synthetic databases.

In the first experiment, the nspRule algorithm runs with $min\_sup = 0.006$ and $min\_nor\_conf = 0.2$ on different data sizes: from 5 to 20 times of DS1. The scale of DS1*5-DS1*20 is varied from 21.3M to 85.4M. In the second experiment, the nspRule algorithm runs with $min\_sup = 0.06$ and $min\_nor\_conf = 0.2$ on different data sizes: from 5 to 20 times of DS3. The scale of DS3*5-DS3*20 is varied from 13.7M to 54.6M. As shown in Fig. 6(a)(b), we can see that nspRule's running time and maximum memory usage grows linearly with the size of $|S|$.

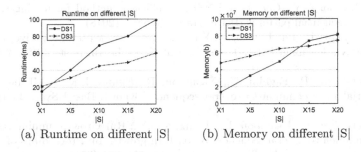

(a) Runtime on different $|S|$ (b) Memory on different $|S|$

Fig. 6. The scalability experiment on DS1.1–DS1.4 and DS3.1–DS3.4

6   Conclusion

Very few algorithms have been proposed to mine NSR and all only mine NSR from PSPs rather than from NSPs, which may result in important information being missing. NSPs can provide more comprehensive information from both positive and negative aspects, which cannot be replicated by only analysing PSPs. In this paper, we proposed a novel algorithm named nspRule to mine NSRs based on NSPs. We conducted experiments on four datasets to evaluate the nspRule algorithm's performance. The experiment results show that under the condition of small memory usage and runtime, many PNSRs can be efficiently mined by the nspRule algorithm.

There is no denying that some of rules are in conflict (e.g., $<ab> \Rightarrow \neg<cd>$ and $<ab> \Rightarrow <c\neg d>$) and cannot be used to make decisions simultaneously, i.e., the decision made by $<ab> \Rightarrow \neg<cd>$ may not be correct. So, we see various opportunities to further work on NSRs. One important issue is to discover actionable NSRs which can enable action-taking for decision-making. Our further work is to find actionable NSRs because many NSRs obtained from NSPs are in conflict (e.g., $<a\neg b> \Rightarrow <cd>$ and $<a\neg b> \Rightarrow <\neg cd>$, $<a\neg b> \Rightarrow <cd>$ and $<a\neg b> \Rightarrow <c\neg d>$, etc.).

References

1. Cao, L., Dong, X., Zheng, Z.: e-NSP: efficient negative sequential pattern mining. Artif. Intell. **235**, 156–182 (2016)
2. Dalmas, B., Fournier-Viger, P., Norre, S.: TWINCLE: a constrained sequential rule mining algorithm for event logs. Procedia Comput. Sci. **112**, 205–214 (2017)
3. Dong, X., Gong, Y., Cao, L.: F-NSP+: a fast negative sequential patterns mining method with self-adaptive data storage. Pattern Recogn. **84**, S0031320318302310 (2018)
4. Dong, X., Gong, Y., Cao, L.: e-RNSP: an efficient method for mining repetition negative sequential patterns. IEEE Trans. Cybern. (2018)
5. Dong, X., Qiu, P., Lü, J., Cao, L., Xu, T.: Mining top-k useful negative sequential patterns via learning. IEEE Trans. Neural Netw. Learn. Syst. **30**(9), 2764–2778 (2019)
6. Fournier-Viger, P., Faghihi, U., Nkambou, R., Nguifo, E.M.: CMRules: mining sequential rules common to several sequences. Knowl. Based Syst. **25**(1), 63–76 (2012)
7. Fournier-Viger, P., Nkambou, R., Tseng, V.S.M.: RuleGrowth: mining sequential rules common to several sequences by pattern-growth. In: Proceedings of the 2011 ACM Symposium on Applied Computing, pp. 956–961 (2011)
8. Gao, X., Gong, Y., Xu, T., Lü, J., Zhao, Y., Dong, X.: Toward to better structure and constraint to mine negative sequential patterns. IEEE Trans. Neural Netw. Learn. Syst. (2020)
9. Guyet, T., Quiniou, R.: NegPSpan: efficient extraction of negative sequential patterns with embedding constraints. Data Min. Knowl. Disc. **34**(2), 563–609 (2020)
10. Hsueh, S.C., Lin, M.Y., Chen, C.L.: Mining negative sequential patterns for e-commerce recommendations. In: IEEE Asia-Pacific Services Computing Conference (2008)
11. Qiu, P., Gong, Y., Zhao, Y., Cao, L., Zhang, C., Dong, X.: An efficient method for modeling nonoccurring behaviors by negative sequential patterns with loose constraints. IEEE Trans. Neural Netw. Learn. Syst. (2021)
12. Setiawan, F., Yahya, B.N.: Improved behavior model based on sequential rule mining. Appl. Soft Comput. S1568494618300413 (2018)
13. Tran, M.-T., Le, B., Vo, B., Hong, T.-P.: Mining non-redundant sequential rules with dynamic bit vectors and pruning techniques. Appl. Intell. **45**(2), 333–342 (2016). https://doi.org/10.1007/s10489-016-0765-3
14. Wang, W., Cao, L.: VM-NSP: vertical negative sequential pattern mining with loose negative element constraints. ACM Trans. Inf. Syst. (TOIS) **39**(2), 1–27 (2021)
15. Zhang, M., Xu, T., Li, Z., Han, X., Dong, X.: e-HUNSR: an efficient algorithm for mining high utility negative sequential rules. Symmetry **12**(8), 1211 (2020)
16. Zhao, Y., Zhang, H., Cao, L., Zhang, C., Bohlscheid, H.: Efficient mining of event-oriented negative sequential rules. In: 2008 IEEE/WIC/ACM International Conference on Web Intelligence, WI 2008, 9–12 December 2008, Sydney, NSW, Australia, Main Conference Proceedings (2008)
17. Zhao, Y., Zhang, H., Cao, L., Zhang, C., Bohlscheid, H.: Mining both positive and negative impact-oriented sequential rules from transactional data. In: Theeramunkong, T., Kijsirikul, B., Cercone, N., Ho, T.-B. (eds.) PAKDD 2009. LNCS (LNAI), vol. 5476, pp. 656–663. Springer, Heidelberg (2009). https://doi.org/10.1007/978-3-642-01307-2_65

18. Zheng, Z., Zhao, Y., Zuo, Z., Cao, L.: An efficient GA-based algorithm for mining negative sequential patterns. In: Zaki, M.J., Yu, J.X., Ravindran, B., Pudi, V. (eds.) PAKDD 2010. LNCS (LNAI), vol. 6118, pp. 262–273. Springer, Heidelberg (2010). https://doi.org/10.1007/978-3-642-13657-3_30
19. Zheng, Z., Zhao, Y., Zuo, Z., Cao, L.: Negative-GSP: an efficient method for mining negative sequential patterns. In: Proceedings of the Eighth Australasian Data Mining Conference, vol. 101. pp. 63–67 (2009)

CrossIndex: Memory-Friendly and Session-Aware Index for Supporting Crossfilter in Interactive Data Exploration

Tianyu Xia[1,2], Hanbing Zhang[1,2], Yinan Jing[1,2(✉)], Zhenying He[1,2], Kai Zhang[1,2], and X. Sean Wang[1,2,3]

[1] School of Computer Science, Fudan University, Shanghai, China
{tyxia19,hbzhang17,jingyn,zhenying,zhangk,xywangCS}@fudan.edu.cn
[2] Shanghai Key Laboratory of Data Science, Shanghai, China
[3] Shanghai Institute of Intelligent Electronics and Systems, Shanghai, China

Abstract. Crossfilter, a typical application for interactive data exploration (IDE), is widely used in data analysis, BI, and other fields. However, with the scale-up of the dataset, the real-time response of crossfilter can be hardly fulfilled. In this paper, we propose a memory-friendly and session-aware index called CrossIndex, which can support crossfilter-style queries with low latency. We first analyze a large number of query workloads generated by previous work and find that queries in the data exploration workload are inter-dependent, which means these queries have overlapped predicates. Based on this observation, this paper defines the inter-dependent queries as a session and builds a hierarchical index that can be used to accelerate crossfilter-style query processing by utilizing the overlapped property of the session to reduce unnecessary search space. Extensive experiments show that CrossIndex outperforms almost all other approaches and meanwhile keeps a low building cost.

Keywords: Crossfilter · Index · Data analysis

1 Introduction

Interactive data exploration (IDE) [3,4,9] is widely used in both scientific research and daily business activities. As the most representative application of IDE, crossfilter is a user interface containing several two-dimensional or three-dimensional charts that reveal the aggregation results of the raw dataset, allowing users to perform interactive operations such as brushing and linking on these charts. It not only lowers the threshold of data analysis, enabling users with varying degrees of skills in the field of data science to manipulate, explore, and analyze data but also shows the results in the form of visualization, helping analysts find insights faster and then draw a conclusion [20].

© The Author(s), under exclusive license to Springer Nature Switzerland AG 2022
A. Bhattacharya et al. (Eds.): DASFAA 2022, LNCS 13245, pp. 476–492, 2022.
https://doi.org/10.1007/978-3-031-00123-9_38

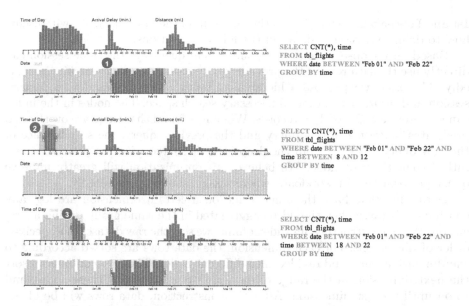

Fig. 1. Queries of Square Crossfilter (https://square.github.io/crossfilter/) in an exploration process.

For example, Fig. 1 shows the process when an airline analyst is exploring part of the flights[1] dataset. He first selects the date from February 1st to February 22, and then drags the brush on the time dimension to find how does airline pattern change between mornings and nights. Through linking and brushing [9,20], a great number of queries will be generated [3]. In order to keep a good interactive experience, crossfilter needs to respond to these queries caused by user interactions in real-time [18]. To deal with the challenge, A variety of methods have been proposed including approximate query processing [1,6,7,25], prefetching [8,10,26], progressive aggregation [11,13,27], data cube [17], and index [12]. However, these approaches more or less suffer as the scale of available data increases, and they ignore the valuable relation hidden in workloads which was recognized as ad-hoc [4] previously. It can be found that there is a relation between the queries during this exploration process. As shown in Fig. 1, the second and third query share the same date predicate with the first query.

To utilize such a characteristic of crossfilter workloads, we first analyze the user interaction logs of exploration on three real-world datasets, and find that periodically a series of queries have overlapped predicates. Such phenomenon occurs when the user discovers something interesting, adds more brushes to views, and performs further detailed queries. We define these queries as a session, and according to data provenance [21], this relation means that the results of these queries are from the same data rows. In the example of Fig. 1, the results of the second and third query come from data rows whose date is between February

---

[1] https://community.amstat.org/jointscsg-section/dataexpo/dataexpo2009.

1st and February 22, which is exactly the result of the first query. We will discuss how to define a session in detail in the following sections.

Based on the observation above, subsequent queries in the same session can directly use the data rows of the previous query result, avoiding searching repeatedly. Therefore, we propose a hierarchical index called CrossIndex to capture session and utilize it to prune unnecessary search space. The nodes in the index can represent a range of data records. When it is found that there are overlapped predicates between the new query and the previous query, the search space of the new query can start from the deepest node shared with the previous query rather than the root node. We believe this optimization will greatly improve query performance for workloads where session shows up repeatedly.

Data rows that have the same value on a certain dimension are logically continuous because they should be aggregated in the same group during aggregation. When constructing our index offline, we sort the raw dataset in a specified order of dimensions. For each dimension, the data rows are sorted according to the dimension value and divided into different nodes. Then we continue to sort the next dimension in the range of data rows represented by these nodes, and so on until the last dimension. After the construction, data rows will be physically continuous, enabling efficient sequential scan access compared to random access [16]. Besides, because queries in a session often scan the same data rows, the nodes in CrossIndex use an array to store the nodes of the next layer so that the CPU cache can be utilized to speed up index search.

In summary, this paper contributes mainly in three aspects:

- We analyze and summarize the characteristics of workloads in the user study conducted by Battle et al. [3], and introduce the concept of the session.
- We propose CrossIndex, a hierarchical index that enables sequential memory access and utilizes sessions for pruning unnecessary search space to speed up crossfilter-style query processing.
- Extensive experiments on the crossfitler benchmark [3] demonstrate that CrossIndex achieves better results than all other approaches, doubling response rate compared to GPU database methods while keeping the same query latency.

The rest of this paper is organized as follows. We summarize the characteristics of workloads and define the problem in Sect. 2. We elaborate on the construction of the proposed CrossIndex and query processing on it in Sect. 3. Section 4 shows the experiments and discussion of results. We review related work in Sect. 5 before discussing how to update the index in Sect. 6 and concluding with Sect. 7.

2   Preliminaries

2.1   Characterizing Workloads

Compared with the traditional OLTP and OLAP context, the query in the crossfilter application has two main characteristics. First, a drag interaction will generate many queries [4], which requires high real-time performance. Second, user

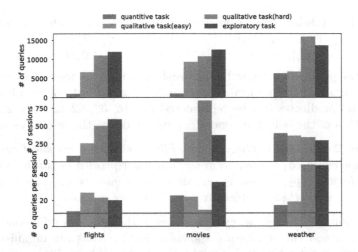

Fig. 2. The number of queries, sessions, and queries per session in the workloads of crossfilter. Generally, the complexity increases from quantitative task to exploratory and more complex task involves more queries.

exploration is unpredictable. Based on the above two points, existing database benchmarks such as TPC-H[2], TPC-DS[3], and SSB[4] are not suitable for IDE scenarios represented by crossfilter because the workloads they represent cannot reflect the real ones generated by user interaction.

However, if we start from the user's point of view, we will realize they are exploring with certain goals. When they explore, the speed of the brush shows a pattern from fast to slow. At the beginning of the process, it always comes with many trials and errors. After determining a certain area of interest, they adjust the range of the brush more carefully. If the range meets the goals, they will continue to explore, otherwise restart the process above. Battle et al.'s analysis [3] on user study designed for 4 types of tasks on 3 different datasets also prove that.

We carefully analyze the user interaction logs and find that the user always continuously adjusts brush range to get an area of interest, and then switches to another view to repeat. Falcon [20] also mentions that users always focus on one view at a time which is called active view, while other views are called passive views. The current query generated by brush on the active view is often based on the previous query results with further filter conditions added.

We call these queries a session. Before explaining how to define a session, we first introduce an operator $\in$ as follows:

---

[2] http://www.tpc.org/tpch/.

[3] http://www.tpc.org/tpcds/.

[4] https://www.cs.umb.edu/~poneil/StarSchemaB.PDF.

Definition 1 (Belonging operator $\in$). *Given two interval sets* $A = \{[v_{s1}^a, v_{e1}^a], [v_{s2}^a, v_{e2}^a], \cdots, [v_{sn}^a, v_{en}^a]\}$ *and* $B = \{[v_{s1}^b, v_{e1}^b], [v_{s2}^b, v_{e2}^b], \cdots, [v_{sn}^b, v_{en}^b]\}$. *If* $[v_{si}^a, v_{ei}^a] \in [v_{si}^b, v_{ei}^b]$ *for every* $i \in [1, n]$, *interval set* $A \in B$.

Since the predicate brought by the brush is an interval, we will use an interval list to represent all the predicates of a query. Take the last query in Fig. 1 as an example, its predicates can be represented as $\{[02\text{–}02, 02\text{–}22], [18, 22]\}$. With the definition of the belonging operator, we can define the session as follows:

Definition 2 (Session). *Given a crossfilter scenario, a workload comprises a series of queries* $q_1, q_2, \cdots, q_n$ *in order, and their predicates can be represented as interval sets* $p_1, p_2, \cdots, p_3$. *If several adjacent queries* $q_i, q_{i+1}, \cdots, q_j$ *satisfy* $p_k \in p_i (i < k \leq j)$, *these queries are said to become a session.*

We summarize the frequency of sessions in the user study [4] conducted by Battle et al., and the results are shown in Fig. 2. Tasks are organized into 4 main classes: quantitative tasks, qualitative task(easy), qualitative task(hard), and exploratory task. Generally, each session contains at least about 10 queries, which means if the overlapped query results are utilized, the performance of a large part of workloads would improve.

2.2 Problem Statement

Based on the characteristics of crossfilter workloads, the crossfilter problem in IDE can be defined as follows:

Problem Statement. *Given a crossfilter scenario, a user explores a certain dataset* R *by crossfilter and generates a series of queries. The order of these queries can be expressed as* $q_1, q_2, \cdots, q_n$. *A session contains at least one query, so the original query sequence can be divided into multiple sessions* $s_1, s_2, \cdots, s_m (m \leq n)$. *The corresponding response time of the query is* $t_1, t_2, \cdots, t_n$. *For a response time threshold* θ *which represents the maximum latency the user can accept, it is guaranteed that the average response time of all queries is less than* θ, *which can be expressed as the following formula:*

$$\frac{1}{n} \sum_{t=1}^{n} t_i \leq \theta (1 \leq i \leq n)$$

We will explain how to build and use CrossIndex to solve this problem in the following sections, and θ will be set to 100ms by default which is a reasonable latency for crossfilter [3].

3   Accelerating Crossfilter by CrossIndex

3.1   CrossIndex Construction

The construction of CrossIndex requires the metadata of the raw dataset, including the type of dimension that users are interested in and the bin width of the visualization. These metadata can uniquely determine a CrossIndex.

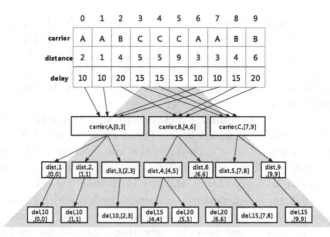

Fig. 3. The construction process of CrossIndex on an example dataset. Triples in the rectangle show the details of CrossIndex nodes. After the end of the construction process, the original dataset will be sorted into the order represented by CrossIndex

Dimension Representation. Suppose the raw dataset is R, and dimensions of interest is $D = \{d_1, \ldots, d_n\}$, and the distinct value contained in the ith dimension is $V_i = \{v_i^1, \ldots, v_i^m\}$. CrossIndex uses each dimension as a layer to build a hierarchical structure, and each node represents data rows whose current dimension has the same value. In the same node, the meaning of data rows are identical, so the interval $[start, end]$ (indicates the start and end id of rows) can be used to represent their range. Therefore a node can be represented by a triple $(d_i, V_i^j, [start, end])$. Starting from the first dimension, datasets are sorted in the order of V_1, resulting in an array of nodes. For the following dimensions, tree nodes are recursively constructed under the range represented by intervals of previous layer nodes. The root node is represented as $(root, null, [0, len(R) - 1])$. The nodes of each layer need to be sorted, so if the row number of dataset is N, the time complexity of the construction is $O(N * \log_2 N * |D|)$.

Partitioning raw data into hierarchical representations can rapidly skip undesired search spaces during querying. Since each tree node is constructed based on previous layers, the data row will not exceed the range of the previous node. As shown in Fig. 3, the first dimension is the *carrier*, and the dimension *distance* is built based on *carrier* nodes. The node with value B ensures that *distance* nodes under it are in the range of interval $[4, 6]$. After the construction process is completed, because the original dataset has been sorted, its logically continuous data rows become physically continuous. The search result of CrossIndex is a collection of intervals, and each interval represents data rows stored sequentially in memory, which means data access will be faster than previously random access [16].

Binned Aggregation. When data scales up, real-time performance should be ensured by choosing a appropriate resolution of visualized data [19]. Binned aggregation can convey both global and local features, and the bin size can also be flexibly adjusted according to needs. On the other hand, visualizations in Crossfilter are basically binned plots, so CrossIndex uses the distinct value of binned dimension d_i as V_i.

CrossIndex has different binning methods for different types of dimensions, mainly for discrete and continuous types. For discrete types such as categorical dimensions, we can simply treat each value as a bin. For continuous types such as numerical and temporal dimensions, if each value is simply used as a bin, it is easy to cause the bin size to be too large, thereby increasing the construction cost of CrossIndex. Therefore, it is necessary to specify the bin width of the continuous type before construction, so that the continuous type can be converted as the discrete type to process. Some binning algorithms [23] can be used to calculate the bin width so that the bin count is moderate, and the visualization fits well on screen.

Construction Order. The order in which the dimensions of CrossIndex are constructed will affect the efficiency of the query to a certain extent. For example, if a query only has a predicate in the last dimension, it needs to cross the first few layers and will not be filtered until the last layer. When the interval is finally aggregated, it may be time-consuming and can easily become a performance bottleneck. Experiments show that a heuristic way is to construct CrossIndex in order of cardinality from small to large, which leads to skipping more search spaces when the first a few dimensions are filtered. Section 4.4 will compare the performance of CrossIndex built in different orders in detail.

3.2 Crossfilter-Style Query Processing

The query generated by the user's interaction on the crossfilter generally contains a series of predicates caused by brush and a specified dimension (current visualization), which corresponds to the WHERE and GROUP BY clauses of the SQL statement. For such a query, CrossIndex will first sort its predicates in the construction order of dimension. Then we scan and filter the index nodes along the hierarchy until reaching the last dimension with a predicate. At this time, the nodes that meet the conditions are the result of the query, and their intervals are the data row position. Finally, the result intervals need to be aggregated by GROUP BY dimension. If the level of GROUP BY dimension is smaller than the level of result node dimension, intervals should trace back to GROUP BY dimension. If the level of GROUP BY dimension is exactly the level of the result, intervals only need to be aggregated by the current dimension. If the level of GROUP BY dimension is greater than the result node dimension, further traversing will be performed until hitting GROUP BY dimension. As shown in Fig. 4, the GROUP BY dimension is *carrier*, so the final result should be aggregated by *carrier*.

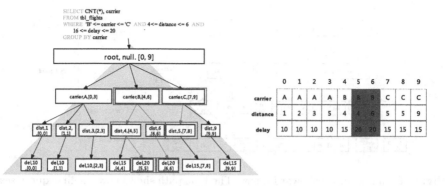

Fig. 4. Example for single query processing in CrossIndex. The red path shows how a query is processed in CrossIndex. The result of this example is $\{B : \{[5,5], [6,6]\}\}$, which means data rows that satisfy the query predicate are in the fifth and sixth row. (Color figure online)

The final result of the CrossIndex query is a set of keys and values. The key represents the grouping value of the GROUP BY dimension, and the value is the interval array. According to the interval array, the actual data row can be found and the corresponding aggregation function calculation is performed. Histograms representing the number of rows of data falling within a certain bin are the most common visualization in crossfilter, which can be calculated simply with $O(1)$ time complexity through interval by $end - start + 1$. Other aggregate functions need to be handed over to the underlying database system for execution.

3.3   Optimization for Crossfilter Workloads

Under crossfilter workloads, searching from the root node every time will bring unnecessary search overhead. Psallidas et al. [21,22] point out that IDE applications can use data provenance. The session defined above can represent the lineage of queries. They are all from the same subset of the raw data, which means they belong to the same subtree in CrossIndex. Because the current query (except for the first one) in one session is based on the previous query, the current query result can be obtained by returning a subset of the previous query result through backward query [22]. The actual search starts from the result nodes of the previous query and finds the result faster than searching from the root node. To implement this optimization, we use an array cache to store result nodes of the previous query in every layer of CrossIndex. For example, the user first filters data with *carrier* between B and C, *distance* between 4 and 6, and *delay* between 16 and 20 on the visualizations. The generated query starts from the CrossIndex root node to find the result. Next, the user adjusts the brush range of *delay* as between 5 and 15, and the query generated at this time can be searched from the previous results node. Figure 5 above shows how two different queries are processed in CrossIndex.

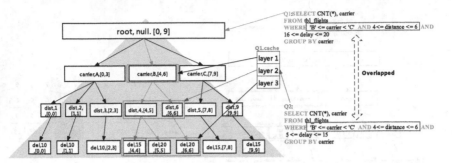

Fig. 5. Example for a backward query. The blue path shows how the first query search in CrossIndex, while the red path shows how the second query search. (Color figure online)

In order to maximize the effect of optimization, it is necessary to find the deepest overlapping node in the CrossIndex hierarchy between the current query and the previous query. If the current query is set as q, and the previous query is set as q_{bak}, the algorithm of backward query is described as follows. First, sort the predicate order of q and q_{bak} query according to the order of CrossIndex construction. Looping from the first dimension, compare two predicates on each dimension and record the current shared deepest layer h. If

- Case 1: the predicates of q and q_{bak} are different, determine whether the predicate of q_{bak} contains the predicate of q. Since the predicate generated by user interaction is in the form of an interval, it is only necessary to compare whether the upper and lower bounds of q_{bak} are greater or smaller than the upper and lower bounds of q. If it contains, update h to the current layer height and exit the loop, otherwise exit the loop directly.
- Case 2: the predicates of q and q_{bak} are the same, update h and continue to the next round.
- Case 3: neither q nor q_{bak} has a predicate, skip and continue to the next round.
- Case 4: either q or q_{bak} has a predicate, exit the loop.

Under the workloads of crossfilter, the backward query can greatly reduce the search time of subsequent queries in a session in CrossIndex. Subsequent experimental in Sect. 4 can also prove the effect of optimization.

4 Experiments

4.1 Setup

Environment. The default experimental environment is a single Ubuntu18.04 LTS system server, with 64G memory, 40 cores (Intel(R) Xeon(R) Gold 5215 CPU @ 2.50 GHz), and more than 3 TB of disk space. The distributed server environment is one Master and four Slaves, all of which are Ubuntu 18.04 LTS operating system. The Master has 128G memory, 16 cores (Intel(R) Xeon(R) Bronze 3106 CPU @ 1.70 GHz), and more than 50 TB of disk space. The Slave has 128G memory, 32 cores (Intel(R) Xeon(R) Silver 4208 CPU @ 2.10 GHz), and more than 7 TB of disk space. The bandwidth of distributed environment is 2000 Mbits/s. The distributed environment is only used for the construction and testing of Kylin data cube[5]. This is because we cannot afford the huge time Kylin is consuming on data cube construction in a single server, and meanwhile Kylin performs better in the distributed environment than in the stand-alone one. Other experiments basically run on a single node server without special specifications.

Datasets and Workloads. We use three real-world datasets [3]: Flights, Movies, and Weather, which are divided into three sizes: 1M rows, 10M rows, and 100M rows. 100M version of three datasets are synthetic, generated by data generation tool provided by IDEBench [9], which uses a statistical model to ensure that the distribution of the synthetic dataset is similar to the actual dataset. The query workloads come from User Study conducted by Battle et al. [3]. There are 128 workflows in total(44 for Flights, 36 for Movies, and 48 for Weather), and each workflow contains all the interactive operations performed by the user in a certain task. We use IDEBench for crossfilter [3] to translate workflows into query workloads tested in experiments.

Metrics. Response rate indicates the proportion of queries that are successfully answered. Queries that are cancelled, discarded, and exceed the time threshold θ are all unsuccessful answered queries. Response rate can be expressed as follows:

$$Response\ Rate = \frac{|Q_a|}{|Q_i|}, \tag{1}$$

where Q_a represents queries that are successfully answered and Q_i represents queries that are issued. We set θ to the same 100 ms as [3].

Mean query duration represents the mean response time of the answered queries, which means that dropped queries are not included. It is calculated as Formula 2.

$$Mean\ Query\ Duration = \frac{\sum_{q \in Q_a} t(q)}{|Q_a|} \tag{2}$$

The two metrics above can directly show how is the throughput of the system and whether the interactive system meets the latency requirement. In addition, we use expansion rate, the ratio of the size of the index to the size of the original dataset, to evaluate the storage cost of CrossIndex offline construction. It is shown as follows:

$$ExpansionRate(Data\ Index) = \frac{|Data\ Index|}{|R|} \tag{3}$$

---

[5] http://kylin.apache.org/.

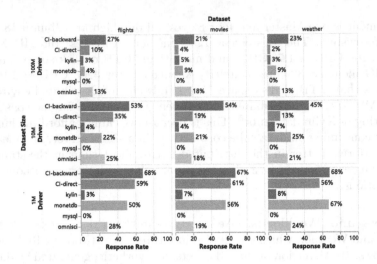

Fig. 6. Response rates of six systems, faceted by dataset and dataset size.

Compared Approaches. We compare our optimized CrossIndex called *CI-backward* and unoptimized version called *CI-direct* with the popular relational database *MySQL*, the in-memory columnar database *MonetDB* [5], the GPU database *Omnisci*[6], and the data cube engine *Kylin*.

4.2   Query Performance

Response Rate. Figure 6 shows the response rate results for all workflows. Overall, as the dataset scales up, basically the performance of each method will be worse, which is the same as our expectations. Larger data leads to more queries failing to reach the response time threshold, so the response rate decreases. MySQL has never been able to return results within 100ms. This is because MySQL is designed mainly for transactional workloads, which leads to poor performance in analytical scenarios such as IDE. Kylin has only slightly improved performance compared to MySQL, but the response rate has never exceeded 10%. MonetDB's response rate decrease rapidly with scale-up, from 50%–67% to 4%–9%. Omnisci's response rate does not exceed 30%, this may be because data needs to be transferred between GPU and CPU. However, its performance is the most stable, enough to prove the role of GPU computing power in queries.

CI-direct can maintain a response rate above 50% in the 1M scenario, but once the amount of data increases, the size of CrossIndex will also increase, and searching from the root node every time will bring huge overhead. So the performance drops rapidly, and the response rate is only 2% to 10% in the case of 100M. CI-backward utilizes the session and outperforms all other methods in all datasets of all sizes. Even in the case of 100M, it can maintain more than a 20% response rate.

---

[6] https://www.omnisci.com/.

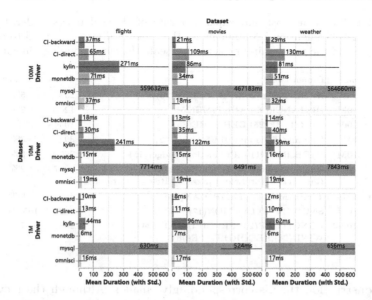

Fig. 7. Mean query duration results of six systems, faceted by dataset and dataset size

Mean Query Duration. Figure 7 shows the mean duration of all answered queries, and the standard deviation is represented by error bars. As the dataset scales up, the mean duration of almost all methods will increase. Kylin has the largest standard deviation due to various orders of predicates in the query workloads. If the order of these predicates conforms to the construction order of the data cube, the response time will be shorter, otherwise the time will be longer. This is also the reason for the large variance of CI-direct. Omnisci has the smallest change in standard deviation and the mean query duration is from 16 ms to 37 ms. MonetDB performs better than CI-direct, with the mean query duration from 6 ms to 71 ms. CI-direct still uses row storage, so even sequential access to rows is not better than column storage. The mean duration of CI-backward has never exceeded 40 ms. However, the standard deviation of CI-backward is larger than that of MonetDB and Omnisci. This is because when the session is switched, it is necessary to search from the CrossIndex root node again, during which the search time may exceed 100 ms. In general, CrossIndex can meet the 100 ms threshold while maintaining a fairly good response rate.

4.3   Offline Cost

The offline construction cost of CrossIndex mainly comes from the grouping and sorting operation. Table 1 shows the construction time and expansion rate in different configurations and sizes. According to the results shown in the table, it is easy to find that the build time increases as the dataset scales up. The number and bin count of the dimensions will also affect the construction time. Compared with the other two datasets, the Flights dataset has fewer dimensions,

Table 1. Offline time cost and expansion rate of three datasets under 1M, 10M, and 100M size. 'Dimensions' column shows the order of dimensions used to construct CrossIndex, and the number in brackets represents the corresponding bin count

Dataset	Size	Dimensions	Time	Expansion rate
Flights	1M	ARR_DELAY(20),DEP_DELAY(20),	25 s	68.5%
	10M	DISTANCE(23),AIR_TIME(25),	126 s	26.2%
	100M	ARR_TIME(25),DEP_TIME(25)	1046 s	7.7%
Movies	1M	Running_Time_min(9),US_Gross(12),	39 s	93.3%
	10M	IMDB_Rating(15),Production_Budget(15),	260 s	55.3%
	100M	US_DVD_Sales(17),Rotten_Tomatoes_Rating(20),Worldwide_Gross(37),Release_Date(50)	2087 s	51.2%
Weather	1M	TMEP_MIN(11),TEMP_MAX(13),	46 s	169.8%
	10M	SNOW(14),ELEVATION(17),	292 s	105.9%
	100M	LONGITUDE(23),PRECIPIATION(24),WIND(24),RECORD_DATE(48)	2088 s	57.7%

so the construction time is correspondingly smaller. Although the movies and weather data sets have the same number of dimensions, the bin count of weather is relatively larger, which leads to a longer construction time. More dimensions and bins will also cause the size of CrossIndex to become larger, so the expansion rate of Movies and Weather is relatively higher, while that of Flights is lower.

Grouping depends on the cardinalities of each dimension involved. The more groups these dimensions have, the larger the CrossIndex would be. Suppose the group numbers of each dimensions are $G = \{g_1, \ldots, g_n\}$, the space complexity of construction is $O(\prod_{i=1}^n g)$

4.4   Effect of Construction Order

We evaluate the effect of construction order for the performance of CrossIndex on the 10M datasets. We use three different construction orders: based on the cardinality of dimension from low to high (ASC), from high to low (DESC), and random(RANDOM). In order to avoid queries in individual workflows that are conducive to certain build order, we experiment on all workflows and focus on the mean query duration. Figure 8 shows that the ASC order is more efficient than random and DESC. The lower the cardinality of dimension(the higher the selectivity), the more data can be filtered through this dimension. The former such dimension is in the order of CrossIndex, the more it can use the session to reduce unnecessary search space, so ASC performs better. DESC is sometimes good or bad, because high cardinality dimensions filter out less data than lower cardinality dimensions, leaving more search space in the subsequent search process of CrossIndex. Query latency on Weather is better than the other datasets because query workloads of Weather contain more sessions, therefore CrossIndex optimization can be fully utilized.

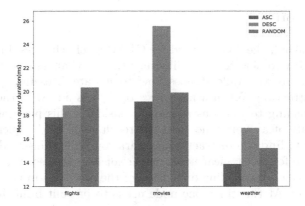

Fig. 8. Mean query duration under different construction orders for 10M dataset size.

5   Related Work

The database and visualization communities have proposed many approaches for crossfilter, and we categorize these approaches as *data cube and indexing, column stores, approximation,* and *prefetching.* For the approximation and prefetching approaches, their results are not 100% accurate and are not suitable for crossfilter scenarios [3], so we have not compared our CrossIndex with them.

Data Cube and Indexing. Data cube and index calculate the aggregation results of the original dataset offline, and directly use these aggregation results when querying. Nanocubes [17] builds an index that can fit in memory for spatiotemporal datasets. Liu et al.'s imMens [19] utilizes dense data structure to overcome the exponential increase in cost due to the number of dimensions. Falcon by Moritz et al. [20] focuses on the index on the active view and calculates the index under the new condition online when switching the active view.

Column Stores. For OLAP workloads, compared with row-stores databases, column-stores always achieve better performance, which have been adopted in MonetDB [5], SeeDB [24], and Profiler [15].

Approximation. In order to fill the gap between data volume and interactivity, many works [1,6,7,25] utilize approximate query processing (AQP) to speed up data processing. Others support progressive data visualization such as incremental sampling-based works [11], and range-based binning works [19,20].

Prefetching. Prefetching can be divided into two categories, namely based on currently explored visualizations and historical data. Approaches based on currently explored visualizations are exemplified by imMens [19] and SW [14]. For the latter, examples are XmdvTool [8], ForeCache [2] and iExplore [26].

6   Discussion

Generally speaking, the dataset used for OLAP rarely changes. In fact, changes may happen. If Crossindex is rebuilt every time it changes, a certain cost will be incurred. Here we would discuss how to update CrossIndex to avoid the cost of reconstruction. When a new piece of data is added to the dataset, the value corresponding to the constructed dimension of this piece of data is used as the predicate of a query, and then the result node of the query is searched in CrossIndex to find where exactly new data row will be inserted. If the result node cannot be found, it means that it cannot be matched in a certain layer. The value of the corresponding dimension of the new data in this layer has not appeared before. At this time, one node needs to be split from the upper layer of the unmatched layer, and continue to create new nodes along the way until the last layer of CrossIndex; if a node in the last layer is found, the right interval value of the node needs to be modified to the original value plus one. Finally, in both cases, it is necessary to modify all the original nodes on the right part of the search path including the nodes on the search path in CrossIndex, and increase their right interval value by one, otherwise the row number of the CrossIndex index will be inaccurate. Deleting or modifying a certain piece of data in the dataset is a similar strategy. Let the number of CrossIndex nodes is n, the time complexity of the update strategy is $O(n)$.

7   Conclusion

In this paper, we propose an index-based method called CrossIndex for crossfilter in data exploration. We first analyze the characteristics of queries of typical IDE applications crossfilter and define the concept of the session. Then, a memory-friendly hierarchical index structure is proposed to handle crossfilter queries. In order to support better real-time performance, we design an algorithm to make CrossIndex aware of sessions and use sessions to prune the search space. The experiments performed on crossfilter IDEBench show that CrossIndex can satisfy a 100ms query latency threshold while maintaining a response rate of more than 20%, which outperforms almost all other approaches.

Acknowledgement. This work is supported by the NSFC (No. 61732004, No. U1836207 and No. 62072113), the National Key R&D Program of China (No. 2018YFB1004404) and the Zhejiang Lab (No. 2021PE0AC01).

References

1. Agarwal, S., Mozafari, B., Panda, A., Milner, H., Madden, S., Stoica, I.: BlinkDB: queries with bounded errors and bounded response times on very large data. In: EuroSys, pp. 29–42 (2013)
2. Battle, L., Chang, R., Stonebraker, M.: Dynamic prefetching of data tiles for interactive visualization. In: SIGMOD, pp. 1363–1375 (2016)

3. Battle, L., et al.: Database benchmarking for supporting real-time interactive querying of large data. In: SIGMOD, pp. 1571–1587 (2020)
4. Battle, L., Heer, J.: Characterizing exploratory visual analysis: a literature review and evaluation of analytic provenance in tableau. In: CGF, vol. 38, pp. 145–159 (2019)
5. Boncz, P.A., Zukowski, M., Nes, N.: Monetdb/x100: hyper-pipelining query execution. In: CIDR, vol. 5, pp. 225–237 (2005)
6. Chaudhuri, S., Ding, B., Kandula, S.: Approximate query processing: no silver bullet. In: SIGMOD, pp. 511–519 (2017)
7. Ding, B., Huang, S., Chaudhuri, S., Chakrabarti, K., Wang, C.: Sample+ seek: approximating aggregates with distribution precision guarantee. In: SIGMOD, pp. 679–694 (2016)
8. Doshi, P.R., Rundensteiner, E.A., Ward, M.O.: Prefetching for visual data exploration. In: DASFAA, pp. 195–202 (2003)
9. Eichmann, P., Zgraggen, E., Binnig, C., Kraska, T.: IDEBench: a benchmark for interactive data exploration. In: SIGMOD, pp. 1555–1569 (2020)
10. Fekete, J., Fisher, D., Nandi, A., Sedlmair, M.: Progressive data analysis and visualization (Dagstuhl seminar 18411). Dagstuhl Rep. 8(10), 1–40 (2018)
11. Fisher, D., Popov, I., Drucker, S., Schraefel, M.: Trust me, I'm partially right: incremental visualization lets analysts explore large datasets faster. In: SIGCHI, pp. 1673–1682 (2012)
12. Gray, J., et al.: Data cube: a relational aggregation operator generalizing group-by, cross-tab, and sub-totals. DMKD 1(1), 29–53 (1997)
13. Hellerstein, J.M., Haas, P.J., Wang, H.J.: Online aggregation. In: SIGMOD, pp. 171–182 (1997)
14. Kalinin, A., Cetintemel, U., Zdonik, S.: Interactive data exploration using semantic windows. In: SIGMOD, pp. 505–516 (2014)
15. Kandel, S., Parikh, R., Paepcke, A., Hellerstein, J.M., Heer, J.: Profiler: integrated statistical analysis and visualization for data quality assessment. In: AVI, pp. 547–554 (2012)
16. Li, L., et al.: BinDex: a two-layered index for fast and robust scans. In: SIGMOD, pp. 909–923 (2020)
17. Lins, L., Klosowski, J.T., Scheidegger, C.: NanoCubes for real-time exploration of spatiotemporal datasets. TVCG 19(12), 2456–2465 (2013)
18. Liu, Z., Heer, J.: The effects of interactive latency on exploratory visual analysis. TVCG 20(12), 2122–2131 (2014)
19. Liu, Z., Jiang, B., Heer, J.: imMens: real-time visual querying of big data. In: CGF, vol. 32, pp. 421–430 (2013)
20. Moritz, D., Howe, B., Heer, J.: Falcon: balancing interactive latency and resolution sensitivity for scalable linked visualizations. In: SIGCHI, pp. 1–11 (2019)
21. Psallidas, F., Wu, E.: Provenance for interactive visualizations. In: HILDA, pp. 1–8 (2018)
22. Psallidas, F., Wu, E.: Smoke: fine-grained lineage at interactive speed. Proc. VLDB Endow. (2018)
23. Satyanarayan, A., Russell, R., Hoffswell, J., Heer, J.: Reactive vega: a streaming dataflow architecture for declarative interactive visualization. TVCG 22(1), 659–668 (2015)
24. Vartak, M., Rahman, S., Madden, S., Parameswaran, A., Polyzotis, N.: SeeDB: efficient data-driven visualization recommendations to support visual analytics. Proc. VLDB Endow. 8(13), 2182–2193 (2015)

25. Wu, Z., Jing, Y., He, Z., Guo, C., Wang, X.S.: POLYTOPE: a flexible sampling system for answering exploratory queries. World Wide Web **23**(1), 1–22 (2019). https://doi.org/10.1007/s11280-019-00685-x
26. Yang, Z., et al.: iExplore: accelerating exploratory data analysis by predicting user intention. In: Pei, J., Manolopoulos, Y., Sadiq, S., Li, J. (eds.) DASFAA 2018. LNCS, vol. 10828, pp. 149–165. Springer, Cham (2018). https://doi.org/10.1007/978-3-319-91458-9_9
27. Zhang, Y., Zhang, H., He, Z., Jing, Y., Zhang, K., Wang, X.S.: Parrot: a progressive analysis system on large text collections. Data Sci. Eng. **6**(1), 1–19 (2021)

GHStore: A High Performance Global Hash Based Key-Value Store

Jiaoyang Li[1,2], Yinliang Yue[1,2(✉)], and Weiping Wang[1,2]

[1] Institute of Information Engineering, Chinese Academy of Sciences, Beijing, China
{lijiaoyang,yueyinliang,wangweiping}@iie.ac.cn
[2] School of Cyber Security, University of Chinese Academy of Sciences,
Beijing, China

Abstract. Log-Structured Merge tree (LSM-tree) has become the mainstream data structure of persistent key-value (KV) stores, but it suffers from serious write and read amplification. In update intensive workloads, repeated and useless compaction of outdated data makes the problem more serious. So we design an efficient global segmented hashmap to record the level of the latest KV pairs, and we present GHStore based on it, which is a key-value store that improves overall performance in write, read and range query simultaneously for update intensive workloads. A read operation of GHStore does not need to search from top to bottom, and a write-induced compaction operation ignores outdated records. The experiments show that for update intensive workloads, compared to widely-used key-value stores (e.g. RocksDB, Wisckey and PebblesDB), GHStore decreases read latency by 10%–50%, range query latency by 15%–60%, while increases write throughput by 4%–55%.

Keywords: Key-value store · LSM-tree · Compaction · Global segmented hashmap

1 Introduction

Persistent key-value stores have become an important part of storage infrastructure in data centers. They are widely deployed in large-scale production environments to serve search engine including e-commerce [9], graph database [1,16], distributed storage [15,17,22], data cache [12], cloud database [18], stream processing [4] and so on.

Log-Structured Merge tree (LSM-tree) [23] has become the main stream data structure of persistent key-value (KV) stores. Various distributed and local stores built on LSM-trees are widely used, such as LevelDB [13], RocksDB [11], Cassandra [19], MongoDB [22] and HBase [14]. The main advantage of LSM-tree over other indexing structures (such as B-trees [6]) is that it maintains sequential access patterns for writes, which is efficient on both solid-state storage devices and hard-disk drives.

A. Bhattacharya et al. (Eds.): DASFAA 2022, LNCS 13245, pp. 493–508, 2022.
https://doi.org/10.1007/978-3-031-00123-9_39

However, LSM-tree remains two challenging problems: write amplification and read amplification. High write amplification caused by the compaction process increases the I/O on storage devices, which reduces write performance. Read amplification is also serious due to multiple levels of search to lookup a KV pair, and it limited the read performance. The existing works make some efforts to solve one of them [10]. For example, some of them focus on improving write performance [2,5,20,21,24,26], or improving read performance [8,28,29]. But few of them improve write, read and range query performance at the same time.

We found that in update intensive workloads, there are multiple versions for a key, but only the latest one is valid. The outdated data causes repeated and useless compaction because they are not to be deleted timely, which makes the write amplification more serious. To address this challenge, we complement the LSM-tree with an additional structure *global segmented hashmap* (GHmap), and design GHStore based on it. GHMap in GHStore records the level of the latest KV pairs, so that a read operation does not need to search from top to bottom, and a write-induced compaction operation ignores outdated records. We evaluate GHStore using YCSB [7] and show that for update intensive workloads, compared to widely-used key-value stores(e.g. RocksDB, Wisckey, PebblesDB), GHStore decreases read latency by 10%–50%, range query latency by 15%–60%, while increases write throughput by 4%–55%. Meanwhile, GHStore reduces read amplification by 3x, write amplification by 1.5x and space amplification by 1.5x.

In summary, the paper makes the following contributions:

- We design an efficient global segmented hashmap to play the role of "God-view", called GHmap. GHmap records the level of the latest KV pairs, to remove outdated KV pairs timely and avoid searching data level by level.
- We design a key-value store based on GHmap called GHStore. In update intensive workloads, GHStore improves overall performance in write, read and range query simultaneously.
- Experiments show that GHStore performs better than several widely used key-value stores (e.g.RocksDB, Wisckey and PebblesDB).

2   Background and Motivation

2.1   Log-Structured Merge Tree

Log-Structured Merge tree (LSM-tree) [23] is a disk-oriented, hierarchical data structure, which defers and batch write requests in memory to exploit the high sequential write bandwidth of storage devices.

RocksDB is a typically key-value store implemented using LSM-tree. It contains multiple levels: two levels in memory, organized data by Memtable and several levels on disk, organized data by SSTable. A Memtable is a skip-list, which could keep the KV pairs in order. An SSTable is a file, divided into data blocks, filter blocks and index blocks, and uses compaction to ensure the ordering of KV pairs.

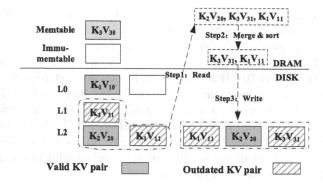

Fig. 1. In the last step of compaction, not all data written to disk is valid.

When L_i triggers compaction [30], RocksDB read some KV pairs from L_i and $L_{(i+1)}$, and then write to $L_{(i+1)}$ after merge and sort. Thus, some KV pairs are read and written many times when they are eventually migrated from L_0 to L_6 through a series of compactions, leading to high write amplification.

To lookup a KV pair, RocksDB needs to check multiple levels, each level may have candidate SSTables. To find a KV pair within an SSTable file, RocksDB needs to read multiple metadata blocks within the file(index block, filter blocks [3], data block). So the reading process requires multiple disk accesses, resulting in the read amplification. Prior studies [21] show that the write amplification could reach up 50x, which read amplification could reach up 300x.

2.2 Motivation

Although Compaction is the fundamental cause of write amplification, it is hard for LSM-tree to avoid it. So it is particularly important to optimize compaction. In the last step of compaction [30], some KV pairs will be written to the disk. However, in the update intensive workloads, not all data written to disk is valid. For example, the latest version of K_3 at Memtable and the latest version of K_1 at L_0. As shown in Fig. 1, after L_1 compaction, not only the valid data K_2V_{20}, but also the outdated data K_3V_{31} and K_1V_{11}, are rewritten to the disk. The outdated

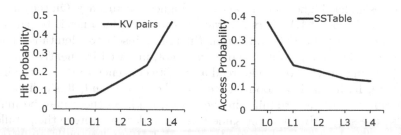

Fig. 2. SSTable access frequence and hit probability of KV pairs at different levels.

data is deleted until the latest one is down to the adjacent level. Before they are deleted, they participate in compaction many times and write to the disk again and again. *So it is necessary to delete outdated data timely.*

On the other hand, key-value stores suffer from severe read amplification. We run experiments on 70 GB RocksDB with 1 million get requests. The Fig. 2 shows that most of KV pairs are located at higher levels, while the SSTables at the lower levels are accessed more frequently. This is because when lookup a KV pair, key-value store needs to check multiple SSTables from the lowest level to the highest level until the key is found or all levels have been checked. This process increases disk access, due to the read amplification. *So avoiding searching data level by level is necessary.*

3   GHStore Design

Figure 3 shows the architecture of GHStore. It consists of two structures: GHmap and LSM-tree. GHmap is the key of GHStore, which plays the role of "God-view". It records the level of the latest KV pairs, to help GHStore to remove outdated data timely and avoid searching level by level from a global perspective. The LSM-tree is divided into R-layer and NR-layer. GHStore only records the level of the latest KV pairs in R-layer into GHmap, to reduce memory cost. In update intensive workloads, with the help of the GHmap, GHStore improves overall performance in write, read and range query simultaneously.

3.1   Global Segmented Hashmap(GHmap)

GHmap records level instead of SSTable, because the memory consumption is more controllable. The number of SSTables increases with the raise of KV pairs, while the number of levels is seven at most in LSM-tree, so only one byte is needed to represent it.

Design: As shown in Fig. 3, GHmap is a hash table that internally is made of an array of N hash tables, which are called submaps. Every submap holds a mutex, costs 8 bytes [25]. Inserting or looking up a record consists of two steps: (1) Calculate the index of the target submap by the key. (2) Decide the index of the target bucket by both the segmentID and the key.

The design has higher performance and higher concurrency. On the one hand, the level of the latest KV pairs is always variable. GHmap needs to participate in all processes of the PUT operation (including insert to Memtable, flush to disk, compaction, etc.), so it would be put, got, updated frequently. Therefore, GHmap should provide very high read and write efficiency to reduce the impact on storage, hashmap is a good choice. On the other hand, flush threads and compaction threads are running in the background. So there may be multiple threads accessing GHmap at the same time. In order to improve the parallelism of it, we split a hashmap into several segments, GHmap internally protects each segment access with its associated mutex.

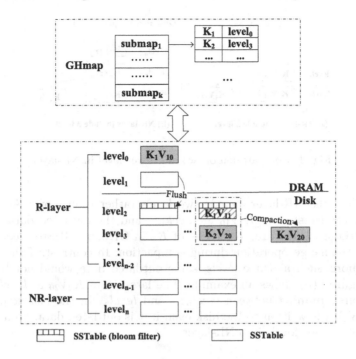

Fig. 3. GHStore architecture

Benifit: GHmap has two benifits: (1) Help to detect the outdated KV pairs and deletes them timely. When the storage updates a KV pair, such as K_1V_{10}, there is a new KV pair put into Memtable. GHmap updates the record to present the level of the K_1's latest version is $level_0$(Memtable), and KV pairs in other levels are out of date. If an outdated KV pair participate in compaction, it would be deleted so that it has no chance to write again. Therefore, the write amplification also will be reduced. (2) Help to locate the level of KV pairs. Compared to search level by level, the storage locates the target level quickly, which is good for reducing read amplification and improving read performance.

3.2 GHStore Optimization

As the global structure, GHStore updates and searches GHmap frequently. So the GHmap is put into memory during GHStore open. Considering that recording all the data of LSM-tree will cost too much memory, we divide the LSM-tree into two parts: the first N-2 levels are R-layer and the last two levels are NR-layer. GHmap only records the latest KV pairs in the R-layer, rather than records in both R-layer and NR-layer.

The capacity of LSM-tree increases level by level. The lower level is 10 times larger than the upper level. The R layer only accounts for 1% of the total, so it greatly reduces the memory consumption.

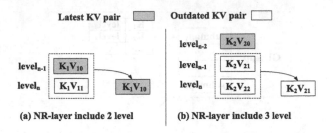

Fig. 4. Comparation of two and three level in NR-layer.

Why does the NR-layer include two levels rather than more? As shown in Fig. 4 (a), the latest data K_1V_{10} at $level_{n-1}$ and the outdated data K_1V_{11} at $level_n$. Although there are no records of K_1 in GHmap, GHStore could delete K_1V_{11} by the merge operation during compaction. In contrast, if the NR-layer includes more, such as three levels, some outdated data would not be deleted timely. Figure 4 (b) shows an example. The latest data K_2V_{20} at $level_{n-2}$, and compaction is running between $level_{n-1}$ and $level_n$. In the last step of compaction, K_2V_{21} is written to the disk even if it is outdated data. So it is better to put the last two levels into NR-layer.

3.3   Efficient GHStore Operations

This section briefly describes how various operations are implemented in GHStore, including four operations: Compaction, Put, Get and Range Query. The Range query operation in GHStore is handled similarly to RocksDB.

Put: The put() operation inserts or updates the mapping from key to value in the GHStore. If the key already exists in the system, its associated value is updated. Firstly, the put() operation writes KV pairs to an in-memory skip-list called the *Memtable*, and insert or update the corresponding record in GHmap. When the Memtable reaches a certain size, it becomes immutable. Then GHStore deletes outdated KV pairs and flushes the latest KV pairs to storage as an SSTable file at $level_2$ (we called memtable as $level_0$ and immutable memtable as $level_1$). Meanwhile, GHmap is updated due to the level of the latest data change. When each level reaches a certain size, it is compacted into the higher level. During the compaction process, GHStore removes outdated KV pairs, only rewrites the latest data to the next level.

Compaction: Compaction is triggered when the actual size of a level exceeds its target size. The compaction of $level_i$ includes three steps: (1) GHStore selects an SSTable randomly, calculates the data boundary (minkey, maxkey) of it. Then it selects the SSTables that overlap with (minkey, maxkey) from the $level_{i+1}$, reads them to the memory too. (2) After that, it merges the duplicate key, deletes outdated KV pairs with GHmap, and sorts the rest of the data. (3) Finally, GHStore puts them into SSTables to $level_{i+1}$, and updates GHmap.

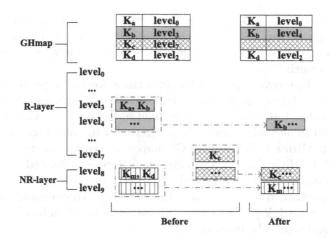

Fig. 5. Compaction

In the second step, GHStore detects outdated data by comparing the datas level with the GHmaps record. If the data come from R-layer, the record must exist in GHmap. As shown in Fig. 5, GHmap tells the storage that the latest KV pair of K_a locates in $level_0$ and the latest KV pair of K_b locates in $level_3$. So K_a at $level_3$ is out of date, while K_b is valid. The output of compaction between $level_3$ and $level_4$ only contains K_b. If the data come from NR-layer, such as K_m and K_d in Fig. 5, the data is outdated if there is a record in GHmap. So the current KV pair of K_d is outdated. For the K_m, there is no record in GHmap, so it is not sure if the current KV pair is the latest one. But the older one is deleted by merging, so the left KV pair must be the latest one and should be rewritten to disk.

In the third step, GHStore updates GHmap after the latest KV pairs are written to the higher level. As shown in Fig. 5, if the KV pair output to R-layer, such as K_b, GHmap would be updated. Otherwise, such as K_c, GHStore deletes the record of it.

Get: The get() operation returns the latest value of the key. The get() operation locates the level of the data by GHmap, and searches the target SSTable within the level. If there is no record in GHmap, the system searches SSTable file in the NR-layer.

Specifically, since there are multiple candidate SSTables in $level_2$, we use bloom filter to avoid reading unnecessary SSTables off storage.

3.4   Crash Consistency

The data in-memory (Memtables, metadata, GHmap) may be lost when the system crashed. For the KV pairs in Memtables, metadata (manifest files), GHStore adopts *Write-ahead Logging* (WAL) for crash recovery, which is also used in RocksDB. For the records in GHmap, GHStore tends to recover them during the compaction process. On the one hand, the GHmap is updated frequently, and thus logging them as soon as the record changes is not a good idea. On the other hand, the recovery of the lost records by scanning the whole storage consumes too much time.

Therefore, the recovery process includes the following steps: (i) Initialize the GHmap with the latest checkpoint when the GHStore reopens. (ii) Recover the data in Memtable, immutable Memtable and metadata by WAL. (iii) Flush the Memtable and immutable Memtable as SSTables to the disk, and record the information in GHmap. (iv) Update GHmap in the compaction process. During the recovery of GHmap, it can also help the GHStore to deleted some outdated data. For example, there is a record K_a, $level_4$ in GHmap. When the $level_i$ triggers compaction, if $level_i$ is lower than $level_4$, GHmap should be updated because the current KV pair is newer. If $level_i$ is higher than $level_4$, the KV pair is outdated and should be deleted.

4   Evaluation

In this section, we run extensive experiments to demonstrate the key accomplishments of GHStore: (1) GHStore performance in various workloads (Sect. 4.2 and Sect. 4.3). (2) The strengths of GHmap (Sect. 4.5). (3) The memory consumption of GHStore (Sect. 4.6).

4.1   Experimental Setup

Our experiments are run on a machine with a 24-core Intel(R) Xeon(R) Platinum 8260M CPU @ 2.40 GHz processor, 64 GB RAM and running CentOS 8.3 LTS with the Linux 4.18.0 kernel and the ext4 file system. The machine has one 1TB HDD and one 128 GB SSD.

We compare GHStore with widely-used key-value stores, such as RocksDB, Wisckey and PebblesDB. We use their default parameters: 64 MB Memtables/SSTables, 256 MB $level_2$ size, AF of 10 and compression algorithm is *Snappy*. We also allow the compared key-value stores to use all available capacity in our disk space, so that their major overheads come from the read and write amplification in the LSM-tree management.

We use db_bench and YCSB [7] to provide a more complete view of GHStore's behavior.

4.2   Performance Comparison

This section evaluates GHStore performance using different micro-benchmarks in update intensive workloads. We evaluate GHStore performance in different dimensions, such as write throughput in different update ratios, read latency in a different dataset, range query latency in different query lengths. We also analyze write and read amplification.

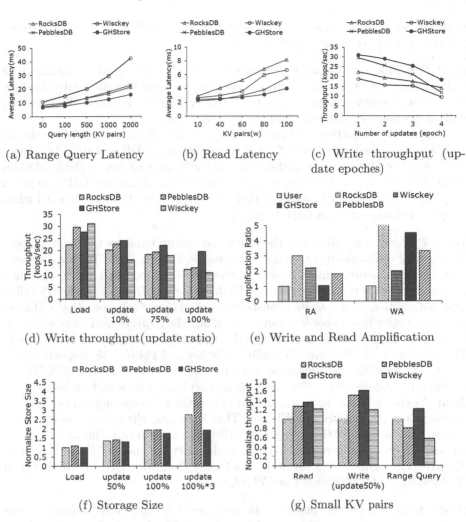

Fig. 6. Micro benchmarks performance

Range Query Performance. We evaluate the range query performance of GHStore and other key-value stores in update intensive workloads. We first load 100 GB of KV pairs into each key-value store. A KV pair consists of a 24B key, and a 1 KB value. We then repeatedly issue updates over the existing 100 GB of KV pairs three times. As shown in Fig. 6(a), in most cases, GHStore does well. The latency of GHStore is 18%–25% lower than RocksDB, 24%—62% lower than Wisckey, and 15%–29% lower than PebblesDB.

For an LSM-based key-value store (without key-value separate), the range query is comprised of seek() operation followed by several next() operations. Range query performance depends mainly on two factors: the number of levels and the number of next() operations. Compared with others, GHStore does fewer next() operations due to a higher density of valid data. We should thank GHmap for its efforts in reducing outdated data.

Read Performance. We do experiments to measure the read performance of GHStore and other key-value stores. We use the same key size and value size as the range query experiment, but the data volume is different. We issue one million get requests and calculate the average latency.

Figure 6(b) shows that GHStore is always doing well compared with RocksDB and Wisckey. The latency is 25%–50% lower than RocksDB and 10%–47% lower than Wisckey. However, the performance of GHStore is similar to PebblesDB on small datasets, like 10 GB. But with the data volume increases, GHStore has a growing advantage. For example, GHStore is 37% higher than PebblesDB when we issue get requests to a 100 GB storage.

Write Performance. We study the impact of update numbers on the write performance of GHStore and other key-value stores. We first load 50 GB of data into each key-value store, which consists of the 24 B key, and 1 KB value, and then repeatedly update them several times. Updates in each time follow a heavy-tailed Zipf distribution with a Zipfian constant of 0.99. Figure 6(c) shows that GHStore performs better than other key-value stores, and the improvement increase as the number of updates rises. Generally, write throughput of GHStore is 30%–48%, 64%–93% and 4%–55% over RocksDB, Wisckey and PebblesDB respectively.

We do another experiment to evaluate the write performance of GHStore in different update ratios. We insert 10 million unique keys and update part of them. As shown in Fig. 6(d), in the load phase, the write throughput of GHStore is lower than PebblesDB and Wisckey. That is because there are fewer duplicate keys in the storage and the strength of GHStore is not obvious. But in the update phases, GHStore performs well. For example, when we update 75% of KV pairs, the write throughput of GHStore is 54% higher than RocksDB, 14% higher than PebblesDB, and 23% higher than Wisckey.

Write Amplification. We measure the *write amplification* (WA) of four systems on the same experiment of randomly writing 40 GB dataset. Figure 6(e) shows the results measured by the ratio of the amount of data written to SSDs and the amount of data coming from users. The WA of GHStore is 1.26x lower than RocksDB. GHStore has the lower WA since it deletes more outdated KV pairs during compaction, which is beneficial to reduce WA. Unfortunately, The WA of GHStore is higher than PebblesDB and Wisckey. Because we only guarantee that outdated KV pairs are not rewritten to disk, the latest KV pairs are still rewritten many times.

Read Amplification. We measure the *read amplification* (RA) of four systems for workloads that randomly read 3 million KV pairs from 40 GB storage. We analyze the AR by calculating the ratio of the amount of SSTables read from the disk and the amount of SSTables that contains target data. As the result present in Fig. 6(e), GHStore has the smallest RA because it avoids searching data level by level.

Space Amplification. In the beginning, we load 100 GB unique KV pairs into each key-value store. The space size of GHStore is near RocksDB, Wisckey, and PebblesDB. However, as shown in Fig. 6(f), after we updated part of the keys several times, GHStore consumes 10%–20% smaller space size than other stores. For instance, after we repeatedly update the existing 100 GB of KV pairs three times (update 100% * 3 workload), the space overhead of PebblesDB is 1.5x that of GHStore.

Small KV Pairs. We insert 50 GB KV pairs into storage (the key is 16bytes and the value is 128 bytes), and update 50% of the existing data five times. As shown in Fig. 6(g), the result is similar to results with large keys. GHStore obtains higher read, write and range query performance.

4.3 YCSB Workloads

Table 1. YCSB core workloads description

Workloads	Description
Load	100% writes
YCSB A	50% reads and 50% updates
YCSB B	95% reads and 5% updates
YCSB C	100% reads
YCSB D	95% reads(latest writes) and 5% updates
YCSB E	95% range queries and 5% updates
YCSB F	50% reads and 50% reads-modify-writes

The industry standard in evaluating key-value stores is the Yahoo Cloud Serving Benchmark [7], a widely used micro-benchmark suite delivered by Yahoo!. The suite has six core workloads (described in Table 1), each representing a different real-world scenario.

We run the YCSB benchmark with Zipfian key distribution. We set the Memtable size as 128 MB, SSTable size as 64 MB, and allow at most 4 threads to run the compaction process. In the beginning, we load a 100 GB dataset with 1 KB KV pairs. Then we evaluate workload A-F with one million KV pairs respectively. Figure 7 presents the results: GHStore outperforms RocksDB, Wisckey and PebblesDB on most workloads except Load.

On write-dominated workloads like Load, since there is only one version of keys, the strength of GHStore is not obvious.

For the read-dominated workloads, such as *workloadB,workloadC* and *workloadD*, GHStore achieves better throughput than other systems. For example, in workloadB, GHStore obtains 1.35x, 1.2x and 1.11x better than RocksDB, Wiscey and PebblesDB respectively.

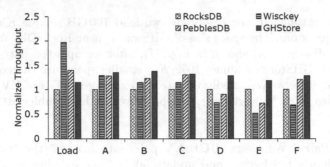

Fig. 7. YCSB Performance

For the range-query-only *workloadE*, GHStore surprisingly achieves 2.31x better throughput than Wisckey, and 1.64x better throughput than PebblesDB. The reason for that is GHStore has a higher density of valid data in SSTables, so the number of next() operations in the range query process is decreased.

The difference between *workloadA* and *workloadF* is workloadF does a get() operation before a put() operation. Although GHStore wins by a nose compared to PebblesDB, GHStore performs much better than RocksDB and Wisckey. GHStore obtains 1.88X and 1.28x better throughput than Wisckey and RocksDB respectively.

4.4   Performance on SSD

To examine the impact of different devices, we conduct experiments with SSD. Similarly, we first load 50 GB of data to each key-value store, each KV pair consists of the 24 B key, and 1 KB value. And then repeatedly update the existing KV pairs five times. We evaluate GHStore performance in write, read and range query dimensions. As the result shown in Fig. 8, in most cases, GHStore does well compared with other key-value stores. Therefore, our system can be adapted to a variety of storage devices.

4.5   GHmap Strengths

We study the impact of submap on the insert latency of GHmap. We vary the number of submap from 2^0 to 2^6. Figure 9 shows that the segmented hashmap has lower insert latency than the normal hashmap, and the number of submaps make some effect on latency. But more submaps do not always better. If we increase the number of submaps, we expect to see more parallelism, but diminishing returns. Because every submap resize will quickly block the other threads until the resize is completed. So in this paper, we choose $N = 16$ as the number of submaps.

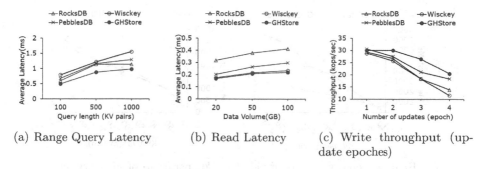

(a) Range Query Latency   (b) Read Latency   (c) Write throughput (up-date epoches)

Fig. 8. GHStore performance on SSD

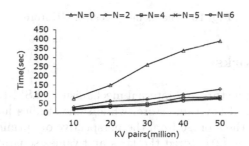

Fig. 9. Submaps influence on GHmap.

4.6 Memory Consumption

As described in Sect. 3.2, to reduce the memory consumption, GHmap only records the level of the latest KV pairs in R-layer, rather than both R-layer and NR-layer. As shown in Fig. 10(a), the optimization can greatly reduce memory consumption.

Then we compare the memory consumption of GHStore and RocksDB. We fix the size of each KV pair, which consists of the 22B key and 1 KB value. As shown in Fig. 10(b), GHStore has similar memory consumption to RocksDB. Specifically, we load 100 million (about 220 GB) of KV pairs, GHStore only has 0.05% higher consumption than RocksDB. RocksDB searches a KV pair from top level to bottom level, and each level has candidate SSTables. To reduce SSTable access, RocksDB adds bloom filter to each SSTable, and caches them to memory. GHStore uses GHmap to index the target level, so we don't need bloom filter except for $level_2$ (only the $level_2$ have multiple candidate SSTables). Therefore, GHStore memory cost is acceptable.

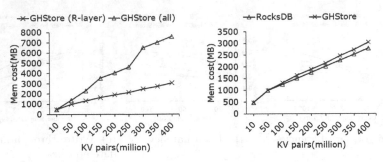

(a) Reduce memory cost by divide LSM-tree into R-layer and NR-layer (b) Comparation of memory cost between GHStore and RocksDB

Fig. 10. The memory cost of GHStore

5   Related Works

Starting from the structure and internal mechanism of LSM-tree, GHStore optimizes LSM-tree based key-value store, while some studies have also optimized the performance of the storage from the perspective of optimizing the structure of LSM-tree. Wisckey [21] stored the keys and values separately to reduce the depth of LSM-tree, thus improving the write performance. However, this mechanism of key-value separation does not support efficient read and range query. PebblesDB [26] mitigates WA by using guards to maintain partially sorted levels. However, the data inside each guard are out of order and duplicated. So the read and range query performance are not good. UniKV [28] grouped the disk data into several boxes and divided intra-box data into two parts: UnsortedStore and SortedStore. UniKV placed the hot data in UnsortedStore and used hash index to speed up the read performance, while other data are stored in SortedStore and the write performance is accelerated by the KV separation, thus UniKV can improve write performance while maintaining read performance as much as possible. By contrast, we have improved both the read performance and the range query performance without losing too much write performance, and thus the overall performance is better.

GHStore timely filters the stale data in the system, which is helpful to compress the system size. Some related work also optimized the performance of the storage system from the perspective of compressing the system size. By changing the storage mode of the index in SSTables, SlimDB [27] relieved the space requirement from data and index, thus reducing the size of the system. However, this kind of optimization method has not removed the outdated KV pairs in the system, so there is still a lot of outdated data in the system.

6   Conclusion

This paper presents GHStore, a key-value store that improves overall performance in write, read and range query simultaneously for update intensive workloads. GHStore consists of two structures: GHmap and LSM-tree. GHmap is the key of GHStore, implemented by a global segmented hashmap. It records the level of the latest KV pairs, to help GHStore to remove outdated data timely and avoid searching level by level from a global perspective. The LSM-tree is divided into R-layer and NR-layer. GHStore only records the level of the latest KV pairs in R-layer into GHmap to reduce memory cost. GHStore outperforms widely-used stores such as RocksDB, Wisckey, and PebblesDB on many workloads. In the future, we will apply GHStore to various production environments.

Acknowledgements. We would like to thank the reviewers for their comments. This work was partially supported by BMKY2020B10.

References

1. Baidu: Hugegraph. https://github.com/hugegraph/hugegraph (2019)
2. Balmau, O., et al.: Triad: Creating synergies between memory, disk and log in log structured key-value stores. In: Proceedings of the 2017 USENIX Conference on Usenix Annual Technical Conference, pp. 363–375. USENIX ATC 2017, USENIX Association, USA (2017)
3. Bloom, B.H.: Space/time trade-offs in hash coding with allowable errors. Commun. ACM **13**(7), 422–426 (1970)
4. Carbone, P., Ewen, S., Fóra, G., Haridi, S., Richter, S., Tzoumas, K.: State management in apache flink®: consistent stateful distributed stream processing. Proc. VLDB Endowment **10**(12), 1718–1729 (2017)
5. Chan, H.H., et al.: Hashkv: Enabling efficient updates in {KV} storage via hashing. In: 2018 {USENIX} Annual Technical Conference ({USENIX}{ATC} 18), pp. 1007–1019 (2018)
6. Comer, D.: Ubiquitous b-tree. ACM Comput. Surv. (CSUR) **11**(2), 121–137 (1979)
7. Cooper, B.F., Silberstein, A., Tam, E., Ramakrishnan, R., Sears, R.: Benchmarking cloud serving systems with YCSB. In: Proceedings of the 1st ACM Symposium on Cloud Computing, pp. 143–154 (2010)
8. Dayan, N., Athanassoulis, M., Idreos, S.: Monkey: optimal navigable key-value store. In: Proceedings of the 2017 ACM International Conference on Management of Data, pp. 79–94 (2017)
9. DeCandia, G., et al.: Dynamo: amazon's highly available key-value store. ACM SIGOPS Oper. Syst. Rev. **41**(6), 205–220 (2007)
10. Dong, S., Kryczka, A., Jin, Y., Stumm, M.: Rocksdb: evolution of development priorities in a key-value store serving large-scale applications. ACM Trans. Stor. (TOS) **17**(4), 1–32 (2021)
11. Facebook: Rocksdb. http://RocksDB.org (2017)
12. Gade, A.N., Larsen, T.S., Nissen, S.B., Jensen, R.L.: Redis: a value-based decision support tool for renovation of building portfolios. Build. Environ. **142**, 107–118 (2018)
13. Ghemawat, S., Dean, J.: Leveldb. https://github.com/google/LevelDB (2011)

14. Harter, T., et al.: Analysis of {HDFS} under hbase: a facebook messages case study. In: 12th {USENIX} Conference on File and Storage Technologies ({FAST} 14), pp. 199–212 (2014)
15. Huang, D., et al.: TIDB: a raft-based HTAP database. Proc. VLDB Endowment **13**(12), 3072–3084 (2020)
16. Jain, M.: Dgraph: synchronously replicated, transactional and distributed graph database. Birth (2005)
17. cockroach Labs: Cockroachdb. https://github.com/cockroachdb/cockroach (2017)
18. Lai, C., Jiang, S., Yang, L., Lin, S., Cong, J.: Atlas: Baidu's key-value storage system for cloud data. In: Symposium on Mass Storage Systems & Technologies, pp. 1–14 (2015)
19. Lakshman, A., Malik, P.: Cassandra: a decentralized structured storage system. ACM SIGOPS Oper. Syst. Rev. **44**(2), 35–40 (2010)
20. Lin, Z., Kai, L., Cheng, Z., Wan, J.: Rangekv: An efficient key-value store based on hybrid dram-nvm-SSD storage structure. IEEE Access **8**, 154518–154529 (2020)
21. Lu, L., Pillai, T.S., Gopalakrishnan, H., Arpaci-Dusseau, A.C., Arpaci-Dusseau, R.H.: Wisckey: separating keys from values in SSD-conscious storage. ACM Trans. Storage (TOS) **13**(1), 1–28 (2017)
22. MongoDB: Mongodb. https://github.com/mongodb/mongo (2017)
23. O'Neil, P., Cheng, E., Gawlick, D., O'Neil, E.: The log-structured merge-tree (lsm-tree). Acta Inform. **33**(4), 351–385 (1996)
24. Pan, F., Yue, Y., Xiong, J.: dcompaction: Delayed compaction for the lsm-tree. Int. J. Parallel Program. **45**(6), 1310–1325 (2017)
25. Popovitch, G.: parallel-hashmap. https://github.com/greg7mdp/parallel-hashmap (2020)
26. Raju, P., Kadekodi, R., Chidambaram, V., Abraham, I.: Pebblesdb: building key-value stores using fragmented log-structured merge trees. In: Proceedings of the 26th Symposium on Operating Systems Principles, pp. 497–514 (2017)
27. Ren, K., Zheng, Q., Arulraj, J., Gibson, G.: Slimdb: a space-efficient key-value storage engine for semi-sorted data. Proc. VLDB Endowment **10**(13), 2037–2048 (2017)
28. Zhang, Q., Li, Y., Lee, P.P., Xu, Y., Cui, Q., Tang, L.: Unikv: toward high-performance and scalable kv storage in mixed workloads via unified indexing. In: 2020 IEEE 36th International Conference on Data Engineering (ICDE), pp. 313–324. IEEE (2020)
29. Zhang, W., Xu, Y., Li, Y., Zhang, Y., Li, D.: Flamedb: a key-value store with grouped level structure and heterogeneous bloom filter. IEEE Access **6**, 24962–24972 (2018)
30. Zhang, Z., et al.: Pipelined compaction for the LSM-tree. In: 2014 IEEE 28th International Parallel and Distributed Processing Symposium, pp. 777–786. IEEE (2014)

Hierarchical Bitmap Indexing for Range Queries on Multidimensional Arrays

Luboš Krčál[1]([✉]) [ID], Shen-Shyang Ho[2] [ID], and Jan Holub[1] [ID]

[1] Department of Computer Science, Czech Technical University in Prague,
Prague, Czech Republic
{lubos.krcal,jan.holub}@fit.cvut.cz
[2] Department of Computer Science, Rowan University, Glassboro, NJ, USA
hos@rowan.edu

Abstract. Bitmap indices are widely used in commercial databases for processing complex queries, due to their efficient use of bit-wise operations. Bitmap indices apply natively to relational and linear datasets, with distinct separation of the columns or attributes, but do not perform well on multidimensional array scientific data.

We propose a new method for multidimensional array indexing that considers the spatial component of multidimensional arrays. The hierarchical indexing method is based on sparse n-dimensional trees for dimension partitioning, and bitmap indexing with adaptive binning for attribute partitioning. This indexing performs well on range queries involving both dimension and attribute constraints, as it prunes the search space early. Moreover, the indexing is easily extensible to membership queries.

The indexing method was implemented on top of a state of the art bitmap indexing library Fastbit, using tables partitioned along any subset of the data dimensions. We show that the hierarchical bitmap index outperforms conventional bitmap indexing, where an auxiliary attribute is required for each dimension. Furthermore, the adaptive binning significicantly reduces the amount of bins and therefore memory requirements.

1 Introduction

Research in many areas produces large scientific datasets, which are stored in multidimensional arrays of arbitrary size, dimensionality and cardinality, such as QuikSCAT [10] or RapidScat [13] satellite data.

Majority of the current systems rely on linearization of the array data, enabling many one-dimensional access methods to be used. Others, such as array databases [2,16], work natively with multidimensional arrays.

Very efficient method of indexing arbitrary data is bitmap indexing. Bitmap indices leverage hardware support for fast bit-wise operations and are space-efficient. For higher-cardinality attributes, this efficiency is achieved by sophisticated multi-level and multi-component indices. Bitmap indices are used in majority of commercial relational databases [6,7,14].

A. Bhattacharya et al. (Eds.): DASFAA 2022, LNCS 13245, pp. 509–525, 2022.
https://doi.org/10.1007/978-3-031-00123-9_40

Major disadvantage of bitmap indices for multidimensional array data indexing is their linear nature. Even with index compression such as WAH [21] (variation of run-length compression), this only partially suppresses the issue.

Our major contribution is a new method of bitmap indexing designed natively
for multidimensional array data that overcomes the dimensionality-induced inef-
ficiencies. The method is based on n-dimensional sparse trees for dimension partitioning, and on attribute partitioning using adaptively binned indices.

We demonstrate the performance on range queries involving both dimension
and attribute constraints on large partitioned datasets.

Our algorithm fundamentally differs from standard spatial indexing methods,
such as cSHB [11] or grid-based bitmap index from [12]. We focus on multidimensional arrays, with data points directly addressable with a set of integral
coordinates, and each coordinate maps to exactly one data point.

2   Related Work

Traditional indexing methods like B-trees and hashing are not effectively applicable to index multiple attributes in a single index, being replaced by multidimensional indexing methods, such as R-trees [8], R*-trees [3], KD-trees, n-
dimensional trees (quadtrees, octrees, etc.).

The drawbacks of traditional indexing algorithms led to bitmap indices [4]
and their applications for scientific data [15]. Bitmap indices are naturally based
on linear data, ideal for relational databases. Space filling curves, such as Z-order
curve and Hilbert curves were used for linearization and subsequent querying of
multidimensional data. Hilbert curves were used in [9], while Z-order curves were
used in [11], which is a system for querying spatial data (not arrays).

The boom of multidimensional, scientific array data gave birth to open-source
multidimensional array-based data management and analytics systems, namely
RasDaMan [2] and SciDB [16]. These databases work natively with multidimensional arrays, but lack some of the effective query processing methods implemented in other databases.

3   Preliminaries

3.1   Array Data Model

An array $\mathcal{A}$ consists of *cells* with *dimensions* indexed by $d_1, \ldots, d_n$. Each
cell is a tuple of several *attributes* $a_1, \ldots, a_m$. We assume the structure of
the attributes is the same for all cells in the array. The array is denoted as
$\mathcal{A}\langle a_1, \ldots, a_m\rangle[d_1, \ldots, d_n]$.

We form a *query* on arrays based on *constraints* on dimensions and
attributed. Figure 1 shows a query on array $\mathcal{A}\langle a_1, \ldots, a_m\rangle[d_1, \ldots, d_n]$ that has a
constraint on an attribute a and a constraint on dimension d_2.

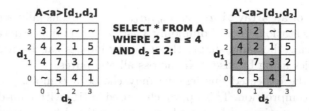

Fig. 1. An example of a range query on a two-dimensional array. The result is on the right, with the selected values highlighted.

3.2 Distributed Arrays

Due to the large size of scientific data, it is often necessary to split the data into subarrays called *chunks*, or *partitions* in conventional databases.

There are two commonly used strategies. Regularly gridded chunking, where all chunks are of equal shape and do not overlap. This array data model is known in SciDB as MAC (Multidimensional Array Clustering) [16]. The second strategy is irregularly gridded chunking, which is one of the chunking option in RasDaMan [2].

3.3 Bitmap Indexing

Bitmap indices, originally introduced in [4], were shown to be very efficient for read-only or append-only data, and are used in many relational databases and for scientific data management.

The structure of bitmaps is determined by a *binning* strategy. For high cardinality attributes, binning is the essential minimum to keep the size of the index reasonable [22]. Binning effectively reduces the overall number of bitmaps required to index the data, but increases the number of cells that have to be later verified. This is called a *candidate check*. Two most common binning strategies are *equi-width* binning and *equi-depth* binning.

Another aspect of bitmap indexing is *encoding* [4]. *Range encoding* uses $B-1$ bitmaps, each bitmap R_i encodes a range of bins $[B_1, B_i]$. *Interval encoding* [5] uses $\frac{|B|}{2}$ bitmaps, each bitmap I_i is based on two range encoded bitmaps.

Binary run-length compression algorithms are usually applied on bitmap indices to reduce the overall size. There are two representative compression algorithms, namely Byte-aligned Bitmap Code – BCC [1] and Word-Aligned Hybrid (WAH) compression [21].

4   Hierarchical Bitmap Array Index

We now briefly discuss a common way of indexing multidimensional arrays using additional bitmap indexes for each dimension. Then we describe the structure of our hierarchical bitmap array index.

Attributes of an array $\mathcal{A}\langle a_1, \ldots, a_m \rangle [d_1, \ldots, d_n]$ are usually stored in a linearized representation, most commonly C-style row-major ordering. With all attributes having the same shape, these binary indices can be used to execute selection queries using bitwise AND across all attributes.

Based on the expected queries, we may choose a combination of binning, encoding and compression. This approach is used in [19] with equi-depth binning or in [18] with v-optimized binning based on v-optimal histograms and C-style row-major linearization in [17].

Another option is to use either Z-order or Hilbert space filling curves to further increase locality of the dimensions. We chose to use the concepts of n-dimensional space partitioning to structure our bitmaps and avoid the need to explicitly enumerate bitmaps for the array dimensions.

4.1 Partitioning of Arrays

We partition the array $\mathcal{A}\langle a_1, \ldots, a_m \rangle [d_1, \ldots, d_n]$ into a set of regularly gridded chunks C in the *Multidimensional Array Clustering* fashion described in Sect. 3.2, such that each chunk C is defined on a hyperrectangle with the two opposite corners defined by the points $(o_1, \ldots o_n)$ and $(e_1, \ldots, e_n)$:

$$C_A[o_1, \ldots o_n, e_1, \ldots, e_n] = \mathcal{A}\langle a_1, \ldots, a_m \rangle [o_1 \leq d_1 < e_1, \ldots, o_n \leq d_n < e_n]$$

All chunks in our data model are of the same shape, i.e., for all chunks C, C' from array A, it holds that $C_A[e_k] - C_A[o_k] = C'_A[e_k] - C'_A[o_k]$ for all dimensions k. Chunks are not overlapping and completely cover the array A. By chunking the array, we limit the domain of both attributes and dimensions per partition.

We choose to use bitmap indexing on attributes and auxiliary bitmap indexing for dimensions of the chunk. Note that the dimension indices are the same for all chunks of the same shape in the array, since for each chunk, we can simply subtract its offset from the dimensions query constraints. Therefore these auxiliary dimension indices are only stored in memory for the lifetime of the index.

We propose a unified solution that solves both the problem with dimension attributes and with synopsis of array chunks. Our solution is in a form of hierarchical bitmap index on top of a n-dimensional tree with variable binning for each node in the tree. This allows our index to discard invalid or completely matches nodes. The hierarchical index also allows for a smooth transition between index levels and the leaf indices, due to the current matching state being passed further down the tree during query execution.

4.2 Structure of the Array Chunk Index

Each chunk $C[o_1, \ldots, o_n]$ of array $\mathcal{A}$ is associated with exactly one leaf $N_\ell(o_1, o_2, \ldots, o_n)$. Independently, each leaf uses an equi-depth binning index with a total of at most BINS bins, where bin boundaries $bins(N_\ell)$ of the index are based on the distribution of exact chunk values. The leaf's dimension boundaries correspond to its associated chunk's boundaries.

Accounting for empty values is done using a special bitmask, known as *empty bitmask*. For each chunk, we thus have a total of BINS + 1 indices.

Except for very narrow dimension range queries, a dimension query will either cover the whole span of a leaf node, or result in a one-sided dimension range query once the query processing reaches a single chunk. Thus, the ideal encodings for chunks are *range* and *interval* encodings. We have chosen interval encoding as our default encoding since it uses half the memory range encoding does.

4.3   Construction of the Hierarchical Bitmap Array Index

To deal with the higher level index, we create a special composite index on tree similar to n-dimensional tree. Each internal node of the index has at most F children, where F is called a *fanout*. Our bitmap indices are based on the fanout and we want to utilize binary operations as much as possible. For this reason, the fanout F should be a multiple of the processor word size W, or as close to it as possible.

The index tree construction works in a bottom-up fashion, where the leaf nodes are indexed at first. This allows both data appending and modification. Each internal node is constructed from at most F direct children and with at most BINS attribute bins, with one additional index for empty bitmask.

Let $B = (min(N_1), max(N_1)), \ldots, (min(N_F), max(N_F))$ be the set of all intervals ranging from the minimum to the maximum value of the indexed attribute α among all the child nodes N_i. The set B is the set of bins – the individual interval boundaries are delimiters, where the attribute's value spans a different subset of child nodes. Meaning anytime we cross any bin threshold from B, at least one child either becomes a new partial or full match or is no longer a partial or full match. Formally, $nodesin(a) \subset N_i$ is a function of the attribute value $a \in \alpha$, which returns a subset of child nodes, such that $N_i \in nodesin(a) \iff min(N_i) \leq a \leq max(N_i)$.

4.4   Bin Boundaries Merging in Parent Nodes

The number of bins B from all F child nodes is higher than BINS for majority of the internal nodes N, therefore it is necessary to reduce the size of the set of bins B. There are several strategies to choose the parent bins R such that $|R| =$ BINS. An example of such binning reduction is in Fig. 2.

The first strategy is to use an equi-width distribution of the bins. This is the ideal choice assuming the attribute part of the query is uniformly distributed or when there is no prior knowledge about the attribute query.

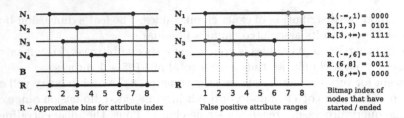

Fig. 2. Example of merging $|B| = 8$ bin boundaries to $|R| = 4$ bin boundaries for 4 child nodes $N_1, \ldots, N_4$. False positive ranges are marked in red. Two sided range encoded bitmaps are generated for R. (Color figure online)

The second strategy is to use equi-depth binning. This is ideal if the attribute distribution of the child nodes is skewed. It is possible to maintain the weights of the bins for leaf nodes. However, internal nodes can only make estimates about the weight of merged bins. In each internal node and leaf, we store the weight estimate $w(b)$, where $b \in B$. The weighted square error $wse(b)$ of a bin b and weighted sum square error $wsse(B)$ are:

$$wse(b) = \left| w(b) - \frac{w(B)}{\text{BINS}} \right|^2 \qquad wsse(B) = \sum_{b \in B} wse(b)$$

To estimate the weight of merged bin $r \in R \subset B$, we assume uniform distribution of values over the intervals of bins $b \in B$. Estimated weight of r is:

$$w(r) = \sum_{b \in B} w(b) \cdot |b \cap r|$$

We cannot use the trivial algorithm for equi-depth binning, because we can only iterate by bins of variable weight, instead of iterating by single data points. This is why we need to approximate the equi-depth using a simple iterative algorithm. Details on selecting $R \subset B$ approximately equi-depth bins are shown in Algorithm 1. We first start with equi-width binning (line 1). Then, we generate sets of all possible bin splits and merges (lines 2–3), setup two priority queues and evaluate all possible splits and merges in terms of weighted sum square error (lines 4–11). After that, we perform one valid split and one merge on the binning as long as this leads to an improvement of the overall binning (lines 14–18).

Input: set of bins B, set of weights $w(b)$, $b \in B$, number of output bins BINS
Result: approx equi-depth bins $R \subset B$, $|R| = BINS$

```
1  E ← eq-width bins from B, |B| =BINS;
2  S ← all possible split bins of E;
3  M ← all possible merged bins of E;
4  Q_S ← priority_queue();
5  Q_M ← priority_queue();
6  for s ∈ S do // bins to split
7  |   add (s, Δwse(s)) to Q_SPLIT;
8  for (m, m') ∈ M do // bins to merge
9  |   add ((m, m'), Δwse((m, m')) to Q_MERGE;
   // split that decreases wsse the most
10 (s, Δwse(s) ← min(Q_S);
   // merge that increases wsse the least
11 ((m, m'), Δwse((m, m'))) ← min(Q_M);
12 while Δwse((m, m')) > Δwse(b) do
13 |   split b;
14 |   merge (b, b');
15 |   update R, S, M, Q_M, Q_S;
```

Algorithm 1: Iterative equi-depth binning approximation of parent bins.

4.5 Double Range Encoding of Bitmap Indices in Internal Nodes

Unlike in bitmap indexing in leaves where one encodes positions of individual values, we encode sets of child nodes $nodesin(a)$ for attribute values a in the internal nodes.

We will now describe an effective bitmap encoding of $nodesin(a)$, for $a \in r \in R$. Let's have two adjacent intervals $r \in R$ and $r' \in R$, such that $r_h = r'_\ell$. Note that since $R \subset B$, we have $nodesin(r) \neq nodesin(r')$. If $nodesin(r') \supset nodesin(r)$, then r' corresponds to a bin, where nodes are added, and we add r' to a set R_+. Else, if $nodesin(r') \subset nodesin(r)$, then nodes are removed in set $nodesin(r')$, and we add r' to set R_-. Otherwise, some nodes are added and some are removed and we add r' to both R_+ and R_-. In our example in Fig. 2, $R_+ = \{[1, 3), [3, 6)\}$ and $R_- = \{(3, 6], (6, 8]\}$.

There is no guarantee that $|R_+| = |R_-|$. If we wanted, we could run Algorithm 1 separately on boundaries B_+ and B_- (likewise defined) and with $\frac{BINS}{2}$ bins, but then we'd lose the equi-width approximation.

Now, we encode $|R_+| + 1$ bitmaps using range encoding, so that the index for each bin $r_+ \in R_+$ corresponds to children, whose attribute range minimum $min(N_i)$ is less or equal to the upper boundary of the interval r_+. In our example, bitmap corresponding to $r = [1, 3) \in R_+$ is 0101, indicating that N_1 and N_3 have started in or before this interval. Similarly, we encode $|R_-| + 1$ bitmaps for values r_- using inverse range encoding, i.e., children, whose attribute range maximum $max(N_i)$ is greater than r_- are encoded by 0 in the bitmap, representing children that have already ended before or in the interval r_-.

Any two bitmaps then allow evaluation of partial and complete matches (see Sect. 5.1) using only two bitmap reads and one logical operation for both partial and complete query.

4.6 Locality of the Hierarchical Index

In order to preserve locality of the data during queries, we store the whole index in a locality preserving linearization of an n-dimensional tree. The index data consist of bin boundaries, weight estimates and bitmap indices.

We use the Hilbert space-filling curve to linearize the node's children index. Hilbert curve has perfect locality, but it does not preserve dimensions ordering. This means we precompute bitmaps for dimension constraints for each block of Hilbert curve separately.

5 Querying Dimensions and Attributes

In this work, we focus on selection queries over both dimensions and single attribute of an array. Such query consists of a set of dimension constraints and attribute constraints. Let's specify a query Q over an array $\mathcal{A}\langle a_1, \ldots, a_m \rangle [d_1, \ldots, d_n]$ as a set of ranges over attributes Q_A and dimensions Q_D.

$$Q = Q_A \cup Q_D = \{(a, a_\ell, a_h), \ldots\} \cup \{(d, d_\ell, d_h), \ldots\}$$

where (a, a_ℓ, a_h) is a triple specifying attribute's lower bound and its (exclusive) upper bound; same goes for dimensions. It is possible for a query to not specify constraints for some dimensions, in which case we fill all remaining dimensions to get a complete query.

The core of the query algorithm is a breadth-first descent through the index tree. At each level, the search space is pruned according to both dimension and attribute values. Let N be the currently searched node, N_i be its child nodes, where $0 \leq i < F$. Throughout the query processing, we maintain a queue of partially matched nodes P and a set of completely matched nodes C. We start at a root node N_r, setting $P = \{N_r\}$

Let p, p_D, p_A and c, c_D, c_A be zero bitmaps of size F; the bitmaps p indicates partial attribute matches among the children of node N, p_D indicated partial dimensions matches, p_A indicates partial matches, similarly the vectors c, c_D, c_A indicate complete matches. We will now set these vectors according to the query Q for the first node in the queue P. The partial and complete matches bitmap computation is also described in Algorithm 2 and in Fig. 3.

Input: query $Q = \{(a, a_\ell, a_h), (d, d_\ell, d_h), \ldots\}$; current node N; node's children
$\quad\quad N_1, \ldots, N_F$; dimension boundaries $N[d]_\ell, N[d]_h$ for N and all N_i;
Result: partial matches p; complete matches c;

```
1  𝒫_A, 𝒞_A ← load attribute index for node N;
2  𝒫_D, 𝒞_D ← precomputed dimension index;
3  p_D ← {0}^F, p_A ← {0}^F, p;
4  c_D ← {1}^F, c_A ← {1}^F, c;
5  if a_h < min(N) or a_ℓ > max(N) then
6  |  return p ← {0}^F, c ← {0}^F ;                    // no matches
7  c_A = c_A & 𝒞_A(a_ℓ, a_h) ;                        // complete attribute match
8  p_A = p_A | 𝒫_A(a_ℓ, a_h) & ~c ;                   // partial attribute match
9  for (d, d_ℓ, d_h) in Q_D do
10 |  if d_h < N[d]_ℓ or d_ℓ > N[d]_h then
11 |  |  return P ← {0}^F, C ← {0}^F ;               // no matches
12 |  if d_ℓ > N[d]_ℓ then
13 |  |  p_D = p_D | 𝒫_D(d_ℓ) ;                       // partial dimension match
14 |  if d_h < N[d]_h then
15 |  |  p_D = p_D | 𝒫_D(d_h) ;                       // partial dimension match
16 |  c_D = c_D & 𝒞_D(d_ℓ, d_h) ;                     // complete dimension match
17 p_D ← p_D & c_D;
18 c_D ← c_D & ~p_D;
19 c ← c_A & c_D;
20 p ← (p_A | c_A) & (p_D | c_D) & ~c;
21 return p, c
```

Algorithm 2: Evaluation of partial and complete match bitmaps for a node.

5.1 Attribute Based Matches

In this subsection, we explain how attribute bitmask is set. This subsection further describes lines 5–8 in Algorithm 2.

If $a_h < min(N)$, or $a_\ell > max(N)$, there are neither partial nor complete attribute matches and we terminate processing of the current node.

Let $\mathcal{P}_A(a_\ell, a_h)$ be a *partial attribute match* interval-encoded bitmask specific to node N for an array, with bits set to one corresponding to children N_i so that the intersection $[a_\ell, a_h) \cap [min(N_i), max(N_i)) \neq \emptyset$.

$$\mathcal{P}_{B|A}(a)[i] = 1 \iff min(N_i) \leq a$$
$$\mathcal{P}_{E|A}(a)[i] = 1 \iff max(N_i) \geq a$$
$$\mathcal{P}_A(a_\ell, a_h)[i] = 1 \iff \mathcal{P}_{B|A}(a_h)[i] \wedge \neg\mathcal{P}_{E|A}(a_\ell)[i]$$

The first expression describes bitmap set to 1 for children that have started before or at value a, the second expression describes children that have ended at or after a. The third expression then combines both.

To evaluate partial matches using $\mathcal{P}_A(a_\ell, a_h)$, we first use binary search on the set of attribute boundaries from R_+ (corresponding to starting bitmaps) and set of attribute boundaries from R_- (corresponding to ending bitmaps). To find two bins $L \in R_+$ and $H \in R_-$ such that $a_\ell \in L$ and $a_h \in H$. These bins L and H mark the attribute boundary bins. Then, $\mathcal{P}_{B|A}(a_h)$ is identical to $R_+[H]$ and $\neg \mathcal{P}_{E|A}(a)$ is identical to $R_-[L]$. The bitmap indices R_+ and R_-, each queried for a single bin, are described in Sect. 4.3. Then we combine $\mathcal{P}_A(a_\ell, a_h)$ to p using bitwise OR.

Now, we process complete candidates in a similar fashion. Let $\mathcal{C}_A(a_\ell, a_h)$ be a *complete attribute match* bitmask specific to node N for array of shape $\mathcal{S}$, so that the intersection $[a_\ell, a_h] \cap [min(N_i), max(N_i)] = [a_\ell, a_h]$.

$$\mathcal{C}_A(a_\ell, a_h)[i] = 1 \iff \mathcal{P}_{B|A}(a_\ell)[i] \wedge \neg \mathcal{P}_{E|A}(a_h)[i]$$

This expression is very similar to $\mathcal{P}_A(a_\ell, a_h)$, describing children that have started at or before a_ℓ and have not ended at or before a_h. To evaluate $\mathcal{C}_A(a_\ell, a_h)$, we query $R_+[L]$ and $R_-[H]$. Then, we add the result to c using bitwise OR and remove those from p, i.e., $p = p \wedge \neg c$.

5.2   Dimension Based Matches

Next, we explain how the dimension masks are set. This subsection further describes lines 9–17 in Algorithm 2.

If for a dimension d we have $d_h < N[d]_\ell$ or $d_\ell > N[d]_h$, there are neither partial nor complete dimension matches and we terminate processing the current node.

Unlike attribute query, the evaluation of dimension query is the same for all nodes N, so all the bitmaps for processing dimensions queries are *precomputed*.

Let $\mathcal{P}_d(d_\ell, d_h)$ be a *partial dimension match*, where d is a dimension in the query constraint (d, d_ℓ, d_h). The partial dimension match indicates child nodes N_i such that the intersection $[N_i[d]_\ell, N_i[d]_h) \cap [d_\ell, d_h) \neq \emptyset$.

Let's fix a dimension d for which we evaluate partial matches $\mathcal{P}_d(d_\ell, d_h)$:

$$\mathcal{P}_d(d_\ell)[i] = 1 \iff d_\ell \in N_i[d] \wedge d_\ell \neq N_i[d]_\ell$$
$$\mathcal{P}_d(d_h)[i] = 1 \iff d_h \in N_i[d] \wedge d_h \neq N_i[d]_h$$
$$\mathcal{P}_d(d_\ell, d_h)[i] = 1 \iff \mathcal{P}_d(d_\ell)[i] \vee \mathcal{P}_d(d_h)[i]$$
$$\mathcal{C}_D(d_\ell, d_h)[i] = \bigcap_{1 \leq d \leq \text{DIMS}} \mathcal{P}_d[i]$$

The first expression describes which children N_i have dimension d range such that the query limit d_ℓ falls inside the range, but it is not equal to the lower limit of that range. The second expression is similar, but for d_h. Third expression combines the partial matches over the previous query limits. And the fourth expression combines all dimensions.

Partial dimension matches are evaluated using one precomputed bitmap index corresponding to

$$\mathcal{P}_d(b)[i] = 1 \iff b = N_i[d]$$

where b is a bucket corresponding to the chunking of the array $\mathcal{A}$. There are a total of F_d such buckets along dimension d, resulting in a total of $F_d \cdot d$ bitmaps of size F. We query these bitmaps for all dimensions and combine them using bitwise OR.

There is a special case of false negative dimension result. If d_ℓ or d_h is equal to the d'th dimension range border of a child node N_i, and at the same time the other end of d_ℓ or d_h causes the dimension to be fully covered in N_i, i.e. $d_\ell = N_i[d]_\ell$ and $d_h \geq N_i[d]_h$ or $d_h = N_i[d]_h$ and $d_\ell \leq N_i[d]_\ell$, the query is evaluated as partial match for N_i and dimension d, while in fact dimension d contributes to complete matches. A check for this scenario requires comparing the dimension ranges of child nodes to the query range, and was ignored on purpose, as it complicates and slows down the query process.

For complete candidates, we will slightly modify the definition of $\mathcal{C}_A$ used for attributes. Let $\mathcal{C}_d(d_\ell, d_h)$ be a *complete dimension match*. The complete dimension match indicates which child nodes N_i are *partially or fully* covered by interval $[d_\ell, d_h]$. Despite the semantics indicating partially matches should not be included, we later trim the complete dimension match bitmap accordingly.

$$\mathcal{C}_d(d_\ell, d_h)[i] = 1 \iff [d_\ell, d_h] \cap N_i[n] \neq \emptyset$$
$$\mathcal{C}_D(d_\ell, d_h)[i] = \bigcap_{1 \leq d \leq \text{DIMS}} \mathcal{C}_d[i]$$

Complete dimension matches are evaluated using two precomputed bitmap indices corresponding to

$$\mathcal{C}_{B|d}(b)[i] = 1 \iff b \leq N_i[d]$$
$$\mathcal{C}_{E|d}(b)[i] = 1 \iff b \geq N_i[d]$$

similarly to bitmaps used for partial matches. There is a total of $2 \cdot F_d \cdot d$ bitmaps of size F for complete matches. We query these bitmaps for all dimensions and combine them using AND.

We now combine the partial dimension matches bitmap $\mathcal{C}_D(d_\ell, d_h)$ with $\mathcal{C}_D(d_\ell, d_h)$, such that $\mathcal{P}_D(d_\ell, d_h) = \mathcal{P}_D(d_\ell, d_h) \wedge \neg \mathcal{C}_D(d_\ell, d_h)$. During the evaluation of dimension matches, we used a total of $3 \cdot d$ index queries. An example of dimension query is displayed in the top row in Fig. 3.

5.3   Partial and Complete Matches

Now that we have both attribute and dimension, and both partial and complete candidates, we may proceed to merging the candidates and generating a bitmap representing the set of complete node children matches $\mathcal{C}$ and a bitmap representing the set of partial node children matches $\mathcal{P}$ that will be recursively explored. This subsection further describes lines 20–22 in Algorithm 2.

The $\mathcal{C}$ bitmap is easier to obtain, as it is the intersection of both complete bitmaps without partial candidates bitmaps.

$$\mathcal{C} = \mathcal{C}_A \wedge \mathcal{C}_D$$

We obtain the set of partial candidates $\mathcal{P}$ by joining the dimension-based partial candidates with the attribute-based candidates and clipping both by complete candidates

$$\mathcal{P} = (\mathcal{P}_A \vee \mathcal{C}_A) \wedge (\mathcal{P}_D \vee \mathcal{C}_D) \wedge \neg \mathcal{C}$$

We then iterate through the results, adding child nodes from $\mathcal{C}$ to the result set and the partial candidates $\mathcal{P}$ into the queue to be processed subsequently. This process is done on top of Hilbert curve indices, as it is trivial to generate Hilbert curve indices corresponding to nodes in the lower levels. The Hilbert curve ordering of the inner nodes and breadth-first traversal also ensures single traversal through the index.

5.4   Implementation and Fastbit Integration

Fastbit [20] is an open source library that implements several state of the art algorithms for bitmap indexing. It's not a complete database management system, rather a data processing tool, as its main purpose is to facilitate selection queries and estimates. Fastbit's key technological features are WAH bitmap compression, multi-component, and multi-level indices with many different combinations of encoding and binning schemes.

The implementation of our ArrayBit algorithm is built directly on top of Fastbit library. We use Fastbit's partitions to setup the lowest level of our indices (leaves), and base our binning indices on Fastbit's single-level binning index. This allows our hierarchical bitmap index to smoothly transition into the leaf nodes by using existing bitmask before executing range queries on the leaves.

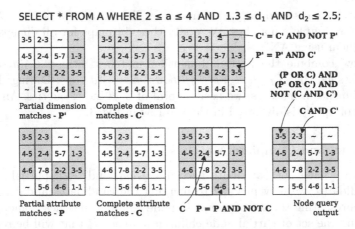

Fig. 3. Processing of a query in a single node of the hierarchical index. Top row represents dimension constraints, bottom row represents attribute constraints. Blue nodes represent partial matches and green represent complete matches. (Color figure online)

This approach requires preprocessing of the data into evenly shaped partitions, generating empty bitmasks and shape metadata. The same shape metadata is also used for mesh queries in Fastbit. During partition indexing, we inject more metadata into the partition to be able to rebuild ArrayBit index only at any stage, without affecting any backward compatibility with Fastbit's query execution. The table is then indexed as described in Sect. 4.

During query execution, only ArrayBit hierarchical index and the leaf indices are resident in memory. Selection queries require disk access to load raw data for all partially matched leaf nodes.

6   Experimental Evaluation

We have tested our implementation on large multidimensional arrays, running queries with both dimension and attribute constraints. Since our solution is based on bitmap indexing, we compared our implementation with state of the art bitmap indexing implementations from Fastbit library.

We measured the time efficiency for each individual query, i.e. the index construction time and query execution time, and space requirements for the index. Timing was measured as an average of 100 runs with preconstructed and preloaded index in the main memory.

We use the same library from Fastbit for CPU and wall time measuring. We tried to not include non-essential steps in the time measurements, such as query string parsing, by using the low level API of Fastbit. Space requirements were measured based on the disk space required to store the bitmap index together with all relevant metadata.

The experiments were run on a single machine – AMD Ryzen Threadripper 1920X @ 3.50 GHz, 4×16 GB RAM, NVMe Samsung SSD 970 EVO Plus 1TB (for data only, different from system drive); running Ubuntu 18.04 (4.20.3 kernel).

6.1   Datasets

This section describes the synthetic and real world datasets in detail. A summary of the dataset properties is available in Table 1.

DGauss. We use a synthetic dataset generated from a *sum of multidimensional gaussian distribution* DGAUSS. Its only attribute $a1$ of `double` type is a sum of 32 randomly initialized Gaussian distributions in D dimensions.

For our experiments, we use a 2D version of the dataset partitioned into 1024 evenly distributed partitions of shape $(1024, 1024)$. Each partition is decorated with metadata indicating its shape and position in within the global array.

RapidScat. As a representative of a large real world dataset, we use a complete Level 2B Ocean Wind Vectors data from the RapidScat instrument mounted on the International Space Station from 2014 to 2016.

The data has 3 native dimensions: *along_track*, *cross_track*, and *ambiguities* (only used for some attributes) and one auxiliary dimension: *orbit*. We use the dimensions *along_track*, *cross_track*, and *orbit* to model 3 dimensional data array.

The total compressed RapisScat data is over 100 GB, uncompressed raw NetCDS is almost 500 GB, but after extracting the only attribute (and keeping all dimensional data) and transforming in Fastbit's binary format, we got to total size of 119 GB including a bitmap index of size 2.6 GB.

Table 1. Overview of datasets used for experimental evaluation.

Dataset	Size	Index size	Dimensions	Attributes	Partitions
DGauss	26 GB (1 attribute)	796 MB	2 (x, y)	1 (a1)	1024
RapidScat	119 GB (1 attribute)	2.6 GB	3 (orbit, along, cross)	22, 1 used (retrieved_wind_speed)	6930

6.2   Bitmap Indexing Methods

We compare ArrayBit with state of the art bitmap indexing library Fastbit, including their implementation of spatial queries.

FastBit represents a naive algorithm, where the dimensions in the datasets are indexed using bitmap indices the same way the attributes are. The index is using 32 equi-depth binned indices, interval encoding and WAH compression. The queries for all dimensions and attributes are executed via logical *AND*.

FastBit::MeshQuery uses an extension of the Fastbit library designed for executing spatial queries on designated attributes. When mesh tags are specified in the dataset, and the query contains only constraints from those dimensions, a mesh query will be executed. Mesh query is able to handle regular meshes, treating cells of meshes as connected regions in space. The configuration is the same as in the naive Fastbit: using 32 binned indices, range encoding and WAH compression on all attributes. For mesh queries, Fastbit generates larger index.

ArrayBit represents our hierarchical multidimensional index. We use 16 equi-depth binned indices, range encoding and WAH compression to index the partitions. For the hierarchical index, we use 16 approximately equi-depth binned indices (described in Sect. 4.4) with two sided range encoding and no compression. We only use half of the bins for the bitmaps, this compensates the required space for the hierarchical index.

6.3   Range Queries

In our work, we focus on mixed attribute and dimension queries. Regardless of the dataset, we describe the queries based on *selectivity*, i.e. the overall ratio of the query result to the size of the entire array size.

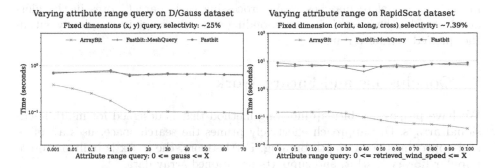

Fig. 4. Comparison of `FastBit`, `FastBit::MeshQuery`, and `ArrayBit` on a set of queries with fixed dimension ranges and varying attribute range.

Figure 4 shows the execution time dependency based on the selectivity of the queried attribute range. We can see that on both datasets, both `FastBit` and `FastBit::MeshQuery` have a constant execution time. This is due to the inability to effectively filter out partitions outside of the queried attribute range. `ArrayBit` effectively discards all partitions which cannot yield an attribute match. The number of such partitions increases with higher attribute range, hence the downward trend in the plot.

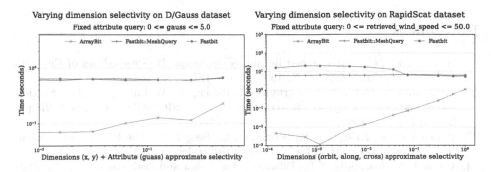

Fig. 5. Comparison of `FastBit`, `FastBit::MeshQuery`, and `ArrayBit` on a set of queries with fixed attribute range and varying dimension range.

Figure 5 shows the dependency of execution time based on the selectivity of the dimension query constraints. The attribute range in the queries is set such that vast majority of the partitions qualifies as having potential results. From both subplots, we can see that with increased dimension range, the query execution time increases almost linearly. This correlates with the number of partitions that have to be loaded and have queries executed within them.

In the last query from the RapidScat dataset, where ArrayBit is still a bit faster than the reference FastBit algorithm. This is due to some partitions marked as complete matches when processing their parent notes in the hierarchical index, because their corresponding attribute range is completely within the queried attribute range.

7   Conclusions and Future Work

We have proposed a bitmap indexing method that is designed for multidimensional arrays. Our approach effectively prunes the search space, uses adaptive and approximate equi-depth binning. Furthermore, the index is built on top of popular library Fastbit, and supports partitioned array data.

Our experimental results show that the proposed bitmap indexing method outperforms standard bitmap approaches for mixed attribute and dimension range query processing. We have tested our algorithm on large synthetic and real world datasets, with queries of varying dimension and attribute selectivity.

Future work includes several interesting topics. It is possible to adapt the tree structure dynamically based on dimensions, such as adaptive mesh refinement widely used in physical simulations. We may want to explore the possibility of using index structures other than n-dimensional trees for mapping out and combining the bitmap index and different binning strategies. A robust templated implementation of these parameters would yield a good ground for further indexing experiments. Implementation-wise, a step forward would be to integrate out algorithm as part of the core Fastbit library.

References

1. Antoshenkov, G.: Byte-aligned bitmap compression. In: Proceedings of the Data Compression Conference. DCC 1995, p. 476. IEEE (1995)
2. Baumann, P., Dehmel, A., Furtado, P., Ritsch, R., Widmann, N.: The multidimensional database system RasDaMan. In: ACM SIGMOD Record, vol. 27, pp. 575–577. ACM (1998)
3. Beckmann, N., Kriegel, H.P., Schneider, R., Seeger, B.: The R*-tree: an efficient and robust access method for points and rectangles, vol. 19. ACM (1990)
4. Chan, C.Y., Ioannidis, Y.E.: Bitmap index design and evaluation. In: ACM SIGMOD Record, vol. 27, pp. 355–366. ACM (1998)
5. Chan, C., Ioannidis, Y.: An efficient bitmap encoding scheme for selection queries. ACM SIGMOD Record (1999). http://dl.acm.org/citation.cfm?id=304201
6. Chou, J., et al.: Parallel index and query for large scale data analysis. In: Proceedings of 2011 International Conference for High Performance Computing, Networking, Storage and Analysis, p. 30. ACM (2011)

7. Gosink, L., Shalf, J., Stockinger, K., Wu, K., Bethel, W.: HDF5-FastQuery: accelerating complex queries on HDF datasets using fast bitmap indices. In: 18th International Conference on Scientific and Statistical Database Management, pp. 149–158. IEEE (2006)

8. Guttman, A.: R-trees: a dynamic index structure for spatial searching, vol. 14. ACM (1984)

9. Lawder, J.K., King, P.J.H.: Querying multi-dimensional data indexed using the Hilbert space-filling curve. ACM SIGMOD Rec. **30**(1), 19–24 (2001)

10. Lungu, T., Callahan, P.S.: QuikSCAT science data product user's manual: overview and geophysical data products. D-18053-Rev A, version 3, p. 91 (2006)

11. Nagarkar, P., Candan, K.S., Bhat, A.: Compressed spatial hierarchical bitmap (cSHB) indexes for efficiently processing spatial range query workloads. Proc. VLDB Endow. **8**(12), 1382–1393 (2015)

12. Park, K.: A hierarchical binary quadtree index for spatial queries. Wirel. Netw. **25**(4), 1913–1929 (2018). https://doi.org/10.1007/s11276-018-1661-z

13. SeaPAC: Rapidscat level 2b ocean wind vectors in 12.5km slice composites version 1.1. In: NASA Physical Oceanography DAAC (2015). https://doi.org/10.5067/RSX12-L2B11

14. Sinha, R.R., Winslett, M.: Multi-resolution bitmap indexes for scientific data. ACM Trans. Database Syst. (TODS) **32**(3), 16 (2007)

15. Stockinger, K.: Bitmap indices for speeding up high-dimensional data analysis. In: Hameurlain, A., Cicchetti, R., Traunmüller, R. (eds.) DEXA 2002. LNCS, vol. 2453, pp. 881–890. Springer, Heidelberg (2002). https://doi.org/10.1007/3-540-46146-9_87

16. Stonebraker, M., Brown, P., Zhang, D., Becla, J.: SciDB: a database management system for applications with complex analytics. Computing in Science and Engineering **15**(3), 54–62 (2013). https://doi.org/10.1109/MCSE.2013.19

17. Su, Y., Wang, Y., Agrawal, G.: In-situ bitmaps generation and efficient data analysis based on bitmaps. In: Proceedings of the 24th International Symposium on High-Performance Parallel and Distributed Computing, pp. 61–72. ACM (2015)

18. Wang, Y., Su, Y., Agrawal, G.: A novel approach for approximate aggregations over arrays. In: Proceedings of the 27th International Conference on Scientific and Statistical Database Management, p. 4. ACM (2015)

19. Wang, Y., Su, Y., Agrawal, G., Liu, T.: SciSD: novel subgroup discovery over scientific datasets using bitmap indices. In: Proceedings of Ohio State CSE Technical report (2015)

20. Wu, K., et al.: FastBit: interactively searching massive data. J. Phys. Conf. Seri. **180**, 012053 (2009). IOP Publishing

21. Wu, K., Otoo, E.J., Shoshani, A.: Optimizing bitmap indices with efficient compression. ACM Trans. Database Syst. (TODS) **31**(1), 1–38 (2006)

22. Wu, K.L., Yu, P.S.: Range-based bitmap indexing for high cardinality attributes with skew. In: Proceedings of the Twenty-Second Annual International COMPSAC 1998, pp. 61–66. IEEE (1998)

Membership Algorithm
for Single-Occurrence Regular
Expressions with Shuffle and Counting

Xiaofan Wang[✉]

State Key Laboratory of Computer Science, Institute of Software,
Chinese Academy of Sciences, Beijing 100190, China
wangxf@ios.ac.cn

Abstract. Since shuffle introduced into regular expressions makes the membership problem NP-hard, and an efficient membership algorithm facilitates processing many membership-based applications, it is an essential work to devise an efficient membership algorithm for regular expressions which can support shuffle. In this paper, we focus on the membership problem for single-occurrence regular expressions with shuffle and counting (SOREFCs). First, we define *single-occurrence finite automata with shuffles and counters* (SFA(&,#)s), which can recognize the languages defined by SOREFCs. We prove that the membership problem for SFA(&,#)s is decidable in polynomial time. Then, we devise a membership algorithm for SOREFCs by constructing equivalent SFA(&,#)s. Experimental results demonstrate that our algorithm is efficient in membership checking.

1 Introduction

Shuffle (&) [18] applied to any two strings returns the set of all possible interleavings of the symbols in the two strings. For example, the shuffle of ab and cd is $ab\&cd = \{abcd, acbd, acdb, cdab, cadb, cabd\}$. Shuffle has been applied in different fields, such as modeling and verification of concurrent systems [14], modeling workflow [22,24], modeling XSD schemata and Relax NGs [12,16,26,31] in XML database systems. Counting [20,31] is used to express repeated patterns, such as $a^{[m,n]}$, which represents m to n consecutive repetitions of a. Counting introduced into regular expressions mainly models XSD schemata [31] in XML database systems, and regular expressions with shuffle and counting also models the crucial aspects of other schema languages [19,23].

However, shuffle introduced into regular expressions makes the membership problem NP-hard [27], and the membership problem for regular expressions with shuffle and counting is PSPACE-complete [16]. Since the tractable membership problem for automata facilitates diverse applications based on membership queries, such as synthesizing models [1,28], learning automata [3,8,25], detecting bugs and then designing possible fixes [2,30], etc., it is an essential work to devise efficient membership algorithm for regular expressions which can support shuffle. Furthermore, for 431,885 regular expressions extracted from XSD files, which

A. Bhattacharya et al. (Eds.): DASFAA 2022, LNCS 13245, pp. 526–542, 2022.
https://doi.org/10.1007/978-3-031-00123-9_41

were collected from Open Geospatial Consortium (OGC) XML Schema repository[1], and 761,278 regular expressions extracted from Relax NG files, which were searched from Maven[2] and GitHub[3]. Table 1 shows that the proportions of single-occurrence regular expressions (SOREs) and single-occurrence regular expressions with shuffle and counting (SOREFCs), and indicates that SORE-FCs have significant practicability. Thus, in this paper, we focus on membership algorithm for SOREFCs.

Table 1. Proportions of SOREs and SOREFCs.

Subclasses	SOREFCs	SOREs
% of XSDs	96.72	62.42
% of Relax NGs	98.34	60.23

For restricted regular expressions with shuffle and counting, Ghelli et al. [13,17] proposed the corresponding linear time membership algorithms. However, Kleene star is only applied to symbol disjunctions, no alphabet symbol appears twice and counting is only applied to single alphabet symbol. These restrictions imply that the above membership algorithms are difficult to apply to SOREFCs. Although there are some classic automata which can recognize the languages defined by the regular expressions supporting shuffle [4,5,9,16,21,29], the membership problem[4] for each of them is either NP-complete [4,5,9,21,29] or PSPACE-complete [16]. The above finite automata supporting shuffle are as follows: parallel finite automaton (PFA) [29], shuffle automaton (SA) [21], non-deterministic finite automaton supporting shuffle (NFA(&)) [16], concurrent finite-state automaton (CFSA) [4,5] and partial derivative automaton (PDA) [9]. For PFAs, SAs and NFA(&)s, they have many ε-transitions, which can lead to unnecessarily non-deterministic recognitions. A PDA is a plain deterministic finite automaton (DFA), which can result in an exponential blow up of the size of DFA [9]. Additionally, for the non-deterministic finite automata supporting shuffle and counting (NFA(&,#)s) [16], which can recognize the languages defined by regular expressions with shuffle and counting, the membership problem is also PSPACE-complete [16]. Recently, finite automata with shuffle (FA(&)s) [32] are proposed to model XML schemata or workflows, although the membership problem for FA(&)s can be decidable in polynomial time [32], FA(&)s do not support counting, and there are many shuffle markers in an FA(&).

Therefore, for solving above problems, it is an essential work to devise an efficient membership algorithm for SOREFCs. Different from existing works, we propose more succinct and polynomial decidable (for membership problem) finite automata: single-occurrence finite automata with shuffles and counters

---

[1] http://schemas.opengis.net/.

[2] https://mvnrepository.com/.

[3] https://github.com/topics/.

[4] In this paper, the mentioned membership problem is the uniform version that both the string and a representation of the language are given as inputs.

(SFA(&,#)s), which recognize the languages defined by SOREFCs. We devise a membership algorithm for SOREFCs by constructing equivalent SFA(&,#)s. We can ensure that our algorithm is efficient in membership checking. The main contributions of this paper are as follows.

- We introduce a new type of automata with shuffles and counters: SFA(&,#)s, for which the membership problem is decidable in polynomial time. An SFA(&,#) recognizes the language defined by a SOREFC.
- We devise a membership algorithm for SOREFCs by constructing equivalent SFA(&,#)s. We prove that the membership problem for SOREFCs is also decidable in polynomial time.
- We provide evaluations on our algorithm in terms of the time performance for membership checking. Experimental results demonstrate that our algorithm is efficient in membership checking.

The rest of this paper is organized as follows. Section 2 gives the basic definitions. Section 3 describes the SFA(&,#) and provides an example of such an automaton. Section 4 presents the membership algorithm for SOREFCs by constructing equivalent SFA(&,#)s. Section 5 presents experiments. Section 6 concludes the paper.

2   Preliminaries

Let Σ be a finite alphabet of symbols. A standard regular expression over Σ is inductively defined as follows: ε and $a \in \Sigma$ are regular expressions, for any regular expressions r_1, r_2 and r_3, the disjunction $(r_1|r_2)$, the concatenation $(r_1 \cdot r_2)$, and the Kleene-star r_1^* are also regular expressions. Usually, we omit writing the concatenation operator in examples. The regular expressions with shuffle and counting are extended from standard regular expressions by adding the shuffle operator: $r_1 \& r_2$, and the counting $r_1^{[m,n]}$ where $m \in \mathbb{N}$, $n \in \mathbb{N}_{/1}$, $\mathbb{N} = \{1, 2, 3, \cdots\}$, $\mathbb{N}_{/1} = \{2, 3, 4, ...\} \cup \{+\infty\}$, and $m \leq n$. Note that, r^+, r^* and $r?$ are used as abbreviations of $r^{[1,+\infty]}$, $r^{[1,+\infty]}|\varepsilon$ and $r|\varepsilon$, respectively.

The language $\mathcal{L}(r)$ is defined in the following inductive way: $\mathcal{L}(\varepsilon) = \{\varepsilon\}$; $\mathcal{L}(a) = \{a\}$; $\mathcal{L}(r_1|r_2) = \mathcal{L}(r_1) \cup \mathcal{L}(r_2)$; $\mathcal{L}(r_1 r_2) = \mathcal{L}(r_1)\mathcal{L}(r_2)$; $\mathcal{L}(r_1^*) = \mathcal{L}(r_1)^*$; $\mathcal{L}(r_1^{[m,n]}) = \{w_1 \cdots w_i | w_1, \cdots, w_i \in \mathcal{L}(r_1), m \leq i \leq n\}$; $\mathcal{L}(r_1 \& r_2) = \mathcal{L}(r_1) \& \mathcal{L}(r_2) = \bigcup_{s_1 \in \mathcal{L}(r_1), s_2 \in \mathcal{L}(r_2)} s_1 \& s_2$. The shuffle operation & is defined inductively as follows: $u \& \varepsilon = \varepsilon \& u = \{u\}$, for $u \in \Sigma^*$; and $au \& bv = \{az | z \in u \& bv\} \cup \{bz | z \in au \& v\}$, for $u, v \in \Sigma^*$ and $a, b \in \Sigma$. & also obeys the associative law, that is $\mathcal{L}(r_1 \& (r_2 \& r_3)) = \mathcal{L}((r_1 \& r_2) \& r_3)$. We specify that all expressions of form $(r_1 \& r_2) \& r_3$ or $r_1 \& (r_2 \& r_3)$ are rewritten as $r_1 \& r_2 \& r_3$, and let $\mathcal{L}(r_1 \& r_2 \& r_3) = \mathcal{L}((r_1 \& r_2) \& r_3)$. For a regular expression r, $|r|$ denotes the length of r, which is the number of symbols and operators occurring in r plus the size of the binary representations of the integers [15]. For any two strings $u, v \in \Sigma^+$, if a string $s \in \Sigma^+$ and $s \in u \& v$, then s is a shuffled string. For a directed graph (digraph) $G(V, E)$, $G. \succ (v)$ $(v \in G.V)$ denotes the set of all direct successors of v in G. $G. \prec (v)$ denotes the set of all direct predecessors of v in G. For space consideration, all omitted proofs can be found at https://github.com/GraceXFun/MemSFA.

2.1 SOREs, SOREFCs, MDS and MDC

SORE is defined as follows.

Definition 1 (SORE [6,7]). *Let Σ be a finite alphabet. A single-occurrence regular expression (SORE) is a standard regular expression over Σ in which every alphabet symbol occurs at most once.*

Since $\mathcal{L}(r^*) = \mathcal{L}((r^+)?)$, a SORE does not use the Kleene-star operation (*). SOREFC extending SORE with shuffle and counting is defined as follows.

Definition 2. *[SOREFC] Let Σ be a finite alphabet. A single occurrence regular expression with shuffle and counting (SOREFC) is a regular expression with shuffle and counting over Σ in which every alphabet symbol occurs at most once.*

A SOREFC also does not use Kleene-star (*) operators, and an iteration operator ($^+$) in an SOREFC is written as the counting operator ($^{[1,+\infty]}$). SOREFCs are deterministic regular expressions [11]. SOREs is a subclass of SOREFCs. In this paper, a SOREFC forbids immediately nested counters, and expressions of the forms $(r?)?$ and $(r?)^{[m,n]}$ for regular expression r.

Example 1. $a\&b$, $(c^{[1,2]}|d)^{[3,4]}$, $a?b(c^{[1,2]}\&d?)(e^{[1,+\infty]})?$, and $(a?b)\&(c|d\&e)^{[3,4]}f$ are SOREFCs, while $a(b|c)^+a$ is not a SORE, therefore not a SOREFC. However, the expressions $((a\&b)^{[3,4]})^{[1,2]}$, $((a^{[3,4]})?)^{[1,2]}$, and $((a^{[3,4]})?)?$ are forbidden.

Maximum nesting depth of shuffle (MDS) and maximum nesting depth of counting (MDC) are defined as follows.

Definition 3 (MDS). *For regular expressions $r_1, \cdots, r_k$ $(k \geq 2)$,*
$MDS(\varepsilon) = MDS(a) = 0$ $(a \in \Sigma)$, $MDS(r_1^{[m,n]}) = MDS(r_1)$, $MDS(r_1 r_2) = MDS(r_1|r_2)$
$= \max(MDS(r_1), \quad MDS(r_2))$, *and* $MDS(r_1\&r_2\&\cdots\&r_k) = \max(MDS(r_1),$
$MDS(r_2), \cdots, MDS(r_k)) + 1$.

Definition 4 (MDC). *For regular expressions r_1 and r_2, $MDC(\varepsilon) = MDC$
$(a) = 0$ $(a \in \Sigma)$, $MDC(r_1 r_2) = MDC(r_1|r_2) = MDC(r_1\&r_2) = \max(MDC(r_1),$
$MDC(r_2))$, and $MDC(r_1^{[m,n]}) = MDC(r_1) + 1$.*

Note that, r_1^+ can be rewritten as $r_1^{[1,+\infty]}$.

Example 2. Let regular expressions $r_1 = a\&(b\&d)e\&c$ and $r_2 = (a\&(b\&d^{[1,2]})e)^+$
$\&c^+$, then $MDS(r_1) = 2$, $MDC(r_1) = 0$, $MDS(r_2) = 3$ and $MDC(r_2) = 2$.

3 Single-Occurrence Finite Automata with Shuffles and Counters

3.1 Shuffle Markers, Counters and Update Instructions

For recognizing the language defined by a SOREFC r, and for the ith subexpression of the form $r_i = r_{i_1}\&r_{i_2}\&\cdots\&r_{i_k}$ $(i, k \in \mathbb{N}, k \geq 2)$ in r, there is a shuffle mark $\&_i$ in an SFA($\&$,#) for starting to recognize the strings derived by r_i. For each subexpression r_{i_j} $(1 \leq j \leq k)$, there is a concurrent marker $\|_{ij}$ in an SFA($\&$,#) for starting to recognize the symbols or strings derived by r_{i_j}. Since $\&$ is associative, there are at most $\lceil \frac{|\Sigma|-1}{2} \rceil$ shuffle markers and at most $|\Sigma|$ concurrent markers in an SFA($\&$,#) (see Theorem 1). Let $\mathbb{D}_\Sigma = \{1, 2, \cdots, \lceil \frac{|\Sigma|-1}{2} \rceil\}$ and $\mathbb{P}_\Sigma = \{1, 2, \cdots, |\Sigma|\}$.

An SFA($\&$,#) runs on a given finite sample, first, there are counters which count the numbers of the strings or substrings that are repeatedly matched by the SFA($\&$,#) each time. Then, update instructions are used to compute the minimum and maximum of the values obtained by the counters.

Counter variables are presented as follows. Let $H(V, E)$ denote the node transition graph of an SFA($\&$,#). A loop marker $+_k$ $(k \in \mathbb{N})$ is a node in H marking a strongly connected component (excluding singleton) in H. There are at most $2|\Sigma|-1$ loop markers (see Theorem 1). Let $\mathbb{B}_\Sigma = \{1, 2, \cdots, 2|\Sigma|-1\}$. There are corresponding counter variables for the nodes with self-loop, the marker $\&_i$ $(i \in \mathbb{D}_\Sigma)$ and the marker $+_k$ $(k \in \mathbb{B}_\Sigma)$. Let $V_c(H) = \{v | v \in H. \succ (v), v \in H.V\} \cup \{\&_i | i \in \mathbb{D}_\Sigma\} \cup \{+_k | k \in \mathbb{B}_\Sigma\}$. Let $\mathcal{C}$ denote the set of counter variables, and let $c(v) \in \mathcal{C}$ $(v \in V_c(H))$ denote a counter variable. The mapping $\theta \colon \mathcal{C} \mapsto \mathbb{N}$ is the function assigning a value to each counter variable in $\mathcal{C}$. θ_1 denotes that $c(v) = 1$ for each $v \in V_c(H)$.

Update instructions are introduced as follows. Let partial mapping $\beta \colon \mathcal{C} \mapsto \{\mathbf{res}, \mathbf{inc}\}$ ($\mathbf{res}$ for reset, $\mathbf{inc}$ for increment) represent an update instruction for each counter variable. β also defines mapping g_β between mappings θ. For each $v \in V_c(H)$, if $\beta(c(v)) = \mathbf{res}$, then $g_\beta(\theta) = 1$; if $\beta(c(v)) = \mathbf{inc}$, then $g_\beta(\theta) = \theta(c(v)) + 1$. Let $l(v)$ and $u(v)$ denote lower bound and upper bound variables for counter variable $c(v)$ $(v \in V_c(H))$, respectively. Let $T = \{(l(v), u(v)) | v \in V_c(H)\}$. We define mapping $\gamma \colon T \mapsto \mathbb{N} \times \mathbb{N}$ as a function assigning values to lower bound and upper bound variables: $l(v)$ and $u(v)$. Let $\mathbf{Min} \colon V_c(H) \mapsto \mathbb{N}$ and $\mathbf{Max} \colon V_c(H) \mapsto \mathbb{N}$. $(\mathbf{Min}, \mathbf{Max}) \models \gamma$ holds if and only if $\mathbf{Min}(v) \leq l_v \leq \mathbf{Max}(v)$ or $\mathbf{Min}(v) \leq u_v \leq \mathbf{Max}(v)$ for any $(l_v, u_v) = \gamma(l(v), u(v))$ $(v \in V_c(H))$. γ_∞ denotes all upper bound variables that are initialized to $-\infty$ and all lower bound variables that are initialized to $+\infty$. Let partial mapping $\alpha \colon T \mapsto (\min(\{T.l(v) | v \in V_c(H)\} \times \mathcal{C}), \max(\{T.u(v) | v \in V_c(H)\} \times \mathcal{C}))$ be an update instruction for $(l(v), u(v))$. $\alpha(l(v), u(v)) = (\min(l(v), c(v)), \max(u(v), c(v)))$. α also defines the partial mapping $f_\alpha \colon \gamma \times \theta \mapsto \gamma$ such that $f_\alpha(\gamma, \theta)((l(v), u(v)), c(v)) = (\min(\pi_1^2(\gamma(l(v), u(v))), \theta(c(v))), \max(\pi_2^2(\gamma(l(v), u(v))), \theta(c(v))))$. Both $\alpha = \emptyset$ and $\beta = \emptyset$ denote empty instructions. $g_\emptyset(\theta) = \theta$ and $f_\emptyset(\gamma, \theta) = \gamma$.

3.2   Single-Occurrence Finite Automata with Shuffles and Counters

Definition 5 (SFA($\&$, #)). *A single-occurrence finite automaton with shuffles and counters (SFA($\&$,#)) is a tuple $(V, Q, \Sigma, q_0, q_f, H, \mathbf{Min}, \mathbf{Max}, \Phi)$, where the members are described as follows:*

- *Σ is a finite and non-empty set of alphabet symbols.*
- *q_0 and q_f: q_0 is the unique initial state, q_f is the unique final state.*

- $V = \Sigma \cup V_1$, where $V_1 \subseteq \{+_i, \&_j, ||_{jk}\}_{i\in\mathbb{B}_\Sigma, j\in\mathbb{D}_\Sigma, k\in\mathbb{P}_\Sigma}$.
- $Q = \{q_0, q_f\} \cup V_2$, where $V_2 \subseteq 2^V$. A state $q \in Q \setminus \{q_0, q_f\}$ is a set of nodes in V.
- $H(V_h, E, R, \mathcal{C}, T)$ is a node transition graph.
  - $H.V_h = V \cup \{q_0, q_f\}$.
  - $H.R : \{+_i, \&_j\}_{i\in\mathbb{B}_\Sigma, j\in\mathbb{D}_\Sigma} \mapsto 2^\Sigma$.
  - $H.\mathcal{C}$ is a set of counter variables. $H.\mathcal{C} = \{c(v) | v \in V_c(H)\}$. For recognizing a string by an SFA(&,#), $c(v)$ ($v \in \Sigma$) is used to count the number of the symbol v repeatedly matched by SFA(&,#) each time. $c(+_i)$ (resp. $c(\&_j)$) is used to count the number of the strings where the first letters are in $H.R(+_i)$ (resp. the first letters are in $H.R(\&_j)$) repeatedly matched by SFA(&,#) each time.
  - $H.T = \{(l(v), u(v)) | v \in V_c(H)\}$. $l(v)$ and $u(v)$ are respectively lower bound variable and upper bound variable for the counter variable $c(v)$.
- **Min** : $V_c(H) \mapsto \mathbb{N}$ and **Max** : $V_c(H) \mapsto \mathbb{N}$. For $v \in V_c(H)$ and $c(v) \in H.\mathcal{C}$, **Min**(v) (resp. **Max**(v)) denotes the lower bound (resp. the upper bound) of counter variable $c(v)$.
- $\Phi(H, X, z)$ where $X \subseteq V_c(H)$ and $z \in H.V_h$ is a function returning the tuple consisting of the partial mapping of α and the partial mapping of β (α and β are update instructions) for each node in X transiting to the node z in H. $\Phi(H, X, z) = (A, B)$, where
  - $A = \{\emptyset\} \cup \{H.T \mapsto (\min(H.T.l(x), H.\mathcal{C}.c(x)), \max(H.T.u(x), H.\mathcal{C}.c(x))) | (x \in H.\succ(x) \wedge z \neq x) \vee (\exists x' \in H.\succ(x) : x' \in \{+_i, \&_j\}_{i\in\mathbb{B}_\Sigma, j\in\mathbb{D}_\Sigma} \wedge z \notin H.R(x')), x\in V_c(H) \wedge x \in X\}$,
  - $B = \{\emptyset\} \cup \{c(x) \mapsto \mathbf{res} | (x \in H.\succ(x) \wedge z \neq x) \vee (\exists x' \in H.\succ(x) : x' \in \{+_i, \&_j\}_{i\in\mathbb{B}_\Sigma, j\in\mathbb{D}_\Sigma} \wedge z \notin H.R(x')), x\in V_c(H) \wedge x \in X\} \cup \{c(x) \mapsto \mathbf{inc} | (x \in H.\succ(x) \wedge z = x) \vee (\exists x' \in H.\succ(x) : x' \in \{+_i, \&_j\}_{i\in\mathbb{B}_\Sigma, j\subset\mathbb{D}_\Sigma} \wedge z \in H.R(x')), x\in V_c(H) \wedge x \in X\}\}$.

Essentially, SFA(&,#)s are finite automata supporting counting and shuffle (FACF) [33] that every symbol labels at most one node in the node transition graph. The configuration of an SFA(&,#) is defined as follows.

Definition 6 (Configuration of an SFA(&, #)). A configuration of an SFA(&, #) is a 3-tuple (q, γ, θ), where $q \in Q$ is the current state, γ: $A.H.T \mapsto \mathbb{N} \times \mathbb{N}$ and θ: $A.H.\mathcal{C} \mapsto \mathbb{N}$. The initial configuration is $(q_0, \gamma_\infty, \theta_1)$, and a configuration is final if and only if $q = q_f$.

For an SFA(&,#) $\mathcal{A}$, we specify that $\mathcal{A}.H.R(+_i)$ ($i\in\mathbb{B}_\Sigma$) is a set of alphabet symbols, where an alphabet symbol is the first letter of the string that can be repeatedly matched by SFA(&,#) $\mathcal{A}$ from the state including the node $+_i$. $\mathcal{A}.H.R(\&_j)$ ($j\in\mathbb{D}_\Sigma$) is also a set of alphabet symbols, where an alphabet symbol is the first letter of the shuffled string that can be recognized by SFA(&,#) $\mathcal{A}$ from the state including the node $\&_j$. Then, the transition function of an SFA(&,#) is defined as follows.

Definition 7 (Transition Function of an SFA(&, #)). The transition function δ of an SFA(&,#) $(V, Q, \Sigma, q_0, q_f, H, \mathbf{Min}, \mathbf{Max}, \Phi)$ is defined for any configuration (q, γ, θ) and the symbol $y \in \Sigma \cup \{\dashv\}$, where $\dashv$ denotes the end symbol of a string.

1. $q = q_0$ or q is a set, where $q = \{v\}$ or $\{+_i\}$ $(v \in \Sigma, i \in \mathbb{B}_\Sigma)$:
   - $y \in \Sigma$: $\delta((q, \gamma, \theta), y) = \{(\{z\}, \gamma', g_\beta(\theta)) | z \in H. \succ (x) \wedge (z = y \vee y \in H.R(z)), x \in \{q_0, v, +_i\}, z \in \{y\} \cup \{+_j\}_{j \in \mathbb{B}_\Sigma}, \gamma' = f_\alpha(\gamma, \theta) \wedge (\mathbf{Min}, \mathbf{Max}) \models \gamma', (\alpha, \beta) = \Phi(H, \{x\}, z)\}$.
   - $y = \dashv$: $\delta((q, \gamma, \theta), y) = \{(p, \gamma', g_\beta(\theta)) | p \in H. \succ (x) \wedge p = q_f, x \in \{q_0, v\}, \gamma' = f_\alpha(\gamma, \theta) \wedge (\mathbf{Min}, \mathbf{Max}) \models \gamma', (\alpha, \beta) = \Phi(H, \{x\}, p)\}$.
2. q is a set and $q = \{\&_i\}$ $(i \in \mathbb{D}_\Sigma)$: $\delta((q, \gamma, \theta), y) = \{(p, f_\alpha(\gamma, \theta), g_\beta(\theta)) | p = H. \succ (\&_i), y \in H.R(\&_i), (\alpha, \beta) = (\emptyset, \emptyset)\}$.
3. q is a set and $|q| \geq 2$:
   - $y \in \Sigma$: $\delta((q, \gamma, \theta), y) = \bigcup_{1 \leqslant t \leqslant 3} \delta_t((q, \gamma, \theta), y)$.
     - $\delta_1((q, \gamma, \theta), y) = \{((q \backslash \{x\}) \cup \{z\}, \gamma', g_\beta(\theta)) | z \in H. \succ (x), z = y \vee (\exists i \in \mathbb{D}_\Sigma : z = \&_i \wedge y \in H.R(z)), x \in q, \gamma' = f_\alpha(\gamma, \theta) \wedge (\mathbf{Min}, \mathbf{Max}) \models \gamma', (\alpha, \beta) = \Phi(H, \{x\}, z)\}$.
     - $\delta_2((q, \gamma, \theta), y) = \{((q \backslash \{\&_i\}) \cup H. \succ (\&_i), f_\alpha(\gamma, \theta), g_\beta(\theta)) | \&_i \in q \wedge y \in H.R(\&_i), (\alpha, \beta) = (\emptyset, \emptyset), i \in \mathbb{D}_\Sigma\}$.
     - $\delta_3((q, \gamma, \theta), y) = \{((q \backslash W) \cup \{z\}, \gamma', g_\beta(\theta)) | \exists i \in \mathbb{D}_\Sigma \forall x \in W : \{z, \&_i\} \subseteq H. \succ (x) \wedge x \in H. \prec (\&_i) \wedge ((z = y \wedge y \notin R(\&_i)) \vee (z = \&_i \wedge y \in R(\&_i))) \wedge |W| = |H. \succ (\&_i)|, W \subseteq q, \gamma' = f_\alpha(\gamma, \theta) \wedge (\mathbf{Min}, \mathbf{Max}) \models \gamma', (\alpha, \beta) = \Phi(H, W, y)\}$.
   - $y = \dashv$: $\delta((q, \gamma, \theta), y) = \{(q_f, \gamma', g_\beta(\theta)) | \exists i \in \mathbb{D}_\Sigma \forall x \in W : x \in H. \prec (\&_i) \wedge |W| = |H. \succ (\&_i)|, W \subseteq q, \gamma' = f_\alpha(\gamma, \theta) \wedge (\mathbf{Min}, \mathbf{Max}) \models \gamma', (\alpha, \beta) = \Phi(H, q, q_f)\}$.

Definition 8 (Deterministic SFA(&, #)). *An SFA(&,#) is deterministic if and only if* $|\delta((q, \gamma, \theta), y)| \leq 1$ *for any configuration* (q, γ, θ) *and the symbol* $y \in \Sigma \cup \{\dashv\}$.

Example 3. Let $\Sigma = \{a, b, c, d, e\}$ and $V = \Sigma \cup \{\&_1, ||_{11}, ||_{12}, \&_2, ||_{21}, ||_{22}, +_1\}$. Let $Q = \{q_0, q_f, \{\&_1\}, \{||_{11}, ||_{12}\}, \{||_{11}, \&_2\}, \{||_{11}, e\}, \{+_1, ||_{12}\}, \{a, ||_{12}\}, \{b, ||_{12}\}, \{b, \&_2\}, \{b, e\}, \{+_1, e\}, \{a, e\}, \{||_{11}, ||_{21}, ||_{22}\}, \{||_{11}, c, ||_{22}\}, \{||_{11}, ||_{21}, d\}, \{||_{11}, c, d\}, \{b, ||_{21}, ||_{22}\}, \{b, c, ||_{22}\}, \{b, ||_{21}, d\}, \{b, c, d\}, \{a, c, ||_{22}\}, \{a, c, d\}, \{a, d, ||_{11}\}, \{b, d, ||_{11}\}\}$. Figure 1 shows SFA(&,#) $\mathcal{A} = (V, Q, \Sigma, q_0, q_f, H, \mathbf{Min}, \mathbf{Max}, \Phi)$ recognizing regular language $\mathcal{L}((((ab)^+)?\&(c\&d)^+e)^+)$. The node transition graph $\mathcal{A}.H$ is the directed graph illustrated in Fig. 1. $\mathcal{A}.H.R(+_1) = \{a\}$, $\mathcal{A}.H.R(\&_1) = \{a, c, d\}$, and $\mathcal{A}.H.R(\&_2) = \{c, d\}$. For each $v \in \{+_1, \&_1, \&_2\}$, $(\mathbf{Min}(v), \mathbf{Max}(v))$ lists as follows. $(\mathbf{Min}(+_1), \mathbf{Max}(+_1)) = (1, 2)$, $(\mathbf{Min}(\&_2), \mathbf{Max}(\&_2)) = (1, 2)$ and $(\mathbf{Min}(\&_1), \mathbf{Max}(\&_1)) = (1, +\infty)$. The update instructions specified by Φ are also presented in Fig. 1. $\mathcal{A}$ is a deterministic SFA(&,#) (see Definition 8).

Theorem 1. *For a SOREFC r, if there exists an SFA(&,#) $\mathcal{A}$ such that $\mathcal{L}(\mathcal{A}) = \mathcal{L}(r)$, then the SFA(&,#) $\mathcal{A}$ has at most $\lceil \frac{|\Sigma| - 1}{2} \rceil$ shuffle markers, at most $|\Sigma|$ concurrent markers and at most $2|\Sigma| - 1$ loop markers.*

Theorem 2. *SFA(&,#)s recognize the languages defined by SOREFCs. For any given string $s \in \Sigma^*$, and an SFA(&,#) $\mathcal{A}$, it can be decided in $\mathcal{O}(|s||\Sigma|^3)$ time whether $s \in \mathcal{L}(\mathcal{A})$.*

Theorem 1 and Theorem 2 follow Theorem 5.4 and Theorem 6.3 in [33], respectively.

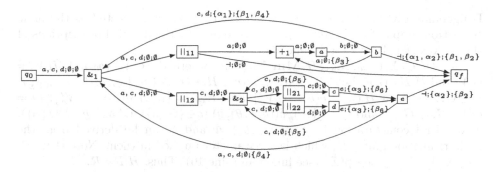

Update instructions:

$$\alpha_1 : (l(+_1), u(+_1)) \mapsto (\min(l(+_1), c(+_1)), \max(u(+_1), c(+_1))).$$
$$\alpha_2 : (l(\&_1), u(\&_1)) \mapsto (\min(l(\&_1), c(\&_1)), \max(u(\&_1), c(\&_1))).$$
$$\alpha_3 : (l(\&_2), u(\&_2)) \mapsto (\min(l(\&_2), c(\&_2)), \max(u(\&_2), c(\&_2))).$$
$$\beta_1 : c(+_1) \mapsto \mathbf{res}; \ \beta_2 : c(\&_1) \mapsto \mathbf{res}; \ \beta_3 : c(+_1) \mapsto \mathbf{inc}; \ \beta_4 : c(\&_1) \mapsto \mathbf{inc};$$
$$\beta_5 : c(\&_2) \mapsto \mathbf{inc}; \ \beta_6 : c(\&_2) \mapsto \mathbf{res}.$$

Fig. 1. The SFA(&,#) $\mathcal{A}$ for regular language $\mathcal{L}((((ab)^{[1,2]})?\&(c\&d)^{[1,2]}e)^{[1,+\infty]})$. The label of the transition edge is $(y; A_i; B_j)$ $(i, j \in \mathbb{N})$, where $y \in \Sigma \cup \{\dashv\}$ is the current symbol and A_i (resp. B_j) is the set of the update instructions from α (resp. β). $\alpha_m \in A_i$ ($m \in \{1, 2, 3\}$) is an update instruction for the lower bound and upper bound variables of the counting operator, and $\beta_n \in B_j$ ($n \in \{1, 2, 3, 4, 5, 6\}$) is an update instruction for the counter variable.

4   Membership Algorithm for SOREFC

We devise membership algorithm for SOREFCs by constructing equivalent SFA-(&,#)s. For any given SOREFC r, and a string $s \in \Sigma^*$, first, we construct an equivalent SFA(&,#) $\mathcal{A}$ for r. Then, the SFA(&,#) $\mathcal{A}$ recognizes the string s to check whether $s \in \mathcal{L}(\mathcal{A})$. Algorithm 1 presents the membership algorithm for SOREFCs.

Constructing an Equivalent SFA(&,#) for an SOREFC. We present how to construct an SFA(&,#) for a given SOREFC. Since an SFA(&,#) has shuffle marks $\&_i$, loop marks $+_k$ and concurrent marks $||_{ij}$ $(i, j, k \in \mathbb{N})$, we can construct an SFA(&,#) $\mathcal{A}$ by introducing the above marks into a given SOREFC r. First, we introduce shuffle marks, loop marks and concurrent marks into the given SOREFC r, a new expression r' is obtained. Then, we construct Glushkov automaton G for r', and G is converted to the node transition graph H of an SFA(&,#) $\mathcal{A}$. Finally, we present the detailed descriptions of the SFA(&,#) $\mathcal{A}$. Algorithm 2 shows how to construct an SFA(&,#) $\mathcal{A}$ for a given SOREFC r. Theorem 3 shows that the constructed SFA(&,#) $\mathcal{A}$ is equivalent to r (i.e., $\mathcal{L}(\mathcal{A}) = \mathcal{L}(r)$).

In Algorithm 2, initially, $i = 1$ (line 3), $j = 1$ (line 9). For a regular expression r, $first(r)$ denotes the set of the first letters of the strings derived by r. The Glushkov automaton G is constructed for r' by using the method proposed by

Brüggemann-Klein [10]. The finally obtained G can be respected as the node transition graph of an SFA(&,#). Then, we present the detailed descriptions of the SFA(&,#) $\mathcal{A}$ obtained in line 15.

$\mathcal{A} = (V, Q, \Sigma, G.q_0, G.q_f, H, \mathbf{Min}, \mathbf{Max}, \Phi)$, where $V = G.V \setminus \{q_0, q_f\}$, $H.V_h = G.V$ and $H.E = G.E$. Let $V'_c = \{v | (v \in \Sigma \wedge v \in H. \succ (v)) \vee (v \in \{+_i, \&_j\}_{i \in \mathbb{B}_\Sigma, j \in \mathbb{D}_\Sigma} \wedge v \in G.V)\}$, then $H.\mathcal{C} = \{c(v) | v \in V'_c\}$ and $H.\mathcal{T} = \{(l(v), u(v)) | v \in V'_c\}$. $Q = Q' \cup \{G.q_0, G.q_f\}$ and $Q' = \bigcup_q \bigcup_y \delta((q, \gamma, \theta), y)$ $(q \in \{G.q_0\} \cup Q'$ and $y \in \Sigma \cup \{\dashv\})$. The initial configuration is $(G.q_0, \gamma_\infty, \theta_1)$. Φ and δ can be derived from the node transition graph H, which is a parameter implied in them. Note that, $\mathcal{R}$: $\{+_i, \&_j\}_{i \in \mathbb{B}_\Sigma, j \in \mathbb{D}_\Sigma} \mapsto \wp(\Sigma)$ (see line 6 and line 10). Thus, $H.R = \mathcal{R}$.

Algorithm 1. $Membership_{\&}^{\#}$

Input: An SOREFC r and a string s;
Output: *true* if $s \in \mathcal{L}(r)$ or *false* otherwise;
1: Let SFA(&, #) $\mathcal{A} = ConsEquSFA_{\&}^{\#}(r)$;
2: **if** $Recognize(\mathcal{A}, s)$ **then return** *true*;
3: **return** *false*;

Algorithm 2. $ConsEquSFA_{\&}^{\#}$

Input: An SOREFC r;
Output: An SFA(&, #) $\mathcal{A}$: $\mathcal{L}(\mathcal{A}) = \mathcal{L}(r)$;
1: For each subexpression r_b in r:
2:    **if** there exists subexpressions $r_1, r_2, \cdots, r_k$ $(k \geq 2)$ in r: $r_b = r_1 \& r_2 \& \cdots \& r_k$
   then
3:       Let $r_b = \&_i(||_{i1}r_1| \, ||_{i2}r_2| \cdots | \, ||_{ik}r_k)$; $(\mathbf{Min}(\&_i), \mathbf{Max}(\&_i)) = (1, 1)$;
4:       **if** $r_b^{[m,n]}$ $(m \in \mathbb{N}, n \in \mathbb{N}_{/1})$ is a subexpression of r **then**
5:          Let $r_b^{[m,n]} = r_b^+$; $(\mathbf{Min}(\&_i), \mathbf{Max}(\&_i)) = (m, n)$;
6:          $\mathcal{R}(\&_i) = first(r_1) \cup first(r_2) \cup \cdots \cup first(r_k)$; $i = i + 1$;
7:    **if** there exists subexpression r_1 in r: $r_b = r_1^{[m,n]}$ $(m \in \mathbb{N}, n \in \mathbb{N}_{/1})$ **then**
8:       **if** $r_1 \neq a \in \Sigma$ and there does not exist subexpressions $e_1, e_2, \cdots, e_k$ $(k \geq 2)$ in r_1: $r_1 = e_1 \& e_2 \& \cdots \& e_k$ **then**
9:          Let $r_b = (+_j r_1)^+$; $(\mathbf{Min}(+_j), \mathbf{Max}(+_j)) = (m, n)$;
10:         $\mathcal{R}(+_j) = first(r_1)$; $j = j + 1$;
11:      **if** $r_1 = a \in \Sigma$ **then** Let $r_b = r_1^+$; $(\mathbf{Min}(a), \mathbf{Max}(a)) = (m, n)$;
12: $r' = r$; Construct Glushkov automaton G for r';
13: Add a node q_f in G; add edges $\{(v, q_f) | v$ is final state of $G\}$ in G;
14: Let q_f denote the final state of G;
15: SFA(&, #) $\mathcal{A} = (V, Q, \Sigma, G.q_0, G.q_f, H, \mathbf{Min}, \mathbf{Max}, \Phi)$;
16: **return** $\mathcal{A}$;

Example 4. For a SOREFC $r = (((ab)^{[1,2]})?\&(c\&d)^{[1,2]}e)^{[1,+\infty]}$, we introduce shuffle marks, loop markers and concurrent markers into r, a new expression $r' = (\&_1(||_{11}((+_1ab)^+)?| \, ||_{12}(\&_2(||_{21}c| \, ||_{22}d))^+e))^+$ is obtained. The constructed Glushkov automaton is showed in Fig. 2(a). The node transition graph H of an SFA(&,#) constructed for r' is illustrated in Fig. 2(b). The finally obtained SFA(&,#) is presented in Fig. 1.

Theorem 3. *For a given SOREFC r, ConsEquSFA$_{\&}^{\#}$ can construct an equivalent SFA($\&$,$\#$) $\mathcal{A}$ in $\mathcal{O}(\frac{81}{4}|\Sigma||r|)$ time.*

Recognizing. The above constructed SFA($\&$,$\#$) $\mathcal{A}$ recognizes the string s to check whether $s \in \mathcal{L}(\mathcal{A})$. Let *Recognize* denote the process of SFA($\&$,$\#$) $\mathcal{A}$ recognizing the string s. For SFA($\&$,$\#$) $\mathcal{A}$ and the string s as input, *Recognize* outputs *true* or *false*. If SFA($\&$,$\#$) $\mathcal{A}$ can recognize each symbol in the string s, and the state $\mathcal{A}.q_f$ can be reached when the symbol $\dashv$ is read, then *Recognize* returns *true*, s can be accepted by $\mathcal{A}$; Otherwise, *Recognize* returns *false*, s cannot be accepted by $\mathcal{A}$. The rules for SFA($\&$,$\#$) $\mathcal{A}$ identifying each symbol in s are specified by the transition function $\mathcal{A}.\delta$, and the initial configuration is $(q_0, \gamma_\infty, \theta_1)$.

Theorem 4. *For any given SOREFC r and a string $s \in \Sigma^*$ as inputs, algorithm Membership$_{\&}^{\#}$ takes $\mathcal{O}(\frac{81}{4}|\Sigma||r| + |s||\Sigma|^3)$ time to chech whether $s \in \mathcal{L}(r)$.*

According to Theorem 2 and Theorem 3, we can prove Theorem 4.

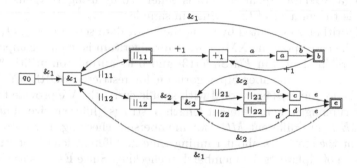

(a) The Glushkov automaton G.

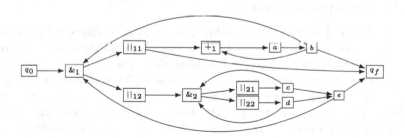

(b) The node transition graph H.

Fig. 2. (a) is the Glushkov automaton G for $r' = (\&_1(||_{11}((+_1 ab)^+)?|\,||_{12}(\&_2(||_{21}c|\,||_{22}d))^+ e))^+$. (b) is the node transition graph H of the SFA($\&$,$\#$) $\mathcal{A}$ (see Fig. 1).

5   Experiments

In this section, we evaluate our algorithm on XML data (including positive data and negative data) in time performance of membership checking.

We searched Relax NG files from above repositories, and then extracted 1000 diverse SOREFCs with average $MDS = 2$ and average $MDC = 2$ from Relax NG files for each alphabet size in $\{10, 20, \cdots, 100\}$. Let Q_1 denote the set of the above 1000 SOREFCs with alphabet size 20. Let Q_2 ($Q_2 \supset Q_1$) denote the set of the above 10000 SOREFCs with the alphabet size ranging from 10 to 100. Additionally, we extracted 1000 diverse SOREFCs with alphabet size 20 and average $MDC = 2$ from Relax NG files for each $MDS \in \{1, 2, \cdots, 10\}$. Let Q_3 denote the set of 10000 SOREFCs with MDS ranging from 1 to 10. We also extracted 1000 diverse SOREFCs with alphabet size 20 and average $MDS = 2$ from Relax NG files for each $MDC \in \{1, 2, \cdots, 10\}$. Let Q_4 denote the set of 10000 SOREFCs with MDC ranging from 1 to 10. For each target expression in Q_i ($i \in \{1, 2, 3, 4\}$), the random sample in experiments, which is a finite set of strings, is extracted from the corresponding XML data, which is either positive data or negative data. Negative data is generated by using ToXgene[5]. The size of sample is the number of the strings in sample.

Our algorithm is evaluated by using the above data sets (Q_1, Q_2, Q_3 and Q_4) including the corresponding XML data. Our algorithm is mainly compared with the membership algorithm *FlatStab* [13] and brics automaton utilities[6] (BAU). *FlatStab* is a linear membership algorithm for restricted SOREFCs [13], and BAU can be extremely fast to deal with shuffle currently. We provide the statistics about running time in different length of strings, different size of alphabets, different MDS and different MDC for membership checking. For automata, we also present the statistics about running time in different length of strings and different size of alphabets for membership checking. Since FA($\&$)s are more efficient in membership checking than other automata supporting shuffle (including BUA) [32], our automaton model SFA($\&$,$\#$) is mainly compared with NFA($\&$,$\#$) and FA($\&$) in membership checking.

Positive Data. For each membership algorithm or utilities, which inputs a target expression e with its derived string s ($s \in \mathcal{L}(e)$ is extracted from positive data), we record the corresponding running time. For each one of 1000 target expressions in Q_1, we extracted the corresponding 1000 strings (forming a set S_1) with fixed length, which ranges from 10^3 to 10^4. In Fig. 3(a), the running time for a given length of string is the average of the corresponding recorded 10^6 (1000 * 1000) running times. For each alphabet size in $\{10, 20, \cdots, 100\}$, and for each one of 1000 target expressions with that alphabet size in Q_2, we also extracted the 1000 strings (forming a set S_2) with fixed lengths of 5000. In Fig. 3(b), the running time for a given alphabet size is the average of the corresponding recorded 10^6 running times.

---

[5] http://www.cs.toronto.edu/tox/toxgene/.
[6] https://www.brics.dk/automaton/.

For each $MDS \in \{1, 2, \cdots, 10\}$, and for each one of 1000 target expressions with that MDS in Q_3, we also extracted the 1000 strings with fixed lengths of 5000. In Fig. 3(c), the running time for a given MDS is the average of the corresponding recorded 10^6 running times. For each $MDC \in \{1, 2, \cdots, 10\}$, and for each one of 1000 target expressions with that MDC in Q_4, we also extracted the 1000 strings with fixed lengths of 5000. In Fig. 3(d), the running time for a given MDC is the average of the corresponding recorded 10^6 running times.

Figure 3(a)–Figure 3(d) show that the running times for $membership_{\&}^{\#}$ are lower than that for BAU and $FlatStab$. Figure 3(a) presents that the running time for $membership_{\&}^{\#}$ is less than 0.17 s, when the length of string is not over 10^4. Figure 3(b) illustrates that the running time for $membership_{\&}^{\#}$ is less than 0.365 s, when the alphabet size is not over 100. Especially, Fig. 3(c) demonstrates that the running time for $membership_{\&}^{\#}$ is less than 0.5 s, when MDS is not over 10. Figure 3(d) demonstrates that the running time for $membership_{\&}^{\#}$ is less than 0.3 s, when MDC is not over 10. Thus, for positive data, the time performance of $membership_{\&}^{\#}$ shows that $membership_{\&}^{\#}$ is efficient in membership checking.

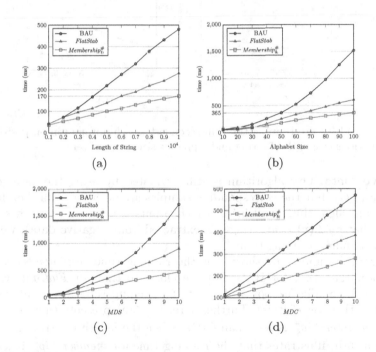

Fig. 3. (a), (b), (c) and (d) are running times for each algorithm or utilities on positive data as the functions of length of string, alphabet size, MDS and MDC, respectively.

For each automaton recognizing each string, we record the corresponding running time. For each one of 1000 target expressions in Q_1, we extracted the corresponding 1000 strings in S_1. In Fig. 4(a), the running time for a given length of string is the average of the corresponding recorded 10^6 $(1000 * 1000)$ running times. For each alphabet size in $\{10, 20, \cdots, 100\}$, and for each one of 1000 target expressions with that alphabet size in Q_2, we also extracted the corresponding 1000 strings in S_2. In Fig. 4(b), the running time for a given alphabet size is the average of the corresponding recorded 10^6 running times. Note that, for each target expression $r \in Q_1$ or $r \in Q_2$, each automaton is equivalently transformed from r.

Figure 4(a) and Fig. 4(b) present that the running times for SFA(&,#) are more lower than that for NFA(&,#), but are closer to that for FA(&). Although the running time for SFA(&,#) is higher than that for FA(&), FA(&) does not support counting, the running time for SFA(&,#) is less than 0.2 s, when the length of string is not over 10^4. The running time for SFA(&,#) is less than 0.4 s, when the alphabet size is not over 100. For positive data, SFA(&,#) is also efficient in membership checking.

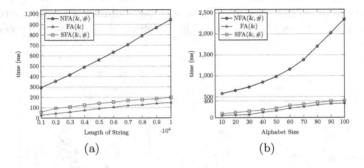

Fig. 4. (a) and (b) are running times in seconds for each automaton on positive data as the functions of length of string and alphabet size, respectively.

Negative Data. Our algorithm is still evaluated by using the data sets Q_1, Q_2, Q_3 and Q_4, but the corresponding samples are negative data. I.e., for each membership algorithm or utilities, which inputs a target expression e with a string s that $s \notin \mathcal{L}(r)$ (the string s is extracted from negative data), we record the corresponding running time.

Figure 5(a)–Figure 5(d) show that the running times for $membership^{\#}_{\&}$ are more lower than that for BAU, but are closer to that for *FlatStab*. Even for any given *MDS*, Fig. 5(c) presents that the running time for $membership^{\#}_{\&}$ is lower than that for *FlatStab*. Furthermore, Fig. 5(a) presents that the running time for $membership^{\#}_{\&}$ is less than 0.06 s, when the length of string is not over 10^4. Figure 5(b) illustrates that the running time for $membership^{\#}_{\&}$ is less than 0.85 s, when the alphabet size is not over 100. Especially, Fig. 5(c) demonstrates that the running time for $membership^{\#}_{\&}$ is less than 0.1 s, when *MDS* is not over 10. Figure 5(d) demonstrates that the running time for $membership^{\#}_{\&}$ is less than 0.126 s, when *MDC* is not over 10. Thus, for given negative data, the time performance of $membership^{\#}_{\&}$ still demonstrates that $membership^{\#}_{\&}$ is efficient in membership checking.

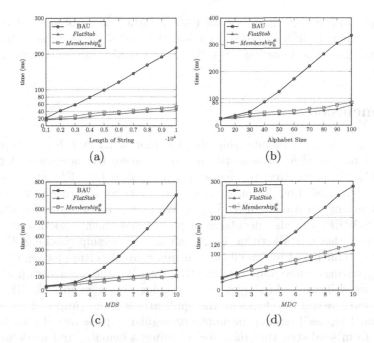

Fig. 5. (a), (b), (c) and (d) are running times for each algorithm or utilities on negative data as the functions of length of string, alphabet size, *MDS* and *MDC*, respectively.

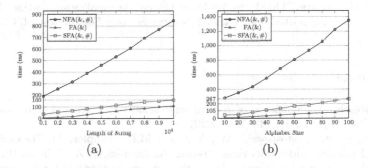

Fig. 6. (a) and (b) are running times in seconds for each automaton on negative data as the functions of length of string and alphabet size, respectively.

For each automaton and each string from negative data, we also provide the statistics about running time in different length of strings and different size of alphabets for membership checking. We still evaluate each automaton by using regular expressions from Q_1 and Q_2.

Both Fig. 6(a) and Fig. 6(b) present that the running times for SFA($\&,\#$) are more lower than that for NFA($\&,\#$). In Fig. 6(a), the running time for SFA($\&,\#$) is closer to that for FA($\&$). The running time for SFA($\&,\#$) is less than 0.16 s, when the length of string is not over 10^4. In Fig. 6(b), although the running time for SFA($\&,\#$) is higher than that for FA($\&$), FA($\&$) does not support counting, and the running time for SFA($\&,\#$) is less than 0.267 s, when the alphabet size is not over 100. For negative data, SFA($\&,\#$) is also efficient in membership checking.

6   Conclusion

This paper proposed a membership algorithm for SOREFCs. First, we proposed automata model: SFA ($\&,\#$)s, which can recognize the languages defined by SOREFCs. We prove that the membership problem for SFA ($\&,\#$)s is decidable in polynomial time. Then, we devise membership algorithm for SOREFCs by constructing equivalent SFA ($\&,\#$)s. We prove that the membership problem for SOREFCs is also decidable in polynomial time. Experimental results demonstrate that our algorithm is efficient in membership checking. For future work, we can further study an efficient membership algorithm for regular expressions supporting shuffle and counting (RE ($\&,\#$)s), the membership algorithm needs more delicate technologies for processing non-deterministic RE ($\&,\#$)s. Furthermore, we can also focus on the applications of our proposed membership algorithm. Such as learning automata or regular expressions (by membership queries) from semi-structure data for modeling schemata, and modeling workflows from streaming data for managing streaming database.

Acknowledgements. Thanks for professor George Barmpalias supporting this work, which was also supported by National Nature Science Foundation of China (No. 11971501).

References

1. Ade-Ibijola, A.: Synthesis of regular expression problems and solutions. Int. J. Comput. Appl. **42**(8), 748–764 (2020)
2. Aichernig, B.K., Mostowski, W., Mousavi, M.R., Tappler, M., Taromirad, M.: Model learning and model-based testing. In: Bennaceur, A., Hähnle, R., Meinke, K. (eds.) Machine Learning for Dynamic Software Analysis: Potentials and Limits. LNCS, vol. 11026, pp. 74–100. Springer, Cham (2018). https://doi.org/10.1007/978-3-319-96562-8_3
3. Becerra-Bonache, L., Dediu, A.H., Tîrnăucă, C.: Learning DFA from correction and equivalence queries. In: Sakakibara, Y., Kobayashi, S., Sato, K., Nishino, T., Tomita, E. (eds.) ICGI 2006. LNCS (LNAI), vol. 4201, pp. 281–292. Springer, Heidelberg (2006). https://doi.org/10.1007/11872436_23
4. Berglund, M., Björklund, H., Björklund, J.: Shuffled languages representation and recognition. Theoret. Comput. Sci. **489**, 1–20 (2013)
5. Berglund, M., Björklund, H., Högberg, J.: Recognizing shuffled languages. In: Dediu, A.-H., Inenaga, S., Martín-Vide, C. (eds.) LATA 2011. LNCS, vol. 6638, pp. 142–154. Springer, Heidelberg (2011). https://doi.org/10.1007/978-3-642-21254-3_10

6. Bex, G.J., Neven, F., Schwentick, T., Tuyls, K.: Inference of concise DTDs from XML data. In: International Conference on Very Large Data Bases, Seoul, Korea, September, pp. 115–126 (2006)
7. Bex, G.J., Neven, F., Schwentick, T., Vansummeren, S.: Inference of concise regular expressions and DTDs. ACM Trans. Database Syst. **35**(2), 1–47 (2010)
8. Björklund, J., Fernau, H., Kasprzik, A.: Polynomial inference of universal automata from membership and equivalence queries. Inf. Comput. **246**, 3–19 (2016)
9. Broda, S., Machiavelo, A., Moreira, N., Reis, R.: Automata for regular expressions with shuffle. Inf. Comput. **259**, 162–173 (2018)
10. Brüggemann-Klein, A.: Regular expressions into finite automata. Theoret. Comput. Sci. **120**(2), 197–213 (1993)
11. Brüggemann-Klein, A., Wood, D.: One-unambiguous regular languages. Inf. Comput. **142**(2), 182–206 (1998)
12. Clark, J., Makoto, M.: Relax NG Tutorial. OASIS Committee Specification (2001). http://www.oasis-open.org/committees/relax-ng/tutorial-20011203.html
13. Colazzo, D., Ghelli, G., Sartiani, C.: Linear time membership in a class of regular expressions with counting, interleaving, and unordered concatenation. ACM Trans. Database Syst. (TODS) **42**(4), 24 (2017)
14. Garg, V.K., Ragunath, M.: Concurrent regular expressions and their relationship to petri nets. Theoret. Comput. Sci. **96**(2), 285–304 (1992)
15. Gelade, W., Gyssens, M., Martens, W.: Regular expressions with counting: weak versus strong determinism. SIAM J. Comput. **41**(1), 160–190 (2012)
16. Gelade, W., Martens, W., Neven, F.: Optimizing schema languages for XML: numerical constraints and interleaving. SIAM J. Comput. **38**(5), 2021–2043 (2009)
17. Ghelli, G., Colazzo, D., Sartiani, C.: Linear time membership in a class of regular expressions with interleaving and counting. In: Proceedings of the 17th ACM Conference on Information and Knowledge Management, pp. 389–398. ACM (2008)
18. Ginsburg, S., Spanier, E.H.: Mappings of languages by two-tape devices. J. ACM (JACM) **12**(3), 423–434 (1965)
19. Harris, S., Seaborne, A.: SPARQL 1.1 query language. W3C Recommendation **21**(10), 778 (2013)
20. Hume, A.: A tale of two greps. Softw. Pract. Exp. **18**(11), 1063–1072 (1988)
21. Jedrzejowicz, J., Szepietowski, A.: Shuffle languages are in P. Theoret. Comput. Sci. **250**(1–2), 31–53 (2001)
22. Kougka, G., Gounaris, A., Simitsis, A.: The many faces of data-centric workflow optimization: a survey. Int. J. Data Sci. Anal. **6**(2), 81–107 (2018). https://doi.org/10.1007/s41060-018-0107-0
23. Libkin, L., Martens, W., Vrgoč, D.: Querying graphs with data. J. ACM (JACM) **63**(2), 14 (2016)
24. Lou, J.G., Fu, Q., Yang, S., Li, J., Wu, B.: Mining program workflow from interleaved traces. In: Proceedings of the 16th ACM SIGKDD International Conference on Knowledge Discovery and Data Mining, pp. 613–622 (2010)
25. Maler, O., Mens, I.-E.: A generic algorithm for learning symbolic automata from membership queries. In: Aceto, L., Bacci, G., Bacci, G., Ingólfsdóttir, A., Legay, A., Mardare, R. (eds.) Models, Algorithms, Logics and Tools. LNCS, vol. 10460, pp. 146–169. Springer, Cham (2017). https://doi.org/10.1007/978-3-319-63121-9_8
26. Martens, W., Neven, F., Niewerth, M., Schwentick, T.: BonXai: combining the simplicity of DTD with the expressiveness of XML Schema. ACM Trans. Database Syst. (TODS) **42**(3), 15 (2017)
27. Mayer, A.J., Stockmeyer, L.J.: The complexity of word problems-this time with interleaving. Inf. Comput. **115**(2), 293–311 (1994)

28. García Soto, M., Henzinger, T.A., Schilling, C., Zeleznik, L.: Membership-based synthesis of linear hybrid automata. In: Dillig, I., Tasiran, S. (eds.) CAV 2019. LNCS, vol. 11561, pp. 297–314. Springer, Cham (2019). https://doi.org/10.1007/978-3-030-25540-4_16

29. Stotts, P.D., Pugh, W.: Parallel finite automata for modeling concurrent software systems. J. Syst. Softw. **27**(1), 27–43 (1994)

30. Tappler, M., Aichernig, B.K., Bloem, R.: Model-based testing IoT communication via active automata learning. In: 2017 IEEE International Conference on Software Testing, Verification and Validation (ICST), pp. 276–287. IEEE (2017)

31. Thompson, H., Beech, D., Maloney, M., Mendelsohn, N.: W3C XML Schema Definition Language (XSD) 1.1 Part 1: Structures (2012)

32. Wang, X.: Learning finite automata with shuffle. In: Karlapalem, K., et al. (eds.) PAKDD 2021. LNCS (LNAI), vol. 12713, pp. 308–320. Springer, Cham (2021). https://doi.org/10.1007/978-3-030-75765-6_25

33. Wang, X.: Research on learning algorithms for extended regular expressions and their automata. Ph.D. thesis, University of Chinese Academy of Sciences (2021). (in Chinese)

(p, n)-core: Core Decomposition in Signed Networks

Junghoon Kim[1] and Sungsu Lim[2](✉)

[1] Nanyang Technological University, Singapore 639798, Singapore
junghoon001@e.ntu.edu.sg
[2] Chungnam National University, Daejeon 34134, South Korea
sungsu@cnu.ac.kr

Abstract. Finding cohesive subgraphs is a key and fundamental problem in network science. k-core is a widely used cohesive subgraph model having many applications. In this paper, we study core decomposition in the signed networks named (p, n)-core that combines k-core and signed edges. (p, n)-core preserves internal sufficient positive edges and deficient negative edges simultaneously to get high-quality cohesive subgraphs. We prove that finding exact (p, n)-core is NP-hard. Therefore, we propose two algorithms with pruning techniques. Finally, we demonstrate the superiority of our proposed algorithms using seven real-world networks.

Keywords: Signed network analysis · Core decomposition

1 Introduction

With proliferation of development of the mobile and communication technology, nowadays, many people use social networking services anytime and anywhere. Since analyzing social networks helps us understand human activity and relationship, there are many research studies to capture the characteristics of the social networks [3,5,7,10]. These days, many social network services consist of a social network and meta-data information named attributes such as a user or link information. One of the representative attributed social networks is a signed network [2,5]. The signed network contains a set of nodes, and a set of positive and negative edges. Note that edge sign presents the relationship between two users. Signed network mining has many applications including friend recommendation and finding marketing targets.

In this paper, we study a core decomposition problem [10] in the signed networks. Specifically, we extend the classic k-core decomposition on signed networks by considering signed edges. Note that we do not extend other models such as k-truss and k-ecc owing to the simple and intuitive structure of the k-core. There are many variations of the k-core [7] including attributed k-core, distance-generalized k-core, radius-bounded k-core, etc. Recently, Giatsidis et al. [2] propose directed signed core decomposition. However, the problem [2] does not consider internal negative edges of the resultant subgraphs since it focuses on

A. Bhattacharya et al. (Eds.): DASFAA 2022, LNCS 13245, pp. 543–551, 2022.
https://doi.org/10.1007/978-3-031-00123-9_42

external negative edges. It implies that a resultant subgraph cannot be directly utilized for the applications such as group recommendation as each subgraph may contain many negative edges.

To handle the abovementioned issue, we formulate (p, n)-core by considering two key components: (1) *Considering positive and negative edge constraints*: we consider both positive and negative edge constraints to guarantee the sufficient internal positive edges and deficient internal negative edges; and (2) *Maximality*: we aim to maximize the size of nodes.

2   Related Work

Giatsidis et al. [2] study the signed (l^t, k^s)-core problem in signed directed networks. Specifically, given a signed directed network G, and two parameters k and l, it aims to find (l^t, k^s)-core which is a maximal subgraph H of G of which each node $v \in H$ has $deg_{in}^s(v, H) \geq k$ and $deg_{out}^t(v, H) \geq l$. Note that $s, t \in \{+, -\}$. As we have discussed before, it does not consider the internal negative edges, and thus the resultant (l^t, k^s)-core may contain many negative internal edges which lead to a meaningless result. Li et al. [6] study the (α, k)-clique problem. Specifically, given α, k, and r, it aims to enumerate all maximal (α, k)-cliques and find top r maximal cliques where (α, k)-clique is a clique satisfying negative and positive constraints. Wu et al. [11] study the signed (k, r)-truss problem. They present balanced and unbalanced triangles to model a (k, r)-truss. Specifically, given a signed network G and two positive integers k and r, a signed (k, r)-truss is a subgraph S of G which satisfies (1) $sup^+(e, S) \geq k$; (2) $sup^-(e, S) \leq r$; and (3) maximality constraint. Support $sup^+(e, S)$ (or $sup^-(e, S)$) indicates that the number of balanced (or unbalanced) triangles contains the edge e in S. The author defines that a triangle is balanced if it contains odd number of positive edges, otherwise, the triangle is unbalanced.

Compared with both (k, r)-truss [11] and (α, k)-clique [6] problems, our proposed (p, n)-core has less cohesiveness level [1]. However, our model is much simpler to understand and intuitive. In addition, selecting a parameter is relatively not challengeable compared with a (k, r)-truss and a (α, k)-clique to end users. In the (k, r)-truss, an end-user may be required to understand the details of the problem for selecting proper k and r. Similarly, the end-user may be overwhelmed to select proper parameters α, k, and r in the (α, k)-clique problem.

3   Problem Statement

A signed network is modeled as a graph $G = (V, E^+, E^-)$ with a set of nodes V, a set of positive edges E^+, and a set of negative edges E^-. We denote a positive graph (or a negative graph) G^+ (or G^-) to represent the induced subgraph consisting of positive(or negative) edges, and we denote V^+(or V^-) to represent a set of nodes in the positive or negative graph. In this paper, we consider that G is undirected. Given a set of nodes $C \subseteq V$, we denote $G[C]$ as the induced subgraph of G which takes C as its node set and $E[C] = \{(u, v) \in E | u, v \in C\}$

as its edge set. Note that we allow that any pair of nodes can have positive and negative edges together since a positive edge represents public relationship, while a negative edge may present private-banned relationship. In many social networks, we can hide or block friends due to personal issue. To introduce our problem, we present some basic definitions and k-core problem.

Definition 1. *(Positive edge constraint). Given a signed network $G = (V, E^+, E^-)$ and positive integer p, a subgraph H of G satisfies the positive edge constraint if any nodes in H have at least p positive edges in H, i.e., $\delta(H) \geq p$.*

Definition 2. *(Negative edge constraint). Given a signed network $G = (V, E^+, E^-)$ and positive integer n, a subgraph H of G satisfies the negative edge constraint if any nodes in H have negative edges less than n in H, i.e., $\gamma(H) < n$.*

Problem 1. (k-core [10]). Given a graph G and positive integer k, k-core, denoted as H, is a maximal subgraph where each node has at least k neighbors in H.

We denote a subgraph H is a p-core graph if $H = G[D]$ where D is p-core.

Problem 2. ((p, n)-core). Given a signed network G, positive integer p, and n, (p, n)-core, denoted as C, is a maximal subgraph of which every node satisfies the positive and negative edge constraints in C.

Theorem 1. *Finding a solution of (p, n)-core is NP-hard.*

Proof. To show the hardness, we utilize the k-clique problem which is a classic problem and NP-hard. First, suppose that we have an instance of k-clique : $I_{KC} = (G = (V, E), k)$. We then show a reduction from I_{KC} to an instance of our problem. We first construct a signed network $S = (V, E^+, E^-)$ where $E^+ = E$, and we generate $|V|(|V| - 1)$ negative edges in E^-, and then we set $n = k + 1$ and $p = k - 1$. It implies that the size of the solution must be larger than or equal to k, and less than $k + 1$, i.e., the size is k and its minimum degree is $k - 1$. It indicates that finding a solution of $I_{pncore} = (S, k - 1, k + 1)$ is to find a k-clique in I_{KC}. From the abovementioned, we show a reduction from an instance I_{KC} to the instance I_{pncore}. Therefore, (p, n)-core is also NP-hard.

In this paper, we consider a signed network as a two-layer graph: $G^+ = (V^+, E^+)$ and $G^- = (V^-, E^-)$ for making better understanding.

4   Algorithms

4.1   Follower-Based Algorithm (**FA**)

As we aim to identify a maximal subgraph, we are required to remove a set of nodes to satisfy the constraints. When removing a node, we notice that a set of nodes can be removed together in a cascading manner due to the positive edge constraint. Thus, we first define a concept named *follower*.

Definition 3. (*Followers*). *Given a positive graph H and node $v \in H$, $F(v)$ consists of (1) node v (2) the nodes that are deleted cascadingly owing to the positive edge constraint when node v removed.*

Property 1. Suppose that we have a signed graph H with $\delta(H) \geq p$. When we remove a node $u \in H^+$ with $|F(u)| = 1$, the only node u in H^- is removed, that is, there are no additional nodes to be removed from H^-. However, when we remove a node u from H^-, a set of nodes which contains the node u can be removed together owing to the positive edge constraint.

Owing to the Property 1, we notice that removing the multiple nodes is determined by the positive graph. Thus, we design our FA algorithm.

Strategy 1. Suppose that we have a signed graph H with $\delta(H) \geq p$. First, we find a set of nodes in H^- of which each has at least n negative edges. We call them key nodes. Next, we find candidate nodes which are a union of the key nodes and neighbors of the key nodes. For every candidate, we compute all followers and find a node which has the minimum followers. We then remove the node and its followers. This procedure is repeated until the remaining graph satisfies both edge constraints.

Complexity. The time complexity of FA is $O(|V|^2(|V|+|E^+|))$ where $|V|(|V|+|E^+|)$ is to compute all followers and the maximum number of iterations is $|V|$.

4.2 Disgruntled Follower-Based Algorithm (DFA)

In FA, we observe that it suffers from efficiency and effectiveness issues as it only focuses on the size of followers to find proper nodes to be removed.

To consider the negative edge constraint, we introduce a definition named *disgruntlement* which is an indicator that describes how helpful it is to satisfy the negative edge constraint when we remove a node. Note that a node having the large disgruntlement is preferred to be removed since it helps to satisfy the negative edge constraint. The definition of disgruntlement can be checked in Definition 4. Note that we prefer the large disgruntlement and the small number of followers, that is, we aim to maximize the following function: $O^*(.) = \frac{D(.)}{|F(.)|}$ where $D(.)$ is a disgruntlement value and $F(.)$ is a set of nodes in followers. Thus, we design an algorithm by iteratively removing a node which has the maximum $O^*(.)$. However, a major concern is *"how to compute $|F(.)|$ efficiently"*. To answer the question, we propose a pruning strategy to improve efficiency.

Definition 4. (*Disgruntlement*). *Given a negative graph $H = (V, E)$, node $u \in V$, and negative edge threshold n, the disgruntlement of node u is as follows.*

$$D(u) = D^{self}(u) + D^{neib}(u)$$

$$D^{self}(u) = \begin{cases} 0, & \text{if } u \text{ is not a key node,} \\ |N(u, H)| - n + 1, & \text{if } u \text{ is a key node} \end{cases} \quad (1)$$

$$D^{neib}(u) = \sum_{w \in N(u,H)} 1, \forall N(w, H) \geq n$$

Computing a Lower-Bound. In a positive graph, we discuss the lower-bound of the size of followers. First, we define VD-node (verge of death).

Definition 5. *(VD-node). Given an integer p, graph $H = (V, E)$ with $\delta(H) \geq p$, a node $v \in V$ is VD-node if its $|N(v, H)|$ and coreness[1] are p.*

The VD-node implies that the node will be deleted after we remove any neighbor nodes of the VD-node since the degree of the VD-nodes is exactly p.

Definition 6. *(VD-cc). Given a graph H, and a set of VD-nodes V', we denote a set of connected components of V' as VD-ccs.*

Strategy 2. When any node in a VD-cc is removed, all nodes in the VD-cc are removed together. Thus, given a node v, if any neighbor node $w \in N(v, G)$ is VD-node, we find a VD-cc of node w. After combining all VD-ccs of $N(v, G)$, we compute the lower bound node v by summarizing all sizes of the VD-ccs.

Computing an Upper-Bound. In this Section, we present an approach to find an upper-bound by incorporating the hierarchical structure of the k-core [3,10].

Definition 7. *(CCNode). Given a positive graph H and integer k, we define a set of connected subgraphs of induced subgraphs by the k-core as a set of CCNodes, i.e., each connected subgraph of k-core is a CCNode.*

We can consider that any pair of CCNodes is not connected and the number of maximal CCNode is $\frac{|V^+|}{(k+1)}$. By utilizing the CCNode, we define CCTree.

Definition 8. *(CCTree). CCTree is a tree consisting of a dummy root node and a set of CCNodes. CCNodes in the same tree level imply that they belong to the same x-core, and a pair of a parent and a children of the CCNodes which are connected in CCTree implies that the parent CCNode (in x'-core) contains the children CCNode (in $(x' + 1)$-core).*

Strategy 3. Initially, we find p-core graph H. Then, We construct a CCTree by making a dummy node and add it to the CCTree as a root node r. We next find all the connected components in H. The connected subgraphs can be the CCNodes and they can be the children of r with level 1 of the CCTree. Then, for each CCNode C_i, we check whether C_i contains a p'-core with $p' = p + l$ where l is the level of the C_i. If the p'-core exists, we add a new CCNode C_j by computing the connected components of the p'-core then make C_i be the parent of C_j if $C_j \subseteq C_i$. This process is repeated. After constructing the CCTree, we are ready to compute $UB(.)$. We first find a starting CCNode of node v by finding a CCNode cur which contains node v while the tree-level of the cur is smaller than other CCNodes containing node v. We then add two upper-bounds $CI(.)$ and $RI(.)$. We find a size of the node sets which are not deleted after removing node v since they are already sufficiently connected or are in the different connected component. Each upper-bound can be computed as follows:

---

[1] Coreness of node v is q if node v belongs to the q-core but not to the $(q + 1)$-core.

- *Children immutable size $CI(.)$*: The sum of the size of the children's of *cur* is the $CI(.)$. Since the children of *cur* do not contain node v, and their coreness value is larger than *cur*;
- *Root immutable size $RI(.)$*: We traverse the CCTree from the *cur* to the root by iteratively traversing the parent. After visiting a parent, we add children's size of the connected components of each parent except for the children which is visited previously. The sum of the size of children is $RI(.)$.

By summarizing the $CI(.)$ and $RI(.)$, we can find the upper-bound of a node. More specifically, $UB(u) = |V| - (CI(u) + RI(u))$.

Algorithmic Procedure. Here we combine the lower-bound, upper-bound, and disgruntlement to design DFA.

Strategy 4. Instead of computing all nodes' followers, we aim to compute a few followers for efficiency. We firstly compute following three values : (1) disgruntlement score $D(.)$; (2) lower-bound $LB(.)$; and (3) upper-bound $UB(.)$. Next, we can get the following result.

$$LB(.) \leq |F(.)| \leq UB(.) \Rightarrow \frac{D(.)}{LB(.)} \geq \frac{D(.)}{|F(.)|} \geq \frac{D(.)}{UB(.)} \Rightarrow UB^*(.) \geq O^*(.) \geq LB^*(.)$$

In our DFA algorithm, we utilize the above inequality. Given two nodes u and v, if $LB^*(v) > UB^*(u)$, intuitively we do not need to compute the followers of node u. Hence, we firstly compute the node having the largest $LB^*(.)$ and then find a set of candidate nodes to compute the followers. To update coreness in every iteration, we adopt a Purecore algorithm [9].

Complexity. The time complexity of DFA is $O(|V|(|V|(|V| + |E^+|) + |E^-|))$ since it may be required to compute all nodes' followers $O(|V|(|V| + |E^+|))$ and computing disgruntlement takes $O(|E^-|)$ in the maximum iterations $|V|$.

5   Experiments

Dataset. Table 1 reports the statistics of the real-world networks. As we do not consider the weighted temporal networks, we ignore the edge weights and keep the most recent edges with signs for OTC and Alpha datasets [4].

Algorithms. To the best of our knowledge, there is no direct competitor in the previous literature. Since (k, r)-truss and (α, k)-clique are different problem, we do not include them for fair comparison. Thus, we only report our algorithms. Note that we report the results if the algorithms are terminated within 24 h.

Table 1. Real-world network datasets

Dataset	$|V|$	+	−	Max. coreness	# of triangles
Alpha [4]	3,783	12,759	1,365	18	16,838
OTC [4]	5,881	18,250	3,242	19	23,019
Epinions (EP) [5]	132K	590,466	120,744	120	3,960,165
Slashdot090211 (SD0211) [5]	82,140	382,167	118,314	54	418,832
Slashdot090216 (SD0216) [5]	81,867	380,078	117,594	54	414,903
Slashdot811106 (SD1106) [5]	77,350	354,073	114,481	53	395,289
Wiki-interaction (WI) [8]	138,587	629,523	86,360	53	2,599,698

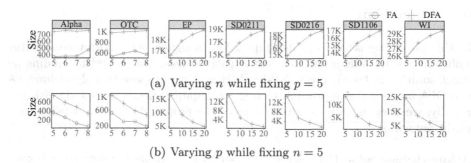

(a) Varying n while fixing $p = 5$

(b) Varying p while fixing $n = 5$

Fig. 1. Experimental results on real-world networks

Experiments on Real-World Networks. Figure 1 reports the size of resultant (p, n)-core of our algorithms. In Fig. 1a, we fix $p = 5$ then vary the negative edge threshold n from 5 to 20 (for Alpha and OTC, we vary the values from 5 to 8 as the sizes of both graphs are small). When the parameter n increases, we observe that the (p, n)-core returns larger solution as the large n implies that we allow more negative edges. We observe that for all cases, DFA returns the best result. Unfortunately, the proposed FA returns small-sized (p, n)-core as it does not consider the disgruntlement when it removes the nodes. In Fig. 1b, we fix $n = 5$ then vary the positive edge threshold p. We identify that when the p value increases, proposed algorithms consistently return small-sized (p, n)-core as the larger p implies that the resultant subgraphs become more cohesive.

Case Study: Community Search. One of the famous community search models is to maximize the minimum degree [1]. By setting the query node, we identified a community by fixing $n = 5$ in OTC and Alpha datasets by utilizing our proposed DFA. In Fig. 2, positive edges are grey-coloured, and the negative edges are red-coloured. We notice that the query nodes are densely connected to the other nodes in the community and there are a few negative edges.

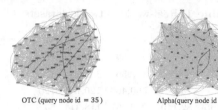

OTC (query node id = 35)   Alpha(query node id = 7)

Fig. 2. Identifying a community

6   Conclusion

In this paper, we study core decomposition in signed networks by considering sufficient positive edges and deficient negative edges. We prove that finding an exact solution of (p, n)-core is NP-hard. Hence, we propose two algorithms: FA and DFA by designing a pruning strategy to improve effectiveness and efficiency. Finally, we demonstrate the superiority of the proposed algorithms using real-world networks.

Acknowledgements. This work was supported by Institute of Information & communications Technology Planning & Evaluation (IITP) grant funded by the Korea government (MSIT) (No. 2020-0-01441, Artificial Intelligence Convergence Research Center (Chungnam National University)) and supported by the National Research Foundation of Korea (NRF) grant funded by the Korea government (MSIT) (No. 2019R1F1A1063231).

References

1. Fang, Y., et al.: A survey of community search over big graphs. VLDBJ **29**(1), 353–392 (2020)
2. Giatsidis, C., Cautis, B., Maniu, S., Thilikos, D.M., Vazirgiannis, M.: Quantifying trust dynamics in signed graphs, the s-cores approach. In: SDM, pp. 668–676 (2014)
3. Kim, J., Guo, T., Feng, K., Cong, G., Khan, A., Choudhury, F.M.: Densely connected user community and location cluster search in location-based social networks. In: SIGMOD, pp. 2199–2209 (2020)
4. Kumar, S., Spezzano, F., Subrahmanian, V., Faloutsos, C.: Edge weight prediction in weighted signed networks. In: ICDM, pp. 221–230. IEEE (2016)
5. Leskovec, J., Huttenlocher, D., Kleinberg, J.: Signed networks in social media. In: SIGCHI, pp. 1361–1370 (2010)
6. Li, R.H., et al.: Efficient signed clique search in signed networks. In: ICDE, pp. 245–256. IEEE (2018)
7. Malliaros, F.D., Giatsidis, C., Papadopoulos, A.N., Vazirgiannis, M.: The core decomposition of networks: theory, algorithms and applications. VLDBJ **29**(1), 61–92 (2020)
8. Maniu, S., Abdessalem, T., Cautis, B.: Casting a web of trust over Wikipedia: an interaction-based approach. In: WWW, pp. 87–88 (2011)

9. Sariyüce, A.E., Gedik, B., Jacques-Silva, G., Wu, K.L., Çatalyürek, Ü.V.: Streaming algorithms for k-core decomposition. VLDB **6**(6), 433–444 (2013)
10. Seidman, S.B.: Network structure and minimum degree. Soc. Netw. **5**(3), 269–287 (1983)
11. Wu, Y., Sun, R., Chen, C., Wang, X., Zhu, Q.: Maximum signed (k, r)-truss identification in signed networks. In: CIKM, pp. 3337–3340 (2020)

TROP: Task Ranking Optimization Problem on Crowdsourcing Service Platform

Jiale Zhang[1], Haozhen Lu[1], Xiaofeng Gao[1(✉)], Ailun Song[2], and Guihai Chen[1]

[1] MoE Key Lab of Artificial Intelligence, Department of Computer Science and Engineering, Shanghai Jiao Tong University, Shanghai, China
{zhangjiale100,haozhen.lu,gaoxiaofeng,chen-gh}@sjtu.edu.cn
[2] Tencent Inc., Shenzhen, China
ailunsong@tencent.com

Abstract. Crowdsourcing has the potential to solve complex problems, especially tasks that are easy for humans but difficult for computers. Service providers of emerging crowdsourcing platforms hope that crowdsourcing tasks on their service platforms can be executed as much as possible in available time. We consider from the perspective of the crowdsourcing service platform and study how to rank tasks to minimize the maximum timeout of tasks. We first formalize the Task Ranking Optimization Problem (TROP) and study its offline version. In the case of online scenario, we propose an Iterative Hungarian Algorithm for Task Ranking Optimization Problem, considering task deadline and click transfer rate with ranking amplification. Experiments on a real crowdsourcing service platform and the simulations based on real datasets demonstrate the superiority of proposed algorithm.

Keywords: Ranking optimization · Scheduling · Crowdsourcing platforms · Hungarian algorithm · Online algorithm

1 Introduction

Crowdsourcing [5] is getting more and more attention since it has the potential to solve complex problems. Crowdsourcing means outsourcing to the crowd, whose applications include crowdsourcing database [2], data anotation for machine learning [6] and so on. Task recommendation (or allocation) is a basic problem in the research field of crowdsourcing. A typical crowdsourcing system works in the following way: Task requesters design their tasks and release them on a crowdsourcing platform. Workers look for tasks on the platform, fulfill some and get

This work was supported by the National Key R&D Program of China [2019YFB2102200]; the National Natural Science Foundation of China [61872238, 61972254] and Shanghai Municipal Science and Technology Major Project [2021SHZDZX0102].

the corresponding rewards. In this process, the crowdsourcing platform needs to recommend or allocate tasks to workers, check the quality of their jobs and give out rewards. Task allocation or recommendation is proposed to select workers for a task or recommend a list of tasks to workers. It aims to match workers with suitable tasks to make the crowdsourcing tasks finished quickly and properly.

Task recommendation is more promising and practical than task allocation. Traditional task allocation scheme has the implicit hypothesis that all the workers would finish the allocated jobs without loss of quality. It simplifies the interaction too much. Interaction phases of session between worker and crowdsourcing platform include exploration, receiving, submission and reward: The platform display the tasks to workers on the interface; Workers choose what they want to do according to their willingness; Workers finish their packages and submit the result; The platform assesses worker's answers and delivers them rewards. Task recommendation is a non-mandatory form of allocation that respects the worker's liberty.

Task ranking optimization is essential in task recommendation since the display ranking of tasks can greatly affect how much time it takes for a certain task to be finished [4, 7]. From June to July 2019, we tracked and recorded the display rankings and total submissions of each crowdsourced task on a crowdsourcing platform every second. In addition, we requested the crowdsourcing platform to perform topping operations on specific tasks every 3 to 5 days to observe the impact of different task display rankings on task submission speed. Figure 1 shows a representative record of 8 tasks issued almost in the same period. The x-axis is the timeline with unit of $1\,s$. The y-axis is the number of answers submitted. Different colored lines represent different tasks. Among them, $task_1$ to $task_4$ are released at time 0, and $task_5$ to $task_8$ are released at time 100,000. The blue vertical line represents the topping operation, which fixes $task_5$ to $task_8$ and $task_1$ to $task_4$ in the top 1 to 8 of the display list respectively.

The curves in Fig. 1 illustrate that display rankings of tasks have a significant impact on the speed of task submissions. Submission speed of a task would slow down gradually after the task is published. The reason is that a new-coming task would occupy the premier position on the interface by default and the old ones would not get enough exposure. After the topping operation, the total submissions of the 8 tasks increases quickly. Additionally, the order of task finishing time is almost in accordance with the position order of the 8 tasks (which is the same with the order in legend). By displaying the tasks at the premier position on the interface, one task can be soon finished. Therefore, exposure is all we need (Fig. 2).

Our main contribution includes:

1. We clarify the difference between traditional crowdsourcing platforms and emerging crowdsourcing platforms. Based on our practical observations, we propose the Task Ranking Optimization Problem (TROP), which considers task deadline and click transfer rate with ranking amplification.
2. We design an online task-ranking algorithm based on Hungarian bipartite matching algorithm. We propose an iterative method for the case of different ranking for different groups.

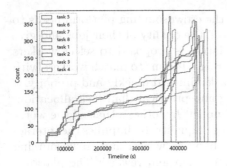
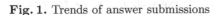

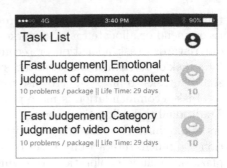

Fig. 1. Trends of answer submissions   **Fig. 2.** Task display list

3. We compare our scheme with existing ranking schemes and verify the effectiveness of our scheme.

2   Problem Statement

In this section, we formulate the Task Ranking Optimization Problem (TROP). The crowdsourcing platform has a worker set W of m workers and a task set T of n tasks. The characteristics for each task T_i include ranking $r(T_i)$, quantity of questions q_i, beginning time b_i, consumption time $c(i, r_i)$, and deadline d_i. Note that $c(i, r_i)$ is a function of task i and its ranking r_i. We need to give out a display ranking scheme $r = (r_1, r_2, \cdots, r_n)$ to minimize the maximal overdue time. r is an n-tuples to represent the display of of n tasks so that $\{r_1, r_2, \cdots, r_n\} = \{1, 2, \cdots, n\}$. For instance, if there are $n = 3$ tasks and we give a scheme $r = (2, 3, 1)$. It means we put task T_2 first, task T_3 second and task T_1 last. It is obvious that there are $n!$ possible rankings in total.

The minimization of maximal overdue time problem TROP is formulated as the following:

Definition 1 (Task Ranking Optimization Problem).

$$\min_{r} \max_{i} \max (0, b_i + c(i, r_i) - d_i), \tag{1}$$

$\max (0, b_i + c(i, r_i) - d_i)$ is the overdue time of task T_i. We want to design a ranking scheme r to minimize the maximal overdue time of task $T_i (i = 1, 2, \cdots, n)$. It is not practical to calculate Eq. 2 by enumerating of r since the complexity would be $O(n \cdot n!)$ in that case.

3   Offline Task Ranking Optimization

We further formulate the offline version of TROP as an integer programming. In the offline version of TROP, the input includes n tasks T_i and their quantity of questions q_i, beginning time b_i, deadline d_i, and the relation between ranking and the working speed p_{ijt} (It can be used to calculate the consumption time function $c(i, r_i)$.) We propose an integer programming of offline task ranking to minimize the maximal overdue time, which is denoted as δ here.

Definition 2 (Offline Task Ranking Optimization Problem).

$$
s.t. \begin{cases}
\qquad\qquad\qquad \min \delta \\
\sum\limits_{j \in R, t \in [0,T]} x_{ijt} p_{ijt} \geq q_i, & T_i \in \mathbf{T}, \\
\sum\limits_{T_i \in T} x_{ijt} = 1, & j \in R, t \in [0, D], \\
\qquad x_{ijt} = 0, & \forall i, t \leq b_i, t \geq d_i + \delta, \\
\qquad x_{ijt} \in \{0, 1\}, & T_i \in \mathbf{T}, j \in R, t \in [0, D],
\end{cases}
$$

where the indicator x_{ijt} represents whether the task i would be allocated to the ranking j at round t while p_{ijt} represents the number of submitted answers by allocated task T_i to ranking j at round t. The first constraint promises that there are enough submitted answers for each task. The second one promises that each ranking would have only one task in each round. Also, task allocation should be occurred during its beginning time and ending time. The objective δ is the maximal overdue time.

Problem 2 is NP-hard but we need to get the exact solution as a lower bound for the online version of the problem. Programming solvers can tackle small-scale instances while we can use heuristic search such as Genetic Algorithm (GA) [1] to get approximation for large-scale instances.

4 Online Task Ranking Optimization

In this section, we propose an online scheme to optimize the task ranking without the information of the number of submitted answers p_{ijt}.

Definition 3 (Online Task Ranking Optimization Problem).

$$
\min_r \max_i \max \left(0, b_i + c(i, r_i) - d_i \right), \tag{2}
$$

The deadline d_i is not known until round $t = b_i$ and the ranking and the working speed relation p_{ijt} is not given as input.

Online version of Task Ranking Optimization is a practical problem to model the real application but unfortunately the methods to solve the offline TROP are not effective here. First, we cannot get full information of begin time and deadlines of all task so that the integer programming before is not a proper formulation of the online problem. Second, we cannot calculate the consumption time $c(i, r_i)$ because the working speed p_{ijt} is not given. We tackle the second problem first by estimating $c(i, r_i)$ with a more detailed model and then design online ranking method based on bipartite graph matching and Hungarian Algorithm.

We estimate the consumption time $c(i, r_i)$ as follow. Denote the raw click transfer rate (CTR) vector as A, where $A(i)$ represents the CTR of workers on the task T_i. The click transfer rate can be predicted by user's historical data and the similarity of word embedding of task description by using the technique like DeepFM [3]. The ranking of tasks on the interface, r, would also affect the CTR. Here, we suppose that the amplification function of ranking, $f(r(T_i))$, is non-increasing. The CTR of workers on the task T_i after amplification is

$$A_r(T_i) = A(T_i) \times f(r(T_i)) \tag{3}$$

So for the task T_i with remaining quantity q_i', the consumption time $c(i, r_i)$ is estimated as

$$c(i, r_i) = \frac{q_i'}{m A_r(T_i)} \tag{4}$$

We construct a bipartite graph where the tasks T with remaining quantity q', beginning time of the current status b' and deadline d are on the left; the rankings r are on the right. The weight of edge (T_i, r_j) is defined as $\max(0, b_i + c(i, r_i) - d_i)$. Hungarian Algorithm is utilized to find a minimum weighted matching of bipartite graph which optimizes our objective, the sum of edges.

The ranking should be updated when changes have been made under the following situations.

1. Sticky Operation (Active): Choose a task and move it to the top, i.e. swap it with the first task on the ranking list. The rest of tasks should be rearranged.
2. New Task Coming (Passive): New tasks would be appended to the tail of ranking list. The initial position is not important since we would rearrange the ranking.
3. Old Task Finishing (Passive): When the task whose ranking is j is finished, every task after it would advance one rank in the ranking which brings them the augmentation of CTR.

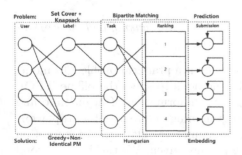

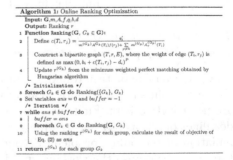

Fig. 3. Illustration of process flow   **Fig. 4.** Algorithm for online ranking

With no capacity requirements, microtask can be done by any worker. It is the case that we discussed previously. In the case of macrotask, like translating French to Chinese, the platform would offer entrance only for the qualified workers. To manage the users and the tasks, label system is proposed. Users can get a capacity label when they pass the corresponding examination. Then, some tasks would be available for the workers with the specific label. The whole process is illustrated in Fig. 3. Greedy Algorithm ($O(\log n)$-approximated) is used for the set cover problem to get the label representatives. Specially, the tasks with no requirement are already covered by the blank label. With these labels, workers can be partitioned into groups. To realize the personalization, we can set each individual worker as a single group (Fig. 4).

Given the same pool of tasks $\mathbf{T}$ and different groups $G_k \in \mathbf{G}$ of workers, the model is slightly changed as workers are split into several groups with different transfer rate matrix. The estimated consumption time $c(i, r_i)$ would modified as

$$c(i, r_i) = \frac{q_i'}{\sum_{\mathbf{G}} m^{(G_k)} A_r^{(G_k)}(T_i)}. \tag{5}$$

where, $m^{(G_k)}$ represents the number of workers in G_k, and $A_r^{(G_k)}(T_i)$ represents the CTR of workers in G_k on the task T_i after amplification.

To satisfy the objective of Eq. (2), we first run the Hungarian Algorithm for each group individually. We note the rankings are $r_1, \ldots, r_k$. Then, r_j is recalculated in a round-robin way by Eq. (5). The convergence represents that a local minimal is achieved. The algorithm is given as Algorithm 4, where P is a large constant to promise the domination of the maximal $\max(0, b_i + c(T_i, r_j) - d_i)$ in the sum of weights.

5 Experiments

In this section, we report experiments on a crowdsourcing platform and simulations with datasets from Tencent crowdsourcing platform to offer insights into the effectiveness of the proposed model. As the workers' personal information is classified information, we cannot obtain workers' profile. Among worker-related information, we can only use workers' submission records for different tasks, including: worker id, task id, question id, whether the answer is correct, the time to receive the question, and the time to submit the answer.

5.1 The Effectiveness of CTR Vector Prediction

To observe the prediction accuracy of the CTR β of existing workers on the platform, we treat each task in datasets as a *new task*, and other tasks as existing tasks. $\hat{\beta}$ is the predicted click rate of this *new task*, and the actual β is divided by the predicted $\hat{\beta}$ for comparison. In Fig. 5, the histogram of ratio of real value and prediction value is indicated. Most cases falls in the range $[0.9, 1.3]$. The result indicates that we can basically predict β.

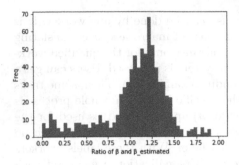
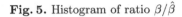
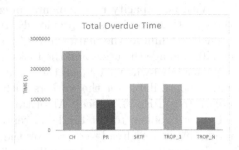

Fig. 5. Histogram of ratio $\beta/\hat{\beta}$

Fig. 6. Comparison of total overdue time

5.2 Performance Comparison

In this subsection, the effectiveness of our online scheme is verified by simulations on datasets from the crowdsourcing platform. We compare the overdue time of the following ranking strategies:

- CH: Just let the task display on the interface chronologically;
- PR: The task ranking for each user is arranged according to his/her preference for each task in descending order. The calculation of workers' preferences for different tasks is based on workers' historical records and the similarity matrix of tasks.
- SRTF: Shortest Remaining Time First strategy is utilized while the remaining time is also estimated by $c(i, r_i)$.
- TROP: Result calculated by our online algorithm. It can be categorized into several cases according to the group division, including
  1. $TROP_1$: All the workers are in the same group;
  2. $TROP_N$: One individual worker is treated as a group.

We simulate 100 times and the average overdue time of each algorithm is shown in Fig. 6. In each simulation, available new task information is randomly generated, including the number of new tasks, as well as the word embedding vector, number of questions, and deadline for each new task. At the same time, the click transfer rate vector A is also randomly generated. In addition, in each simulation, 200 workers who submitted more than 50 of the existing tasks will be randomly selected from datasets. We set $f(r(T_i)) = \frac{1}{r(T_i)}$ in simulations. Such a simplification is equivalent to considering the decelerating effect of the lower task ranking on the task submission speed.

Figure 6 shows the results, it is indicated that $TROP_1$ performs the best in macroscopic allocation scheme. $TROP_N$ is better than recommendation only considering users' preference. Numeric result shows that $TROP_N$ decreases about 80% of consumption time for the original scheme CH. $TROP_N$ decreases about 50% of consumption time for the PR scheme. This result also reflects that display rankings have a significant acceleration effect on task submissions. The

PR scheme only considers the preferences of workers, and ignores the impact of task display rankings. Therefore, although the effect of the PR scheme is improved over the original scheme, it is not the best solution. In addition, $TRPO_1$ is equivalent to only considering the impact of task display rankings, while ignoring the preferences of workers. However, $TROP_N$ not only considers the preferences of workers, but also considers the impact of task display rankings, so $TROP_N$ has the best result.

6   Conclusion

Considering that display rankings of tasks will affect the submission speed of tasks, we optimize the task display ranking to minimize the maximum overdue time of tasks. We propose the Task Ranking Optimization Problem, and use genetic algorithm to find the lower bound of the maximum overdue time of tasks in offline task allocation. Then, to realize online task ranking optimization, Iterative Hungarian Algorithm is proposed to find the local minimal for ranking optimization. Experiments on the real crowdsourcing platform verify that the task display ranking has a significant impact on the task submission speed. Simulation results verify the effectiveness of our proposed algorithm.

References

1. Balin, S.: Non-identical parallel machine scheduling using genetic algorithm. Expert Syst. Appl. **38**(6), 6814–6821 (2011)
2. Chai, C., Fan, J., Li, G., Wang, J., Zheng, Y.: Crowdsourcing database systems: overview and challenges. In: IEEE International Conference on Data Engineering (ICDE), pp. 2052–2055 (2019)
3. Guo, H., Tang, R., Ye, Y., Li, Z., He, X.: DeepFM: a factorization-machine based neural network for CTR prediction. In: International Joint Conference on Artificial Intelligence (IJCAI), pp. 1725–1731 (2017)
4. Haas, D., Wang, J., Wu, E., Franklin, M.J.: Clamshell: speeding up crowds for low-latency data labeling. Proc. VLDB Endow. **9**(4), 372–383 (2015)
5. Howe, J.: The rise of crowdsourcing. Wired Mag. **14**(6), 1–4 (2006)
6. Huang, K., Yao, J., Zhang, J., Feng, Z.: Human-as-a-service: growth in human service ecosystem. In: IEEE International Conference on Services Computing (SCC), pp. 90–97 (2016)
7. Zeng, Y., Tong, Y., Chen, L., Zhou, Z.: Latency-oriented task completion via spatial crowdsourcing. In: IEEE International Conference on Data Engineering (ICDE), pp. 317–328

HATree: A Hotness-Aware Tree Index with In-Node Hotspot Cache for NVM/DRAM-Based Hybrid Memory Architecture

Gaocong Liu[1], Yongping Luo[1], and Peiquan Jin[1,2($\boxtimes$)]

[1] University of Science and Technology of China, Hefei, China
jpq@ustc.edu.cn
[2] Key Laboratory of Electromagnetic Space Information, CAS, Hefei, China

Abstract. The emerging of Non-Volatile Memory (NVM) has changed the traditional DRAM-only memory system. Compared to DRAM, NVM has the advantages of non-volatility and large capacity. However, as the read/write speed of NVM is still lower than that of DRAM, building DRAM/NVM-based hybrid memory systems is a feasible way of adding NVM into the current computer architecture. This paper aims to optimize the well-known B$^+$-tree for hybrid memory. We present a new index called *HATree (Hotness-Aware Tree)* that can identify and maintain hot keys with an in-node hotspot cache. The novel idea of HATree is using the unused space of the parent of leaf nodes (PLNs) as the hotspot data cache. Thus, no extra space is needed, but the in-node hotspot cache can efficiently improve query performance. We present the new node structures and operations of HATree and conduct experiments on synthetic workloads using real Intel Optane DC Persistent Memory. The comparative results with three existing state-of-the-art indices, including FPTree, LBTree, and BaseTree, suggest the efficiency of HATree.

Keywords: Hybrid memory · B$^+$-tree · Hotspot · In-node cache

1 Introduction

In recent years, Non-Volatile Memory (NVM) has emerged as an alternative to the next-generation main memories [3,9]. Although NVM has the advantages of non-volatility, byte addressability, the current NVM is much slower than DRAM. Thus, a better choice is to use both DRAM and NVM in the memory architecture, forming a hybrid memory architecture.

This paper aims to develop a new index structure toward the hybrid memory architecture involving DRAM and NVM. Previous indices for hybrid memory, such as FPTree [7] and LBTree [4], have proposed using unsorted leaf nodes on NVM. However, the unsorted leaf nodes are not friendly to search operations because we have to traverse an unsorted leaf node to find the matched records.

© The Author(s), under exclusive license to Springer Nature Switzerland AG 2022
A. Bhattacharya et al. (Eds.): DASFAA 2022, LNCS 13245, pp. 560–568, 2022.
https://doi.org/10.1007/978-3-031-00123-9_44

Thus, we propose to develop a new index for the hybrid memory that can offer high search performance but keep similar updating performance as FPTree and LBTree. The new index proposed is named *HATree* (*Hotness-Aware Tree*). HATree employs the similar write-optimized techniques of FP-tree and LBTree but proposes a novel idea of using the unused space of the parent of leaf nodes as the hotspot data cache. Thus, no extra space is needed, but the in-node hotspot cache can efficiently improve query performance. Briefly, we make the following contributions in this study:

(1) We notice that the parent of leaf nodes (denoted as PLNs) in the B^+-tree contain about 30% unused space. Following this observation, we propose to use the unused space of the PLNs to cache hotspot entries and improve the search performance of the tree index without extra DRAM caches.
(2) Motivated by the in-node hotspot caching idea, we propose HATree, a Hotness-Aware B^+-tree for the DRAM/NVM-based hybrid memory architecture. We detail the new node structures of HATree and present the operations on HATree (Sect. 3).
(3) We conduct experiments on real Intel Optane DC Persistent Memory with synthetic trace and compare HATree with three hybrid-memory-oriented tree indices, including FPTree [7], LBTree [4], and BaseTree (the logless FPTree). The results show that HATree outperforms all competitors (Sect. 4) .

2   Related Work

The B^+-tree has been widely used in modern DBMSs. So far, most indexes for hybrid memory are based on the B^+-tree, such as FPTree [7] and LBTree [4]. There are also studies on tree indices toward NVM-only memory architecture [5, 6], which are orthogonal to this study.

To take full advantage of DRAM and NVM, most NVM-aware indices adopt the selective persistence model [7,10], i.e., leaf nodes are stored on NVM, and inner nodes are stored on DRAM. In addition, current NVM-based indices usually adopt the idea of unsorted leaf nodes [1,5,7], meaning that the entries are not ordered within a leaf node. However, when updating an entry inside a leaf node, the index has to append the new entry at the end of all the entries and mark the old entry invalid.

The existing FPTree and LBTree indices adopt the same searching strategy as the conventional B^+-tree. When searching a key, they start at the root node and traverse a path of inner nodes till a leaf node residing in NVM is reached. To this end, at least one NVM access and multiple DRAM accesses will be caused during a searching operation. One recent work called TLBtree [5] proposed to use a read-optimized structure for inner nodes and write-optimized sub-indices for leaf nodes. However, TLBtree is presented for the NVM-only memory architecture and does not utilize any DRAM in its implementation.

On the other hand, many database accesses are skewed, indicating that most requests will focus on a small portion of data. Therefore, it could be helpful

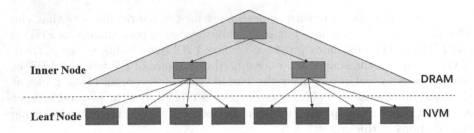

Fig. 1. The high-level index architecture of HATree.

for improving the searching performance if the frequently accessed hot data can be cached on DRAM. One possible way to cache hot data is to use an extra DRAM space to store hotspot entries [8]. However, this approach consumes a large amount of additional DRAM space, which is not space efficient.

3   Hotness-Aware B⁺-tree

HATree is a Hotness-Aware B⁺-tree without using an extra DRAM cache. Inspired by a previous study discussing the space utilization of the 2–3 tree [11], we notice that the inner nodes in the B⁺-tree are not fully filled, meaning that there is unused space in the parent nodes of leaf nodes, i.e., PLNs. Thus, we propose to utilize the unused space of PLNs as the in-node hotspot cache. With such a mechanism, HATree can answer queries with the in-node hotspot cache but does not consume additional DRAM space.

3.1   Index Structure of HATree

The high-level index structure of HATree is shown in Fig. 1. HATree also adopts the common idea of current tree indices on hybrid memory, i.e., putting all leaf nodes on NVM to make the index persistent and recoverable and maintaining all inner nodes on DRAM to accelerate search performance. Such an index structure has also been used in previous work, including FPTree [7] and NV-tree [10].

Figure 2 shows the leaf-node structure of HATree. The metadata field contains the same information as FPTree, such as *bitmap* and *fingerprints*. The two shadow *siblings* are used to implement log-less splitting [4,5]. The size of the leaf nodes is preferably a multiple of 256B (the row size of the Intel Optane DC Persistent Memory) to reduce read/write amplification [5,9].

As HATree focuses on the in-node cache in the PLN nodes, it proposes a new inner-node structure, as shown in Fig. 3. The major difference between the HATree's inner-node structure and the previous one is summarized as follows: (1) HATree uses the unused space of the PLN node as the in-node hotspot cache. (2) HATree adds some new metadata in the metadata field. However, we ensure that the size of the added new metadata, together with the old metadata, will not exceed one cacheline size (64 B). Therefore, the new metadata will not cause extra cacheline reads.

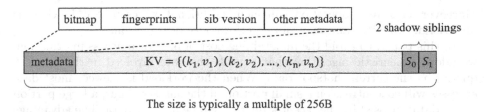

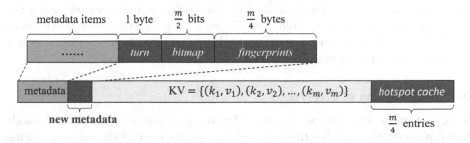

Fig. 2. The leaf-node structure of HATree.

Fig. 3. The inner-node structure of HATree in comparison with the previous design.

Note that although the PLN nodes in the B^+-tree have about 30% unused space, we remain the 25% space of a PLN node as the in-node hotspot cache. As shown in Fig. 3, let m be the total number of entries, the last $\frac{m}{4}$ entries are remained as the in-node hotspot cache.

The new metadata includes three fields. *turn* records the starting position of the next scan of the cache replacement algorithm (we adopt a second-chance replacement algorithm in the implementation). *fingerprints* maintain a one-byte hash value for each cached entry, which is used to accelerate searching in the in-node hotspot cache, i.e., we can scan the *fingerprints* to know whether a key is within the hotspot cache quickly. *bitmap* contains two bits for each cached entry: one *status* bit is used to indicate whether the entry is used or not and the other *access* bit is used as the access bit of the entry. Note that the data entries maintained in the in-node hotspot cache are not ordered because the sorting of entries will introduce additional DRAM writes. As the number of the cached entries is determined, we use the *status* bit in the *bitmap* to reflect the current status (used or not used) of each cached entry. The *access* bit in the *bitmap* is used by the second-chance cache replacement algorithm. As a consequence, the *bitmap* in the new metadata can be any of the following four values: 00, 01, 10, and 11. When operating on the hotspot cache, we will check the *bitmap* value to tell the status of cached entries.

3.2 Hotspot Identification

A simple approach to identifying hotspot data is maintaining historical access information. However, maintaining statistics has to consume extra space.

Moreover, each read operation will incur additional writes to NVM, which is not friendly to NVM and may cause performance degradation.

In this paper, to avoid the maintaining cost of historical access information, we adopt a heuristic algorithm that regards the data involved in the recent k queries as the current hotspot data. When the workload is skewed, most data accesses will concentrate on a small portion of the dataset. Thus, a large portion of the data accessed by the recent k queries will be hotspot data. The advantage of such a heuristic strategy is that no additional space is required and each read operation only needs to amortize $\frac{1}{k}$ write operations. Its disadvantage is that the accuracy rate may be lower than statistics-based strategies, especially when the recent queries involve many cold data. To avoid the long existence of cold data in the in-node hotspot cache, we need to design an effective replacement policy to move the cold data out of the cache.

3.3 Operations of HATree

This section details the operations of HATree with the support of the in-node hotspot cache. We first describe the read and update operations and then discuss the PLN node splitting and merging.

Read. To answer a read request on HATree, we first look up the key in the hotspot cache. If it is not found, we check the leaf node. Before searching in the cache, we first perform a prefetching operation on the PLN node. We assume that the keys to be queried for every K read operations are hot keys, and they need to be cached in the PLN cache. Since normal entries can also be stored in the node, we need to find the real starting position of the cache. We check the *status* bit in the *bitmap* to determine whether the entry is used and look up the *fingerprints* to see whether the searched key is in the cache. If the key is found in the cache, we update the corresponding *access* bit in the *bitmap* to indicate that the entry has been accessed recently.

Insert. Because the hotspot cache is a read cache, the insert operation is almost unaffected. Only when the leaf node splits so that the PLN node needs to insert a new entry and this entry occupies the cache space, we need to change the corresponding state bits of the entry to 00. We first check whether the PLN node is full, and return if it is full. Otherwise, if the hotspot cache in the PLN node is full, we perform the second-chance cache replacement. If the cache is not full, we find a free entry to insert the entry.

Update and Delete. In order to ensure consistency, in addition to updating the leaf nodes, the corresponding hotspot cache must also be updated. Note that we can prefetch the PLN node before updating the hotspot cache in the PLN node. The delete operation is similar to the update operation. When performing the delete operation in the cache, we need to update the *bitmap* or the *fingerprints* to indicate that the entry has been deleted.

Merging and Splitting of PLN Nodes. When a PLN node splits, its bitmap must be 0; thus we need not perform additional operations. When PLN nodes are merged, we only keep the cache of the merged node, and no additional operations are required.

4 Performance Evaluation

All experiments were run on a server equipped with two 36-core Intel Xeon Gold CPUs. The server contains 256 GB DRAM and 512 GB Optane DC Persistent Memory distributed upon two sockets. To avoid the NUMA effect on the experimental results, we run all experiments using the CPU and the Optane within the same socket. The operating system on the server is Ubuntu with the 5.4.0 kernel. We configure all the Optane into the App-Direct mode and utilize PMDK 1.8 to manage the files on the Optane [2].

We compare HATree with three tree-like indices that were all proposed for DRAM/NVM-based hybrid memory architecture, including FPTree [7], LBTree [4], and BaseTree (the logless FPTree). All the indexes have the same node size. The default node sizes of a leaf node and an inner node are 256 B and 512 B, respectively.

In all experiments, We use an 8-byte key and an 8-byte value as the entry, which is compatible with the configuration in LBTree. For each workload, we first perform 32M random insert operations and then run the workload.

4.1 Search Performance

We first generate nine read-only workloads with different skewnesses to evaluate the search performance of four indexes, each with 32M operations. The results are shown in Fig. 4(a). HATree outperforms the other three indices when the skewness of the workload is greater than 0.5. In particular, when the skewness is 0.9, HATree improves the search performance by 27.9%, 20.6%, and 13.4%, compared to LBTree, FPTree, and BaseTree, respectively. When the workload is less skewed, the search performance of HATree drops at most 2.3%, compared to the best performing BaseTree. This is because the benefits of the hotspot cache are decreasing when the requests become less skewed.

To compare the performance of HATree under random-access and highly-skewed workloads, we further generate three read-only workloads, denoted as *Random*, *Hotspot5*, and *Hotspot9*. The *Random* workload consists of randomly distributed requests. The *Hotspot5* and *Hotspot9* workloads simulate the highly-skewed workload [1]. In *Hotspot5*, 50% of accesses concentrate on 1% of the data entries, and in *Hotspot9*, 90% of accesses only touch 1% of the data. Figure 4(b) shows the search performance of all indexes under the three workloads. For the *Random* workload, all indexes show similar performance. For the *Hotspot5*

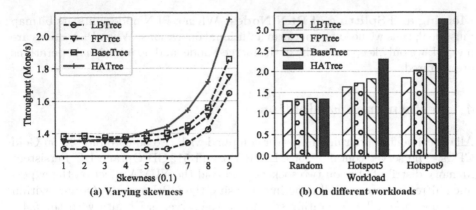

Fig. 4. Comparison of search performance.

workload, HATree improves the search performance by 40.1%, 32.8%, and 25.2%, compared to LBTree, FPTree, and BaseTree, respectively. And for the *Hotspot9* workload, HATree improves the search performance by 76.2%, 60.9%, and 48.8%, compared to LBTree, FPTree, and BaseTree, respectively. As a result, we can see that HATree performs best on highly-skewed workloads.

4.2   Updating Performance

Next, we measure the updating performance of HATree. Many search-optimized indices are not write-optimized [5]. As HATree adopts a similar method as FPTree and LBTree to deal with data updating, we expect that HATree can achieve comparable updating performance with other indices.

In this experiment, we generate four workloads consisting insert and delete operations with skewness of 0.8. We vary the number of operations and evaluate the execution time of all indices. Note that the inserted key does not exist in the index and the deleted key must be in the tree. Before evaluating HATree, we perform 32M search operations to fill up the in-node hotspot cache. Figure 5 summarizes the insert and delete performance of the four indexes. The insertion performance of HATree is basically the same as that of BaseTree, which is 16.1%–17.1% lower than the insertion performance of LBTree and 121.3%–122.6% better than that of FPTree. The delete performance of HATree is 1.2%–2.2% lower than that of BaseTree. To sum up, HATree exhibits similar updating performance with BaseTree and LBTree. This result is understandable because the in-node hotspot cache proposed in HATree only benefits search operations. There is no new optimization used in HATree, and we remain the improving of the updating performance of HATree as one of our future work.

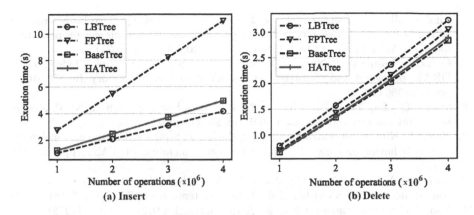

Fig. 5. Comparison of updating performance.

5   Conclusions and Future Work

In this paper, we presented a new index called HATree for DRAM/NVM-based hybrid memory. The novel idea of HATree is using the unused space of the leaf nodes' parents as the in-node hotspot cache, which maintains the hot entries within the parent nodes. Such an approach does not need extra memory space but can efficiently accelerate query performance. We develop detailed node structures and operations for HATree and conducted experiments on synthetic workloads using real Intel Optane DC Persistent Memory. The comparison with three state-of-the-art indices showed that HATree has the best search performance and comparable updating performance on both workloads.

In the future, we will consider to propose a learned secondary index with in-node caches [13]. Also, we will investigate the feasibility of using in-node caches to optimize other index structures like the Radix Tree [12].

Acknowledgments. This paper is supported by the National Science Foundation of China (grant no. 62072419).

References

1. Chen, J., et al.: HotRing: a hotspot-aware in-memory key-value store. In: FAST, pp. 239–252 (2020)
2. Intel: Intel optane DC persistent memory. https://www.intel.com/content/www/us/en/architecture-and-technology/optane-dc-persistent-memory.html
3. Kim, D., Choi, W.G., Sung, H., et al.: A scalable and persistent key-value store using non-volatile memory. In: SAC, pp. 464–467 (2019)
4. Liu, J., Chen, S., Wang, L.: LB+-Trees: optimizing persistent index performance on 3DXPoint memory. Proc. VLDB Endow. **13**(7), 1078–1090 (2020)
5. Luo, Y., Jin, P., Zhang, Q., Cheng, B.: TLBtree: a read/write-optimized tree index for non-volatile memory. In: ICDE, pp. 1889–1894 (2021)

6. Luo, Y., Jin, P., Zhang, Z., Zhang, J., Cheng, B., Zhang, Q.: Two birds with one stone: boosting both search and write performance for tree indices on persistent memory. ACM Trans. Embed. Comput. Syst. **20**(5s), 1–25 (2021)

7. Oukid, I., Lasperas, J., Nica, A., Willhalm, T., Lehner, W.: FPTree: a hybrid SCM-DRAM persistent and concurrent B-tree for storage class memory. In: SIGMOD, pp. 371–386 (2016)

8. Wang, Q., Lu, Y., Li, J., Shu, J.: Nap: a black-box approach to NUMA-aware persistent memory indexes. In: OSDI, pp. 93–111 (2021)

9. Yang, J., Kim, J., Hoseinzadeh, M., Izraelevitz, J., Swanson, S.: An empirical guide to the behavior and use of scalable persistent memory. In: FAST, pp. 169–182 (2020)

10. Yang, J., Wei, Q., Chen, C., Wang, C., Yong, K.L., He, B.: NV-tree: reducing consistency cost for NVM-based single level systems. In: FAST, pp. 167–181 (2015)

11. Yao, A.C.C.: On random 2–3 trees. Acta Informatica **9**(2), 159–170 (1978)

12. Zhang, J., Luo, Y., Jin, P., Wan, S.: Optimizing adaptive radix trees for NVM-based hybrid memory architecture. In: BigData, pp. 5867–5869 (2020)

13. Zhang, Z., Jin, P., Wang, X., Lv, Y., Wan, S., Xie, X.: COLIN: a cache-conscious dynamic learned index with high read/write performance. J. Comput. Sci. Technol. **36**(4), 721–740 (2021)

A Novel Null-Invariant Temporal Measure to Discover Partial Periodic Patterns in Non-uniform Temporal Databases

R. Uday Kiran[1]([⊠])[iD], Vipul Chhabra[2][iD], Saideep Chennupati[2][iD], P. Krishna Reddy[2][iD], Minh-Son Dao[3][iD], and Koji Zettsu[3][iD]

[1] The University of Aizu, Aizuwakamatsu, Fukushima, Japan
udayrage@u-aizu.ac.jp
[2] IIIT-Hyderabad, Hyderabad, Telangana, India
{vipul.chhabra,saideep.c}@research.iiit.ac.in, pkreddy@iiit.ac.in
[3] NICT, Koganei, Tokyo, Japan
{dao,zettsu}@nict.go.jp

Abstract. "Rare item problem" is a fundamental problem in pattern mining. It represents the inability of a pattern mining model to discover the knowledge about frequent and rare items in a database. In the literature, researchers advocated the usage of null-invariant measures as they disclose genuine correlations without being influenced by the object co-absence in the database. Since the existing null-invariant measures consider only an item's *frequency* and disregard its temporal occurrence information, they are inadequate to address the *rare item problem* faced by the partial periodic pattern model. This paper proposes a novel null-invariant measure, called *relative periodic-support*, to find the patterns containing both frequent and rare items in non-uniform temporal databases. We also introduce an efficient pattern-growth algorithm to find all desired patterns in a database. Experimental results demonstrate that our algorithm is efficient.

Keywords: Pattern mining · Rare item problem · Null-invariant measure

1 Introduction

Partial periodic patterns [5] are an important class of regularities that exist in a temporal database. The basic model of partial periodic pattern involves discovering all patterns in a temporal database that satisfy the user-specified *minimum period-support* (*minPS*) constraint. *MinPS* controls the minimum number of periodic recurrences of a pattern in a database. Since only one *minPS* is used for the entire database, this model also implicitly assumes that all items in the database have similar occurrence behavior, and thus, suffers from the dilemma known as the *rare item problem* [6,7,13]. That is, we either miss the partial

© The Author(s), under exclusive license to Springer Nature Switzerland AG 2022
A. Bhattacharya et al. (Eds.): DASFAA 2022, LNCS 13245, pp. 569–577, 2022.
https://doi.org/10.1007/978-3-031-00123-9_45

periodic patterns containing rare items at high $minPS$ or produce too many patterns at low $minPS$. This paper makes an effort to address this problem.

The contributions of this paper are as follows. First, we propose a generic model of a partial periodic pattern by introducing a new measure known as *relative periodic-support* (RPS). This measure determines the periodic interestingness of a pattern by taking into account the frequencies of its items. The proposed measure allows the user to specify a high periodic-support threshold value for a pattern containing only frequent items and a low periodic-support threshold value for a pattern containing rare items. Consequently, enabling the user to find partial periodic patterns containing frequent and rare items without producing too many patterns. Second, we show that our measure satisfies both *null-invariant* [3] and *convertible anti-monotonic* [10] properties. The null-invariant property facilitates our model to address the rare item problem by disclosing genuine correlations without influencing the items co-absence in the database. It is the first null-invariant temporal measure to the best of our knowledge. The convertible anti-monotonic property facilitates our model to be practicable on massive real-world databases. Third, we introduce the concept of *irregularity pruning* and present an efficient pattern-growth algorithm to find all desired patterns in a database. Experimental results demonstrate that our algorithm is not only memory and runtime efficient but also highly scalable. Finally, we present two case studies where our model was applied to discover helpful information in air pollution and traffic congestion databases.

The organization of the paper is as follows. Section 2 describes the related work on frequent pattern mining, rare item problem, and periodic pattern mining. Section 3 presents the proposed model of partial periodic pattern. Section 4 describes the mining algorithm. Section 5 reports the experimental results. Section 6 concludes the paper with future research directions.

2   Related Work

Several alternative measures of *support*, such as χ^2 [2], *all-confidence* [8] and *Kulczynski* [3], have been described in the literature to address the problem. Each measure has a selection bias that justifies the significance of one pattern over another. As a result, there exists no universally acceptable best measure to judge the interestingness of a pattern for any given database or application. Tan et al. [10] introduced the *null-invariance* property and several other properties to aid the user in selecting a measure. Since then, several studies [3,8] recommended the usage of measures that satisfy the *null-invariance* property. It is because this property discloses genuine correlation without being influenced by the object co-absence in a database. Unfortunately, existing null-invariant frequency-based are inadequate to address the rare item problem in partial periodic pattern mining. Thus, this paper explores a new null-invariant temporal measure to address the rare item problem in partial periodic pattern model.

Most previous studies [4,9,11,12] extended the frequent pattern model to discover periodic-frequent patterns in a database. As a result, these studies required too many input parameters and discovered only full (or perfect) periodically occurring frequent patterns. Uday et al. [5] tackled these two problems by proposing the model of partial periodic pattern that may exist in a temporal database. Unfortunately, this model suffers from the rare item problem. This paper addresses this problem using a null-invariant temporal measure.

3 Proposed Model

Let $I = \{i_1, i_2, \cdots, i_n\}$, $n \geq 1$, be a set of items. Let $X \subseteq I$ be a pattern (or an itemset). A pattern containing k number of items is called a k-pattern. A transaction $t_{tid} = (tid, ts, Y)$, where $tid \geq 1$ represents the transaction identifier, $ts \in R^+$ represents the timestamp and $Y \subseteq I$ is a pattern. An (irregular) **temporal database** TDB is a collection of transactions. That is, $TDB = \{t_1, t_2, \cdots, t_m\}$, $1 \leq m \leq |TDB|$, where $|TDB|$ represents the size of database. If a pattern $X \subseteq Y$, it is said that X occurs in transaction t_{tid}. The timestamp of this transaction is denoted as ts_{tid}^X. Let $TS^X = \{ts_a^X, ts_b^X, \cdots, ts_c^X\}$, $a, b, c \in (1, |TDB|)$, denote the set of all timestamps in which the pattern X has appeared in the database. The *support* of X in TDB, denoted as $sup(X)$, represents the number of transactions containing X in TDB. That is, $sup(X) = |TS^X|$.

Example 1. Let $I = \{abcdefg\}$ be the set of items. The temporal database of the items in I is shown in Table 1. The set of items a and c, i.e., $\{a, c\}$ (or ac, in short) is a pattern. It is a 2-pattern because it contains two items. The pattern ac appears in the transactions whose timestamps are 1, 3, 4, 12, 13, 15 and 16. Therefore, $ts_1^{ac} = 1$, $ts_2^{ac} = 3$, $ts_3^{ac} = 4$, $ts_{12}^{ac} = 12$, $ts_{12}^{ac} = 13$, $ts_{13}^{ac} = 15$ and $ts_{14}^{ac} = 16$. The complete set of timestamps at which ac has occurred in Table 1, i.e., $TS^{ac} = \{1, 3, 4, 12, 13, 15, 16\}$. The *support* of ac, i.e., $sup(ab) = |TS^{ac}| = 7$.

Table 1. Temporal database

tid	ts	items	tid	ts	items	tid	ts	items	tid	ts	items	tid	ts	litems
1	1	acd	4	5	def	7	8	adf	10	11	ae	13	15	abcg
2	3	abce	5	6	deg	8	9	bcd	11	12	abcf	14	16	abcd
3	4	abcd	6	7	adg	9	10	adf	12	13	abcd			

Definition 1. (*Periodic appearance of itemset X.*) Let ts_j^X, $ts_k^X \in TS^X$, $1 \leq j < k \leq m$, denote any two consecutive timestamps in TS^X. The time difference between ts_k^X and ts_j^X is referred as **an inter-arrival time** of X, and denoted as iat^X. That is, $iat^X = ts_k^X - ts_j^X$. Let $IAT^X = \{iat_1^X, iat_2^X, \cdots, iat_k^X\}$, $k = sup(X) - 1$, be the set of all inter-arrival times of X in TDB. An inter-arrival time of X is said to be **periodic** if it is no more than the user-specified maximum inter-arrival time (*maxIAT*). That is, a $iat_i^X \in IAT^X$ is said to be **periodic** if $iat_i^X \leq maxIAT$.

Example 2. The pattern ac has consecutively appeared in the transactions whose timestamps are 1 and 3. The difference between these two timestamps gives an inter-arrival time of ac. That is, $iat_1^{ac} = 3 - 1 = 2$. Similarly, other inter-arrival times of ac are: $iat_2^{ac} = 4 - 3 = 1$, $iat_3^{ac} = 12 - 4 = 8$, $iat_4^{ac} = 13 - 12 = 1$, $iat_5^{ac} = 15 - 13 = 2$ and $iat_6^{ac} = 16 - 15 = 1$. Therefore, the set of all inter-arrival times of ac in Table 1, i.e., $IAT^{ac} = \{2, 1, 8, 1, 2, 1\}$. If the user-specified $maxIAT = 2$, then iat_1^{ac}, iat_2^{ac}, iat_4^{ac}, iat_5^{ab} and iat_6^{ac} are considered as the periodic occurrences of ac in the data. On the contrary, iat_3^{ac} is considered as an irregular occurrence of ac because $iat_3^{ac} \nleq maxIAT$.

Definition 2. (Period-support of pattern X [5].) *Let $\widehat{IAT^X} \subseteq IAT^X$ be the set of all inter-arrival times that have value no more than $maxIAT$. That is, $\widehat{IAT^X} \subseteq IAT^X$ such that if $\exists iat_k^X \in IAT^X$: $iat_k^X \leq maxIAT$, then $iat_k^X \in \widehat{IAT^X}$. The period-support of X, denoted as $PS(X) = \frac{|\widehat{IAT^X}|}{|TDB|-1}$, where $|TDB| - 1$ denote the maximum number of inter-arrival times a pattern may have in TDB.*

Example 3. Continuing with the previous example, $\widehat{IAT^{ac}} = \{2, 1, 1, 2, 1\}$. Therefore, the *period-support* of 'ac,' i.e., $PS(ac) = \frac{|\widehat{IAT^{ac}}|}{|TDB|-1} = \frac{|\{2,1,1,2,1\}|}{13} = 0.38$.

Definition 3. (Relative period-support of pattern X.) *The relative period-support of pattern X, denoted as $RPS(X) = \frac{|\widehat{IAT^X}|}{min(sup(i_j)|\forall i_j \in X)-1}$, where $min(sup(i_j)|\forall i_j \in X) - 1$ represents the minimum number of inter-arrival times of a least frequent item in X.*

Example 4. Continuing with the previous example, the *relative period-support* of ac, i.e., $RPS(ac) = \frac{|\widehat{IAT^{ac}}|}{min(sup(a),sup(c))-1} = \frac{|\{2,1,1,2,1\}|}{min(11,8)-1} = \frac{5}{7} = 0.714 \ (= 71.4\%)$. It can be observed that though the pattern ac has low *period-support* in the entire data, it has a high *relative period-support*. Thus, this measure facilitates us to find patterns containing both frequent and rare items.

We now define the partial periodic pattern using the proposed measure.

Definition 4. (Partial periodic pattern X.) *A pattern X is said to be a partial periodic pattern if $PS(X) \geq minPS$ and $RPS(X) \geq minRPS$, where $minPS$ and $minRPS$ represent the user-specified minimum period-support and minimum relative period-support, respectively. Our model employs $minPS$ constraint to prune the patterns that have very less number of periodic occurrences in the database.*

Example 5. If the user-specified $minPS = 28\%$ and $minRPS = 60\%$, then ac is a partial periodic pattern because $PS(ac) \geq minPS$ and $RPS(ac) \geq minRPS$.

Definition 5. (Problem definition.) *Given a temporal database (TDB) and the user-specified maximum inter-arrival time (maxIAT) minimum period-support (minPS), and minimum relative period-support (minRPS), the problem of partial periodic pattern mining is to find all patterns in TDB that satisfy the minPS and minRPS constraints. The generated patterns satisfy the* null-invariant *(see Property 1) and* convertible anti-monotonic *(see Property 2)* properties.

Property 1. (**Null-invariance property** [10].) A binary measure of association is null-invariant if $O(M + C) = O(M)$, where M is a 2×2 contingency matrix, $C = [0\ 0; 0\ k]$ and k is a positive constant.

Property 2. (**The convertible anti-monotonic property.**) Let $Y = \{i_1.i_2,$ $\cdots, i_k\}, k \geq 1$ be an ordered pattern such that $sup(i_1) \leq sup(i_2) \leq \cdots \leq sup(i_k)$. If Y is a partial periodic pattern, then $\forall X \subset Y$, $X \neq \emptyset$ and $i_1 \in X$, X is also a partial periodic pattern.

4 Generalized Partial Periodic Pattern-Growth (G3P-Growth)

The G3P-growth algorithm involves the following two steps: (*i*) compress the database into Generalized Partial Periodic Pattern-tree (G3P-tree) and (*ii*) recursively mine the tree to discover the complete set of partial periodic patters. Algorithm 1 describes the procedure to find PPIs using G3P-list. Algorithms 2 describes the procedure for constructing the prefix-tree. Algorithm 3 describes the procedure for finding partial periodic patterns from the G3P-tree.

Algorithm 1. G3P-List (*TDB*: temporal database, *I*: set of items, *per*: period, *minPS*: minimum period-support and *minRPS*: minimum relative period-support)

1: Let id_l be a temporary array that records the *timestamp* of the last appearance of each item in S. Let ts_{cur} denote the current timestamp of a transaction. Let $[ts_a, ts_b]$ denote the last time interval recorded in *p-list*.
2: **for** each transaction $t \in TDB$ **do**
3:    **for** each item $i \in t$ **do**
4:       **if** i exists in G3P-list **then**
5:          $++s(i)$.
6:          **if** $ts_{cur} - id_l(i) \leq per$ **then**
7:             $++ps(i)$
8:             Set the last list of *p-list(i)* as $[ts_a, ts_{cur}]$.
9:          **else**
10:             **if** $ts_a == ts_b$ **then**
11:                Replace the previous ts_a value in the last list of *p-list(i)* with $[ts_{cur}, ts_{cur}]$.
12:             **else**
13:                Add another list entry into the *p-list(i)* with $[ts_{cur}, ts_{cur}]$.
14:       **else**
15:          Add i to the G3P-list with $S(i) = 1$, $ps(i) = 0$ and $id_l(i) = ts_{cur}$.
16:          Add an entry, $[ts_{cur}, ts_{cur}]$, into the *p-list(i)*.
17: Prune all uninteresting items from the list with *period-support* less than *minPS*.

---

5   Experimental Evaluation

Since there exists no algorithm to find partial periodic patterns that may exist in a spatiotemporal database, we only evaluate the proposed G3P-growth algorithm to show that it is not only memory and runtime efficient, but also highly scalable as well. We also demonstrate the usefulness of our model with two case studies.

Algorithm 2. G3P-Tree (TDB, G3P-list)
***
1: Create the root node in G3P-tree, $Tree$, and label it *"null"*.
2: **for** each transaction $t \in TDB$ **do**
3:    Select the PNIs in t and sort them in L order. Let the sorted list be $[e|E]$, where e is the first item and E is the remaining list. Call $insert\_tree([e|E], ts_{cur}, Tree)$ [5].
4: call G3P-growth ($Tree$, *null*);

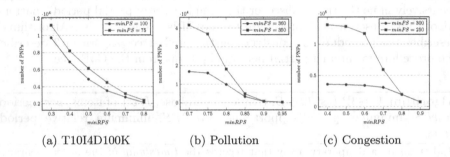

(a) T10I4D100K   (b) Pollution   (c) Congestion

Fig. 1. Number of patterns generated at different $minPS$ and $minRPS$ values

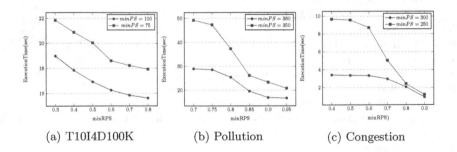

(a) T10I4D100K   (b) Pollution   (c) Congestion

Fig. 2. Runtime requirements of G3P-growth

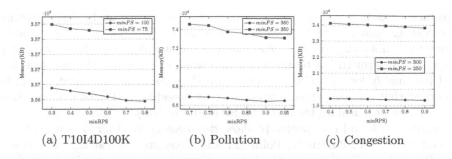

(a) T10I4D100K   (b) Pollution   (c) Congestion

Fig. 3. Memory used for the G3P-tree construction

Algorithm 3. G3P-growth $(Tree, \alpha)$

1: **while** items in the header of $Tree$ **do**
2:     Generate pattern $\beta = i \cup \alpha$. Traverse $Tree$ using the node-links of β, and construct an array, TS^{β}, which represents the list of timestamps in which β has appeared periodically in TDB. Construct β's conditional pattern base and β's conditional G3P-tree $Tree_{\beta}$ if period-support is greater than or equal to $minPS$, relative period-support is greater than or equal to $minRPS$ and the distance between all of its is no more than the user-specified $maxDist$.
3:     **if** $Tree_{\beta} \neq \emptyset$ **then**
4:         call PP-growth $(Tree_{\beta}, \beta)$;
5:     Remove i from the $Tree$ and push the i's ts-list to its parent nodes.

The G3P-growth algorithm was written in Python 3 and executed on a machine with 2.5 GHz processor and 8 GB RAM. The experiments have been conducted on synthetic (**T10I4D100K**) and real-world (**Pollution and Congestion**) databases. The **T10I4D100K** [1] is a widely used synthetic database for evaluating frequent pattern mining algorithms. This transactional database is converted into a temporal database by considering *tids* as timestamps. This database contains 870 items and 100,000 transactions. The **Pollution** database contains 1600 items and 720 items. The minimum, average and maximum transactions lengths are 11, 460 and 971, respectively. The **Congestion** database contains 1782 items and 1440 transactions. The minimum, average and maximum transactions lengths are 11, 66.25 and 267, respectively.

Figure 1a, b and c respectively show the number of partial periodic patterns generated in T10I4D100K, Congestion and Pollution databases at different $minRPS$ and $maxPS$ values. The $maxIAT$ value in T10I4D100K, Congestion and Pollution databases is set to 5000, 5 min and 5 min, respectively. It can be observed that increase in $minRPS$ and/or $minPS$ have negative effect on the generation of patterns. It is because many patterns fail to satisfy the increased $minPS$ and $minRPS$ values.

Figure 2a, b and c respectively show the runtime requirements of G3P-growth in T10I4D100K, Congestion and Pollution databases at different $minPS$ and $minRPS$ values. The $maxIAT$ value in T10I4D100K, Congestion and Pollution databases has been set to 5000, 5 min and 5 min, respectively. The following observations can be drawn from these figures: (i) Increase in $minPS$ and/or $minRPS$ values may decrease the runtime requirements of G3P-growth. It is because G3P-growth has to find fewer patterns. (ii) As per as the database size and reasonably low $minPS$ and low $minRPS$ values are concerned, it can be observed that mining patterns from the corresponding G3P-tree is rather time efficient (or practicable) for both synthetic and real-world databases.

Figure 3a, b and c respectively show the memory used for the construction of G3P-tree in T10I4D100K, Pollution and Congestion databases at different $minPS$ and $minRPS$ values. The following two observations can be drawn from these figures: (i) increase in $minPS$ may decrease the memory requirements of G3P-tree construction. It is because many items fail to be periodic items. (ii) increase in $minRPS$ has very little effect on the memory requirements of G3P-tree. It is because G3P-tree constructed by periodic items.

6   Conclusions and Future Work

Rare item problem is a major problem encountered by the partial periodic pattern model. This paper tackled this challenging problem of great importance by proposing a novel null-invariant temporal measure known as *relative period-support*. We have also shown that finding the patterns using our measure is practicable on big data as the generated patterns satisfy the *convertible anti-monotonic property*. Furthermore, this paper proposed an efficient algorithm to discover the desired patterns. By conducting experiments, we have shown the efficiency of our algorithm. As a part of future work, we like to investigate the models to find patterns in dynamic graphs and data streams.

References

1. Agrawal, R., Srikant, R., et al.: Fast algorithms for mining association rules. VLDB. **1215**, 487–499 (1994)
2. Brin, S., Motwani, R., Silverstein, C.: Beyond market baskets: generalizing association rules to correlations. SIGMOD Rec. **26**(2), 265–276 (1997)
3. Kim, S., Barsky, M., Han, J.: Efficient mining of top correlated patterns based on null-invariant measures. In: Gunopulos, D., Hofmann, T., Malerba, D., Vazirgiannis, M. (eds.) ECML PKDD 2011. LNCS (LNAI), vol. 6912, pp. 177–192. Springer, Heidelberg (2011). https://doi.org/10.1007/978-3-642-23783-6_12
4. Uday Kiran, R., Krishna Reddy, P.: Towards efficient mining of periodic-frequent patterns in transactional databases. In: Bringas, P.G., Hameurlain, A., Quirchmayr, G. (eds.) DEXA 2010. LNCS, vol. 6262, pp. 194–208. Springer, Heidelberg (2010). https://doi.org/10.1007/978-3-642-15251-1_16
5. Kiran, R.U., Shang, H., Toyoda, M., Kitsuregawa, M.: Discovering partial periodic itemsets in temporal databases. In: SSDBM, pp. 30:1–30:6 (2017)
6. Kiran, R.U., Venkatesh, J.N., Fournier-Viger, P., Toyoda, M., Reddy, P.K., Kitsuregawa, M.: Discovering periodic patterns in non-uniform temporal databases. In: Kim, J., Shim, K., Cao, L., Lee, J.-G., Lin, X., Moon, Y.-S. (eds.) PAKDD 2017. LNCS (LNAI), vol. 10235, pp. 604–617. Springer, Cham (2017). https://doi.org/10.1007/978-3-319-57529-2_47

7. Liu, B., Hsu, W., Ma, Y.: Mining association rules with multiple minimum supports. In: Proceedings of the Fifth ACM SIGKDD International Conference on Knowledge Discovery and Data Mining, pp. 337–341. ACM (1999)
8. Omiecinski, E.R.: Alternative interest measures for mining associations in databases. IEEE TKDE **15**(1), 57–69 (2003)
9. Özden, B., Ramaswamy, S., Silberschatz, A.: Cyclic association rules. In: International Conference on Data Engineering, pp. 412–421 (1998)
10. Tan, P.N., Kumar, V., Srivastava, J.: Selecting the right interestingness measure for association patterns. In: SIGKDD, pp. 32–41 (2002)
11. Tanbeer, S.K., Ahmed, C.F., Jeong, B.-S., Lee, Y.-K.: Discovering periodic-frequent patterns in transactional databases. In: Theeramunkong, T., Kijsirikul, B., Cercone, N., Ho, T.-B. (eds.) PAKDD 2009. LNCS (LNAI), vol. 5476, pp. 242–253. Springer, Heidelberg (2009). https://doi.org/10.1007/978-3-642-01307-2_24
12. Venkatesh, J.N., Uday Kiran, R., Krishna Reddy, P., Kitsuregawa, M.: Discovering periodic-correlated patterns in temporal databases. In: Hameurlain, A., Wagner, R., Hartmann, S., Ma, H. (eds.) Transactions on Large-Scale Data- and Knowledge-Centered Systems XXXVIII. LNCS, vol. 11250, pp. 146–172. Springer, Heidelberg (2018). https://doi.org/10.1007/978-3-662-58384-5_6
13. Weiss, G.M.: Mining with rarity: a unifying framework. SIGKDD Explor. Newsl. **6**(1), 7–19 (2004)

Utilizing Expert Knowledge and Contextual Information for Sample-Limited Causal Graph Construction

Xuwu Wang[1], Xueyao Jiang[1], Sihang Jiang[1], Zhixu Li[1],
and Yanghua Xiao[1,2(✉)]

[1] Shanghai Key Lab. of Data Science, School of Computer Science, Fudan University,
Shanghai, China
{xwwang18,xueyaojiang19,zhixuli,shawyh}@fudan.edu.cn
[2] Fudan-Aishu Cognitive Intelligence Joint Research Center, Shanghai, China

Abstract. This paper focuses on causal discovery, which aims at inferring the underlying causal relationships from observational samples. Existing methods of causal discovery rely on a large number of samples. So when the number of samples is limited, they often fail to produce correct causal graphs. To address this problem, we propose a novel framework: Firstly, given an expert-specified causal subgraph, we leverage contextual and statistical information of the variables to expand the subgraph with positive-unlabeled learning. Secondly, to ensure the faithfulness of the causal graph, with the expanded subgraph as the constraint, we resort to a structural equation model to discover the entire causal graph. Experimental results show that our method achieves significant improvement over the baselines, especially when only limited samples are given.

Keywords: Causal graph · Causal discovery · Knowledge graph

1 Introduction

Causal graph can be treated as a special kind of knowledge graph that only consists of causal relations [3]. The edge from node v_i to node v_j represents that the occurrence of the cause variable v_i triggers the occurrence of the effect variable v_j.

Causal discovery [5] is one of the most effective methods to construct a directed acyclic graph (DAG) as the causal graph. Pearl et al. [8] propose detailed definitions of the causal graph and corresponding causal model, which makes it applicable to downstream tasks that require high interpretability. Existing causal discovery methods mainly rely on statistical analysis to infer the causal graph from observed samples (see Fig. 1(a) for example), which could be roughly divided into two lines: The first line of works is constraint-based method [13–15,17], which utilizes conditional independence tests over the samples to find the graph skeleton first and then decides the direction of edges. The second is the

A. Bhattacharya et al. (Eds.): DASFAA 2022, LNCS 13245, pp. 578–586, 2022.
https://doi.org/10.1007/978-3-031-00123-9_46

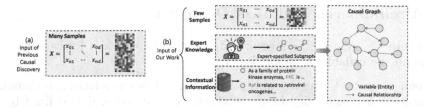

Fig. 1. Comparison of the information sources of different methods. The example causal graph is a consensus protein signaling causal graph. The samples are expression levels of the proteins and phospholipid components.

score-based method [14,17], which assigns a score to each graph in terms of the samples and then searches for the graph with the highest score. Overall, these methods have produced appealing results from large-scale samples. However, collecting large-scale samples is *time-consuming* and *labor intensive* in practice. For example, collecting samples may need manual measurement of the variables in the field of biology [11] or long-term follow-up visiting in pathophysiology relevant research [12]. As a result, it is a usual case to collect only a limited number of samples.

To make up for the lack of samples, we propose to resort to other sources of information that commonly exist and are easily accessible in practice. The new sources of information include 1) *expert knowledge*: some already known causal relations provided by human experts, which can be treated as a causal subgraph; and 2) *contextual information*: the textual description of each variable obtained from either domain documentations, knowledge graphs or even the Internet. As the example shown in Fig. 1(b), while the expert knowledge can be used to provide weak supervision, the contextual information provides plentiful contexts about the variables as the supplement to the samples.

In this paper, we propose a novel causal graph construction framework with two key phases. The expert-specified causal subgraph is usually small since it is difficult for human experts to identify many causal relations based on their experience and knowledge. So in **Phase 1**, to expand the expert-specified subgraph, we train a classifier that can predict new causal relations based on the annotations of the subgraph. The realization of this classifier mainly benefits from two aspects: 1) The contextual information of the variables and the samples; 2) Positive-Unlabeled (PU) learning [9]. These newly-predicted relations form a so-called predicted subgraph. However, this expanded subgraph may break the faithfulness condition [7] that is the crucial basis of causal discovery. To tackle this problem, in **Phase 2**, we use a structural equation model to search for the DAG that is the most likely to generate the observed samples. The predicted subgraph is utilized as the constraint to guide the search.

2   Preliminaries and Task Definition

Variable. A variable v_i represents an object to be explored for causality. The set of all the variables is denoted as $V = \{v_i\}_{i=1}^d$.

Causal Graph. A causal graph $\mathcal{G}$ recording the causal relationships among d variables is a *directed acyclic graph* (DAG) with d nodes. $\mathcal{G}$ is defined by the adjacency matrix $\mathbf{A}$. Following previous works [16], instead of operating on the discrete space $\mathbf{A} \in \{0,1\}^{d \times d}$, we operate on the continuous space $\mathbf{A} \in \mathbb{R}^{d \times d}$: $\mathbf{A}_{ij} \neq 0$ indicates v_i is the cause of v_j.

Observational Samples. $\mathbf{X} \in \mathbb{R}^{n \times d}$ is a matrix of n i.i.d. observational samples from the joint distribution of the d variables. $\mathbf{X}_{ij}$ is the jth variable's value in the ith sample. We abbreviate observational samples as samples.

Expert-Specified Causal Subgraph. The expert-specified causal subgraph $\hat{\mathcal{G}}$ is a spanning subgraph of the entire causal subgraph $\mathcal{G}$. That is to say, $\hat{\mathcal{G}}$ includes all nodes of $\mathcal{G}$ and a subset of edges of $\mathcal{G}$. The subgraph with $\alpha\%$ of the edges in $\mathcal{G}$ is denoted as $\hat{\mathcal{G}}^{\alpha\%}$. In the following text, when α is not explicitly stated, we represent it with $\hat{\mathcal{G}}$. In real applications, experts often only specify whether there is a causal relationship between variables, but can rarely provide the detailed strength of the causal relationship. So the adjacency matrix of $\hat{\mathcal{G}}$ is simply defined as: $\hat{\mathbf{A}} \in \{0,1\}^{d \times d}$, where $\hat{\mathbf{A}}_{ij} = 1$ represents v_i is the cause of v_j.

Task Definition. Suppose there are d variables to be explored for causality. Given the samples $\mathbf{X} \in \mathbb{R}^{n \times d}$, we need to infer the causal graph $\mathcal{G}$ that is faithful to the joint distribution behind the observed samples $\mathbf{X}$. Specially, we focus on how to infer the $\mathcal{G}$ when *limited* samples of $\mathbf{X}$ are given, i.e., n is small. And contextual information and an expert-specified subgraph are provided to tackle this problem.

3   Methodology

Our method of constructing a causal graph consists of two key phases.

3.1   Phase 1: PU Causal Classifier

In the first phase, with the expert-specified subgraph $\hat{\mathcal{G}}$ as the labelling criterion, we train a binary classifier $\phi : (v_i, v_j) \rightarrow \{0,1\}, v_i, v_j \in V$, which predicts whether the variable v_i is the cause of v_j. The framework of the PU causal classifier is presented in Fig. 2(a) and elaborated below.

Preprocess. Before training, the dataset used to train the classifier is built as follows. Given $\hat{\mathcal{G}}$ with the adjacency matrix of $\hat{\mathbf{A}}$, for every instance $(v_i, v_j) \in V \times V$, if $\hat{\mathbf{A}}_{ij} = 1$, v_i is the cause of v_j, and (v_i, v_j) is labeled as 1; otherwise (v_i, v_j) is unlabeled. This positive-unlabeled setting is solved in the optimization procedure later.

Design of the Classifier ϕ. We use two kinds of features to enhance the classifier ϕ. The first kind of features is the textual contexts of the variables, which provide the *semantic* information for causality. Specifically, for each variable v_i, we retrieve its context from the database or the Web: s_i. Then the semantic feature vector of v_i is obtained through a BERT [2] encoder: $\mathbf{se}(\mathbf{v}_i) = \mathrm{BERT}(s_i)$.

The second kind of features is the observational samples, which provides the *statistical* information for causality. Considering only limited samples are provided, we propose to enhance a variable's sample information by aggregating its correlated variables' sample information. To achieve it, we first transform samples in matrix form into a fully connected sample graph $\mathcal{O}$ with the adjacency matrix $\mathbf{B}$. The nodes of $\mathcal{O}$ are variables and the edge weight reflects the correlation between variables. Assuming that the distributions of correlated variables over the samples are similar, the edge weight $\mathbf{B}_{ij}$ is defined as the reciprocal of the Wasserstein distance [6] between $\mathbf{X}_{.i}$ and $\mathbf{X}_{.j}$: $\mathbf{B}_{ij} = \frac{1}{W(\mathbf{X}_{.i}, \mathbf{X}_{.j})}$[1]. Here $\mathbf{X}_{.i}$ represents variable v_i's distribution over the samples. Then the graph convolutional network (GCN) [4] is used to pass information between correlated variables over the sample graph $\mathcal{O}$. The sample feature for every variable is thus generated as: $\mathbf{sa}(\mathbf{v}_i) = \mathrm{GCN}(\mathbf{B})$

Then we concatenate the semantic and statistical features of v_i and v_j:

$$\mathbf{v}_{ij} = \mathbf{se}(\mathbf{v}_i) \oplus \mathbf{sa}(\mathbf{v}_i) \oplus \mathbf{se}(\mathbf{v}_j) \oplus \mathbf{sa}(\mathbf{v}_j) \tag{1}$$

and feed $\mathbf{v}_{ij}$ to a multilayer perceptron (MLP) and a softmax layer. The probability that v_i is the cause of v_j is finally calculated as: $f(v_i, v_j) = \mathrm{softmax}(\mathrm{MLP}(\mathbf{v}_{ij}))$.

Optimization Objective. As mentioned earlier, the training data consists of a set of positive (P) instances and a set of unlabeled (U) instances. To predict new causality from the unlabeled instances, we resort to PU learning. Following [9], the optimization objective of our PU classifier is defined as a bounded nonnegative PU loss:

$$L = \underbrace{\frac{\pi_p}{n_p} \sum_{k=1}^{n_p} l\left(f\left(x_k^p\right), 1\right)}_{(\clubsuit)} + \max\left(0, \underbrace{\frac{1}{n_u} \sum_{k=1}^{n_u} l\left(f\left(x_k^u\right), 0\right)}_{(\spadesuit)} - \underbrace{\frac{\pi_p}{n_u} \sum_{k=1}^{n_p} l\left(f\left(x_k^p\right), 0\right)}_{(\heartsuit)} \right)$$
$$\tag{2}$$

where x^p and x^u represent positive instance and unlabeled instance respectively. $l(logit, label)$ is the CrossEntropy loss function. n_p and n_u represent the number of positive and unlabeled instances respectively. π_p is a hyper-parameter indicating the estimated percentage of the positive instances. By weakening the penalty for labeling unlabeled instances as negative instances (i.e. $(\spadesuit) - (\heartsuit)$), the classifier is able to identify some positive instances from the unlabeled instances.

---

[1] Other metric functions such as JS divergence, co-occurrence frequency can also be utilized to generate $\mathbf{B}_{ij}$.

Postprocess. Once ϕ is trained, its prediction can be transformed into a pre-dicted causal subgraph $\bar{\mathcal{G}}$ with the adjacency matrix $\bar{\mathbf{A}}$. Specifically, we use the logit of $f(v_i, v_j)$ as the adjacency matrix's entry when the logit is higher than 0.5. Otherwise, the entry is set as 0.

3.2 Phase 2: SEM with Subgraphs

In the second phase, based on $\bar{\mathcal{G}}$, we use SEM to search for the $\mathcal{G}$ with the highest likelihood to generate $\mathbf{X}$. Three requirements are achieved as illustrated below.

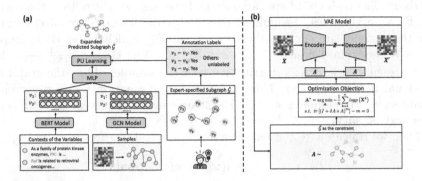

Fig. 2. Overview of the proposed method. Phase (a) corresponds to the PU causal classifier. Phase (b) corresponds to the SEM with the constraint of the subgraph.

R1) $\mathcal{G}$ is Faithful to the Joint Distribution Behind $\mathbf{X}$. To search for such a $\mathcal{G}$, we model it as a continuous optimization problem to find the adjacency matrix $\mathbf{A}^*$ by minimizing the negative log likelihood of obtaining $\mathbf{X}$:

$$\mathbf{A}^* = \arg\min_{\mathbf{A}} -\frac{1}{n} \sum_{k=1}^{n} \log p\left(\mathbf{X}^k\right) \tag{3}$$

To model this generative procedure, based on the Markov condition and causal sufficiency assumption, $\mathbf{X}$ is generated by the linear structural equation model (SEM): $\mathbf{X} = \mathbf{A}^\top \mathbf{X} + \mathbf{Z} \quad \rightarrow \quad \mathbf{X} = \left(\mathbf{I} - \mathbf{A}^\top\right)^{-1} \mathbf{Z}$. Here $\mathbf{Z} \in \mathbb{R}^{n \times d}$ is the noise matrix. $\mathbf{I}$ is an identity matrix.

To instantiate the above SEM model with neural networks, we adopt the variational autoencoder (VAE) model proposed in [14]: as shown in Fig. 2(b), the encoder of the VAE takes the samples $\mathbf{X}$ as input, and outputs the hidden variable $\mathbf{Z}$:

$$g(\mathbf{Z}) = \left(\mathbf{I} - \mathbf{A}^T\right) \mathrm{MLP}\left(\mathbf{X}\right). \tag{4}$$

where $\mathbf{A}$ is learnable together with other parameters. $g(\mathbf{Z})$ represents the variance and bias of $\mathbf{Z}$. Then the decoder takes $g(\mathbf{Z})$ as the input to recover $\mathbf{X}$:

$$\mathbf{X}' = \mathrm{MLP}\left(\left(\mathbf{I} - \mathbf{A}^T\right)(g(\mathbf{Z}))\right) \tag{5}$$

With this VAE model, the log likelihood of $\mathbf{X}$ can be transformed into

$$\frac{1}{n}\sum_{k=1}^{n}\log p\left(\mathbf{X}^{k}\right) = \frac{1}{n}\sum_{k=1}^{n}\log\int p(\mathbf{X}^{k}|\mathbf{Z})p(\mathbf{Z})d\mathbf{Z} \tag{6}$$

R2) $\mathcal{G}$ is a DAG. As analyzed in [14], the requirement that $\mathbf{A} \in DAGs$ can be quantified as $tr\left[(I + \delta A \circ A)^{m}\right] - m = 0$. Here $tr(\cdot)$ represents the trace of a matrix. δ is a hyper-parameter. After adding this DAG constraint with the augmented Lagrangian, the optimization objective is transformed into:

$$\mathbf{A}^{*} = \underset{\mathbf{A}}{\arg\min} - \frac{1}{n}\sum_{k=1}^{n}\log p\left(\mathbf{X}^{k}\right) + \lambda_{1}l\left(\mathbf{A}\right) + \frac{c}{2}|l\left(\mathbf{A}\right)|^{2} \tag{7}$$

where $l\left(\mathbf{A}\right) = tr\left[(I + \delta A \circ A)^{m}\right] - m$. λ_{1} is the Lagrange multiplier. c is the penalty term.

R3) $\mathcal{G}$ is Similar with $\bar{\mathcal{G}}$. We extend the above model by adding minimizing the difference between $\mathbf{A}$ and $\bar{\mathbf{A}}$ as another optimization objective:

$$\mathbf{A}^{*} = \underset{\mathbf{A}}{\arg\min} - \frac{1}{n}\sum_{k=1}^{n}\log p\left(\mathbf{X}^{k}\right) + \lambda_{1}l\left(\mathbf{A}\right) + \frac{c}{2}|l\left(\mathbf{A}\right)|^{2} + \lambda_{2}h\left(\mathbf{A}, \bar{\mathbf{A}}\right) \tag{8}$$

$h\left(\mathbf{A}, \bar{\mathbf{A}}\right)$ is defined as the element-wise difference between $\mathbf{A}$ and $\bar{\mathbf{A}}$:

$$h\left(\mathbf{A}, \bar{\mathbf{A}}\right) = \sum_{i=1}^{d}\sum_{j=1}^{d}\left(||\mathbf{A}_{ij}| - |\bar{\mathbf{A}}_{ij}|| \times \mathbb{1}(\bar{\mathbf{A}}_{ij})\right) \tag{9}$$

Here $\mathbb{1}(\bar{\mathbf{A}})$ is an indicator function which returns 1 when $\bar{\mathbf{A}}_{ij} \neq 0$ and returns 0 when $\bar{\mathbf{A}}_{ij} = 0$. We apply $|\cdot|$ to the matrix entry for we only care about the existence of a causal relationship instead of its polarity (positive or negative), which is the same in [14].

4   Experiment

4.1   Experimental Setups

Datasets. We perform experiments on 2 real datasets: ALARM, SACHS [1,11]. The ALARM is used to infer the causal relationships between device alarms, where the problem of limited samples is prominent. It has 61 samples, 41 variables and 1307 edges in the causal graph. Specially, due to the characteristics of this application, the ground-truth graph does not satisfy the DAG constraint. So we remove the DAG constraint from all the methods. The SACHS is a dataset used to analyze the protein signal network with 853 observational samples, 11 variables and 17 edges in the causal graph, which is also used in [17].

Table 1. Comparison with the baselines. The bold numbers indicate the best results. '-' indicates that some datasets do not support the experimental setting, or the model failed to make predictions. Since SACHS only contains 17 causal relations, we set $\beta \in \{5, 10, 20, 40\}$ instead.

Dataset	Method	Experimental settings							
	β	0.5		1		2		–	
	Metric	TPR	SHD	TPR	SHD	TPR	SHD	TPR	SHD
ALARM	PC	0	785	1.53	765	1.53	765	–	–
	CAM	0	785	0	785	8.72	671	–	–
	NOTEARS	1.45	769	1.76	767	1.07	772	–	–
	DAGGNN	10.94	757	22.26	593	27.09	523	–	–
	CDRL	2.21	757	19.36	580	33.28	465	–	–
	ours ($\hat{\mathcal{G}}^{10\%}$)	81.41	302	80.95	**307**	83.09	293	–	–
	ours ($\hat{\mathcal{G}}^{20\%}$)	**82.63**	**226**	**83.63**	523	**85.23**	**291**	–	–
	β	5		10		20		40	
	Metric	TPR	SHD	TPR	SHD	TPR	SHD	TPR	SHD
SACHS	PC	–	–	–	–	–	–	–	–
	CAM	11.76	18	17.65	15	19.61	16	21.57	15
	NOTEARS	**47.06**	19	35.29	16	23.53	17	29.41	16
	DAGGNN	41.18	21	41.18	20	21.76	16	35.29	19
	CDRL	11.76	16	27.45	**14**	23.53	14	37.25	**14**
	ours ($\hat{\mathcal{G}}^{10\%}$)	**47.06**	18	**52.94**	16	35.29	17	47.06	16
	ours ($\hat{\mathcal{G}}^{20\%}$)	**47.06**	14	**52.94**	14	**47.06**	10	**52.94**	15

Limited Samples Settings. To study the model performance of different sample sizes, we randomly select samples of different sizes from the original dataset. We use $\beta = \frac{n}{d}$ to quantify the sample size. A smaller β indicates fewer samples, especially when $\beta < 1$.

Baselines. We compare with various effective baselines including PC [13], CAM [10], NOTEARS [16], DAGGNN [14], CDRL [17].

Implementation. We randomly sample $\alpha\%$ gold causal relations as the expert-specified subgraph. For the first phase, an Adam optimizer with the learning rate of $1e-4$ is used to optimize the classifier for 200 epochs. For datasets that cannot retrieve the corresponding contextual information, we use a one-hot vector instead. For the second phase, the batch size is set as 100 at each iteration. We pick 64 as the hidden dimension of the encoder and the decoder. Other hyper-parameters are set the same as [14].

Evaluation Metrics. We report the performance of two metrics: true positive rate (TPR) and structural hamming distance (SHD) [14]. A higher TPR and a lower SHD indicate that the generated causal graph is better.

4.2   Experimental Results

We present the comparison between our method with $\hat{\mathcal{G}}^{10\%}/\hat{\mathcal{G}}^{20\%}$ and the baselines in Table 1. Since there are not many causal relationships in the causal graph, $\hat{\mathcal{G}}^{10\%}$ and $\hat{\mathcal{G}}^{20\%}$ are very small, which provide only weak supervision. Besides, to test the model performance with limited samples, we set $\beta \in \{0.5, 1, 2, 4\}$. This is different from the experimental settings of existing methods, that usually requires β to be larger than 50. All the results are the mean of 3 runs performed on the same device. We can see that our method achieves significant improvement compared with all the baselines under most settings. This phenomenon is even more pronounced for TPR. This result shows that our method has a huge advantage in recalling more causal relationships.

5   Conclusion

In this paper, towards the real problem of few samples in causal discovery, we propose a framework that infers causal relationships with the supervision of an expert-specified subgraph. To expand the small given subgraph, a PU classifier is first used to predict new causality. Second, we use a structural equation model with the given and predicted subgraphs as initialization or constraints to search for the entire causal graph. Experimental results on two real datasets and two synthetic datasets show that our method achieves both effectiveness and efficiency.

Acknowledgement. This research was supported by the National Key Research and Development Project (No. 2020AAA0109302), National Natural Science Foundation of China (No. 62072323), Shanghai Science and Technology Innovation Action Plan (No. 19511120400) and Shanghai Municipal Science an Technology Major Project (No. 2021SHZDZX0103).

References

1. Barabási, A.L., Albert, R.: Emergence of scaling in random networks. Science **286**(5439), 509–512 (1999)
2. Devlin, J., Chang, M.W., et al.: BERT: pre-training of deep bidirectional transformers for language understanding. In: NAACL, pp. 4171–4186, June 2019
3. Heindorf, S., Scholten, Y., et al.: Causenet: towards a causality graph extracted from the web. In: Proceedings of CIKM, pp. 3023–3030 (2020)
4. Kipf, T.N., Welling, M.: Semi-supervised classification with graph convolutional networks. In: Proceedings of the 5th International Conference on Learning Representations, ICLR 2017 (2017)
5. Lu, N., Zhang, K., et al.: Improving causal discovery by optimal bayesian network learning. In: AAAI (2021)
6. Olkin, I., Pukelsheim, F.: The distance between two random vectors with given dispersion matrices. Linear Algebra Appl. **48**, 257–263 (1982)
7. Pearl, J.: Probabilistic reasoning in intelligent systems: networks of plausible inference. Elsevier (2014)

8. Pearl, J., Mackenzie, D.: The book of why: the new science of cause and effect. Basic books (2018)
9. Peng, M., Xing, X., et al.: Distantly supervised named entity recognition using positive-unlabeled learning. In: Proceedings of ACL, pp. 2409–2419 (2019)
10. Ramsey, J., Glymour, M., Sanchez-Romero, R., Glymour, C.: A million variables and more: the fast greedy equivalence search algorithm for learning high-dimensional graphical causal models, with an application to functional magnetic resonance images. Int. J. Data Sci. Anal. 3(2), 121–129 (2017)
11. Sachs, K., Perez, O., et al.: Causal protein-signaling networks derived from multi-parameter single-cell data. Science 308, 523–529 (2005)
12. Shen, X., Ma, S., et al.: challenges and opportunities with causal discovery algorithms: application to Alzheimer's pathophysiology. Sci. Rep. 10, 1–12 (2020)
13. Spirtes, P., Glymour, C.N., Scheines, R., Heckerman, D.: Causation, prediction, and search. MIT press (2000)
14. Yue Yu, J.C., et al.: Dag-gnn: dag structure learning with graph neural networks. In: Proceedings of the 36th International Conference on Machine Learning (2019)
15. Zhang, K., Peters, J., et al.: Kernel-based conditional independence test and application in causal discovery. arXiv preprint arXiv:1202.3775 (2012)
16. Zheng, X., Aragam, B., et al.: Dags with no tears: Continuous optimization for structure learning. N (2018)
17. Zhu, S., Ng, I., Chen, Z.: Causal discovery with reinforcement learning. In: International Conference on Learning Representations (2020)

A Two-Phase Approach for Recognizing Tables with Complex Structures

Huichao Li[1], Lingze Zeng[1], Weiyu Zhang[1], Jianing Zhang[1], Ju Fan[2], and Meihui Zhang[1($\boxtimes$)]

[1] Beijing Institute of Technology, Beijing, China
{3120191013,3220201116,meihui_zhang}@bit.edu.cn
[2] Renmin University of China, Beijing, China
fanj@ruc.edu.cn

Abstract. Tables contain rich multi-dimensional information which can be an important source for many data analytics applications. However, table structure information is often unavailable in digitized documents such as PDF or image files, making it hard to perform automatic analysis over high-quality table data. Table structure recognition from digitized files is a non-trivial task, as table layouts often vary greatly in different files. Moreover, the existence of spanning cells further complicates the table structure and brings big challenges in table structure recognition. In this paper, we model the problem as a cell relation extraction task and propose T2, a novel two-phase approach that effectively recognizes table structures from digitized documents. T2 introduces a general concept termed *prime relation*, which captures the direct relations of cells with high confidence. It further constructs an alignment graph and employs message passing network to discover complex table structures. We validate our approach via extensive experiments over three benchmark datasets. The results demonstrate T2 is highly robust for recognizing complex table structures.

Keywords: Data mining · Table structure recognition · Message passing networks

1 Introduction

Table, as a widely used format for data organization in documents, often contains rich structured information, which is an important source to build knowledge repositories or perform data analytics tasks in many important applications. However the table structure information is often unavailable in documents such as PDF or image files. To analyze the tabular data residing in the documents, it is crucial to conduct a preprocessing step recognizing the structure of the tables.

Nevertheless, table structure recognition is a non-trivial task on account of the following difficulties. Tables are not organized with unified layout. For instance, some tables contain full border lines while others are only partially segmented by lines. Furthermore, many tables have complex header structure, where spanning cells occupy multiple columns or rows.

A. Bhattacharya et al. (Eds.): DASFAA 2022, LNCS 13245, pp. 587–595, 2022.
https://doi.org/10.1007/978-3-031-00123-9_47

Early works [4,7,10] recognize table structures mainly by extracting hand-crafted features and employing heuristics based algorithms, which highly rely on layout hints and thus are not able to generalize well in practice. Recent works [1,5,9] model the problem as a cell relation prediction task, and propose graph-based models to capture the relations (i.e., vertical/horizontal alignment or no relation) between cells. Some work [9] detects all pair-wise cell relations, while others [1,5] detect relations only between a cell and its k nearest neighbors. We argue that on the one hand it is unnecessary to perform sophisticated graph-based computation for all cell pairs, yielding a high computational complexity. On the other hand, it is inadequate to compute only local relations between k nearest neighbors, which is insufficient to reconstruct the table structure.

We observe that in most cases cells that belong to the same column/row tend to have overlap in the vertical/horizontal direction, except for the cases where spanning cells are present. Based on the observation, we propose T2, a two-phase table structure recognition approach, where we introduce a general concept termed *prime relation* based on cell overlap, which captures the strong evidence of cells belonging to the same column/row. Then we develop an effective method to generate prime rows and columns with prime relation. To discover the complex table structure with spanning cells, we design a novel graph-based alignment model, where we model the prime column/row as a prime node with carefully designed features and learn the alignment relationship between prime columns/rows and spanning cells through message passing network. Comparing to prior works, our proposed approach is pure data oriented without relying on any external information. Further, it is able to completely recognize complex table structure and save a lot computation cost with the help of prime notion.

2   The T2 Framework

T2 takes a table as input, and employs some tools, e.g., Tabby [10] and CascadeTabNet [8], to detect cell locations of the table. The obtained cell locations are further fed into the following two-phase method to discover the relations between cells and finally output columns and rows of the table. In phase one, T2 detects the *prime relations* and generates *prime columns and rows*. In phase two, T2 takes as input the generated prime columns and rows, and builds an alignment graph and learns the column/row alignment relation through message passing network. For ease of presentation, we only present the process of column relation discovery. We omit 'column/row' when the context is clear.

2.1   Phase One: Prime Relation Generation

A *cell* is denoted as a vector $c_i = (x_{i1}, x_{i2}, y_{i1}, y_{i2})$, where $x_{i1}, x_{i2}, y_{i1}, y_{i2} \in R$, represent the coordinates of the cell bounding box. Specifically, (x_{i1}, y_{i1}) and (x_{i2}, y_{i2}) are the coordinates of the lower-left corner and upper-right corner respectively. We observe that in most cases the cells that have column/row relation tend to overlap in vertical/horizontal direction with each other except for the cases where complex structure presents (i.e., spanning cells). Based on this observation, we define *prime relation*, to capture the strong evidence of cells that ought to be aligned. Formally, two cells are defined to have prime column/row relation if and only if the two cells overlap in vertical/horizontal direction. Subsequently, prime column/row is defined as a set of cells, such that there exists at least one cell (called *pivot cell*) that has prime column/row relation with all other cells in the set.

Next, we introduce how prime columns are generated. We take the identified cell locations as input and compute the width of each cell. We start with the cell that has the minimum width to act as the pivot cell. When multiple cells have the same minimum width, we break the ties randomly. We then detect the prime relation between the pivot cell and all other cells based on the cell overlap, and group all cells that have prime column relation w.r.t. the pivot cell to generate a prime column. Next, we select the pivot (i.e. cell with the minimum width) from all the cells that are not yet associated with any prime column and generate the prime column. The process is repeated until the generated prime columns cover all cells in the table.

Note that each time we choose the cell with minimum width to be the pivot. This is because tables may contain wide spanning cells that overlap with multiple columns. If we choose such cell to be the pivot, we will end up generating a giant prime column with mixed data from multiple columns. Ideally, we expect the prime column to have high *purity* in the sense that it only contains the cells from a single column.

2.2   Phase Two: Graph-Based Alignment Model

In this phase, we aim to discover relations that are not detected in phase one, mainly due to the existence of spanning cells. Detailed steps are as follows:

Alignment Graph Construction. Based on the prime columns PC and prime rows PR detected previously, we construct an *alignment graph*, which is defined as an undirected graph $\mathcal{G} = (C \cup PC, E_{cp} \cup E_{cc})$ with a set C of *cell nodes* and a set PC of *prime-column nodes*. There are two kinds of edges between the nodes: (i) E_{cp} contains a set of **c2p edges** that connect all cell nodes in C and all prime-column nodes in PC; (ii) E_{cc} contains a set of **c2c edges**, where each edge $e_{cc} = (c_i, c_j)$ indicates that cells c_i and c_j are adjacent horizontally in the same prime row. We consider a c2p edge $e_{cp} = (c_i, pc_j)$ is pre-determined if cell c_i is already in prime column pc_j; otherwise, the edge e_{cp} is called undetermined. Figure 1(b) shows an example alignment graph for the table in Fig. 1(a).

Alignment Graph Representation. Given the alignment graph $\mathcal{G}$, our task is to perform a classification on each c2p edge (c_i, pc_j) to predict whether c_i has column relation with all cells in pc_j. To support the classification task, we first present our design for node/edge features in the alignment graph $\mathcal{G}$.

(1) Features of Prime-Column Nodes: Intuitively, we would like to include the *horizontal location* information of cells and prime columns as their features. The

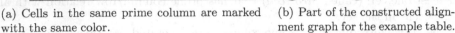

	pc_5		pc_3 pc_1		pc_4 pc_2	
pr_1	c_1 System		c_6 Dev		c_9 Test	
pr_2			c_7 EM c_8 F1		c_{10} EM c_{11} F1	
pr_3	c_2 Top Leaderboard Systems (Dec 10th, 2018)					
pr_4	c_3 Human		86.3 89.0		86.9	89.5
pr_5	c_4 #1 Single - MIR-MRC (F-Net)				74.8	78.0
pr_6	c_5 #2 Single - nlnet				74.2	77.1

(a) Cells in the same prime column are marked with the same color.

(b) Part of the constructed alignment graph for the example table.

Fig. 1. Generated prime columns and rows as well as the constructed alignment graph of the example table.

idea is that, a cell c_i that is very close to a prime column pc_j in the horizontal direction is more likely to have relation with pc_j, comparing to a prime column pc_k that is far from c_i. A straightforward way is to define the closeness on absolute coordinates. However, this may not be effective as tables may have various layouts and sizes. Instead, we use *relative orderings* of prime columns to represent their location features. Specifically, we use $pc_j.x_1$ and $pc_j.x_2$ to denote the x-coordinates of prime column pc_j, which are computed by averaging x-coordinates of cells in pc_j. Then, we sort the prime columns in ascending order of $pc_j.x_1$ ($pc_j.x_2$) and use r_{j1} (r_{j2}) to denote the *rank* of pc_j in the obtained ordering, and present feature vector $\vec{pc}_j$ for prime column pc_j as

$$\vec{pc}_j = (r_{j1}, r_{j2}) \tag{1}$$

(2) Features of Cell Nodes: We also utilize horizontal location information as features to represent cell nodes. We use "range" of cells' prime columns' ranks (because of the existence of spanning cells that may belong to multiple prime columns) to represent the horizontal location. Formally, we use PC_{c_i} to represent the prime columns that cell c_i belongs to, and represent feature vector $\vec{c}_i$ of c_i as

$$\vec{c}_i = (\min_{pc_k \in PC_{c_i}} \{r_{k1}\}, \max_{pc_k \in PC_{c_i}} \{r_{k2}\}) \tag{2}$$

For example, as cell c_2 in Fig. 1(a) belongs to 4 prime columns with min and max rank $1, 4$. Thus, we obtain its feature vector $\vec{c}_2 = (1, 4)$.

(3) Initial c2p Edge Features: We denote the initial features of c2p edge e_{ij} as a d-dimensional vector $\vec{e}_{ij}^{(0)}$. We consider the type of edge e_{ij}, i.e., pre-determined or undetermined, which is denoted as $e_{ij}.\texttt{type}$. Based on the type, we obtain the initial feature vector $\vec{e}_{ij}^{(0)}$ as,

$$\vec{e}_{ij}^{(0)} = \texttt{M}_1(e_{ij}.\texttt{type}) \tag{3}$$

where $\texttt{M}_1$ is a learnable embedding function that maps from the edge type code to a d-dimensional vector.

Message Passing Framework. We utilize a message passing framework to iteratively update c2p edge features, so as to incorporate more information of neighbors. In the t-th iteration, we use the current edge feature vector $\vec{e}_{ij}^{(t)}$ to compute $\vec{e}_{ij}^{(t+1)}$ by considering the following three components.

(1) Intrinsic Features of Edge: The first component considers intrinsic features of edge e_{ij}, i.e., whether locations of its corresponding cell node c_i and prime-column node pc_j are close. Intuitively, the smaller the difference of the corresponding ranks between c_i and pc_j is, the more likely c_i has relation with pc_j.

However, the above intuition may not be true for spanning cells. For instance, cell c_2 has relation with pc_5, although their feature vectors have significant differences, i.e., $(1,4)$ vs. $(1,1)$. To alleviate this issue, we further consider the number of cells in the corresponding prime row. Let $|PC|$ denote the number of prime columns in the table, and $|pr_{c_i}|$ be the number of cells in the prime row pr_{c_i} that cell c_i belongs to. We take the difference between $|PC|$ and $|pr_{c_i}|$ as a "compensation". The idea is that, when this difference is significant, it implies $|pr_{c_i}|$ is small, which indicates spanning cells may be present. As such, large rank differences may be acceptable. To formalize the above ideas, we introduce the intrinsic feature of a c2p edge $e_{ij} = (c_i, pc_j)$ as

$$
\begin{aligned}
f_{ij}^{\mathrm{I}} = \mathtt{F}_1 \{ (\mathtt{M}_2(\vec{c}_i.r_{i1}) - \mathtt{M}_2(\vec{pc}_j.r_{j1})) \oplus (\mathtt{M}_2(\vec{c}_i.r_{i2}) - \mathtt{M}_2(\vec{pc}_j.r_{j2})) \\
\oplus (\mathtt{M}_3(|PC|) - \mathtt{M}_3(|pr_{c_i}|) \}
\end{aligned}
\tag{4}
$$

where $\mathtt{M}_2$ and $\mathtt{M}_3$ are embedding functions like $\mathtt{M}_1$ but with different learnable parameters, $\oplus$ is the concatenation operation, and $\mathtt{F}_1$ is a fully connected feed forward network (FFN) consisting of two linear layers followed by $\mathtt{ReLU}$ [6] as activation function. The output dimension of $\mathtt{F}_1$ is d, i.e., the same as $\vec{e}_{ij}^{(t)}$.

(2) Neighboring Features of Edge: Given edge $e_{ij} = (c_i, pc_j)$, the second component considers whether neighboring cells of c_i is close to prime-column pc_j. Note that "neighboring cells" of c_i are the cells having c2c edges with c_i, e.g., c_1 and c_9 are neighbors of c_6. The intuition is that, given pc_j is close to c_i and its neighbors, if neighbors of c_i have less evidence to have relation with pc_j, then c_i would be more likely to have relation with pc_j. Formally, we first define an aggregated function for e_{ij} as

$$
\phi_{ij} = \mathtt{M}_4(\mathbf{abs}(\vec{c}_i.r_{i1} - \vec{pc}_j.r_{j1})) \oplus \mathtt{M}_4(\mathbf{abs}(\vec{c}_i.r_{i2} - \vec{pc}_j.r_{j2})) \oplus \vec{e}_{ij}^{(t)}
\tag{5}
$$

where $\mathtt{M}_4$ is an embedding functions and $\mathbf{abs}$ is the absolute value function. Based on the aggregated function, we consider the neighbor set $\mathcal{N}_i$ of cell c_i and compute the neighboring edge features as

$$
f_{ij}^{\mathrm{N}} = \sum_{u \in \mathcal{N}_i} \mathtt{F}_2(\phi_{ij} \oplus \phi_{uj})
\tag{6}
$$

where $\mathtt{F}_2$ is another FFN that has the same structure with $\mathtt{F}_1$.

(3) Row Features of Cell: We also find the difference of cell numbers across prime rows is an important indicator of the column structure of the table. Suppose a

table has 4 prime rows, and the number of cells in each row is $\{2,4,4,4\}$. It is very likely that the first prime row contains column-spanning cells that occupy multiple columns. Based on this, we design the third component to capture row features of cell c_i. More specifically, we consider all prime rows in the generated PR and arrange the cell numbers in these prime rows, i.e., $\{|pr_1|, |pr_2|, \ldots, |pr_{|PR|}|\}$, as a sequence. Then, we use the sequence to train a Bi-directional GRU [2] to capture the differences among cell numbers across prime rows. Finally, we obtain row feature of cell c_i as $f_{ij}^R = h_{pr_{c_i}}$, where $h_{pr_{c_i}}$ is the hidden state of our GRU model corresponding to the prime row that c_i belongs to.

Overall, by considering all the three components mentioned above, we introduce the message passing equation as below.

$$\vec{e}_{ij}^{(t+1)} = \vec{e}_{ij}^{(t)} + f_{ij}^I + f_{ij}^N + f_{ij}^R \tag{7}$$

Relation Determination via Classification. Based on our message passing framework, we obtain $\vec{e}_{ij}^{(T)}$ after T iterations over the alignment graph $\mathcal{G}$. Then, we feed $\vec{e}_{ij}^{(T)}$ to a Multilayer Perceptron (MLP) to obtain a probability distribution over 2 classes. According to the final prediction result of each c2p edge $e_{ij} = (c_i, pc_j)$, we determine the column relations as follows. If the predicted result is 1, we determine that c_i has column relations with all cells in prime column pc_j. Otherwise, we remove the column relations between c_i and pc_j. Based on these, we reconstruct the columns of the table.

3 Experiments

3.1 Experimental Setup

Datasets. We use three datasets. ICDAR-2013 [3] contains 156 tables with manual annotations. SciTSR [1] is constructed automatically. We filter out some erroneous samples and finally it contains 10961 and 275 tables for training and testing. SciTSR-COMP is a subset of SciTSR and only contains tables with spanning cells. It has 635 tables for testing after the filtering process.

Baselines. We compare with three state-of-the-art approaches. GraphTSR [1] utilizes graph attention network to predict relations between cells and their k nearest neighbors. GFTE [5] utilizes GCN to retrieve relations between cells and their k nearest neighbors. Apart from cell position features, it also takes extra features such as image feature and textual feature. DGCNN* [9] combines position feature extracted by DGCNN [11] and image feature to classify relations between all cell pairs. For fair comparison, we only use cell position as input feature in GFTE and DGCNN*.

3.2 Evaluation

The graph-based alignment model is trained on SciTSR training set, and tested on SciTSR testing set, SciTSR-COMP and ICDAR-2013. In the rest of this section, we focus on the results of column relation alignment. This is because (1) Most of the spanning cells are column-spanning cells; (2) We find there are many mislabeled row relations in the dataset, largely because many cells with muti-line text are wrongly split into several vertically adjacent cells.

Table 1. Overall comparison results

Method	SciTSR-test			SciTSR-COMP			ICDAR-2013		
	Precision	Recall	F1	Precision	Recall	F1	Precision	Recall	F1
GraphTSR	**0.994**	0.982	0.988	**0.989**	0.949	**0.969**	0.910	0.779	0.839
DGCNN*	0.455	0.909	0.606	0.405	0.863	0.552	0.420	0.641	0.507
GFTE	0.723	0.778	0.750	0.710	0.397	0.509	0.791	0.826	0.808
Ours	0.993	**0.990**	**0.992**	0.976	**0.961**	**0.969**	**0.952**	**0.970**	**0.961**

We compare the overall results of our framework to the state-of-the-art approaches. Since GraphTSR and GFTE only detect the relations between cells and their k nearest neighbours, we adopt the same strategy used in [3]. Specifically, we only consider the relations between cells and their adjacent neighbors and evaluate the approaches in terms of precision, recall and F1 only for these adjacent relations.

The results is shown in Table 1. We can see that we achieve the best F1 score over all datasets. Although the precision score of our method is slightly lower than that of GraphTSR on the SciTSR and SciTSR-COMP datasets, we achieve higher recall on all the datasets. Notice that our method greatly outperforms all other methods on ICDAR-2013 dataset. This is partly because our alignment model takes cells' ranks rather than the bounding box coordinates as input. Thus, our approach is more robust to the absolute locations of cells and achieve better generalization ability.

3.3 Ablation Study

We perform ablation study to investigate the effectiveness of the proposed three components in message passing framework. We believe the first component f^I which considers the location difference between cells and prime columns is an essential part, for model to correctly recognize the relations, and thus we take f^I as the base model. We subsequently incorporate the second component f^N and the third component f^R, and evaluate their effectiveness.

Experiments are conducted on SciTSR-COMP dataset, which is the most challenging dataset containing many tables with complex structure. As discussed previously, due to the spanning cells, phase one may miss many relations. In particular, according to the statistics, there are 54326 relations that are not found in phase one. When comparing different models, we evaluate how many missing relations are predicted correctly and wrongly. The results are reported in Table 2. For ease of comparison, we also present the recall (i.e., the proportion of the correct relations w.r.t. all the missing relations) and the error rate (i.e., the proportion of the wrong relations w.r.t. all found relations).

Table 2. Additional relations found in phase two.

Model	#correct	#wrong	Recall (%)	Error (%)
$f^I(a)$	26900	4322	49.52	**13.84**
$+f^N(b)$	31956	6012	58.82	15.83
$+f^N+f^R(c)$	32216	5258	**59.30**	14.03

From the results, it can be observed that: (1) With phase two, we can correctly retrieve many more additional relations. Even the base model achieves recall 49.52%. (2) Comparing (a) and (b), we can conclude that propagating information between neighboring cells is very helpful in discovering more indirect relations. With f^N, we manage to increase the recall to 58.82% with slightly higher error rate. (3) When combining all three components, we find more correct relations with the best recall 59.30%. Meanwhile, with the help of the GRU model, we reduce the number of wrong relations and the error rate.

4   Conclusion

In this work, we propose a novel two-phase approach to tackle table structure recognition problem. We introduce a notion called prime relation and propose a graph-based alignment model to efficiently detect the complex table structure. Our approach is proved to be effective and robust on real-world datasets.

Acknowledgement. This work is supported by NSF of China (62072461).

References

1. Chi, Z., Huang, H., Xu, H.D., Yu, H., Yin, W., Mao, X.L.: Complicated table structure recognition. arXiv preprint arXiv:1908.04729 (2019)
2. Chung, J., Gulcehre, C., Cho, K., Bengio, Y.: Empirical evaluation of gated recurrent neural networks on sequence modeling. arXiv preprint arXiv:1412.3555 (2014)
3. Göbel, M., Hassan, T., Oro, E., Orsi, G.: ICDAR 2013 table competition. In: ICDAR, pp. 1449–1453 (2013)
4. Kieninger, T., Dengel, A.: The T-Recs table recognition and analysis system. In: DAS, pp. 255–270 (1998)
5. Li, Y., Huang, Z., Yan, J., Zhou, Y., Ye, F., Liu, X.: GFTE: graph-based financial table extraction. In: ICPR Workshops, pp. 644–658 (2020)
6. Nair, V., Hinton, G.E.: Rectified linear units improve restricted boltzmann machines. In: ICML, pp. 807–814 (2010)

7. Oro, E., Ruffolo, M.: PDF-TREX: an approach for recognizing and extracting tables from PDF documents. In: ICDAR, pp. 906–910 (2009)
8. Prasad, D., Gadpal, A., Kapadni, K., Visave, M., Sultanpure, K.: CascadeTabNet: an approach for end to end table detection and structure recognition from image-based documents. In: CVPR Workshops, pp. 572–573 (2020)
9. Qasim, S.R., Mahmood, H., Shafait, F.: Rethinking table recognition using graph neural networks. In: ICDAR, pp. 142–147 (2019)
10. Shigarov, A., Mikhailov, A., Altaev, A.: Configurable table structure recognition in untagged pdf documents. In: DocEng, pp. 119–122 (2016)
11. Wang, Y., Sun, Y., Liu, Z., Sarma, S.E., Bronstein, M.M., Solomon, J.M.: Dynamic graph CNN for learning on point clouds. ACM TOG **38**(5), 146:1–146:12 (2019)

Towards Unification of Statistical Reasoning, OLAP and Association Rule Mining: Semantics and Pragmatics

Rahul Sharma[1]([✉])(ID), Minakshi Kaushik[1](ID), Sijo Arakkal Peious[1](ID), Mahtab Shahin[1](ID), Amrendra Singh Yadav[2](ID), and Dirk Draheim[1](ID)

[1] Information Systems Group, Tallinn University of Technology, Tallinn, Estonia
{rahul.sharma,minakshi.kaushik,sijo.arakkal,mahtab.shahin,
dirk.draheim}@taltech.ee
[2] Vellore Institute of Technology - VIT Bhopal, Bhopal, India

Abstract. Over the last decades, various decision support technologies have gained massive ground in practice and theory. Out of these technologies, statistical reasoning was used widely to elucidate insights from data. Later, we have seen the emergence of online analytical processing (OLAP) and association rule mining, which both come with specific rationales and objectives. Unfortunately, both OLAP and association rule mining have been introduced with their own specific formalizations and terminologies. This made and makes it always hard to reuse results from one domain in another. In particular, it is not always easy to see the potential of statistical results in OLAP and association rule mining application scenarios. This paper aims to bridge the artificial gaps between the three decision support techniques, i.e., statistical reasoning, OLAP, and association rule mining and contribute by elaborating the semantic correspondences between their foundations, i.e., probability theory, relational algebra, and the itemset apparatus. Based on the semantic correspondences, we provide that the unification of these techniques can serve as a foundation for designing next-generation multi-paradigm data mining tools.

Keywords: Data mining · Association rule mining · Online analytical processing · Statistical reasoning

1 Introduction

Nowadays, decision-makers and organizations are using a variety of modern and old decision support techniques (DSTs) with their specific features and limited scope of work. However, in the era of big data and data science, the huge volume and variety of data generated by billions of internet devices demand advanced

---

This work has been partially conducted in the project "ICT programme" which was supported by the European Union through the European Social Fund.

DSTs that can handle a variety of decision support tasks. Currently, no single DST can fulfill this demand. Therefore, to provide advanced decision support capabilities, this paper contributes by elaborating the semantic correspondences between the three popular DSTs, i.e., statistical reasoning (SR) [13], online analytical processing (OLAP) [3] and association rule mining (ARM) [1,11]. These correspondences between SR, ARM and OLAP, and vice versa, appear to be easy, but none of these have been implemented in practice, nor they have been discussed in the state of the art. However, substantial research has been done over the years to enhance OLAP, data warehousing, and data mining approaches [7]. In particular, in data mining, Kamber et al. [8], Surjeet et al. [2] have presented different ways to integrate OLAP and ARM together. Later, Han et al. [5] have proposed DBMiner for interactive mining. In the state of the art, the adoption of concepts in between OLAP and ARM (and vice versa) are referred to as automatic OLAP [14] and multi-dimensional ARM [8]. We appraise all approaches for the integration of the OLAP and ARM. However, the concept of semantic correspondences between DSTs is yet to be elaborated in the state-of-the-art. To establish semantic correspondences between the three DSTs, we use probability theory and conditional expected values (CEVs) at the center of our considerations. CEVs correspond to *sliced average aggregates* in OLAP and would correspond to potential *ratio-scale confidences* in a generalized ARM [4]. Elaborating these concepts between DSTs will enable decision-makers to work with cross-platform decision support tools [6,10] and check their results from different viewpoints.

The paper is structured as follows: In Sect. 2, we elaborate semantic mapping between the SR and ARM, i.e., between probability theory and itemset apparatus. In Sect. 3, we discuss the semantic mapping between the SR and OLAP, i.e., between probability theory and relational algebra. Conclusion is given in Sect. 4.

2   Semantic Mapping Between SR and ARM

We stick to the original ARM concepts and notation provided by Agrawal et al. [1]. However, ARM is also presented for numerical data items as quantitative ARM [12], numerical ARM [9].

In classical ARM, first, there is a *whole itemset* $\mathfrak{I} = \{I_1, I_2, \ldots, I_n\}$ consisting of a *total number* n of items $I_1, I_2, \ldots, I_n$. A subset $X \subseteq \mathfrak{I}$ of the whole itemset is called an *itemset*. We then introduce the concept of a *set of transactions* T (*that fits the itemset* $\mathfrak{I}$) as a relation as follows:

$$T \subseteq TID \times \underbrace{\{0,1\} \times \cdots \times \{0,1\}}_{n-\text{times}} \tag{1}$$

Here, TID is a finite set of transaction identifiers. For the sake of simplicity, we assume that it has the form $TID = \{1, \ldots, N\}$. In fact, we must impose a uniqueness constraint on TID, i.e., we require that T is right-unique, i.e., a function given as,

$$T \in TID \longrightarrow \underbrace{\{0,1\} \times \cdots \times \{0,1\}}_{n-\text{times}} \qquad (2)$$

Given (2), we have that N in $TID = \{1, \ldots, N\}$ equals the size of T, i.e., $N = |T|$. Henceforth, we refer to T interchangeably both as a relation and as a function, according to (1) resp. (2). For example, we use $t = \langle i, i_1, \ldots i_n \rangle$ to denote an arbitrary transaction $t \in T$; similarly, we use $T(i)$ to denote the i-th transaction of T more explicitly etc. Given this formalization of the transaction set T, it is correct to say that T is a binary relation between TID and the whole itemset. In that, $I_1, I_2, \ldots, I_n$ need to be thought of as column labels, i.e., there is exactly one bitmap column for each of the n items in $\mathfrak{I}$, compare with (1) and (2). Similarly, Agrawal et al. have called the single transaction a bit vector and introduced the notation $t[k]$ for selecting the value of the transaction t in the k-th column of the bitmap table (in counting the columns of the bitmap table, the TID column is omitted, as it merely serves the purpose of providing transaction identities), i.e., given a transaction $\langle tid, i_1, \ldots i_n \rangle \in T$, we define $\langle tid, i_1, \ldots i_n \rangle[k] = i_k$. Less explicit, with the help of the usual tuple projection notation π_j, we can define $t[k] = \pi_{k+1}(t)$. Let us call a pair $\langle \mathfrak{I}, T \rangle$ of a whole itemset $\mathfrak{I}$ and a set of transaction T that fits $\mathfrak{I}$ as described above an ARM $frame$. Henceforth, we assume an ARM frame $\langle \mathfrak{I}, T \rangle$ as given.

A transaction, as previously stated, is a bit vector. For the sake of simplicity, Let's start with some notation that makes it possible to treat a transaction as an itemset. Given a transaction $t \in T$ we denote the set of all $items$ $that$ $occur$ in t as $\{t\}$ and we define it as follows:

$$\{t\} = \{I_k \in \mathfrak{I} \mid t[k] = 1\} \qquad (3)$$

The $\{t\}$ notation provided by (3) will prove helpful later because it allows us to express transaction properties without having to use bit-vector notation, i.e., without having to keep track of item numbers k of items I_k.

Given an $I_j \in \mathfrak{I}$ and a transaction $t \in T$, Agrawal says [1] that I_j is $bought$ by t if and only if $t[j] = 1$. Similarly, we can say that t $contains$ I_j in such case. Next, given an itemset $X \subseteq \mathfrak{I}$ and a transaction $t \in T$, Agrawal says that t $satisfies$ X if and only if $t[j] = 1$ for all $I_j \in X$. Similarly, we can say that t $contains$ all of the $items$ of X in such case. Next, we can see that t satisfies X if and only if $X \subseteq \{t\}$. Henceforth, we use $X \subseteq \{t\}$ to denote that t satisfies X.

Given an itemset $X \subseteq \mathfrak{I}$, the relative number of all transactions that satisfy X is called the $support$ of X and is denoted as $Supp(X)$, i.e., we define:

$$Supp(X) = \frac{|\{t \in T \mid X \subseteq \{t\}\}|}{|T|} \qquad (4)$$

It's perfectly reasonable to discuss an itemset's support once more. X as the relative number of all transactions that each contain all of the items of X.

An ordered pair of itemsets $X \subseteq \mathfrak{I}$ and $Y \subseteq \mathfrak{I}$ is called an $association$ $rule$, and is denoted by $X \Rightarrow Y$. Now, the relative number of all transactions that satisfy Y among all of those transactions that satisfy X is called the *confidence of* $X \Rightarrow Y$, and is denoted as $Conf(X \Rightarrow Y)$, i.e., we define:

$$Conf(X \Rightarrow Y) = \frac{|\{t \in T \mid Y \subseteq \{t\} \wedge X \subseteq \{t\}\}|}{|\{t \in T \mid X \subseteq \{t\}\}|} \tag{5}$$

Usually, the confidence of an association rule is introduced via supports of itemsets as follows:

$$Conf(X \Rightarrow Y) = \frac{Supp(X \cup Y)}{Supp(X)} \tag{6}$$

It can easily be checked that (5) and (6) are equivalent.

2.1 Semantic Mapping Between Association Rule Mining and SR (Probability Theory)

Here, we compare probability theory to the concepts defined in ARM. Given an ARM frame $F = \langle \mathfrak{I}, T \rangle$. next we map the concepts defined in ARM to probability space $(\Omega_F, \Sigma_F, \mathsf{P}_F)$. First, we define the set of outcomes Ω_F to be the set of transactions T. Next, we define Σ_F to be the power set of Ω_F. Finally, given an event $X \in \Sigma_F$, we define the probability of X as the relative size of X, as follows:

$$\Omega_F = T \tag{7}$$
$$\Sigma_F = \mathbb{P}(T) \tag{8}$$
$$\mathsf{P}_F(X) = \frac{|X|}{|T|} \tag{9}$$

In the sequel, we drop the indices from Ω_F, Σ_F, and P_F, i.e., we simply use Ω, Σ, and P to denote them, but always keep in mind that we actually provide a mapping from ARM frames F to corresponding probability spaces $(\Omega_F, \Sigma_F, \mathsf{P}_F)$. The idea is simple. Each transaction is modeled as an outcome and, as usual, also a basic event. Furthermore, each set of transactions is an event.

We step forward with item and itemsets. For each item $I \in \mathfrak{I}$ we introduce the *event that item I is contained in a transaction*, and we denote that event as $[\![I]\!]$. Next, for each itemset $X \subseteq \mathfrak{I}$, we introduce the *event that all of the items in X are contained in a transaction* and we denote that event as $[\![X]\!]$. We define:

$$[\![I]\!] = \{t \mid I \in \{t\}\} \tag{10}$$
$$[\![X]\!] = \bigcap_{I \in X} [\![I]\!] \tag{11}$$

As usual, we identify an event $[\![I]\!]$ with the characteristic random variable $[\![I]\!] : \Omega \longrightarrow \{0,1\}$ and use $\mathsf{P}([\![I]\!])$ and $\mathsf{P}([\![I]\!]{=}1)$ as interchangeable.

2.2 Formal Mapping of ARM Support and Confidence to Probability Theory

Based on the mapping provided by (7) through (11), we can see how ARM *Support* and *Confidence* translate into probability theory.

Lemma 1 (Mapping ARM *Support* to Probability Theory) *Given an itemset $X \subseteq \mathfrak{I}$, we have that:*

$$Supp(X) = \mathsf{P}(\llbracket X \rrbracket) \tag{12}$$

Proof. According to (11), we have that $\mathsf{P}(\llbracket X \rrbracket)$ equals

$$\mathsf{P}(\underset{I \in X}{\cap} \llbracket I \rrbracket) \tag{13}$$

Due to (10), we have that (13) equals

$$\mathsf{P}\left(\underset{I \in X}{\cap} \{t \in T \mid I \in \{t\}\}\right) \tag{14}$$

We have that (14) equals

$$\mathsf{P}(\{t \in T \mid \underset{I \in X}{\wedge} I \in \{t\}\}) \tag{15}$$

We have that (15) equals

$$\mathsf{P}(\{t \in T \mid X \subseteq \{t\}\}) \tag{16}$$

According to (9), we have that (16) equals

$$\frac{|\{t \in T \mid X \subseteq \{t\}\}|}{|T|} \tag{17}$$

According to (4), we have that (17) equals $Supp(X)$ □

Lemma 2 (Mapping ARM *Confidence* to Probability Theory) *Given an itemset $X \subseteq \mathfrak{I}$, we have that:*

$$Conf(X \Rightarrow Y) = \mathsf{P}(\llbracket Y \rrbracket \mid \llbracket X \rrbracket) \tag{18}$$

Proof. Omitted.

With these mappings, we provide that a set of items in ARM $\mathfrak{I} = \{I_1, I_2, \ldots, I_m\}$ are equivalent to the set of events $\mathfrak{I} = \{I_1 \subseteq \Omega, \ldots, I_m \subseteq \Omega\}$ in probability theory. Transactions T in ARM are equivalent to the set of outcomes Ω in probability space $(\Omega, \Sigma, \mathsf{P})$. Support of an itemset X in ARM is equivalent to the relative probability of the itemset X. Confidence of an association rule $X \Rightarrow Y$ is equivalent to the conditional probability of Y in the presence of X.

3   Semantic Mapping Between SR and OLAP

As per our findings, conditional operations on bitmap (Binary) columns correspond to conditional probabilities, whereas conditional operations on numerical columns correspond to conditional expected values, e.g., we model a sample OLAP Table 1 in probability theory. We consider that Table 1 is equivalent to the set of outcomes Ω in probability space (Ω, Σ, P), a row r is an element of Ω, i.e. $r \in \Omega$ and each column c is equivalent to a random variable $\mathbb{R}$. We consider numerical columns as *finite real-valued* random variables (For Example: Salary $\in \Omega \subseteq \mathbb{R}$) and bitmap columns are considered as events (For Example: Freelancer $\subseteq \Omega$). The following is a probabilistic interpretation of the OLAP Table 1.

Table 1. A sample OLAP table.

City	Profession	Education	Age group	Freelancer	Salary
New York	Lawyer	Master	25–30	0	3.800
Seattle	IT	Bachelor	18–25	1	4.200
Boston	Lawyer	PhD	40–50	1	12.700
L.A	Chef	High School	30–40	0	3.700
...	...	...	...	...	...

3.1   Semantic Mapping Between OLAP Averages and SR

Generally, decision-makers use SQL queries to interact with OLAP [3]. Therefore, we use OLAP queries to be mapped with SR, i.e., probability theory. We have a simple OLAP average query; (SELECT AVG(Salary) FROM Table 1). If the number of rows of Table 1 is represented by $|\Omega|$ and the number of rows that contain a value i in column C are equivalent to $\#_C(i)$ then AVG(Salary) FROM Table 1 will compute the average of all the salaries, i.e., a fraction of the sum of the column (Salary) and the total number of rows in the table. In probability theory, the *average* of a random variable X is the *Expected Value* of $X = \mathsf{E}[X]$. We compare the expected value of X, i.e., $\mathsf{E}(X)$ with the output of the AVG query in OLAP. We have:

$$OLAP - Query \ (SELECT \ AVG \ (Salary) \ FROM \ Table \ 1) \tag{19}$$

$$\text{Expected Value: } \mathsf{E}(\text{Salary}) = \sum_{i \in I_{\text{Salary}}} i \cdot \mathsf{P}(\text{Salary} = i) \tag{20}$$

$$= \sum_{i \in I_{\text{Salary}}} i \cdot \frac{\#_{\text{Salary}}(l)}{|\Omega|} = \frac{\sum_{r \in \Omega} \text{Salary}(r)}{|\Omega|} \tag{21}$$

As per Eq. 20 and Eq. 21, the average of a random variable X in probability theory and simple averages of an OLAP query provides the same outcome. Hence, we say that an average query in OLAP corresponds to expected values in probability theory. The conditional average queries in OLAP calculate averages of a column with a WHERE clause. For example, we have an average SQL query with some conditions where the target column is numerical and conditional variables have arbitrary values. We have: SELECT AVG(Salary) FROM Table 1 WHERE City $= Seattle$ AND Profession$=IT$;. In probability theory, we compute the conditional average of a random number using its conditional expectation. Therefore, the conditional expectation of a random number Y with condition X is given as:

$$E(Y|X) = \sum_{n=0}^{\infty} i_n \cdot P(Y = i_n | X) \tag{22}$$

$$f(i) = E(Y = i_n | X) \tag{23}$$

Here, the value $E(Y = i_n | X)$ is dependent on the value of i. Therefore, we say that $E(Y = i_n | X)$ is a function of i, which is given in Eq. 23. We compare the conditional expected value of $E(Y = i_n | X)$ with the output of the conditional AVG query in OLAP. We have:

$$OLAP\ Query: SELECT\ AVG(Salary)\ FROM\ Table\ 1$$
$$WHERE\ City = Seattle\ AND\ Profession = IT; \quad (24)$$

$$\text{Conditional Expected Value: } E(Salary \mid City = Seattle \cap Profession = IT) \quad (25)$$

$$E(Y|X) = \sum_{i \in I_C} i \cdot P(Y = i \mid X) \tag{26}$$

As per Eq. 25 and Eq. 26, the average of a random variable Y with condition X (Conditional Expected values) and the conditional average of an OLAP query provides the same outcome. Hence, we can say that a conditional average query in OLAP corresponds to the conditional expected values in probability theory. Based on these mappings in OLAP, conditional averages on binary columns correspond to conditional probability and they also correspond to confidence in ARM.

4   Conclusion

In this paper, we elaborated semantic correspondences between the three DSTs, i.e., SR, OLAP and ARM. We identify that SR, OLAP, and ARM operations complement each other in data understanding, visualization, and making individualized decisions. In the proposed mappings, it is identified that OLAP and ARM have common statistical reasoning, exploratory data analysis methods and offer similar solutions for decision support problems. Based on these findings, we can review current obstacles in each of SR, OLAP and ARM. Furthermore, the semantic correspondences between the three DSTs will be helpful in designing certain next-generation hybrid decision support tools.

References

1. Agrawal, R., Imieliński, T., Swami, A.: Mining association rules between sets of items in large databases. ACM SIGMOD Rec. **22**(2), 207–216 (1993). https://doi.org/10.1145/170036.170072
2. Chaudhuri, S., Dayal, U.: Data warehousing and olap for decision support. In: Proceedings of the 1997 ACM SIGMOD International Conference on Management of Data, SIGMOD 1997, pp. 507–508. Association for Computing Machinery, New York (1997). https://doi.org/10.1145/253260.253373
3. Codd, E.F.: Providing olap (on-line analytical processing) to user-analysts: An it mandate. Available from Arbor Software's web site-http://www.arborsoft.com/papers/coddTOC.html (1993)
4. Hartmann, S., Küng, J., Chakravarthy, S., Anderst-Kotsis, G., Tjoa, A.M., Khalil, I. (eds.): DEXA 2019. LNCS, vol. 11706. Springer, Cham (2019). https://doi.org/10.1007/978-3-030-27615-7
5. Han, J., Fu, Y., Wang, W., Chiang, J., Zaïane, O.R., Koperski, K.: DBMiner: interactive mining of multiple-level knowledge in relational databases. In: Proceedings of SIGMOD'96 - the 1996 ACM SIGMOD International Conference on Management of Data, p. 550. Association for Computing Machinery (1996). https://doi.org/10.1145/233269.280356
6. Heinrichs, J.H., Lim, J.S.: Integrating web-based data mining tools with business models for knowledge management. Decis. Support Syst. **35**(1), 103–112 (2003). https://doi.org/10.1016/S0167-9236(02)00098-2
7. Imieliński, T., Khachiyan, L., Abdulghani, A.: Cubegrades: generalizing association rules. Data Min. Knowl. Disc. **6**(3), 219–257 (2002)
8. Kamber, M., Han, J., Chiang, J.: Metarule-guided mining of multi-dimensional association rules using data cubes. In: Proceedings of VLDB'1994 - the 20th International Conference on Very Large Data Bases, KDD 1997, pp. 207–210. AAAI Press (1997)
9. Kaushik, M., Sharma, R., Peious, S.A., Shahin, M., Yahia, S.B., Draheim, D.: A systematic assessment of numerical association rule mining methods. SN Comput. Sci. **2**(5), 1–13 (2021)
10. Arakkal Peious, S., Sharma, R., Kaushik, M., Shah, S.A., Yahia, S.B.: Grand reports: a tool for generalizing association rule mining to numeric target values. In: Song, M., Song, I.-Y., Kotsis, G., Tjoa, A.M., Khalil, I. (eds.) DaWaK 2020. LNCS, vol. 12393, pp. 28–37. Springer, Cham (2020). https://doi.org/10.1007/978-3-030-59065-9_3
11. Sharma, R., Kaushik, M., Peious, S.A., Yahia, S.B., Draheim, D.: Expected vs. unexpected: selecting right measures of interestingness. In: Song, M., Song, I.-Y., Kotsis, G., Tjoa, A.M., Khalil, I. (eds.) DaWaK 2020. LNCS, vol. 12393, pp. 38–47. Springer, Cham (2020). https://doi.org/10.1007/978-3-030-59065-9_4
12. Srikant, R., Agrawal, R.: Mining quantitative association rules in large relational tables. SIGMOD Rec. **25**(2), 1–12 (1996)
13. Stigler, S.M.: The History of Statistics: The Measurement of Uncertainty Before 1900. Harvard University Press (1986)
14. Zhu, H.: On-line analytical mining of association rules. In: Master's thesis. Simon Fraser University, Burnaby, Brithish Columbia, Canada (1998)

A Dynamic Heterogeneous Graph Perception Network with Time-Based Mini-Batch for Information Diffusion Prediction

Wei Fan, Meng Liu, and Yong Liu[✉]

Heilongjiang University, Harbin, China
{2201792,2191438}@s.hlju.edu.cn, liuyong123456@hlju.edu.cn

Abstract. Information diffusion prediction is an important task to understand how information spreads among users. Most previous studies either only focused on the use of diffusion sequences, or only used social networks between users to make prediction, but such modeling is not sufficient to model the diffusion process. In this paper, we propose a novel Dynamic Heterogeneous Graph Perception Network with Time-Based Mini-Batch (DHGPNTM) that can combine dynamic diffusion graph and social graph for information diffusion prediction. First, we propose a Graph Perception Network (GPN) to learn user embedding in dynamic heterogeneous graphs, and combine temporal information with user embedding to capture users' dynamic preferences. Then we use a multi-head attention to generate users' context-dependence embedding, and design a fusion gate to selectively integrate users' dynamic preferences and context-dependence embedding. The extensive experiments on real datasets demonstrate the effectiveness and efficiency of our model.

Keywords: Social network · Information diffusion prediction · Graph Perception Network · Multi-head attention

1 Introduction

Analyzing information diffusion data to explore information diffusion mechanism has gradually become a hot research topic, which receive great attention in data mining. The information diffusion prediction task aims to study how information is transmitted between users and predict who will be infected in the future, which plays an important role in many applications, e.g., rumor detection [1], epidemiology [2], viral marketing [3], media advertising [4] and the spread of news and memes [5,6].

Traditional methods usually rely on explicit features, such as temporal information [7], user characteristics [8], content [9] and interaction between users [10]. Although they have significantly improved the performance of diffusion prediction, the feature engineering process requires much manual effort and extensive domain knowledge. Recent work attempts to use deep learning to address this

© The Author(s), under exclusive license to Springer Nature Switzerland AG 2022
A. Bhattacharya et al. (Eds.): DASFAA 2022, LNCS 13245, pp. 604–612, 2022.
https://doi.org/10.1007/978-3-031-00123-9_49

problem. Some researchers propose models based on diffusion sequences [11–14]. There are also some researchers explore the interaction of users in social networks [8,15,16]. They widely use graph neural networks to aggregate user embedding in social network for predicting information diffusion [17–21].

However, there are several challenges with the recent work. On one hand, the above works either only focus on the use of diffusion path, or only use social network, but do not consider two important factors. On the other hand, although graph neural networks have been used to model graph data, existing GNNs have still the problem of over smoothing.

In order to overcome the above challenges, we propose a novel $\underline{D}$ynamic $\underline{H}$eterogeneous $\underline{G}$raph $\underline{P}$erception $\underline{N}$etwork with $\underline{T}$ime-Based $\underline{M}$ini-Batch (DHG-PNTM) for Information Diffusion Prediction. First, we construct heterogeneous graphs composed of the social graph and dynamic diffusion graphs. Second, we propose a graph neural network model called Graph Perception Network (GPN) to capture user structural embedding. Then we embed the temporal information into the user embedding through the self-attention mechanism to learn users' dynamic preferences. Afterwards, a multi-head attention mechanism is used to capture users' context-dependence embedding. Finally, we design a fusion gate to selectively integrate users' dynamic preferences and context-dependence embedding. To further enhance the training speed of model, we also propose a time-based mini-batch method to construct mini-batch input for training model. The experimental results show that DHGPNTM significantly outperformes several baseline methods. **The source code can be found at** https://github.com/DHGPNTM/DHGPNTM.

2 Related Work

2.1 Diffusion Path Based Methods

Diffusion path based methods interpret the interpersonal influence based on observed diffusion sequences. In some work [11–13], deep learning is used to automatically learn the embedding of the diffusion path. For example, Topo-LSTM [11] constructs a dynamic directed acyclic graph (DAG) composed of diffusion paths, and uses topology-aware node embedding to extend the LSTM mechanism to learn the DAG structure. DeepDiffuse [13] uses the infection timestamp and attention mechanism to predict when and who will be infected based on the previously observed diffusion paths.

2.2 Social Graph Based Methods

Social graph based methods utilizes social networks to explain the interpersonal influence on diffusion prediction and improve prediction accuracy in some work [8,14–16]. For example, [14] uses the structural features between neighbors in social network, and uses a framework based on RNN to model the diffusion sequence. [16] proposes a multi-scale information diffusion prediction model. They use a combination of RNN and reinforcement learning to predict the next infected user and the total number of infected users.

3   Problem Definition

A social graph can be defined as $G = (V, E)$, where V and E represent the user set and the edge set, respectively. If there is a link between u and v, then $e_{u,v} = 1$. In addition, the diffusion of message x_i is recorded as $x_i = \{(v_{i,1}, t_{i,1}), (v_{i,2}, t_{i,2}), ..., (v_{i,N_c}, t_{i,N_c})\}$, where the element $(v_{i,j}, t_{i,j})$ indicates user $v_{i,j}$ was infected at time $t_{i,j}$ in the i-th cascade, and N_c indicates how many people have spread this message. The information diffusion prediction task aims to predict the next infected user v_{i,N_c+1}. That is to say, we need a model to learn the conditional probability $P(v_{i,N_c+1}|\{(v_{i,1}, t_{i,1}), (v_{i,2}, t_{i,2}), ..., (v_{i,N_c}, t_{i,N_c})\})$.

We use a diffusion cascade x_i to construct a diffusion graph $G_{x_i} = (V_{x_i}, E_{x_i})$, where $V_{x_i} \subset V$. If user u reposts or comments on message posted by user v, then $e_{v,u} = 1$. We use diffusion cascade set $X = \{x_1, x_2, ..., x_M\}$ to construct the diffusion graph set $G_X = \{G_{x_1}, G_{x_2}, ..., G_{x_M}\}$, where M is the total number of messages. Then we combine small diffusion graphs in G_X to construct a large diffusion graph G_M. Finally, we divide the diffusion graph G_M into n time intervals. Thus the diffusion graph G_M can be defined as $G_M = \{G_M^1, G_M^2, ..., G_M^n\}$, where each G_M^i denotes the dynamic diffusion graph in the i-th time interval. Please note that G_M^i also contains dynamic diffusion action before the i-th time interval. Thus we have $G_M^{i-1} \subset G_M^i$.

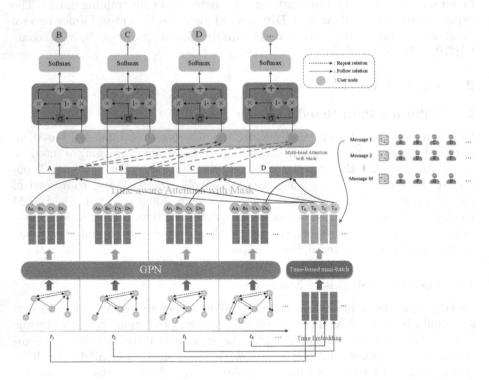

Fig. 1. The architecture of dynamic heterogeneous graph perception network with time-based mini-batch.

4   Method

The overview architecture of our model is shown in Fig. 1. First, we construct dynamic heterogeneous graphs based on a social graph and dynamic diffusion graphs. Second, we design a graph perception network (GPN) that can learn user embedding. Third, we construct a mini-batch input called time-based mini-batch and design a time-aware attention to learn users' dynamic preferences. Following this, the multi-head attention with mask is used to learn context dependence, and the fusion gate is used to fuse users' dynamic preferences and dependence-aware embedding to predict the user that will be infected.

4.1   Heterogeneous Graph Construction

We use a social graph $G = (V, E)$ and the dynamic diffusion graph $G_M^i = (V_M^i, E_M^i)$ in the i-th time interval to construct the dynamic heterogeneous graph $G_H^i = (V, E_H^i)$ in the i-th time interval, where $E_H^i = E \cup E_M^i$. Finally, the dynamic heterogeneous graph set is defined as $G_H = \{G_H^1, G_H^2, ..., G_H^n\}$, where each G_H^i is the heterogeneous graph in the i-th time interval.

4.2   Graph Perception Network (GPN)

Each GPNConv layer in GPN performs graph convolution operations and feature aggregation of neighbors, the embedding h_v^l of user v in the l-th layer is updated through the following formula:

$$e_{uv} = ReLU((f_{MLP}^l(h_v^{l-1}))^\mathsf{T} W_a^l f_{MLP}^l(h_u^{l-1})), \qquad (1)$$

$$\alpha_{uv} = Softmax(e_{uv}) = \frac{exp(e_{uv})}{\sum_{k \in N(v) \cup \{v\}} exp(e_{kv})}, \qquad (2)$$

$$h_v^l = ReLU(\sum_{u \in N(v) \cup \{v\}} \alpha_{uv} \cdot f_{MLP}^l(h_u^{l-1})), \qquad (3)$$

where f_{MLP}^l is a two-layer perception network in the l-th layer, $W_a^l \in R^{d \times d}$ is a learnable matrix parameter where d is the dimension size, and $N(v)$ is the neighbors of node v. The final output embedding of GPN can be defined as follows:

$$h_v = ReLU(f_P(h_v^0) + f_P(h_v^L)), \qquad (4)$$

where f_P is a single-layer perceptual network for residual connection.

4.3   User Dynamic Preferences Based on Mini-Batch

In order to speed up the training of our model, we propose a time-based mini-batch method. In a batch, we divide the sequences of all messages into multiple groups according to the length of $step\_len$. Then we use the latest timestamp in each group as the timestamp for all users in the same group. Through this Mini-Batch process, we obtain the converted timestamp of all users. Each converted timestamp will be assigned to one of n time intervals. Each time interval correponds to a temporal embedding. Finally, we can obtain the dynamic preferences $\widetilde{v}_{ij} \in R^d$ of user j in message i through the following formula:

$$\widetilde{v}_{ij} = Softmax(\frac{h_j^T t'}{\sqrt{d}}) \cdot h_j, \quad t' = Lookup(t_{ij}), \tag{5}$$

where t_{ij} is the converted timestamp of user j in message i, $t' \in R^d$ is a temporal embedding corresponding to t_{ij}, $Lookup()$ converts a time timestamp into a temporal embedding, and $h_j \in R^d$ is an embedding of user j obtained by GPN.

4.4 Dependency-Aware User Embedding

We use the learned user preferences $\widetilde{V} = \{\widetilde{v}_{11}, \widetilde{v}_{12}, ..., \widetilde{v}_{MN_c}\} \in R^{M \times N_c \times d}$ to capture the context dependency between users through a multi-head attention mechanism. This process can be defined as follows:

$$Attention(Q, K, S) = Softmax(\frac{QK^T}{\sqrt{d_k}} + C)S,$$
$$h_i = Attention(\widetilde{V}W_i^Q, \widetilde{V}W_i^K, \widetilde{V}W_i^S), \tag{6}$$
$$Z = [h_1 : h_2 : ... : h_{Head}]W^O,$$

where $W_i^Q, W_i^K, W_i^S \in R^{d \times d_k}$, $W^O \in R^{d \times d}$ are learnable paremeters, and $Head$ is the number of heads of multi-head attention. $d_k = d/Head$. $Z \in R^{M \times N_c \times d}$ represents dependency-aware user embedding in all messages. The mask $C \in R^{M \times N_c \times N_c}$ is used to turn off the attention weight for future time.

4.5 Fusion Gate

In order to integrate dependency-aware user embedding with dynamic preferences, we design a fusion gate to produce final user embedding as follows:

$$F = sigmoid(\widetilde{V}W_f^1 + ZW_f^2 + b_f), A = F \odot \widetilde{V} + (1 - F) \odot Z, \tag{7}$$

where $W_f^1, W_f^2 \in R^{d \times d}$ and $b_f \in R^{M \times N_c \times d}$ are learnable paremeters. After we obtain the final user embedding $A \in R^{M \times N_c \times d}$, we calculate the diffusion probability $\widehat{Y}$:

$$\widehat{Y} = Softmax(W_2(Relu(W_1 A^T + b_1))^T + b_2), \tag{8}$$

where $W_1 \in R^{|V| \times d}, W_2 \in R^{M \times M}, b_1 \in R^{|V| \times N_c \times M}$ and $b_2 \in R^{M \times N_c \times |V|}$ are the learnable parameters. Finally, we use the minimized cross-entropy loss to train model, which is formulated as follows:

$$\mathcal{L}(\theta) = -\sum_{i=1}^{M}\sum_{j=2}^{N_c}\sum_{k=1}^{|V|} y_{ijk} \log(\widehat{y}_{ijk}), \tag{9}$$

where θ denotes all the parameters that will be learned, y_{ijk} is the groundtruth label, and $y_{ijk} = 1$ means that the user k is the j-th user that forwards message i, otherwise $y_{ijk} = 0$.

5   Experiments

5.1   Experimental Settings

Datasets. Like previous works [16,22], we conduct extensive experiments on three datasets Memetracker [5], Twitter [23] and Douban [24] to validate the proposed model.

Baselines. In order to evaluate the performance of DHGPNTM, we compare our model with the following state-of-the-art baselines: TopoLSTM [11], Deep-Diffuse [13], NDM [25], SNIDSA [14], FOREST [16] and DyHGCN [22].

Evaluation Metrics. We use two widely used ranking metrics for evaluation: Mean Average Precision on top k (Map@k) and Hits score on top k (Hits@k).

Parameter Setups. Our model is implemented by PyTorch. The parameters are updated by Adam optimizer. The batch size is set to 16. The dimension size of user embedding and temporal embedding are both 64. We use GPN composed of a layer of GPNConv to learn user embedding. The *step_len* is set to 5. The *head* number of multi-head attention is set to 8. The number of time intervals n is set to 8.

5.2   Experimental Results

The experimental results on three datasets are shown in Table 1. Compared with TopoLSTM, DeepDiffuse and NDM, DHGPNTM has a great improvement in terms of hits@k and map@k. These baseline methods model the diffusion path as a sequence, and they do not use social network. However, social network can reflect the relationships between users, facilatating information flow between users. The experimental results show that the social network has a positive influence on information diffusion prediction.

Compared with SNIDSA and FOREST, DHGPNTM also has a great improvement in terms of hits@k and map@k. Although SNIDSA and FOREST use social network, they do not take diffusion structure into consideration. DHG-PNTM not only uses social network, but also exploits diffusion graphs to model diffusion behavior, which brings a significant performance improvement.

Compared with DyHGCN, DHGPNTM has a clear improvement as well. Although DyHGCN uses both social network and diffusion graphs to model diffusion behavior, it uses GCN to capture structural information. As shown in literature [26], when multiple GCN layers are applied to learn node embedding, all nodes will converge to the same semantics. Therefore, we design a novel GPN network to replace GCN to alleviate over smoothness. Furthermore, We propose a fusion gate to combine users' dynamic preferences and context-dependence embedding to further enhance the performance. Hence, DHGPNTM is superior to DyHGCN in terms of hits@k and map@k.

Table 1. Experimental results of all models on three datasets. Due to the lack of social graph in Memetracker, we ignore the TopoLSTM and SNIDSA model in Memetracker.

Datasets	Model	hits@10	hits@50	hits@100	map@10	map@50	map@100
Twitter	DeepDiffuse	4.57	8.80	13.39	3.62	3.79	3.85
	TopoLSTM	6.51	15.48	23.68	4.31	4.67	4.79
	NDM	21.52	32.23	38.31	14.30	14.80	14.89
	SNIDSA	23.37	35.46	43.49	14.84	15.40	15.51
	FOREST	26.18	40.95	50.39	17.21	17.88	18.02
	DyHGCN	28.10	47.17	58.16	16.86	17.73	17.89
	DHGPNTM	**29.68**	**48.65**	**59.86**	**18.13**	**18.99**	**19.15**
Douban	DeepDiffuse	9.02	14.93	19.13	4.80	5.07	5.13
	TopoLSTM	9.16	14.94	18.93	5.00	5.26	5.32
	NDM	10.31	18.87	24.02	5.54	5.93	6.00
	SNIDSA	11.81	21.91	28.37	6.36	6.81	6.91
	FOREST	14.16	24.79	31.25	7.89	8.38	8.47
	DyHGCN	15.92	28.53	36.05	8.56	9.12	9.23
	DHGPNTM	**17.86**	**31.32**	**38.87**	**10.37**	**10.98**	**11.09**
Memetracker	DeepDiffuse	13.93	26.50	34.77	8.14	8.69	8.80
	NDM	25.44	42.19	51.14	13.57	14.33	14.46
	FOREST	29.43	47.41	56.77	16.37	17.21	17.34
	DyHGCN	29.74	48.45	58.39	16.48	17.33	17.48
	DHGPNTM	**30.70**	**50.48**	**60.63**	**18.01**	**18.92**	**19.06**

6   Conclusion

In this paper, we investigate the problem of information diffusion prediction. We propose a novel dynamic heterogeneous graph perception network with time-based mini-batch, called DHGPNTM, to model the social graph and dynamic diffusion graphs. In DHGPNTM, we design a graph perception network (GPN) to learn user embedding, and design a fusion gate to selectively integrate users' dynamic preferences and context-dependence embedding. The experimental results show that our model is better than the state of the art baselines.

Acknowledgement. This work was supported by the Natural Science Foundation of Heilongjiang Province in China (No. LH2020F043), the Innovation Talents Project of Science and Technology Bureau of Harbin in China (No. 2017RAQXJ094), the Foundation of Graduate Innovative Research of Heilongjiang University in China (No. YJSCX2021-076HLJU).

References

1. Takahashi, T., Igata, N.: Rumor detection on twitter. In: SCIS&ISIS, pp. 452–457. IEEE (2012)
2. Wallinga, J., Teunis, P.: Different epidemic curves for severe acute respiratory syndrome reveal similar impacts of control measures. Am. J. Epidemiol. **160**(6), 509–516 (2004)
3. Leskovec, J., Adamic, L.A., Huberman, B.A.: The dynamics of viral marketing. ACM Trans. Web (TWEB) **1**(1), 5-es (2007)
4. Li, H., Ma, X., Wang, F., Liu, J., Xu, K.: On popularity prediction of videos shared in online social networks. In: CIKM, pp. 169–178. ACM (2013)
5. Leskovec, J., Backstrom, L., Kleinberg, J.: Meme-tracking and the dynamics of the news cycle. In: SIGKDD, pp. 497–506. ACM (2009)
6. Vosoughi, S., Roy, D., Aral, S.: The spread of true and false news online. Science **359**(6380), 1146–1151 (2018)
7. Cheng, J., Adamic, L.A., Dow, P.A., Kleinberg, J.M., Leskovec, J.: Can cascades be predicted? In: WWW, pp. 925–936 (2014)
8. Yang, Y., et al.: RAIN: social role-aware information diffusion. In: AAAI (2015)
9. Tsur, O., Rappoport, A.: What's in a hashtag?: content based prediction of the spread of ideas in microblogging communities. In: WSDM, pp. 643–652. ACM (2012)
10. Goyal, A., Bonchi, F., Lakshmanan, L.V.: Learning influence probabilities in social networks. In: WSDM, pp. 241–250. ACM (2010)
11. Wang, J., Zheng, V.W., Liu, Z., Chang, K.C.C.: Topological recurrent neural network for diffusion prediction. In: ICDM, pp. 475–484. IEEE (2017)
12. Wang, Y., Shen, H., Liu, S., Gao, J., Cheng, X.: Cascade dynamics modeling with attention-based recurrent neural network. In: IJCAI, pp. 2985–2991 (2017)
13. Islam, M.R., MUTHIAHS, A.: DeepDiffuse: predicting the 'Who' and 'When' in Cascades. In: ICDM, pp. 1055–1060. IEEE (2018)
14. Wang, Z., Chen, C., Li, W.: A sequential neural information diffusion model with structure attention. In: CIKM, pp. 1795–1798. ACM (2018)
15. Bourigault, S., Lamprier, S., Gallinari, P.: Representation learning for information diffusion through social networks: an embedded cascade model. In: WSDM, pp. 573–582. ACM (2016)
16. Yang, C., Tang, J., Sun, M., Cui, G., Liu, Z.: Multi-scale information diffusion prediction with reinforced recurrent networks. In: IJCAI, pp. 4033–4039 (2019)
17. Kipf, T.N., Welling, M.: Semi-supervised classification with graph convolutional networks. arXiv preprint arXiv:1609.02907 (2016)
18. Wang, H., Li, J., Luo, T.: Graph semantics based neighboring attentional entity alignment for knowledge graphs. In: ICIC, pp. 355–367 (2021)
19. Zhang, M., Cui, Z., Neumann, M., Chen, Y.: An end-to-end deep learning architecture for graph classification. In: AAAI (2018)
20. Zhang, X., Zhang, T., Zhao, W., Cui, Z., Yang, J.: Dual attention graph convolutional networks. In: IJCNN, pp. 238–251 (2019)
21. Xu, K., Hu, W., Leskovec, J., Jegelka, S.: How powerful are graph neural networks? arXiv preprint arXiv:1810.00826 (2018)
22. Yuan, C., Li, J., Zhou, W., Lu, Y., Zhang, X., Hu, S.: DyHGCN: a dynamic heterogeneous graph convolutional network to learn user's dynamic preferences for information diffusion prediction. In: PKDD, pp. 347–363 (2020)

23. Hodas, N.O., Lerman, K.: The simple rules of social contagion. Sci. Rep. **4**(1), 1–7 (2014)
24. Zhong, E., Fan, W., Wang, J., Xiao, L., Li, Y.: ComSoc: adaptive transfer of user behaviors over composite social network. In: SIGKDD, pp. 696–704. ACM (2012)
25. Yang, C., Sun, M., Liu, H., Han, S., Liu, Z., Luan, H.: Neural Diffusion Model for Microscopic Cascade Prediction. arXiv preprint arXiv:1812.08933 (2018)
26. Huang, Z., Wang, Z., Zhang, R.: Cascade2vec: learning dynamic cascade representation by recurrent graph neural networks. IEEE Access **7**, 144800–144812 (2019)

Graphs

Cascade-Enhanced Graph Convolutional Network for Information Diffusion Prediction

Ding Wang[1,2], Lingwei Wei[1,2], Chunyuan Yuan[3], Yinan Bao[1,2],
Wei Zhou[1(✉)], Xian Zhu[1,2], and Songlin Hu[1,2]

[1] Institute of Information Engineering, Chinese Academy of Sciences, Beijing, China
{wangding,weilingwei,baoyinan,zhouwei,zhuxian,husonglin}@iie.ac.cn
[2] School of Cyber Security, University of Chinese Academy of Sciences,
Beijing, China
[3] JD.com, Beijing, China

Abstract. Information diffusion prediction aims to estimate the probability of an inactive user to be activated next in an information diffusion cascade. Existing works predict future user activation either by capturing sequential dependencies within the cascade or leveraging rich graph connections among users. However, most of them perform prediction based on user correlations within the current cascade without fully exploiting diffusion properties from other cascades, which may contain beneficial collaborative patterns for the current cascade. In this paper, we propose a novel Cascade-Enhanced Graph Convolutional Networks (CE-GCN), effectively exploiting collaborative patterns over cascades to enhance the prediction of future infections in the target cascade. Specifically, we explicitly integrate cascades into diffusion process modeling via a heterogeneous graph. Then, the collaborative patterns are explicitly injected into unified user embedding by message passing. Besides, we design a cascade-specific aggregator to adaptively refine user embeddings by modeling different effects of collaborative features from other cascades with the guidance of user context and time context in the current cascade. Extensive experiments on three public datasets demonstrate the effectiveness of the proposed model.

Keywords: Information diffusion prediction · Graph neural network · Heterogeneous graph

1 Introduction

The progress of online media has made it convenient for people to post and share information online, triggering large information diffusion cascades. As a

---

D. Wang and L. Wei—Both are first authors with equal contributions.

A. Bhattacharya et al. (Eds.): DASFAA 2022, LNCS 13245, pp. 615–631, 2022.
https://doi.org/10.1007/978-3-031-00123-9_50

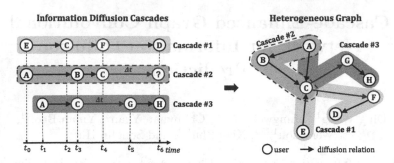

Fig. 1. A motivation example of exploiting collaborative patterns from other cascades. Suppose there are three cascades available and the current cascade is Cascade #2.

huge number of cascades are tracked and recorded, researchers have been motivated to investigate the pattern of information dissemination on the Internet via the task of information diffusion prediction. This novel task plays an increasingly significant role in many areas, such as modeling user behavior for recommendation [25] and influence prediction [14,27].

To address the task of diffusion prediction, researchers [1,21,23,26,27,29] have proposed a plethora of approaches. Prior studies [9,21,22] try to understand the diffusion pattern via sequential models. They regard the diffusion cascade as a sequence and predict the future user activation by capturing linear correlations within the current cascade. Although these models have achieved remarkable results, they are incapable of capturing the complex user dependencies due to over-simplified modeling for the diffusion process. Recently, with the observation that most of the diffusion cascade is created in the online social network and the promising result of graph representation in several areas, some works [16,18, 23,27,29] leverage rich connections among users to model the diffusion process, demonstrating remarkable prediction performances.

However, previous works mainly focus on leveraging information within the current cascade for prediction without fully exploiting collaborative diffusion patterns provided by other cascades. As shown in Fig. 1, when predicting the potential diffusion trend in Cascade #2 after timestamp t_4, collaborative patterns from the other cascades might provide helpful suggestions, e.g., new pairwise user $C \rightarrow F$ from Cascade #1 and user $C \rightarrow G$ from Cascade #3. Moreover, it can be observed that distinguishing the effect of diffusion patterns from different cascades could enhance predicting accuracy for the current cascade. For example, when predicting the next infected users after user C in Cascade #2 at timestamp t_6, the property from Cascade #3 may offer more helpful information than that from Cascade #1. Cascade #3 not only shares more infected users with the current Cascade #2 but also exhibits a similar temporal pattern between user C and user G, indicating that user G is more likely to be infected next in Cascade #2. Therefore, it is beneficial but challenging to effectively exploit relevant collaborative patterns to promote the task of diffusion prediction.

In this paper, we propose a novel Cascade-Enhanced Graph Convolutional Network (CE-GCN), which effectively exploits collaborative patterns over cascades to enhance the learning of user representation for the current cascade. Specifically, we first assign nodes for all cascades and users, and exploit multiple relations to construct the heterogeneous graph to comprehensively model the diffusion process. Then, we design a novel message passing mechanism on the constructed heterogeneous graph to fully explore rich collaborative patterns among different cascades and generate representations for both users and cascades. After that, we design a cascade-specific aggregator to distinguish the impact of different collaborative patterns. The module automatically refines the user representation by focusing on more relevant features within the current cascade. Finally, the refined user representations are leveraged to predict the diffusion process by retrieving potential users via a multi-head self-attention module.

To evaluate the proposed model, we conduct experiments on three public datasets. The experimental results demonstrate that CE-GCN consistently outperforms several strong baselines, which shows the effectiveness of the proposed model.

The contributions of this paper are three-fold:

- We propose a novel Cascade-Enhanced Graph Convolutional Network (CE-GCN) to enhance the learning of user embeddings by exploiting collaborative patterns over all cascades.
- We design a cascade-specific aggregator to adaptively refine user representations by distinguishing different effects of collaborative patterns from other cascades based on user context and time context in the current cascade.
- We conduct extensive experiments on three public datasets. CE-GCN significantly outperforms state-of-the-art models on the information diffusion prediction task.

2 Related Work

In this section, we review related studies on information diffusion prediction and graph neural networks.

2.1 Information Diffusion Prediction

Information diffusion prediction explores to forecast the diffusion trend based on the historical infected users. Early researchers model the diffusion process by assuming an underlying diffusion model, such as the independent cascade model or linear threshold model [10], which requires extensive manual analysis for the social network and diffusion cascades.

With the promising achievement of deep learning in sequence prediction, researchers have adopted different models for the task. Prior researchers [7,9,15, 21,26] model the diffusion cascade as a sequence and explore to detect the diffusion tendency by capturing linear dependencies within the cascade. For example, Wang et al. [21] model the diffusion process as a directed acyclic graph and explore user dependencies in the cascade through an improved Long Short-Term Memory (LSTM) module. Yang et al. [26] propose to capture long-term dependency in the cascade via self-attention mechanism and convolution neural networks (CNN). Islam et al. [9] employ the embedding technique and attention mechanism to incorporate the infected timestamp with user correlation for prediction.

Recently, as works [4,6] have found that rich social connections among users could show potential dependencies, some researchers [18,23,24,29] improved user representations by leveraging rich graph connections. For example, Wang et al. [23] fuse the target diffusion paths and social graph into Recurrent Neural Network (RNN) to construct user embeddings for prediction. Yang et al. [27] propose a comprehensive diffusion prediction model. They employ reinforcement learning and social connections to jointly predict the next infected user and estimate the total size of infected users, which enhances the performance on both microscopic and macroscopic diffusion prediction tasks. Sankar et al. [16] propose a fully generative model via a variational graph autoencoder to capture complex correlations within users for prediction. Su et al. [18] model the diffusion process as a heterogeneous graph and incorporate meta-path properties and text information to predict social contagion adoption. Yuan et al. [29] utilize diffusion and social relations to construct user representations based on a discrete-time dynamic graph neural network.

However, most previous works learn user embeddings based on user dependencies in the current cascade, failing to fully utilize the collaborative diffusion pattern from other cascades. To tackle the challenge, this paper investigates collaborative patterns over all cascades to enhance the learning of user embeddings.

2.2 Graph Neural Networks

Recently, Graph Neural Network (GNN) has been widely applied to analyze graph structures in many tasks [28] due to its convincing performance and high interpretability. As the most commonly used model, graph convolutional networks (GCNs) [12] have been broadly applied to capture graph level dependency via message passing mechanism between nodes. Unlike standard neural networks, GCNs retain node states that can represent the node from its neighborhood with arbitrary depth, which could build more expressive node representations for downstream tasks. Moreover, variants of graph neural network [17,20] have been designed for different tasks with a convincing performance.

For our model, with the graph embedding technique and our novel message passing mechanism, users in the heterogeneous graph will leave traits related to cascade-aware collaborative information and influential neighbors in their representations, leading our model to discover potential diffusion trends in the social network.

3   Problem Statement

Formally, we define the input of our problem as a set of cascades $\mathcal{C} = \{c_1, c_2, ..., c_{|\mathcal{C}|}\}$ on a user set $\mathcal{U} = \{u_1, u_2, ..., u_{|\mathcal{U}|}\}$. Each cascade $c \in \mathcal{C}$ is recorded

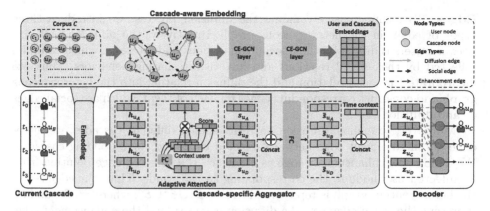

Fig. 2. The architecture of the proposed CE-GCN. The corpus $\mathcal{C}$ contains all observed cascades in the training set. The **FC** denotes the fully-connected neural network layer.

as a series of user behaviors, *i.e.*, $c = \{(u_1, t_1), (u_2, t_2), ..., (u_{|c|}, t_{|c|})\}$, where the element (u_i, t_i) denotes that user u_i is infected in this cascade at time t_i. Here, infection means this user takes part in the cascade process (e.g., resharing a content in social media). The infected users are ordered by time, thus $t_{i-1} < t_i$.

The task of **information diffusion prediction** aims to predict user u_{i+1}, who will be infected next, based on the given user behavior sequence. The goal is to build a model to learn the function of the conditional probability $p(u_{i+1}|\{(u_1, t_1), (u_2, t_2), ..., (u_i, t_i)\})$.

4 The Proposed Model

In this section, we introduce our novel Cascade-Enhanced Graph Convolutional Network (CE-GCN) for the task of information diffusion prediction. The overview architecture of CE-GCN is shown in Fig. 2.

4.1 Cascade-Aware Embedding

To jointly model users and cascades, we first represent the diffusion process as a heterogeneous graph. Then, we design a novel message passing mechanism on the constructed heterogeneous graph to capture the collaborative diffusion patterns among different cascades to enhance user and cascade representations.

Constructing Heterogeneous Graph. For the given cascade set $\mathcal{C}$ and the user set $\mathcal{U}$, we construct a heterogeneous graph to represent the diffusion process, which is $\mathcal{G} = \{\mathcal{V}, \mathcal{E}\}$, where $\mathcal{V}$ and $\mathcal{E}$ are node set and edge set, respectively.

Nodes. To capture collaborative diffusion patterns among cascades, we first explicitly incorporate cascades into the heterogeneous graph by constructing corresponding vertices. Therefore, the heterogeneous graph contains two types of vertices, *i.e.*, $\mathcal{V} = \mathcal{V}_u \cup \mathcal{V}_c$, where $\mathcal{V}_u = \{v_u | u \in \mathcal{U}\}$ denotes the user node set and $\mathcal{V}_c = \{v_c | c \in \mathcal{C}\}$ denotes the cascade node set.

Edges. To comprehensively model the diffusion process, we introduce three types of edges in the heterogeneous graph. Intuitively, the diffusion behavior between users is driven by their social connections and interests, which could be reflected by their social relations and historical resharing behaviors. Thus, we incorporate the social relation and the diffusion relation into the heterogeneous graph and construct corresponding edges for each of them, which are social edges e^s and diffusion edges e^d.

Moreover, to exploit diffusion patterns among different cascades, we introduce an enhancement edge between infected users and corresponding cascades. We denote the enhancement edge as e^e, which is a directed edge pointing from an infected user node to the corresponding cascade node. Thus, the edge set in the heterogeneous graph is represented as $\mathcal{E} = \mathcal{E}^d \cup \mathcal{E}^s \cup \mathcal{E}^e$, where $\mathcal{E}^d, \mathcal{E}^s$, and $\mathcal{E}^e$ represent the social edge set, the diffusion edge set, and the enhancement edge set, respectively.

Learning Unified Embedding. In this section, we introduce our layer-wise message passing mechanism in our CE-GCN. We define different message-passing mechanisms on the heterogeneous graph for each type of node to capture collaborative features to construct enhanced representations.

For user node v_u at $(l+1)^{th}$ CE-GCN layer, it shares connection with both user nodes and cascade nodes. Thus, we divide the neighbors by their node types and separately aggregate their contextual information, which is formulated as,

$$
\begin{aligned}
a_u^{\mathcal{N}_u^u(l+1)} &= f\big(\frac{1}{|\mathcal{N}_u^u|} \sum_{u_i \in \mathcal{N}_u^u} W_{u_i}^{u(l+1)} h_{u_i}^{(l)}\big), \\
a_c^{\mathcal{N}_c^u(l+1)} &= f\big(\frac{1}{|\mathcal{N}_c^u|} \sum_{c_i \in \mathcal{N}_c^u} W_{c_i}^{u(l+1)} h_{c_i}^{(l)}\big),
\end{aligned}
\tag{1}
$$

where $\mathcal{N}_c^u$ and $\mathcal{N}_u^u$ are neighbors that share edges with focal user node v_u with different node types. $h_{u_i}^{(l)}$ and $h_{c_i}^{(l)}$ are corresponding neighbor node representations from the last layer. $W_{u_i}^{u(l+1)}$ and $W_{c_i}^{u(l+1)}$ are learnable weight matrices to aggregate contextual features from different neighbors. The function $f(\cdot)$ contains normalize, dropout, and activate operations.

Then, we concatenate the above contextual information and apply an MLP layer to filter useful information for the focal user node v_u, which is formulated as follows,

$$
h_u^{(l+1)} = \text{MLP}\big(\big[a_u^{\mathcal{N}_u^u(l+1)}; a_c^{\mathcal{N}_c^u(l+1)}; h_u^{(l)}\big]\big),
\tag{2}
$$

where $h_u^{(l)}$ and $h_u^{(l+1)}$ are the output embedding for user node v_u in the lth CE-GCN layer and $(l+1)^{th}$ CE-GCN layer, respectively.

For cascade node v_c at $(l+1)^{th}$ CE-GCN layer, it only connects with users that are infected in the corresponding cascade. Thus, the message passing for the cascade node is to aggregate the contextual information from all infected users. We describe the process as,

$$a_c^{\mathcal{N}_u^c(l+1)} = f(\frac{1}{|\mathcal{N}_u^c|} \sum_{u_i \in \mathcal{N}_u^c} W_{u_i}^{c(l+1)} h_{u_i}^{(l)}),$$

$$(3)$$

$$h_c^{(l+1)} = \text{MLP}([a_c^{\mathcal{N}_u^c(l+1)}; h_c^{(l)}]),$$

where $\mathcal{N}_u^c$ is a set of users neighbors for the corresponding cascade node. $W_{c_i}^{u(l+1)}$ is a learnable weight matrix to aggregate contextual features from different users.

For the input of our first CE-GCN layer, we utilize a normal distribution [5] to randomly initialize the user and cascade representations $h_u^{(0)}$ and $h_c^{(0)}$. Then, we stack the CE-GCN layer L times and collect the output user and cascade representation h_u and h_c from the last CE-GCN layer as the output of our Cascade-aware Embedding module.

4.2   Cascade-Specific Aggregator

To better perform prediction for the current cascade, we further design a cascade-specific aggregator to refine the generated embeddings for users in the current cascade with the consideration of user context and time context. The core idea of this module is to differentiate the effect of collaborative patterns from different cascades to enhance user representation and diffusion prediction.

Fusing User Context. We employ an adaptive attention mechanism to focus on user context in the current cascade. The core idea is to capture the diffusion dependencies among input cascade users. Specifically, the attention score between user $u_j \in \{u_1, u_2, ..., u_j, ..., u_i\}$ and its context user $u_k \in \{u_1, ..., u_{j-1}\}$ can be computed as,

$$\beta_{kj} = \frac{\exp\left(\langle\tanh(W_h^c h_{u_k}), \tanh(W_h^t h_{u_j})\rangle\right)}{\sum_{r=1}^{j-1} \exp\left(\langle\tanh(W_h^c h_{u_r}), \tanh(W_h^t h_{u_j})\rangle\right)},$$

$$(4)$$

where W_h^c and W_h^t are transformation matrices for the context user and the target user respectively. These transformation matrices could differentiate the collaborative features by the given user when predicting different cascades. $\langle\cdot, \cdot\rangle$ represents the inner product operation.

Then, the cascade-specific user representation s_{u_j} of user u_j is calculated as,

$$s_{u_j} = \sum_{k=1}^{j-1} \beta_{kj} h_{u_k}.$$

$$(5)$$

Finally, we apply a residual connection to fuse the user embedding h_{u_j} and the cascade-specific representation s_{u_j} to generate user vectors $\tilde{s}_{u_j}$ for each user u_j by a fully-connected layer. That is,

$$\tilde{s}_{u_j} = \sigma([h_{u_j}; s_{u_j}]W_s + b_s),$$

$$(6)$$

where W_s and b_s are trainable parameters.

Fusing Time Context. Previous studies [2,3] have shown that temporal diffusion pattern within the cascade is affected by historical user influences, especially by the last infected user. Moreover, this type of user influence decays as time passes. Inspired by them, we capture the temporal diffusion patterns within the cascade via their context infected timestamps and employ a neural function to represent the user influence.

Give the time sequence $\{t_{u_1}, t_{u_2}, ..., t_{u_i}\}$ of the current cascade, we represent time context of user $u_j \in \{u_1, ..., u_i\}$ as $\Delta t_{u_j} = t_{u_{j+1}} - t_{u_j}$ to indicate the changing influence from previous users. Then, we discretize it as a one-hot vector, denoted as $\mathbf{t}_{u_j}$, where $\mathbf{t}_{u_j}^n = 1$ if $t_{n-1} < \Delta t_{u_j} < t_n$. t_{n-1} and t_n are defined by splitting the time range $(0, T_{max}]$ into $\mathbf{T}$ intervals, $i.e.$, $\{(0, t_1], ..., (t_{n-1}, t_n], ..., (t_{T-1}, T_{max}]\}$, where T_{max} refers to the max observation time in the given cascade set $\mathcal{C}$.

Then, we map $\mathbf{t}_{u_j}$ to time vector $\boldsymbol{\mu}_{u_j}$ to express the time-decay effect of the previous influence of users via a fully connected layer:

$$\boldsymbol{\mu}_{u_j} = \sigma(\mathbf{W}_t \mathbf{t}_{u_j} + \mathbf{b}_t), \tag{7}$$

where $\mathbf{W}_t$ and $\mathbf{b}_t$ are learnable parameters.

After that, we make concatenation of the above two vectors to produce the final user representation $\mathbf{z}_{u_j}$, $i.e.$,

$$\mathbf{z}_{u_j} = [\tilde{\mathbf{s}}_{u_j}; \boldsymbol{\mu}_{u_j}]. \tag{8}$$

4.3 Diffusion Prediction

To predict future infections, we first reconstruct the current cascade representation based on the final user representations, denoted as $\mathbf{Z} = [\mathbf{z}_{u_1}, \mathbf{z}_{u_2}, ..., \mathbf{z}_{u_i}]$. Then, we apply a masked multi-head self-attention module [19] to retrieve the potential infections at each given timestamp. As it could parallelly attend to each position in sequence modeling, the module is much faster and easier to capture the context information than RNNs when processing the cascade sequence. Therefore, the infected user representation at each timestamp for current cascade c is predicted as the following,

$$\text{Attention}(\mathbf{Q}, \mathbf{K}, \mathbf{V}) = \text{softmax}\left(\frac{\mathbf{Q}\mathbf{K}^T}{\sqrt{d_k}} + \mathbf{M}\right)\mathbf{V},$$

$$o_i^d = \text{Attention}\left(\mathbf{Z}\mathbf{W}_i^Q, \mathbf{Z}\mathbf{W}_i^K, \mathbf{Z}\mathbf{W}_i^V\right), \tag{9}$$

$$C = [o_1^d; ...; o_H^d]\mathbf{W}^O,$$

where $\mathbf{W}_i^Q, \mathbf{W}_i^K, \mathbf{W}_i^V, \mathbf{W}^O$ are learnable parameters. H is the number of heads in the multi-head self-attention module. d_k is the scaling factor and $\mathbf{M}$ refers to a mask matrix to mask the future information to avoid leaking labels, which is calculated as:

$$M_{ij} = \begin{cases} 0 & i \leq j, \\ -\infty & \text{otherwise.} \end{cases} \tag{10}$$

Then, based on the sequence of predicted user representations C, we apply a two-layer fully-connected neural network to decode infected probabilities for all users at each time step, $i.e.$,

$$\hat{y} = W_2\mathbf{ReLU}(W_1 C^T + b_1) + b_2,\qquad(11)$$

where W_1, W_2, b_1, and b_2 are learnable parameters.

Finally, we train the model with the cross-entropy loss. The loss function is defined as,

$$\mathcal{L}(\theta) = -\sum_{i=2}^{L}\sum_{j=1}^{|U|} y_{ij} \log(\hat{y}_{ij}),\qquad(12)$$

where $y_{ij} = 1$ denotes that the predicted user j is infected at position i, otherwise $y_{ij} = 0$. θ denotes all parameters needed to be learned in CE-GCN.

5   Experimental Setups

In this section, we briefly describe experimental setups including datasets, comparison methods, evaluation metrics, and parameter settings.

5.1   Datasets

Following the previous works [27,29], we evaluate our model on three public real-world datasets, $i.e.$, Twitter, Douban, and Memetracker datasets. The detailed statistics of the datasets are presented in Table 1.

1) **Twitter** dataset [8] records the tweets containing URLs during October 2010 on Twitter[1]. Each URL is viewed as a diffusion item spreading among users. The social relations are pre-defined by the follow relation on Twitter. 2) **Douban** dataset [30] is collected from a Chinese social website named douban[2], where people share their book reading statuses. Each book is considered as a diffusion item, and a user is infected if he or she reads it. Social relations in this dataset are pre-defined by co-occurrence relations. If two users take part in the same discussion more than 20 times, they are considered as friends to each other. 3) **Memetracker**[3] dataset [13] tracks the most frequent quotes and phrases, $i.e.$ memes, to analyze the migration of information online. Each meme is considered as a diffusion item and each URL is treated as a user. The dataset has no underlying social network.

For all datasets, following previous works [23,27,29], we use 80% of cascades for training, 10% for validation, and the rest for testing.

---

[1] https://www.twitter.com.
[2] https://www.douban.com.
[3] http://memetracker.org/.

Table 1. Statistics of three datasets.

Datasets	# Users	# Links	# Cascades	Avg. Length
Twitter	12,627	309,631	3,442	32.60
Douban	23,123	348,280	10,602	27.14
Memetracker	4,709	–	12,661	16.24

5.2 Comparison Methods

We compare the proposed **CE-GCN** with the following methods:

TopoLSTM [21] views the diffusion cascade as a dynamic directed acyclic graph and extends the standard LSTM module to improve the prediction performance. This model relies on the pre-defined social graph to make predictions.

DeepDiffuse [9] employs the embedding technique and attention mechanism to incorporate the infected time with user correlation. Given the previously observed cascade, the model can predict when and who is going to be infected.

NDM [26] aims to capture the long-term dependency in the cascade sequence by leveraging self-attention mechanism and convolution neural networks to construct user representations.

SNIDSA [23] employs a recurrent network with a user embedding layer and incorporates the social relation and diffusion paths by structure attention. The model requires pre-defined social network for prediction.

FOREST [27] is a multi-scale diffusion prediction model based on reinforcement learning, which both employed RNN and reinforcement learning to jointly predict the next infected user in a microscopic view and estimate the total size of infected users in a macroscopic view.

DyHGCN [29] utilizes the discrete-time dynamic graph neural network to discover user dynamic preference and learn user representations. It achieves state-of-the-art performance in the task of information diffusion prediction.

5.3 Evaluation Metrics

Following the previous works [27, 29], we consider the information diffusion prediction task as an information retrieval task by ranking all the uninfected users according to their infected probabilities. We evaluate the performance of our model with other baselines in terms of two metrics, *i.e.*, Mean Average Precision on top K (*MAP@K*) and HITS scores on top K (*HITS@K*). The higher Hits@k and MAP@k indicate better performance.

5.4 Parameter Settings

The parameters are updated by Adam algorithm [11]. The parameters in Adam, β_1 and β_2, are 0.90 and 0.98 respectively. The learning rate is set to 0.001. The batch size of the samples in the training set is 16. The dimensionality of user embedding is set to $d = 64$. The time interval T in Sect. 4.2 is selected from $\{50, 100, 500, 1000, 2500, 5000, 7500, 10000\}$ and the best setting is 10000. The dimension of time context embeddings d_t is set to 8. We apply a two-layer GCN with kernel size 128 to learn the underlying diffusion pattern. The number of GCN layers is selected from $\{1, 2, 3, 4, 5\}$, and the best setting is 2. The number of heads H in a multi-head attention module is chosen from $\{2, 4, 6, 8, 10, 12, 14, 16, 18, 20\}$ and the best setting is 14. All optimal hyper-parameters are selected by Grid Search algorithm according to performance on the validation set. Then we report the performance on the test set.

6   Results and Analysis

In this section, we report experimental results and make further analysis.

6.1   Experimental Results

The experimental results are shown in Table 2 and Table 3. From the results, CE-GCN consistently achieves the best performance on all datasets at all evaluation metrics, which proves the effectiveness of the proposed model.

(1) **TopoLSTM**, **DeepDiffuse**, and **NDM** mainly focus on diffusion dependencies in the current cascade for diffusion prediction. **SNIDSA**, **FOREST**, **DyHGCN**, and **CE-GCN** incorporates social relations in the diffusion cascade to obtain better performance. It reveals that both diffusion relations and social relations play vital roles in the task of information diffusion prediction.

(2) Both **SNIDSA** and **FOREST** model diffusion cascade sequences with RNN-based methods. They achieve worse performance than graph-based models like **DyHGCN** and our model **CE-GCN** which jointly encodes diffusion cascades and social networks. The results show the powerful learning abilities of graph neural networks to capture complex user dependencies in multiple diffusion cascades.

(3) Compared with **DyHGCN**, our model **CE-GCN** consistently outperforms on three datasets. These results demonstrate the superiority of considering collaborative diffusion patterns from other cascades to enhance user embedding construction for diffusion prediction.

6.2   Ablation Study

To study the relative importance of each module in the CE-GCN, we conduct ablation studies over the different parts of the model as follows:

– **w/o CA:** We remove the cascade-specific aggregator and directly predict diffusion cascades based on the unified user representations.

Table 2. Results on Twitter and Douban datasets (%). Improvements of CE-GCN are statistically significant with $p < 0.01$ on paired t-test.

Models	Twitter						Douban					
	HITS@K			MAP@K			HITS@K			MAP@K		
	$K = 10$	$K = 50$	$K = 100$	$K = 10$	$K = 50$	$K = 100$	$K = 10$	$K = 50$	$K = 100$	$K = 10$	$K = 50$	$K = 100$
DeepDiffuse	4.57	8.80	13.39	3.62	3.79	3.85	9.02	14.93	19.13	4.80	5.07	5.13
TopoLSTM	6.51	15.48	23.68	4.31	4.67	4.79	9.16	14.94	18.93	5.00	5.26	5.32
NDM	21.52	32.23	38.31	14.30	14.80	14.89	10.31	18.87	24.02	5.54	5.93	6.00
SNIDSA	23.37	35.46	43.39	14.84	15.40	15.51	11.81	21.91	28.37	6.36	6.81	6.91
FOREST	26.18	40.95	50.39	17.21	17.88	18.02	14.16	24.79	31.25	7.89	8.38	8.47
DyHGCN	28.98	47.89	58.85	17.46	18.30	18.45	16.34	28.91	36.13	9.10	9.67	9.78
CE-GCN	**31.48**	**50.87**	**61.12**	**19.31**	**20.20**	**20.35**	**20.28**	**34.11**	**41.77**	**11.70**	**12.33**	**12.44**
Improve (%)	8.63	6.22	3.45	10.60	11.86	10.30	24.11	17.82	14.59	28.57	27.51	27.20

Table 3. Results on Memetracker dataset (%). For fair comparison, TopoLSTM and SNIDSA are excluded due to the absence of social graph in the Memetracker dataset. Improvements of CE-GCN are statistically significant with $p < 0.01$ on paired t-test.

Models	Memetracker					
	HITS@K			MAP@K		
	$K = 10$	$K = 50$	$K = 100$	$K = 10$	$K = 50$	$K = 100$
DeepDiffuse	13.93	26.50	34.77	8.14	8.69	8.80
NDM	25.44	42.19	51.14	13.57	14.33	14.46
FOREST	29.43	47.41	56.77	16.37	17.21	17.34
DyHGCN	29.90	48.30	58.43	17.64	18.48	18.63
CE-GCN	**37.01**	**56.04**	**65.09**	**21.54**	**22.43**	**22.56**
Improve (%)	23.78	14.88	10.73	22.11	21.37	21.10

- **w/o CE:** We omit cascade nodes in the heterogeneous graph and learn user embeddings in the graph where nodes only refer to users and edges are built by social relations and diffusion relations.
- **w/o CE & CA:** We remove both the cascade-aware embedding and cascade-specific aggregator. This ablated model predicts diffusion process by capturing the time-decay patterns based on the diffusion cascade sequence.

The experimental results of ablation studies are presented in Table 4. Referring to the results, we have the following findings:

(1) When removing the cascade-specific aggregator (**CE-GCN w/o CA**), the performance drops to some extent, which shows the effectiveness of the CA module. With the guidance of related user context and time context in the current cascade, CA can generate an enhanced user representation for prediction.
(2) When removing explicit modeling of cascade nodes in the heterogeneous graphs, **CE-GCN w/o CE** mainly learns explicit dependencies among users. The worse performance implies that exploiting additional property from other cascades is indeed beneficial for information diffusion prediction.

Table 4. Ablation study on three datasets (%).

Models	Twitter				Douban				Memetracker			
	HITS@K		MAP@K		HITS@K		MAP@K		HITS@K		MAP@K	
	$K = 50$	$K = 100$	$K = 50$	$K = 100$	$K = 50$	$K = 100$	$K = 50$	$K = 100$	$K = 50$	$K = 100$	$K = 50$	$K = 100$
CE-GCN	**50.87**	**61.12**	**20.20**	**20.35**	**34.11**	**41.79**	**12.33**	**12.44**	**56.04**	**65.09**	**22.43**	**22.56**
- w/o CA	50.53	60.83	19.52	19.67	31.29	39.15	10.97	11.08	50.86	60.76	19.32	19.46
- w/o CE	49.93	59.75	20.12	20.26	31.90	39.49	11.67	11.77	54.11	63.20	21.40	21.53
- w/o CE & CA	20.13	27.43	5.93	6.04	19.13	24.56	7.38	7.46	31.22	40.84	11.16	11.29

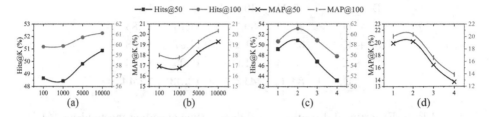

Fig. 3. Parameter analysis on the Twitter dataset. (a) and (b) denotes the results against the number of time intervals; (c) and (d) denotes the results against the number of GCN layers.

(3) When we only predict diffusion based on the diffusion cascade sequence by capturing the time-decay patterns (**CE-GCN w/o CE & CA**), the performance drops significantly. It reflects the complexity of information diffusion caused by user participation.

6.3 Parameter Analysis

In this part, we analyze the impact of two key parameters in CE-GCN, *i.e.*, the number of time intervals, and the number of GCN layers.

Figure 3(a) and (b) show the performance against the number of time intervals T on the Twitter dataset. When T increases, the performance emerges an overall upward trend. The trend implies that fusing time context into user representations is helpful for information diffusion prediction in the current cascade. Larger T indicates that the time range is split into more pieces, expressing more fine-grained time-decay patterns from the infected time difference to assist diffusion prediction.

Figure 3(c) and (d) show the performance against different numbers of GCN layers. As the number of GCN layers increases, the performance increases first and then decreases. When applying deeper GCNs, the performance drops significantly due to the over-fitting and over-smoothing problem. A two-layer GCN is the most suitable for models to capture user dependencies.

6.4 Further Study

In this part, we conduct experiments on the Douban dataset to further analyze how CE-GCN improves the performance of information diffusion prediction.

Table 5. Results against different graphs on Douban dataset (%). S., D., and E. denotes that we construct the heterogeneous graph by social, diffusion, and enhancement edge.

Models	HITS@K		MAP@K	
	$K = 50$	$K = 100$	$K = 50$	$K = 100$
S	31.88	39.48	11.36	11.41
D	31.70	39.13	11.66	11.76
E	32.24	40.06	11.42	11.53
S. + D	32.42	39.78	11.75	11.85
S. + E	33.93	41.24	11.97	12.08
D. + E	33.56	41.06	12.05	12.13
S. + D. + E	**34.11**	**41.77**	**12.33**	**12.44**

Table 6. Experimental results over different time encoding mechanisms(%).

Models	HITS@K		MAP@K	
	$K = 50$	$K = 100$	$K = 50$	$K = 100$
CE-GCN	**34.11**	**41.79**	**12.33**	**12.44**
CE-GCN$_{Slice}$	33.36	41.16	11.43	11.54
CE-GCN$_{Relative}$	33.15	40.93	11.39	11.50
CE-GCN$_{None}$	32.53	40.03	10.83	10.94

The Effect on User Context. We extend several variant models by leveraging different graphs to encode user representations. The result is shown in Table 5, where S., D., and E. indicate that we construct the heterogeneous graph based on social, diffusion, and enhancement edges, respectively. Firstly, the best results are obtained by combining the three edges, which denotes that considering relations across cascades provides both social and propagation patterns in the user context. Moreover, combining the enhancement graph with other graphs could effectively promote the model performance, which shows that the enhancement graph could capture collaborative diffusion patterns among different cascades and promote user representations.

The Effect on Time Context. We conduct experiments against different time encoding mechanisms to analyze the effect on time context. The results are shown in Table 6. CE-GCN$_{Slice}$ generates time context by splitting time range into slices. CE-GCN$_{Relative}$ leverages relative position embedding as time context. CE-GCN$_{None}$ is a variant without any time context. When omitting the time encoder or using the relative position embedding as time context, the inferior performance reveals that the infected timestamp could bring beneficial information for prediction. Moreover, compared with CE-GCN$_{Slice}$, our model show better performance. The fact demonstrates that time difference could reflect user influence in the cascade more precisely, bringing more accurate prediction result.

7 Conclusion

In this paper, we have proposed Cascade-Enhanced Graph Convolutional Network (CE-GCN) to effectively exploit collaborative patterns over all cascades and enhance the learning of user representation for the task of information diffusion prediction. Specifically, CE-GCN models the diffusion process as a heterogeneous graph and explicitly injects the collaborative patterns into user embeddings via message passing. After that, we design a cascade-specific user encoder to adaptively refine the user representation by distinguish different effects of features from other cascades based on user context and time context. We conduct experiments on three public real-world datasets. The extensive results show that CE-GCN obtains the state-of-the-art performance, which demonstrates the effectiveness of our model.

References

1. Bourigault, S., Lamprier, S., Gallinari, P.: Representation learning for information diffusion through social networks: an embedded cascade model. In: Proceedings of the Ninth ACM International Conference on Web Search and Data Mining (WSDM), pp. 573–582 (2016)
2. Cao, Q., Shen, H., Cen, K., Ouyang, W., Cheng, X.: DeepHawkes: bridging the gap between prediction and understanding of information cascades. In: Proceedings of the 2017 ACM on Conference on Information and Knowledge Management, CIKM, pp. 1149–1158 (2017)
3. Chen, X., Zhang, K., Zhou, F., Trajcevski, G., Zhong, T., Zhang, F.: Information cascades modeling via deep multi-task learning. In: Proceedings of the 42nd International ACM SIGIR Conference on Research and Development in Information Retrieval (SIGIR), pp. 885–888 (2019)
4. Cheng, J., Adamic, L.A., Dow, P.A., Kleinberg, J.M., Leskovec, J.: Can cascades be predicted? In: 23rd International World Wide Web Conference (WWW), pp. 925–936. ACM (2014)
5. Glorot, X., Bengio, Y.: Understanding the difficulty of training deep feedforward neural networks. In: Proceedings of the Thirteenth International Conference on Artificial Intelligence and Statistics, AISTATS 2010, vol. 9, pp. 249–256 (2010)
6. Gomez-Rodriguez, M., Balduzzi, D., Schölkopf, B.: Uncovering the temporal dynamics of diffusion networks. In: Proceedings of the 28th International Conference on Machine Learning (ICML) pp. 561–568 (2011)
7. Gomez-Rodriguez, M., Leskovec, J., Krause, A.: Inferring networks of diffusion and influence. In: Proceedings of the 16th ACM SIGKDD International Conference on Knowledge Discovery and Data Mining (KDD), pp. 1019–1028 (2010)
8. Hodas, N., Lerman, K.: The simple rules of social contagion. Scientific reports 4 (2014)
9. Islam, M.R., Muthiah, S., Adhikari, B., Prakash, B.A., Ramakrishnan, N.: Deepdiffuse: predicting the 'who' and 'when' in cascades. In: IEEE International Conference on Data Mining, (ICDM), pp. 1055–1060 (2018)
10. Kempe, D., Kleinberg, J.M., Tardos, É.: Maximizing the spread of influence through a social network. In: Proceedings of the Ninth ACM SIGKDD International Conference on Knowledge Discovery and Data Mining (KDD), pp. 137–146 (2003)

11. Kingma, D.P., Ba, J.: Adam: a method for stochastic optimization. In: 3rd International Conference on Learning Representations (ICLR) (2015)
12. Kipf, T.N., Welling, M.: Semi-supervised classification with graph convolutional networks. In: 5th International Conference on Learning Representations (ICLR) (2017)
13. Leskovec, J., Backstrom, L., Kleinberg, J.M.: Meme-tracking and the dynamics of the news cycle. In: Proceedings of the 15th ACM SIGKDD International Conference on Knowledge Discovery and Data Mining (KDD), pp. 497–506 (2009)
14. Qiu, J., Tang, J., Ma, H., Dong, Y., Wang, K., Tang, J.: DeepInf: social influence prediction with deep learning. In: Proceedings of the 24th ACM SIGKDD International Conference on Knowledge Discovery & Data Mining (KDD), pp. 2110–2119 (2018)
15. Saito, K., Ohara, K., Yamagishi, Y., Kimura, M., Motoda, H.: Learning diffusion probability based on node attributes in social networks. In: Foundations of Intelligent Systems - 19th International Symposium (ISMIS), pp. 153–162 (2011)
16. Sankar, A., Zhang, X., Krishnan, A., Han, J.: Inf-VAE: A variational autoencoder framework to integrate homophily and influence in diffusion prediction. In: Caverlee, J., Hu, X.B., Lalmas, M., Wang, W. (eds.) The Thirteenth ACM International Conference on Web Search and Data Mining (WSDM), pp. 510–518 (2020)
17. Schlichtkrull, M.S., Kipf, T.N., Bloem, P., van den Berg, R., Titov, I., Welling, M.: Modeling relational data with graph convolutional networks. In: The Semantic Web - 15th International Conference (ESWC), pp. 593–607 (2018)
18. Su, Y., Zhang, X., Wang, S., Fang, B., Zhang, T., Yu, P.S.: Understanding information diffusion via heterogeneous information network embeddings. In: Li, G., Yang, J., Gama, J., Natwichai, J., Tong, Y. (eds.) DASFAA 2019. LNCS, vol. 11446, pp. 501–516. Springer, Cham (2019). https://doi.org/10.1007/978-3-030-18576-3_30
19. Vaswani, A., et al.: Attention is all you need. In: Advances in Neural Information Processing Systems 30: Annual Conference on Neural Information Processing Systems 2017 (NIPS), pp. 5998–6008 (2017)
20. Velickovic, P., Cucurull, G., Casanova, A., Romero, A., Liò, P., Bengio, Y.: Graph attention networks. In: 6th International Conference on Learning Representations (ICLR) (2018)
21. Wang, J., Zheng, V.W., Liu, Z., Chang, K.C.: Topological recurrent neural network for diffusion prediction. In: 2017 IEEE International Conference on Data Mining (ICDM), pp. 475–484 (2017)
22. Wang, Z., Chen, C., Li, W.: Attention network for information diffusion prediction. In: Companion of the Web Conference 2018 on the Web Conference 2018, WWW 2018, pp. 65–66 (2018)
23. Wang, Z., Chen, C., Li, W.: A sequential neural information diffusion model with structure attention. In: Proceedings of the 27th ACM International Conference on Information and Knowledge Management, (CIKM), pp. 1795–1798 (2018)
24. Wang, Z., Chen, C., Li, W.: Information diffusion prediction with network regularized role-based user representation learning. ACM Trans. Knowl. Discov. Data 13(3), 29:1–29:23 (2019)
25. Xian, Y., Fu, Z., Muthukrishnan, S., de Melo, G., Zhang, Y.: Reinforcement knowledge graph reasoning for explainable recommendation. In: Proceedings of the 42nd International ACM SIGIR Conference on Research and Development in Information Retrieval (SIGIR), pp. 285–294 (2019)
26. Yang, C., Sun, M., Liu, H., Han, S., Liu, Z., Luan, H.: Neural diffusion model for microscopic cascade study. IEEE Trans. Knowl. Data Eng. 33(3), 1128–1139 (2021)

27. Yang, C., Tang, J., Sun, M., Cui, G., Liu, Z.: Multi-scale information diffusion prediction with reinforced recurrent networks. In: Proceedings of the Twenty-Eighth International Joint Conference on Artificial Intelligence (IJCAI), pp. 4033–4039 (2019)
28. Yao, L., Mao, C., Luo, Y.: Graph convolutional networks for text classification. In: The Thirty-Third AAAI Conference on Artificial Intelligence (AAAI), pp. 7370–7377 (2019)
29. Yuan, C., Li, J., Zhou, W., Lu, Y., Zhang, X., Hu, S.: DyHGCN: a dynamic heterogeneous graph convolutional network to learn users' dynamic preferences for information diffusion prediction, pp. 347–363 (2020)
30. Zhong, E., Fan, W., Wang, J., Xiao, L., Li, Y.: Comsoc: adaptive transfer of user behaviors over composite social network. In: The 18th ACM SIGKDD International Conference on Knowledge Discovery and Data Mining (KDD), pp. 696–704 (2012)

Diversify Search Results Through Graph Attentive Document Interaction

Xianghong Xu[1], Kai Ouyang[1], Yin Zheng[2], Yanxiong Lu[2],
Hai-Tao Zheng[1,3(✉)], and Hong-Gee Kim[4]

[1] Shenzhen International Graduate School, Tsinghua University, Shenzhen, China
{xxh20,oyk20}@mails.tsinghua.edu.cn
[2] Department of Search and Application, Weixin Group, Tencent, Beijing, China
alanlu@tencent.com
[3] Pengcheng Laboratory, Shenzhen 518055, China
zheng.haitao@sz.tsinghua.edu.cn
[4] Seoul National University, Seoul, South Korea
hgkim@snu.ac.kr

Abstract. The goal of search result diversification is to retrieve diverse documents to meet as many different information needs as possible. Graph neural networks provide a feasible way to capture the sophisticated relationship between candidate documents, while existing graph-based diversification methods require an extra model to construct the graph, which will bring about the problem of error accumulation. In this paper, we propose a novel model to address this problem. Specifically, we maintain a document interaction graph for the candidate documents of each query to model the diverse information interactions between them. To extract latent diversity features, we adopt graph attention networks (GATs) to update the representation of each document by aggregating its neighbors with learnable weights, which enables our model not dependent on knowing the graph structure in advance. Finally, we simultaneously compute the ranking score of each candidate document with the extracted latent diversity features and the traditional relevance features, and the ranking can be acquired by sorting the scores. Experimental results on TREC Web Track benchmark datasets show that the proposed model outperforms existing state-of-the-art models.

Keywords: Search result diversification · Graph attention networks · Document interaction

1 Introduction

Research shows that most queries provided by users are short and could be ambiguous [12, 23, 24]. Search result diversification aims to satisfy different information needs of users under ambiguous and short queries. For example, a query

---

X. Xu and K. Ouyang—Equal contribution.

A. Bhattacharya et al. (Eds.): DASFAA 2022, LNCS 13245, pp. 632–647, 2022.
https://doi.org/10.1007/978-3-031-00123-9_51

about "apple" may contain two different subtopics: one is about the famous company or brand, the other is about the fruit. Search result diversification methods need to retrieve a list of documents that covers different subtopics in this scenario.

Most of the traditional search result diversification models used manually craft functions with empirically tuned parameters [1, 2, 7, 11, 22]. In recent years, many researchers have tried to use machine learning methods in order to learn a ranking function automatically. Existing approaches to search result diversification can be mainly categorized into three classes [25]: explicit approaches [6, 7, 11, 13, 22] model subtopic coverage (explicit features) of the ranking results, implicit methods [2, 25, 30–32, 34] model the novelty of documents (implicit features), and ensemble models [17, 20] use both explicit and implicit features. The previous studies [7, 11, 13, 20] have demonstrated that models which use subtopic information always perform better than those that do not. However, subtopic mining itself is a sophisticated challenge in search result diversification [28], though there are some representative subtopic mining methods [8, 18, 21, 33], a more commonly used method in practice is collecting subtopics from search suggestions (such as Google Suggestions) [11, 13]. Some models [13, 17] even need subtopic information at the inference stage, it is not realistic in real scenarios for most search engines. Therefore, implicit models have attracted much attention [10], the most recent implicit model [25] leverages graph convolutional networks [15] (GCNs) to model the greedy selection process, it even outperforms existing explicit models. However, it requires an extra classifier to construct the graph, which will lead to classification errors and cause error accumulation in the diversification model [25]. Besides, though the ranking model is implicit, it requires the subtopic information to train the classifier.

To tackle the issues above, we propose a novel graph-based implicit diversification model. Inspired by the characteristics of graph attention networks [27] (GATs) that the representation of nodes can be updated with the information from their neighbors in different weights, and without the need to construct graph structure upfront, we adopt GAT to avoid training an extra model to construct the graph in our model. Similar to the previous graph-based model [25], we construct a graph for the candidate documents of each query, where each node in the graph can have interaction with its neighbors. Specifically, in order to capture latent diversity features, we elaborate graph attentive layers to update the representation of each node, which empowers each node can discriminatively exploit the information from its neighbors. Then the extracted diversity features are used by a neural diversity scorer to calculate the diversity scores. For simplicity, we calculate relevance scores with traditional relevance features by a neural relevance scorer. Finally, a parameter λ is used to balance the diversity and relevance of each document. In addition, the neural scorers can take the entire candidate document set as input, and generate the diversity score and relevance score of each document simultaneously. Besides, the graph solely needs the document representation as input, it does not require subtopic information. Since we adopt GAT to model document interaction for the search result diversification task, our model is denoted as GAT-DIV.

Overall, the main contributions of this paper are summarized as follows:

- We propose a novel graph-based model for search result diversification, which has an end-to-end structure, rather than the two-stage pipeline structure that the existing graph-based diversification model uses. It does not require additional models to construct the graphs, so it alleviates the error accumulation problem.
- We maintain a candidate document interaction graph and adapt GAT to extract the latent diversity features from the graph. The proposed method does not require the subtopic information, which is easier to generalize in real scenarios.
- Experimental results show that the proposed model outperforms the state-of-the-art models on diversity test benchmark TREC Web Track datasets.

2   Related Work

2.1   Search Result Diversification

From the perspective of using implicit and explicit features, existing approaches to search result diversification can be categorized into implicit, explicit, and ensemble models.

Implicit Models. Most implicit models follow the framework of MMR [2]:

$$S_{\text{MMR}}(q, d, C) = \lambda S^{rel}(d, q) - (1 - \lambda) \max_{d_j \in C} S^{sim}(d, d_j), \qquad (1)$$

where λ is the parameter to balance relevance and novelty, S^{rel} is the relevance score of d to the query q, S^{sim} is the similarity score of d to d_j, and C is the candidate document set. SVM-DIV [32] aimed to use structural SVM to model the diversity of the documents. R-LTR [34] considered search result diversification as a sequential selection process, and used likelihood loss to optimize the learning process. The neural tensor network (NTN) [31] considered diversity as the relationship of documents and can learn the relation function automatically. PAMM [30] proposed to optimize evaluation metrics directly instead of a loosely related loss function. Graph4DIV [25] used a document intent classifier based on BERT [9] to judge whether two documents cover the same intent, in order to construct a graph to model the document selection process.

Explicit Models. xQuAD [22] exploited query reformulation to discover different aspects of the query. PM2 [7] suggested retrieving documents by proportionality corresponding to each subtopic of the query. These two models are representative of traditional explicit methods, and researchers did further studies based on these models. TxQuAD and TPM2 [6] characterized user intents by terms coverage. A hierarchical structure was introduced in HxQuAD and HPM2 [11]. Supervised models can learn the parameters automatically instead of manually empirical tuning. A representative supervised method is DSSA [13], and many further studies are based on it. DSSA proposed a novel list-pairwise diversity ranking framework. It utilized recurrent neural networks (RNNs) to model the selected documents, and used the attention mechanism to capture the subtopic coverage based on the selected list.

Ensemble Models. Recently, some researchers have suggested that leveraging both explicit features and implicit features is helpful for getting diversity ranking. DESA [20] used Transformer [26] encoder and self-attention to model subtopic coverage and document novelty. DVGAN [17] generated training data by using explicit and implicit features based on a generative adversarial network structure.

2.2   Graph in Search Result Diversification

Graph is a common data structure to represent nonlinear relationships, it is widely used in IR. For example, PageRank [19] is a representative algorithm to model the importance of web pages. In search result diversification, Graph4DIV [25] is the first method that models the greedy selection process in search result diversification by graph neural networks.

3   Proposed Model

3.1   Problem Definition

Given a query q, the candidate document set of q is $\mathcal{D}$, and there are n candidate documents in $\mathcal{D}$. The search result diversification task aims to retrieve a new ranked list $\mathcal{R}$ based on candidate documents set $\mathcal{D}$, where the top-ranked documents can cover as many subtopics as possible. For the retrieved documents, both the relevance to the query and the diversity between documents are crucial for the task.

3.2   Architecture

For a query q and the associated candidate documents, the model can calculate the diversity score and the relevance score for each document simultaneously. The relevance features $\mathbf{R}_i$ of document d_i is generated by traditional methods, and the latent diversity features $\mathbf{H}$ is extracted from the graph attentive layers. In addition, the width of the edges in the graph represents the different attention coefficients, and there are K independent attention heads for each layer, which will be described in detail in Sect. 3.3. Figure 1 shows the architecture of GAT-DIV and the detailed process of calculating the ranking score of one document, please note that the ranking score of each candidate document can be calculated simultaneously. Then the final ranking list $\mathcal{R}$ can be generated by

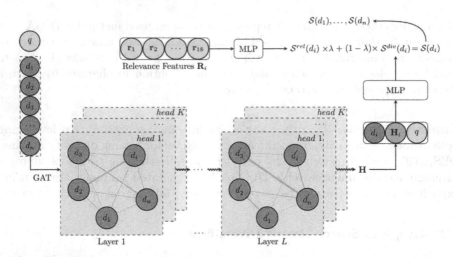

Fig. 1. Architecture of GAT-DIV, where $\mathbf{r}_j$ is the element of $\mathbf{R}_i$, $\mathbf{R}_i$ and $\mathbf{H}_i$ are the relevance features and latent diversity features of the i-th document, respectively. Lines of different thicknesses represent different weights.

ranking the scores of the candidate documents. In sum, the ranking scores $\mathcal{S}_\mathcal{D}$ of the candidate documents in $\mathcal{D}$ can be calculated by:

$$\mathcal{S}_\mathcal{D} = \underbrace{\lambda\mathcal{S}^{rel}(q,\mathcal{D})}_{\text{relevance}} + \underbrace{(1-\lambda)\mathcal{S}^{div}(\mathcal{D})}_{\text{diversity}}, \tag{2}$$

where λ is the parameter to balance the relevance scores $\mathcal{S}^{rel}(q,\mathcal{D})$ and the diversity scores $\mathcal{S}^{div}(\mathcal{D})$. Equation (2) is the common format of search result diversification methods that considering both the relevance and the diversity.

To calculate the relevance scores $\mathcal{S}^{rel}(q,\mathcal{D})$, we use the same relevance features $\mathbf{R}$ as the previous methods [13,17,20,25]. The simplified process of calculating the relevance score of a candidate document d_i can be described as:

$$\mathcal{S}^{rel}(q,d_i) = \text{MLP}(\mathbf{R}_i), \tag{3}$$

where MLP indicates a multi-layer perceptron layer, d_i is the i-th document in the candidate set $\mathcal{D}$, and $\mathbf{R}_i$ is the relevance features of d_i. For each document d_i, $\mathbf{R}_i$ has 18 features that illustrated in detail in Table 1. We use traditional relevance methods, rather than using pre-trained models to generate relevance features.

To calculate the diversity scores $\mathcal{S}^{div}(\mathcal{D})$, we solely use the implicit features of the candidate documents. The computation process can roughly divide into two stages: first, extracting the latent diversity features $\mathbf{H}$ of the documents through the document interaction graph. Then calculating the diversity scores of the candidate set $\mathcal{D}$ with the latent features $\mathbf{H}$, the representation of query and

Table 1. The relevance features of a document. Each of the first 3 features is applied to 5 parts of the document: body, title, URL, anchor, and the whole document, respectively.

Name	Description	#Features
TF-IDF	TF-IDF model	5
BM25	BM25 with default parameters	5
LMIR	LMIR with Dirichlet smoothing	5
PageRank	PageRank score	1
#inlinks	Number of inlinks	1
#outlinks	Number of outlinks	1

the documents. The process of computing the diversity score of one document is demonstrated as follows:

$$\mathbf{H}(d_i) = \mathcal{F}(\mathbf{E}(d_i), \mathcal{G}_\mathcal{D}), \tag{4}$$

$$\mathcal{S}^{div}(d_i) = \text{MLP}(q, \mathbf{E}^{(0)}(d_i), \mathbf{H}(d_i)), \tag{5}$$

where $\mathbf{H}(d_i)$ indicates the latent diversity features of document d_i, $\mathbf{E}(d_i)$ is the total representation of d_i in the graph, $\mathbf{E}^{(0)}(d_i)$ is the initial representation of d_i, $\mathcal{F}$ is the function that describes how to extract latent diversity features of document d_i, $\mathcal{G}_\mathcal{D}$ is the corresponding interaction graph for the candidate document set $\mathcal{D}$ of the query q, MLP indicates another multi-layer perceptron layer which is different from that used in calculating the relevance score.

The process of computing diversity scores $\mathcal{S}^{div}(\mathcal{D})$ is the focus of our model, there are two key components of the computation process. (1) Graph construction and the node aggregation algorithms. Similar to the previous graph-based method [25], we build an interaction graph for each query. Different from the previous method, we do not require an extra model to judge whether the nodes in the graph are connected, it is a complete graph that each node connects with other nodes. The node aggregation process will consider learnable weights of different neighbors, and the latent diversity features can be extracted from the graph. (2) Diversity scoring based on the features in the graph. The latent diversity representation will be utilized in the scoring process. The initial representation of the nodes is $\mathbf{E}^{(0)} = [\mathbf{E}^{(0)}(d_1), \ldots, \mathbf{E}^{(0)}(d_n)]$, it will be updated on the l-th graph attentive layer as $\mathbf{E}^{(l)} = [\mathbf{E}^{(l)}(d_1), \ldots, \mathbf{E}^{(l)}(d_n)]$. The latent diversity features $\mathbf{H}$ can be computed with the total representation $\mathbf{E} = [\mathbf{E}^{(0)}, \ldots, \mathbf{E}^{(L)}]$ through the function $\mathcal{F}$. Finally, input the initial representation, the latent features, and the query into the MLP diversity scorer.

3.3 Diversity Scoring

To calculate the diversity score $\mathcal{S}^{div}(d_i)$ for the document d_i, latent diversity features $\mathbf{H}_i$ are needed. In this section, we introduce the graph construction, node aggregation, and latent diversity features generation algorithms for diversity scoring.

Graph Construction. Given the current query q and the corresponding candidate document set $\mathcal{D}$, we construct a document interaction graph $\mathcal{G}_{\mathcal{D}}$ by considering each document in $\mathcal{D}$ as a node in $\mathcal{G}_{\mathcal{D}}$, and the total number of nodes in $\mathcal{G}_{\mathcal{D}}$ is $n = |\mathcal{D}|$. Therefore, the adjacent matrix $\mathbf{A}$ for the interaction graph $\mathcal{G}_{\mathcal{D}}$ can be defined as:

$$\mathbf{A}[i,j] = \begin{cases} 1, \text{ if } d_i \text{ needs to interact with } d_j; \\ 0, \text{ else.} \end{cases} \tag{6}$$

where $\mathbf{A} \in \mathbb{R}^{n \times n}$, $\mathbf{A}[i,j]$ is the i-th row and j-th column element of $\mathbf{A}$, which indicates the information interaction need between d_i and d_j. Based on the perspective that the information from all the documents in $\mathcal{D}$ is helpful to generate the latent diversity features of one document, each element in $\mathbf{A}$ is 1, which means that the document interaction graph is a complete graph.

Node Aggregation. First, we will briefly introduce the graph attentive layer adapted in this paper. The initial representation of the nodes in $\mathcal{G}_{\mathcal{D}}$ denote as $\mathbf{E}^{(0)} = [\mathbf{E}^{(0)}(d_1), \ldots, \mathbf{E}^{(0)}(d_n)]$, and $\mathbf{E}^{(0)}(d_i) \in \mathbb{R}^F$ where F is the number of initial features of each node. A graph attentive layer can generate a new feature presentation for each node $\mathbf{E}^{(1)} = [\mathbf{E}^{(1)}(d_1), \ldots, \mathbf{E}^{(1)}(d_n)]$ and $\mathbf{E}^{(1)}(d_i) \in \mathbb{R}^{F'}$. In order to get sufficient expressive power of the graph, we adapt a learnable weight matrix $\mathbf{W}^{(0)} \in \mathbb{R}^{F' \times F}$ to transform the initial features to the high-level features. Then attention mechanism is applied on the nodes:

$$e_{ij}^{(0)} = \mathcal{A}\left[\mathbf{W}^{(0)}\mathbf{E}^{(0)}(d_i), \mathbf{W}^{(0)}\mathbf{E}^{(0)}(d_j)\right], \tag{7}$$

$$e_{ij}^{(l)} = \mathcal{A}\left[\mathbf{W}^{(l)}\mathbf{E}^{(l)}(d_i), \mathbf{W}^{(l)}\mathbf{E}^{(l)}(d_j)\right], \tag{8}$$

where $\mathbf{W}^{(0)} \in \mathbb{R}^{F' \times F}$, $\mathbf{W}^{(l)} \in \mathbb{R}^{F' \times F'}, l > 0$, $\cdot^{(l)}$ indicates the element at layer l, $\mathcal{A} : \mathbb{R}^{F'} \times \mathbb{R}^{F'} \to \mathbb{R}$ is a shared function to compute the attention score for each layer.

To make attention coefficients easily comparable across different nodes, we can normalize them through:

$$\alpha_{ij}^{(l)} = \frac{\exp(e_{ij}^{(l)})}{\sum_{k \in \mathcal{N}_i} \exp(e_{ik}^{(l)})}, \tag{9}$$

where $\mathcal{N}_i$ is the set of neighbors of node i.

The attention score in this paper can be calculated as follows:

$$\mathcal{A}\left[v_i^{(l)}, v_j^{(l)}\right] = \text{MLP}(v_i^{(l)}) + \text{MLP}(v_j^{(l)}), \tag{10}$$

where $v_i^{(l)}$ and $v_j^{(l)}$ is the transformed representation $\mathbf{W}^{(l)}\mathbf{E}^{(l)}(d_i)$ and $\mathbf{W}^{(l)}\mathbf{E}^{(l)}(d_j)$ at layer l for node i and node j, respectively.

Algorithm 1: Node Aggregation at layer l

Input: document representation $\mathbf{E}^{(l)}$ at l-th layer, adjacent matrix $\mathbf{A}$ for $\mathcal{G}_\mathcal{D}$.
Output: document representation $\mathbf{E}^{(l+1)}$ at $(l+1)$-th layer.

1   $V^{(l)} \leftarrow \emptyset$
2   **for** $k : 0 \rightarrow K - 1$ **do**
3   **for** $i : 0 \rightarrow n - 1$ **do**
4       $\lfloor \; v_i \leftarrow \mathbf{W}_k^{(l)} \mathbf{E}^{(l)}(d_i)$
5   **for** $i : 0 \rightarrow n - 1$ **do**
6   **for** $j : 0 \rightarrow n - 1$ **do**
7         $\lfloor \; e_{ij} \leftarrow \mathcal{A}(v_i, v_j)$
8       $\alpha_{k,ij} \leftarrow \mathrm{softmax}_j(e_{ij})$
9       $Agg_i \leftarrow \sum_{j \in \mathcal{N}_i} \alpha_{k,ij}^{(l)} v_j$
10    $\lfloor \; V^{(l)} \leftarrow [V^{(l)} \| Agg]$ //multi-head concatenation
11   $V^{(l+1)} \leftarrow \mathbf{O}^{(l)} V^{(l)} + \mathbf{b}^{(l)}$
12   **return** $V^{(l+1)}$

In order to stabilize the attention mechanism learning process, we adapt multi-head attention similar to [26,27]. Specifically, if K independent attention mechanism is applied, the output of the l-th layer is:

$$h_i^{(l)} = \mathbf{O}^{(l)} \left(\mathop{\Big\|}_{k=1}^{K} \left[\sum_{j \in \mathcal{N}_i} \alpha_{k,ij}^{(l)} \mathbf{W}_k^{(l)} \mathbf{E}^{(l)}(d_j) \right] \right) + \mathbf{b}^{(l)}, \tag{11}$$

where $h_i^{(l)}$ is the output representation of i-th node in the l-th layer, $\mathbf{O}^{(l)} \in \mathbb{R}^{F' \times KF'}$ and $\mathbf{b}^{(l)} \in \mathbb{R}^{F'}$ are learnable linear transformation and bias in order to compute the output of the concatenation of multi-head attention mechanism. $\cdot_k$ indicates the k-th independent attention mechanism.

According to Eq. (10) and (11), the node aggregation algorithm for node i at layer l can be briefly described as:

$$\mathbf{E}^{(l+1)}(d_i) = h_i^{(l)}, \tag{12}$$

where $\mathbf{E}^{(l+1)}(d_i)$ is the representation of d_i at $(l+1)$-th layer, $h_i^{(l)}$ is the output of l-th layer.

Therefore, the representation of a document at layer l can interact with the information from other documents and generate a new representation by Eq. (12). The node aggregation algorithm at layer l is summarized as Algorithm 1.

Latent Diversity Features Generation. Empirically, stacking layers can give the model more capacity to model the document representation with the information from other documents, and the latent diversity features for the documents is the output of the last layer. The process of generating latent diversity features $\mathbf{H}$ is summarized as Algorithm 2.

Algorithm 2: Generating Latent Diversity Features

    Input: inital document representation $\mathbf{E}^{(0)}$.
    Output: latent diversity features $\mathbf{H}$.
1 $\mathbf{H} \leftarrow \emptyset$
2 $\mathbf{E} \leftarrow \mathbf{E}^{(0)}$
3 **for** $l : 0 \rightarrow L - 1$ **do**
4     $\lfloor$ $\mathbf{E} \leftarrow NodeAggregation(\mathbf{E})$
5 $\mathbf{H} \leftarrow \mathbf{E}$
6 **return** $\mathbf{H}$

Computing the Diversity Score. According to Eq. (5), to calculate the diversity score S^{div} of d_i, the representation of q, the initial representation of the document, and the latent diversity features of the document are needed. The key component have described in Algorithm 1 and 2.

3.4 Optimization and Ranking

Given a context sequence $\mathcal{C}$ and a pair of positive and negative documents d_i and d_j associated with a query, a training sample $(\mathcal{C}, d_i, d_j)$ can transform into two ranking sequences $r_1 = [\mathcal{C}, d_i]$ and $r_2 = [\mathcal{C}, d_j]$. Then constructing a document interaction graph for each list to extract the latent diversity features, the attention coefficients are reflected in the width of the edges in the figure. Finally, calculating the ranking score and the metric for each ranking sequence based on the latent features, and computing the loss to train the model.

The List-Pairwise Data Sampling. Because of the dataset limitation of search result diversification task, we used the list-pairwise sampling method as the previous models [13,17,20,25] did to generate enough training data.

    The sampling process is demonstrated as follows: given a query and the corresponding candidate documents, the context sequence $\mathcal{C}$ is the subsequence that selected in random lengths from the optimal ranking or random ranking. Then sample two documents d_i and d_j in the ranking list apart from $\mathcal{C}$ to generate two sequence $r_1 = [\mathcal{C}, d_1]$ and $r_2 = [\mathcal{C}, d_2]$. And the metric of positive ranking sequence $M(r_1)$ should be better than the negative sequence $M(r_2)$.

The Loss Function. Based on list-pairwise data sampling, the loss function is defined as:

$$\mathcal{L} = \sum_{q \in \mathcal{Q}} \sum_{s \in S_q} |\Delta M| \left[y_s \log(P(r_1, r_2)) + (1 - y_s) \log(1 - P(r_1, r_2)) \right], \quad (13)$$

where s is a training sample, S_q is all the training samples of q, $\Delta M = M(r_1) - M(r_2)$ represents the weight of the sample, $y_s = 1$ for the weight is positive, otherwise $y_s = 0$. $P(r_1, r_2) = \sigma(s_{r_1} - s_{r_2})$ indicates the probability of the sample being positive, where s_{r_i} is the ranking score of r_i calculated by the model.

The Ranking Process. Different from greedy selection models, our model does not require maintaining a selected list to select the best document one by one. The model will take the entire candidate document set as input, and jointly return the ranking scores with the consideration of relevance and diversity of all candidate documents. More details can be found in [20].

4   Experimental Settings

4.1   Data Collections

We conducted experiments on TREC Web Track dataset[1] from 2009 to 2012, which is the same as the previous works [13,17,20,25]. There are 200 queries in the dataset in total, however 2 queries (#95 and #100) have no subtopic judgment, so there are 198 queries in the experiment. Each of the queries has 3 to 8 subtopics, and the relevance judgment of corresponding candidate documents are given in binary at the subtopic level by the TREC assessors. Explicit models will use subtopic information, and implicit models will not. Our model does not require subtopic information.

4.2   Evaluation Metrics

We use TREC official diversity evaluation metrics of α-nDCG [4] and ERR-IA [3]. Besides, we also take NRBP [5] into consideration. The parameter α is set to 0.5 as the default settings given by TREC evaluation program. Consistent with previous search result diversification models [13,20,25,30,31,34] and TREC Web Track, we adopt the top 50 results of Lemur[2] for diversity re-ranking. These metrics are computed by official evaluation program on the top 20 documents of retrieved ranking lists.

4.3   Baseline Models

We compare our models with various models which can be categorized into 4 classes as follows:

Non-diversified Models. Lemur: For a fair comparison, we use the same result as [13,25]. **ListMLE** [29] is a representative learning-to-rank method without counting diversity.

---

[1] https://boston.lti.cs.cmu.edu/Data/clueweb09/.
[2] Lemur service: http://boston.lti.cs.cmu.edu/Services/clueweb09_batch.

Explicit Models. xQuAD [22], **PM2** [7], **TxQuAD, TPM2** [6], **HxQuAD, HPM2** [11] and **DSSA** [13]. These models are some representative unsupervised explicit baseline models for comparison. DSSA is a state-of-the-art supervised explicit method that model selected documents using RNNs and select documents with greedy strategy from the candidate documents using subtopic attention. All these 7 models use the parameter λ to balance the relevance and diversity of candidate documents linearly. HxQuAD and HPM2 use an additional parameter α to control the weights of subtopics in different hierarchical layers.

Implicit Models. R-LTR [34], **PAMM** [30], **NTN** [31] and **Graph4DIV** [25]. These models are some representative supervised implicit baseline models for comparison. For PAMM, α-nDCG@20 is used as the optimization metrics and the number of positive rankings l^+ and negative rankings l^- is tuned for each query. The neural tensor network (NTN) can be used on both R-LTR and PAMM, so the two models are denoted as R-LTR-NTN and PAMM-NTN, respectively. Graph4DIV is a state-of-the-art implicit method that uses graph convolutional networks to model the relationship of documents and the query, it selects the best candidate document greedily.

Ensemble Models. DESA [20] and **DVGAN** [17]. These two methods are state-of-the-art models that associate both explicit features and implicit features to get ranking results. DESA uses an encoder-decoder structure. DVGAN has a generative adversarial network structure.

4.4   Implementation Details

We conduct 5-fold cross-validation experiments on the pre-processed version dataset with the same data subset split as in [13, 20, 25]. We use doc2vec [16] embeddings as the initial queries and documents with the dimension of 100, which is the same as the previous state-of-the-art explicit model [13], implicit model [25], and ensemble model [20]. We use the following optimizer and model configuration settings: the optimizer is Adam [14] with learning rate $\eta = 0.001$. The neural diversity scorer has 2 hidden layers with dimensions 256 and 64, followed by a one-dimension full connection output layer. The neural relevance scorer has 2 hidden layers with dimensions 18 and 8, followed by a one-dimension full connection output layer. The parameter λ is fixed as 0.5, same as the previous studies [13, 20, 25].

5   Experimental Results

5.1   Overall Results

Table 2 shows the overall results of all baseline models and our model. GAT-DIV outperforms all the baseline methods in terms of α-nDCG, ERR-IA, and NRBP.

Table 2. Performance comparison of all models. The best result overall is in bold, the underline is the best result of all baselines.

	ERR-IA	α-nDCG	NRBP
Lemur	.271	.369	.232
ListMLE	.287	.387	.249
xQuAD	.317	.413	.284
TxQuAD	.308	.410	.272
HxQuAD	.326	.421	.294
PM2	.306	.401	.267
TPM2	.291	.399	.250
HPM2	.317	.420	.279
R-LTR	.303	.403	.267
PAMM	.309	.411	.271
R-LTR-NTN	.312	.415	.275
PAMM-NTN	.311	.417	.272
DSSA (2017)	.356	.456	.326
DESA (2020)	.363	.464	.332
DVGAN (2020)	.367	.465	.334
Graph4DIV (2021)	.370	.468	.338
GAT-DIV (Ours)	**.380**	**.476**	**.351**
Improv	2.9%	1.8%	3.8%

Compared with non-diversified and the conventional representative diversification models, GAT-DIV consistently outperforms these models. Because non-diversified solely consider the relevance between the query. Some conventional models use handcrafted ranking functions, and others directly model the similarity between documents as the diversity.

Compared with the state-of-the-art models, DSSA is the state-of-the-art explicit model which leverages RNNs, attention mechanism, and greedy selection strategy to generate the diversity ranking list. Graph4DIV is the state-of-the-art implicit method that uses a document intent classifier based on BERT to judge whether two documents covering the same subtopic, and applies GCN to model the greedy selection process. As for the two state-of-the-art ensemble methods, DESA has a global optimal diversity ranking framework and DVGAN uses generative adversarial network architecture. To sum up, our model can outperform all the state-of-the-art explicit, implicit, and ensemble methods. Compared with the best performing baseline model, our method achieves statistically significant improvements on α-nDCG, ERR-IA, and NRBP by about 1.8%, 2.9%, and 3.8%, respectively.

Graph4DIV is an impressive implicit graph-based model, because it outperforms many models which use subtopic information and demonstrates that

Table 3. Performance of GAT-DIV with different settings.

(L, K, F')	ERR-IA	α-nDCG	NRBP
(1,3,256)	.365	.464	.335
(1,4,256)	**.380**	**.476**	**.351**
(1,5,256)	.367	.465	.334
(2,3,256)	.368	.467	.337
(2,4,256)	.374	.472	.344
(2,5,256)	.371	.469	.340
(3,4,256)	.361	.461	.328
(1,4,512)	.367	.466	.335
(1,4,128)	.376	.471	.346

appropriate modeling structures can alleviate the lack of subtopic information. Our graph-based model does not require additional models to construct the graphs and enables each document discriminatively has interaction with other documents.

5.2 Discussion and Ablation Study

Our model is implicit and concise, the effectiveness of GAT-DIV is the joint efforts of the global ranking strategy and the latent diversity feature extraction structure. We conducted further experiments to investigate the effects of different GAT-DIV settings and the effects of different latent diversification feature extraction structures.

Effects of GAT Settings. The number of graph attentive layers, the number of independent attention heads of each layer, and the number of hidden features F' are crucial in our model. We conduct a series of experiments to investigate the effects of these parameters, and the experimental results of different settings of the number of layers L, heads K, and hidden features F' are shown in Table 3. Based on the results, we can discover that stacking graph attentive layers may not improve the ability to capture the latent diversity features. One possible reason is the graph in the proposed model is a complete graph, which empowers each node in the graph can directly have interaction with all other documents. Therefore, the number of layers, heads, and hidden features in this paper are set as 1, 4, and 256, respectively.

Effects of Different Feature Extraction Structures. In order to explore the performance of different feature extraction structures, we conduct experiments on using GCNs of the different number of layers with the same adjacent matrix as GAT-DIV to capture latent diversity features, denoted as n-layer GCN, where n is the number of layers. For a comprehensive and fair comparison, we study

Table 4. Performance of different diversity feature extraction structures.

	ERR-IA	α-nDCG	NRBP
Graph4DIV	.370	.468	.338
1-layer GCN	.367	.465	.337
2-layer GCN	.372	.471	.342
3-layer GCN	.375	.471	.344
4-layer GCN	.374	.471	.344
GAT-DIV	**.380**	**.476**	**.351**

n-layer GCN because Graph4DIV [25] uses GCN to generate diversified ranking results either. Specifically, n is set from 1 to 4, the adjacent matrix and the number of hidden features are the same as in GAT-DIV, and the experimental results are shown in Table 4. On the one hand, GAT-DIV outperforms n-layer GCN models, because the latter treat the importance of each node's neighbors equally, while the former discriminatively exploit the information from other documents. It demonstrates the effectiveness of the elaborated feature extraction structure. On the other hand, n-layer GCN models outperform Graph4DIV, the latter uses a greedy selection strategy, and the former are non-greedy ranking methods. Graph4DIV uses pre-trained language models to construct the graphs, however, n-layer GCN models slightly outperform Graph4DIV. So it illustrates that we do not require extra models to construct the graphs to some extent. Furthermore, n-layer GCN models and GAT-DIV outperforms DESA, though these graph-based methods use complete graphs, the empirical studies show that they are more effective than DESA. It also demonstrates the effectiveness of our feature extraction structures.

To sum up, the experimental results in Table 3 and 4 demonstrate that the non-greedy ranking strategy and the elaborate latent diversity feature extraction structure are both beneficial to the search result diversification task.

6 Conclusion

In this paper, we proposed a novel graph-based implicit method for the search result diversification task. Our model first constructs a document interaction graph for each query, then models the interaction between each document and its neighbors by adopting the graph attention network. To extract latent diversity features, we design node aggregation and latent diversity feature extraction algorithms. The proposed model can take the whole candidate document set as input, then respectively calculates the diversity and relevance score of each document simultaneously by the extracted diversity features and traditional relevance features, finally jointly returns the ranking score of each document. Experimental results on TREC diversity test benchmark datasets demonstrate the superiority of our model compared to existing state-of-the-art models. In the future, we plan to explore more effective diversity feature extraction structures, and apply our diversity feature extraction structure to the explicit search result diversification model.

Acknowledgement. This research is supported by National Natural Science Foundation of China (Grant No. 6201101015), Beijing Academy of Artificial Intelligence (BAAI), Natural Science Foundation of Guangdong Province (Grant No. 2021A1515012640), the Basic Research Fund of Shenzhen City (Grand No. JCYJ20210324120012033 and JCYJ20190813165003837), Overseas Cooperation Research Fund of Tsinghua Shenzhen International Graduate School (Grant No. HW2021008), and research fund of Tsinghua University - Tencent Joint Laboratory for Internet Innovation Technology.

References

1. Agrawal, R., Gollapudi, S., Halverson, A., Ieong, S.: Diversifying search results. In: WSDM, pp. 5–14 (2009)
2. Carbonell, J., Goldstein, J.: The use of MMR, diversity-based reranking for reordering documents and producing summaries. In: SIGIR, pp. 335–336 (1998)
3. Chapelle, O., Metlzer, D., Zhang, Y., Grinspan, P.: Expected reciprocal rank for graded relevance. In: CIKM, pp. 621–630 (2009)
4. Clarke, C.L., et al.: Novelty and diversity in information retrieval evaluation. In: SIGIR, pp. 659–666 (2008)
5. Clarke, C.L.A., Kolla, M., Vechtomova, O.: An effectiveness measure for ambiguous and underspecified queries. In: Azzopardi, L., et al. (eds.) ICTIR 2009. LNCS, vol. 5766, pp. 188–199. Springer, Heidelberg (2009). https://doi.org/10.1007/978-3-642-04417-5_17
6. Dang, V., Croft, B.W.: Term level search result diversification. In: SIGIR, pp. 603–612 (2013)
7. Dang, V., Croft, W.B.: Diversity by proportionality: an election-based approach to search result diversification. In: SIGIR, pp. 65–74 (2012)
8. Dang, V., Xue, X., Croft, W.B.: Inferring query aspects from reformulations using clustering. In: CIKM, pp. 2117–2120 (2011)
9. Devlin, J., Chang, M.W., Lee, K., Toutanova, K.: BERT: pre-training of deep bidirectional transformers for language understanding. In: NAACL, pp. 4171–4186 (2019)
10. Goswami, A., Zhai, C., Mohapatra, P.: Learning to diversify for e-commerce search with multi-armed bandit. In: SIGIR Workshop (2019)
11. Hu, S., Dou, Z., Wang, X., Sakai, T., Wen, J.R.: Search result diversification based on hierarchical intents. In: CIKM, pp. 63–72 (2015)
12. Jansen, B.J., Spink, A., Saracevic, T.: Real life, real users, and real needs: a study and analysis of user queries on the web. Inf. Process. Manag. **36**(2), 207–227 (2000)
13. Jiang, Z., Wen, J.R., Dou, Z., Zhao, W.X., Nie, J.Y., Yue, M.: Learning to diversify search results via subtopic attention. In: SIGIR, pp. 545–554 (2017)
14. Kingma, D.P., Ba, J.: Adam: a method for stochastic optimization. In: 3rd International Conference on Learning Representations, ICLR 2015, San Diego, CA, USA, 7–9 May 2015, Conference Track Proceedings (2015)
15. Kipf, T.N., Welling, M.: Semi-supervised classification with graph convolutional networks. In: ICLR (2017)

16. Le, Q., Mikolov, T.: Distributed representations of sentences and documents. In: International Conference on Machine Learning, pp. 1188–1196. PMLR (2014)

17. Liu, J., Dou, Z., Wang, X., Lu, S., Wen, J.R.: DVGAN: a minimax game for search result diversification combining explicit and implicit features. In: SIGIR, pp. 479–488 (2020)

18. Nguyen, T.N., Kanhabua, N.: Leveraging dynamic query subtopics for time-aware search result diversification. In: de Rijke, M., Kenter, T., de Vries, A.P., Zhai, C.X., de Jong, F., Radinsky, K., Hofmann, K. (eds.) ECIR 2014. LNCS, vol. 8416, pp. 222–234. Springer, Cham (2014). https://doi.org/10.1007/978-3-319-06028-6_19

19. Page, L., Brin, S., Motwani, R., Winograd, T.: The pagerank citation ranking: bringing order to the web. Technical report, Stanford InfoLab (1999)

20. Qin, X., Dou, Z., Wen, J.R.: Diversifying search results using self-attention network. In: CIKM, pp. 1265–1274 (2020)

21. Rafiei, D., Bharat, K., Shukla, A.: Diversifying web search results. In: WWW, pp. 781–790 (2010)

22. Santos, R.L., Macdonald, C., Ounis, I.: Exploiting query reformulations for web search result diversification. In: WWW, pp. 881–890 (2010)

23. Silverstein, C., Marais, H., Henzinger, M., Moricz, M.: Analysis of a very large web search engine query log. In: ACM SIGIR Forum, vol. 33, pp. 6–12. ACM New York (1999)

24. Song, R., Luo, Z., Wen, J.R., Yu, Y., Hon, H.W.: Identifying ambiguous queries in web search. In: WWW, pp. 1169–1170 (2007)

25. Su, Z., Dou, Z., Zhu, Y., Qin, X., Wen, J.R.: Modeling intent graph for search result diversification. In: SIGIR (2021)

26. Vaswani, A., et al.: Attention is all you need. In: NeurIPS, pp. 5998–6008 (2017)

27. Velickovic, P., Cucurull, G., Casanova, A., Romero, A., Liò, P., Bengio, Y.: Graph attention networks. In: ICLR (2018)

28. Wang, C.J., Lin, Y.W., Tsai, M.F., Chen, H.H.: Mining subtopics from different aspects for diversifying search results. Inf. Retrieval 16(4), 452–483 (2013)

29. Xia, F., Liu, T.Y., Wang, J., Zhang, W., Li, H.: Listwise approach to learning to rank: theory and algorithm. In: ICML, pp. 1192–1199 (2008)

30. Xia, L., Xu, J., Lan, Y., Guo, J., Cheng, X.: Learning maximal marginal relevance model via directly optimizing diversity evaluation measures. In: SIGIR, pp. 113–122 (2015)

31. Xia, L., Xu, J., Lan, Y., Guo, J., Cheng, X.: Modeling document novelty with neural tensor network for search result diversification. In: SIGIR, pp. 395–404 (2016)

32. Yue, Y., Joachims, T.: Predicting diverse subsets using structural SVMs. In: ICML, pp. 1224–1231 (2008)

33. Zheng, W., Fang, H., Yao, C.: Exploiting concept hierarchy for result diversification. In: CIKM, pp. 1844–1848 (2012)

34. Zhu, Y., Lan, Y., Guo, J., Cheng, X., Niu, S.: Learning for search result diversification. In: SIGIR, pp. 293–302 (2014)

On Glocal Explainability of Graph Neural Networks

Ge Lv[1]($\boxtimes$) iD, Lei Chen[1] iD, and Caleb Chen Cao[2] iD

[1] The Hong Kong University of Science and Technology, Hong Kong, China
{glvab,leichen}@cse.ust.hk
[2] Huawei Technologies, Hong Kong, China
caleb.cao@huawei.com

Abstract. Graph Neural Networks (GNNs) derive outstanding performance in many graph-based tasks, as the model becomes more and more popular, explanation techniques are desired to tackle its black-box nature. While the mainstream of existing methods studies instance-level explanations, we propose Glocal-Explainer to generate model-level explanations, which consumes local information of substructures in the input graph to pursue global explainability. Specifically, we investigate faithfulness and generality of each explanation candidate. In the literature, fidelity and infidelity are widely considered to measure faithfulness, yet the two metrics may not align with each other, and have not yet been incorporated together in any explanation technique. On the contrary, generality, which measures how many instances share the same explanation structure, is not yet explored due to the computational cost in frequent subgraph mining. We introduce adapted subgraph mining technique to measure generality as well as faithfulness during explanation candidate generation. Furthermore, we formally define the *glocal explanation generation problem* and map it to the classic weighted set cover problem. A greedy algorithm is employed to find the solution. Experiments on both synthetic and real-world datasets show that our method produces meaningful and trustworthy explanations with decent quantitative evaluation results.

Keywords: Graph neural network · Explainability · XAI

1 Introduction

Graph neural networks (GNNs) have received significant attention due to their outstanding performance in various real-life applications, such as social networks [12], citation graphs [6], knowledge graphs [33], recommendation systems [14] and so on. GNNs generally adopt a message propagation strategy to recursively aggregate neural information from neighboring nodes and links [10,23,26] to capture both node features and graph topology. Owing to the broad usage of graph data and promising effectiveness, GNNs become increasingly important and popular. Nonetheless, the model shares the same black-box nature with

A. Bhattacharya et al. (Eds.): DASFAA 2022, LNCS 13245, pp. 648–664, 2022.
https://doi.org/10.1007/978-3-031-00123-9_52

other deep learning techniques, practitioners can hardly understand their decision mechanism, so they cannot fully trust the model. Hence human-intelligible explanation on how GNNs make prediction is intensively desired so that the models can be inspected and possibly amended by domain experts before being deployed in crucial applications.

In recent years, extensive research efforts have been devoted to studying the explainability of GNNs [16,18,24,29,30]. The task is nontrivial since, different from images and texts, graphs do not follow certain grid-like or sequential structure, and encompass rich relational information, which blocks existing explaining techniques for images and texts to be directly applied. Specifically, explaining a pretrained GNN is to determine a substructure[1] of an input graph that is critical for the model to make a prediction [18,29]. The main stream of exiting methods studies instance-level explainability, which generates input-specific explanations [18,24,29]. While global (model-level) explanation is much less explored, which outputs a set of input-independent explanations for a target class [16,30]. As a matter of fact, both levels of explainability are important because instance-level techniques are more precise locally and specific for a target instance, while model-level methods can unveil high-level knowledge that the model learned from the training set.

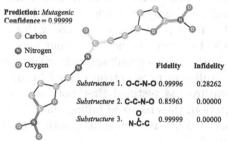

Fig. 1. An example from the MUTAG dataset showing the two metrics fidelity and infidelity may not align with each other.

For both categories of works, the fundamental challenge lies in how to qualify explainability. In the literature, **faithfulness** has been constantly focused on to measure explainability, which examines if an explanation faithfully explains the model's behavior [25,31]. A number of tools from various theories are borrowed to measure faithfulness, including Mutual Information [16,29], conditional probability [24], Input Optimization [30], Shapley Value [32] and so on. In general, these metrics either measure fidelity [18,31], which equals the model's predicted confidence drop when removing a candidate substructure [32]; or infidelity [13,31], which equals the model's confidence change when solely keeping the substructure [16,24,29,30]. Though both metrics are decent tools for validating faithfulness, they may not align with each other in pursuing optimal explainability even though their way of perturbing the input graph complement one another. To give an example, Fig. 1 shows an instance from the MUTAG dataset[2], which consists of molecular structures classified by

---

[1] In this work, we explicitly focus on explanation for topology structure of the input graph since feature explanation in GNNs is analogous to that in non-graph based neural networks, which has been widely studied [25].

[2] The MUTAG dataset [9] is widely used to study the explainability of GNN on graph classification task in many existing works [7,29,30,32].

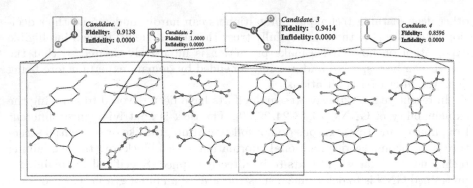

Fig. 2. A set of selected examples from the MUTAG dataset with a set of explanation candidates, whose explained instances are boxed in different colors. (Color figure online)

their mutagenic effect. The instance is predicted *Mutagenic* by a pretrained GNN with a confidence of 0.999999. Three candidate substructures of the instance are highlighted by bold edges in different colors (red, blue and yellow, respectively), their fidelity and infidelity values are also shown. While *Substructures* 1 and 2 retain superiority in either fidelity or infidelity only, *Substructure* 3 achieves the best of both worlds, which makes it the most promising candidate with the highest faithfulness to the pretrained GNN. However, to the best of our knowledge, none of the existing works consider both metrics simultaneously given the difficulty that the two measurements from different perspectives cannot be simply added up to optimize. The rich literature has testified that both metrics are vital to the qualifying an ideal explanation, thus to avoid falling short from either perspective, a technique to construct an objective concerning both fidelity and infidelity of explanations is in great need.

On the contrary of faithfulness being intensively studied, **generality** of explanations has not yet been explored. An explanation is a subgraph of the input graph by definition, an explanation for multiple instances is, hence, a common subgraph of instances being explained by the light of nature. Intuitively, an explanation shared by more instances should be considered with higher global explainability, as it represents the general topological characteristics of the group; we term this property "generality" of an explanation. Shown in Fig. 2 is a set of instances from the MUTAG dataset with a number of explanation candidates that can potentially "explain" them[3], whose explained instances are boxed in corresponding colors. Based on the number of instances a substructure can explain, *Candidate* 3 is the best since it has the highest generality in addition to high fidelity and infidelity scores. In contrast, though *Candidate* 2 has better measurement regarding faithfulness, it can convey local explainability on one single instance only. Similarly, *Candidate* 1 and 4 are not optimal either. Higher generality implies a substructure is truly the high-level knowledge learned by the GNN from the training set. As illus-

---

[3] For the ease of demonstration, assume the set of instances one candidate can explain is known for the moment.

trated in the example, generality connects the explanation of each single instance to the discriminative characteristic of the class captured by the GNN. The main issue that hinders the development in investigating the generality of explanations is the computational cost of finding optimal common substructures for multiple graphs [29]. Tackling the issue requires subgraph isomorphism checking, which is NP-hard and embedding-based methods cannot ensure topological isomorphism. Nonetheless, measuring generality reveals the true correlation between instance- and model- level explanation, which is essential to finding global explanation.

To build a unified system that qualifies explanations based on both fidelity and infidelity metrics as well as generality, we propose Glocal-Explainer, a novel framework for finding model-level explanation. Glocal-Explainer is the first method designed to interpret a GNN model globally using the local information given by individual instances. The contribution of our work is summarized as below:

- We introduce subgraph mining technique adapted to the GNN explanation task for measuring the generality of candidates, and define domination set based on fidelity and infidelity as a metric to evaluate faithfulness of a candidate.
- We formally define the *glocal explanation generation* problem, and propose a novel framework Glocal-Explainer to incorporate faithfulness and generality so as to solve the problem with a theoretical guarantee of $log(k)$ approximation to the optimal solution.
- We conduct experiments on both synthetic and real-world datasets and show that our method outputs meaningful and trustworthy explanations with decent quantitative evaluation results for pretrained GNN models.

2   Related Work

2.1   Local Explanation of GNNs

Local explanability refers to instance-level explanation techniques, which aim to acquire explanations for a target instance by identifying parts of the data that are critical to the model's decision. Recently, instance-level technique has been present main stream of GNN explanation. GNNExplainer [29] is the ground-breaking work for GNN explanation, which uses a mask to screen out important part of the input, then feed it into the trained GNN for evaluating and updating the mask. Many of the following works [16,21] employ the same technique and even though different in mask and objective function design, they share the same high-level idea. Other existing works can be categorized into surrogate methods [24], gradients-based methods [18] and decomposition methods [22]. While instance-level methods are precise locally, it falls short in giving high-level explanation of a GNN.

2.2   Global Explanation of GNNs

Global explanation methods are termed model-level methods, which output general and input-independent explanations that interpret the overall behavior of

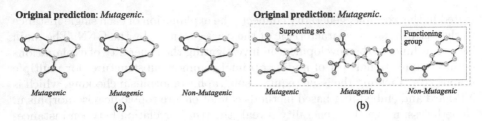

Fig. 3. (a) Examples of dispensable and indispensable substructures. **(b)** An example to show the difference between supporting set and functioning group of a substructure s^*; functioning group of a candidate is always a subset of its supporting set. (Color figure online)

the model. Such explanations not only provide high-level reasoning of the GNN's prediction, they also offer easy generalization to an inductive setting which is the nature of many GNN applications [16]. To the best of our knowledge, the only existing model-level method is XGNN [30]. It proposes to explain GNNs via graph generation using reinforcement learning. However, the two problems in its output are: first, none of the explanations exists in the original data, thus they cannot truly "explain" any class; second, the structure violates chemical rules, which will decrease the user's trust in the model. Thus we aim to develop a new and trustworthy model-level explanation method.

3   On the Perspective of Generality

GNN explanation serves human understanding, thus should be exploited in a natural and intuitive fashion. In specific, the presence of some substructure in an instance is regarded as an identification of membership to some class in the eyes of a GNN. Based on the intuition, we proposed Glocal-Explainer to find model-recognize genetic substructures for each interested class. For illustrative convenience, we introduce the generality perspective first.

3.1   Counterfactual Qualification

Our goal is to bridge the gap between global and local explainability for measuring the generality of a candidate. To start with, a candidate for explanations needs to be defined locally. The idea of counterfactual [2,11,25] is frequently used in explanation techniques for non-graph tasks and introduced to GNNs recently [15,17,20,29], it helps users understand black-box models in human reasoning fashion: a subtractive counterfactual investigates the situation "if it had not been ... then the outcome would not have been...", which performs better than the additive one on subsequent logical reasoning tasks [2]. We follow this idea to perturb the input by removing a substructure and check if prediction from the GNN changes to determine if the substructure is critical for the model.

Definition 1. *Given a pretrained GNN $\phi(\cdot)$, a substructure s of a graph g is said to be indispensable to g, denoted by $s \rightsquigarrow_\phi g$, if removing s from g results in GNN changes its prediction on g.*

Figure 3(a) shows an instance originally predicted as *Mutagenic* by the pretrained GNN. Three substructures are highlighted in separated copies. Removing the first two substructures (in yellow) does not lead to change of GNN's prediction, hence they are not critical to the model making a decision on this instance. Whereas, if the third substructure (in red) is removed, a new prediction (*Nonmutagenic*) is given, hence the last substructure is *indispensable* to the instance. Subsequently, substructures with such a property can be well considered as counterfactual candidates of an explanation.

3.2 Candidate Generation

A few existing methods attempt to generate a model-level explanation by finding a locally optimal one first then generalize it to the entire class [16,29]. However, a better way to capture generality and pursue globally ideal explainability is to generate explanation candidates while computing their generality along the way. In this regard, frequent subgraph mining comes into play naturally and adapted techniques for facilitating mining of GNN explanation candidates need to be designed to encode GNN's identifying candidates into the workflow. Luckily, in the field of graph mining there is a rich research body offering many decent tools for mining frequent subgraphs [4,8,27,28], which allow us to explore the candidate space while determining the generality of a candidate at the same time. Take the gSpan algorithm [27] as an example, its high-level idea of the mining procedure is pattern growing: starting from size-one candidates and growing the pattern one edge at a time, it performs a depth-first search on a tree-shaped search space of subgraphs (the search space is also referred as an enumeration tree). As shown in the center of Fig. 4, each node of the tree corresponds to a subgraph with one edge extended compared to that in its parent node. In other words, a subgraph contained in a node is always a supergraph of the one in its parent node. Due to this anti-monotone property of subgraph isomorphism over the search space, the size of supporting set of subgraphs decreases monotonously along a path from the root to a leaf on the enumeration tree, as the supporting set of a graph is always larger or equal to that of its supergraph. Owing to subgraph mining techniques, we are now ready to introduce our candidate mining details.

3.3 Mining Strategy

The mining algorithms aim at all subgraphs with frequency higher than the minimum support parameter, yet it brings unnecessary computational cost. Instead of searching exhaustively, a pruning strategy needs to be introduced. Based on the definition of a counterfactual explanation candidate, we further define the functioning group of a substructure in the eyes of a GNN as below:

Definition 2. *Given a pretrained GNN $\phi(\cdot)$ and its corresponding training set $\mathcal{D}$, functioning group of a substructure s, denoted by $\mathfrak{F}_{\phi(\cdot)}(s)$, is the set of instances from the training set that share the same predicted label with s and to which s is indispensable. Formally,*

$$\mathfrak{F}_{\phi(\cdot)}(s) = \{g \mid \hat{y}'_{\phi(g)} = \hat{y}_{\phi(s)} \wedge s \rightsquigarrow_\phi g, \forall g \in \mathcal{D}\} \tag{1}$$

Functioning group reflects one substructure's scope of effectiveness; meanwhile, it acts as a direct measurement of generality for a candidate. The reason why supporting set of the candidate is not used is that the presence of some subgraph may not be recognized by the GNN so that the underlying instance would be predicted as a specific class. Shown in Fig. 3(b) is an example to illustrate the difference between supporting set and functioning group of a substructure s^*. All three graphs contain s^*, i.e., they are all in the supporting set of s^*. However, for only the last one, s^* is indispensable since removing it changes GNN's prediction while the other two do not share this property. Therefore, we believe that the GNN recognizes some other pattern instead of s^* for them. Thus, the first two instances are not included in the functioning group of the substructure s^*. In conclusion, the intrinsic relation between supporting set and functioning group is that the latter is always a subset of the former, i.e., the size of the supporting set of one subgraph is the upper bound of its functioning group.

As mentioned in Sect. 3.2, the size of supporting set of mined subgraphs will decrease monotonously along a path from root to leaves on the searching tree. On the contrary, size of functioning group might increase constantly. The rational is that one substructure is indispensable to an instance proves that it contains critical information for the model to predict its class, hence its supergraph contains the information as well, which will also be indispensable to the same instance. Due to the black-box nature of GNN, this assumption can not be proved theoretically for the moment, yet it is observed in our experimental study. According to this philosophy, we design the pruning strategy for mining candidates as the following: the mining should stop, i.e., stop growing a mined pattern when its functioning group equals its supporting set. The idea is straightforward, if the mining were to continue, the size of functioning group might increase but the size of supporting set will decrease. However, the latter is the upper bound of the former, thus continually growing will not mine new candidates with a larger functioning group, i.e., higher generality. As a smaller substructure is preferable to a lager one [18,31], we stop exploring. In summary, when the functioning group equals the supporting set, the mining will be stopped since no more promising candidates with higher generality can be found.

4   On the Perspective of Faithfulness

Besides generality, faithfulness is the other goal of finding optimal explanations. As introduced in Sect. 1, assume a candidate substructure is truly critical, fidelity is to measure model's confidence drop when being removed, where a higher value indicates a better explanation result [32]. On the contrary, infidelity is for measuring the confidence change keeping only the substructure, where a lower value implies a better quality of explanation [16, 29]. To incorporate the two, we refer to the 2-D skyline operator problem [1]. The definition of *Skyline* is the set of data points out of the database that are not dominated by any other point [1]. Inspired by the idea of comparing "quality" of instances among data points themselves, we borrow the concept of *domination set* and apply it using fidelity and infidelity as the two dimensions. Formally, we define the two metrics for one instance first. Assume a candidate substructure s is predicted as label c when inputting it into the pretrained GNN, then we define:

Definition 3. *Individual fidelity of s on an instance g is calculated as:*

$$fid_{\phi(\cdot)}(s, g) = \phi(g)_{\hat{y}=c} - \phi(g/s)_{\hat{y}=c}$$

where g/s refers to removing the substructure s from the instance g.

The higher the fidelity, the more significant s is to g.

Definition 4. *Individual infidelity of s on an instance g is calculated as:*

$$inf_{\phi(\cdot)}(s, g) = \phi(g)_{\hat{y}=c} - \phi(s)_{\hat{y}=c}$$

The lower the infidelity, the less informative the removed part is, hence s is of better explainability.

Subsequently, the fidelity and infidelity of a candidate are measured among its functioning group, the overall *faithfulness* of a candidate is defined as below:

Definition 5. *Fidelity of a substructure s measures how much information s takes away when it is removed, which is calculated as:*

$$Fid_{\phi(\cdot)}(s) = \frac{1}{\|\mathfrak{F}_{\phi(\cdot)}(s)\|} \sum_{\forall g \in \mathfrak{F}_{\phi(\cdot)}(s)} fid_{\phi(\cdot)}(s, g)$$

Definition 6. *Infidelity of a substructure s evaluates how confident the GNN is to make the same prediction as the original instance, which is computed as:*

$$Inf_{\phi(\cdot)}(s) = \frac{1}{\|\mathfrak{F}_{\phi(\cdot)}(s)\|} \sum_{\forall g \in \mathfrak{F}_{\phi(\cdot)}(s)} inf_{\phi(\cdot)}(s, g)$$

The two faithfulness metrics serve as the two dimensions for the set of candidates. We are interested in how many other candidates one can dominate, i.e., this candidate is certainly better than how many others considering both fidelity and infidelity. Formally, the domination set of s is formally defined as:

$$\mathfrak{D}_{\phi(\cdot)}(s) = \{s' | Fid_{\phi(\cdot)}(s') \leq Fid_{\phi(\cdot)}(s) \wedge Inf_{\phi(\cdot)}(s') \geq Inf_{\phi(\cdot)}(s)\}$$

An illustrating example of domination set is shown on the right in Fig. 4. Candidates are plotted as data points with infidelity and 1-fidelity in two axes[4]. The

---

[4] To cater to the plotting convention of skyline problem so that the domination set locates in the upper right corner.

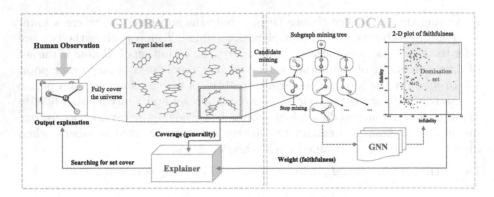

Fig. 4. The workflow of Glocal-Explainer.

more others one candidate can dominate, the better its quality is. Such design has two advantages: firstly, it consider two metrics equally and allow us to aim for the best of both worlds; secondly, instead of comparing the absolute values, qualifying candidates by contrast of one another scales the metrics value, thus avoiding potential problems of handling skewed distribution in any metrics.

5   The Proposed Glocal-Explainer

We aim for mode-level explanation of GNN's general behavior with both high generality in the data and faithfulness to the model. Furthermore, we take the total number of explanations into consideration. In practice, users prefer fewer explanations to conclude knowledge learned by the GNN that is sufficient for understanding. Hence, we propose to aim for a set of explanations, whose generality is large enough to cover all instances in the training set with the target label. In this fashion, high-level knowledge is summarized by incorporating faithfulness in instances, hence achieving the best regarding both metrics. This immediately leads to the weighted set cover problem with direct mapping detailed in below.

Given a pretrained GNN $\phi(\cdot)$ and its training set $\mathcal{D}$, the set of instances from $\mathcal{D}$ with predicted labels as the target class c^* is considered the *universe* with instances as *elements* to be covered, mathematically,

$$\mathcal{U}_{\phi(\cdot),c^*}(\mathcal{D}) = \{g \in \mathcal{D} \mid \phi(g) = c^*\} \qquad (2)$$

The reason why predicted label is used instead of the ground truth label is for capturing the knowledge that the GNN learned but not fed in. During training, GNN might not be able to achieve a training accuracy of 100%, the difference between the universe defined by Eq. (2) and the set of all instances with true label c^* distinguishes between GNN's recognized pattern and real characteristics among instances within the target class. In summary, we are interested in what GNNs see instead of what humans see. Each explanation candidate is a *set* that can possibly be picked to form a set cover, furthermore, functioning group defined by Eq. (1) serves as the *coverage* of a candidate. Finally, the weight of each candidate is calculated by measuring its faithfulness via domination set:

$$\omega(s) = 1 - \frac{\|\mathfrak{D}_{\phi(\cdot)}(s)\|}{\|\mathcal{C}\|}, \tag{3}$$

where $\mathcal{C}$ is the set of all candidates. Thus a candidate with the highest faithfulness that can dominate all the others will have the weight 0, while a candidate that can dominate fewer others will be assigned a larger weight. In this fashion, faithfulness of candidates is evaluated internally. Now we are ready to define the *glocal explanation generation* problem formally in the following:

Definition 7. *Given a GNN $\phi(\cdot)$ trained on a dataset $\mathcal{D}$, a target class c^* and a group of candidates $\mathcal{C}$ mined from $\mathcal{D}$, the glocal explanation generation problem is to find a subset $\mathcal{C}'$ of $\mathcal{C}$ s.t.:*

$$\underset{\mathcal{C}'}{\mathrm{argmin}}\{\sum_{s \in \mathcal{C}'} \omega(s) \mid \bigcup \mathfrak{F}_{\phi(\cdot)}(s) = \mathcal{U}_{\phi(\cdot),c^*}(\mathcal{D})\}$$

where $\mathfrak{F}_\phi(\cdot)$, $\mathcal{U}_{\phi,c^*}(\mathcal{D})$ and $\omega(s)$ are defined by Eqs. (1), (2) and (3) respectively.

As a result, the classic greedy algorithm to approximate the optimal weighted set cover with a $log(k)$ factor [3] can be used as the solution. Since all mappings of settings between the two problems are clear and straightforward, we do not go into details and include the algorithm in this paper.

The workflow of the proposed method is shown in Fig. 4. Locally, subgraph mining technique is used to mine candidates from the instances while computing the fidelity, infidelity and generality; the pruning strategy operates when the functioning group size equals the supporting set of a candidate; faithfulness of the candidates are calculated by investigating their domination set. Globally, with generality and faithfulness serving as coverage and weight, the explainer outputs final explanations as a set cover over the universe. Such a framework consumes local information in the input graph to explain GNNs globally, hence we give it the term "Glocal".

6   Experimental Evaluation

We evaluate our explainer on using both synthetic and real-world data, i.e., isAcyclic dataset and MUTAG dataset in graph classification task. The design of the experiments attempts to answer two questions: 1) how does our model perform in an identical setting compared to the existing model-level explainer? 2) how does our model perform when measured by various metrics?

6.1   Datasets and Experimental Setup

Table 1 summarizes the properties of the two datasets in use; in the following we describe them in detail.

Synthetic Dataset. isAcyclic datasets is a synthetic dataset built by Hao et al. [30] specifically for evaluating model-level explanations, where the ground truth explanations are prepared. Each instance is labeled either *Cyclic* or *Acyclic* according to if there exists any cycle in its graph structure. The former class consists of graphs with grid-like, circle, wheel or circular ladder structures; while the latter contains star-like, binary tree, path and full rary tree graphs.

MUTAG Dataset. The dataset consists of molecule structures of chemical compounds with 7 chemical elements, where nodes represent atoms and edges represent chemical bonds. Each instance is classified according to its mutagenic effect on a bacterium [5]. The dataset is widely used in GNN explanation works, since it has golden knowledge in the underlying domain: three or more fused rings and nitrogen dioxide (-NO_2 structure) are known to be mutagenic [5]. For fair comparison with the existing technique [30], we also ignore edge label.

Table 1. Statistics of the datasets and the corresponding GNN accuracy.

Dataset	Classes	*Avg.* nodes	*Avg.* edges	*Acc.*[a]
isAcyclic	2	30.04	28.46	0.964
MUTAG	2	17.93	19.79	0.978

[a]For fair comparison, we attempt to use the same pretrained GNNs as the compared method [30]. However, official implementation of the work is open to the public on MUTAG dataset only. Therefore we build a GNN for isAcyclic dataset according to the structure described in [30] and train it to achieve comparable accuracy. For evaluation, we report the official results with GNN scores for XGNN from the original paper and use scores from our own trained GNN for the proposed method.

GNN Models to Explain. In this work, we focus on graph classification task for the purpose of demonstration. Without loss of generality, node-level and link-level tasks can be handled easily by using the target's computational graph as input. GCN models [10] are used as explained networks on the two datasets, both models follow the design in the compared method [30] for the sake of parallel testing. We describe the GNNs briefly here. For the synthetic dataset isAcyclic, node degrees are used as the input node features. The network consists of two GCN layers with output dimensions as 8 and 16 accordingly and global mean pooling is performed to obtain the graph representations, followed by a fully-connected layer as the final classifier. *Sigmoid* is utilized as the non-linear activation function. For the real-world MUTAG dataset, one-hot feature encoding is used as input node feature. The network structure consists of three GCN layers with input dimension as 7 and hidden dimensions as {32, 48, 64}, respectively. Global mean pooling and fully-connected layer classifier are also employed. *ReLU* is utilized as the non-linear activation function. We trained the models to reasonable training accuracy (reported in Table 1) to ensure the GNNs have captured the knowledge from the training set.

6.2   Compared Method

We compare our method with XGNN [30], which is, to the best of our knowledge, the only existing work that generates global explanation. XGNN borrows the idea of Input Optimization from explanation task in text and image field, by which, GNN's predicted confidence is the only optimization goal and evaluation metric, while the generality of an explanation is not considered. It employs reinforcement learning to find explanations by graph generation, thus the output is independent of all instances in the dataset, and checking generality needs extra effort to carry out subgraph isomorphism, which is NP-hard. Hence we mainly compare predicted confidence between our proposed Glocal-Explainer and XGNN. As a matter of fact, the experimental results show that, generality is barely measurable for the output of XGNN. The training hyper-parameters are all set to be the same as the optimized ones in the original paper. The reinforcement learning framework requires user-set maximum number of nodes in the output explanations and initial node for the graph generation. We set the maximum number of nodes to be the average number of nodes of all outputs from our proposed Glocal-Explainer, and then test all possible initial nodes to give a comprehensive comparison.

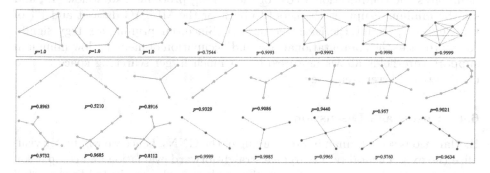

Fig. 5. Visualization of results on isAcyclic dataset, the first row represents *Cyclic* class, the second and the third rows correspond to *Acyclic* class. The output of Glocal-Explainer and XGNN are shown with blue and red nodes respectively, GNN predicted scores (see Table footnote a) for explanations are reported below each structure. (Color figure online)

6.3   Candidate Mining Algorithm

As discussed in Sect. 3.2, subgraph mining technique is employed in Glocal-Explainer to generate candidates of explanations. Specifically in this paper, we choose to use the gSpan algorithm [27], yet other mining techniques based on Depth-First Searching on tree-structure searching space are also eligible to be plugged in. Hence, superior graph mining algorithms developed later can be adopted to improve the performance of Glocal-Explainer further.

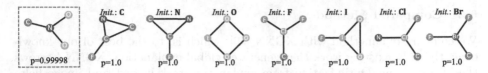

Fig. 6. Results of *Mutagenic* class in MUTAG dataset with corresponding GNN scores. Shown in the green dashed box is the output of Glocal-Explainer and the other structures are the outputs of XGNN of the same size using different initial nodes. (Color figure online)

There are three parameters to set for mining the candidates, the minimum and maximum number of nodes in the subgraph mined and the support threshold (the lowest frequency). Since one single edge is not informative enough to deliver insights as an explanation and overlarge substructures are difficult for humans to understand, we set the minimum number of nodes to be 3 and maximum to be 7, following [30]. It is worth noting that Glocal-Explainer actually automatically decides how many nodes are optimal in the explanation according to their generality (coverage) and faithfulness (weight) to the GNN.

Regarding the setting of the support parameter in subgraph mining, which dominates the computational cost of the mining procedure, we follow [19] and set the minimum support threshold to be 1 for the MUTAG dataset and 1% of the mining space for the isAcyclic dataset. Since our pruning strategy significantly reduces the computational cost and occupation of memory, low minimum support thresholds allows us to explore a much larger searching space to find candidates of lower frequency.

6.4   Result and Discussion

Explanations serve human understanding of the GNNs, hence we will firstly visualize the experimental results to give straightforward evaluation and comparison between methods. Secondly, quantitative analysis is also conducted to give a full picture of how our explainer performs measured by various metrics.

Qualitative Analysis for isAcyclic Dataset. We report the results of our explainer as well as XGNN in Fig. 5. For our method, blue nodes are used. Due to official implementation of the compared method not being available to the public (see Table footnote a), we present the official results from the original paper. Explanations generated by Glocal-Explainer are all consistent with the rules which builds up the synthetic dataset. Meanwhile, all explanations except the second one for *Acyclic* class receive high GNN scores which is comparable to XGNN's, showing that the results are highly recognized by the model. Though very few of the final explanations retain an unsatisfactory score, it may be the result of considering both faithfulness and generality and find the model-level explanations as an entirety, which do not guarantee individual optima. In addition, all structures for *Cyclic* class found by XGNN other than the triangle

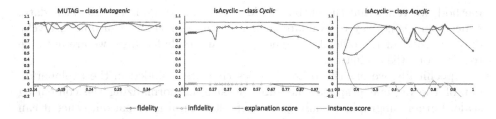

Fig. 7. Plots of fidelity, infidelity and GNN confidence of both the explanations and the respective instances of Glocal-Explainer's outputs; horizontal axis represents the ratio of nodes that are included in the explanation out of the original graph.

do not exist in the dataset, the method suffers from serious out-of-distribution (OOD) problem. For the *Acyclic* class, our explainer includes 4 out of 5 outputs of XGNN while providing a richer repository for summarizing the knowledge that GNN learned for this class. Yet unlike XGNN, our method does not require any user-defined parameters such as explanation size to shape the output.

Qualitative Analysis for MUTAG Dataset. The outputs of *Mutagenic* class in MUTAG dataset together with GNN corresponding scores are shown in Fig. 6. We use the official implementation of XGNN to generate explanations which have the same size as our final output and use all possible atoms as the initial node. The output of Glocal-Explainer is shown in the left dashed box in green with a predicted confidence 0.99998, which indicates that this structure faithfully explains the model's decision. It also verifies that the pretrained GNN captures the domain knowledge. Recall that, as introduced in Sect. 6.1, chemistry experts found that nitrogen dioxide (NO_2 structure) attached to carbon rings are mutagenic [5]. The explanation correctly captures the symbolic structure of nitrogen dioxide.

From the perspective of generality, the algorithm outputs one single substructure that covers the whole universe, which means it is indispensable to all instances predicted as mutagenic. Such explanation is of highest generality, thus represents common characteristics for the target class and top-level knowledge that GNN learned from training. On the contrary, we have verified that the explanations output by XGNN do not exist in any of the graphs from the dataset in *either* class (see Fig. 2 for reference), which means they have <u>zero</u> generality. XGNN outputs irrational structures that violate the fundamental rules in the chemical area, such as the bound of chemical valence of the atoms is broken.

Quantitative Analysis of the Results . Although we aim to generate model-level explanation, we are still interested in whether the explanations maintain outstanding faithfulness locally. Hence we measure fidelity and infidelity of Glocal-Explainer's output by averaging the scores among instances in the respective functioning groups for quantitative study. Whereas for the compared method, we verify that the output for MUTAG dataset do not exist in the data, hence there is no way to evaluate their faithfulness locally; regarding isAcyclic dataset, without the official implementation and node features, we are not able to carry out the evaluation.

Specifically, we use the *ratio* of nodes that are included in the explanation out of the original graph to measure the size for normalizing the comparing scale. Larger subgraphs usually carry more structural information, hence it can affect the faithfulness metrics and smaller substructures tend to be less important (low GNN score, low fidelity and high infidelity). We plot the curves of fidelity, infidelity and GNN scores of both the explanations and the respective instances for reference against size ratio in Fig. 7. Over the two datasets, Glocal-explainer achieves high fidelity and low infidelity with a high GNN score for the generated explanation. It is interesting that some infidelity score is negative, which results from the extremely high conference score of the output substructure. Such phenomenon indicates that the GNN is even more confident about the prediction, given the explanation substructure. In very rare cases, the output is not very promising. Considering it may be the trade off between faithfulness and generality, the method aims for model-level explanation, we believe that is reasonable.

Overall, Glocal-Explainer can produce meaningful and trustworthy explanation with outstanding algorithmic evaluation, thus it can help humans to understand the model, increase their trust and improve the GNN.

7   Conclusion

Graph neural networks are widely employed owing to their outstanding performance. Yet users cannot understand their decision making mechanism. In this work, we propose Glocal-Explainer to generate model-level explanations. We adapt subgraph mining technique to compute the generality of candidates, and introduce domination set to measure their faithfulness. We further define the *glocal explanation generation* problem and employ a greedy algorithm to find the solution. Experimental results show that Glocal-Explainer outputs trustworthy explanations with superior quantitative evaluation results.

Acknowledgment. This work is partially supported by National Key Research and Development Program of China Grant No. 2018AAA0101100, the Hong Kong RGC GRF Project 16209519, CRF Project C6030-18G, C1031-18G and C5026-18G, AOE Project AoE/E-603/18, RIF Project R6020-19, Theme-based project TRS T41-603/20R, China NSFC No. 61729201, Guangdong Basic and Applied Basic Research Foundation 2019B151530001, Hong Kong ITC ITF grants ITS/044/18FX and ITS/470/18FX, Microsoft Research Asia Collaborative Research Grant, HKUST-NAVER/LINE AI Lab, Didi-HKUST joint research lab, HKUST-Webank joint research lab grants.

References

1. Börzsönyi, S., Kossmann, D., Stocker, K.: The skyline operator. In: ICDE, pp. 421–430. IEEE Computer Society (2001)
2. Byrne, R.M.: Counterfactuals in explainable artificial intelligence (XAI): evidence from human reasoning. In: IJCAI, pp. 6276–6282. ijcai.org (2019)
3. Chvátal, V.: A greedy heuristic for the set-covering problem. Math. Oper. Res. 4(3), 233–235 (1979)
4. Cohen, M., Gudes, E.: Diagonally subgraphs pattern mining. In: DMKD, pp. 51–58. ACM (2004)
5. Debnath, A.K., Lopez de Compadre, R.L., Debnath, G., Shusterman, A.J., Hansch, C.: Structure-activity relationship of mutagenic aromatic and heteroaromatic nitro compounds. correlation with molecular orbital energies and hydrophobicity. J. Med. Chem. 34(2), 786–797 (1991)
6. Huan, Z., Quanming, Y., Weiwei, T.: Search to aggregate neighborhood for graph neural network. In: ICDE, pp. 552–563. IEEE (2021)
7. Jang, E., Gu, S., Poole, B.: Categorical reparameterization with gumbel-softmax. In: ICLR. OpenReview.net (2017)
8. Jiang, C., Coenen, F., Zito, M.: A survey of frequent subgraph mining algorithms. Knowl. Eng. Rev. 28(1), 75–105 (2013)
9. Kersting, K., Kriege, N.M., Morris, C., Mutzel, P., Neumann, M.: Benchmark data sets for graph kernels (2016). http://graphkernels.cs.tu-dortmund.de
10. Kipf, T.N., Welling, M.: Semi-supervised classification with graph convolutional networks. In: ICLR. OpenReview.net (2017)
11. Koh, P.W., Liang, P.: Understanding black-box predictions via influence functions. In: ICML, vol. 70, pp. 1885–1894. PMLR (2017)
12. Li, Z., et al.: Hierarchical bipartite graph neural networks: towards large-scale e-commerce applications. In: ICDE, pp. 1677–1688. IEEE (2020)
13. Liang, J., Bai, B., Cao, Y., Bai, K., Wang, F.: Adversarial infidelity learning for model interpretation. In: SIGKDD, ACM (2020)
14. Liu, B., Zhao, P., Zhuang, F., Xian, X., Liu, Y., Sheng, V.S.: Knowledge-aware hypergraph neural network for recommender systems. In: Jensen, C.S., et al. (eds.) DASFAA 2021. LNCS, vol. 12683, pp. 132–147. Springer, Cham (2021). https://doi.org/10.1007/978-3-030-73200-4_9
15. Lucic, A., ter Hoeve, M., Tolomei, G., de Rijke, M., Silvestri, F.: CF-GNNExplainer: counterfactual explanations for graph neural networks. arXiv preprint arXiv:2102.03322 (2021)
16. Luo, D., et al.: Parameterized explainer for graph neural network. In: NIPS (2020)
17. Numeroso, D., Bacciu, D.: Explaining deep graph networks with molecular counterfactuals. CoRR abs/2011.05134 (2020)
18. Pope, P.E., Kolouri, S., Rostami, M., Martin, C.E., Hoffmann, H.: Explainability methods for graph convolutional neural networks. In: CVPR. IEEE Computer Society (2019)
19. Saigo, H., Nowozin, S., Kadowaki, T., Kudo, T., Tsuda, K.: gBoost: a mathematical programming approach to graph classification and regression. Mach. Learn. 75(1), 69–89 (2009)
20. Sanchez-Lengeling, B., et al.: Evaluating attribution for graph neural networks. In: NIPS (2020)
21. Schlichtkrull, M.S., Cao, N.D., Titov, I.: Interpreting graph neural networks for NLP with differentiable edge masking. CoRR abs/2010.00577 (2020)

22. Schnake, T., et al.: Higher-order explanations of graph neural networks via relevant walks. arXiv preprint arXiv:2006.03589 (2020)
23. Velickovic, P., et al.: Graph attention networks. In: ICLR. OpenReview.net (2018)
24. Vu, M.N., Thai, M.T.: PGM-explainer: probabilistic graphical model explanations for graph neural networks. In: NIPS (2020)
25. Xiao-Hui, L., et al.: A survey of data-driven and knowledge-aware explainable AI. TKDE (2020)
26. Xu, K., Hu, W., Leskovec, J., Jegelka, S.: How powerful are graph neural networks? In: ICLR. OpenReview.net (2019)
27. Yan, X., Han, J.: gSpan: graph-based substructure pattern mining. In: ICDM, pp. 721–724. IEEE Computer Society (2002)
28. Yan, X., Han, J.: CloseGraph: mining closed frequent graph patterns. In: SIGKDD, pp. 286–295. ACM (2003)
29. Ying, Z., Bourgeois, D., You, J., Zitnik, M., Leskovec, J.: GNNExplainer: generating explanations for graph neural networks. In: NIPS (2019)
30. Yuan, H., Tang, J., Hu, X., Ji, S.: XGNN: towards model-level explanations of graph neural networks. In: SIGKDD, pp. 430–438. ACM (2020)
31. Yuan, H., Yu, H., Gui, S., Ji, S.: Explainability in graph neural networks: a taxonomic survey. CoRR abs/2012.15445 (2020)
32. Yuan, H., Yu, H., Wang, J., Li, K., Ji, S.: On explainability of graph neural networks via subgraph explorations. In: ICML. PMLR (2020)
33. Zhang, J., Liang, S., Deng, Z., Shao, J.: Spatial-temporal attention network for temporal knowledge graph completion. In: Jensen, C.S., et al. (eds.) DASFAA 2021. LNCS, vol. 12681, pp. 207–223. Springer, Cham (2021). https://doi.org/10.1007/978-3-030-73194-6_15

Temporal Network Embedding with Motif Structural Features

Zhi Qiao, Wei Li[✉], and Yunchun Li

Beijing Key Lab of Network Technology,
School of Computer Science and Engineering, Beihang University, Beijing, China
{zhiqiao,liw,lych}@buaa.edu.cn

Abstract. Temporal network embedding aims to generate a low-dimensional representation for the nodes in the temporal network. However, the existing works rarely pay attention to the effect of meso-dynamics. Only a few works consider the structural identity of the motif, while they do not consider the temporal relationship of the motif. In this paper, we mainly focus on a particular temporal motif: the temporal triad. We propose the Temporal Network Embedding with Motif Structural Features (MSTNE), a novel temporal network embedding method that preserves structural features, including structural identity and temporal relationship of the motif during the evolution of the network. The MSTNE samples the neighbor node based on the temporal triads and models the effects of different temporal triads using the Hawkes process. To distinguish the importance of different structural and temporal triads, we introduce the attention mechanism. We evaluate the performance of MSTNE on four real-world data sets. The experimental results demonstrate that MSTNE achieves the best performance compared to several state-of-the-art approaches in different tasks, including node classification, temporal link prediction, and temporal node recommendation.

Keywords: Temporal network embedding · Structural features · Motif · Attention · Hawkes process

1 Introduction

Network embedding aims to generate a low-dimensional representation for the nodes in the network while preserving a similar relationship between all nodes in the network [5]. It has recently attracted much attention from academia and industry. The traditional works of network embedding were proposed in the 2000s, which aim to reduce the dimensionality of the data [1,17]. These methods preserve the local geometry structure between nodes by constructing an affinity matrix and generating a low-dimensional representation. However, these methods have higher computational costs and do not suitable for large-scale networks. With the continuous development of deep learning, DeepWalk [14] was proposed. DeepWalk designs a random walk strategy so that each node obtains its context in the network and uses skip-gram to learn effective representations. Based on

Deepwalk, many other network embedding methods are proposed later, such as LINE [18], Node2vec [6], Stru2vec [16]. However, the methods mentioned above are only suitable for static networks. Nevertheless, in the real world, the network structure is evolving. At present, most methods do not pay attention to the Meso-dynamics during the evolution of the network.

Meso-dynamics is widely used in mining the network structure and evolution [7]. Meso-dynamics focus on the interaction between a group of nodes and edges, thus preserving neighborhood similarity at the group level. Usually, meso-dynamic performs in the formation process of the small subgraph pattern, small subgraph pattern in the network is also called the network motif, which is crucial for understanding the evolution network [5]. For example, the nodes that constitute a specified motif usually have close relationships and are more likely to belong to the same category or related. In this article, we mainly focus on a particular temporal motif: the temporal triad [13]. The reason why we choose triad is that the triad closure process is a fundamental mechanism in the formation and evolution of temporal networks [3]. At present, few works focus on the triad. For instance, DynamicTriad [20] considers the influence of subgraphs evolving from open triad to closed triad. MTNE [8] considers the triad evolution process and preserves motif-aware high-order proximities. However, they ignore the influence of the temporal relationship between the edges of the triad.

To show the influence of temporal relationships in the triads, we take an example of message delivery. Figure 1 shows the evolution of two triads G^1 and G^2 when $\delta = 2\,\mathrm{s}$, $6\,\mathrm{s}$, $10\,\mathrm{s}$, where δ denotes the timestamp. Suppose A gets a new message when $\delta = 2\,\mathrm{s}$, and wants to deliver the message to B and C. The blue circle means this node has got the new message, and the green circle means this node has not got the new message. Due to $G^1_{A,B}$ established earlier than $G^1_{B,C}$, when $G^1_{A,B}$ is established, B has got the new message, and since there will occur a message delivery between B and C when $\delta = 6\,\mathrm{s}$, A can let B delivery the new message to C. However, when edge $G^2_{B,C}$ is established, B has not got the message, so that B can not deliver the message to C. To deliver the message to C, the established probability of $G^2_{A,C}$ is much higher than $G^1_{A,C}$. When $\delta = 6\,\mathrm{s}$, the structure identities of the two temporal triads are the same, but edge $G^1_{A,B}$ and edge $G^1_{B,C}$ has different temporal relationships. The method that does not consider the temporal relationships (e.g. Deepwalk, Node2vec) will generate similar representations for C because they have the same neighbor node B. However, the evolution patterns of two triads are different, and different triads will lead to different results when $\delta = 10\,\mathrm{s}$ which $G^2_{A,C}$ established but $G^1_{A,C}$ do not. From Fig. 1 we observe that different temporal triads will affect the evolution of the network. Thus, it is essential to consider the temporal relationship and structural identities of the motif in the evolution of the network.

In addition, most of the methods focus on motifs exploring the whole network [12,19], which have high computation costs in preprocessing, and storing a large amount of motif also wastes much memory.

In this paper, to address the above challenges, we propose Temporal Network Embedding with Motif Structural Features (MSTNE). MSTNE samples the

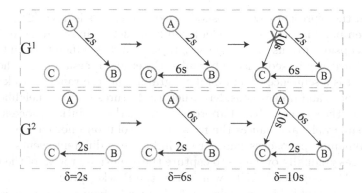

Fig. 1. A toy example of the different temporal motif evolution, G^1 and G^2 are two different motif, δ represents the timestamp.

neighbors and finds the neighbor nodes that form the temporal triad for selected edges. Then measure the impact factor of temporal motif use the conditional intensity function based on Hawkes process [9]. MSTNE considers structural features that include both the structural identity and temporal relationship of the triad to generate an ideal vector representation for the nodes in the network. In summary, the contributions of our work are as follows:

- To find neighbor nodes without exploring the whole network while preserving the temporal relationship and structural identities as much as possible. We propose a novel neighbor node sampling method based on the temporal motif called T-Motif Sampling, which finds neighbor nodes that can form a temporal triad for selected edges. Through the T-Motif Sampling method, our MSTNE model can preserve both the structural identity and temporal relationship of the motif.
- To measure the impact factor of temporal motif sampled by T-Motif sampling. We propose a novel conditional intensity function based on the Hawkes process, which considers the influence of the open triad and closed triad. In addition, to distinguish the importance of different temporal triads, we propose an attention mechanism for 24 kinds of triads with different structural identities and temporal relationships.
- The experimental results on four real data sets demonstrate that the MSTNE achieves the best performance over the existing approaches in the tasks of node classification, temporal link prediction, and temporal recommendation.

2   Related Work

The approaches for temporal network embedding can be categorized into two groups, the snapshot-based approach and the dynamic approach. The snapshot is a subgraph sequence that divides the temporal network into different moments.

One of the representative snapshot-based approaches is TNE [21]. They proposed a temporal latent space model for link prediction in social networks. TNE does not consider motifs in a single subgraph, and it is difficult to capture local structural features in sequence subgraphs. Du et al. [4] assume that the temporal network is a sequence of temporal subgraphs and use random walk to select neighbor nodes and do not consider structural features between neighbor nodes. In addition, the snapshot-based approach ignores the evolution process between different subgraphs and causes the miss of the evolution information.

The dynamic approaches consider the influence of different events. The influence changes gradually overtime to capture the evolution process of the network. Zuo et al. [22] proposed HTNE, which uses the Hawkes process to measure the probability of establishing an edge. HTNE mainly focuses on considering the temporal relationship between neighbor nodes, ignoring the structural features between neighbor nodes. Therefore, many works pay attention to the effect of microscopic dynamics. Lu et al. [10] proposed MMDNE, and they consider both micro-dynamics and macro-dynamics in the temporal network that are mutually evolved and alternately affected. HNIP proposed by Qiu et al. [15] captures the highly non-linear structure in the temporal network.

However, the above methods ignore the evolution of the meso-dynamics. Zhou et al. [20] proposed DynamicTriad, which considered the influence of subgraphs evolving from open triad to closed triad. Huang et al. [8] proposed MTNE. Although they consider the meso-dynamics, they still did not consider the temporal relationship of the motif, which will affect the evolution of the network, and different temporal triads will affect the evolution of the network.

Therefore, we propose a temporal network embedding method MSTNE, which includes a novel neighbor node sampling method: T-Motif Sampling based on the temporal motif and a novel conditional intensity function based on Hawkes process to measure the impact factor of the temporal motif.

3   Problem Formulation

Definition 1: Temporal Network. Temporal network is defined as $G = (V, E, T)$, where V represents the set of nodes, E represents the set of edges, and T represents the set of timestamps. In the temporal network, $e_{v_i,v_j}^t (v_i, v_j \in V, t \in T)$ denotes edges, which means at time t, an edge is established between the two nodes v_i and v_j. Note that the edges between nodes v_i and v_j may be repeatedly established. There will be many temporal motifs with different numbers of nodes in the temporal network. We mainly focus on a particular temporal motif composed of three nodes, which are called the temporal triad.

Definition 2: Temporal Triad. Given a set of three nodes $\Psi = (v_s, v_h, v_t)$. v_s, v_h, v_t is three nodes in the network. If there exist at least non-repeated two edges between three nodes, we call it the temporal triad, for example $\exists e_{v_s,v_h}^{t_1}, e_{v_h,v_t}^{t_2} \in E$. As shown in Fig. 2, assume that edge $e_{A,C}$ is the last to be established, different type of the edge direction and temporal relationship

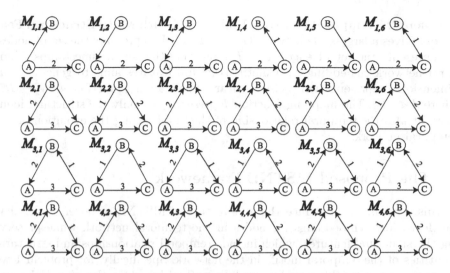

Fig. 2. Different types of temporal triad, M indicates the type of temporal triad.

between $e_{B,C}$ and $e_{A,B}$ will affect the establish probability of edge $e_{A,C}$. Thus there are 24 different temporal triads, which include 18 different closed triads and 6 open triads. The numbers on edges indicate the temporal relationship between the three edges, 1 is the earliest edge, and 3 is the last edge. The bi-directional edge, such as the right edge of $M_{3,3}$, represents a bi-directional edge established between two points at the same time. A closed triad refers to a triad that can form a closed loop, such as $M_{2,1} - M_{4,6}$. The open triad refers to the triad that does not form a closed loop, such as the $M_{1,1} - M_{1,6}$.

Problem 1: Temporal Triad Sampling. The purpose of temporal triad sampling is to find neighbor nodes without exploring the whole network while preserving temporal relationship and structural identities as much as possible. First, given an edge $e^t_{v_i,v_j}$, which is last to be established in the sampled temporal triad, temporal triad sampling aims to find a sequence of neighbor nodes.

$$\omega = \{(v_1, t_{l_1}, t_{r_1}, m_1), (v_2, t_{l_2}, t_{r_2}, m_2), \ldots, (v_h, t_{l_h}, t_{r_h}, m_h)\}$$
$$t_{l_h}, t_{r_h} < t, m_h \in [1, 24], h \leq H \tag{1}$$

where H represents the number of neighbor nodes need to be found, all v_h can be combined with v_i, v_j to form the temporal triad in Fig. 2. v_h, v_i and v_j preserve the structural identities between the three nodes, which indicates that the three nodes can form the temporal triad. t_{l_h}, t_{r_h} are the timestamp of the other two edges e_{v_i,v_h} and e_{v_h,v_j}, which preserve temporal relationship of neighbor nodes. m_h are the types of the triad. In the case of $e^t_{v_i,v_j}$ is a bi-directional edge, it is sampled as two edges with different directions. When selecting neighbor nodes, we preserve the structural identity and temporal relationship of neighbor nodes.

Problem 2: Temporal Network Embedding with Motif Structural Features. Gives a large-scale network $G = (V, E, T)$, V represents the set of nodes, E represents the set of edges, and T represents the set of timestamps. Temporal network embedding with motif structural features aims to generate a d-dimensional representation vector, by learning a mapping function: $\Phi : V \leftarrow R^d$, where $d \ll |V|$. The mapping function Φ preserves not only the structural identity but also the temporal relationship of the motif during the evolution of the temporal network.

4   The Proposed MSTNE Framework

In this section, we introduce the architecture of MSTNE, a novel model that can learn ideal embeddings for nodes in the temporal network while preserving the structural features, which include temporal relationships and structural identities of the temporal motifs in the network. Specifically, for problem 1 we propose a novel neighbor node sampling method based on the temporal motif called T-Motif Sampling. For Problem 2 we propose a novel conditional intensity function based on the Hawkes process and an attention mechanism to measure the impact factor of temporal motif sampled by T-Motif sampling. And then, use maximum likelihood estimation and stochastic gradient descent to optimize the impact factor and then generate the ideal representation.

4.1   Neighbor Node Sampling Method Based on the Temporal Motif

Most of the previous works, such as DeepWalk [14] and LINE [18], optimize the sampling strategy of neighbor nodes based on random walk. However, they did not consider the structural identity and temporal relationship between neighbor nodes. In addition, methods focus on motifs exploring the whole network [12,19], which will lead to colossal resource consumption, and the relationship between the edges in motifs which has a large time interval is not close and can be ignored. Thus, we propose a neighbor node sampling method called T-Motif Sampling.

T-Motif Sampling samples based on the edge, The input of Algorithm 1 $G = (V, E, T)$ represents the list of all temporal edges. $h = e_{v_i,v_j}^t$ denotes the edge that needs to find neighbor nodes. Too many nodes or edges in the network result in higher complexity of the neighbor node sampling method, and the influence of the edge, which has a large time interval with h is very small. So we specify the size W of a history window. H refers to the number of neighbor nodes that need to be selected.

In Algorithm 1, the $findidx$ method (line 3) finds $neighbours$ which means W edges connected to the node v_i and v_j, that established before timestamp t. Then randomly select an edge h_γ from $neighbours$, and check whether another edge h_β can form a temporal triad in $neighbours$ except edge h_γ. The $ifclosedtriad$ method (line 6) is used to determine whether the current three edges can form a closed temporal triad as $M_{2,1} - M_{4,6}$ shown in Fig. 2, if not h_γ and h is considered to be an open temporal triad as $M_{1,1} - M_{1,6}$ shown in Fig. 2. m is the type of

Algorithm 1. T-Motif Sampling(G, W, H, h)

Input: Temporal Network $G = (V, E, T)$; Select edge: $h = e^t_{v_i,v_j}$; Window size: W; Neighbor nodes size: H;

Output: sequence of neighbor nodes: ω

1: initialize: $triad = \{\}$, $\omega = \{\}$
2: **while** Length of $\omega < H$ **do**
3:     $neighbours = findidx(G, W, h)$ ▷ $neighbours$ is W historical edges
4:     Random Sample h_γ from $neighbours$
5:     **for** each $h_\beta \in neighbours$ except h_γ **do**
6:         **if** $ifclosedtriad(h_\gamma, h_\beta, h)$ is True **then**
7:             Add (h_γ, h_β, m) to $triad$ ▷ m is type of closed triad (h_γ, h_β, h)
8:     Add (h_γ, m) to $triad$ ▷ m is type of open triad (h_γ, h)
9:     Random Sample h_t from $triad$ ▷ h_t means (h_γ, h_β, m) or (h_γ, m)
10:    **if** h_t is closed triad **then**
11:        t_γ, t_β is the timestamp of edge h_γ, h_β
12:        Add $(v_h, t_\gamma, t_\beta, m)$ to ω ▷ v_h is the intersection of edge h_β and h_γ
13:        **if** $t_\gamma < t_\beta$ **then**
14:            $h = h_\gamma$
15:        **else**
16:            $h = h_\beta$
17:    **else**
18:        Add $(v_h, t_\gamma, 0, m)$ to ω ▷ v_h is the nodes of edge h_γ expect node v_i
19:        $h = h_\gamma$
        return ω

---

triad. To line 8, the *triad* list contains all neighbor nodes that can form an open and closed temporal triad with h_γ. Then, randomly select h_t from the list of *triad* (line 9). If h_t is a closed temporal triad, we will add $(v_h, t_\gamma, t_\beta, m)$ to ω (line 12). In order to ensure temporal relationships between the neighbor nodes, we will choose the edge with the smallest timestamp to continue the iteration. For the open temporal triad, there is only one timestamp. To ensure that each tuple in ω has the same dimension, mark the other timestamp to 0 as shown in line 18. Then continue sampling based on this edge until the number of neighbor nodes reaches H, as shown in lines 9 to 19 of Algorithm 1. The time complexity of the sampling algorithm is $O(H \times W)$. The time complexity of the general technique, which explores the whole network, has minimum time complexity $O(\frac{D}{2} \times |E|)$ [12], where $|E|$ is the counts of edges D is the maximum degree of all nodes. Our method performs best when H,W is 3,10 (see Sect. 5.5 for details), both degree D and counts of edges $|E|$ are much greater than 10. So the time complexity of T-Motif sampling is much less than the general technique.

Figure 3 shows an example of T-Motif sampling. First, we find a neighbor node B that can form a closed temporal triad and add it in the sequence ω. 11 is the timestamp of the edge $e_{A,B}$, 10 is the timestamp of the edge $e_{B,C}$, $M_{2,5}$ is the type of the triad as shown in Fig. 3. Since the timestamp of $e_{B,C}$ is smaller than that of $e_{A,B}$, we select edge $e_{B,C}$ and continue the iteration. The remaining two iterations were not found closed temporal triad, so we select

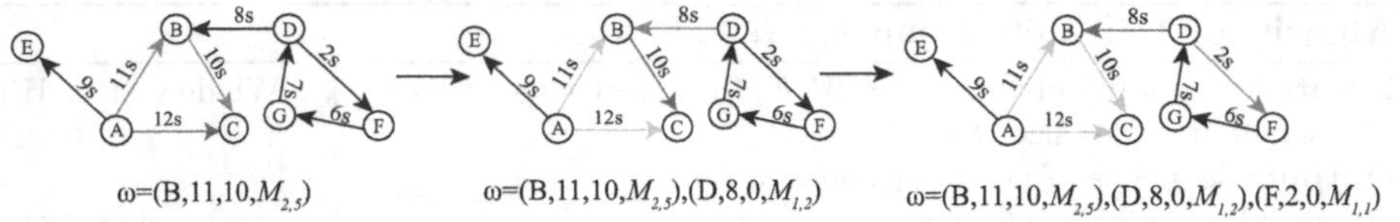

ω=(B,11,10,$M_{2,5}$) ω=(B,11,10,$M_{2,5}$),(D,8,0,$M_{1,2}$) ω=(B,11,10,$M_{2,5}$),(D,8,0,$M_{1,2}$),(F,2,0,$M_{1,1}$)

Fig. 3. An example of T-Motif sampling, the red arrows indicate the edges which need to find the neighbor node, the blue arrows indicate the closed temporal triads or open temporal triads found, the gray arrows indicate edges that have been sampled, and M is the type of the temporal triads. (Color figure online)

neighbor nodes D and F, which can form an open temporal triad, and mark 0 to the second timestamp. $M_{1,2}$ and $M_{1,1}$ are the type of the two open temporal triads. It can be seen from Fig. 3 that our sampling algorithm is continuously expanding outward while preserving the structural identity of the high-order neighbor nodes, which can form a triad. In addition, our sampling method is based on the window before the selected edge and always selects the edge with the smallest timestamp for iteration. So our sampling method fully preserves the temporal relationship and the structural identity between neighbor nodes.

4.2   Impact Factor Measure of Temporal Triad

The Hawkes process is a point process with self-motivation influence. It considers the influence of historical events that occurred in the past period on current events. Here we use conditional intensity function to measure the impact factor of temporal triads. In the impact factor measure, we think that both the open triad and the closed triad will influence the evolution of the temporal network. Thus, we propose a novel conditional intensity function is as follows:

$$\widetilde{\lambda}_{x,y}(t) = \mu_{x,y} + \sum_{t_h < t} \alpha_{\gamma_h} * mask + \sum_{t_h < t} \alpha_{\theta_h}, \quad \lambda_{x,y}(t) = exp(\widetilde{\lambda}_{x,y}(t)) \quad (2)$$

where $\widetilde{\lambda}_{x,y}(t)$ represents the impact factor of $e_{x,y}$ at timestamp t, which indicates the probability of establishing an edge between nodes x and y. γ_h is the neighbor node set of $e_{x,y}$, which can form the closed temporal triad. θ_h is the neighbor nodes that can form the open temporal triad. We denote the impact factor of them as α_{γ_h} and α_{θ_h}. t_h denotes the occurrence time of historical events. $\mu_{x,y}$ represents the base rate, which indicates the similarity between node x and y. $\mu_{x,y}$ usually expressed as the negative squared Euclidean distance: $-||x - y||^2$. Since the impact factor should be positive, we regard $\lambda_{x,y}(t)$ as an impact factor by an exponential function, and the value range of it is between 0 and 1. For some special nodes, all of their neighbor nodes can only form an open temporal triad. Here we use a mask to distinguish the types of neighbor nodes. If the closed temporal triad can be formed, the neighbor node mask is 1. Otherwise, the mask is 0.

Impact Factor of Closed Temporal Triad. The influence of the closed temporal triad is produced during the process of forming the closed temporal triad from the open temporal triad. As shown in Fig. 1, the two edges established at $\delta = 6$ s will influence the establish probability of last edge $G^1_{A,C}$ and $G^2_{A,C}$ at $\delta = 10$ s which is the influence of closed triad. Here we give a closed triad(x, h, y), the impact factor of (x, h, y) is as follows:

$$\alpha_{\gamma h} = \alpha_{x,h} + \alpha_{h,y}, \quad \alpha_{x,h} = \sum_{t_{x,h} < t} \mu_{x,h} * e^{-\delta(t - t_{x,h})} \tag{3}$$

where h is the node that is adjacent to x and y and can form a closed temporal triad. $\alpha_{x,h}$ measures the impact factor of the edge $e_{x,h}$ on the $e_{x,y}$, $\alpha_{h,y}$ measures the impact factor of the edge $e_{h,y}$ on $e_{x,y}$. Where t is the timestamp of the edge $e_{x,y}$, $t_{x,h}$ is the timestamp of the edge $e_{x,h}$. The influence should be decreasing over time, $e^{-\delta(t - t_{x,h})}$ represents the time decay effect, where δ is a variable used to measure the decay of influence over time, which can be obtained during training. $\alpha_{h,y}$ can also be expressed as the $\alpha_{x,h}$.

Impact Factor of Open Temporal Triad. In addition to the influence of the closed temporal triad, the open temporal triad also influences the formation process. The influence of the open temporal triad is produced during the process of forming the open temporal triad from the single edge. As shown in Fig. 1, the edge established when $\delta = 2$ s: $G^1_{A,B}$ and $G^2_{B,C}$, will influence the establish probability of edge $G^1_{B,C}$ and $G^2_{A,B}$ at $\delta = 6$ s which is the influence of open triad. The impact factor of the open temporal triad can be expressed in the following exponential form:

$$\alpha_{\theta h} = \sum_{t_{x,h}, \, t_{h,y} < t} |\mu_{x,h} - \mu_{y,h}| * e^{-\delta(t_{y,h} - t_{x,h})} \tag{4}$$

The impact factor between the open temporal triad can be obtained by subtracting the base rate of the two edges and multiplying by the time difference between the two edges.

4.3 Attention Mechanisms for Triad with Different Structural Identity and Temporal Relationship

Different open temporal triads have different probabilities for turning into a closed temporal triad, and different single edges have different probabilities turning into open temporal triads. To distinguish the impact factor of 18 closed temporal triads and 6 open temporal triads, we add an attention mechanism. The attention mechanism formula is shown as follows:

$$\omega_{\gamma h}(c_n) = \frac{e^{c_n}}{\sum_{c'_n} e^{c'_n}}, \quad \omega_{\theta h}(o_n) = \frac{e^{o_n}}{\sum_{o'_n} e^{o'_n}} \tag{5}$$

where c_n represents the type of the closed triads. $\omega_{\gamma h}$ represents the weights of different closed temporal triads. We take c_n as input and then use a SoftMax function to compute the final attention weight. c'_n indicates the other closed temporal triad types except c_n. o_n represents the type of open triad, ω_{θ_h} represents the weights of different open temporal triads. o'_n indicates the other open temporal triad types except o_n. Therefore, we can reformulate the conditional intensity function as:

$$\widetilde{\lambda}_{x,y}(t) = \mu_{x,y} + \sum_{t_h < t} \omega_{\gamma h}\alpha_{\gamma h} * mask + \sum_{t_h < t} \omega_{\theta h}\alpha_{\theta h} \tag{6}$$

4.4   Loss Function

By considering candidate triads $H_\gamma(t)$ selected according to the T-Motif Sampling, the probability of edge $e_{x,y}$ establish at time t can calculate by the maximum likelihood estimation:

$$p(x,y|H_\gamma(t)) = \frac{\lambda_{x,y}(t)}{\sum_{y'} \lambda_{y'|x}(t)}, \quad \log L = \sum_{(e^t_{x,y}) \in E} \log p(x,y|H_\gamma(t)) \tag{7}$$

where y' denotes all the nodes in the network except v_x. After taking a log function, the likelihood of all edges in the network is $\log L$.

Then we use SDG to calculate and update the gradient of the embedding. But when we update the gradients, we need to compute the gradients of all nodes in the network, resulting in a high computation cost. Therefore, we use the negative sampling algorithm to reduce the amount of calculation. According to the degree distribution of nodes: $P_n(v) \propto d_v^{\frac{3}{4}}$, to sample negative samples, the optimized the loss formula is as follows,

$$-\log \sigma(\widetilde{\lambda}_{x,y}(t)) - \sum_{k=1}^{K} E_{v^k \sim P_n(v)} \log \sigma(-\widetilde{\lambda}_{x,k}(t)) \tag{8}$$

where K represents the number of negative samples according to the $P_n(v)$ distribution, σ represents the Sigmoid function: $\frac{1}{1+e^{-x}}$.

5   Experimental Results

5.1   Datasets

We compare the performance of the proposed MSTNE framework with seven other state-of-the-art methods on four data sets of different sizes. We first briefly introduce the four data sets.

- **DBLP**[1] is a co-author network, where the authors come from ten different fields.

---

[1] https://dblp.uni-trier.de/.

- **WikiTalk**[2] is a temporal network, and each edge represents Wikipedia users editing each other's talk page.
- **School**[3] is a temporal network between students in a high school. Each edge represents the connection between two students. Students come from three different classes.
- **Eucore**[4] is an email data of a European research institution. Each edge represents an email sent between members of the institution

The statistics of the 4 datasets are summarized in Table 1.

5.2   Comparison Approaches

Next, we describe the 7 network embedding approaches for comparison.

- **DeepWalk** [14] uses random walks and skip-gram to generate the ideal representation of the nodes. DeepWalk does not consider the temporal relationships between nodes in the network.
- **Node2vec** [6] is an extension method of DeepWalk. The transition probability during each walk is affected by the edge weight. Same as DeepWalk, Node2vec also does not consider temporal relationships.
- **HTNE** [22] uses the Hawkes process to model the influence between events in the temporal network, which does not capture the process of network evolution.
- **MMDNE** [10] is an extension HTNE, which considers the influence of Micro- and Macro-dynamics based on HTNE.
- **MTNE** [8] is also an extension of HTNE, which uses the Hawkes process to model the influence of motif and generate the representation of the node. But it does not consider the temporal relationships of the motif.
- **Dynamictriad** [20] considers the influence of triad in the process of network evolution and captures the dynamic information in the network.
- **HNIP** [15] is a temporal random walk method, which captures the highly non-linear structural identity in the temporal network.

5.3   Parameter Settings

We set the embedding length d to 64 in all methods. In our approach MSTNE, the batch size is 500, the learning rate of SDG is 0.01, the number of positive neighbor size H is 3, the number of negative neighbor size K is 3, and the window size W is 10. For a fair comparison, we also use the optimal parameter settings in the comparison approaches. For each group of experiments, we performed the experiments 10 times and took the average value as the final results.

---

[2] http://snap.stanford.edu/data/wiki-talk-temporal.html.
[3] http://www.sociopatterns.org/datasets/high-school-dynamic-contact-networks.
[4] http://snap.stanford.edu/data/email-Eu-core-temporal.html.

Table 1. Dataset description

Dataset	Nodes	Static edges	Temporal edges	Timesteps	Labels
DBLP	28085	162451	236894	27	10
WikiTalk	1140149	7833140	3309592	2320	0
School	178	9846	18648	331	3
Eucore	986	24929	332334	526	0

5.4   Performance Evaluation

Evaluation of Node Classification. For node classification tasks, we use two labeled datasets: the DBLP and the School. We use embedding as input to train a logistic regression classifier. We adjust the ratio of the training set from 20% to 80% and compare the Macro-F1 and Micro-F1 of different methods. As shown in Table 2, the approaches that consider temporals, including HTNE, MMDNE, HNIP, Dynamictriad, MTNE, and MSTNE, are significantly better than Deep-Walk and Node2vec, in terms of Macro-F1 and Micro-F1. In all methods that consider temporal, MSTNE achieves the best performance. We think this is due to the fact that we consider the structural identity of different temporal triads, the nodes that can form temporal triads are more likely to have the same label. Compared with MTNE that only considers the structural identity of temporal triads, the proposed MSTNE achieves better effects, the highest increase in Micro-F1 is 1.43%, Macro-F1 is 1.02%. We believe this is because we consider the influence of triads with different temporal relationships.

Evaluation of Temporal Link Prediction. For the task of temporal link prediction, we selected three data sets of different sizes: DBLP, Eucore, and WikiTalk. We hope to judge whether there is an edge between two nodes at time t_k based on the data before time t_k which t_k in the middle position for each dataset. For edge e_{v_i,v_j}^t, we take $|v_i - v_j|$ as input and train a logistic regression classifier. All edges with a timestamp less than t_k are the training set, and those edges established at timestamp t_k are the positive edges in the test set, then add the unconnected edges with the same number of positive edges in the test set to be the negative edges. When adding negative edges, too many node pairs lead to too much computation cost. So we randomly sample about 1%, 1%, and 0.1% node pairs in three datasets for evaluation. Here use precision, recall, and F1 score to evaluate the system performance. As shown in Table 3, same as the node classification result, the methods considering the temporal are better than DeepWalk and Node2vec. In addition, the performances of HTNE, MTNE, MMDNE, and MSTNE are better than DynamicTriad and HNIP. We think that this is due to the use of Hawkes process modeling because the Hawkes process is modeled based on the established probability of the edge. Among these approaches, our MSTNE achieves the best performance. Compared with MTNE, the precision of MSTNE has increased by 1.20%, 0.64%, and 1.07%, respectively.

Table 2. Evaluation of node classification

Dataset	Methods	TrainRatio							
		20%		40%		60%		80%	
		Macro-F1	Micro-F1	Macro-F1	Micro-F1	Macro-F1	Micro-F1	Macro-F1	Micro-F1
DBLP	DeepWalk	0.5716	0.5927	0.5779	0.5952	0.5836	0.5999	0.5818	0.5989
	Node2vec	0.5898	0.5994	0.5998	0.6047	0.6033	0.607	0.6013	0.6065
	HTNE	0.6536	0.6645	0.6475	0.6746	0.6507	0.6765	0.653	0.6786
	MMDNE	0.6379	0.6496	0.6475	0.6551	0.6507	0.6574	0.653	0.6573
	HNIP	0.6414	0.6631	0.6538	0.670	0.6610	0.6724	0.6702	0.6735
	DynamicTriad	0.6017	0.6470	0.6230	0.6513	0.6483	0.6680	0.6642	0.6695
	MTNE	0.6505	0.6723	0.6691	0.6749	0.6753	0.6811	0.6771	0.6816
	MSTNE	**0.6589**	**0.6767**	**0.6759**	**0.6826**	**0.6757**	**0.6840**	**0.6788**	**0.6959**
School	DeepWalk	0.7951	0.7973	0.8044	0.8086	0.9090	0.9090	0.9340	0.9380
	Node2vec	0.8547	0.8627	0.8857	0.8869	0.9080	0.9090	0.9342	0.9401
	HTNE	0.9265	0.9215	0.9036	0.9090	0.9085	0.9145	0.9486	0.9486
	MMDNE	0.9200	0.9207	0.9201	0.9210	0.9476	0.9477	0.9483	0.9495
	HNIP	0.9023	0.9043	0.9208	0.9217	0.9324	0.9412	0.9653	0.9732
	DynamicTriad	0.9203	0.9280	0.9119	0.9201	0.9305	0.9399	0.9654	0.9579
	MTNE	0.9260	0.9210	0.9190	0.9195	0.9428	0.9435	0.9640	**0.9743**
	MSTNE	**0.9273**	**0.9281**	**0.9211**	**0.9217**	**0.9479**	**0.9480**	**0.9742**	**0.9743**

Table 3. Evaluation of temporal link prediction

Dataset	Metric	DeepWalk	Node2vec	HTNE	MMDNE	HNIP	DynamicTriad	MTNE	MSTNE
DBLP	Precision	0.7250	0.7545	0.8115	0.8370	0.7873	0.7760	0.8421	**0.8541**
	Recall	0.7199	0.7401	0.8085	0.8280	0.7769	0.7451	0.8356	**0.8484**
	F1	0.7123	0.7249	0.7945	0.8003	0.7420	0.7340	0.8150	**0.8202**
WikiTalk	Precision	0.7738	0.7702	0.7890	0.7910	0.7814	0.7795	0.8089	**0.8153**
	Recall	0.7642	0.7607	0.7815	0.7830	0.7783	0.7645	0.8001	**0.8051**
	F1	0.7339	0.7432	0.7685	0.7689	0.7580	0.7485	0.7935	**0.7994**
Eucore	Precision	0.7448	0.7577	0.8066	0.8157	0.8041	0.7665	0.8272	**0.8379**
	Recall	0.7407	0.7480	0.7976	0.8128	0.8038	0.7620	0.8241	**0.8354**
	F1	0.7226	0.7424	0.7950	0.8086	0.7900	0.7569	0.8166	**0.8236**

Table 4. Evaluation of temporal node recommendation

Dataset	Metric	DeepWalk	Node2vec	HTNE	MMDNE	HNIP	DynamicTriad	MTNE	MSTNE
DBLP	Pre.@10	0.0630	0.0524	0.0782	0.0821	0.0745	0.0619	0.0843	**0.0927**
	Recall	0.1426	0.1245	0.1527	0.1691	0.1495	0.1332	0.1717	**0.1734**
School	Pre.@10	0.0835	0.1003	0.1449	0.1529	0.1346	0.1246	0.1606	**0.1766**
	Recall	0.1130	0.1254	0.1856	0.1892	0.1674	0.1537	0.1982	**0.2002**

It can be seen that considering the influence of triads with different structural identities and temporal relationships improves the effect of link prediction.

Evaluation of Temporal Node Recommendation. The task of temporal node recommendation uses the network before time t_k for training. Then we generate the embedding of nodes. We sort them according to the similarity between nodes to predict the top-k possible neighbors at time t_k. Same as [2], for Deepwalk and Node2vec, we use the inner product as the similarity because they

(a) DeepWalk   (b) HTNE   (c) MSTNE

Fig. 4. 2d t-SNE visualization for DBLP Dataset. Each point represents a node, and each color represents the label of the node. Green is Data mining, and dark blue is Computer vision, light blue is Computer network. (Color figue online)

use the inner product as the optimization target during the training process. The remaining methods used the negative Euclidean distance vector as the similarity between nodes. Here we choose two datasets, the School and the DBLP for evaluation. Since there are too many pairs of nodes in DBLP, we randomly select 1% of them for evaluation. Same as temporal link prediction, we also set t_k in the middle position. Our experimental results are shown in Table 4. Table 4 shows two indicators, Pre.@10 and Recall, Pre.@10 represents the accuracy of the recommended top-10 nodes with the highest similarity. It can be seen that the methods considering temporal are still better than DeepWalk and Node2vec. In the embedding methods considering temporal, our MSTNE achieves the best performance, which is 1.6% higher of Pre.@10 than MTNE in the School data set. We think the reason is that compared with MTNE we capture the temporal relationship of the same structural identity triad.

Network Visualization. Figure 4 shows the visualization result of reducing the dimensionality of the multi-dimensional embedding into two-dimensional embedding using t-SNE [11]. As shown in Fig. 4(a), DeepWalk does not separate the three categories of nodes. Figure 4(b) shows the embedding result of HTNE. Although HTNE separates dark blue nodes from the other two categories of nodes, light blue and green are not clearly separated. Figure 4(c) shows the embedding result of MSTNE. Compared with HTNE and DeepWalk, MSTNE effectively distinguishes the three types of nodes, which illustrates the necessity of modeling the structural identity and temporal relationship of the motif.

5.5   Designation of Parameters

There are three important parameters in MSTNE, the size of positive neighbors H, the size of negative neighbors K, and the window size W for sampling. We conduct experiments on the DBLP data set for the effects of different values of these three parameters. Figure 5 shows the relationship between the size of positive and negative neighbors and Micro-F1. In Fig. 5(a), we fixed the size of negative neighbors to 5 to test the value of positive neighbors. When the size

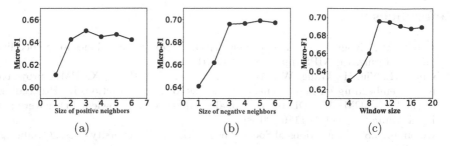

Fig. 5. Impacts of positive and negative neighbors in DBLP. The x-axis of (a) represents the size of the positive neighbors, the x-axis of (b) represents the size of the negative neighbors, the x-axis of (c) represents the size of the window. The y-axis of both (a), (b), and (c) is the value of Micro-F1 in DBLP with the task of node classification.

of positive neighbors is less than 3, Micro-F1 gradually increases. It reaches the peak at 3 while the Micro-F1 begins to drop after 3. In Fig. 5(b), we fixed the size of positive neighbors to 3. When the number of negative neighbors is less than 3, Micro-F1 gradually increases. When it is greater than 3, the increase is not significant. We also compare the effects of different window sizes. Figure 5(c) shows the relationship between the sampled window size W and the Micro-F1. It can be seen that when the window size is 10, Micro-F1 reaches its peak and then decreases gradually. We think that is because when the window size is large, the sampled neighbor nodes will occur at a larger time interval, the influence will be small, resulting in the decrease of Micro-F1. Considering both the efficiency and the accuracy, we think that the size of positive neighbors, size of negative neighbors, window size are 3, 3, 10, the method achieves the best performance, so we set it as the default parameter settings.

6   Conclusion

In this paper, we proposed the MSTNE, a novel temporal network embedding method. We use the neighbor node sampling method to capture both the structural identity and temporal relationship of the motif during the evolution of the network and use the Hawks process to measure the impact factor of the temporal motif. Compared with other temporal network embedding methods, we further explored the influence of the temporal triad on the network evolution process. Since the MSTNE considers both the structural identity and temporal relationship of the motif, the experimental results show that the MSTNE achieves the best performance on the four real-world data sets. In the future, we will expand our model on other types of motifs and integrate the attributes of temporal edges into our model.

Acknowledgment. This work is supported by the National Key Research and Development Program of China (Grant No. 2016YFB1000304) and National Natural Science Foundation of China (Grant No. 1636208).

References

1. Belkin, M., Niyogi, P.: Laplacian eigenmaps and spectral techniques for embedding and clustering. In: NIPS, pp. 585–591 (2001)
2. Chen, H., Yin, H., Wang, W., Wang, H., Nguyen, Q.V.H., Li, X.: PME: projected metric embedding on heterogeneous networks for link prediction. In: Proceedings of the 24th ACM SIGKDD International Conference on Knowledge Discovery and Data Mining, pp. 1177–1186 (2018)
3. Coleman, J.S.: Foundations of Social Theory. Harvard University Press, Cambridge (1994)
4. Du, L., Wang, Y., Song, G., Lu, Z., Wang, J.: Dynamic network embedding: an extended approach for skip-gram based network embedding. In: IJCAI, vol. 2018, pp. 2086–2092 (2018)
5. Goyal, P., Ferrara, E.: Graph embedding techniques, applications, and performance: a survey. Knowl.-Based Syst. **151**, 78–94 (2018)
6. Grover, A., Leskovec, J.: node2vec: Scalable feature learning for networks, pp. 855–864 (2016)
7. Huang, H., Dong, Y., Tang, J., Yang, H., Chawla, N.V., Fu, X.: Will triadic closure strengthen ties in social networks? ACM Trans. Knowl. Discov. Data (TKDD) **12**(3), 1–25 (2018)
8. Huang, H., Fang, Z., Wang, X., Miao, Y., Jin, H.: Motif-preserving temporal network embedding. In: Proceedings of the Twenty-Ninth International Joint Conference on Artificial Intelligence, IJCAI-2020, pp. 1237–1243 (2020)
9. Laub, P.J., Taimre, T., Pollett, P.K.: Hawkes processes. arXiv preprint arXiv:1507.02822 (2015)
10. Lu, Y., Wang, X., Shi, C., Yu, P.S., Ye, Y.: Temporal network embedding with micro-and macro-dynamics. In: Proceedings of the 28th ACM International Conference on Information and Knowledge Management, pp. 469–478 (2019)
11. Van der Maaten, L., Hinton, G.: Visualizing data using T-SNE. J. Mach. Learn. Res. 9(11) (2008)
12. Meira, L.A., Máximo, V.R., Fazenda, Á.L., Da Conceiçao, A.F.: ACC-Motif: accelerated network motif detection. IEEE/ACM Trans. Comput. Biol. Bioinf. **11**(5), 853–862 (2014)
13. Paranjape, A., Benson, A.R., Leskovec, J.: Motifs in temporal networks. In: Proceedings of the Tenth ACM International Conference on Web Search and Data Mining, pp. 601–610 (2017)
14. Perozzi, B., Al-Rfou, R., Skiena, S.: DeepWalk: online learning of social representations, pp. 701–710 (2014)
15. Qiu, Z., Hu, W., Wu, J., Liu, W., Du, B., Jia, X.: Temporal network embedding with high-order nonlinear information, vol. 34, no. 04, pp. 5436–5443 (2020)
16. Ribeiro, L.F., Saverese, P.H., Figueiredo, D.R.: struc2vec: Learning node representations from structural identity, pp. 385–394 (2017)
17. Roweis, S.T., Saul, L.K.: Nonlinear dimensionality reduction by locally linear embedding. Science **290**(5500), 2323–2326 (2000)
18. Tang, J., Qu, M., Wang, M., Zhang, M., Yan, J., Mei, Q.: Line: large-scale information network embedding, pp. 1067–1077 (2015)
19. Wernicke, S., Rasche, F.: FANMOD: a tool for fast network motif detection. Bioinformatics **22**(9), 1152–1153 (2006)
20. Zhou, L., Yang, Y., Ren, X., Wu, F., Zhuang, Y.: Dynamic network embedding by modeling triadic closure process, vol. 32, no. 1 (2018)

21. Zhu, L., Guo, D., Yin, J., Ver Steeg, G., Galstyan, A.: Scalable temporal latent space inference for link prediction in dynamic social networks, vol. 28, pp. 2765–2777. IEEE (2016)
22. Zuo, Y., Liu, G., Lin, H., Guo, J., Hu, X., Wu, J.: Embedding temporal network via neighborhood formation. In: Proceedings of the 24th ACM SIGKDD International Conference on Knowledge Discovery Data Mining, pp. 2857–2866 (2018)

Learning Robust Representation Through Graph Adversarial Contrastive Learning

Jiayan Guo[1(✉)], Shangyang Li[2(✉)], Yue Zhao[3], and Yan Zhang[1]

[1] School of Artificial Intelligence, Peking University, Beijing, China
{guojiayan,zhyzhy001}@pku.edu.cn
[2] Peking-Tsinghua Center for Life Sciences, IDG/McGovern Institute for Brain
Research, Academy for Advanced Interdisciplinary Studies, Peking University,
Beijing, China
syli@pku.edu.cn
[3] Academy for Advanced Interdisciplinary Studies, Peking University, Beijing, China
zhaoyue@stu.pku.edu.cn

Abstract. Existing studies show that node representations generated by graph neural networks (GNNs) are vulnerable to adversarial attacks, such as unnoticeable perturbations of adjacent matrix and node features. Thus, it is requisite to learn robust representations in graph neural networks. To improve the robustness of graph representation learning, we propose a novel **Graph Adversarial Contrastive Learning** framework (GraphACL) by introducing adversarial augmentations into graph self-supervised learning. In this framework, we maximize the mutual information between local and global representations of a perturbed graph and its adversarial augmentations, where the adversarial graphs can be generated in either supervised or unsupervised approaches. Based on the Information Bottleneck Principle, we theoretically prove that our method could obtain a much tighter bound, thus improving the robustness of graph representation learning. Empirically, we evaluate several methods on a range of node classification benchmarks and the results demonstrate GraphACL could achieve comparable accuracy over previous supervised methods.

Keywords: Graph neural network · Graph adversarial attack · Robust representation learning

1 Introduction

Graph neural networks (GNNs) have enabled significant advances on graph-structured data [9,16] and are widely used in many applications like node classification, graph classification, and recommendation systems. However, existing works show that they are vulnerable towards adversarial attacks [26–28] like

---

J. Guo, S. Li and Y. Zhao—Equal contribution.

A. Bhattacharya et al. (Eds.): DASFAA 2022, LNCS 13245, pp. 682–697, 2022.
https://doi.org/10.1007/978-3-031-00123-9_54

unnoticeable perturbations, which is still a critical challenge in employing GNNs in safety-critical applications.

Albeit various studies have been proposed to ensure the robustness of the graph neural networks against adversarial attacks [8,13,20,24], the significance of adversarial augmentations has been ignored, especially under unsupervised learning setting. Recently, self-supervised learning has achieved remarkable performances on graph-structured data, like DGI [17], GraphCL [23], etc. These works use pairs of augmentations on unlabeled graphs to define a classification task for pretext learning of graph representations. Also, GraphCL [23] has found that contrastive learning with randomly generated graph augmentations can somehow increase the robustness; however, we argue that such randomly generated samples are not the optimal choice to achieve the robustness of representations and adversarial augmentations can perform provably much better.

Thereby, we present a novel adversarial self-supervised learning framework to learn robust graph representations. We introduce adversarial samples into the input. Primairly, both supervised and unsupervised approaches can be used to generate adversarial samples. For example, Metattack [28], a supervised adversarial attack method, can be directly applied. Besides, we further propose an unsupervised method to generated adversarial graphs, which uses unsupervised contrastive loss as the target of Metattack to generete adversarial samples. After generating perturbed graphs, we maximize the similarity between representations of the clean graph and the adversarial attacked graph to suppress distortions caused by adversarial perturbations. This will result in representations that are robust against adversarial attacks.

We refer to this novel adversarial self-supervised graph representation learning method as **Graph Adversarial Contrastive Learning** (GraphACL). To the best of our knowledge, this is the first attempt to use adversarial samples to increase the robustness of graph representations based on contrastive learning. We also build a theoretical framework to analyze the robustness of graph contrastive learning based on the Information Bottleneck Principle. To verify the effectiveness of GraphACL, we conduct experiments on public academic dataset, Cora, Citeseer, Pubmed under both targeted attack (i.e., Netattack) and global attack (i.e., Metattack). Experimental results suggest that GraphACL outperforms DGI and other baselines significantly, thus proving our method can learn robust representations under various graph adversarial attacks.

In summary, our contributions are as follows:

- We propose GraphACL, a general framework to use self-supervised graph contrastive learning with adversarial samples to learn robust graph representations.
- We theoretically prove that our method could improve the robustness of graph representation learning from the perspective of information theory.
- We present an unsupervised graph adversarial attack method that use meta-gradient to poison the graph structure to maximize the contrastive loss between clean and perturbed graphs.

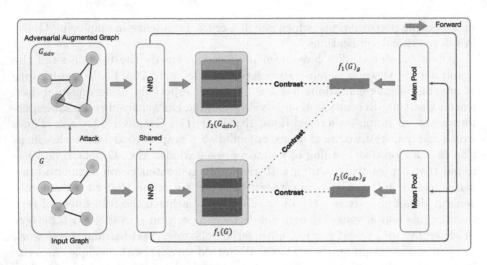

Fig. 1. Graph adversarial contrasive learning framework

- We conduct extensive experiments to demonstrate the effectiveness of our proposed GraphACL under various types of adversarial attacks, which indicates that GraphACL can significantly improve the performance of previous methods both in evasive and poisoning settings.

2   Methodologies

2.1   Graph Adversarial Attack

In this subsection, we will formulate the classic optimization problem of graph adversarial attack. Let $G = (A, X)$ be an attribute graph with adjacency matrix $A \in \{0, 1\}^{N \times N}$ and attribute matrix $X \in \mathbb{R}^{N \times D}$, where N is the number of nodes and D is the dimension of the node feature vector. Considering a semi-supervised node classification task, where labels of the nodes $\mathcal{V}_L \in \mathcal{V}$ are given. Each node is assigned as one class in $\mathcal{C} \in \{c_1, ..., c_k\}$. The goal of adversarial attack can be mathematically formulated as a bilevel optimization problem

$$\max_{G_{attack} \in \Phi(G)} \mathcal{L}(f_{\theta^*}(G_{attack}))$$
$$s.t. \quad \theta^* = \underset{\theta}{argmin}\, \mathcal{L}(f_\theta(G)) \tag{1}$$

where $\Phi(G)$ is the space of perturbation on the input graph, $\mathcal{L}$ is the cross entropy by default and $f_\theta(\cdot)$ is the surrogate model.

Based on whether to re-train the model on the attacked graph, the attack type is categorized by poisoning attack and evasive attack. Poisoning attack requires re-training while evasive attack does not.

2.2   Graph Adversarial Contrastive Learning Framework

As illustrated in Fig. 1, we now present our framework to learn robust representations via adversarial contrastive training. Firstly, we conduct adversarial generation on the perturbed graph. Then, we use the input graph and an adversarial augmented graph as different views of the same graph. A shared encoder like GCN encodes multi-views of the graph and then outputs respective local representations $f_1(G)$ and $f_2(G_{adv})$, where G is the input graph and G_{adv} is the adversarial augmentation. $f_1(\cdot)$ and $f_2(\cdot)$ are encoders that can be the same or different with their unshared projection layers. Crossed local-global information maximization is implemented by maximizing the information between local representations of the input graph and global representations of the adversarial graph, vice versa. The GraphACL framework is modified on DGI framework by additionally introducing an adversarial augmented view of the input graph. The other omitted settings are the same with DGI, and negative samples are also used. Therefore, the improvement of GraphACL over DGI is of our concern.

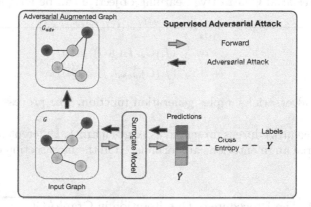

Fig. 2. Generation of graph adversarial augmentations under supervised loss.

When several labels are known, adversarial augmentation G_{adv} could be obtained by supervised generation method. The process is illustrated in Fig. 2 Then we use the contrastive learning objective to maximize the similarity between input examples G and their instance-wise adversarial augmentation G_{adv}. Then we can formulate our Graph Adversarial Contrastive Learning objective as follow:

$$\mathcal{L}_{GACL}^{sup} = \min_{f_1, f_2}(L_{cl}^{self}(f_1(G), f_1(G)_{global})$$
$$+ \alpha L_{cl}^{adv}(f_1(G), f_2(G_{adv})_{global}) \qquad (2)$$
$$+ \beta L_{cl}^{adv}(f_1(G)_{global}, f_2(G_{adv}))),$$

where L_{cl} is contrastive loss that is negative mutual information essentially and α balances between contrastive loss L_{cl}^{self} and L_{cl}^{adv}. Similar to DGI [17], $f(\cdot)_{global}$ is the global representation of the whole graph.

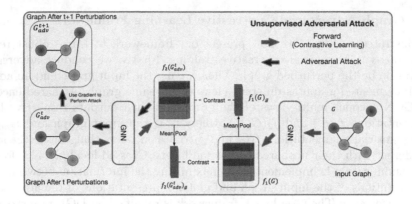

Fig. 3. Generation of graph adversarial augmentations under unsupervised loss.

If no label information is given, the unsupervised adversarial training strategy of Graph Adversarial Contrastive Learning objective can be formulated as:

$$
\begin{aligned}
\mathcal{L}_{GACL}^{unsup} = \min_{f_1, f_2} \max_g (&L_{cl}^{self}(f_1(G), f_1(G)_{global}) \\
&+ \alpha L_{cl}^{adv}(f_1(G), f_2(g(G))_{global}) \\
&+ \beta L_{cl}^{adv}(f_1(G)_{global}, f_2(g(G)))),
\end{aligned}
\tag{3}
$$

where $g(\cdot)$ is adversarial samples generation function. The process is illustrated in Fig. 3.

The detailed procedure is presented as **Algorithm 1**. Different ways of generating adversarial augmentations are formulated in the next section of theoretical analysis.

Algorithm 1. The Procedure of One Iteration in GraphACL

Require: Input Graph $G = (A, X)$;
Ensure: $f_1(\cdot)$, $f_2(\cdot)$: graph encoders;
1: **if** use supervised adversarial augmentation **then**
2:     generate adversarial graph $\hat{G}$ based on Eq. (13);
3: **else**
4:     generate adversarial graph $\hat{G}$ based on Eq. (17);
5: **end if**
6: Generate node representations of input graph $f_1(G)$;
7: Generate node representations of adversarial augmented graph $f_2(G_{adv})$;
8: Generate global representation of input graph $f_1(G)_g$ by mean pooling;
9: Generate global representation of adversarial augmented graph $f_2(G_{adv})_g$ by mean pooling;
10: Compute contrastive loss by Eq. (2) or Eq. (3);
11: Back propagate gradients and update $f_1(\cdot)$ and $f_2(\cdot)$;
12: **return** $f_1(\cdot)$ and $f_2(\cdot)$;

---

3   Theoretical Analysis on Graph Adversarial Contrastive Learning

In this section, we first formulate the Information Bottleneck Principle in graph self-supervised learning and achieve the related objective function. Then, we illustrate the generation of adversarial augmentations in Figure ??. Finally, we derive desirable lower bounds of the information bottleneck and formulate the objective function according to supervised and unsupervised adversarial augmentations.

3.1   Information Bottleneck Principle for Graph Self-supervised Learning

The Information Bottleneck (IB) [14,15] provides an essential principle for representation learning from the perspective of information theory, which is an optimal representation need to contain minimal yet sufficient information for downstream tasks. It encourages the representation to involve as much information about the target as possible to obtain high prediction accuracy, and discard redundant information that is irrelevant to the target. In graph representation learning, each graph $G(A, X)$ contains information of both the graph structure $A \in \mathbb{R}^{N \times N}$ and node features $X \in \mathbb{R}^{N \times d}$. Applying IB to graph self-supervised learning, we desire to learn an optimal graph representation Z, which is informative about the original graph $G \in \mathcal{G}$, but invariant to its augmentations $\widehat{G} \in \widehat{\mathcal{G}}$. This principle can be formulated as follows:

$$\mathcal{L}_{\text{IB}} \triangleq \beta I(Z, \widehat{\mathcal{G}}) - I(Z, \mathcal{G}), \tag{4}$$

where $I(\cdot, \cdot)$ denotes mutual information between variables and $\beta > 0$ is a hyperparameter to control the trade-off between preserving information and being invariant to distortions.

We use $\widehat{G}$ to represent different views of the corresponding graph G. For the first term of $\mathcal{L}_{\text{IB}}$, we utilize an upper bound proved in [4] to derive a tractable bound of the mutual information between Z and $\widehat{G}$:

$$I(Z, \widehat{\mathcal{G}}) \leq \sum_{z \in Z} \sum_{\widehat{G} \in \widehat{\mathcal{G}}} p(\widehat{G}, z) \log(p(z \mid \widehat{G}))$$
$$- \sum_{z \in Z} \sum_{\widehat{G} \in \widehat{\mathcal{G}}} p(\widehat{G}) p(z) \log(p(z \mid \widehat{G})), \tag{5}$$

Also, the mutual information between Z and $\mathcal{G}$ can be written as

$$I(Z, \mathcal{G}) = \sum_{z \in Z} \sum_{G \in \mathcal{G}} p(z, G) \log \frac{p(z, G)}{p(z) p(G)}$$
$$= \sum_{z \in Z} \sum_{G \in \mathcal{G}} p(z, G) \log p(G|z) + H(\mathcal{G}). \tag{6}$$

The entropy term $H(\mathcal{G})$ could be dropped which results in

$$I(Z,\mathcal{G}) \geq \sum_{z \in Z} \sum_{G \in \mathcal{G}} p(z,G) \log p(G|z). \tag{7}$$

By combining Eq. (4), Eq. (5) and Eq. (7), we can minimize the upper bound of IB by:

$$\begin{aligned}
\hat{\mathcal{L}}_{\text{IB}} = \beta \frac{1}{NM} \sum_{i=1}^{N} \sum_{j=1}^{M} [\log p(z_j^i \mid \widehat{G}_j^i) \\
- \frac{1}{M} \sum_{k=1}^{M} \log p(z_k^i \mid \widehat{G}_j^i)] \\
- \frac{1}{NM} \sum_{i=1}^{N} \sum_{j=1}^{M} \log p(G^i \mid z_j^i),
\end{aligned} \tag{8}$$

where N is the number of original graphs, M is the number of augmentations of each original input graph. In graph contrastive learning, $p(z_j^i \mid \widehat{G}_j^i)$ can be viewed as an encoder $f_\theta : \widehat{\mathcal{G}} \to Z$.

We assume $p(z_k^i \mid z_j^i, \widehat{G}_j^i) = p(z_k^i \mid z_j^i)$, which means the representation z_k^i of an augmented graph cannot depend directly on another augmented graph $\widehat{G}_j^i$. Also, since the function f_θ is deterministic, we have

$$\begin{aligned}
p(z_k^i \mid \widehat{G}_j^i) = \sum p(z_k^i \mid z_j^i, \widehat{G}_j^i) p(z_j^i \mid \widehat{G}_j^i) \\
= p(z_k^i \mid z_j^i) p(z_j^i \mid \widehat{G}_j^i) = p(z_k^i \mid z_j^i)
\end{aligned} \tag{9}$$

Further, Eq. (8) can be written as:

$$\begin{aligned}
\hat{\mathcal{L}}_{\text{IB}} = \beta \frac{1}{NM} \sum_{i=1}^{N} \sum_{j=1}^{M} [-\frac{1}{M} \sum_{k=1}^{M} \log p(z_k^i \mid z_j^i)] \\
- \frac{1}{NM} \sum_{i=1}^{N} \sum_{j=1}^{M} \log p(G^i \mid z_j^i),
\end{aligned} \tag{10}$$

where $p(z_k^i|z_j^i)$ could be viewed as a similarity measurement between representations of different augmentations. Eventually, we formulate the problem with IB and obtain a general objective function of graph self-supervised learning as Eq. (10). Intuitively, the objective function motivates GNN to increase the averaged similarity of representations between different augmentations, thus making the learned representations invariant and robust to various different views.

3.2 Generation of Supervised Graph Adversarial Augmentations

The generation of supervised graph adversarial augmentation is schematically shown in Fig. 2, which utilizes previous graph adversarial attack methods like

Metattack [28]. Primarily, a surrogate model is applied to the perturbed graph to generate predictions. Then the supervised loss is computed by cross entropy. Finally, we use the gradient to modify the structure of the original graph to generate adversarial samples.

Suppose G is the original graph and its node labels are Y, we consider a softmax regression layer between Z and Y. The posterior class probabilities can be written as:

$$
\begin{aligned}
P_{Y|Z}(y \mid z) &= \frac{e^{w_y^T z}}{\sum_k e^{w_k^T z}} \\
&= \frac{e^{w_y^T f_\theta(G)}}{\sum_k e^{w_k^T f_\theta(G)}}
\end{aligned}
\tag{11}
$$

where $\mathcal{W} = \{w_y\}_{y=1}^k$ is the vector of classification parameters for class y and θ is the parameter of the encoder $f_\theta(\cdot)$, which are learned by minimizing the cross-entropy loss

$$
L_{ce}(G, Y; \mathcal{W}, \theta) = -\log \frac{e^{w_y^T f_\theta(G)}}{\sum_k e^{w_k^T f_\theta(G)}}.
\tag{12}
$$

Given the learned encoder and classifier, an optimal perturbation for G is generated by maximizing the cross-entropy loss:

$$
\begin{aligned}
G_{adv*} &= \arg \max_{G_{adv}} L_{ce}(G_{adv}, y; \mathcal{W}, \theta) \\
&\text{s.t. } G_{adv} \in \Phi(G),
\end{aligned}
\tag{13}
$$

where $\Phi(G)$ means the space of perturbation on the original graph. Then we can further formulate a constrained optimization problem as following

$$
\begin{aligned}
G_{adv} &\in \Phi(X) \\
&\text{s.t. } \mathcal{Q}(G, G_{adv}) < \epsilon
\end{aligned}
\tag{14}
$$

where $\mathcal{Q}(\cdot)$ represents a distance measurement function, ϵ is a parameter for imperceptible perturbation evaluation.

$$
\begin{aligned}
&\frac{1}{K} \sum_{k=1}^M \log p[f_\theta(\widehat{G}) \mid f_\theta(G)] \\
&= \frac{1}{K} \sum_{k=1}^M \log \sum p[f_\theta(\widehat{G}) \mid f_\theta(G), y) p(y \mid f_\theta(G)] \\
&= \frac{1}{K} \sum_{k=1}^M \log \sum p[f_\theta(\widehat{G}) \mid y) p(y \mid f_\theta(G))] \\
&> \log[\sum p(f_\theta(G_{adv*}) \mid y) p(y \mid f_\theta(G))] \\
&= \log p[f_\theta(G_{adv*}) \mid f_\theta(G)],
\end{aligned}
\tag{15}
$$

Since similarity between clean graph representations and adversarial augmentation representations becomes a lower bound of the averaged similarity between representations on the original graph and all augmentations.

3.3   Generation of Unsupervised Graph Adversarial Augmentations

The unsupervised graph adversarial augmentation generation is schematically in Fig. 3. Graph i is mapped into an example pair $(\widehat{G}_k^i, \widehat{G}_j^i)$. Graph contrastive learning is performed through maximizing the agreement between an positive pair. Equation (10) tells us that if we want to get a more robust representation, we need to increase $\frac{1}{K}\sum_{k=1}^{M}\log p(z_k^i|z_j^i)$. Similar to supervised situation, there is a lower bound $\frac{1}{K}\sum_{k=1}^{M}\log p[f_\theta(\widehat{G}_k^i)|f_\theta(\widehat{G}_j^i)] > \log p[f_\theta(G_{adv*}^i)|f_\theta(\widehat{G}_j^i)]$.

The choice of G_{adv*} could be formulate as a two-stage alternative optimization problem: one is self-supervised learning, the other is adversarial attack or generation of adversarial augmentations. A generative function $g(\cdot)$ is introduced to denote the generation of adversarial samples G_{adv*}. For example, the generation function $g(\cdot)$ can be the same as it in Metattack. In the first stage, adversarial samples can be generated by $G_{adv} = g(G)$, which is further considered as an augmentation or a different view of the perturbed G. Hence, self-supervised learning is conducted to maximize the mutual information between different views by optimizing corresponding encoders f_1 and f_2. Then, given the encoders, we can optimize $g(\cdot)$ by using adversarial attacks to minimize self-supervised loss and obtain a new adversarial graph. Still using Metattack as an example, $g(\cdot)$ is optimized by attacking the gradient of the self-supervised loss.

Finally, we formulate the two-stage of unsupervised training strategy – adversarial attack and self-supervised learning as an underlying min-max objective function in the following:

$$\min_{g}\max_{f_1,f_2} I(f_1(G), f_2(g(G))) \tag{16}$$

In practice, as shown in Fig. 3, the adversarial unsupervised training strategy is modified on DGI framework:

$$\min_{g}\max_{f_1,f_2}\{I[f_1(G)_{global}, f_2(g(G))] \\ + I[f_1(G), f_2(g(G))_{global}]\}. \tag{17}$$

4   Experiments

4.1   Experimental Settings

To evaluate the robustness of different models against adversarial attacks, we conduct experiments on the following benchmarks with Netattack [1] and Metattack [28], where adversarial augmentations are generated by supervised or unsupervised contrastive loss, respectively. Netattack works based on boolean features; therefore, the features in each dataset are preprocessed to be 0 or 1. We follow the experimental settings in Netattack [1, 21, 23] exactly: We test the classification accuracy of the 40 selected target nodes: 10 nodes with the highest margin of classification, which is most likely to be classified correctly; 10 nodes with the lowest margin of classification but still classified correctly, which may be easily attacked; 20 other random nodes. Each perturbation denotes a filp on a boolean feature or a modification on an edge related to the node. Robustness experiments are evaluated on a clean graph and corrupted graphs with a number of perturbations from 1 to 4. For Metattack, we use the standard Metattack setting with a perturbation rate of 0.05 and 0.2 on clean graphs to generate modified graphs. Then we test the node classification accuracy on the modified graph. In this experiment, the adversarial augmentations are generated by our proposed unsupervised contrastive loss.

We evaluate two types of robustness tasks, including evasive and poisoning. We include the baselines such as GCN, RGCN, GAT; the results are cited from [21]. Previous work has not included the pre-trained models in attack experiments, and DGI is now considered to be compared with since we desire to evaluate the impact of adversarial augmentations. GIB [21] is one of the previous SOTA on these experiments; however, it is not related to our comparison on whether to use adversarial augmentations nor unsupervised pretraining. To verify the impact of introducing the adversarial augmentation, we focus on the improvements of GraphACL over DGI and GCN. For Metattack, we only evaluate the model's performance on the evasive task.

Datasets are summarized in the supplementary materials. The results in Table 1 and 2 denote averaged classification accuracy and standard deviation over 5 random seeds. GACL is short for GraphACL.

We denote DGI and GraphACL as pre-trained methods, which are unsupervised pre-trained with only the graph and features and without any other information in downstream tasks. No previous study includes pre-trained methods in robustness experiments; however, we find it effective to defend unknown attacks with pretraining. Our main hypothesis is that the adversarial augmentations will help the model learn more robust representations, which is confirmed by comparing DGI and GraphACL. Thereby, we highlight the best results in DGI, GCL and GraphACL in bold, surpassing all other results except few special cases.

4.2   Robustness Evaluation Under Netattack

Netattack. For experiments on Cora, GraphACL with pretraining and adversarial augmentations, outperforms all previous methods like GCN remarkably. In evasive experiments, GraphACL surpasses GCN by 16.0%, RGCN by 18.0%, GAT by 19.5% and DGI by 5.5% on average on the task with one perturbation. Also, GraphACL achieves 17.0% and 2.0% improvements over GCN and DGI respectively, when being poisoned on one perturbation case. When the number of perturbations gets larger, the averaged results of GraphACL are still a bit higher.

Table 1. Classification accuracy(%) under netattack over 5 random seeds

	Model	Clean	Evasive				Poisoning			
			1	2	3	4	1	2	3	4
Cora	GCN	80.0±7.87	51.5±4.87	38.0±6.22	31.0±2.24	26.0±3.79	47.5±7.07	39.5±2.74	30.0±5.00	26.5±3.79
	RGCN	80.0±4.67	49.5±6.47	36.0±5.18	30.5±3.25	25.5±2.09	46.5±5.75	35.5±3.70	29.0±3.79	25.5±2.73
	GAT	77.8±3.97	48.0±8.73	39.5±5.70	36.5±5.48	32.5±5.30	50.5±5.70	38.0±5.97	33.5±2.85	26.0±3.79
	DGI	82.5±4.33	62.0±4.81	46.0±3.79	34.0±5.18	27.5±3.06	62.5±3.54	43.5±3.79	31.5±6.75	26.5±4.18
	GCL	64.4±4.27	53.8± 5.20	38.8±8.54	25.6±3.15	18.8±3.23	41.9±6.25	33.1±8.00	28.8±7.77	23.1±5.54
	GACL	82.0±3.26	**67.5±5.00**	**46.0±5.76**	**35.5±4.81**	**29.0±6.75**	**64.5±5.70**	**44.0±7.42**	**33.5±6.02**	**27.5±5.30**
Citeseer	GCN	71.8±6.94	42.5±7.07	27.5±6.37	18.0±3.26	15.0±2.50	29.0±7.20	20.5±1.12	17.5±1.77	13.0±2.09
	RGCN	73.5±8.40	41.5±7.42	24.5±6.47	18.5±6.52	13.0±1.11	31.0±5.48	19.5±2.09	13.5±2.85	5.00±1.77
	GAT	72.3±8.38	49.0±9.12	33.0±5.97	22.0±4.81	18.0±3.26	38.0±5.12	23.5±4.87	16.5±4.54	12.0±2.09
	DGI	78.5±5.76	64.0±4.18	49.5±4.47	36.5±5.18	30.5±5.97	57.5±4.68	40.0±7.70	31.0±2.24	25.5±5.70
	GCL	70.0±7.36	59.4±6.57	47.5±5.40	36.3±6.29	32.5±4.08	50.6±5.54	38.3±6.77	38.1±4.27	26.9±8.00
	GACL	77.5±3.06	**66.0±3.79**	**53.0±5.70**	**46.0±4.87**	**37.0±1.12**	**63.5±4.18**	**41.0±3.79**	**40.0±9.19**	**30.5±5.97**
Pubmed	GCN	82.6±6.98	39.5±4.81	32.0±4.81	31.0±5.76	31.0±5.76	36.0±4.18	32.5±6.37	31.0±5.76	28.5±5.18
	RGCN	79.0±5.18	39.5±5.70	33.0±4.80	31.5±4.18	30.0±5.00	38.5±4.18	31.5±2.85	29.5±3.70	27.0±3.70
	GAT	78.6±6.70	41.0±8.40	33.5±4.18	30.5±4.47	31.0±4.18	39.5±3.26	31.0±4.18	30.0±3.06	35.5±5.97
	DGI	79.0±7.20	40.5±5.86	31.0±4.54	29.5±3.71	28.0±2.74	40.0±4.81	31.0±3.79	28.5±4.18	28.0±4.68
	GCL	67.5±7.07	**45.3±1.77**	**35.8±6.29**	28.3±5.10	28.1±4.26	40.25±1.77	33.25±6.61	**30.3±1.44**	19.4±3.75
	GACL	83.0±5.42	43.0±5.42	34.0±5.18	**30.0±3.06**	**28.5±4.18**	**41.0±3.79**	**34.0±2.85**	29.5±3.71	**28.5±4.18**

GraphACL also achieves significant improvements on both evasive and poisoning experiments on Citeseer. Note that many nodes in Citeseer have few degrees, thus making the attack much harder to defend. When the number of perturbations is 1, GraphACL surpasses DGI and GCN by 2.0% and 23.5% on evasive tasks, 3.5% and 34.5% on poisoning tasks. Especially, when the number of perturbations is 3, GraphACL surpasses DGI by 9.5% on evasive tasks and 9.0% on poisoning tasks. All results of GraphACL surpass DGI and GCN a lot on both evasive and poisoning tasks. We attribute the success to the added views of adversarial augmentations, which makes the model more defensive to the unseen attacks on graphs like Citeseer.

Additionally, GraphACL achieves the best on the clean graph and obtains similar results to GraphCL on Pubmed both in evasive setting and poisoning setting. When the number of perturbations is 1, GraphACL improves the averaged accuracy on GCN, RGCN, GAT, and DGI by 3.5%, 3.5%, 2.0%, and 2.5%, respectively on evasive tasks.

4.3   Robustness Evaluation Under Metattack

Metattack. We also evaluate the performances of DGI and GraphACL on Cora and Citeseer under Metattack in Table 2. We use our proposed unsupervised attack method in Fig. 3 to generate graph adversarial samples for GraphACL. On the second row, clean denotes evaluation on the clean graph after Metattack, while 0.05 and 0.2 denotes the perturbed rate on the graph. The first column denotes the the perturbed rate in GraphACL training, which is related to adversarial generation. When this rate is 0.000, the method is indeed DGI. While the rate increases, the perturbation gets stronger. As information bottleneck demonstrated, there would be a desired representation containing sufficient information with less nuisance to get more robust performance. The results in Table 2 prove the same idea. The performance gets higher first, achieves a peak, and then goes down when the perturbed rate increases. Surprisingly, we find that the optimal results are all related to GraphACL with 0.030 rate of perturbation. If the perturbation rate is much larger, the graph is corrupted too much to maintain sufficient information, thus resulting in poor performance. The best performances of GraphACL all outperform DGI. When the evalutaion graph is more perturbed, which means the robust representation is much more needed, the improvements get much higher. In Cora, the best GraphACL achieves 1.4%, 2.2% and 3.5% higher in performance than DGI on clean, 0.05-perturbed, 0.2-perturbed graph respectively. In Citeseer, the best GraphACL achieves 2.0%, 5.4% and 3.2% higher in performance than DGI on clean, 0.05-perturbed, 0.2-perturbed graph respectively.

Table 2. Classification Accuracy(%) under Metattack over 5 random seeds

Dataset	Cora			Citeseer		
Rate\Task	Clean	0.05	0.2	Clean	0.05	0.2
DGI	75.2 ± 2.71	73.8 ± 2.48	71.5 ± 2.77	67.5 ± 3.44	64.9 ± 4.18	65.9 ± 4.10
GraphACL-0.001	75.4 ± 2.71	74.7 ± 3.23	73.1 ± 3.56	69.1 ± 1.47	69.0 ± 1.56	68.9 ± 1.63
GraphACL-0.010	76.6 ± 2.21	75.7 ± 2.47	74.1 ± 2.55	70.2 ± 1.90	69.9 ± 2.01	68.8 ± 2.04
GraphACL-0.020	76.2 ± 2.22	74.8 ± 2.93	74.9 ± 3.84	68.6 ± 3.59	68.0 ± 3.83	66.4 ± 3.28
GraphACL-0.030	$\mathbf{76.8 \pm 0.82}$	$\mathbf{76.0 \pm 1.40}$	$\mathbf{75.0 \pm 1.70}$	$\mathbf{70.5 \pm 1.88}$	$\mathbf{70.3 \pm 1.87}$	$\mathbf{69.1 \pm 2.40}$
GraphACL-0.040	73.7 ± 3.95	73.3 ± 3.90	72.4 ± 3.96	69.2 ± 2.50	68.8 ± 2.54	67.3 ± 3.21
GraphACL-0.050	74.4 ± 4.11	73.8 ± 4.67	72.8 ± 5.55	69.8 ± 1.70	69.5 ± 1.88	68.2 ± 2.35

4.4 Perturbation Rate Sensitivity for Adversarial Samples

In Fig. 4, we evaluate GraphACL with different perturbation rates for adversarial samples on Cora, Citeseer and Pubmed. Experiments are conducted over 5 random seeds. The solid line denotes GraphACL with zero perturbation rate served as a baseline, which means no augmentation is included, i.e., DGI. The dotted line denotes GraphACL with the least positive perturbation rate, which usually performs the worst within GraphACL. The other three lines in dash-dot style related to GraphACL with different suitable perturbation rates. Their performances are similar, which means our method is not sensitive to the perturbation rate in a reasonable range. The difference is mainly based on the dataset. Within the range, there is a best perturbation rate for GraphACL to improve the performance over baselines up to 8%.

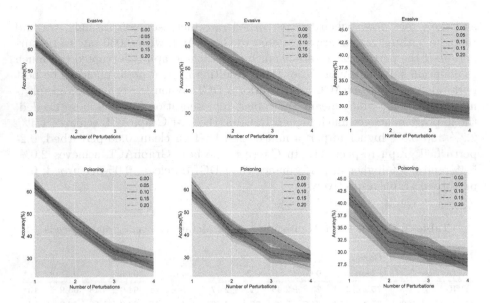

Fig. 4. Results on different perturbation rates for adversarial samples

5   Related Work

5.1   Adversarial Attack and Defense on Graph Data

The first graph adversarial attack is proposed by Zugner et al. to generate adversarial graph data using an efficient greedy search method [26]. The generated graph can be used to fool GNN or other traditional graph learning methods. Then some methods are proposed to attack the topological structure by adding or removing edges according to the gradient of a surrogate model. Xu et al. proposed an optimization-based attack method based on the gradient of the surrogate model [22]. Zugner et al. presented to use meta-gradient to guide the perturbation of graph adjacency matrix [28]. Wu et al. argued that integrated gradients can better reflect the effect of perturbing certain features or edges [20]. Also, Bojchevski et al. took DeepWalk [11] as base method using eigen-decomposition and genetic algorithm based strategy to attack the network embedding [1].

To ensure the robustness under adversarial attack, many methods have been proposed to defense GNN models [2,5,18,19,21,24]. Wang et al. thought the vulnerabilities of graph neural networks are related to the aggregation layer and the perceptron layer [18]. To address these two disadvantages, they propose an adversarial training framework with a modified GNN model to improve the robustness of GNNs. Chen et al. proposed different defense strategies based on adversarial training for target and global adversarial attack with smoothing distillation and smoothing cross-entropy loss function [2]. Feng et al. proposed a method of adversarial training for the attack on node features with a graph adversarial regularizer which encourages the model to generate similar predictions on the perturbed target node and its connected nodes [5]. Wang et al. pointed out that the values of perturbation could be continuous or even negative [19]. Zhu et al. proposed to use Gaussian distribution to increase the robustness of Graph Convolutional Network [24]. Wu et al. applied the information bottleneck principle on semi-supervised learning settings [21] to defense the targeted node attack [26].

5.2 Self-supervised Graph Representation Learning

Self-supervised contrastive learning [3,7] showed significant performance on graph-structured data. The contrastive learning approach usually needs to generate augmented graph sample pairs of the original graph. Then the similarity between the representation of augmented graph pairs is minimized to learn graph representations. Veličković et al. proposed to maximize the information between local and global graph representations to learn node representations [17]. Zhu et al. proposed various augmentation strategies to generate augmented graph samples [25]. Hassani et al. introduced multi-view contrastive learning [6] that maximizes the information between graph and its diffusion versions [10]. Qiu et al. used an anonymous random walk to generate augmented subgraphs from a large graph and minimize the similarity between the paired subgraphs and maximize the similarity between subgraphs and negative samples [12]. Although various methods have been proposed to use self-supervised contrastive learning to learn graph representations, few works considered the quality of augmentation samples.

6   Conclusion and Discussion

To summarize, we introduce adversarial augmentations into graph self-supervised representation learning and propose a novel Graph Adversarial Contrastive Learning (GraphACL) framework. Theoretically, we obtain an upper bound of the Information Bottleneck loss function for graph contrastive learning. With adversarial augmentations, our method could result in a much tighter bound and more robust representations. Based on the theoretical analysis, we formulate the GraphACL framework and present relative objective functions in both supervised and unsupervised settings. To verify the empirical performance, we conduct experiments on classic benchmarks attacked by Netattack or Metattack. GraphACL outperforms DGI and other baselines on both evasive and poisoning tasks, thus proving itself a more robust way of graph representation learning. The analysis of different perturbation rates also indicates that our method is not sensitive to the rate. Albeit our model is built on top of Deep graph infomax (DGI), our theory can be easily extended to other models by combining adversarial learning and graph self-supervised learning together.

References

1. Bojchevski, A., Günnemann, S.: Adversarial attacks on node embeddings via graph poisoning. In: Proceedings of the 36th International Conference on Machine Learning, pp. 695–704. PMLR, 09–15 June 2019
2. Chen, J., Wu, Y., Lin, X., Xuan, Q.: Can adversarial network attack be defended? CoRR abs/1903.05994 (2019)
3. Chen, T., Kornblith, S., Norouzi, M., Hinton, G.: A simple framework for contrastive learning of visual representations. In: Proceedings of the 37th International Conference on Machine Learning, pp. 1597–1607 (2020)
4. Cheng, P., Hao, W., Dai, S., Liu, J., Gan, Z., Carin, L.: Club: a contrastive log-ratio upper bound of mutual information. In: ICML 2020: 37th International Conference on Machine Learning, vol. 1, pp. 1779–1788 (2020)
5. Feng, F., He, X., Tang, J., Chua, T.S.: Graph adversarial training: dynamically regularizing based on graph structure. IEEE Trans. Knowl. Data Eng. **6**, 2493–2504 (2021)
6. Hassani, K., Khasahmadi, A.H.: Contrastive multi-view representation learning on graphs. In: Proceedings of the 37th International Conference on Machine Learning, pp. 4116–4126. Proceedings of Machine Learning Research (2020)
7. He, K., Fan, H., Wu, Y., Xie, S., Girshick, R.: Momentum contrast for unsupervised visual representation learning. In: Proceedings of the IEEE/CVF Conference on Computer Vision and Pattern Recognition, pp. 9729–9738 (CVPR), June 2020
8. Jin, W., Ma, Y., Liu, X., Tang, X., Wang, S., Tang, J.: Graph structure learning for robust graph neural networks. In: Proceedings of the 26th ACM SIGKDD International Conference on Knowledge Discovery & Data Mining, pp. 66–74 (2020)
9. Kipf, T.N., Welling, M.: Semi-supervised classification with graph convolutional networks. arXiv preprint arXiv:1609.02907 (2016)
10. Klicpera, J., Weiß enberger, S., Günnemann, S.: Diffusion improves graph learning. Adv. Neural Inf. Process. Syst. **32** (2019)
11. Perozzi, B., Al-Rfou, R., Skiena, S.: Deepwalk: online learning of social representations. In: Proceedings of the 20th ACM SIGKDD International Conference on Knowledge Discovery and Data Mining, pp. 701–710. KDD 2014 (2014)
12. Qiu, J., et al.: GCC: Graph contrastive coding for graph neural network pre-training. In: Proceedings of the 26th ACM SIGKDD International Conference on Knowledge Discovery & Data Mining, pp. 1150–1160 (2020)
13. Tang, X., Li, Y., Sun, Y., Yao, H., Mitra, P., Wang, S.: Transferring robustness for graph neural network against poisoning attacks. In: Proceedings of the 13th International Conference on Web Search and Data Mining, pp. 600–608 (2020)
14. Tishby, N., Pereira, F.C.N., Bialek, W.: The information bottleneck method. In: Proceedings 37th Annual Allerton Conference on Communications, Control and Computing, 1999, pp. 368–377 (2000)
15. Tishby, N., Zaslavsky, N.: Deep learning and the information bottleneck principle. In: 2015 IEEE Information Theory Workshop (ITW), pp. 1–5 (2015)
16. Veličković, P., Cucurull, G., Casanova, A., Romero, A., Lio, P., Bengio, Y.: Graph attention networks. arXiv preprint arXiv:1710.10903 (2017)
17. Veličković, P., Fedus, W., Hamilton, W.L., Liò, P., Bengio, Y., Hjelm, R.D.: Deep graph infomax. arXiv preprint arXiv:1809.10341 (2018)
18. Wang, S., et al.: Adversarial defense framework for graph neural network (2019)
19. Wang, X., Liu, X., Hsieh, C.: Graphdefense: Towards robust graph convolutional networks. CoRR abs/1911.04429 (2019)

20. Wu, H., Wang, C., Tyshetskiy, Y., Docherty, A., Lu, K., Zhu, L.: Adversarial examples for graph data: deep insights into attack and defense. In: Proceedings of the Twenty-Eighth International Joint Conference on Artificial Intelligence, IJCAI-19, pp. 4816–4823 (2019)
21. Wu, T., Ren, H., Li, P., Leskovec, J.: Graph information bottleneck. arXiv preprint arXiv:2010.12811 (2020)
22. Xu, K., et al.: Topology attack and defense for graph neural networks: an optimization perspective. In: Proceedings of the Twenty-Eighth International Joint Conference on Artificial Intelligence, IJCAI-19, pp. 3961–3967 (2019)
23. You, Y., Chen, T., Sui, Y., Chen, T., Wang, Z., Shen, Y.: Graph contrastive learning with augmentations. Adv. Neural Inf. Process. Syst. **33**, 5812–5823 (2020)
24. Zhu, D., Zhang, Z., Cui, P., Zhu, W.: Robust graph convolutional networks against adversarial attacks. In: Proceedings of the 25th ACM SIGKDD International Conference on Knowledge Discovery & Data Mining. ACM (2019)
25. Zhu, Y., Xu, Y., Yu, F., Liu, Q., Wu, S., Wang, L.: Deep graph contrastive representation learning. arXiv preprint arXiv:2006.04131 (2020)
26. Zügner, D., Akbarnejad, A., Günnemann, S.: Adversarial attacks on neural networks for graph data. In: Proceedings of the 24th ACM SIGKDD International Conference on Knowledge Discovery & Data Mining, pp. 2847–2856 (2018)
27. Zügner, D., Borchert, O., Akbarnejad, A., Günnemann, S.: Adversarial attacks on graph neural networks: Perturbations and their patterns. ACM Trans. Knowl. Discov, Data **14**(5), 1–31 (2020)
28. Zügner, D., Günnemann, S.: Adversarial attacks on graph neural networks via meta learning. In: International Conference on Learning Representations (ICLR) (2019)

What Affects the Performance of Models? Sensitivity Analysis of Knowledge Graph Embedding

Han Yang[1], Leilei Zhang[1], Fenglong Su[2], and Jinhui Pang[3(✉)]

[1] Peking University, Beijing, China
{hyang001,zhang_leilei}@pku.edu.cn
[2] National University of Defense Technology, Changsha, Hunan, China
sufenglong18@nudt.edu.cn
[3] Beijing Institute of Technology, Beijing, China
pangjinhui@bit.edu.cn

Abstract. Knowledge graph (KG) embedding aims to embed entities and relations into a low-dimensional vector space, which has been an active research topic for knowledge base completion (KGC). Recent researchers improve existing models in terms of knowledge representation space, scoring function, encoding method, etc., have achieved progressive improvements. However, the theoretical mechanism behind them has always been ignored. There are few works on **sensitivity analysis** of embedded models, which is extremely challenging. The diversity of KGE models makes it difficult to consider them uniformly and compare them fairly. In this paper, we first study the internal connections and mutual transformation methods of different KGE models from the generic group perspective, and further propose a unified KGE learning framework. Then, we conduct an in-depth sensitivity analysis on the factors that affect the objective of embedding learning. Specifically, in addition to the impact of the embedding algorithm itself, this article also considers the structural features of the dataset and the strategies of the training method. After a comprehensive experiment and analysis, we can conclude that the Head-to-Tail rate of datasets, the definition of model metric function, the number of negative samples and the selection of regularization methods have a greater impact on the final performance.

Keywords: Knowledge graph embedding · Group theory · Sensitivity analysis

1 Introduction

Knowledge Graphs (KGs) have emerged as a core abstraction for incorporating human knowledge into intelligent systems, which become increasingly popular in various downstream tasks including semantic search [2,28], question answering [1,8], and recommendation system [26,31]. In general, a KG can be seen as a collection of triple facts

---

H. Yang and L. Zhang—Contributed equally to this research. This work was supported by the National Key RD Program of China under Grant No. 2020AAA0108600.

A. Bhattacharya et al. (Eds.): DASFAA 2022, LNCS 13245, pp. 698–713, 2022.
https://doi.org/10.1007/978-3-031-00123-9_55

in the triple format, expressed as *(head entity, relation, tail entity)* also abbreviated as *(h, r, t)*, e.g., (Donald Trump, presidentOf, USA). Knowledge graph embedding aims to compress both relation and entity into continuous low-dimensional embedding spaces while preserving the intrinsic graph properties and its underlying semantic information. These approaches provide a way to perform reasoning in KGs with simple numerical computation in continuous spaces.

Recent years witnessed tremendous research efforts on the KGE models, which can be roughly divided into translation-based models, bilinear models, and other neural network models [25]. These models are dedicated to transferring to more complex representation spaces, designing different scoring functions or loss functions, thus making up for the shortcomings of previous works and improving the performance of embedded learning. Although more and more tailored models have shown promising performance on this task, the theoretical mechanism behind them has been much less well-understood to date. There is still a lack of comprehensive study to explore the influencing factors that lead to the improvements of the results [7, 14], which can be helpful to enhance the interpretability of KGE models. In this paper, we focus on the sensitivity analysis of KGE for the first time, which is quite challenging. For one thing, the heterogeneity between different models impedes the proposal of a unified KGE abstract representation, making it difficult to compare existing models and discover which modules of the embedding algorithm lead to progress. Besides, the performance of KGE is also affected by model training, such as regularization and negative sampling method, etc. This makes the independent analysis of factors affecting model performance more complicated, especially when results are reproduced from prior studies that used a different experimental setup. Moreover, the performance of the same KGE model on different datasets may be very different, which also attracts us to analyze how the structural characteristics of the dataset influence the embedded learning objective.

To overcome the aforementioned challenges, we express the popular KGE models into a unified form, that is, a metric space based on the Abelian group. Based on group isomorphism, we further analyzed typical KGE models such as TorusE [6], RotatE [20], DisMult [29], ComplEx [23], and proposed that they can all be regarded as variants of TransE [3] in terms of metrics, and completely different types of KGE learning algorithms can also be converted to each other. We choose circle group in Sect. 3.3 to illustrate the relation between them in detail and intuitively.

Moreover, we conduct a systematical sensitivity analysis of KGE from three aspects: dataset characteristics, model architecture, and model training. Through statistics and analysis of the structural characteristics of the dataset, we innovatively pointed out that the Head-to-Tail rate will have a significant impact on the effect of KGE models in the knowledge graph completion task. For the KGE model architecture itself, we quantitatively proved the limitations of the commonly used Euclidean metric function, and discussed the impact of model hyperparameters based on the unified KG representation learning framework. We also conduct a large number of experiments by changing the training strategy in common experimental settings to quantify and summarize the impact of different training methods on model performance. Surprisingly, we discovered that the number of negative samples, whether to perform regularization and the choice of the regularization method are important to the embedding effect, but the negative sampling method does not matter. The conclusion of KGE sensitivity analysis is quite helpful to improve the existing KGE models.

To sum up, the highlights of the paper can be summarized as:

(1) To the best of our knowledge, we are the first work focusing on the sensitivity analysis of the knowledge graph embedding models, which is of great significance for improving the interpretability of representation learning.
(2) We innovatively provide a unified framework for several popular KGE models, and explored the theoretical and conversion methods between these models, which helps to fairly analyze the influence of various factors of the embedded algorithm from a new perspective.
(3) We define a variety of dataset structural features to better analyze how dataset features affect the goal of embedding learning and pointed out the impact of different training strategies on model performance.

2 Preliminaries: Knowledge Graph Embedding

Various KGE models have been proposed for the KG completion task in recent years. For a more intuitive discussion, we only review the methods that are directly related to our work, without considering the multi-modal embedding with external information.

2.1 General Architecture

Knowledge graph embedding models learn to encode a collection of factual triplets from a knowledge graph $G = \{(h, r, t)\} \subseteq \mathcal{E} \times \mathcal{R} \times \mathcal{E}$ into low dimensional, continuous vectors $(\mathbf{h}, \mathbf{r}, \mathbf{t})$, where $\mathbf{h}, \mathbf{t} \in \mathbb{R}^k$ and $\mathbf{r} \in \mathbb{R}^d$. Typical KGE approaches follow a clear workflow that consist of four component:

(1) *Random Initialization.* Randomly initialize the entity and relation vectors, which generally uses an embedding lookup table to convert the sparse discrete one-hot vectors into dense distributed representations;
(2) *Scoring Function.* Define a scoring function to measure the plausibility of facts. The scoring function $s : \mathcal{E} \times \mathcal{R} \times \mathcal{E} \to \mathbb{R}$ takes form $s(h, r, t) = f(h, r, t)$ and assigns scores to all potential triples $(h, r, t) \in \mathcal{E} \times \mathcal{R} \times \mathcal{E}$, where f may be either a fixed function or a parameterized function;
(3) *Interaction Mechanism.* Design the interaction mechanism to model the interactions of entities and relations to compute the matching score of a triple. The most popular interaction mechanisms include linear or bilinear models, factorization models, and neural networks. This is the main component of a model;
(4) *Training Strategy.* Training the KGE model by maximizing the confidence of triples, with training strategies such as negative sampling and regularization.

2.2 KGE Models

Based on the scoring function and adopted interaction mechanism, we roughly divide previous work into translation-based models, bilinear models and other models.

Translation-Based Model. These models are known for their simplicity and efficiency, which measure the plausibility of a triple as the distance between the head entity and the tail entity. The scoring functions of translation-based models usually adopt L^1 or L^2 distance as the distance metric. Taking TransE [3] as an example, the scoring function is:

$$f_r(h, t) = -\|\mathbf{h} + \mathbf{r} - \mathbf{t}\|_p = -\left(\sum_d^D |\mathbf{h}_d + \mathbf{r}_d - \mathbf{t}_d|^p\right)^{1/p}, \tag{1}$$

where $\mathbf{h}, \mathbf{r}, \mathbf{t}$ are the embeddings of h, r, t, respectively, p is the order of Minkowski metric, such as taxicab distance is 1 and Euclidean distance is 2. D is the dimension size of the embedding space, $x = (x_1, x_2, ..., x_D)$ is a point in D-dimensional space.

TransE is the seminal work for translation-based model, which interprets relation as a translation vector $\mathbf{r}$ so that entities can be connected, formally as $\mathbf{h} + \mathbf{r} \approx \mathbf{t}$. The follow-up variants of TransE are proposed to overcome the flaws of TransE in dealing with 1-to-N, N-to-1, and N-to-N relations, such as TransH [12] and TransR [12]. TransD [9] and TranSparse [10] simplify the projection matrices, while TorusE [5] and ManifoldE [27] introduce other representation spaces. RotatE [20] defines each relation as a rotation from the head entity to the tail entity in the complex vector.

Bilinear Models. These models, also known as semantic models, use the scoring function in the form of trilinear product between entities and relations to measure the semantic similarity.

The most classical representative method is the RESCAL [18] model, which represents KG as a three-way tensor, then DistMult [29] simplifies RESCAL by restricting relation matrices to be diagonal, HolE [17] further combines the expressive power of RESCAL with the efficiency and simplicity of DistMult. ComplEx [23] entends HolE to the complex space so as to better model asymmetric relations. The analogical embedding framework [13] restricts the embedding dimension and scoring function, thus it can recover or equivalently obtain several models.

Other Models. Traditional translation-based and bilinear models cannot meet the requirements of KGE, there are some works proposed to obtain better and more effective entity and relation embeddings. QuatE [32] takes advantage of quaternion representations to enable rich interactions between entities and relations. ConvE [4] is the first work to use the convolutional neural network (CNN) framework for KG completion. In addition, we notice that some models are significantly better than other KGE models, such as ConvKB [16], CapsE [24], KBAT [15], etc. Recent work [21] has pointed out that their outstanding performances are caused by containing a large number of identical scores. Coper-ConvE [19] only conducted tail entity prediction experiments, which are simpler than head entity prediction, thus it is unfair to them with other models. They will not be discussed in this article.

3   A Unified Knowledge Graph Embedding Framework

In this section, we present a detailed theoretical analysis of the popular and typical KG representation learning models, such as TransE, TorusE, DisMult, ComplEx, RotatE, which have achieved competitive results. We summarized the above five models into the metric space based on the Abelian group, and further discussed the influence of metric methods and group operations on the KGE model performance in Sect. 5.3.

3.1   Abelian Group and Metric Space

Abelian Group: An abelian group, also called a commutative group, is a set, G, together with an operation $*$ that combines any two elements a and b of G to form another element of G, denoted $a * b$. For all a, b in an abelian group, the set and operation, $(G, *)$,

$$a * b = b * a. \tag{2}$$

Metric Space: A metric space is an ordered pair (G, d) where G is a set and d is a metric on G, i.e., a function $d : G \times G \to \mathbb{R}$ such that for any $x, y, z \in G$, the following holds:

1. $d(x, y) \geq 0$
2. $d(x, y) = 0 \iff x = y$
3. $d(x, y) = d(y, x)$
4. $d(x, z) \leq d(x, y) + d(y, z)$.

3.2   Group Representation of KGE Models

Following the definition of Abelian group and metric space, we transfer the process of knowledge graph embedding models into a three-state workflow on the group space:

(1) The group operation of the head entity h and the relation r on the abelian group G, aiming at generating a target characteristic $\tilde{t}$ in the group:

$$\tilde{t} = h * r, h, r \in G. \tag{3}$$

(2) Calculate the distance between the generated target characteristic $\tilde{t}$ and the ground-truth tail entity t on the metric space $< G, * >$.

$$d(\tilde{t}, t), d : G \times G \to \mathbb{R}. \tag{4}$$

(3) Design the loss function $F(d)$ and use it to train the whole KGE model.

We discuss the characteristics of the several selected typical models and further summarize their group representations in Table 1. TransE interprets relation as a translation vector r, formally, it calculates the distance between the characteristics of entity $\tilde{t}$ and the tail entity t in the metric space $< R^n, *, d >$, where d is the Euclidean distance. ComplEx and RotatE belong to the semantic model and the translation-based rotation model respectively, but they are highly similar from the perspective of group representation, both of them act in the high-dimensional complex number field $\mathbb{C}^n$ and perform

Table 1. The transformations and metrics of several typical KGE models

Models	Scoring function	Group operation	Metric	Transformation function
TransE	$-\|\mathbf{h}+\mathbf{r}-\mathbf{t}\|$	+	Euclidean	–
TorusE	$-\min_{(x,y)\in([h]+[r])\times[t]}\|\mathbf{x}-\mathbf{y}\|_i$	+(Lie Group)	Euclidean	$0.5*\sin(2*\pi*(\mathbf{h}+\mathbf{r}-\mathbf{t})+1.5\pi)$ $+0.5\sim\sin(\mathbf{h}+\mathbf{r}-\mathbf{t})$
DisMult	$-\sum_{k=1}^{K}\mathbf{h}_k\mathbf{r}_k\mathbf{t}_k$	*	Dot product	$e^{(\mathbf{h}+\mathbf{r}-\mathbf{t})}$
ComplEx	$-\operatorname{Re}\left(\sum_{k=1}^{K}\mathbf{r}_k\mathbf{h}_k\bar{\mathbf{t}}_k\right)$	* (Complex)	Dot product	$\cos(\mathbf{h}+\mathbf{r}-\mathbf{t})$
RotatE	$-\|\mathbf{h}\circ\mathbf{r}-\mathbf{t}\|$	* (Complex)	Euclidean	$0.5*\sin(0.5(\mathbf{h}+\mathbf{r}-\mathbf{t}))$

group operations such as complex multiplication. The difference between them is the metric function in the metric space. ComplEx applies the inner product of two vectors, while RotatE uses Euclidean distance. TorusE performs Lie group addition operation in the multidimensional torus $\mathbb{T}^n$, then computes the distance between the characteristic entity and the real entity by Euclidean distance. DisMult performs the Hadamard product of vectors in $\mathbb{R}^n$, and then measures the characteristic entity and the tail entity through the inner-product operation of vectors.

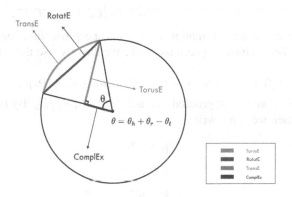

Fig. 1. The unification of KGE models.

3.3   Model Transformation and Unification

We have shown that there is a conditional isomorphism of models, i.e. a kind of mapping relation that maps models to a uniform representation space. In order to enhance the interpretability of the model and facilitate the sensitivity analysis in the following section, we choose the most commonly used circle group to summarize different KGE models. As shown in Fig. 1, we compare different KGE models in the circle group more vividly. The details of the transformation function are described in Table 1, and we give the proof as follow,

Theorem 1. *TransE can be represented as an angle with the size of θ, or as a arc segment(the bleu arc in Fig. 1), then TorusE, ComplEx, RotatE and DisMult can all be regarded as transformations of TransE based on trigonometric functions.*

Proof. (1) **TorusE:** By setting $d = \mathbf{h} + \mathbf{r} - \mathbf{t}$, we get $d - \lfloor d \rfloor \in [0,1)^n$. Then,

$$
\begin{aligned}
f(\mathbf{h}, \mathbf{r}, \mathbf{t}) &= \frac{1}{4} \|(2 - 2\cos(2\pi(\min(d - \lfloor d \rfloor, 1 - d + \lfloor d \rfloor))))\| \\
&= 0.5\|\sin(2\pi d + 1.5\pi)\| \\
&\sim \|\sin(\mathbf{h} + \mathbf{r} - \mathbf{t})\|.
\end{aligned}
\tag{5}
$$

(2) **ComplEx:** By further restricting $|h_i| = |r_i| = |t_i| = C$, we can rewrite $\mathbf{h}, \mathbf{r}, \mathbf{t}$ by

$$
\begin{aligned}
\mathbf{h} &= Ce^{i\theta_h} = C\cos\theta_h + iC\sin\theta_h \\
\mathbf{r} &= Ce^{i\theta_r} = C\cos\theta_r + iC\sin\theta_r \\
\mathbf{t} &= Ce^{i\theta_t} = C\cos\theta_t + iC\sin\theta_t.
\end{aligned}
\tag{6}
$$

Then, we can get

$$
\begin{aligned}
f(\mathbf{h}, \mathbf{r}, \mathbf{t}) &= \|\mathrm{RE}(\mathbf{h} \circ \mathbf{r} \circ \bar{\mathbf{t}})\| = C \left\| \mathrm{RE}\left(e^{i(\theta_h + \theta_r)} \circ e^{-i\theta_t}\right) \right\| \\
&= C \left\| \mathrm{RE}\left(e^{i(\theta_h + \theta_r - \theta_t)}\right) \right\| \\
&= C \|\cos(\theta_h + \theta_r - \theta_t)\| \\
&\sim \|\cos(\theta_h + \theta_r - \theta_t)\|.
\end{aligned}
\tag{7}
$$

(3) **RotatE:** Since the transformation of the trigonometric function calculation for RotatE [20] has been given in previous work, we directly use the conclusion in this paper.

$$
f(\mathbf{h}, \mathbf{r}, \mathbf{t}) = \|\mathbf{h} \circ \mathbf{r} - \mathbf{t}\| \sim \|\sin(\theta_h + \theta_r - \theta_t)\|.
\tag{8}
$$

(4) **DistMult:** Prior work [30] proofed $\mathrm{Trans}\,\mathrm{E} \cong \mathrm{DistMult}\,/\mathbb{Z}_2$. By further restricting $h, r, t > 0$, and then we can rewrite

$$
\begin{aligned}
\mathbf{h} &= e^{\theta_h} \\
\mathbf{r} &= e^{\theta_r}, \\
\mathbf{t} &= e^{\theta_t}
\end{aligned}
$$

$$
f(\mathbf{h}, \mathbf{r}, \mathbf{t}) = \|\mathbf{h} \cdot \mathbf{r} \cdot \mathbf{t}\| = \left\|e^{\theta_h + \theta_r + \theta_t}\right\|
\tag{9}
$$

Let $\theta_t' = -\theta_t$, then we can get

$$
f(\mathbf{h}, \mathbf{r}, \mathbf{t}) = \left\|e^{\theta_h + \theta_r + \theta_t}\right\| \sim \left\|e^{\theta_h + \theta_r - \theta_t'}\right\|.
\tag{10}
$$

$\square$

4   Influencing Factors of Knowledge Graph Models

Sensitivity Analysis quantitatively studies the uncertainty in the output of a black-box model or system, thus can greatly enhance the interpretability of neural networks, which is still blank in the field of knowledge graph embedding. Since the performance of a KGE model is not only determined by the embedding algorithm itself, but also affected by the structural features of the experimental dataset, and various strategies adopted in the training method. We will first review these influencing factors follow, and then analyze and discuss them in detail in the next section.

4.1   Dataset Structural Features

Table 2. Definition of the structural features of KGE datasets

Definition	Description
First-level absolute features	
Number of entities	The number of entities and their proportion to the total entities
Number of relations	The number of relations and their proportion to the total entities
Entity category	The entity is divided into four categories: $1-1$, $1-n$, $n-1$, $n-n$
Relation category	The relation is divided into four categories: $1-1$, $1-n$, $n-1$, $n-n$
Secondary-level absolute features	
Head-to-Tail Rate	The ratio between the number of head entity category and the number of tail entity category
Test-train relative features	
Head-In Rate	Given a relation r in the test dataset, the proportion of its head entities appearing as head entities for the same relation r in the training dataset
Tail-In Rate	Given a relation r in the test dataset, the proportion of its tail entities appearing as tail entities for the same relation r in the training dataset
Avg-In Rate	Given a relation r in the test dataset, the proportion of its head or tail entities that occur with the same relation in the training dataset

The same KGE model performs quite differently on different datasets. It is difficult for a model with better performance on the benchmark dataset to maintain its superiority in the new dataset, which greatly limits the popularization and application of knowledge graph representation learning model in downstream tasks. This paper pioneered a variety of characteristic indicators to describe the structure of the knowledge graph dataset to further study how the dataset affects the performance of the KGE models. We introduced and described the definition of the structural characteristics of KGE datasets in Table 2. The dataset structural features include two types: the absolute structure characteristics of the dataset itself and the relative characteristics describing the relation between the test dataset and the training dataset. Among them, the absolute characteristics include not only the first-level features that can be directly obtained by statistics, such as the number of unique entities, relations, etc., but also the secondary-level features calculated based on the first-level features, such as Head-to-Tail rate, that is the rate of the number of types of head entities h to the number of types of tail entities t for a given relation r.

As for the relative characteristics describing the relation between the training dataset and the test dataset, we give three conceptual definitions: Head-In Rate, Tail-In Rate, and Avg-In Rate. Given a specific triple (h_i, r_i, t_i) which appears in the test dataset, Head-In (Tail-In) Rate refers to the probability of the head entity h_i (tail entity t_i) and relation r_i appear in the triple that contains this relation in the training test set at the same time. Avg-In Rate is the average of Head-In rate and Tail-In rate.

4.2   Embedding Algorithm

We have unified some typical embedding learning models into the form of transE with trigonometric functions. Specifically, the KGE model can be expressed in the form of the trigonometric function $A \sin(\omega * (\theta) + b)$. In this paper, we construct a new model based on the unified framework Sin E as follows:

$$f_{\sin E} = A \sin(\omega * (h + r - t) + b). \tag{11}$$

Then, we analyze the effect of three elements of the trigonometric function including amplitude, frequency, and phase on the entity linking task in Sect. 5.3.

4.3 Model Training

The most commonly used strategies for training the KE model are *negative sampling* and *regularization*.

Negative Sampling Methods: We analyzed not only the method of negative sampling, but also the influence of the number of negative sampling on the objective of embedding learning. For the negative sampling method, we mainly study the impact of three methods including uniform sampling, self-adversarial sampling, and NSCaching sampling [33] on the performance of KGE models. Uniform sampling refers to randomly selecting negative sample entities from all candidate entities. On the basis of uniform sampling, self-adv increases the weight of samples with higher scores in the same batch. NSCaching considers negative examples to be good or bad, and uses a caching mechanism to obtain high-quality negative examples. Moreover, we also analyze the impact of the number of negative samples on the performance of the representation model.

Regularization Method: Regularization method constrains the parameters to be optimized, which helps prevent over-fitting. Prior work [11] has studied the regularization method of the embedded models, and proposed the N3 regularization method. This paper further analyzes the effects of five common matrix norms on the KGE model, and compares them with the pure model that does not use regularization.

1-Norm:

$$\|A\|_1 = \max_{1 \leq j \leq n} \sum_{i=1}^{m} |a_{ij}|. \tag{12}$$

∞-Norm:

$$\|A\|_\infty = \max_{1 \leq i \leq m} \sum_{j=1}^{n} |a_{ij}|. \tag{13}$$

2-Norm:

$$\|A\|_2 = \sqrt{\lambda_{\max}(A^T A)}. \tag{14}$$

Nuclear Norm(Nuc-Norm):

$$\|A\|_* = \mathrm{tr}\left(\sqrt{A^T A}\right) = \mathrm{tr}(\Sigma)$$
$$A = U \sum V^T. \tag{15}$$

Fro Regularization:

$$\|A\|_F = \sqrt{\mathrm{tr}(A^T A)} = \sqrt{\sum_{i=1}^{m} \sum_{j=1}^{n} a_{ij}^2}. \tag{16}$$

5   Sensitivity Analysis of the Influencing Factors in KGE Models

In this section, we perform experiments to revisit the contribution of the various influencing factors in KGE learning models, and hope to answer the following questions through an objective sensitivity analysis of the experimental results.

Q1: How does the dataset structure affect knowledge graph embedding learning?
Q2: How does the model architecture influence the objective of embedding learning?
Q3: What are the effects of different training strategies on the KGE models?

We first introduce the relevant datasets and evaluation tasks used in the sensitivity analysis experiment, and then discuss them in detail.

5.1   Experimental Settings

Table 3. Statistics of the datasets used in this paper.

Dataset	Train	Valid	Test	Ent	Rel
FB15k	483,1442	50,000	59071	14951	1345
FB15k-237	272,115	17,535	20,466	14,541	237
WN18	141,442	5000	5000	40,943	18
WN18RR	86,835	3,3034	3,3134	40,943	11

Experimental Dataset: We conducted experiments on some common datasets including FB15k, WN18 [3], FB15k-237 [22] and WN18RR [4]. FB15k is a subset of Freebase, a large-scale knowledge graph containing general knowledge facts. WN18 is extracted from WordNet3, where words are interlinked by means of conceptual-semantic and lexical relations. FB15k-237 and WN18RR are their corresponding updated version, with inverse relations removed. Statistics of the datasets are provided in Table 3.

Evaluation Protocol. We evaluate the performance of KGE models on the entity linking task, which predicts the missing entity in the triple by minimizing the loss function, that is, given $(h, r, ?)$, predict the tail entity ?.

$$t = \operatorname*{argmin}_{e} \cdot f(h, r, e). \tag{17}$$

We employ three popular metrics to evaluate the performance of link prediction, including Mean Rank (MR), Mean Reciprocal Rank (MRR), and Hit ratio with cut-off values n = 1, 3, 10. MR measures the average rank of all correct entities with a lower value representing better performance. MRR is the average inverse rank for correct entities. Hit@n measures the proportion of correct entities in the top n entities.

5.2 Sensitivity Analysis of Dataset Structural Features

Table 4. The link prediction results of some relations

Relation	HeadInRate	tailInRate	AvgInRate	HeadToTailRate	ErorrRate
dated_money_value/currency	0.00000	1.00000	0.03226	553.91670	0.00000
location_of_ceremony	1.00000	0.01888	0.03704	0.00197	0.88679
food/diet/followers	1.00000	0.00000	0.11764	0.02173	0.46667
athlete_salary/team	0.25000	0.25000	0.25000	1.37500	0.25000

We perform sensitivity analysis of dataset structural characteristics on the FB15K dataset, because it has the largest amount of training triplets and contains various types of relations. As shown in Table 4, we list the characteristic statistical values and error rate of missing triples corresponding to some specific relations in the FB15K dataset. Here we use the hit@10 indicator to represent the error rate. Figure 2 shows the scatter plots of Hit@10 error rate with Avg-In rate and Head-to-Tail rate, respectively. It shows that Hit@10 error rate and Head-to-Tail rate are negatively correlated, while Avg-In Rate has little effect on Hit@10 error rate.

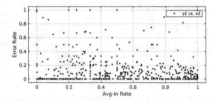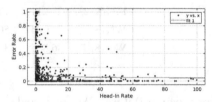

Fig. 2. Error Rate to Avg-In Rate, Head-In Rate, respectively.

Table 5. Error Rate to Head-to-Tail Rate.

HeadTailRate	Hit10 ErrorRate	NumberOfTriples	HeadTailRate	Hit10 ErrorRate	NumberOfTriples
0−0.5	0.171	568	3−3.5	0.058	44
0.5−1	0.142	218	3.5−4	0.089	29
1−1.5	0.161	273	4−4.5	0.035	37
1.5−2	0.099	66	4.5−5	0.025	42
2−2.5	0.098	97	≥5	0.029	523
2.5−3	0.084	25			

The Head-to-Tail Rate is not a uniform distribution, in order to analyze the relations between the two more intuitively, we segment the Head-to-Tail Rate with a step size of 0.5, and calculate the average error rate of all relations in the segment to make the histogram. Table 5 and Fig. 3 further confirm the conclusion that the larger the Head-to-Tail Rate, the lower the Hit@10 error rate.

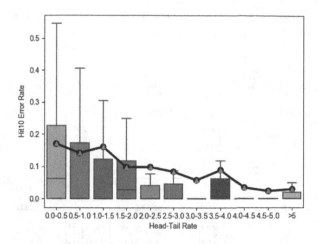

Fig. 3. Error Rate to Head-to-Tail Rate.

5.3 Sensitivity Analysis of KGE Model Architecture

We analyzed the influence of period, margin, amplitude, and phase of SinE. The results are shown in Fig. 4. Among the four factors, the period has the greatest impact. ω can be regarded as the reciprocal of the cycle. With the increase of ω, the performance of the KGE model greatly improves until it reaches the peak, then stabilizes or decreases slowly. The margin has a similar impact, but it is not as obvious as the period. The impact of amplitude on the dataset is relatively small. At first, a large range of amplitude has minimal impact on model performance, and then as the amplitude increases, the model performance gradually decreases. Phase is the parameter that has the least impact on the performance of the model among the four factors.

5.4 Sensitivity Analysis of Model Training Strategies

Table 6. Sensitivity analysis of the number of NS, where NS represents the negative sampling.

Number of NS	1	3	5	10
MRR/Hit@1	0.633/0.510	0.696/0.581	0.721/0.611	0.752/0.654
Number of NS	50	150	200	256
MRR/Hit@1	0.788/0.723	0.795/0.741	0.791/0.738	0.791/0.739

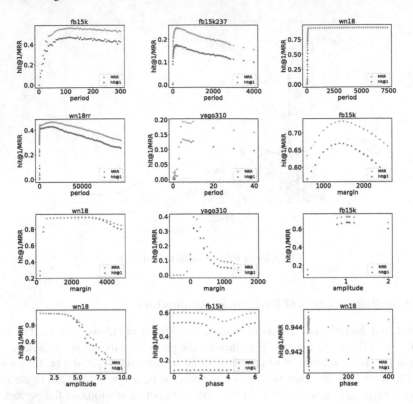

Fig. 4. Sensitivity Analysis of period, margin, amplitude, and phase of SinE.

In order to verify whether the number of negative samples has an impact on the objective of embedding learning, we use RotatE as the basic model, and adopt uniform, self-adversarial, and NSCaching negative sampling methods to train the KGE model on the FB15k dataset. At the same time, we also set the number of negative samples is 1, 3, 5, 10, 50, 100, 150, 200, and 256 to observe the influence of the number of negative samples. The experimental results are shown in Table 6. We can conclude that compared to the negative sample sampling method, the number of samples has a greater impact on downstream tasks.

In addition, We conduct experiments on the selection of negative sampling methods, and the results are shown in Table 7. We can conclude that the impact of negative sampling methods on model performance depends on the dataset. For example, in WN18RR, the uniform sampling performs better, but the performance of uniform sampling is worse in the FB15k-237 dataset. NSCaching is the opposite, with the worst effect on WN18RR, but it has the best performance on FB15k-237 among the three negative sampling methods. The performance of self-adversarial negative sampling is relatively stable on the two datasets. Therefore, this article believes that the choice of negative sampling method should be determined by the structural characteristics of the dataset. Different datasets should be equipped with different negative sampling methods.

Table 7. Sensitivity analysis of NS methods, where NS represents negative sampling.

NS method	WN18RR				FB15K237			
	MRR	Hit@1	Hit@3	Hit@10	MRR	Hit@1	Hit@3	Hit@10
ComplEx + random	0.449	0.387	0.486	**0.556**	0.272	0.191	0.295	0.432
ComplEx + self-adversarial	**0.470**	0.429	0.484	0.555	0.322	0.229	0.352	**0.511**
ComplEx + NSCaching	0.302	–	–	0.481	**0.446**	–	–	0.509

Table 8. Sensitivity analysis of regularization methods on FB15K and WN18 datasets.

Model	FB15K				WN18			
	MRR	H@1	H@3	Hit@10	MRR	H@1	H@3	Hit@10
ComplEx-None	0.83	0.79	0.86	0.90	0.95	0.95	0.95	0.95
ComplEx-FRO	0.84	0.80	0.86	0.90	0.95	0.95	0.95	0.96
ComplEx-N3	0.84	0.79	0.86	0.91	0.95	0.95	0.95	0.96
ComplEx-1	0.83	0.80	0.86	0.90	0.95	0.94	0.95	0.95
ComplEx-2	0.83	0.80	0.86	0.90	0.95	0.94	0.95	0.95
ComplEx-∞	0.83	0.80	0.86	0.90	0.95	0.94	0.95	0.95
ComplEx-nuc	0.83	0.79	0.86	0.90	0.95	0.94	0.95	0.95

Table 9. Sensitivity analysis of regularization methods on FB15k-237 and WN18RR datasets.

Model	FB15k-237				WN18RR			
	MRR	H@1	H@3	Hit@10	MRR	H@1	H@3	Hit@10
ComplEx-None	0.34	0.25	0.37	0.51	0.46	0.43	0.47	0.52
ComplEx-FRO	0.33	0.24	0.36	0.52	0.45	0.42	0.46	0.51
ComplEx-N3	0.35	0.26	0.38	0.53	0.47	0.43	0.48	0.53
ComplEx-1	0.34	0.25	0.37	0.51	0.46	0.43	0.47	0.51
ComplEx-2	0.33	0.25	0.37	0.51	0.45	0.43	0.46	0.51
ComplEx-∞	0.31	0.22	0.34	0.49	0.41	0.39	0.41	0.44
ComplEx-nuc	0.33	0.25	0.37	0.51	0.45	0.43	0.46	0.51

We also implement the six regularization methods mentioned in Sect. 4.3 to train the ComplEx separately, and the experimental results are shown in Table 8 and 9. After analysis, it can be found that (1). The choice of the regularization method is important to the performance of the KGE models. The N3 regularization method has achieved excellent performance on all datasets. Especially for the WN18RR dataset, except for N3, the other regularization methods lead to reduces in the performance of ComplEx, which shows that the regularization method cannot be used arbitrarily. (2). The structural features of KGE datasets also affect the influence of regularization methods. In the denser FB15K and WN18 datasets, whether to use the regularization method has less impact on the model. In the sparse FB237 and WN18 datasets, the opposite is true, which also proves that when the dataset is small, the use of regularization can prevent overfitting.

6   Conclusion

This paper first conducts sensitivity analysis to improve the interpretability of the KGE models. We give a unified representation of several typical KGE models based on TransE + trigonometric functions, and further analyze the transformation methods between them. On this basis, we concluded that the different parameters of the trigonometric function have a significant impact on the objective of embedding learning. Moreover, we discussed the effect of features of the data structure, different model implementation strategies on the KGE models. we found that the Head-to-Tail rate of datasets, the definition of model metric function, the number of negative samples and the selection of regularization methods have a greater impact on the final performance.

References

1. Abujabal, A., Yahya, M., Riedewald, M., Weikum, G.: Automated template generation for question answering over knowledge graphs. In: Proceedings of the 26th International Conference on World Wide Web (2017)
2. Bhagdev, R., Chapman, S., Ciravegna, F., Lanfranchi, V., Petrelli, D.: Hybrid search: effectively combining keywords and semantic searches. In: European Semantic Web Conference (2008)
3. Bordes, A., Usunier, N., Garcia-Duran, A., Weston, J., Yakhnenko, O.: Translating embeddings for modeling multi-relational data. In: (NIPS) (2013)
4. Dettmers, T., Minervini, P., Stenetorp, P., Riedel, S.: Convolutional 2D knowledge graph embeddings. In: AAAI (2018)
5. Ebisu, T., Ichise, R., Torus, E.: Knowledge graph embedding on a lie group. AAAI, Toruse (2018)
6. Ebisu, T., Ichise, R.: TorusE: knowledge graph embedding on a lie group. In: Proceedings of the AAAI Conference on Artificial Intelligence, vol. 32 (2018)
7. Gesese, G.A., Biswas, R., Alam, M., Sack, H.: A survey on knowledge graph embeddings with literals: which model links better literally? Semantic Web (2021)
8. Hu, S., Zou, L., Yu, J.X., Wang, H., Zhao, D.: Answering natural language questions by · subgraph matching over knowledge graphs. IEEE Trans. Knowl. Data Eng. **30**, 824–837 (2017)
9. Ji, G., He, S., Xu, L., Liu, K., Zhao, J.: Knowledge graph embedding via dynamic mapping matrix. In: ACL (2015)
10. Ji, G., Liu, K., He, S., Zhao, J.: Knowledge graph completion with adaptive sparse transfer matrix. In: AAAI (2016)
11. Lacroix, T., Usunier, N., Obozinski, G.: Canonical tensor decomposition for knowledge base completion. In: International Conference on Machine Learning (2018)
12. Lin, Y., Liu, Z., Sun, M., Liu, Y., Zhu, X.: Learning entity and relation embeddings for knowledge graph completion. In: AAAI (2015)
13. Liu, H., Wu, Y., Yang, Y.: Analogical inference for multi-relational embeddings. In: International conference on machine learning. PMLR (2017)
14. Mohamed, S.K., Nováček, V., Vandenbussche, P.Y., Muñoz, E.: Loss functions in knowledge graph embedding models. In: DL4KG@ ESWC (2019)

15. Nathani, D., Chauhan, J., Sharma, C., Kaul, M.: Learning attention-based embeddings for relation prediction in knowledge graphs. arXiv preprint arXiv:1906.01195 (2019)
16. Nguyen, D.Q., Nguyen, T.D., Nguyen, D.Q., Phung, D.: A novel embedding model for knowledge base completion based on convolutional neural network. arXiv preprint arXiv:1712.02121 (2017)
17. Nickel, M., Rosasco, L., Poggio, T.: Holographic embeddings of knowledge graphs. In: AAAI (2016)
18. Nickel, M., Tresp, V., Kriegel, H.P.: A three-way model for collective learning on multi-relational data. In: ICML (2011)
19. Stoica, G., Stretcu, O., Platanios, E.A., Mitchell, T., Póczos, B.: Contextual parameter generation for knowledge graph link prediction. In: AAAI (2020)
20. Sun, Z., Deng, Z.H., Nie, J.Y., Tang, J.: Rotate: Knowledge graph embedding by relational rotation in complex space. In: International Conference on Learning Representations (2018)
21. Sun, Z., Vashishth, S., Sanyal, S., Talukdar, P., Yang, Y.: A re-evaluation of knowledge graph completion methods. In: ACL (2020)
22. Toutanova, K., Chen, D.: Observed versus latent features for knowledge base and text inference. In: Proceedings of the 3rd Workshop on Continuous Vector Space Models and their Compositionality (2015)
23. Trouillon, T., Welbl, J., Riedel, S., Gaussier, É., Bouchard, G.: Complex embeddings for simple link prediction. In: International Conference on Machine Learning. PMLR (2016)
24. Vu, T., Nguyen, T.D., Nguyen, D.Q., Phung, D., et al.: A capsule network-based embedding model for knowledge graph completion and search personalization. In: NAACL (2019)
25. Wang, Q., Mao, Z., Wang, B., Guo, L.: Knowledge graph embedding: A survey of approaches and applications. IEEE Trans. Knowl. Data Eng. 29, 2724–2743 (2017)
26. Wang, X., Wang, D., Xu, C., He, X., Cao, Y., Chua, T.S.: Explainable reasoning over knowledge graphs for recommendation. In: AAAI (2019)
27. Xiao, H., Huang, M., Zhu, X.: From one point to a manifold: knowledge graph embedding for precise link prediction. arXiv preprint arXiv:1512.04792 (2015)
28. Xiong, C., Power, R., Callan, J.: Explicit semantic ranking for academic search via knowledge graph embedding. In: WWW (2017)
29. Yang, B., Yih, W.T., He, X., Gao, J., Deng, L.: Embedding entities and relations for learning and inference in knowledge bases. arXiv preprint arXiv:1412.6575 (2014)
30. Yang, H., Liu, J.: Knowledge graph representation learning as groupoid: unifying TransE, RotatE, QuatE, ComplEx. In: Proceedings of the 30th ACM International Conference on Information & Knowledge Management (2021)
31. Zhang, F., Yuan, N.J., Lian, D., Xie, X., Ma, W.Y.: Collaborative knowledge base embedding for recommender systems. In: SIGKDD (2016)
32. Zhang, S., Tay, Y., Yao, L., Liu, Q.: Quaternion knowledge graph embeddings. Adv. Neural Inf. Process. Syst. 32 (2019)
33. Zhang, Y., Yao, Q., Shao, Y., Chen, L.: NSCaching: simple and efficient negative sampling for knowledge graph embedding. In: 2019 IEEE 35th International Conference on Data Engineering (ICDE), pp. 614–625. IEEE (2019)

CollaborateCas: Popularity Prediction of Information Cascades Based on Collaborative Graph Attention Networks

Xianren Zhang[1,3], Jiaxing Shang[2,3(✉)], Xueqi Jia[2,3], Dajiang Liu[2,3], Fei Hao[4], and Zhiqing Zhang[1]

[1] CQU-UC Joint Co-op Institute, Chongqing University, Chongqing, China
zhangxr2000@foxmail.com, zqzhang@cqu.edu.cn
[2] College of Computer Science, Chongqing University, Chongqing, China
{shangjx,liudj}@cqu.edu.cn
[3] Key Laboratory of Dependable Service Computing in Cyber Physical Society,
Ministry of Education, Chongqing University, Chongqing, China
[4] School of Computer Science, Shaanxi Normal University, Xian, China
fhao@snnu.edu.cn

Abstract. In recent years, with the prosperity of online social media platforms, cascade popularity prediction has attracted much attention from both academia and industry. Due to the recent advance in graph representation learning technologies, many state-of-the-art prediction methods utilize graph neural network to predict the cascade popularity. However, a significant disadvantage shared by these methods is that they treat each cascade independently, while the collaborations among different cascades are ignored. Therefore, in this paper we propose a novel deep learning model **CollaborateCas** which utilizes collaborations among different cascades to learn node and cascade embeddings directly and simultaneously. To this end, we first construct a heterogeneous user-message bipartite graph where different cascades are indirectly connected by common participants. To further capture temporal interdependence among users within each cascade, we construct homogeneous cascade graphs where temporal information is modeled as edge features. Experimental results on two real-world datasets show that our approach achieves significantly higher prediction accuracy compared with state-of-the-art approaches.

Keywords: Information diffusion · Cascade popularity prediction · Graph neural network · Heterogeneous graph · Deep learning

1 Introduction

Recent years have witnessed the prosperity of various online social media platforms which allow users to generate and share various online contents through comments, likes, or retweets. Consequently, the investigation of information diffusion over online social media has attracted much attention [18]. It finds application in a lot of important scenarios such as viral marketing [9], rumor detection

© The Author(s), under exclusive license to Springer Nature Switzerland AG 2022
A. Bhattacharya et al. (Eds.): DASFAA 2022, LNCS 13245, pp. 714–721, 2022.
https://doi.org/10.1007/978-3-031-00123-9_56

[16], etc. Among many of the research topics related to information diffusion, cascade popularity prediction [3], which aims to predict the future popularity of online contents based on their early diffusion patterns, is a key issue.

To address the cascade popularity prediction problem, a lot of research efforts have been devoted. Recently, deep learning techniques have shown their superiority in automatically capturing valuable information from cascades and predicting cascade popularity in an end-to-end manner [12]. Some approaches [2,12] represented cascades as multiple node sequences and then fed them into Recurrent Neural Network (RNN) models [5,10]. To extract underlying diffusion patterns, some researches applied Graph Neural Network (GNN) models [1,6] on cascade graphs [4] or social networks [3,11,14].

Motivation. Although GNN-based approaches have shown high prediction accuracy, a significant disadvantage shared by them is that they treat each cascade independently, while the collaborations among different cascades are ignored. In fact, according to the research of Myers et al. [13], when there are multiple messages spreading over the online social media, these messages will implicitly interact with each other, including both competition and cooperation effects among different cascades. On the one hand, messages with similar content and topics would have a higher chance to be shared by users if they are exposed multiple times to the same user. On the other hand, each user has limited attention with respect to tremendous online contents, thus different messages would implicitly compete with each other [17]. Therefore, it is worthwhile to consider the implicit interactions among different cascades.

Challenges. There are two key challenges in predicting the popularity of cascades when considering the aforementioned factors. The first challenge is how to capture collaborations among different cascades. To this end, instead of treating each cascade independently, multiple cascades should be considered comprehensively and fine-grained user-message interactions should be included into the learning model to get informative cascade embeddings. The second challenge is how to effectively merge temporal and structural information within each cascade. Temporal information can describe the influence of message and predecessors on users' diffusion behavior. Most current methods model temporal information as a chain and use RNN to capture the memory effects. However, modeling temporal information as a chain cannot capture the inter-dependence in tree-like cascade graphs.

To address the above challenges, we propose a novel deep learning model named **CollaborateCas**, which utilizes collaborations among different cascades to learn node and cascade embeddings directly. Specifically, for the first challenge, a heterogeneous user-message bipartite graph is built where users and cascades are represented as two types of nodes and the interactions between users and cascades are taken as edges. Then a type-ware Graph Attention Network (GAT) [15] model is designed to learn representations for the two types of nodes. To deal with the second challenge and based on the observation that users would have different reaction time for different early adopters, we take the difference of infection time as users' edge features in the homogeneous cascade graphs. The proposed approach is tested on two real world datasets and results show that our model significantly outperforms state-of-the-art baselines in terms of prediction accuracy.

In general, the main contributions of our work are as follows:

- For the cascade popularity prediction problem, we make the first attempt to model user-message interactions as a heterogeneous bipartite graph and design a type-aware GAT model to learn user and cascade embeddings simultaneously. Our model is able to capture collaborations among different cascades by learning from the fine-grained user-message interactions.
- Time differences of early adopters and later users are taken as temporal information and encoded into edge features in homogeneous cascade graphs. The temporal and structural information within each cascade graph are used to capture the inter-dependence and attractiveness among different users.
- The proposed approach is evaluated on two real-world datasets. Experimental results indicate that CollaborateCas significantly outperforms state-of-the-art baselines and the average prediction error is reduced by 9.01% and 5.68% respectively on the two datasets.

2   Problem Formulation

We first introduce some preliminaries and basic definitions to formulate the investigated problem.

Definition 1 (Cascade Set). *The data can be represented as a cascade set* $\mathcal{C}^T = \{C_c^T | c \in \mathcal{M}\}$ *which contains cascades with respect to the set of messages* $\mathcal{M}$ *within the observation time window T. Each cascade C_c^T can be represented as a set of tuples $\{(u, v, t)|t \leq T\}$, where (u, v, t) indicates that user v retweeted the message c from user u at time t within the observation time T.*

The purpose of our model is to predict the incremental size of cascade based on observations within a specific time window. Therefore, we define incremental size as follows:

Definition 2 (Incremental Size). *The incremental size of a cascade C_c^T with observation time T after a given time interval Δt is defined as $\Delta S_c = |C_c^{T+\Delta t}| - |C_c^T|$, where $|C_c^T|$ indicates the total number of retweeting behaviors with respect to this cascade by time T.*

Based on the aforementioned definitions, we define the cascade popularity prediction problem as follows:

Definition 3 (Cascade Prediction Problem). *Give a cascade $C_c^T \in \mathcal{C}^T$ within the observation time window T, the cascade popularity prediction problem aims to learn a function $f(\cdot)$ that maps the homogeneous cascade graph $G_c(V, E)$ and heterogeneous bipartite graph $\mathcal{G}(\mathcal{V}, \mathcal{E})$ to $\Delta S_c = |C_c^{T+\Delta t}| - |C_c^T|$.*

3   Methodology

This section will give detailed illustration about our CollaborateCas model. The overall architecture of our deep learning model is shown in Fig. 1.

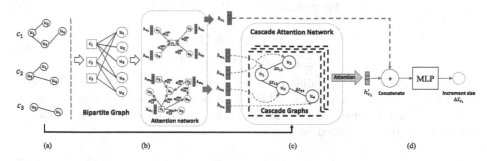

Fig. 1. Overview of CollaborateCas: (a) Input: a cascades set $\mathcal{C}^T$ within observation time T; (b) A heterogeneous bipartite graph is built based on observed cascades and embeddings are learned with a type-aware attention mechanism; (c) User embeddings learned in the previous step are fed into local cascade graph and temporal information is taken as edge features; (d) Both embeddings are concatenated together and then fed into MLP for final prediction.

3.1   Heterogeneous Bipartite Graph Learning

Based on observed cascades, we construct a global user-message graph to explicitly show relationships between messages and users. Since our model involves two different types of nodes, we design a type-aware attention mechanism and use different weights, i.e., W_{um} and W_{mu} to make a distinction between two different information gathering directions. Let

$$\theta_{ij}^{um} = \vec{a}_{um}^T[W_{um}\vec{h}_{c_i}||W_{um}\vec{h}_{u_j}] \tag{1}$$

$$\theta_{ij}^{mu} = \vec{a}_{mu}^T[W_{mu}\vec{h}_{u_i}||W_{mu}\vec{h}_{c_j}] \tag{2}$$

Where $\vec{a}_{um}$ and W_{um} are weights from user to message. $\vec{a}_{mu}$ and W_{mu} are weights from message to user. Then, θ^{um} and θ^{mu} are used to generate attention coefficients by softmax function. The embeddings are upadated as follows:

$$\vec{h}_{c_i} = f(\sum_{j \in N_i} \alpha_{ij}^{um} W_u \vec{h}_{u_j}) \tag{3}$$

$$\vec{h}_{u_i} = f(\sum_{j \in N_i} \alpha_{ij}^{mu} W_m \vec{h}_{c_j}) \tag{4}$$

Where N_i is the set of neighbors in bipartite graph. h_{c_i} and h_{u_i} are embeddings after updating.

3.2   Homogeneous Cascade Graph Learning

In our work,a modified attention mechanism is designed to incorporate temporal information into the graph attention network model. Specifically, we have:

$$\theta_{ij} = f_{mlp}(\Delta t_{ij}) \cdot [W\vec{h}_{u_i} || W\vec{h}_{u_j}] \tag{5}$$

$$\alpha_{ij} = \frac{\exp(LeakyReLU(\theta_{ij}))}{\sum_{k \in N_i} \exp(LeakyReLU(\theta_{ik}))} \tag{6}$$

where Δt_{ij} is the time difference between user i and user j, c is the corresponding cascade, $f_{mlp}()$ is a MLP which is used to project time difference scalar to higher dimensional embedding. Then the cascade embedding is obtained through an attention-based pooling:

$$\vec{h}'_c = \sum_{i \in c} \alpha_i \vec{h}_{u_i} \tag{7}$$

where α_i is the output attention coefficient.

3.3   Cascade Prediction and Loss Function

After embeddings from both heterogeneous bipartite and homogeneous cascade graphs are obtained, they are concatenated and fed into a MLP:

$$\hat{y}_i = MLP([\vec{h}_{c_i} || \vec{h}'_{c_i}]) \tag{8}$$

To optimize parameters of this deep learning model, the loss function is defined as the mean squared error:

$$L = \frac{\sum_i (y_i - \hat{y}_i)^2}{n} \tag{9}$$

Similar to [7], the label is defined as logarithm of incremental size, i.e., $y_i = \log(\Delta S_i + 1)$, where ΔS_i is the incremental size.

4   Evaluation

In this section, we evaluate the performance of our proposed model CollaborateCas by comparing it with several state-of-the-art approaches. Some variants of CollaborateCas are also considered for experimental study.We evaluate our model on two real-world datasets including Sina Weibo dataset [2] and HEP-PH dataset [8].We adopt two commonly used metrics, i.e., MSE [4] (Mean Square Error) and RMSPE [7] (Root Mean Square Percentage Error).

4.1   Baselines

To show the superiority of our approach, we select 5 state-of-the-art approaches and 3 variants as baselines.

- **Feature-linear & Feature-Deep**: We feed some selected features into a linear regression model (Feature-linear) and a MLP (Feature-deep).
- **Node2Vec**: Node2Vec [10] learns node embeddings from cascade graphs.
- **DeepCas**: DeepCas [12] applys GRU neural network to sequences generated from cascade graph.
- **CasCN**: CasCN [4] combines graph convolutional network with LSTM.
- **Deepcon_str**: Deepcon_str [7] regards each cascade as a node and builds two cascade-level graphs.
- **CollaborateCas-bipartite**: CollaborateCas-bipartite removes the part of homogeneous cascade graphs.
- **CollaborateCas-cascade**: CollaborateCas-cascade removes bipartite graph.
- **CollaborateCas-mean**: The attention mechanism at the output of cascade graph is replaced with mean operation.

Table 1. Overall performance between different approaches on the Sina Webo dataset.

T	1 h		2 h		3 h	
Metric	MSE	RMSPE	MSE	RMSPE	MSE	RMSPE
Deeplinear	1.5100	0.3112	1.6455	0.3960	1.7128	0.4979
Deepfeature	1.3116	0.3428	1.5293	0.4485	1.4847	0.5659
Node2Vec	2.1966	0.2500	2.2902	0.4625	2.2107	0.4891
DeepCas	1.0759	0.2229	1.3887	0.3924	1.3003	0.3868
CasCN	1.3336	0.2147	1.4956	0.4131	1.2786	0.4527
Deepcon_str	1.0709	0.2087	1.5049	0.3949	1.4794	0.3776
CollaborateCas	**0.9149**	**0.2019**	**1.2603**	**0.3446**	1.2374	**0.3487**

Table 2. Overall performance between different approaches on the HEP-PH dataset.

T	3 years		5 years		7 years	
Metric	MSE	RMSPE	MSE	RMSPE	MSE	RMSPE
Deeplinear	2.3738	0.5465	2.6249	0.6064	2.9796	0.6908
Deepfeature	2.3973	0.6134	2.2252	0.7486	2.6035	0.7773
Node2Vec	3.4308	0.6675	3.7664	0.8605	3.4933	0.8380
DeepCas	3.0613	0.6102	3.3759	0.9842	3.4412	1.1956
CasCN	2.5551	0.6544	2.1644	0.7142	2.3033	0.7311
Deepcon_str	2.7794	0.6993	2.5188	0.6890	2.7712	0.7880
CollaborateCas	**2.3197**	**0.5351**	**2.1500**	**0.4950**	1.9729	**0.6849**

4.2 Performance Comparison

The experimental results of our proposed model and various baselines are shown in Table 1 and Table 2. CollaborateCas achieves significantly lower MSE and RMSPE than all the baselines. For feature engineering-based methods, feature-linear and feature-deep show similar predictability on this task. Node2Vec and DeepCas have relative lower accuracy than other deep learning models.

CasCN performs worse than Deepcon_str and our model because it treats each cascade independently. Deepcon_str has overall better performance than other deep learning-based baselines. However, this method ignores detailed interactions between users and cascades. CollaborateCas has achieved better results than baselines in all three observation time windows, indicating that our unified modeling of heterogeneous bipartite graph and homogeneous cascade graphs can significantly improve the performance of cascade popularity prediction.

We also compare the performance of different variants of our model, as shown in Table 3. In general, CollaborateCas still performs better than other variants. The most competitive variant is CollaborateCas-bipartite, which means that the heterogeneous bipartite graph is an essential part for cascade prediction.

Table 3. Overall performance between variants of CollaborateCas.

T	1 h		2 h		3 h	
Metric	MSE	RMSPE	MSE	RMSPE	MSE	RMSPE
CollaborateCas-bipartite	0.9809	0.2083	1.2796	0.3851	**1.2281**	0.3601
CollaborateCas-cascade	1.7285	0.2634	2.2310	0.4485	2.2790	0.4649
CollaborateCas-mean	0.9982	**0.1957**	1.3773	0.3850	1.2532	0.3644
CollaborateCas	**0.9149**	0.2019	**1.2603**	**0.3446**	1.2374	**0.3487**

5 Conclusion

To address the cascade popularity problem, we proposed a novel deep learning model called CollaborateCas, which can capture collaborations among different cascades. To this end, we constructed a heterogeneous bipartite graph based on fine-grained user-message interactions and homogeneous cascade graphs incorporating temporal information as edge features. Experiments results demonstrate that CollaborateCas can achieve higher accuracy than state-of-the-art baselines.

Acknowledgements. This work was supported in part by: National Natural Science Foundation of China (Nos. 61966008, U2033213, 61804017).

References

1. Bruna, J., Zaremba, W., Szlam, A., LeCun, Y.: Spectral networks and locally connected networks on graphs. arXiv preprint arXiv:1312.6203 (2013)

2. Cao, Q., Shen, H., Cen, K., Ouyang, W., Cheng, X.: Deephawkes: bridging the gap between prediction and understanding of information cascades. In: Proceedings of the 2017 ACM on Conference on Information and Knowledge Management, pp. 1149–1158 (2017)
3. Cao, Q., Shen, H., Gao, J., Wei, B., Cheng, X.: Popularity prediction on social platforms with coupled graph neural networks. In: Proceedings of the 13th International Conference on Web Search and Data Mining, pp. 70–78 (2020)
4. Chen, X., Zhou, F., Zhang, K., Trajcevski, G., Zhong, T., Zhang, F.: Information diffusion prediction via recurrent cascades convolution. In: 2019 IEEE 35th International Conference on Data Engineering (ICDE), pp. 770–781. IEEE (2019)
5. Cho, K., Van Merriënboer, B., Gulcehre, C., Bahdanau, D., Bougares, F., Schwenk, H., Bengio, Y.: Learning phrase representations using RNN encoder-decoder for statistical machine translation. arXiv preprint arXiv:1406.1078 (2014)
6. Defferrard, M., Bresson, X., Vandergheynst, P.: Convolutional neural networks on graphs with fast localized spectral filtering. Adv. Neural Inf. Process. Syst. **29**, 3844–3852 (2016)
7. Feng, X., Zhao, Q., Liu, Z.: Prediction of information cascades via content and structure proximity preserved graph level embedding. Inf. Sci. **560**, 424–440 (2021)
8. Gehrke J, Ginsparg P, K.J.: Overview of the 2003 KDD cup. In: Acm Sigkdd Explor. Newslett. **5**(2), 149–151 (2003)
9. Gong, Q., et al.: Cross-site prediction on social influence for cold-start users in online social networks. ACM Trans. Web (TWEB) **15**(2), 1–23 (2021)
10. Grover, A., Leskovec, J.: node2vec: scalable feature learning for networks. In: Proceedings of the 22nd ACM SIGKDD International Conference on Knowledge Discovery and Data Mining, pp. 855–864 (2016)
11. Jiang, B., Lu, Z., Li, N., Wu, J., Yi, F., Han, D.: Retweeting prediction using matrix factorization with binomial distribution and contextual information. In: Li, G., Yang, J., Gama, J., Natwichai, J., Tong, Y. (eds.) DASFAA 2019. LNCS, vol. 11447, pp. 121–138. Springer, Cham (2019). https://doi.org/10.1007/978-3-030-18579-4_8
12. Li, C., Ma, J., Guo, X., Mei, Q.: Deepcas: an end-to-end predictor of information cascades. In: Proceedings of the 26th International Conference on World Wide Web, pp. 577–586 (2017)
13. Myers, S.A., Leskovec, J.: Clash of the contagions: cooperation and competition in information diffusion. In: 2012 IEEE 12th International Conference on Data Mining, pp. 539–548. IEEE (2012)
14. Su, Y., Zhang, X., Wang, S., Fang, B., Zhang, T., Yu, P.S.: Understanding information diffusion via heterogeneous information network embeddings. In: Li, G., Yang, J., Gama, J., Natwichai, J., Tong, Y. (eds.) DASFAA 2019. LNCS, vol. 11446, pp. 501–516. Springer, Cham (2019). https://doi.org/10.1007/978-3-030-18576-3_30
15. Veličković, P., Cucurull, G., Casanova, A., Romero, A., Liò, P., Bengio, Y.: Graph attention networks. In: International Conference on Learning Representations (2018)
16. Vosoughi, S., Roy, D., Aral, S.: The spread of true and false news online. Science **359**(6380), 1146–1151 (2018)
17. Weng, L., Flammini, A., Vespignani, A., Menczer, F.: Competition among memes in a world with limited attention. Sci. Rep. **2**(1), 1–9 (2012)
18. Zhou, F., Xu, X., Trajcevski, G., Zhang, K.: A survey of information cascade analysis: models, predictions, and recent advances. ACM Comput. Surv. (CSUR) **54**(2), 1–36 (2021)

Contrastive Disentangled Graph Convolutional Network for Weakly-Supervised Classification

Xiaokai Chu[1,2], Jiashu Zhao[3], Xinxin Fan[1], Di Yao[1], Zhihua Zhu[1], Lixin Zou[4], Dawei Yin[4], and Jingping Bi[1(✉)]

[1] Institute of Computing Technology, Chinese Academy of Sciences, Beijing, China
{chuxiaokai,fanxinxin,yaodi,zhuzhihua,bjp}@ict.ac.cn
[2] University of Chinese Academy of Sciences, Beijing, China
[3] Wilfrid Laurier University, Waterloo, Canada
jzhao@wlu.ca
[4] Baidu Inc., Beijing, China
yindawei@acm.org

Abstract. Node classification on graph-structured data plays an important role in many machine learning applications. Recently, Graph Convolutional Networks (GCNs) have shown remarkable success in the node classification task, due to the ability to aggregate neighborhood information and propagate supervised signals over the graph. However, most GCN-style models require relatively sufficient labeled data, which are not available in many real-world applications. Therefore, we in this paper study the problem of weakly-supervised node classification and propose a *C*ontrastive *D*isentangled *G*raph *C*onvolutional *N*etwork (*CDGCN*) to learn disentangled node representations based on the contrastive learning mechanism. Extensive experimental results show that CDGCN significantly outperforms all baselines on different label sparsities. The code is available at https://github.com/ChuXiaokai/CDGCN.

Keywords: Graph Convolutional Network · Disentangled representation · Contrastive learning · Weakly-supervised node classification

1 Introduction

In recent years, Graph Convolutional Networks (GCNs) [2,3,12] have witnessed a great success on node classification tasks. Nevertheless, the gratifying performance of GCNs and their variants rely greatly on plenty of labeled data [4], which is not always met in reality. In many applications, due to the high cost and potential difficulty in human annotation, we have to predict massive unlabeled nodes (*e.g.*, thousands of nodes) based on sparsely labeled data (*e.g.*, one or two samples per class), *i.e.*, weakly-supervised node classification problem.

A. Bhattacharya et al. (Eds.): DASFAA 2022, LNCS 13245, pp. 722–730, 2022.
https://doi.org/10.1007/978-3-031-00123-9_57

To date, a few of GCN-based approaches [4,11] try to address the weakly-supervised node classification problem. Nevertheless, these models rely on empirical information, which is infective in faced with complicated network data. Moreover, they treat the node's neighborhood as a perceptual whole but ignore the fine-grained differences of the connections between different node pairs, which imply various reasons to build the connections.

With a small number of labeled data, we will need to explore more extra information from the graph in order to learn the relationships between the nodes, and therefore achieve better classification performance. The edges in a graph often imply certain factors, which reflect the reasons that drive interactions between nodes. For example, in a social network, users are connected due to various factors, such as "colleague", "collaboration", "teacher-student", *etc.* Such latent factors often show various properties and context meanings to explain the construction of an edge, which can benefit the clas-

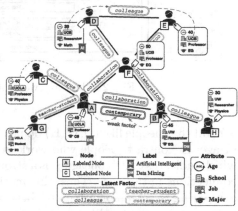

Fig. 1. An toy example of a social network.

sification performance. These factors are not explicitly available in the graph, but can be adaptively extracted via the graph disentangled representation learning [6], which aims to learn a multi-channel representation for a node and each channel describes an aspect of this node that is pertinent to a factor. In graph disentangled representation learning, a neighborhood routing module [6] is performed to analyze the latent factors based on the supervised signals, and assigns nodes to different factor spaces according to the association of the neighborhood regarding each factor.

However, it is hard to directly disentangle factors if the labels are insufficient. When a disentangled model extracts factors directly from a small amount of labeled nodes, it tends to be distracted by some unimportant factors and therefore does not perform well. Figure 1 shows an example of a social network, which includes labeled and unlabeled nodes. Edge (A,B) connects a pair of labeled users A and B, which have the same age value, 45. Thus, a model is very likely to disentangle the factor "contemporary" from this connection based on the supervised signals. However, in this career classification problem, compared to other factors such as "collaboration" and "colleague", "contemporary" is a weaker factor that can influence the classifier, and should be identified and eliminated.

To effectively extract the accurate latent factors in faced with insufficient labeled data, we focus on refining two types of factors: weak factors and coarse-grained factors. For weak factors, such as "contemporary" in the above example,

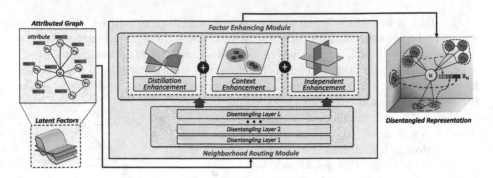

Fig. 2. The framework of the proposed CDGCN.

we find that people like (D, G, H) do not build any connection even they are same in age; in contrast, people who work in the same school are always connected, which indicates that "colleague" is a stronger factor than "contemporary" to drive connections in this case. Thus, this motivates us to identify the weak factors by comparing the common attributes between the connected and unconnected node pairs. For coarse-grained factors, in Fig. 1, the factor "colleague" describes both the social connections of users (A,C) and (D,E), while it has different semantic meanings. For users (A,C), the "colleague" factor is in the context of university UCLA. In terms of (D,E), these two people have the "colleague" factor in the context of university UCB. If we treat these two connections as having exactly the same factor, people in UBC will be recognized to have the same relationship as people in UCLA, while these two universities are two distinct communities. In many classification applications, especially for weakly-supervised problems, the more specific fine-grained context information can further distinguish the differences between nodes and benefit the classification results. Thus, we should distinguish the different contexts to enhance the coarse-grained factors.

Based on the above analysis, in this paper, we propose a novel _C_ontrastive _D_isentangled _G_raph _C_onvolutional _N_etworks (_CDGCN_) to handle the weakly-supervised node classification problem. In CDGCN, two contrastive strategies are designed to enhance the extraction of important factors: (1) *Distillation Enhancement* aims to filter out the weak factors by comparing the attributes between the connected and unconnected nodes; (2) *Context Enhancement* aims to enhance the coarse-grained factor by analyzing the fine-grained differences between different contexts related to this factor. By integrating contrastive learning [8] and neighborhood routing [6], CDGCN can effectively refine high-quality disentangled representations and improve the performance on weakly-supervised node classification significantly, *e.g.*, we always gain 10~20% improvement with only one labeled samples per class available.

2   The Proposed Model

Figure 2 presents the framework of CDGCN: taking an attributed graph as input, which contains multiple latent factors, CDGCN aims to learn the disentangled node representation via two major modules: neighborhood routing module and factor enhancing module. We first introduce some definitions in this paper.

2.1   Preliminaries

Let $G = (\mathcal{V}, \mathcal{E}, \boldsymbol{X}, \boldsymbol{Y})$ be an attributed graph, where $\mathcal{V}$ is the set of nodes, $\mathcal{E}$ denotes the edges, and $(u, v) \in \mathcal{E}$ represents an edge between nodes u and v. $\boldsymbol{X} \in \mathbb{R}^{n \times d_{in}}$ is the attribute matrix, and $\boldsymbol{x}_i$ represents the attribute of node i. $\boldsymbol{Y} \in \mathbb{R}^{|\mathcal{V}| \times C}$ is the label matrix, where C denotes the number of labels. We use $\mathcal{V}^C$ to represent the set of labeled nodes.

We in this paper work on learning disentangled node representation, which assumes that a node representation is composed of K components, each component is corresponding to a factor. In detail, given a node u in an attributed graph G, we aim to refine a K-channel representation $\boldsymbol{z}_u = [\boldsymbol{z}_{u,1}, \boldsymbol{z}_{u,2}, ..., \boldsymbol{z}_{u,K}] \in \mathbb{R}^{K\Delta d}$, each $\boldsymbol{z}_{u,k} \in \mathbb{R}^{\Delta d}$ represents the kth disentangled representation of node u, which describes the aspect that is pertinent to the kth latent factor.

2.2   Neighborhood Routing Module

To obtain the disentangled representation for a node u, we should study K latent factors and assign the neighbors to the related latent factor space. Here we adopt the neighborhood routing mechanism proposed in [6]. In detail, we first project u and all its neighbors into K factor spaces as:

$$z_{o,k} = \sigma(\boldsymbol{W}_k^\top \boldsymbol{x}_o + \boldsymbol{b}_k), \tag{1}$$

where $o \in \{u\} \cup \{v : v \in \mathcal{N}_u\}$, $\mathcal{N}_u$ denotes the neighbors of u, $\boldsymbol{W}_k$ and $\boldsymbol{b}_k$ are parameters to describe the kth latent factor. Then, we apply disentangling layers to identify the latent factor for each neighbor v through iteratively searching for the largest cluster in each factor space k as:

$$z_{u,k}^{(l)} = \frac{z_{u,k}^{(l)} + \sum_{v \in \mathcal{N}_u} p_{v,k}^{(l)} z_{v,k}^{(l)}}{\|z_{u,k}^{(l)} + \sum_{v \in \mathcal{N}_u} p_{v,k}^{(l)} z_{v,k}^{(l)}\|_2}; \qquad p_{v,k}^{(l)} = \frac{exp(z_{v,k}^{(l)\top} z_{u,k}^{(l)})}{\sum_{k=1}^{K} exp(z_{v,k}^{(l)\top} z_{u,k}^{(l)})}. \tag{2}$$

where l represents l^{th} disentangling layer. Each $p_{v,k}^{(l)}$ indicates the probability to assign the neighbor v to k^{th} factor space, by which a neighbor v will send its information in corresponding factor space to node u to construct its output representation $z_u^{(l)} = [z_{u,1}^{(l)}, \cdots, z_{u,K}^{(l)}]$. Equation (2) will be performed for T iterations.

2.3   Factor Enhancing Module

We propose two types of contrastive strategies to ensure the representation quality, *i.e.*, distillation enhancement and context enhancement, as well as a regularization approach to enhance the factors' independence.

Distillation Enhancement. The distillation enhancing strategy aims to identify the strong factors behind the graph while ignore the weak ones. To this end, we introduce a contrastive learning approach by comparing the attributes between the connected nodes and unconnected nodes. In detail, for each neighbor v_i of a node u, we sample a negative node $v_i' \notin \mathcal{N}_u$, which is not connected with node u. Then in each factor space k, a discriminator is adopted to encourage node u to distinguish its true neighbors (which implies the positive factor) from the negative sampled neighbors via the following objective:

$$\ell_d(u, k) = - log\sigma \Big(\mathcal{D}_d \big(\mathbf{z}_{u,k}, \mathcal{R}(\{\mathbf{z}_{v_i,k} : v_i \in \mathcal{N}_u\}) \big) \Big)$$
$$- log\sigma \Big(1 - \mathcal{D}_d \big(\mathbf{z}_{u,k}, \mathcal{R}(\{\mathbf{z}_{v_i',k} : v_i' \notin \mathcal{N}_u\}) \big) \Big), \tag{3}$$

where each vector $\mathbf{z}$ is the output of the last disentangling layer $f^{(L)}(\cdot)$. The function $\mathcal{D}_d(\mathbf{x}, \mathbf{y}) = \mathbf{x}^\top \mathbf{W}_d \mathbf{y} + \mathbf{b}_d$ is a bilinear function, where $\mathbf{W}_d$ and $\mathbf{b}_d$ are the learnable parameters. $\mathcal{R}(\cdot)$ is a readout function to summarize the information from the neighbors, we here consider the following weighted sum to accumulate the information of the positive/negative neighbors according to factor k:

$$\mathcal{R}(\{\mathbf{z}_{v_i,k} : v_i \in \mathcal{N}_u\}) = \sum_{v_i \in \mathcal{N}_u} \overline{p}_{v_i,k} \mathbf{z}_{v_i,k}, \quad \mathcal{R}(\{\mathbf{z}_{v_i',k} : v_i' \notin \mathcal{N}_u\}) = \sum_{v_i' \notin \mathcal{N}_u} \overline{p}_{v_i,k} \mathbf{z}_{v_i',k}, \tag{4}$$

where $\overline{p}_{v_i,k}$ is the mean of the probabilities of a neighbor v_i being assigned to the kth factor in each disentangling layer. Such design is based on an assumption that the assignment probabilities of a neighbor v_i should be consistent in each layer. We use the probability of the neighbor v_i as the probability of the negative neighbor v_i', because we will compare the positive neighbor with the negative neighbor in the same factor space, where they should send the same weight of information for fair comparison.

Context Enhancement. We have discussed that even driven by the same factor, there still exist various reasons to build the connections. By capturing such difference, we can refine more representative and semantically-rich factors. Therefore, our second strategy is to enhance the semantics of a coarse-grained factor by identifying its finer-grained context information. Specifically, given a node u, we randomly select another node a, where $u \neq a$, and consider their respective neighbors $\{v_i : v_i \in \mathcal{N}_u\}$, $\{b_j : b_j \in \mathcal{N}_a\}$. Our purpose is to guide the model to identity and distinguish their context difference in each factor space.

Thus, for each factor k, we propose the following loss function to capture these two contexts:

$$\ell_s(u, k) = -log\sigma\Big(\mathcal{D}_s\big(z_{u,k}, \mathcal{R}(\{z_{v_i,k} : v_i \in \mathcal{N}_u\})\big)\Big)$$
$$-log\sigma\Big(1 - \mathcal{D}_s\big(z_{u,k}, \mathcal{R}(\{z_{b_j,k} : b_j \in \mathcal{N}_a\})\big)\Big). \tag{5}$$

Different from Eq. (3), both the neighborhoods of nodes u and a are observed, thus the readout function of the a's neighbors should be defined as:

$$\mathcal{R}(\{z_{b_j,k} : b_j \in \mathcal{N}_a\}) = \sum_{b_j \in \mathcal{N}_a} \overline{p}_{b_j,k} z_{b_j,k}. \tag{6}$$

By minimizing the loss function (5), the two contexts will be distinct in the kth factor space, leading to the extension of a factor's semantics.

Independence Enhancement. The different components in a disentangled representation should contain non-overlapped information, as it can reflect different views of a node. Thus, inspired by [5], we promote the independence among different factors by performing the empirical version [1] of Hilbert-Schmidt Independence Criterion (HSIC) for each two latent factors as:

$$HSIC(z_{u,i}, z_{u,j}) = (\Delta d - 1)^{-2} tr(K_i H K_j H), \tag{7}$$

where $H = I - 1/\Delta d$ is a matrix with zero mean, K_i and K_j are Gram matrices with inner product as the kernel function, i.e., $K_i = z_{u,i}^{\top} z_{u,i}$.

2.4 Model Optimization

The final objective function of CDGCN contains four parts:

$$\mathcal{L} = \mathcal{L}_c + \alpha\mathcal{L}_d + \beta\mathcal{L}_s + \gamma\mathcal{L}_{HSIC}, \tag{8}$$

where α, β and γ are the hyper-parameters that balance different terms. $\mathcal{L}_c$ is the cross-entropy loss of node classification: $\mathcal{L}_c = -\sum_{u \in \mathcal{V}^C} \sum_{c=1}^{C} y_{u,c} log(\tilde{y}_{u,c})$. Following [6], the predicted vector $\tilde{y}_u$ is obtained via a fully-connected layer with a softmax function on the Lth layer output $z_u^{(L)}$. $\mathcal{L}_d = \sum_{u \in \mathcal{V}} \sum_{k \in K} \ell_d(u, k)$ denotes distillation enhancement to effectively distill the strong factors behind the input graph, while $\mathcal{L}_s = \sum_{u \in \mathcal{V}} \sum_{k \in K} \ell_s(u, k)$ is the total loss of context enhancement. $\mathcal{L}_{HSIC} = \sum_{u \in \mathcal{V}^C} \sum_{i \neq j} HSIC(z_{u,i}, z_{u,j})$ is the HSIC regularization for each labeled node to encourage the independence between different factors.

3 Experiment Evaluation

We conduct experiments on the citation benchmark: Cora, Citeseer and PubMed [10] wherein the nodes indicate the documents and edges denote the citations. We compare with the state-of-the-art self-supervised graph neural networks models, *e.g.*, Co-training, Self-training and their variants [4], M3S [11], DGI [13] and GMI [9]. Furthermore, we also compare with several state-of-the-art node classification approaches, *e.g.*, GCN [3], GAT [12], graph disentangle representation learning models: DisenGCN [6] and IPGDN [5].

Table 1. Node classification accuracy (%) with different labeled samples per class.

	Cora			Citeseer			PubMed		
#Samples/Class	1	2	3	1	2	3	1	2	3
GCN	46.8	50.8	57.0	38.5	43.3	52.5	46.0	51.2	57.2
GAT	49.5	53.0	58.2	41.1	47.3	59.5	53.5	58.4	<u>64.3</u>
DGI	38.0	45.5	59.0	37.5	42.9	53.1	46.4	49.4	52.4
GMI	<u>61.0</u>	<u>63.5</u>	66.7	39.9	45.3	55.7	49.8	51.7	53.2
Co-training	52.4	54.1	55.6	36.9	40.5	49.7	51.0	55.6	61.1
Self-training	55.1	57.0	59.1	47.8	50.2	55.3	50.9	56.4	63.0
Union	53.0	55.5	58.7	45.9	47.3	50.4	52.0	57.3	63.8
Intersection	45.9	50.8	59.2	45.0	50.2	59.7	52.6	55.1	57.4
M3S	58.7	61.7	66.4	<u>52.3</u>	<u>55.1</u>	<u>60.8</u>	<u>55.7</u>	<u>59.3</u>	63.5
DisenGCN	57.8	62.4	<u>70.7</u>	41.5	45.2	52.8	48.8	56.0	63.7
IPGDN	56.2	58.6	62.0	33.9	39.8	52.0	52.4	57.7	64.1
CDGCN	**70.7**	**72.9**	**76.1**	**64.8**	**66.2**	**68.9**	**67.6**	**70.3**	**72.9**
Inc.	15.9%	14.8%	7.6%	23.9%	20.2%	13.3%	21.4%	18.6%	13.4%

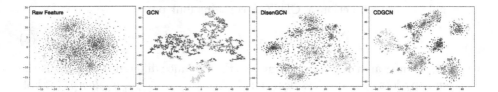

Fig. 3. Visualization of representations learned in Cora with two samples per class.

3.1 Performance Analysis

We examine the model performance with different sparse levels of labeled nodes: from one sample per class to three samples per class for all graphs. We conduct the experiments ten times for each model, where the dataset is randomly re-divided with the given setting for each run. The mean accuracy is reported as the model's performance. The results on different datasets are presented in

Table 1. CDGCN has a great improvement on all datasets, especially when the labels are merely few. For example, when there are only one label per class available for training, CDGCN gains 15.9%, 23.9% and 21.4% improvement than the best performing baselines on each network. Also, when looking at the accuracy *w.r.t.* the number of labels, we can observe that CDGCN is less sensitive to the different sparse levels of labels. The extraordinary improvement indicates that CDGCN can effectively extracts disentangled node representations with the support of distillation enhancement and context enhancement, especially with limited labeled data.

To present a more intuitive comparison, we also show Cora as an example and visualize the learned representations with t-SNE [7] where the model is trained with two samples per class. The result is presented in Fig. 3, where each point represents a node and different colors denote different classes. We observe that with few labeled data, GCN can extract some basic characteristics of the same class from the raw features, still, the different classes are entangled severely and nearly indistinguishable. While DisenGCN performs much better than GCN, the same class nodes tend to be clustered, but different classes are still mixed. CDGCN gains the best performance to make each class tightly cluster together, as well as identifies obvious boundaries between different clusters.

4   Conclusion

We have introduced a novel graph disentangled representation learning model-CDGCN, to handle the weakness of current models on weakly-supervised node classification problem, which incorporates contrastive learning to obtain disentangled node representations for the first time. The multi-facet experimental results exhibit that the proposed CDGCN always achieves the best results and dramatically outperforms the baselines with low label rate settings.

Acknowledgements. This work has been supported by the National Natural Science Foundation of China under Grant No.: 62077044, 61702470, 62002343.

References

1. Gretton, A., Bousquet, O., Smola, A.J., Schölkopf, B.: Measuring statistical dependence with hilbert-schmidt norms. In: ALT (2005)
2. Hamilton, W.L., Ying, Z., Leskovec, J.: Inductive representation learning on large graphs. In: NeurIPS (2017)
3. Kipf, T.N., Welling, M.: Semi-supervised classification with graph convolutional networks. In: ICLR (2017)
4. Li, Q., Han, Z., Wu, X.: Deeper insights into graph convolutional networks for semi-supervised learning. In: AAAI (2018)
5. Liu, Y., Wang, X., Wu, S., Xiao, Z.: Independence promoted graph disentangled networks. In: AAAI (2020)
6. Ma, J., Cui, P., Kuang, K., Wang, X., Zhu, W.: Disentangled graph convolutional networks. In: ICML (2019)

7. van der Maaten, L., Hinton, G.: Visualizing data using t-SNE. J. Mach. Learn. Res. **9**, 2579–2605 (2008)
8. van den Oord, A., Li, Y., Vinyals, O.: Representation learning with contrastive predictive coding. CoRR abs/1807.03748 (2018)
9. Peng, Z., et al.: Graph representation learning via graphical mutual information maximization. In: WWW (2020)
10. Sen, P., Namata, G., Bilgic, M., Getoor, L., Gallagher, B., Eliassi-Rad, T.: Collective classification in network data. AI Mag. **29**(3), 93–106 (2008)
11. Sun, K., Lin, Z., Zhu, Z.: Multi-stage self-supervised learning for graph convolutional networks on graphs with few labeled nodes. In: AAAI (2020)
12. Velickovic, P., Cucurull, G., Casanova, A., Romero, A., Liò, P., Bengio, Y.: Graph attention networks. In: ICLR (2018)
13. Velickovic, P., Fedus, W., Hamilton, W.L., Liò, P., Bengio, Y., Hjelm, R.D.: Deep graph infomax. In: ICLR (2019)

CSGNN: Improving Graph Neural Networks with Contrastive Semi-supervised Learning

Yumeng Song[✉], Yu Gu, Xiaohua Li, Chuanwen Li, and Ge Yu

School of Computer Science and Engineering, Northeastern University, Shenyang,
Liaoning, China
ymsong94@163.com, {guyu,lixiaohua,lichuanwen,yuge}@mail.neu.edu.cn

Abstract. The Graph Neural Network (GNN) is a rising graph analysis
model family that encodes node features into low-dimensional representation vectors by aggregating local neighbor information. Nevertheless, the
performance of GNNs is limited since GNNs are trained only over predictions of the labeled data. Hence, effectively incorporating a great number of unlabeled nodes into GNNs will upgrade the performance of GNNs.
To address this issue, we propose a Contrastive Semi-supervised learning based GNN (CSGNN) that improves the GNN from extra supervision
predicted by contrastive learning. Firstly, CSGNN utilizes multi-loss contrast to learn node representations via maximizing the agreement between
nodes, edges and labels of different views. Then, a semi-supervised fine-
tuner learns from few labeled examples while making the best use of
unlabeled nodes. Finally, we introduce the knowledge distillation based
on label reliability, which further distills the node labels predicted by
contrastive learning into the GNN. Experimentally, CSGNN effectively
improves the classification performance of GNNs and outperforms other
state-of-the-art methods in accuracy over a variety of real-world datasets.

Keywords: Contrastive learning · Semi-supervised learning · Graph
Neural Network

1 Introduction

Graph Neural Networks (GNNs) have aroused more and more attention on
account of the ability to handle the graph-structured data defined on irregular
or non-Euclidean domains. GNNs compute graph node representations through
a propagation process which iteratively aggregates local structural information.
GNNs are clearly superior to traditional graph-based algorithms in quite a few
tasks [6]. Unfortunately, GNNs, as data-driven inference models, are also not free
of the bottleneck when training data is inadequate. The reason is that GNNs are
trained only over predictions of labeled nodes by minimizing the supervised loss,
and predictions of unlabeled nodes do not contribute to the training. In order to
tackle the intrinsic hardness, various researches have emphasized incorporating

© The Author(s), under exclusive license to Springer Nature Switzerland AG 2022
A. Bhattacharya et al. (Eds.): DASFAA 2022, LNCS 13245, pp. 731–738, 2022.
https://doi.org/10.1007/978-3-031-00123-9_58

unlabeled data into GNNs via combining them with self-supervised learning [8] or augmenting topology and attributes of graphs in different ways [12].

Recently, Contrastive Learning (CL) achieves great success in graph representation learning [15,16]. As a popular form of self-supervised learning, CL seeks to maximize the mutual information between the input and its representations by contrasting positive pairs with negative-sampled counterparts. However, CL learns embeddings in a task-agnostic way without using labeled data. This leads us to explore a fusion mechanism of CL and GNNs for graph-based semi-supervised learning. In the latest studies on graph-based CL, [11] proposes a contrastive semi-supervised model CG, which minimizes the contrastive loss, the graph generative loss and the classification loss between graph views together. But CG does not bring out the full power of CL. One of the state-of-the-art researches in computer vision also proposes a contrastive semi-supervised model SIMCLRv2 [1]. SIMCLRv2 distills the generated embeddings of unlabeled data into the downstream student model. Nevertheless, SIMCLRv2 cannot be directly applied to graph-based data.

Based on the above discussion, we propose a Contrastive Semi-supervised learning based GNN (CSGNN) which utilizes knowledge distillation to combine CL with GNNs, with a CL model as the teacher model and a GNN as the student model. To the best of our knowledge, CSGNN is the first research on combining contrastive learning with GNNs through knowledge distillation. Our contributions are summarized as follows:

- This paper provides a contrastive semi-supervised based GNN which could comprehensively leverage the abundant structural and semantic information of unlabeled nodes.
- In the teacher model, a multi-loss contrastive learning method is introduced to learn representations by contrasting positive and negative examples between nodes, edges and labels.
- We design a reliable knowledge distillation method via computing the label reliability based on the Shannon entropy of teacher and student's predictions.
- In experiments, we demonstrate that CSGNN can greatly improve the performance of GNNs in node classification task compared with the state-of-the-art methods on real-world datasets.

2   Related Works

For graph data, graph contrastive learning applies the idea of CL on GNNs. These methods can be categorized based on how the positive and negative samples are constructed. One is to measure the loss of different parts of a graph in latent space by contrasting nodes and the whole graph, nodes and nodes or nodes and subgraphs [5]. The other one uses different data augmentation methods to generate contrastive pairs. GraphCL [13] develops contrastive learning with node dropping, edge perturbation, subgraph sampling and feature masking. MVGRL [3] constructs multiple graph views by sampling subgraphs based

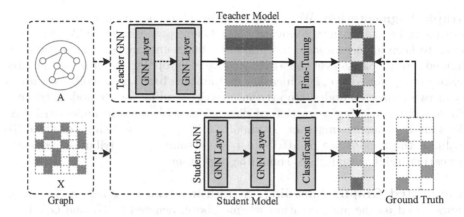

Fig. 1. The overview of CSGNN

on random walks. GCA [16] leverages the network centrality to augment the graph adaptively on both topology and attribute levels. CG [11] augments the graph via localized graph convolution and hierarchical graph convolution and designs a new semi-supervised contrastive loss. Most CL models cannot achieve the best performance for specific tasks through unsupervised learning.

3   Overview

The overall framework of CSGNN is shown in Fig. 1. Given a graph $G = (V, A, X)$ with a node set $V = \{v_1, v_2, ..., v_N\}$, a graph adjacent matrix $A \in R^{N \times N}$ and a node attribute matrix $X \in R^{N \times F}$ where F is the dimension of node attributes, we firstly input A and X of into a trained teacher GNN to generate the node general representation. The teacher GNN is trained via multi-loss contrastive learning, which can measure mutual information in multiple dimensions and obtain the main features of nodes without focusing on the details. The node embedding after fine tuning represents the category probability of each node. For further improving predictive performance and obtaining a compact model, we train the student GNN on the labeled data with ground truths and the unlabeled data with predicted labels from the fine-tuned teacher GNN. Finally, we can generate predictions directly from the student GNN, regardless of the teacher model.

4   Teacher Model with Contrastive Learning

The training process of the teacher model involves the following stages: (1) the adaptive graph augmentation stage, which transforms the original graph into different views; (2) the encoding stage, which generates the node representations via the teacher GNN; (3) the contrasting stage, which contrasts the latent vectors between nodes, edges and labels.

Graph Augmentation. We augment the graph by perturbing possibly unimportant links and features following the schemes proposed in GCA [16] which tends to keep important structures and attributes unchanged. Augmentation is divided into topology-level and attribute-level. On topology-level, we define edge centrality s_{uv}^e for edge e_{uv} to measure its influence based on PageRank centrality of two connected nodes. On attribute-level, we add noises to node attributes via randomly masking a fraction of dimensions with zeros in node attributes. We sample a random mask m_i $Bernoulli(P_{perturbing}^{a_i})$ for each attribute i. To evaluate the importance of attribute i, we assume that attributes frequently appearing in influential nodes should be important.

Augmented Graphs Encoding. At each iteration, we generate two graph views based on the augmentation scheme above, denoted as G_1 and G_2. Then G_1 and G_2 are input into the teacher GNN with shared parameters. The node embeddings are denoted as $U = f(X_1, A_1)$ and $V = f(X_2, A_2)$.

Contrastive Learning. After that, we employ the contrastive loss to train the teacher GNN. We conduct joint contrastive losses between nodes, edges and labels to make the embeddings more conducive to classification. For any node i, its embedding is u_i in view G_1 and v_i in view G_2. The node contrastive loss between a pair of positive examples u_i and v_i is given as follows:

$$\ell_{u_i,v_i} = -\log \frac{\exp(sim(g(u_i), g(v_i))/\tau)}{\sum_{k \neq i} \exp(sim(g(u_i), g(v_k))/\tau) + \sum_{k \neq i} \exp(sim(g(u_i), g(u_k))/\tau)} \tag{1}$$

where $sim(\cdot, \cdot)$ is the cosine similarity, $g(\cdot)$ is a non-linear transformation network, $\sum_{k \neq i} \exp(sim(g(u_i), g(v_k))/\tau)$ is the loss between inter-view negative pairs and $\sum_{k \neq i} \exp(sim(g(u_i), g(u_k))/\tau)$ is the loss between intra-view negative pairs. Since the symmetric among the views, our unsupervised node contrastive loss $\mathcal{L}_{nodes}$ can be presented as:

$$\mathcal{L}_{nodes} = \frac{1}{2N} \sum_{i=1}^{N} (\ell_{u_i,v_i} + \ell_{v_i,u_i}) \tag{2}$$

The goal of the edge contrastive loss is to distinguish between existing edges and non-existing edges within and between views. We reconstruct the adjacency matrix A_1^* and A_2^* based on the node embedding of each view. We also reconstruct the adjacency matrix $A_{1,2}^*$ between two views. We calculate the inner product of node embeddings as the possibility that two nodes have edges for reconstructing the adjacency matrix. Given edge $e_{i,j}$ in graph G, the corresponding edge in A_1^*, A_2^* and $A_{1,2}^*$ are positive examples, and non-existing edges are negative examples. Here, the unsupervised edge contrastive loss can be computed as:

$$\mathcal{L}_{edges} = \frac{1}{3|E|}(\ell_{G_1}^{edges} + \ell_{G_2}^{edges} + \ell_{G_1,G_2}^{edges}) \tag{3}$$

$$\ell_{G_1}^{edges} = -\log \frac{\sum_{e \in E} \exp(A_{1e}^*/\tau)}{\sum_{\hat{e} \notin E} \exp(A_{1\hat{e}}^*/\tau)} \tag{4}$$

where E is the edge set of graph G and A_{1e}^* is the value of edge e in A_1^*. $\ell_{G_2}^{edges}$ and ℓ_{G_1,G_2}^{edges} are similar to Eq. 4.

Our supervised contrastive learning loss will distinguish nodes of the same category and nodes of different categories within and between views, which is defined as:

$$\mathcal{L}_{labels} = \frac{1}{2|L|} \sum_{l \in L} (\ell_l^{labels,G_1} + \ell_l^{labels,G_2}) \tag{5}$$

$$\ell_l^{labels,G_1} = -\log \frac{\sum_{k \in S(l)} \exp(sim(g(l),g(k))/\tau)}{\sum_{\widehat{k} \in Diff(l)} \exp(sim(g(l),g(\widehat{k}))/\tau)} \tag{6}$$

where L is the set of labeled nodes, $S(\cdot)$ is the set of nodes with the same label, $Diff(\cdot)$ is the set of nodes with different labels. ℓ_l^{labels,G_2} is similar to Eq. 6.

By combining node, edge and label contrastive losses, we arrive at the following multi-loss contrastive learning:

$$\mathcal{L} = \mathcal{L}_{nodes} + \lambda_1 \mathcal{L}_{edges} + \lambda_2 \mathcal{L}_{labels} \tag{7}$$

where λ_1 and λ_2 are hyperparameters that control the proportion of the corresponding loss. After training, we input the node embeddings into an L2-regularized logistic regression classifier to generate fine-tuned prediction results.

5   Student Model with Reliable Distillation

5.1   Label Reliability Based on Shannon Entropy

Since the correctness of unlabeled nodes' label predictions is difficult to evaluate, Shannon entropy is used to evaluate the probability of reliable label predictions. However, there are also correct predictions for nodes with high entropy. Therefore, we can compare the prediction results of the student model and the teacher model to enhance the evaluation of label reliability. Formally, we define the label reliability of a node i prediction as follows:

$$R_i = \begin{cases} 1, & \text{if } i \in L \\ 1, & \text{if } t(i) = s(i) \text{ and } H(T_i) < H_{max} \\ \exp(-(H(T_i) + H(S_i))) & \text{if } t(i) = s(i) \text{ and } H(T_i) > H_{max} \\ 0, & \text{if } t(i) \neq s(i) \end{cases} \tag{8}$$

where $H(\cdot)$ computes the Shannon entropy of the vector, $t(\cdot)$ is the label of the teacher's prediction, $s(\cdot)$ is the label of the student's prediction, T is the node prediction vector of the teacher model, S is the prediction vector of student model and H_{max} is the max reliable threshold of entropy.

5.2 Model Training

We train the model based on labeled nodes with ground truths and unlabeled nodes with reliable labels. For each iteration, we update the reliability of the unlabeled nodes. With the improvement of the accuracy of the student model, more and more reliability nodes can be chosen to teach the student GNN. The training loss of the student GNN is defined as:

$$\mathcal{L}_{student} = \frac{1}{|L|} \sum_{l \in L} CE(y_l, \widehat{y}_l) + \frac{1}{|U|} \sum_{u \in U} R(u) CE(\widetilde{y}_u, \widehat{y}_u) \tag{9}$$

where L is the set of labeled nodes, U is the set of unlabeled nodes, y_l is the label of labeled node l, $\widetilde{y}_u$ is the label of unlabeled node u which is learned from the teacher model, $\widehat{y}$ is the prediction of the student GNN and $CE(\cdot)$ is the cross-entropy loss function.

6 Experiments

6.1 Experiment Setting

Five real-world graph datasets are used for the experiments including Cora, Citeseer, Pubmed, Amazon Computers and Amazon Photo [11]. As for baselines, we opt a series of methods including the Label Propagation(LP) [14], Chebyshev [2], GCN [4], GAT [9], DGI [10], GMI [7], MVGRL [3], GCA [16] and CG [11]. For Cora, Citeseer and Pubmed datasets, we use 20 nodes per class as the training set and 30 nodes per class as the validation set. For Amazon Computers and Amazon Photo datasets, we use 30 labeled nodes per class as the training set, 30 nodes per class as the validation set. We report the mean accuracy and the stand derivations of 20 runs. For the hyperparameters of different GNNs, we set them as suggested by their authors. For CSGNN, we set a 2-layer GCN as the teacher GNN and a 2-layer GAT as the student GNN. The hyperparameters of CSGNN are the optimal parameters selected based on experimental results.

6.2 Semi-supervised Classification

The semi-supervised node classification results are reported in Table 1. The results for five datasets exhibit similar trends: CSGNN yields predictions comparable or superior to those of the other contestants. For example, compared to GCN, CSGNN reaches nearly 3.3%, 3.5%, 1.1%, 7.1%, 3.5% gain on five datasets respectively. We also have the following observations: (1) Some unsupervised contrastive learning methods present better performance than baseline semi-supervised learning methods; (2) Two contrastive learning methods GCA and CG are strong competitors for the best performance. They perform well on some datasets, but also fail in some datasets, while CSGNN consistently performs well on all datasets. Hence, we believe that CSGNN can steadily improve GNNs' performance, even better than the state-of-the-art methods.

Table 1. Results of semi-supervised node classification (%)

	Cora	Citeseer	Pubmed	Computers	Photo
LP	68.0	45.3	63.0	70.8 ± 0.0	67.8 ± 0.0
Chebyshev	79.3 ± 1.3	67.4 ± 1.5	75.3 ± 0.5	62.6 ± 0.8	74. 3 ± 0.5
GCN	81.5 ± 0.6	70.7 ± 0.4	79.3 ± 0.2	76.3 ± 0.5	87.3 ± 1.0
GAT	83.1 ± 0.5	72.5 ± 0.7	79.5 ± 0.5	79.3 ± 1.1	86.2 ± 1.5
DGI	81.7 ± 0.6	71.5 ± 0.7	77.3 ± 0.6	75.9 ± 0.6	83.1 ± 0.5
GMI	82.7 ± 0.2	73.0 ± 0.3	80.1 ± 0.2	76.8 ± 0.1	85.1 ± 0.1
MVGRL	82.9 ± 0.7	72.6 ± 0.7	79.4 ± 0.3	79.0 ± 0.6	87.3 ± 0.3
GCA	80.9 ± 0.6	68.1 ± 2.0	80.3 ± 0.9	82.3 ± 0.3	90.4 ± 0.2
CG	83.4 ± 0.7	73.6 ± 0.8	80.2 ± 0.8	79.9 ± 0.6	89.4 ± 0.5
CSGNN	**84.8 ± 1.0**	**74.2 ± 1.2**	**80.8 ± 0.4**	**83.4 ± 1.4**	**90.8 ± 0.1**

6.3   Ablation Study

This section provides an ablation analysis to validate the contributions of different components of CSGNN on three citation datasets. For the variants, we use "T" as the teacher model, "S/R" as the student model without evaluating label reliability, "S" as the student model, "w/o KD" as the GAT without knowledge distillation, "CL-N" as CL with the node loss, "CL-N-E" as CL with node and edge losses, and "CL-ALL" as CL with all losses. The results are summarized in Table 2. It exhibits three interesting patterns: (1) The node, edge and label losses benefit the contrastive learning; (2) Without label reliability, distillation will reduce the performance of the student model, and the performance of "S/R" variants are even lower than the model without distillation; (3) Among different techniques, distillation improves performance more than contrastive learning.

Table 2. Ablation results of semi-supervised node classification(%)

	Cora			CiteSeer			PubMed		
	T	S/R	S	T	S/R	S	T	S/R	S
w/o KD	–	–	83.1	–	–	72.5	–	–	79.5
CL-N	80.9	81.3	84.0	68.1	69.6	72.8	80.3	78.2	80.4
CL-N-E	81.1	83.4	84.2	69.3	71.0	74.2	80.4	77.3	80.7
CL-ALL	81.3	83.6	84.4	69.4	70.8	74.6	80.5	79.5	80.9

7   Conclusion

In this paper, we explore contrastive learning methods for graph-based data and propose a contrastive semi-supervised learning based GNN by knowledge distillation, named CSGNN. CSGNN is able to learn from reliable unlabeled nodes when we distill the predictions of contrastive learning with multi-loss into the down-streaming student model. Extensive experiments demonstrate that CSGNN can consistently outperform the state-of-the-art models in node classification accuracy on real-world datasets.

Acknowledgements. This work is supported by the National Natural Science Fondation of China (62072083 and 61872071).

References

1. Chen, T., Kornblith, S., Swersky, K., Norouzi, M., Hinton, G.: Big self-supervised models are strong semi-supervised learners. arXiv preprint arXiv:2006.10029 (2020)
2. Defferrard, M., Bresson, X., Vandergheynst, P.: Convolutional neural networks on graphs with fast localized spectral filtering. NIPS **29**, 3844–3852 (2016)
3. Hassani, K., Khasahmadi, A.H.: Contrastive multi-view representation learning on graphs. In: ICML, pp. 4116–4126 (2020)
4. Kipf, T.N., Welling, M.: Semi-supervised classification with graph convolutional networks. arXiv preprint arXiv:1609.02907 (2016)
5. Liu, Y., Pan, S., Jin, M., Zhou, C., Xia, F., Yu, P.S.: Graph self-supervised learning: A survey. arXiv preprint arXiv:2103.00111 (2021)
6. Peng, Y., Choi, B., Xu, J.: Graph learning for combinatorial optimization: a survey of state-of-the-art. Data Sci. Eng. **6**(2), 119–141 (2021)
7. Peng, Z., et al.: Graph representation learning via graphical mutual information maximization. In: WWW, pp. 259–270 (2020)
8. Sun, K., Lin, Z., Zhu, Z.: Multi-stage self-supervised learning for graph convolutional networks on graphs with few labeled nodes. In: AAAI, vol. 34, pp. 5892–5899 (2020)
9. Veličković, P., Cucurull, G., Casanova, A., Romero, A., Lio, P., Bengio, Y.: Graph attention networks. arXiv preprint arXiv:1710.10903 (2017)
10. Velickovic, P., Fedus, W., Hamilton, W.L., Liò, P., Bengio, Y., Hjelm, R.D.: Deep graph infomax. ICLR **2**(3), 4 (2019)
11. Wan, S., Pan, S., Yang, J., Gong, C.: Contrastive and generative graph convolutional networks for graph-based semi-supervised learning. In: AAAI, vol. 35, pp. 10049–10057 (2021)
12. Wang, Y., Wang, W., Liang, Y., Cai, Y., Liu, J., Hooi, B.: Nodeaug: semi-supervised node classification with data augmentation. In: KDD, pp. 207–217 (2020)
13. You, Y., Chen, T., Sui, Y., Chen, T., Wang, Z., Shen, Y.: Graph contrastive learning with augmentations. NIPS **33**, 5812–5823 (2020)
14. Zhu, X., Ghahramani, Z., Lafferty, J.D.: Semi-supervised learning using gaussian fields and harmonic functions. In: ICML, pp. 912–919 (2003)
15. Zhu, Y., Xu, Y., Yu, F., Liu, Q., Wu, S., Wang, L.: Deep graph contrastive representation learning. arXiv preprint arXiv:2006.04131 (2020)
16. Zhu, Y., Xu, Y., Yu, F., Liu, Q., Wu, S., Wang, L.: Graph contrastive learning with adaptive augmentation. In: WWW, pp. 2069–2080 (2021)

IncreGNN: Incremental Graph Neural Network Learning by Considering Node and Parameter Importance

Di Wei, Yu Gu$^{(\boxtimes)}$, Yumeng Song, Zhen Song, Fangfang Li, and Ge Yu

School of Computer Science and Engineering, Northeastern University,
Shenyang, Liaoning, China
{guyu,lifangfang,yuge}@mail.neu.edu.cn

Abstract. Graph Neural Network (GNN) has shown powerful learning and reasoning ability. However, graphs in the real world generally exist dynamically, i.e., the topological structure of graphs is constantly evolving over time. On the one hand, the learning ability of the networks declines since the existing GNNs cannot process the graph streaming data. On the other hand, the cost of retraining GNNs from scratch becomes prohibitively high with the increasing scale of graph streaming data. Therefore, we propose an online incremental learning framework IncreGNN based on GNN in this paper, which solves the problem of high computational cost of retraining GNNs from scratch, and prevents catastrophic forgetting during incremental training. Specifically, we propose a sampling strategy based on node importance to reduce the amount of training data while preserving the historical knowledge. Then, we present a regularization strategy to avoid over-fitting caused by insufficient sampling. The experimental evaluations show the superiority of IncreGNN compared to existing GNNs in link prediction task.

Keywords: Graph neural networks · Dynamic graph · Catastrophic forgetting · Incremental learning

1 Introduction

Graphs are ubiquitous in the real world, which have been used for processing deep learning and data mining tasks. Through various analysis of graphs, we can have a deep understanding of complex social relationships and different communication modes. Since graph data is often high-dimensional and difficult to be processed by graph analysis tasks, Graph Embedding (GE) has been widely used as an effective dimensionality reduction technique [6]. As a deep learning-based GE method, GNN gradually shows its advantages in processing graph analysis tasks. The embedding obtained by GNN can not only capture the local neighborhood information, but also enables the embedding to characterize the global neighborhood through multiple iterations.

A. Bhattacharya et al. (Eds.): DASFAA 2022, LNCS 13245, pp. 739–746, 2022.
https://doi.org/10.1007/978-3-031-00123-9_59

At present, most GNNs are designed for static graph data. However, most real-world graphs exist dynamically, that is, as time goes by, new graph streaming data will continue to arrive, which is called the dynamic graph. There are some studies on dynamic graphs [5,10]. However, these dynamic graph embedding methods cannot process the incoming graph data in real time. As a result, such methods must retrain the model from scratch to obtain the embeddings of new nodes. Incremental learning provides ideas for solving this problem, which uses the latest data to update the current model and can prevent catastrophic forgetting. This technology significantly improves the efficiency of the model through reducing the number of nodes involved by retraining, and maintains the similar performance. ContinualGNN [12] is the incremental learning method based on GNN. However, it may reduce the expressive ability of the model and is not suitable for link prediction tasks.

To solve the problems of the existing incremental learning methods, we propose an online incremental learning framework IncreGNN based on GNN, which combines the experience replay-based and regularization-based methods. Motivated by experience replay, we design a sampling strategy based on node importance to sample affected and unaffected nodes separately and learn knowledge of new task on incremental data. Simultaneously, to prevent the historical knowledge from being weaken and overcome the over-fitting problem, we propose a regularization strategy by constraining the modification on the important model parameters. The main contributions of this work are as follows:

- We propose a GNN-based online incremental learning framework IncreGNN, which can efficiently generate node embedding representations in a dynamic environment.
- We design a sampling strategy based on node importance to sample the affected and unaffected nodes separately. At the same time, a regularization strategy based on the importance of model parameters is designed to constrain model parameters.
- We conduct comparative experiments on various datasets, and the results show that IncreGNN can efficiently perform incremental calculations with less accuracy loss.

2 Related Work

Incremental learning is also called continuous learning, or lifelong learning, which is first introduced in Neural Networks to solve multi-task learning problems. The current mainstream incremental learning methods can be divided into three categories including experience replay-based methods [7], regularization-based methods [1,3] and parameter isolation-based methods [4,9]. Due to the complex structure of graphs, most of the current incremental learning methods are applied to image processing and cannot directly process graph data. Continual-GNN [12] is the method to apply incremental learning to graphs. ContinualGNN proposes a novel approximate scoring algorithm to detect the emergence of new patterns. However, ContinualGNN performs sampling by calculating the probability, which will cause the loss of the local structure information of graphs. There are also incremental learning studies [2,13] that focus on improving model accuracy by retaining past knowledge rather than training efficiency, which is different from our concerns.

3   Overview of IncreGNN

Figure 1 shows the architecture of IncreGNN. A new task is defined as training a GNN on incremental data $\triangle G^t$ for the corresponding graph analysis tasks. Historical tasks are defined as training GNNs on historical incremental data $\{\triangle G^1, \triangle G^2, ..., \triangle G^t\}$ before time t to process graph analysis tasks at different time. There is usually only one graph analysis task involved for a GNN during model training. Therefore, we merge the historical incremental data $\{\triangle G^1, \triangle G^2, ..., \triangle G^t\}$ into an intergrated graph G^{t-1}, so that we only need to consider a single historical task when performing incremental learning.

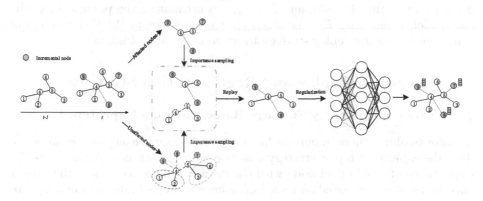

Fig. 1. Overview of IncreGNN

Our goal is to incrementally learn a GNN parameterized by θ with an optimization goal on graph G^t. Based on the above analysis, we explain IncreGNN from the perspective of probability. From this point of view, optimizing the parameters θ is tantamount to finding their most probable values given graph G^t. Therefore, according to Bayes rule, we can obtain the conditional probability by using the prior probability $p(\theta)$ of the parameter and the probability $p(G^t|\theta)$ of graph G^t:

$$log p(\theta|G^t) = log p(G^t|\theta) + log p(\theta) - log p(G^t) \tag{1}$$

Theorem 1. Graph G^t at time t is composed of historical graph G^{t-1} at time $t-1$ and incremental data $\triangle G^t$ at time t. The conditional probability $log p(\theta|G^t)$ can be calculated using historical graph G^{t-1} and incremental data $\triangle G^t$ as conditions:

$$log p(\theta|G^{t-1}, \triangle G^t) = log p(\triangle G^t|\theta) + log p(\theta|G^{t-1}) - log p(\triangle G^t) \tag{2}$$

where the first term $logp(\triangle G^t|\theta)$ is obviously the log-likelihood value of incremental data $\triangle G^t$, that is, the loss function of the learning task at time t is negative $-L_{\triangle G^t}(\theta)$, the second term $logp(\theta|G^{t-1})$ is a posterior distribution of all information learned on the data G^{t-1} related to the task at time $t-1$, and the third term $logp(\triangle G^t)$ is a constant value.

Therefore, according to (2), the total loss function of the IncreGNN at time t can be expressed as:

$$L_{IncreGNN} = L_\triangle + L_{pre} = L_\triangle + L_{olddata} + L_{param} = L_{data} + L_{param} \quad (3)$$

where $L_\triangle$ represents the loss function of the incremental change of the graph at time t, and L_{pre} represents the review of the old knowledge learned on the graph data before time t, so as to avoid catastrophic forgetting. The IncreGNN combines two strategies of experience replay and regularization, where the experience replay strategy is a combination of new and old data. Therefore, L_{pre} can be further decomposed into two parts: $L_{olddata}$, a review of knowledge learned from a part of the old data, and L_{param}, a constraint on the parameters of the old model. Combining $L_\triangle$ and $L_{olddata}$, L_{data} represents the loss function of using the experience replay strategy to process new and old data.

4   Experience Replay and Regularization Strategy

4.1   Experience Replay Strategy Based on Node Importance

In order to obtain more important historical nodes as much as possible, we introduce the experience replay strategy based on node importance. We associate the importance of a historical node with the degree of influence. Then, the experience replay strategy based on node importance is divided into the following two steps:

Step 1: Sampling the important neighbor nodes of the affected nodes in order. The set of incremental data is defined as base change group, where the incremental data includes incremental nodes and edges. Then, the neighbor nodes of the nodes in base change group are regarded as first-order change group, and the neighbor nodes of the nodes in first-order change group are regarded as second-order change group, and so on. It is not difficult to find that the lower the order of the node, the greater the contribution to the incremental node, that is, the greater the impact of the incremental node. So it is necessary to sample as many nodes as possible in low-order change group. Here, we gradually reduce the number of samples as the order increases, and assign the number of node samples at the ratio of $\frac{1}{i}/(1+\frac{1}{2}+...+\frac{1}{K})$ for each order, where K represents a total of K-order change groups of samples, and i represents the i-th change group currently sampled. Node importance of v is measured by the personalized PageRank value between the node and its neighbors in G^{t-1} and denoted as π_v^T.

By calculating node importance, we can get the nodes with high degree of correlation with the incremental nodes according to the number of sampling nodes in each order n_k, which are more important for the knowledge of the old task:

$$I(G^t) = \overset{K}{\underset{k=1}{\cup}} \{v_i | \pi_{v_i}^T > \pi_{v_j}^T, i \in [1, n_k], j \notin [1, n_k]\} \tag{4}$$

Step 2: Performing importance sampling on unaffected nodes. To prevent over-fitting and catastrophic forgetting, it is also necessary to perform partial sampling on unaffected nodes. Since random sampling may sample nodes of the same category or the embeddings are too similar, the incremental learning will be skewed and part of the old knowledge will be covered. Therefore, the unaffected nodes are divided using K-means according to the label or the learned embeddings for the unlabeled nodes in G^{t-1}, and K clusters are obtained. Within each cluster, the degree of the node is used as the criterion to measure node importance. The greater the degree of the node, the higher the node importance within the cluster. Finally, we uniformly sample the same number of k height nodes from each cluster by degree:

$$UI(G^t) = \overset{K}{\underset{k=1}{\cup}} \{v_i | deg_{v_i}^T > deg_{v_j}^T, i \in [1, n_k], j \notin [1, n_k]\} \tag{5}$$

Through node importance sampling strategies, the total number of sampled node sets is M, including the sum of the affected node set $I(G^t)$ and the unaffected node set $UI(G^t)$ sampled in K clusters. These M nodes will be used as the training data for incremental learning of GNN at time t. From the perspective of experience replay, the optimization goal for both retaining historical knowledge and learning new task knowledge is:

$$L_{data} = \sum_{v_i \in I(G^t) \cup UI(G^t)} l(\theta; v_i) \tag{6}$$

4.2   Regularization Strategy Based on Parameter Importance

The purpose of the regularization strategy is to avoid updating the parameters drastically related to the old knowledge. Specifically, a corresponding importance coefficient δ_{ij} is calculated for each parameter θ_{ij}, and the update extent of the parameter is restricted by this importance coefficient. For the parameter θ_{ij} with a large δ_{ij}, the magnitude of its change should be minimized during the gradient descent process, because a large importance coefficient δ_{ij} can indicate that this parameter θ_{ij} is more important to the old knowledge of model learning. Thus this parameter needs to be retained to avoid catastrophic forgetting.

Following Memory Aware Synapses (MAS) [1], for all nodes in the training set, we accumulate the mean value of the gradient calculated on the feature vectors of all nodes to obtain the importance weight δ_{ij} of the parameter θ_{ij}:

$$\delta_{ij} = \frac{1}{N} \sum_{v=1}^{N} ||g_{ij}(x_v)|| \tag{7}$$

where $g_{ij}(x_v) = \frac{\partial(F'(x_v;\theta))}{\partial\theta_{ij}}$ is the partial derivative of the function F' to the parameter θ_{ij}, F' is an approximate function mapping to the real function F, x_v is the feature of each node v and N is the number of nodes in the training set. However, calculating δ requires traversing the entire graph snapshot at the previous time, which will incur a lot of cost to storage and calculation. Therefore, we use the important nodes sampled by experience replay strategy to estimate the importance weight.

Moreover, in order to consolidate more historical knowledge of important parameters, it is necessary to accumulate the parameter importance weight δ calculated by each task at all previous times. Finally, the importance parameter δ_{ij} corresponding to the parameter θ_{ij} at time t is:

$$\delta_{ij} = \frac{1}{tN} \sum_{T=0}^{t-1} \sum_{v=1}^{N} \|g_{ij}(x_v)\| \tag{8}$$

Through approximate calculation to obtain the estimated parameter importance weight, the optimization goal of consolidating historical knowledge from the perspective of constrained model parameters can be obtained:

$$L_{param} = \frac{\lambda}{2} \sum_{i,j} \delta_{ij} (\theta_{ij} - \theta_{ij}^{t-1})^2 \tag{9}$$

where λ is the degree of importance of historical knowledge to new graph.

5 Experiments

5.1 Experiment Setup

We conduct experiments on Enron[1], UCI[2], BC-Alpha[3] and ML-10M[4], which are all divide into 13 graph snapshots following [8]. For link prediction task, 40%, 20% and 40% of the edges are taken as training set, validation set and test set, respectively. We compare the proposed method with five baselines, including EvolveGCN [5], Retrained GAT [11], Pretrained GAT [11], Online GAT [11] and ContinualGNN [12]. Retrained GAT, Pretrained GAT and Online GAT are GATs for retraining, pre-training and dynamic graph online training respectively. The experiment uses all the data at $t = 0$ to train a general model, and then incrementally learns the sampled data at $t = 1...T$. The regularization term λ is (80, 800, 80, 320).

---

[1] https://www.cs.cmu.edu/~./enron/.
[2] http://networkrepository.com/opsahl_ucsocial.php.
[3] http://www.btc-alpha.com.
[4] http://networkrepository.com/ia-movielens-user2tags-10m.php.

Table 1. Performance of link prediction.

Method	Enron			UCI			BC-Alpha			ML-10M		
	AUC	AP	Time(s)	AUC	AP	Time(s)	AUC	AP	Time(s)	AUC	AP	Time(s)
EvolveGCN	65.4	66.0	0.691	77.1	78.9	1.744	79.3	81.0	1.30	85.0	85.6	5.174
RetrainedGAT	**68.9**	**71.5**	0.057	**87.5**	**87.8**	0.677	**91.7**	**91.9**	0.422	**93.1**	**93.9**	3.329
PretrainedGAT	62.5	64.3	0.000	81.9	78.8	0.000	42.0	57.0	0.000	87.6	84.9	0.000
OnlineGAT	44.6	52.2	0.033	58.1	56.2	0.115	64.9	63.0	0.064	61.9	57.3	0.249
ContinualGNN	57.1	64.3	0.006	49.7	52.4	0.061	49.3	53.5	0.147	47.9	49.5	7.469
IncreGNN	**67.3**	**71.2**	0.019	**82.6**	**84.7**	0.143	**90.6**	**90.8**	0.045	89.9	**91.4**	0.070

5.2 Experimental Results

Table 1 shows the average AUC, AP values and training time of all methods. Note that the training time does not include the time to calculate personalized PageRank and K-means, which can be calculated offline before the next snapshot coming. The importance parameter δ_{ij} is updated online. Experimental results show that our proposed algorithm IncreGNN achieves the best performance on both AUC and AP compared to other comparison methods in addition to Retrained GAT. At the same time, IncreGNN can reach experimental results that are similar to the theoretically most accurate Retrained GAT, which shows that the method we propose has high superiority in incremental learning.

Figure 2 shows the performance results of each method across multiple time steps. On the UCI, BCAlpha and ML-10M datasets, our method has reached a very high level and are very stable, indicating that the method we proposed can effectively preserve old knowledge and learn new knowledge. The results of Online GAT on the 4 datasets are very poor and fluctuate greatly. The reason is that Online GAT does not preserve old knowledge, which leads to catastrophic forgetting.

IncreGNN also performs well on the node classification task, but the experiments are omitted due to limited space.

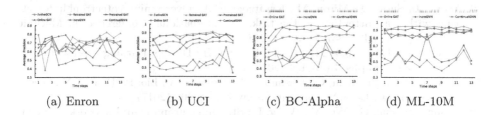

(a) Enron   (b) UCI   (c) BC-Alpha   (d) ML-10M

Fig. 2. Link prediction across multiple time steps

6 Conclusion

In this work, we propose a GNN-based incremental learning framework to process dynamic graphs. We design a sampling strategy based on node importance to sample the affected and unaffected nodes. At the same time, we design a regularization strategy based on the importance of model parameters to constrain the important parameters. Through these two strategies, the problem of catastrophic forgetting in incremental learning can be avoided, and the knowledge of the new task can be learned while preserving the knowledge of the old task. Experimental results verify the effectiveness and efficiency of IncreGNN.

Acknowledgements. This work is supported by the National Natural Science Fondation of China (62072083 and U1811261).

References

1. Aljundi, R., Babiloni, F., Elhoseiny, M., Rohrbach, M., Tuytelaars, T.: Memory aware synapses: Learning what (not) to forget. In: ECCV, pp. 139–154 (2018)
2. Galke, L., Franke, B., Zielke, T., Scherp, A.: Lifelong learning of graph neural networks for open-world node classification. In: IJCNN, pp. 1–8 (2021)
3. Li, Z., Hoiem, D.: Learning without forgetting. IEEE Trans. Pattern Analysis and Machine Intelligence **40**(12), 2935–2947 (2017)
4. Mallya, A., Lazebnik, S.: PackNet: adding multiple tasks to a single network by iterative pruning. In: CVPR, pp. 7765–7773 (2018)
5. Pareja, A., et al.: EvolveGCN: evolving graph convolutional networks for dynamic graphs. In: AAAI, vol. 34, pp. 5363–5370 (2020)
6. Peng, Y., Choi, B., Xu, J.: Graph learning for combinatorial optimization: a survey of state-of-the-art. Data Sci. Eng. **6**(2), 119–141 (2021)
7. Rebuffi, S.A., Kolesnikov, A., Sperl, G., Lampert, C.H.: iCaRL: incremental classifier and representation learning. In: CVPR, pp. 2001–2010 (2017)
8. Sankar, A., Wu, Y., Gou, L., Zhang, W., Yang, H.: DySAT: deep neural representation learning on dynamic graphs via self-attention networks. In: WSDM, pp. 519–527 (2020)
9. Serra, J., Suris, D., Miron, M., Karatzoglou, A.: Overcoming catastrophic forgetting with hard attention to the task. In: International Conference on Machine Learning, pp. 4548–4557. PMLR (2018)
10. Trivedi, R., Farajtabar, M., Biswal, P., Zha, H.: DyRep: learning representations over dynamic graphs. In: ICLR (2019)
11. Veličković, P., Cucurull, G., Casanova, A., Romero, A., Lio, P., Bengio, Y.: Graph attention networks. arXiv preprint arXiv:1710.10903 (2017)
12. Wang, J., Song, G., Wu, Y., Wang, L.: Streaming graph neural networks via continual learning. In: CIKM, pp. 1515–1524 (2020)
13. Xu, Y., Zhang, Y., Guo, W., Guo, H., Tang, R., Coates, M.: GraphSAIL: graph structure aware incremental learning for recommender systems. In: CIKM, pp. 2861–2868 (2020)

Representation Learning in Heterogeneous Information Networks Based on Hyper Adjacency Matrix

Bin Yang and Yitong Wang[✉]

School of Software, Fudan University, Shanghai, China
{yangb20,yitongw}@fudan.edu.cn

Abstract. Heterogeneous information networks(HINs), which usually contain different kinds of nodes and interactions are very common in real world. The richer semantic information and complex relationships have posed great challenges to current representation learning in HINs. Most existing approaches which use predefined meta-paths suffer from high cost and low coverage. In addition, most of the existing methods cannot capture and learn influential high-order neighbors precisely and effectively. In this paper, we attempt to tackle the problem of meta-paths and influential high-order neighbors by proposing an original method HIN-HAM. HIN-HAM captures influential neighbors of target nodes precisely and effectively by generating the hyper adjacency matrix of the HIN. Then it uses convolutional neural networks with weighted multi-channel mechanism to aggregate different types of neighbors under different relationships. We conduct extensive experiments and comparisons on three real datasets and experimental results show the proposed HIN-HAM outperforms the state-of-the-art methods.

Keywords: Heterogeneous information networks · Node embedding · Graph convolutional network

1 Introduction

Heterogeneous information networks(HINs) [10] are ubiquitous in human society such as social networks [2], citation networks [7] and recommendation systems [1]. Different types of nodes and relationships in HINs which contain rich information and complex interactions, has posed great challenges to current research in HINs, in particular the representation learning in HINs.

In the past few years, there have been a series of studies on representation learning in HINs and have achieved quite good results. One of the classic paradigms is to design and use meta-paths, such as Metapath2vec [4] and HIN2vec [5]. In particular, meta-paths are predefined sequence patterns of nodes

This work is supported by the National Key R&D Program of China (No. 2020YFC2008401).

with specific node and relationship types. Recently, some GNNs-based models [13] have been proposed for HINs representation learning, such as HAN [12], HGCN [17] and GTN [14].

However, most existing methods suffer from some limitations: Firstly, meta-paths are designed manually with high cost and low coverage. Secondly, high-order neighbors have not been fully investigated. Finally, different types of nodes/edges should be treated more precisely and effectively rather than cutting connections among them. In view of these limitations, we attempt to tackle the two key challenges by answering the following two questions: 1) how to deal with different types of neighbor nodes/edges to capture influential neighbors (including high-order neighbors) on the target node; 2) how to aggregate influential neighbor nodes to update target node embedding.

We propose an original embedding method for representation learning in Heterogeneous Information Networks based on Hyper Adjacency Matrix (HIN-HAM) to solve the above challenges. Our main contributions are:

1) We introduce a 'hyper adjacency matrix' to precisely capture influential neighbors with different distances from a given target node;
2) We propose a weighted multi-channel mechanism to effectively aggregate information of influential neighbors into the target node;
3) We conduct extensive experiments on three benchmark datasets. Our experimental results show the superiority of HIN-HAM, comparing with the state-of-the-art methods.

2   Related Work

Over the past few years, representation learning [3] has made significant progress. Metapath2vec [4] obtains a series of node sequences through a random walk on a given meta-path. Metagraph2vec [16] extends Metapath2vec using meta-structure-based random walk for sampling. HIN2vec [5] uses shallow neural networks to simultaneously learn the embedding of nodes and relationships in the network. RGCN [9] designs multiple graph convolution layers according to different types of relationships. HAN [12] uses node-level attention to aggregate neighbor information and semantic-level attention to aggregate predefined meta-path information. HetGNN [15] uses restart random walk strategy to sample strongly correlated neighbors and LSTM model to calculate node embeddings of the target node and its neighbors respectively. HGCN [17] uses the GCN [7] model based on HINs to solve collective classification. GTN [14] uses the GNN to learn meta-paths in the graph by identifying multi-hop connections. HGT [6] propose an information aggregation method based on meta-relational learning and heterogeneous self-attention mechanism. While these methods work well in experiments, they still suffer from some limitations: high cost and low coverage of manually-designed meta-paths as well as insufficient to identify influential neighbors.

3 Preliminaries

Definition 1 (Heterogeneous Graph). *A heterogeneous graph can be represented as $\mathcal{G} = (\mathcal{V}, \mathcal{E}, \mathcal{A}, \mathcal{R})$, consists of a node set $\mathcal{V}$ and a edge set $\mathcal{E}$. And it also includes a node type mapping function $\phi : \mathcal{V} \to \mathcal{A}$ and an edge type mapping function $\psi : \mathcal{E} \to \mathcal{R}$, where $\mathcal{A}$ is a set of node types, $\mathcal{R}$ is a set of edge types and $|\mathcal{A}| + |\mathcal{R}| > 2$. A heterogeneous graph can be modeled as a set of adjacency matrices $\{A_k\}_{k=1}^{K}$, where $A_k \in R^{N \times N}$ is a subgraph that contains only type-k edges, $K = |\mathcal{R}|$ and $N = |\mathcal{V}|$. For a heterogeneous graph, it also has a feature matrix $X \in R^{N \times D}$ where D is the feature dimension of each node.*

Definition 2 (Meta-path). *The meta-path is defined as a path $\mathcal{P}$ consisting of node types on the heterogeneous information network mode $T_{\mathcal{G}} = (\mathcal{A}, \mathcal{R})$: $a_1 \xrightarrow{r_1} a_2 \xrightarrow{r_2} \dots \xrightarrow{r_{l-1}} a_l$, where $a_i \in \mathcal{A}$, $r_i \in \mathcal{R}$. The meta-path represents the compound relation $R = r_1 \circ r_2 \dots \circ r_l$ between node a_1 and node a_l, where $\circ$ denotes the composition operator on relations.*

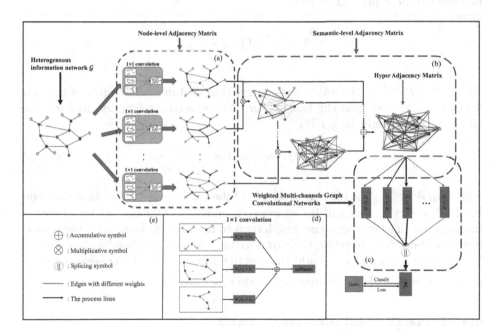

Fig. 1. The overall framework of HIN-HAM. (a) Node-level adjacency matrix. (b) Semantic-level adjacency matrix. (c) Weighted multi-channel graph convolutional networks. (d) The 1×1 convolution in (a). (e) The notations in the framework. The thickness of the edges and arrows reflects their weights.

4   Method

4.1   Overall Framework

Figure 1 shows the architecture of HIN-HAM. The model can be divided into three steps: i) A 1×1 graph convolution layer is used to learn the weights of different types of edges to obtain node-level adjacency matrix; ii) Semantic-level meta-path adjacency matrix is obtained by matrix multiplication and superimposed to obtain hyper adjacency matrix; iii) The weighted multi-channel graph convolutional networks are used to learn node embedding and optimize it.

4.2   Node-Level Adjacency Matrix

In order to express the importance of different types of neighbors under different connections, we introduce node-level adjacency matrix to learn the importance of each type of edge in the heterogeneous graph. We design a specific 1×1 convolution layer to learn the weight of different types of edge. The convolution process can be expressed as follows:

$$\tilde{A} = \sum_{k=1}^{K} (W_k A_k + b_k), \tag{1}$$

where $\tilde{A} \in R^{N \times N}$ is the adjacency matrix that contains the weights of different types of edges and b_k is the bias vector. We use softmax function to normalize the initialized weight to get W_k:

$$W_k = softmax\,(a_k) = \frac{a_k}{\sum_{k=1}^{K} a_k}. \tag{2}$$

where $a_k \in R^{1 \times 1}$ is the initialized weight of the k_{th} subgraph. Note that the node-level adjacency matrix will be asymmetric since the influences of two nodes on one another in a heterogeneous graph tend to be different. Generally, the embeddings of nodes need to retain their own features to prevent complete assimilation by neighbors. Therefore, we add the identity matrix to the heterogeneous graph $\mathcal{G}$, i.e., $A_0 = I$, and $\tilde{A}$ contains each node's own weight.

4.3   Semantic-Level Adjacency Matrix

We propose a novel semantic-level adjacency matrix to learn the importances of different meta-paths. Given the node-level adjacency matrices, we can calculate the adjacency matrix of meta-paths in length l by matrix multiplication.

$$A^{(l)} = \prod_{i=1}^{l} \tilde{A}_i = \prod_{i=1}^{l} \left(\sum_{k=0}^{K} (W_{ik} A_{ik} + b_{ik}) \right), \tag{3}$$

where $A^{(l)} \in R^{N \times N}$ is the meta-path adjacency matrix of the specified length l, $\tilde{A}_i \in R^{N \times N}$ is the node-level adjacency matrix of i_{th} layer, $A_{ik} \in R^{N \times N}$ is the k_{th} subgraph of the heterogeneous graph at i_{th} layer and b_{ik} is the bias vector. W_{ik} is normalized for parameter stability:

$$W_{ik} = softmax\,(a_{ik}) = \frac{a_{ik}}{\sum_{k=0}^{K} a_{ik}}. \tag{4}$$

where $a_{ik} \in R^{1 \times 1}$ is the initialized weight. The meta-path adjacency matrix contains all meta-paths of specified length, and different meta-paths have different weights. Given length L, we sum up the meta-path adjacency matrices with length 1 to L as in Formula (5). The semantic-level adjacency matrix containing all meta-paths no more than length L with different weights can be obtained. We call it the hyper adjacency matrix:

$$G = \sum_{l=1}^{L} \left(A^{(l)} \right). \tag{5}$$

where $G \in R^{N \times N}$. The hyper adjacency matrix can learn all length of meta-paths with length up to L, and assigns different weights to reflect their different importances. So we can capture the influential neighbors with different length more precisely.

4.4   Weighted Multi-channel Graph Convolutional Networks

We applied graph convolutional neural networks to the hyper adjacency matrix and obtained node embedding vector:

$$H = \sigma \left(D^{-1} G X W \right), \tag{6}$$

where D is the degree matrix of hyper adjacency matrix G, $X \in R^{N \times D}$ is the feature matrix and $W \in R^{D \times D}$ is the learnable weight matrix. We try GCN on hyper adjacency matrix G several times and set each output as a channel. Weighted multi-channel mechanism is proposed to aggregate the influential neighbors more effectively and learnt weight coefficients for each channel. We apply GCN to each channel and concatenated multiple node representations as:

$$Z = ||_{i=1}^{C} \left(W_i H_i \right), \tag{7}$$

where H_i is the node embedding vector in the i_{th} channel, W_i is the weight coefficient of channel i, and Z is the final node embedding for node classification. Our loss function is focused on node classification and is defined as a standard cross-entropy on the nodes that have ground truth labels:

$$L = - \sum_{l \in \mathcal{Y}_L} Y^l \ln \left(\Theta \bullet Z^l \right). \tag{8}$$

where Θ is the parameter of the classifier, $\mathcal{Y}_L$ is the set of node indices that have labels, Y^l and Z^l are the labels and embeddings of labeled nodes.

5   Experiment

5.1   Datasets

We conduct extensive experiments on three benchmark datasets: citation networks DBLP and ACM, and a movie dataset IMDB. The details of datasets are described in Table 1.

Table 1. The statistics of the three datasets.

Dataset	Nodes	Edges	Edge types	Features	Training	Validation	Test
DBLP	18405	67946	4	334	800	400	2857
ACM	8994	25922	4	1902	600	300	2125
IMDB	12772	37288	4	1256	300	300	2339

5.2   Baselines

We verify the effectiveness of the proposed HIN-HAM by comparing it with some state-of-the-art baselines, including random walk-based methods and GNNs-based methods. The random walk-based methods includes DeepWalk [8] and Metapath2vec [4]. For the GNNs-based methods, we choose GCN [7] and GAT [11] designed for homogeneous graphs and HAN [12] and GTN [14] designed for heterogeneous graphs.

5.3   Node Classification

We train a shallow neural network as a classifier. Table 2 shows the performance of HIN-HAM compared to other node classification baselines, and HIN-HAM achieved the best performance on all three datasets. GAT performs better than HAN on both DBLP and ACM datasets probably because HAN's use of manually specified meta-paths. GTN performs only second to the proposed HIN-HAM, which illustrates the advantage of learning new graph structures. However, GTN simply use matrix multiplication to define the weights of neighbors of different layers, so the importance of each meta-path cannot be accurately obtained. The proposed HIN-HAM can independently learn the weights of different types and lengths of meta-paths. The design of hyper adjacency matrix could relieve the dependencies of high-order neighbors on lower-order neighbors and learn more rational weights for different neighbors. At the same time, weighted multichannel mechanism can properly balance the results to improve the embedding. Our experimental results prove the effectiveness of proposed HIN-HAM.

5.4   Ablation Study

In this section, we evaluate three different variants of HIN-HAM:

Table 2. The comparison results for node classification on three datasets.

Datasets metrics (%)	DBLP		ACM		IMDB	
	Macro-F1	Micro-F1	Macro-F1	Micro-F1	Macro-F1	Micro-F1
Deep walk	85.89	86.50	70.92	77.69	50.35	54.33
Metapath2vec	90.98	93.01	69.10	75.10	45.15	48.81
GCN	91.29	91.87	79.32	79.15	51.81	54.61
GAT	93.75	94.35	92.23	92.33	52.99	56.89
HAN	92.88	92.99	91.55	91.64	56.77	58.51
GTN	93.68	94.15	92.65	92.52	57.39	58.70
Ours	**94.56**	**95.17**	**93.13**	**93.05**	**59.28**	**60.75**

- HIN-HAM$_{-identity}$. It doesn't add the identity matrix to the heterogeneous graph $\mathcal{G}$.
- HIN-HAM$_{-hyper}$. It doesn't use the hyper adjacency matrix G but only uses a single meta-path adjacency matrix.
- HIN-HAM$_{-weight}$. It doesn't use the weighted multi-channel mechanism but only uses multi-channel splicing.

Comparisons with three variants on DBLP datasets demonstrate that the complete HIN-HAM works the best as shown in Table 3. It can also seen that the techniques proposed in HIN-HAM: hyper adjacency matrix, use of identity matrix, weighted multi-channel mechanism are all useful and contributed to the improvement of HIN-HAM from different aspects. In particular, the performance of HIN-HAM$_{-hyper}$ decreases the most, which further prove the effectiveness of hyper adjacency matrix.

Table 3. Ablation study

Method	Metrics(%)	
	Macro-F1	Micro-F1
HIN − HAM$_{-identity}$	94.17	94.78
HIN − HAM$_{-hyper}$	92.91	93.76
HIN − HAM$_{-weight}$	94.14	94.85
HIN-HAM	**94.56**	**95.17**

6 Conclusion

In this paper, we attempt to solve two basic problems of representation learning in heterogeneous information networks and propose an original node embedding method for representation learning in heterogeneous information networks based on hyper adjacency matrix. The proposed model uses the hyper adjacency matrix to capture the influential neighbors of the target node with different distances and aggregates the neighbor effectively using the convolutional neural network with weighted multi-channel mechanism. HIN-HAM demonstrate its effectiveness by performing classification tasks on three real datasets and outperforming all existing benchmarks. The proposed HIN-HAM is proved to be well interpretable by ablation experiments.

References

1. Berg, R.V.D., Kipf, T.N., Welling, M.: Graph convolutional matrix completion. arXiv preprint arXiv:1706.02263 (2017)
2. Chen, J., Ma, T., Xiao, C.: FastGCN: fast learning with graph convolutional networks via importance sampling. arXiv preprint arXiv:1801.10247 (2018)
3. Cui, P., Wang, X., Pei, J., Zhu, W.: A survey on network embedding. IEEE Trans. Knowl. Data Eng. **31**(5), 833–852 (2018)
4. Dong, Y., Chawla, N.V., Swami, A.: metapath2vec: scalable representation learning for heterogeneous networks. In: The 23rd ACM SIGKDD International Conference (2017)
5. Fu, T.Y., Lee, W.C., Lei, Z.: Hin2vec: explore meta-paths in heterogeneous information networks for representation learning. In: The 2017 ACM (2017)
6. Hu, Z., Dong, Y., Wang, K., Sun, Y.: Heterogeneous graph transformer. In: Proceedings of The Web Conference 2020, pp. 2704–2710 (2020)
7. Kipf, T.N., Welling, M.: Semi-supervised classification with graph convolutional networks. arXiv preprint arXiv:1609.02907 (2016)
8. Perozzi, B., Al-Rfou, R., Skiena, S.: DeepWalk: online learning of social representations. In: Proceedings of the 20th ACM SIGKDD International Conference on Knowledge Discovery and Data Mining, pp. 701–710 (2014)
9. Schlichtkrull, M., Kipf, T.N., Bloem, P., Van Den Berg, R., Titov, I., Welling, M.: Modeling relational data with graph convolutional networks. In: European Semantic Web Conference, pp. 593–607. Springer (2018). https://doi.org/10.1007/978-3-319-93417-4_38
10. Shi, C., Li, Y., Zhang, J., Sun, Y., Philip, S.Y.: A survey of heterogeneous information network analysis. IEEE Trans. Knowl. Data Eng. **29**(1), 17–37 (2016)
11. Veličković, P., Cucurull, G., Casanova, A., Romero, A., Lio, P., Bengio, Y.: Graph attention networks. arXiv preprint arXiv:1710.10903 (2017)
12. Wang, X., et al.: Heterogeneous graph attention network. In: The World Wide Web Conference, pp. 2022–2032 (2019)
13. Yang, C., Xiao, Y., Zhang, Y., Sun, Y., Han, J.: Heterogeneous network representation learning: survey, benchmark, evaluation, and beyond. arXiv e-prints pp. arXiv-2004 (2020)
14. Yun, S., Jeong, M., Kim, R., Kang, J., Kim, H.J.: Graph transformer networks. Adv. Neural. Inf. Process. Syst. **32**, 11983–11993 (2019)

15. Zhang, C., Song, D., Huang, C., Swami, A., Chawla, N.V.: Heterogeneous graph neural network. In: Proceedings of the 25th ACM SIGKDD International Conference on Knowledge Discovery & Data Mining, pp. 793–803 (2019)
16. Zhang, D., Yin, J., Zhu, X., Zhang, C.: MetaGraph2Vec: complex semantic path augmented heterogeneous network embedding. In: Phung, D., Tseng, V.S., Webb, G.I., Ho, B., Ganji, M., Rashidi, L. (eds.) PAKDD 2018. LNCS (LNAI), vol. 10938, pp. 196–208. Springer, Cham (2018). https://doi.org/10.1007/978-3-319-93037-4_16
17. Zhu, Z., Fan, X., Chu, X., Bi, J.: HGCN: a heterogeneous graph convolutional network-based deep learning model toward collective classification. In: Proceedings of the 26th ACM SIGKDD International Conference on Knowledge Discovery and Data Mining, pp. 1161–1171 (2020)

Author Index

Agrawal, Puneet   III-413
Ai, Zhengyang   II-306
Amagata, Daichi   I-224
Ao, Xiang   I-353, I-387, II-166
Appajigowda, Chinmayi   III-527
Araújo, André   III-500
Au, Man Ho   I-404

Bai, Chaoyu   III-272
Bai, Luyi   II-391
Bai, Ting   II-102, II-423
Ban, Qimin   II-85
Bao, Qiaoben   III-238
Bao, Xuguang   III-514, III-518
Bao, Yinan   I-615
Bao, Yuhao   III-509
Bi, Jingping   I-722
Bi, Sheng   I-162
Bi, Wenyuan   I-96
Bian, Shuqing   I-38
Blackley, Suzanne V.   II-673

Cai, Desheng   II-574
Cai, Haoran   III-430
Cai, Xunliang   II-298
Cao, Caleb Chen   I-648
Cao, Shulin   I-107
Cao, Yiming   II-407, III-117
Cao, Zhi   II-489
Carvalho, Arthur   III-500
Chang, Liang   II-248, II-281, III-281,
    III-514, III-518
Chao, Pingfu   I-191
Chatterjee, Ankush   III-413
Chelaramani, Sahil   III-413
Chen, Cen   III-306, III-455
Chen, Guihai   I-552, II-3, II-615, II-706
Chen, Jiajun   II-216
Chen, Jiangjie   III-197
Chen, Lei   I-648
Chen, Lu   III-495
Chen, Qi   III-331
Chen, Siyuan   II-264
Chen, Tongbing   II-590

Chen, Weitong   II-289
Chen, Xiang   II-556
Chen, Xin   II-166, III-377, III-430
Chen, Xingshu   III-133
Chen, Yan   II-681
Chen, Yijiang   II-375
Chen, Yueguo   III-331
Chen, Yunwen   III-36, III-197, III-238,
    III-340
Chen, Yuting   I-341
Chen, Yuxing   I-21
Chen, Zhigang   I-137, III-149
Chen, Zihao   I-309
Chen, Zongyi   III-230
Cheng, Bing   II-298
Cheng, Dawei   III-306, III-455
Cheng, Reynold   III-443
Cheng, Yunlong   II-706
Chennupati, Saideep   I-569
Chhabra, Vipul   I-569
Chi, Jianfeng   I-353, I-387
Ching, Waiki   III-443
Chu, Xiaokai   I-722
Chu, Yuqi   II-574
Couto, Henrique   III-500
Cui, Chuan   II-150
Cui, Hang   III-52
Cui, Lizhen   II-315, II-407, III-117

Damani, Sonam   III-413
Dao, Minh-Son   I-569
Das, Souripriya   I-21
Deng, Sinuo   III-222
Deng, Sucheng   II-556
Ding, Tianyu   III-401
Ding, Yihua   II-590
Dong, Linfeng   I-387
Dong, Qiwen   II-689
Dong, Xiangjun   I-459
Dou, Wenzhou   III-52
Draheim, Dirk   I-596
Drancé, Martin   III-539
Du, Liang   II-681
Du, Ming   III-289

Du, Wei   I-370
Du, Yingpeng   II-19
Duan, Jihang   III-522
Duan, Lei   II-681, III-165
Duan, Zhewen   II-656
Duan, Zhijian   III-389

Fan, Jiangke   II-298
Fan, Ju   I-587
Fan, Wei   I-604
Fan, Xinxin   I-722
Fan, Yu   II-472
Fan, Zhenfeng   II-332
Fan, Zhuoya   I-404
Fang, Chuangxin   III-505
Fang, Junhua   I-191, I-207
Fang, Ruiyu   II-199, III-351
Fang, Shineng   III-197
Feng, Jinghua   I-353, I-387
Feng, Luping   II-118
Feng, Shi   II-256, III-255
Folha, Rodrigo   III-500
Fu, Bin   II-697

Gao, Hanning   II-150
Gao, Peng   III-468
Gao, Shan   II-298
Gao, Xiaofeng   I-552, II-3, II-615, II-706
Gao, Yixu   III-389
Gao, Yuanning   II-615
Gao, Yunjun   III-495
Gao, Zihao   II-281
Goda, Kazuo   I-88
Gong, Zheng   III-213
Gong, Zhiguo   II-556
Grubenmann, Tobias   III-443
Gu, Hansu   II-359
Gu, Ning   I-333, II-359
Gu, Tianlong   III-514, III-518
Gu, Yu   I-325, I-731, I-739
Gudmundsson, Joachim   I-241
Gui, Min   III-297
Gui, Xiangyu   I-122
Guo, Deke   I-441
Guo, Jiayan   I-682
Guo, Shu   II-306
Guo, Tonglei   III-364
Guo, Zhiqiang   II-183
Gupta, Manish   III-413, III-532

Han, Baokun   I-309
Han, Ding   II-523
Han, Donghong   III-255
Han, Tianshuo   III-401
Hao, Fei   I-714
Hao, Jianye   III-389
Hao, Zhifeng   II-556
Hara, Takahiro   I-224
He, Liang   II-85, II-118
He, Ming   III-377, III-401
He, Qing   I-387, II-166
He, Xiaodong   III-425
He, Yi   II-323
He, Yihong   II-455
He, Ying   II-574
He, Zhenying   I-72, I-96, I-476
Ho, Shen-Shyang   I-509
Holub, Jan   I-509
Hong, Yu   III-340
Hou U, Leong   III-68
Hou, Lei   I-107
Hou, Yupeng   I-38
Hu, Huiqi   I-293
Hu, Jun   II-574
Hu, Maodi   II-199, III-351
Hu, Nan   I-162
Hu, Songlin   I-615, II-623
Hu, Wenjin   III-222
Hu, Wenxin   II-85
Hu, Xinlei   III-377
Hu, Xuegang   II-574
Hua, Yuncheng   I-162
Huang, Chen   III-238
Huang, Faliang   II-343
Huang, Hao   III-468
Huang, Junyang   III-238
Huang, Linpeng   I-341
Huang, Qiang   I-232, I-268
Huang, Wei   III-213
Huang, Weiming   II-407
Huang, Xiuqi   II-706
Huang, Yanlong   II-102
Huang, Yanyong   II-656

Ji, Yu   II-118
Jia, Siyu   II-306
Jia, Xueqi   I-714
Jian, Yifei   III-133
Jiang, Qi   I-413
Jiang, Rui   I-333

Jiang, Sihang   I-578
Jiang, Weipeng   III-165
Jiang, Xiaoqi   I-459
Jiang, Xueyao   I-180, I-578
Jiang, Youjia   III-505
Jiao, Pengfei   III-322
Jin, Beihong   I-268
Jin, Cheng   III-314
Jin, Hai   I-122, I-153, I-250
Jin, Peiquan   I-560
Jin, Taiwei   III-364
Jing, Yinan   I-72, I-96, I-476
Johri, Lokesh   III-527
Joshi, Meghana   III-413

Kankanhalli, Mohan   I-232
Kao, Ben   III-443
Kapoor, Arnav   III-532
Kaushik, Minakshi   I-596
Khurana, Alka   III-544
Kim, Hong-Gee   I-632
Kim, Junghoon   I-543
Kiran, R. Uday   I-569
Kitsuregawa, Masaru   I-88
Kou, Yue   III-52
Krčál, Luboš   I-509

Lakkaraju, Kausik   III-527
Lan, Michael   I-55
Lei, Yifan   I-232
Li, Ailisi   I-180, III-36
Li, Anchen   I-171, II-134
Li, Aoran   II-289
Li, Bohan   II 289
Li, Changshu   III-377
Li, Chuanwen   I-325, I-731
Li, Dong   III-389
Li, Dongsheng   II-359, III-3
Li, Fangfang   I-739
Li, Guohui   II-183
Li, Haihong   III-314
Li, Huichao   I-587
Li, Huilin   II-523
Li, Jianjun   II-183
Li, Jiaoyang   I-493
Li, Jingze   II-590
Li, Juanzi   I-107
Li, Kun   II-623
Li, Mengxue   III-364
Li, Peng   II-656

Li, Pengfei   I-191
Li, Renhao   III-165
Li, Ruixuan   II-623
Li, Shangyang   I-682
Li, Shuai   II-298
Li, Shuaimin   III-263
Li, Tao   II-199, III-351
Li, Wei   I-665
Li, Xiang   I-268
Li, Xiaohua   I-731
Li, Xiaoyang   II-689
Li, Xin   I-268
Li, Xinyu   I-404
Li, Xiongfei   III-247
Li, Yexin   II-656
Li, Yingying   I-413
Li, Yongkang   II-298
Li, You   III-505
Li, Yue   I-259
Li, Yunchun   I-665
Li, Zhan   III-481
Li, Zhao   I-459
Li, Zhen   III-117
Li, Zhi   II-183
Li, Zhisong   III-101
Li, Zhixin   II-248, II-281
Li, Zhixu   I-137, I-180, I-578, III-85, III-149,
      III-297
Li, Zonghang   II-455
Liang, Jiaqing   I-180, III-36, III-238
Liang, Yile   II-69
Liang, Yuqi   III-306
Lim, Sungsu   I-543
Lin, Hui   II 85
Lin, Jianghua   III-425
Lin, Junfa   II-264
Lin, Leyu   II-166
Lin, Longlong   I-250
Lin, Meng   II-623
Lin, Yuming   III-505
Lin, Ziyi   I-341
Liu, An   I-137, I-191, I-207, III-85, III-149
Liu, Baichuan   I-425
Liu, Bang   I-180, III-238
Liu, Chang   I-268
Liu, Chengfei   II-472, III-181
Liu, Dajiang   I-714
Liu, Gaocong   I-560
Liu, Haobing   II-53
Liu, Hongtao   III-322

Liu, Hongzhi   II-19, II-697
Liu, Huaijun   II-199
Liu, Jiayv   III-522
Liu, Kuan   II-53
Liu, Lixin   I-404
Liu, Meng   I-604
Liu, Ning   II-407, III-117
Liu, Qi   III-213
Liu, Qingmin   II-3
Liu, Rongke   II-391
Liu, Wei   III-468
Liu, Xiaokai   I-122
Liu, Ximeng   I-413
Liu, Xing   II-631
Liu, Xinyi   III-181
Liu, Yang   I-387
Liu, Yi   II-289
Liu, Yong   I-604, II-232, II-315
Liu, Yudong   III-230
Liu, Zhen Hua   I-21
Liu, Zhidan   II-375
Liwen, Zheng   III-230
Long, Lianjie   II-343
Lu, Aidong   I-370
Lu, Haozhen   I-552
Lu, Jiaheng   I-21
Lu, Jianyun   II-639
Lu, Tun   I-333, II-359
Lu, Xuantao   III-238
Lu, Yanxiong   I-632
Luo, Siqiang   III-455
Luo, Wang   III-468
Luo, Yifeng   III-306, III-455
Luo, Yikai   II-489
Luo, Yongping   I-560
Lv, Fuyu   III-364
Lv, Ge   I-648
Lv, Junwei   II-574
Lv, Xin   I-107

Ma, Denghao   III-331
Ma, Guojie   I-259
Ma, Huifang   II-248, II-281, III-281
Ma, Jianfeng   I-413
Ma, Ling   II-298
Ma, Rui   I-353
Ma, Weihua   II-289
Ma, Xinyu   II-631
Ma, Xuan   II-69
Ma, Yunpu   III-101

Meng, Liu   I-3
Meng, Xiaofeng   I-404
Miao, Chunyan   II-232
Miao, Hang   II-134
Miao, Xiaoye   III-495
Miao, Yinbin   I-413

Narahari, Kedhar Nath   III-413
Ng, Wilfred   III-364
Ni, Jiazhi   I-268
Nie, Tiezheng   III-20, III-52
Ning, Bo   III-181
Niyato, Dusit   II-455

Obradovic, Zoran   II-689
Ouyang, Kai   I-632

Paladi, Sai Teja   III-527
Palaiya, Vinamra   III-527
Pan, Bing   III-481
Pan, Xingyu   I-38
Pang, Jinhui   I-698
Pang, Yitong   II-150
Pavlovski, Martin   II-506, II-689
Pei, Hongbin   III-331
Peious, Sijo Arakkal   I-596
Peng, Qiyao   III-322
Peng, Yuchen   III-495
Pfeifer, John   I-241

Qi, Dekang   II-656
Qi, Guilin   I-162
Qi, Xuecheng   I-293
Qian, Shiyou   I-277
Qian, Tieyun   II-69
Qian, Weining   II-506, II-689, III-306,
    III-455
Qiao, Fan   II-664
Qiao, Zhi   I-665
Qin, Shijun   III-430
Qin, Zhili   II-639
Qu, Jianfeng   I-137, III-85, III-149

Reddy, P. Krishna   I-569
Ren, Ziyao   II-606

Seybold, Martin P.   I-241
Sha, Chaofeng   II-36, III-289
Shahin, Mahtab   I-596
Shang, Jiaxing   I-714
Shang, Lin   II-216
Shang, Mingsheng   II-323

Shao, Junming   II-639
Shao, Kun   III-389
Sharma, Rahul   I-596
Shen, Derong   III-20, III-52
Shen, Fang   III-481
Shen, Qi   II-150
Shen, Shirong   I-162
Shen, Xinyao   III-197
Shen, Zhiqi   II-315
Shi, Bing   II-489
Shi, Chuan   II-199, III-351
Shi, Dan   I-171
Shi, Ge   III-222
Shi, Jiaxin   I-107
Shi, Liye   II-118
Shi, Shengmin   II-590
Shi, Wanghua   I-277
Shi, Xiangyu   III-509
Shi, Yuchen   I-425
Shi, Zhan   III-522
Skoutas, Dimitrios   I-55
Song, Ailun   I-552
Song, Hui   III-289
Song, Kaisong   II-256
Song, Shuangyong   III-425
Song, Weiping   II-298
Song, Xintong   III-509
Song, Xiuting   II-391
Song, Yang   I-38, II-697
Song, Yiping   III-3
Song, Yumeng   I-731, I-739
Song, Zhen   I-739
Srivastava, Biplav   III-527
Su, Fenglong   I-698
Sun, Bo   II-323
Sun, Chenchen   III-20, III-52
Sun, Chuanhou   I-459
Sun, Fei   III-364
Sun, Ke   II-69
Sun, Tao   II-439
Sun, Weiwei   II-590
Sun, Wenya   III-443
Sun, Xigang   II-272
Sun, Yueheng   III-322

Takata, Mika   I-88
Tang, Chunlei   II-673
Tang, Daniel   II-523
Tang, Haihong   II-232
Tang, Jintao   III-3

Tang, Moming   III-306
Tang, Zhihao   III-230
Taniguchi, Ryosuke   I-224
Tao, Hanqing   III-213
Teng, Yiping   III-522
Theodoratos, Dimitri   I-55
Tian, Junfeng   III-297
Tian, Yu   II-590
Tian, Zhiliang   III-3
Times, Valéria   III-500
Tong, Shiwei   III-213
Tong, Yongxin   II-606
Tung, Anthony   I-232

Uotila, Valter   I-21

Van, Minh-Hao   I-370
Viana, Flaviano   III-500
Vinay, M. S.   I-395

Wang, Bin   I-3, III-389
Wang, Binjie   II-664
Wang, Can   II-216
Wang, Changyu   III-331
Wang, Chaoyang   II-183
Wang, Chunnan   III-509
Wang, Chunyang   II-53
Wang, Daling   II-256, III-255
Wang, Ding   I-615
Wang, Dong   II-199, III-351
Wang, Fangye   II-359
Wang, Fei   II-439
Wang, Guoxin   II-232
Wang, Haizhou   III-133
Wang, Hongya   III-289
Wang, Hongzhi   III-509
Wang, Jiaan   III-85, III-149
Wang, Jiahai   II-264
Wang, Jialong   III-481
Wang, Jie   II-272
Wang, Jingyu   I-341
Wang, Jiwen   III-377
Wang, Kai   II-639
Wang, Ke   II-53
Wang, Lei   II-681
Wang, Liping   I-259
Wang, Long   III-514
Wang, Meng   I-162, II-631
Wang, Peng   II-540, II-664
Wang, Pengsen   I-268

Wang, Qi   II-673
Wang, Qiang   III-314
Wang, Sen   II-631
Wang, Sheng   II-298
Wang, Shi   II-523
Wang, Shupeng   II-306
Wang, Wei   II-540, II-664, III-340
Wang, Weiping   I-493
Wang, Wenjun   III-322
Wang, Wentao   III-281
Wang, X Sean   I-72
Wang, X. Sean   I-96, I-476
Wang, Xiaofan   I-526
Wang, Xin   III-481
Wang, Xinpeng   III-101
Wang, Xuwu   I-578, III-297
Wang, Yang   I-325
Wang, Yansheng   II-606
Wang, Yifan   II-298
Wang, Yike   II-248
Wang, Yitong   I-747
Wang, Youchen   I-268
Wang, Yu   III-247
Wang, Yueyi   I-425
Wangyang, Qiming   II-639
Wei, Di   I-739
Wei, Lingwei   I-615
Wei, Xing   I-293
Wei, Xingshen   III-468
Wei, Yunhe   II-248
Wei, Zhihua   II-150
Wei, Zhongyu   III-389
Wen, Ji-Rong   I-38
Wen, Zhihua   III-3
Wu, Bin   II-102, II-423
Wu, Di   II-323
Wu, Han   III-213
Wu, Lifang   III-222
Wu, Lin   II-439
Wu, Longcan   II-256
Wu, Siyuan   III-68
Wu, Wei   I-180
Wu, Wen   II-85, II-118
Wu, Xiaoying   I-55
Wu, Xintao   I-370, I-395
Wu, Yangyang   III-495
Wu, Yiqing   II-166
Wu, Zhen   II-272
Wu, Zhenghao   III-3
Wu, Zhonghai   II-19, II-697

Xia, Tianyu   I-476
Xia, Xiufeng   I-3
Xiang, Ye   III-222
Xiao, Fu   II-648
Xiao, Ning   I-268
Xiao, Shan   III-165
Xiao, Yanghua   I-180, I-578, III-36, III-197,
    III-238, III-297, III-340
Xiao, Yiyong   II-590
Xie, Guicai   III-165
Xie, Rui   I-180
Xie, Ruobing   II-166
Xie, Yi   III-314
Xing, Lehao   III-222
Xing, Zhen   II-375
Xiong, Yun   III-314
Xu, Bo   III-289, III-509
Xu, Chen   I-309
Xu, Feifei   III-101
Xu, Haoran   II-656
Xu, Hongyan   III-322
Xu, Jiajie   I-207, II-472
Xu, Jungang   III-263
Xu, Ke   II-606
Xu, Minyang   II-375
Xu, Ruyao   III-455
Xu, Siyong   II-199
Xu, Tiantian   I-459
Xu, Xianghong   I-632
Xu, Yonghui   II-315, II-407, III-117
Xu, Yongjun   II-439
Xu, Yuan   I-207
Xu, Zenglin   II-455
Xu, Zheng   I-333

Yadav, Amrendra Singh   I-596
Yan, Cheng   I-153
Yan, Ming   III-297
Yang, Bin   I-747
Yang, Bo   I-171, II-134
Yang, Cheng   II-199
Yang, Deqing   I-425
Yang, Fanyi   III-281
Yang, Geping   II-556
Yang, Han   I-698
Yang, Hao   I-353, I-387
Yang, Qinli   II-639
Yang, Shiyu   I-259
Yang, Tianchi   II-199, III-351
Yang, Weiyong   III-468

Yang, Xiaochun   I-3
Yang, Xiaoyu   III-230
Yang, Yifan   II-631
Yang, Yiyang   II-556
Yang, Yonghua   II-315
Yao, Di   I-722
Ye, Jiabo   III-297
Yi, Xiuwen   II-656
Yin, Dawei   I-722
Yin, Hongzhi   II-216
Yin, Jianwei   III-495
Yin, Yunfei   II-343
Yu, Changlong   III-364
Yu, Fuqiang   II-407, III-117
Yu, Ge   I-731, I-739, II-256, III-52
Yu, Han   II-455
Yu, Hongfang   II-455
Yu, Philip S.   III-314
Yu, Runlong   III-213
Yu, Xiaoguang   III-425
Yu, Yonghong   II-216
Yuan, Chunyuan   I-615
Yuan, Lin   I-137, III-149
Yuan, Pingpeng   I-250
Yuan, Shuhan   I-395
Yue, Lin   II-289
Yue, Yinliang   I-493
Yun, Hang   II-69

Zang, Tianzi   II-53
Zang, Yalei   II-289
Zeng, Guanxiong   I-353
Zeng, Li   III-430
Zeng, Lingze   I-587
Zeng, Shenglai   II-455
Zettsu, Koji   I-569
Zhai, Yitao   III-331
Zhang, Aoran   II-216
Zhang, Bolei   II-648
Zhang, Cong   II-423
Zhang, Fusang   I-268
Zhang, Han   II-391
Zhang, Hanbing   I-72, I-96, I-476
Zhang, Heng   III-222
Zhang, Jiale   I-552
Zhang, Jianing   I-587
Zhang, Jiujing   I-259
Zhang, Junbo   II-656
Zhang, Kai   I-72, I-96, I-476
Zhang, Lei   III-117

Zhang, Leilei   I-698
Zhang, Li   II-216
Zhang, Lingzi   II-232
Zhang, Luchen   II-523
Zhang, Luhao   II-199, III-351
Zhang, Meihui   I-587
Zhang, Mengfan   III-522
Zhang, Mi   II-69
Zhang, Ming   II-298
Zhang, Mingming   II-540
Zhang, Nevin L.   III-3
Zhang, Peng   I-333, II-359
Zhang, Qianzhen   I-441
Zhang, Ruisheng   II-606
Zhang, Tao   I-38, II-697
Zhang, Tingyi   I-137, III-149
Zhang, Weiyu   I-587
Zhang, Wenkai   III-101
Zhang, Xi   III-230
Zhang, Xianren   I-714
Zhang, Xiaohui   II-281
Zhang, Xiaoli   III-247
Zhang, Xin   I-268
Zhang, Xingyu   II-36
Zhang, Xu   II-166
Zhang, Yan   I-682
Zhang, Yifei   I-250, II-256, III-255
Zhang, Yiming   II-150
Zhang, Yixin   II-315
Zhang, Yuxiang   III-518
Zhang, Zhao   II-439
Zhang, Zhengqi   III-289
Zhang, Zhiqing   I-714
Zhang, Ziwei   III-481
Zhao, Deji   III-181
Zhao, Fanyou   III-522
Zhao, Feng   I-122, I-153
Zhao, Hang   I-72
Zhao, Hui   II-697
Zhao, Jiashu   I-722
Zhao, Lei   I-137, I-191, I-207, III-85, III-149
Zhao, Long   I-459
Zhao, Mengchen   III-389
Zhao, Pengpeng   I-191, III-85
Zhao, Rongqian   III-430
Zhao, Wayne Xin   I-38
Zhao, Weibin   II-216
Zhao, Wendy   III-481
Zhao, Xiang   I-441
Zhao, Yan   III-281

Zhao, Yue I-682
Zhao, Yuhai I-459
Zheng, Bo III-509
Zheng, Gang III-331
Zheng, Hai-Tao I-632
Zheng, Shengan I-341
Zheng, Wei II-85
Zheng, Yefeng II-631
Zheng, Yin I-632
Zheng, Yu II-656
Zhou, Aoying I-293, I-309
Zhou, Fang II-506, II-689
Zhou, Haolin II-3
Zhou, Jinhua III-430
Zhou, Jinya II-272
Zhou, Rui II-472
Zhou, Shanlin III-101
Zhou, Wei I-615
Zhou, Xiangdong II-375
Zhou, Xin II-232

Zhou, Zimu II-606
Zhu, Rui I-3, III-247
Zhu, Shishun III-468
Zhu, Shixuan II-150
Zhu, Wenwu III-481
Zhu, Xian I-615
Zhu, Yangyong III-314
Zhu, Yanmin II-53
Zhu, Yao II-697
Zhu, Ying III-255
Zhu, Yongchun II-166
Zhu, Zhihua I-722
Zhuang, Fuzhen II-166
Zong, Weixian II-506
Zong, Xiaoning II-315
Zou, Beiqi III-85
Zou, Bo III-425
Zou, Chengming II-332
Zou, Lixin I-722